Crustacean Farming

Ranching and Culture

John F. Wickins

and

Daniel O'C. Lee

Second Edition

Blackwell
Science

Blackwell Science Ltd
Editorial Offices:
Osney Mead, Oxford OX2 0EL
25 John Street, London WC1N 2BS
23 Ainslie Place, Edinburgh EH3 6AJ
350 Main Street, Malden
MA 02148 5018, USA
54 University Street, Carlton
Victoria 3053, Australia
10, rue Casimir Delavigne
75006 Paris, France

Other Editorial Offices:

Blackwell Wissenschafts-Verlag GmbH
Kurfürstendamm 57
10707 Berlin, Germany

Blackwell Science KK
MG Kodenmacho Building
7–10 Kodenmacho Nihombashi
Chuo-ku, Tokyo 104, Japan

Iowa State University Press
A Blackwell Science Company
2121 S. State Avenue
Ames, Iowa 50014–8300, USA

First Edition published 1992
Reprinted 2000
Second Edition published 2002

Set in 9.5/11.5 pt Times
by Sparks Computer Solutions Ltd, Oxford
http://www.sparks.co.uk
Printed and bound in Great Britain by
MPG Books Ltd, Bodmin, Cornwall

DISTRIBUTORS
Marston Book Services Ltd
PO Box 269
Abingdon
Oxon OX14 4YN
(Orders: Tel: 01235 465500
Fax: 01865 465555)

USA and Canada
Iowa State University Press
A Blackwell Science Company
2121 S. State Avenue
Ames, Iowa 50014-8300
(Orders: Tel: 800-862-6657
Fax: 515-292-3348
Web www.isupress.com
email: orders@isupress.com)

Australia
Blackwell Science Pty Ltd
54 University Street
Carlton, Victoria 3053
(Orders: Tel: 03 9347 0300
Fax: 03 9347 5001)

A catalogue record for this title is available from the British Library

ISBN 0-632-05464-6

Library of Congress
Cataloging-in-Publication Data
is available

For further information on
Blackwell Science, visit our website:
www.blackwell-science.com

Contents

Acknowledgements

We are greatly indebted to numerous specialists, researchers and colleagues who provided us with original information, pre-publication manuscripts, photographs, help and encouragement during the preparation of the new edition of *Crustacean Farming*. We also reiterate here our gratitude to those who contributed to the first edition, some of whom have now relocated from the addresses given below:

Hans Ackefors, University of Stockholm, Sweden;
Karl Adamson, North Isles Shellfish, Orkney, Scotland;
Julian Addison, Centre for Environment, Fisheries and Aquaculture Science, Lowestoft, Suffolk;
Dean M. Akiyama, American Soybean Assoc., Singapore;
Adnan Al-Hajj, Consultant, Guayaquil, Ecuador;
Abayomi Alabi, Island Scallops Ltd., Qualicum Beach, British Colombia, Canada;
Geoff Allan, Department of Agriculture, New South Wales, Australia;
Chris Austin, Queensland Institute of Technology, Queensland, Australia;
Mark Ayranto, Aquaculture Specialist, Campbell River, British Columbia, Canada;
Conner Bailey, College of Agriculture, Auburn University, Alabama, USA;
Colin Bannister, Centre for Environment, Fisheries and Aquaculture Science, Lowestoft, Suffolk;
Tony Bart, Fremantle Maritime Centre, Western Australia, Australia;
Adam Body, Darwin, Australia;
John Booth, National Institute of Water and Atmospheric Research Ltd., Wellington, New Zealand;
Claude E. Boyd, Department of Fisheries and Allied Aquacultures, Auburn University, Alabama, USA;
Matt Briggs, Institute of Aquaculture, University of Stirling, Scotland;
James Brock, Aquaculture Veterinarian, Kailua, Hawaii;
Janet Brown, Institute of Aquaculture, University of Stirling, Scotland;
S. and C. Buckhaven, Studham, Bedfordshire;
Craig Burton, Sea Fish Industry Authority, Argyll, Scotland;
Chris Campbell, The Cultured Crustacean Company Ltd., Nanaimo, Canada;
Pedro Cañavate, PEMARES, Cadiz, Spain;
Dave Cannon, Edisto Seafarms Inc., Texas, USA;
Chau-Ling Chan, University of Lancaster, Lancashire;
Paul Clark, Natural History Museum, Kensington, London;
Alan Coghill, OFA, Orkney, Scotland;
Philippe Colivet, Aquaculture Engineer, Nogen sur Eure, France;
W. 'Bill' Cook and Peter Oxford, North Western and North Wales Sea Fisheries Committee, Lancaster University, Lancaster;
Peter Crocos, CSIRO, Cleveland, Australia;
Imre Csavas, FAO Regional Office for Asia and the Pacific, Bangkok, Thailand;
David Currie, Consultant, Perth, Scotland;
John Dallimore, JD & Associates, Hamburg, Germany;
Edwin Derriman, Cornwall Sea Fisheries Committee, Penzance, Cornwall;
M. Esseen, Fisheries Consultant, Anglesey;
Dan Fegan, National Center for Genetic Engineering and Biotechnology, Bangkok, Thailand;
Liu Fengqi, Department of Biology, Nankai University, Tianjin, China;
Tim Flegel, Mahidol University, Bangkok, Thailand;
Andrew Forsythe, Marine Harvest, Canada;
Patrick Franklin, Macallister Elliot and Partners, Lymington, Hampshire;
Peter Fuke, Consultant, Chelmsford, Essex;

Anne Guillaumin, Hatchery & Seafood Specialist, Auray, France;
Dennis Hedgecock, Bodega Bay Laboratory, University of California, USA;
Yves Henocque, Station Marine d'Endoume, Marseille, France;
David Holdich, School of Biological Sciences, University of Nottingham, Nottingham;
D.S. Holker, Marron Growers Association of Western Australia, Australia;
John Hollows, Koura New Zealand Ltd., Dunedin, New Zealand;
Jay Huner, Crawfish Research Centre, University of Louisiana, Lafayette, Louisiana, USA;
Ray Ingle, The Natural History Museum, Kensington, London;
Clive Jones, Freshwater Fisheries and Aquaculture Centre, Walkamin, Queensland, Australia;
David Jones, School of Ocean Sciences, Menai Bridge, Wales;
Li Kangmin and Zhou Xin, Asian Pacific Regional Research and Training Centre for Integrated Fish Farming, Wuxi, China;
Ilan Karplus, Aquaculture Research Organisation, Bet-Dagan, Israel;
Max Keith and Peter Wood, Frippak Feeds, Aberdeen;
Max Keller, Erste Bayerische Satzkrebszucht, Augsberg, Germany;
Tore Kristiansen, Institute of Marine Research, Bergen, Norway;
Ian Laing, Centre for Environment, Fisheries and Aquaculture Science, Weymouth, Dorset;
Craig Lawrence, Fisheries, Western Australian Marine Research Laboratories, North Beach, Western Australia, Australia;
Jean-Francois LeBitoux, Centre Aquacole, Leucate, France;
Phillip G. Lee, National Resource Centre for Cephalopods. University of Texas, Galveston, Texas, USA;
Lionel Letessier, Banggai Sentral Shrimp, Surabaya, Java, Indonesia;
Lewis LeVay, School of Ocean Sciences, Menai Bridge, Wales;
Hervé Lucien-Brun, Sepia International, Paris, France;
Donald J. Macintosh, Centre for Tropical Ecosystems Research, University of Aarhus, Denmark;
Greg Maguire, Tasmanian State Institute of Technology, Tasmania, Australia;
Gay Marsden, Department of Primary Industries, Queensland, Australia;
Greta Martinez, Molokai Sea Farms International, Hawaii;
Ronald D. Mayo, The Mayo Associates, Seattle, Washington, USA;
Douglas McLeod, Marine Resource Consultants, Isle of Skye, Scotland;
Satoshi Mikami, Australian Fresh R&D Corporation, Bribie Island, Queensland, Australia;
Corny Mock, Cornelius Mock and Associates, Galveston, Texas, USA;
Noel Morrissey (retired), Fisheries Department, Western Australian Marine Research Laboratories, North Beach, Western Australia, Australia;
Colin Nash, FAO, Rome, Italy;
Michael New OBE, Aquaculture Development Consultant, Marlow, Buckinghamshire;
Paul Niemeier, National Marine Fisheries Service, Silver Spring, Maryland, USA;
David O'Sullivan, Editor, Austasia Aquaculture Magazine;
Leigh Owens, Department of Microbiology and Immunology, James Cook University, Queensland, Australia;
Stephanie Parkyn, National Institute of Water and Atmospheric Research, Hamilton, New Zealand;
Ian H. Pike, International Association of Fish Meal Manufacturers, Potters Bar, Hertfordshire;
John Portmann, (retired) Centre for Environment, Fisheries and Aquaculture Science, Burnham on Crouch, Essex;
Emanuel Polioudakis, independent anthropologist affiliated to Department of Fisheries and Allied Aquacultures, Auburn University, Alabama, USA;
M.A. Robinson, Senior Fishery Statistician, FAO, Rome, Italy;
R.P. Romaire, Louisiana State University, Agricultural Center, Louisiana, USA;
Bob Rosenberry, Editor, *Aquaculture Digest*, San Diego, USA;
Bill Rowntree, Photographer, School of Ocean Sciences, University College of NorthWales, Wales;
Tony Salisbury, Hatchery Manager, Kimberley College of TAFE, Western Australia, Australia;
Nathan Sammy, Department of Industries and Development, Darwin, Australia;
Rosalie A. Schnick, US Fish and Wildlife Service, Wisconsin, USA;
Alasdair Scott, Centre for Environment, Fisheries and Aquaculture Science, Weymouth, Dorset;
Ephraim Seidman, Kibbutz Ma'agan Michael, Israel;

Robert Shleser, Aquacultural Concepts, San Juan, Puerto Rico;
Nuno Simões, School of Ocean Sciences, Menai Bridge, Wales;
David Smythe and Peter Wilhelmus, New Zealand Clearwater Crayfish (Koura) Ltd., Nelson, New Zealand;
Alan Stewart, University of Stirling, Scotland;
Patrick Sorgeloos, Laboratory of Aquaculture and Artemia Reference Centre, Ghent, Belgium;
Albert Tacon, FAO, Rome, Italy, (currently at The Oceanic Institute, Hawaii);
Mike Timmons, Cornell University, New York, USA;
Len Tong, Ministry of Agriculture and Fisheries, Wellington, New Zealand;
Granvil Treece, Sea Grant Mariculture Specialist, Texas A&M Unversity, Texas, USA;
Susan Utting, Sea Fish Industry Authority, Colwyn Bay, Wales;
Gro I. Van der Meeren, Austevoll Aquaculture Research Station, Norway;
Susan Waddy, Biological Station, St Andrews, New Brunswick, Canada;
Andy C. Watkins, Aqualider, Brazil;
R. Douglas Watson, Department of Biology, University of Alabama, Birmingham, Alabama, USA;
Donald W. Webster, University of Maryland, USA;
Priscilla Weeks, School of Human Sciences and Humanities, University of Houston, Texas, USA;
Dennis M. Weidner, National Marine Fisheries Service, Silver Spring, Maryland, USA;
Philip Wickins, University of Southampton, Southampton, Hampshire;
John F. Wood, Natural Resources Institute, London;
Patrick J. Wood, Euroshrimp, Lubeck, Germany.

We are also grateful to the following colleagues and friends who made time to provide constructive comments on early drafts of individual chapters and sections:

D.J. Alderman, Centre for Environment, Fisheries and Aquaculture Science, Weymouth, Dorset;
R.W. Beales, Overseas Development Administration, London;
T.W. Beard, A.R. Child, S.J. Lockwood and B.E Spencer, Centre for Environment, Fisheries and Aquaculture Science, Conwy, North Wales;
I. Chaston, School of Business Studies, University of Plymouth, Plymouth, Devon;
M. Esseen, Fisheries Consultant, Anglesey;
P. Franklin, Macallister Elliot and Partners, Lymington, Hampshire;
D.M. Holdich, School of Biological Sciences, University of Nottingham, Nottinghamshire;
A.N. Jolliffe, Overseas Development Administration, London;
Clive Jones, Freshwater Fisheries and Aquaculture Centre, Walkamin, Queensland, Australia;
D.A. Jones, School of Ocean Sciences, University College of North Wales, Gwynedd;
G.S. Lee, Inasa, Santo Domingo, Dominican Republic;
G. Parry-Jones, School of Accounting, Banking and Economics, University College of North Wales, Gwynedd;
A. Scott, Centre for Environment, Fisheries and Aquaculture Science, Weymouth, Dorset.

It is a pleasure to thank Dr S.J. Lockwood for providing one of us (DL) with an eminently suitable habitat at the (then) Fisheries Laboratory, Conwy, for the preparation of the first edition of this book. Also, we gratefully acknowledge the help of Sue Walker, David Hyett and Dennis Glasscock of the Centre for Environment, Fisheries and Aquaculture Science in providing facilities and material for the present edition after the closure of the CEFAS Conwy Laboratory and its library facilities in December 1999.

On a more personal note, I (JFW) wish to acknowledge the unreserved support and understanding of my wife Christine and our family, especially since I spent the whole of my first year of retirement working on this volume. I (DL) would like to thank my wife, Gaëlle, for her help with the sections of the book dealing with HACCP, but above all for her patience and encouragement throughout. We are especially grateful for the inspiration provided by Juliette, our baby daughter.

References to proprietary products, Internet sites and organisations do not imply endorsement by the authors nor is any criticism implied of those not mentioned.

Except where specified, all currency units are in US $.

Chapter 1
Introduction

1.1 History

There is no doubt that over the past 50 years the idea of farming shrimp, crayfish, crabs or lobsters has become endowed with considerable 'investor appeal'. Since the early 1950s increasing personal disposable income in Japan and the West has allowed more and more people to explore the delights of eating crustaceans. As a result, consumption has soared and a host of entrepreneurs, businessmen and governmental agencies has rushed to exploit the aquaculture traditions and technologies of the Far East.

For hundreds, perhaps thousands, of years a variety of shrimp, prawn and crab species had been raised as an incidental crop from wild-caught juveniles entering coastal fishponds throughout the Indo-Pacific region. The advent of refrigeration and improved transportation gave the artisan farmers access to high-priced city and international markets and encouraged many to set aside ponds specifically for shrimps and prawns. The hatchery technologies developed by pioneers such as M. Hudinaga of Japan and S.W. Ling of Malaysia allowed much greater control of juvenile supplies. Hatcheries and shrimp farms became widespread during the 1950s and 1960s both throughout the Far East and in the southern USA and Hawaii. Most failed to emulate oriental farming practices successfully during those early days and much money was lost. Nevertheless, valuable lessons were learnt and today shrimp and prawn farms extend throughout most tropical and many subtropical regions of the world and contribute some 25–30% of world supplies.

This level of productivity has not been achieved without considerable social and environmental costs for some countries (Chapter 11), although it is widely recognised that most problems are associated with shrimp farms rather than with freshwater prawn and crayfish production. Widespread construction of marine and brackish-water ponds gives cause for concern because it often involves extensive clearance of mangrove forest (together with all its resources), consequent loss of fish and shrimp nursery areas, coastal erosion and salinisation of coastal lands. Concentrated effluents from large aggregations of shrimp farms have polluted lagoons, estuaries and coastal waters and jeopardised the livelihoods of whole communities. The rising demand for feedstuffs for shrimp and fish farming in particular has put pressure on supplies of low-value fish often consumed directly in developing countries. Ironically, in Taiwan it was the shrimp farmers themselves who first suffered the greatest setbacks when, in 1988, environmental degradation brought about by their own activities resulted in severe disease outbreaks and the near collapse of the industry. Since then, similar catastrophic failures have occurred in most of the major shrimp producing nations of the world, often associated with severe outbreaks of disease (sections 2.5, 8.9, 11.3.3 and 12.2). The estimated economic losses in Taiwan (1987–88) amounted to \$420m, in China (1993) they were \$1bn, while Thailand (1991) lost \$180m in export earnings alone (NACA 1994–95). In 1999–2000, Ecuador, the West's largest producer, lost \$300–500m due to viral disease. Hard lessons have been learnt and, today, much research, technical development and new operating procedures are focused on enhancing the sustainability of the industry.

In western temperate regions there is no long tradition of crustacean farming although since the turn of the century various attempts have been made in Europe and North America to restock natural waters with crayfish (Holdich 1993) and young lobsters (Addison & Bannister 1994). Apart from the restocking programmes in

European inland waters, the only significant freshwater crayfish farming was, until the late 1980s, that practised in the southern USA since about 1950. The last two decades, however, have witnessed an increase in crayfish production, particularly in China where supplies come mainly from the wild but also from incidental polyculture in fishponds. In Europe and Australia crayfish farming has also increased although pond production is small by comparison. The need to export North American crayfish to European markets has wrought significant changes within the production and management practices of the industry (section 1.3.3). As in the early days of shrimp culture, euphoric predictions of crayfish yields often occur in the trade press (Rogers & Holdich 1995). These predictions stimulate entrepreneurs to propose farming projects often based on stocking densities and survival rates that can neither be supported nor refuted because relevant research and pilot studies have not been made. Increasingly however, animals are being produced and this has been sufficient to justify continued research support in countries from the tropics to more temperate zones.

Studies on clawed lobsters in North America between 1965 and 1975 demonstrated that lobsters could be grown to commercial size in only 2 years instead of the 5–7 years taken in the wild, simply by raising the water temperature and by daily feeding. A plethora of commercial culture proposals followed. Many were based on assumptions not fully validated by research and again much money was lost. Perhaps the greatest setback to aquaculture was that reported by Aiken and Waddy (1989) who wrote: 'In both countries (Canada and USA) a productive university–government research effort was extinguished by excessive promotion and premature entrepreneurial interest.' Today, interest focuses on supplementing or extending natural stocks with hatchery-reared juveniles (section 7.8.11).

In the past, spiny lobsters attracted only sporadic aquaculture research effort because of the seemingly intractable difficulties of rearing their delicate larvae. Although fattening of wild-caught juveniles was practised commercially in a few countries, it was not until the late 1980s when researchers in Japan and New Zealand successfully reared a few larvae to the puerulus and juvenile stages that interest was revitalised (Kittaka 2000). Commercial scale larvae culture remains a distant, yet tantalising prospect, while commercial ongrowing and fattening operations are increasing.

Until recently the rearing of crabs appeared to be of little commercial interest except perhaps as a subsistence activity of artisanal farmers. Yet in the last few years, the development of hatchery techniques for mud and mitten crabs has attracted the attention of researchers, and farm production, although still mainly from wild seed, has expanded considerably. Many shrimp farmers have converted their ponds to raise crab in the wake of disease outbreaks.

1.2 Objectives

The recent history of crustacean farming therefore is beset with failures as well as successes, and this originally stimulated the preparation of this book. In the present substantial revision we attempt to provide the technical information required, and to address some of the problems to be faced, by those new to the industry. The information will be relevant not only to all students of aquaculture but also to those who have responsibility for advising or making policy decisions concerning feasibility, investment, financing or implementation of crustacean aquaculture projects. Academic scientific information has generally been kept to a minimum in favour of basic biological and technical descriptions that have direct bearing on the reliability and costs of the various culture options. However, because the first edition of *Crustacean Farming* was so widely used by academic lecturers and students, we have responded by increasing the number of source references in several chapters. Some of the original references are retained for coherence or where little new information is available.

Shrimp farming in particular stands out for its influence on the economies of developing countries and for this reason attention is paid to infrastructure and institutional factors as well as social and environmental impacts. Representatives of all species that are farmed commercially for the table or for restocking, or that are thought to have potential for culture are discussed. A summary of important factors relating to their culture is given in Appendix 1.

Certain other species also deserve mention and include those that form a significant by-catch to the main species being farmed, those captured at a large size and fattened, matured or induced to moult to take advantage of specialist or seasonal markets, and those that are cultivated for bait (sections 7.2.5 and 7.5.9), ornamental display (section 7.4.4.1) or for research purposes (section 7.4.4). No doubt other species exist, which may for one reason or another be worth cultivating. Possible candidates may be found among the crabs, for example the Australian giant crab *Pseudocarcinus gigas* (Gardner

& Northam 1997), the larger, filter-feeding freshwater atyid shrimps (*Atya gabonensis* of West Africa (maximum size 92–124 mm total length), *A. innocous* of the West Indies (21–34 mm carapace length)), or even goose barnacles (*Lepas* spp.; Goldberg & Zabradnik 1984). However, accounts of large-scale culture trials with these novel candidates are either scarce or non-existent.

Several smaller, non-decapod crustaceans are increasingly being cultured as food for the rearing of other organisms and have been included in this revision (section 7.11).

1.3 Current status

By far the greatest tonnage of farmed crustaceans are marine and brackish-water shrimp produced in Southeast Asia and Ecuador (Table 1.1). Although estimates vary considerably, it seems likely that in 1999, almost 815 000 tonnes (mt) were harvested, representing some 25–30% of the world total supply of 2.8×10^6 mt. Farmed freshwater prawns (mainly *Macrobrachium rosenbergii*) now total around 130 000 mt per year. This represents a seven-fold increase in production over the past decade but is primarily due to the inclusion of China and Bangladesh in the statistics (New 2000). Nearly half of the global production emanates from China while Bangladesh contributes around 37%, Taiwan and Thailand about 6% each and India 1%. Vietnam also reports a substantial quantity, the figures for which, unfortunately, are not separated from other farmed freshwater crustaceans. The total annual harvest of wild and extensively farmed freshwater crayfish is probably around 50 000–130 000 mt, of which up to 50 000 mt can come from the southern USA in productive years, but China now produces more than the USA. Reliable figures for the production of cultured spiny lobsters and crabs are not readily available. In the late 1980s aquaculture production of these groups was estimated to be about 1% (6000–7000 mt) of total crustacean production. The 1998 FAO statistics, however, give considerably higher values for crabs alone, possibly because of data included from countries not previously reported. As far as is known, no commercially viable farms for clawed lobsters exist, although ranching and restocking prospects are being assessed with these as well as with some shrimp, prawn, spiny lobster and crab species. Ecological changes, in particular a steady decline in salinity in the Great Salt Lake, Utah, USA in the late 1990s seriously threatened a major global source of *Artemia* cysts used to feed larvae in hatcheries throughout the world. The unexpected, record harvest in late 2000 serves to highlight the vagaries of nature and the potential to affect aquaculture worldwide. Strategies to promote the more efficient use of *Artemia*, for example through nutrient enrichment techniques (section 7.11.2.1), and to extend areas of production are being implemented (Lavens & Sorgeloos 2000).

1.3.1 Marine and brackish-water shrimp

Such is the scale of shrimp farming that in the last two decades it has induced significant changes in market structure and prices (section 3.3.1). For example, an abundance of shrimp in the medium size ranges

Table 1.1 Estimated ranges for crustacean production (mt) from fisheries and aquaculture* for the period 1998–99.

Species/group	Fishery production	Aquaculture production*
Marine shrimps and prawns	2 713 450–2 800 000	600 000–814 000
Freshwater prawns	16 000–21 000	130 000
Freshwater crayfish	36 000–130 000	40 000–70 000
Clawed lobsters	79 146	0
Spiny lobsters	73 575	71
Other lobsters (*Nephrops*, Scyllaridae, Galatheidae)	63 045	0
Crabs (marine and freshwater)	1 284 838	200 660
Artemia	8000**	15–20***

*No attempt has been made to separate yields from restocking programmes.
**Unprocessed cysts, equivalent to 4600 mt saleable product.
***P. Sorgeloos 2001, pers. comm.
Source: Rosenberry 1998, 1999; FAO 2000; Murthy 1998; New 2000; New *et al.* 2000; J. Huner 2001, pers. comm. Values for unspecified species/groups given by FAO 2000 are excluded.

(20–35 g) caused a fall in prices in South-east Asia from \$8.50 to \$4.50 kg^{-1} during the late 1980s. Part of the decline was believed to be the result of a temporary reduction in consumption and consequent increased cold storage of shrimp in Japan. In 1998 US consumption rose 15% but prices fell by 5–20% depending on size (Rosenberry 1998) as the market, surprisingly, sought the smaller sizes of shrimp. Many farmers adapted quickly by producing more crops per year of smaller shrimp. However, such drops in price emphasise the vulnerability of intensive farming methods which have narrower profit margins than many low-cost, extensive and semi-intensive farms (sections 5.2, 10.5 and 10.6.1.5). Global production of farmed marine shrimp reached 733 000 mt in 1994 but fell steadily to 660 000 mt in 1997 before rising again to 814 000 mt in 1999. Production has been adversely affected by a number of factors, notably outbreaks of disease in almost all the major shrimp producing countries (for example the recent outbreaks of white spot virus disease in Central America and Kerala, India, in 1999/2000) and unseasonable fluctuations in the oceanic currents (El Niño and La Niña events) that govern in particular, the availability of seedstock in Ecuador but also impact shrimp farming worldwide (section 11.4.4). The La Niña event of 1983–84 led to an official moratorium on new pond construction in Ecuador from 1984 to 1989 and a widespread shortage of broodstock and juveniles. The El Niño event of 1997–98 produced an abundance of wild post-larvae for stocking that led to the closure of 90% of Ecuadorian hatcheries. It also precipitated violent storms that caused considerable damage to ponds and farm infrastructure elsewhere.

Major consumers of shrimp are the USA and Japan and there has been considerable expansion into European markets with increased sales of value-added product (section 3.3.1). As a result of import bans on tropical shrimp by the EU and USA during the late 1990s for several reasons including food safety (sections 3.2.1 and 11.2.5), Asian producers began to successfully exploit their domestic markets, although progress was temporarily impeded by the Asian financial crisis. A system of quality assurance known as Hazard Analysis, Critical Control Points (HACCP) that focuses on food safety throughout the production and processing stages, is being increasingly adopted by the industry and is helping to increase quality and consumer confidence in crustacean products generally (sections 3.2.2 and 9.6). There is a fear, however, that if the decline in world shrimp fisheries coupled with continued farmed crop failures leads to undersupply of the markets, prices would rise, consumers would buy less and producers' revenue could fall along with foreign investment in farms in developing countries (Keefe & Jolly 2000).

1.3.2 Freshwater prawns

Production of tropical freshwater prawns, though much less than of shrimp, remains of significant interest in many countries. Expensive coastal sites are not required and, because the culture densities employed are lower than those for shrimp, prawn farming is regarded by some as one of the more sustainable forms of crustacean aquaculture. Despite a common perception of generally poorer export marketing opportunities, frozen *Macrobrachium* tails have become a common sight in European, and to a lesser extent American, supermarkets over the last 10 years (section 3.3.2). The heterogeneous growth typical of pond populations remains a major constraint however, and to combat this many farms operate complex stocking and harvesting regimes (section 7.3.7). The benefits of selective stocking from nursery systems, the effect of temperature on sexual maturity and the development of monosex populations are being actively investigated (sections 2.6.3 and 7.3.7). *Macrobrachium rosenbergii* is the prawn preferred by farmers although *M. malcolmsonii* (Table 4.6f) is widely cultivated on the Indian subcontinent. In addition to *M. rosenbergii*, substantial production (15 000 mt) of *M. nipponense* is being reported from China (Wang & Qianghong 1999). Other species, e.g. *M. carcinus*, are grown and trials have been made in Africa with *M. vollenhovenii*. Hybrids have also been studied but none is commercially successful as yet. The growth, yield and marketability of *M. rosenbergii* are now being assessed against that of the Australian redclaw crayfish (*Cherax quadricarinatus*), which grows in a similar culture environment, has no need for brackish water during its life cycle and is less aggressive (section 4.7).

1.3.3 Crayfish

The red swamp crayfish (*Procambarus clarkii*) is the single most important species of this group with most production traditionally (55%) coming from wild and managed stocks in the USA. Significant harvests are now however also being achieved in China (70 000 mt) with Spain, Turkey and Kenya harvesting 3000, 1000 and 500 mt respectively (Skurdal & Taugbøl 2001; Huner 2001). Production from these sources can be highly variable due to changing weather patterns and production statistics seldom distinguish between wild and cultured crayfish. Competition from Chinese product and European demand for larger sizes of crayfish has influenced marketing strategies in the US industry (sections 3.3.3.1 and 7.5.7). Interest in value-added and soft-shell crayfish products in southern USA expanded rapidly in the late 1980s but partially collapsed in the early 1990s, primarily because of high production costs and seasonality of supply. Technology to reduce costs arrived too late for all but a few large producers but nevertheless remains available pending any recovery (Huner 1999; sections 3.3.3.2, 7.5.8 and 12.8.3).

There is increasing commercial interest in crayfish farming in Western Europe and Australia. The potential for culture is good in Europe but fear of crayfish plague, to which native species are susceptible, led to the importation of more resistant North American species, notably signal crayfish (*Pacifastacus leniusculus*), which unfortunately sometimes carried and spread the disease further. Several hundred hatcheries in Europe produce over two million crayfish annually for restocking natural waters (Pérez *et al.* 1997) while in Britain most aquaculture production (7 mt) is for the table (Lewis 2001).

In the late 1980s considerable interest was aroused by claims that 'new' Australian species or strains had particularly good aquaculture potential. The culture of one, redclaw, was consequently implemented both in Australia and abroad despite limited knowledge of culture requirements and performance (Jones 1990; Rouse 1995). However, Australian crayfish are reported to be highly susceptible to plague fungus and although culture outside their native region is spreading, this could be risky. Ecological and commercial disaster could strike if North American or European crayfish from any source were taken into Australia. Today there is widespread commercial interest in three Australian crayfish – redclaw, marron and yabby, which has stimulated the funding of research projects and culture trials in several countries (Table 4.1b). The experiences gained over the next few years in both culture and marketing these species will therefore be critical. In New Zealand, entrepreneurs are now making fresh attempts to farm native crayfish (*Paranephrops* spp.) regardless of pessimistic economic research reports published in the 1970s and 1980s (section 7.7.11).

1.3.4 Clawed lobsters

Catches of North American lobsters have increased from around 48 000 to 80 000 mt over the period from 1984 to 1998 largely due to increased Canadian landings. European lobster catches remained steady at around 2000 mt over the same period. While the culture of both species is technically feasible, the need for individual confinement during ongrowing and the lack of a suitable, cost-effective diet have so far prevented commercial viability. After many years of concerted research effort into the development of battery culture technology and compounded diets, attention has been diverted towards investigating prospects for ranching hatchery-reared juveniles in natural and modified seabed habitats. Research in the UK during the past decade has provided, for the first time, credible evidence that hatchery-reared juveniles survive to enter the fishery and contribute to broodstocks (sections 5.7, 7.8.11 and 8.11).

1.3.5 Spiny lobsters

Australia, New Zealand, South Africa, Cuba, Brazil, Mexico and the USA are the main producers of the 74 000 mt of spiny lobsters derived annually from fisheries. Spiny lobsters are usually sold frozen but the Japanese may pay up to \$100 kg^{-1} for live animals. Recent advances in the culture of larval stages in Japan, New Zealand, Tasmania and elsewhere are encouraging (Kittaka *et al.* 1997; Kittaka 2000) but considerable development of mass culture techniques will be required before commercial culture can succeed. In some areas where juveniles or adults occur naturally, fishermen provide shelters (known as 'casas Cubanas', 'casitas' or 'pesqueros') which modify the seabed habitat and concentrate fished populations (sections 7.9.8 and 8.11.2). Ongrowing or fattening of spiny lobsters in ponds, cages or tanks is possible when adequate supplies of juveniles or undersized lobsters can be obtained (Jeffs & Hooker 2000). Few published details of these operations were found for use in this account, although in New Zealand, initial studies indicate that sea cages could be more cost-effective than land-based systems (Jeffs & James 2000). In New Zealand and Tasmania where the larval ecology of local species is well understood, collection of wild pueruli for commercial ongrowing is allowed under recent fishery quota trade-off arrangements (section 7.9.4).

1.3.6 Crabs

The culture of mud crab (*Scylla* spp.) in South-east Asia and the mitten or river crab (*Eriocheir sinensis*) in China and Korea is now widespread. The majority of farmed crabs come from the extensive ongrowing and fattening of wild-caught juveniles or polyculture operations, while some are raised in semi-intensive monocultures. Stock enhancement of mitten crab is widely practised in China. A variety of farming techniques exist for both species that are suitable for differing economic circumstances. This flexibility has enhanced the mud crab industry's popularity to an extent where the supply of juveniles and adults for stocking has become critical. In recent years advances have been made in mud crab hatchery and feed technology while the prospects for restocking natural populations are being actively investigated in Indonesia and Vietnam (section 7.10.8). In Europe and the USA, however, the mitten crab has become a serious pest wherever populations have become established, through both accidental and intentional translocation. It may be possible to ranch other species (e.g. Japanese swimming crabs, *Portunus* spp.), but the ownership problems that could arise with nomadic species are likely to limit such exercises to public control (section 11.5.3.1). Crab culture has not attracted much commercial interest in the West, although the potential of the king crab *(Mithrax spinosissimus)* has been considered in the Caribbean (Lellis 1992). There is commercial interest in value-added and soft-shell crab products, particularly in the USA (sections 3.3.5 and 7.10.9).

1.4 Advances and constraints

A major research effort has gone into the identification, prevention and treatment of crustacean (especially shrimp) diseases during the past 10 years. New molecular diagnostic tools have been developed but only in the late 1990s was a stable, continuous, albeit transformed, shrimp cell culture line (a fundamental aid to the diagnosis and study of viruses) reported (Tapay & Loh 1999). Experiments are in progress to use retroviral vectors to introduce oncogenes into shrimp cells in the hope of creating immortal cell lines (Shimizu *et al.* 2000) (section 12.3). Specific pathogen free (SPF) and specific pathogen resistant (SPR) stocks of a few important shrimp and crayfish species are now available. Immunostimulants and probiotics (see Glossary) are potentially useful tools in disease prevention and control; however, some claims concerning their effectiveness should be treated with caution (Devresse 1998). Transplantation of wild-caught or cultured species around the world continues to spread diseases despite international recommendations on control of movements (sections 8.9.4 and 11.3.3). Other carriers of pathogenic viruses have been identified including fresh frozen shrimp destined for the table and live bait used by anglers. In addition to transplantations, some animals inevitably escape from farm ponds or tanks and there is a risk of non-endemic species not only becoming established in the wild but also infecting local fished stocks with exotic diseases (Pantoja *et al.* 1999). The record of North American crayfish is particularly poor in this respect. A vaccine against an important disease of clawed lobsters, gaffkaemia, has been developed (section 2.5.5).

While it remains true that great improvements in crustacean productivity can often be gained by upgrading the culture environment, feed composition and husbandry practices, recent advances in the development of precise and specific molecular tools have greatly increased the prospects for developing useful domesticated strains. The genetic variation in stocks and the extent to which key attributes are under genetic control has been assessed in some species, and work has begun on mapping and characterising genes that influence growth, reproduction and disease resistance (sections 2.6.1 and 2.6.2). Large, controlled breeding programmes are prerequisites for further advances in genome studies likely to lead to the development of commercial breeds. Significant progress has been made, for example in Australia, where third and fourth generation *Marsupenaeus japonicus* (see section 2.1 for name changes) and *Penaeus monodon* are demonstrating that reproductive performance of captive broodstock can be comparable to that of wild stocks and that up to 10% increase in growth rates can be obtained from genetically improved lines during commercial production (P. Crocos 1999, pers. comm.; sections 2.6.2 and 12.3).

During the past decade, considerable advances have also been made in understanding nutritional requirements of, and the formulation of specialised compounded diets for, broodstock, larvae and juvenile stages of shrimp (sections 2.4 and 8.8). The differing requirements of a few important species are steadily being defined but even so, much remains to be done concerning the interactions between the dietary components that govern the cost-effectiveness of commercial formulations as well as between those that enhance stress resistance. The value and limitations of live prey as vectors for essential nutrients and medicines for larvae have been established, and methods developed for the culture of a variety of prey organisms (section 7.11). Research into improved diet stability, managing the nutritional contribution from organisms growing naturally in the culture ponds (natural productivity) and into reducing water exchange rates in some specially managed ongrowing systems, has demonstrated potential for more efficient resource utilisation (feed, water), higher yields and for reducing adverse environmental impacts (sections 8.3.6.8, 8.3.7 and 8.8). The dietary needs of several cultivated crayfish (particularly *Cherax* spp.), crab and spiny lobster species are now being more widely investigated, although progress with clawed lobster diets in North America and Great Britain has largely ceased. Global production of manufactured shrimp feeds in 2000 was estimated to be 1.49×10^6 mt (section 3.4.1) and over the current decade, shrimp farming alone is expected to consume around 370 000–480 000 mt of fishmeal annually: about 5–6% of the global supply (Barlow 2000). Research is being conducted on fishmeal substitutes (section 11.5.1.4) and on the preparation of more easily digested and cost-effective formulations that minimise nutrient discharges (e.g. phosphorus) in pond effluents.

The tolerances of the different life cycle stages of several species to toxic levels of metabolites and other substances dissolved in water has been elucidated in part, but considerable work remains to be done especially in the determination of sub-lethal effects on crustacean growth, reproduction and susceptibility to disease (sections 8.5, 8.9 and 12.2). The lack of basic scientific information on pond bottom chemistry and ecology became a key issue on many shrimp farms following commercial crop failures and disease outbreaks in the 1980s. Since then improvements have been made to pond treatment methodology, and new strategies employed for culture and effluent management under low water exchange conditions (sections 7.2.6.6, 8.3.6.8, 8.3.7 and 12.6).

Uncertainties remain regarding changes in global climate patterns and increased risks of pond damage and stock loss by storms and flooding are expected in some areas, which will have an impact upon national coastal zone management strategies (Holmes 1995; Weaver & Green 1998). It is also likely that growth rates and harvesting schedules will be influenced by unpredictable and unseasonable temperature and rainfall patterns (section 11.4.4), for example the 1999 drought seriously reduced crayfish production in Louisiana (Lutz 2000). There is evidence that crustaceans in clear shallow waters (open-air hatcheries, nurseries) could be harmed by ultraviolet radiation-B (UVBR, 280–320 nm), which has increased in some regions (Table 8.3).

The adverse effects on shrimp and fish nurseries resulting from the destruction of mangroves to make farm ponds are now more widely appreciated. However, legislative constraints are increasing, fuelled by rising public concern over the impact of farming practices on the environment including the spread of diseases through stock transfer, escapes of non-endemic strains (and potentially, genetically modified animals) and effluent discharges (sections 11.3.2, 11.3.3 and 11.4.3). Care will be required to ensure that new regulations, designed to protect consumers and ecological systems, are agreed and applied objectively and fairly (Subasinghe 2000). Other factors, which may occasionally handicap progress, include a shortage of skilled technicians and a surfeit of unqualified or inexperienced facilitators and consultants (sections 9.3.6 and 11.5.2.2).

In some areas of crustacean aquaculture, developed countries and multinational corporations now enjoy competitive advantages over many developing countries through the use of advanced technologies and marketing strategies. While this is to be expected, several trade restrictions, duties, quotas and non-tariff barriers are seen to go against the spirit of the World Trade Organization agreement. Similarly, attempts have been made to invoke social, labour and environmental issues to manipulate trade and access to markets in ways which are perceived to be discriminatory in their effect (Subasinghe 2000; sections 11.2.5, 11.5.1.4 and 11.5.3.2). Nevertheless, despite the recent problems related to markets and production, crustacean aquaculture seems likely to retain its momentum and attractiveness to investors for the foreseeable future. Its role in contributing to a healthy diet is well established. However, a careful approach and thorough project appraisal prior to commitment of capital are essential if past mistakes are to be avoided.

1.5 References

Addison J.T. & Bannister R.C.A. (1994) Re-stocking and enhancement of clawed lobster stocks: a review. *Crustaceana*, **67** (2) 131–155.

Aiken D.E. & Waddy S.L. (1989) Culture of the American lobster, *Homarus americanus*. In: *Cold-water Aquaculture in Atlantic Canada* (ed. A.D. Boghen), pp. 79–122. The Canadian Institute for Research on Regional Development, Moncton, Canada.

Barlow S. (2000) Fishmeal and fish oil: sustainable ingredients for aquafeeds. *Global Aquaculture Advocate*, **3** (2) 85–88.

Devresse B. (1998) Nutrition and health: the nutriceutical approach. *International Aquafeed Directory 1997/8*, pp. 51–59.

FAO (2000) http://www.fao.org/waicent/faoinfo/fishery/statist/fisoft/fishplus.htm *(apud* FAO (2000) *FAO yearbook, Fishery statistics, Capture production 1998*. Vol. 86/1 and FAO (2000) *FAO yearbook, Fishery statistics, Aquaculture production 1998*. Vol. 86/2.)

Gardner C. & Northam M. (1997) Use of prophylactic treatments for larval rearing of giant crabs *Pseudocarcinus gigas*

(Lamarck). *Aquaculture*, **158** (3–4) 203–214.

Goldberg H. & Zabradnik J.W. (1984) The feasibility of the gooseneck barnacle *Lepas anatifera* as a candidate for mariculture. *Journal of Shellfish Research*, **4** (1) 110–111.

Holdich D.M. (1993) A review of astaciculture: freshwater crayfish farming. *Aquatic Living Resources*, **6**, 307–317.

Holmes N. (1995) Coastal dynamics and global change: implications for coastal management. In: *Coastal Management in the Asia-Pacific Region: issues and approaches* (eds K. Hotta & I.M. Dutton), pp. 81–93. Japan International Marine Science and Technology Federation, Tokyo, Japan.

Huner J.V. (1999) The fate of the Louisiana soft-shell crawfish. *Aquaculture Magazine*, **25** (3) 46–51.

Huner J.V. (2001) *Procambarus*. In: *Biology of Freshwater Crayfish* (ed. D.M. Holdich), pp. 541–81. Blackwell Science, Oxford, UK.

Jeffs A. & Hooker S. (2000) Economic feasibility of aquaculture of spiny lobsters *Jasus edwardsii*, in temperate waters. *Journal of the World Aquaculture Society*, **31** (1) 30–41.

Jeffs A. & James P. (2000) Cage culture of the spiny lobster *Jasus edwardsii* in New Zealand. In: *Book of abstracts, World Aquaculture '99*, 26 April–2 May 1999, Sydney, Australia, p. 309. World Aquaculture Society, Baton Rouge, LA, USA.

Jones C.M. (1990) *The biology and aquaculture potential of the tropical freshwater crayfish* Cherax quadricarinatus, 109 pp. Information Series QI 90028, Queensland Department of Primary Industry, Australia.

Keefe A.M. & Jolly C.M. (2000) Price flexibility and change in international shrimp supply. *Aquaculture Magazine*, **26** (4) 26–34.

Kittaka J. (2000) Culture of larval spiny lobsters. In: *Spiny Lobsters: fisheries and culture,* 2nd edn (eds B.F. Phillips & J. Kittaka), pp. 508–532. Fishing News Books, Oxford, UK.

Kittaka J., Ono K. & Booth J.D. (1997) Complete development of the green rock lobster, *Jasus verreauxi* from egg to juvenile. *Bulletin of Marine Science*, **61** (1) 57–71.

Lavens P. & Sorgeloos P. (2000) The history, present status and prospects of the availability of *Artemia* cysts for aquaculture. *Aquaculture*, **181** (3–4) 397–403.

Lellis W.A. (1992) A standard reference diet for crustacean nutrition research VI. Responses of postlarval stages of the Caribbean king crab *Mithrax spinosissmus* and the spiny lobster *Panulirus argus*. *Journal of the World Aquaculture Society*, **23** (1) 1–7.

Lewis S.D. (2001) *Pacifastacus*. In: *Biology of Freshwater Crayfish* (ed. D.M. Holdich), pp. 511–40. Blackwell Science, Oxford, UK.

Lutz C.G. (2000) The suffering Louisiana crawfish industry. *Aquaculture Magazine*, **26** (4) 48 54.

Murthy H.S. (1998) Freshwater prawn culture in India. *Infofish International*, (5) 30–36.

NACA (1994–95) *Fish health management in Asia*, 18 pp. (mimeo). Programme proposal to the Office Internationale des Épizooties by Network of Aquaculture Centres in Asia-Pacific.

New M.B. (2000) History and global status of freshwater prawn farming. In: *Freshwater Prawn Culture: the farming of* Macrobrachium rosenbergii (eds M.B. New & W.C. Valenti), pp. 1–11. Blackwell Science, Oxford, UK.

New M.B., Singholka S & Kutty M.N. (2000) Prawn capture fisheries and enhancement. In: *Freshwater Prawn Culture: the farming of* Macrobrachium rosenbergii (eds M.B. New & W.C. Valenti), pp. 411–428. Blackwell Science, Oxford, UK.

Pantoja C.R., Lightner D.V. & Holtschmit K.H. (1999) Prevalence and geographic distribution of infectious hypodermal and hematopoietic necrosis virus (IHHNV) in wild blue shrimp *Penaeus stylirostris* from the Gulf of California, Mexico. *Journal of Aquatic Animal Health*, **11** (1) 23–24.

Pérez J.R., Carral J.M., Celada J.D., Sáez-Royuela M., Muñoz C. & Sierra A. (1997) Current status of astaciculture production and commercial situation of crayfish in Europe. *Aquaculture Europe*, **22** (1) 6–13.

Rogers W.D. & Holdich D.M. (1995) Crayfish production in Britain. In: *Freshwater Crayfish 10* (eds M.C. Geddes, D.R. Fielder & A.M.M. Richardson), pp. 583–595. Louisiana State University, Baton Rouge, LA, USA.

Rosenberry R. (1998) *World Shrimp Farming 1998*, 328 pp. Shrimp News International, **11**, Rosenberry, San Diego, USA.

Rosenberry R. (1999) *World Shrimp Farming 1999*, pp. 320. Shrimp News International, **12**, Rosenberry, San Diego, USA.

Rouse D.B. (1995) Australian crayfish culture in the Americas. *Journal of Shellfish Research*, **14** (2) 569–572.

Shimizu C., Shike H., Dhar A.K., Klimpel K.R. & Burns J.C. (2000) Pantropic retroviral vectors mediate foreign gene expression in shrimp (*Penaeus stylirostris*). In: *Abstracts, Aqua 2000, Responsible aquaculture in the new millennium* (compiled by R. Flos & L. Creswell), p. 646. European Aquaculture Society, Special Publication No. 28.

Skurdal J. & Taugbøl T. (2001) *Astacus*. In: *Biology of Freshwater Crayfish* (ed. D.M. Holdich), pp. 467–510. Blackwell Science, Oxford, UK.

Subasinghe S. (2000) Meeting challenges in aquaculture. *Infofish International*, (2) 3.

Tapay L.M. & Loh P.C. (1999) The antiviral properties of a transformed shrimp lymphoid cell line (OkTr). In: *Book of Abstracts, World Aquaculture '99*, 26 April–2 May 1999, Sydney, Australia, p. 466. World Aquaculture Society, Baton Rouge, LA, USA.

Wang G. & Qianghong S. (1999) Culture of freshwater prawns in China. *Aquaculture Asia*, **4** (2) 14–17.

Weaver A.J. & Green C. (1998) Global climate change: lessons from the past – policy for the future. *Ocean and Coastal Management*, **39**, 73–86.

Chapter 2
Biology

2.1 Terminology

The common names shrimp, prawn, lobster, spiny lobster and crayfish are traditionally applied to different species in different parts of the world. For example, in Britain and Australia, crayfish are freshwater crustaceans but in the USA they are often called crawfish. In Britain, the term crawfish is restricted to members of the Palinuridae but in this book we adopt the more widely used name of spiny lobster for this group. In Australia, spiny lobsters are also known as rock lobsters and in New Zealand as crayfish. The marine and brackish water Penaeidae are called shrimp in the USA, prawns in Australia, India and South Africa while either term may be used in Japan and Taiwan. In Great Britain, small specimens are called shrimp, large specimens, prawns, while in the USA prawn is the name given to the large freshwater carideans of the genus *Macrobrachium*. The FAO convention is to call marine- and brackish- water forms, shrimp; freshwater forms, prawns.

To help clarify the situation, at least as far as readers of this book are concerned, the names and relationships of the main cultivable groups are given in Fig. 2.1. Changes to the scientific names of many farmed penaeid shrimp have been proposed (Pérez Farfante & Kensley 1997) and are being used increasingly in the literature. The proposed changes better convey the interrelationships among penaeids and are used throughout this book (Table 2.1). In addition, the widely cultured mud crab *Scylla serrata* is now believed to embrace some four species (Keenan *et al.* 1999), and for simplicity, will be referred to in this book mainly as *Scylla* spp. Both scientific and common names, as well as key attributes of commercially important species reared for the table, are included in Appendix 1. The majority of species are classified in the order Decapoda of the class Crustacea and are characterised by having five pairs of walking legs, the first often bearing substantial chelae as in the case of clawed lobsters and crayfish. Major features of decapod anatomy relevant to the understanding of their culture biology are shown in Fig. 2.2. A glossary of scientific and technical terms is included at Appendix 3.

Table 2.1 Proposed name changes for farmed shrimp (Pérez Farfante & Kensley 1997).

Proposed name	Present name
Farfantepenaeus aztecus	*Penaeus aztecus*
Farfantepenaeus brasiliensis	*Penaeus brasiliensis*
Farfantepenaeus californiensis	*Penaeus californiensis*
Farfantepenaeus duorarum	*Penaeus duorarum*
Farfantepenaeus notialis	*Penaeus notialis*
Farfantepenaeus paulensis	*Penaeus paulensis*
Farfantepenaeus subtilis	*Penaeus subtilis*
Fenneropenaeus chinensis	*Penaeus chinensis*
Fenneropenaeus indicus	*Penaeus indicus*
Fenneropenaeus merguiensis	*Penaeus merguiensis*
Fenneropenaeus penicillatus	*Penaeus penicillatus*
Litopenaeus occidentalis	*Penaeus occidentalis*
Litopenaeus schmitti	*Penaeus schmitti*
Litopenaeus setiferus	*Penaeus setiferus*
Litopenaeus stylirostris	*Penaeus stylirostris*
Litopenaeus vannamei	*Penaeus vannamei*
Marsupenaeus japonicus	*Penaeus japonicus*
Melicertus kerathurus	*Penaeus kerathurus*
Melicertus latisulcatus	*Penaeus latisulcatus*
Melicertus plebejus	*Penaeus plebejus*
Pleoticus muelleri	*Hymenopenaeus mülleri*
No name change	
Penaeus esculentus	
Penaeus monodon	
Penaeus semisulcatus	

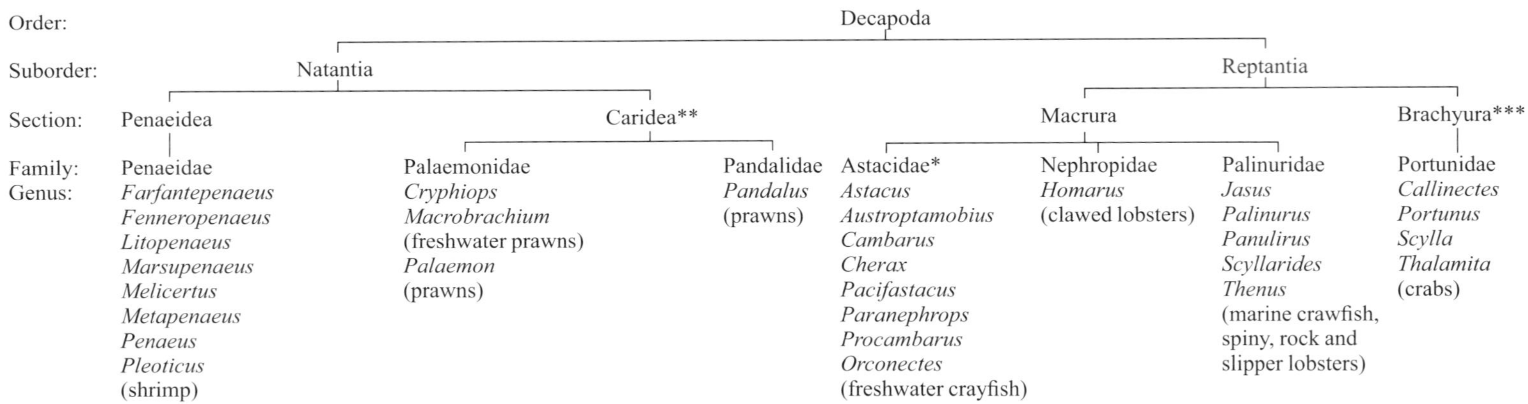

Fig. 2.1 Classification of cultivable decapod Crustacea. *Strictly, cultivable crayfish fall into three families: *Astacidae, Cambaridae* and *Parastacidae*. **Other carideans include *Atya, Lysmata, Sclerocrangon* and *Stenopus*. ***Other brachyurans include *Cancer, Eriocheir, Mithrax* and *Pseudocarcinus*.

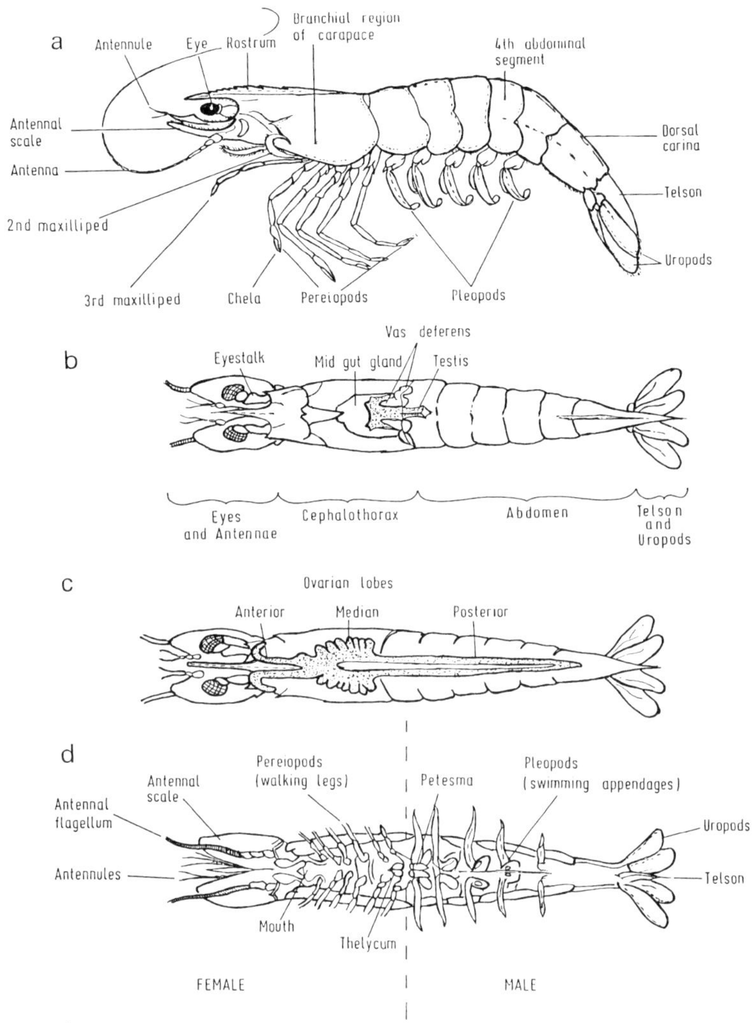

Fig. 2.2 The generalised anatomy of a penaeid shrimp: (a) lateral view; (b) dorsal view of male; (c) dorsal view of female with ripe ovary; (d) ventral view to show the position of the copulatory structures in both males and females (after Wickins 1976).

2.2 Life history

In order to appreciate the different culture conditions required for farmed crustaceans, knowledge of the different life cycles involved is important (Figs. 2.3a–e). The account that follows contains information needed to understand the concepts discussed in later chapters. More detailed treatment of crustacean anatomy, physiology, biochemistry, ecology and behaviour may be found in Bliss (1980–1985) for crustaceans in general; Dall *et al.* (1990) and Fast and Lester (1992) penaeid shrimp; New (1995) and Ismael and New (2000) freshwater prawns; Holdich and Lowery (1988) and Holdich (2001) freshwater crayfish; Cobb and Phillips (1980) and Factor (1995) clawed lobsters; Phillips and Kittaka (2000) for spiny lobsters and Haefner (1985) for crabs.

The sexes are separate in most cultivated decapods, although occasionally individuals in an intersex or hermaphrodite condition are found. A few species, for example the spot prawn, *Pandalus platyceros*, change sex at some time during their lives. Recently a new condition, that of simultaneous hermaphroditism, has been reported for an ornamental shrimp (*Lysmata* spp.) (Bauer & Holt 1998) (section 7.4.4.1). Some crustaceans, for example North American crayfish males (*Procambarus, Orconectes*), exhibit distinct morphological changes during the mating season.

In mature decapods (section 2.3) mating generally occurs when the female is in a soft-shelled condition (i.e.

Fig. 2.3 The generalised life cycles of crustaceans. Typical changes in body form expressed during development of the main cultivated groups are shown. The duration and number of moults vary with species and with temperature but ranges for the species listed in Tables 4.6e–h (reading clockwise from the adult) are shown in these figures.

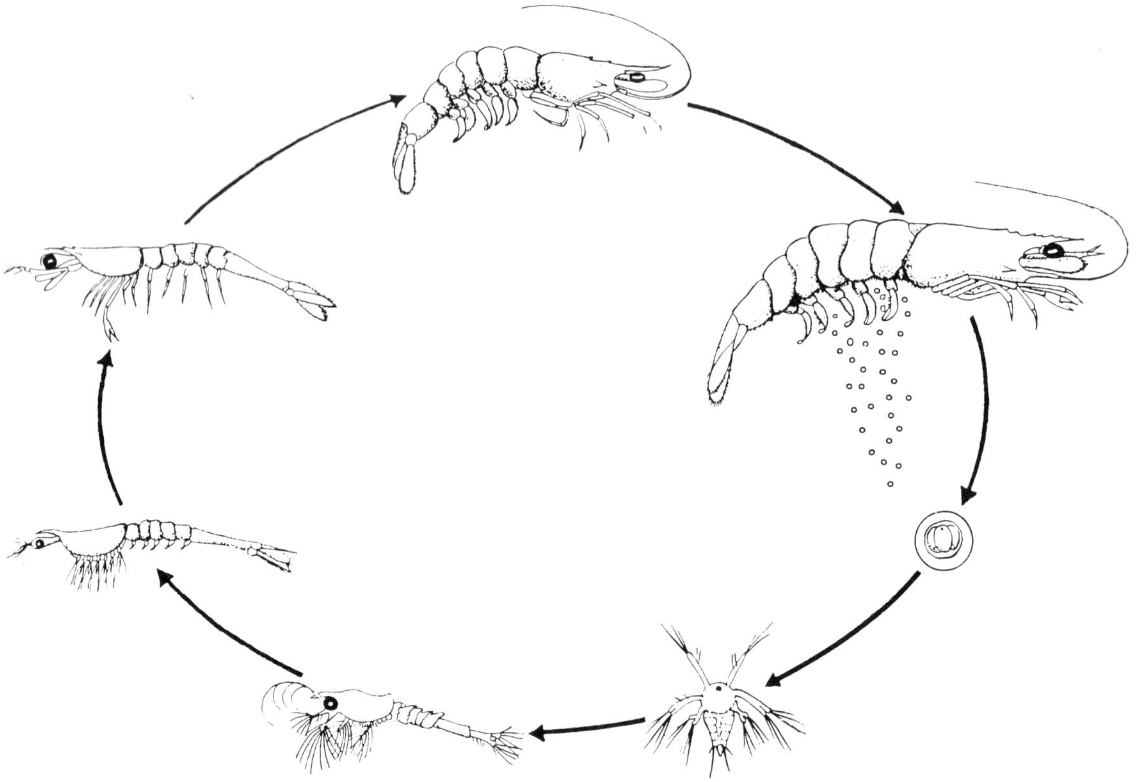

Fig. 2.3a Penaeid shrimp: egg (up to 24 h); nauplius (5–6 instars in 2–3 days); protozoea (3 instars in 3–4 days); mysis (3 instars in 3–5 days); post-larva (3–35 days nursery culture); juvenile to maturity (180–300 days).

newly moulted) and results in the deposition of one or more spermatophores (containing many sperm) in, on or close to the genital openings of the female. Spermatophore deposition occurs during apposition of the ventral surfaces although mating positions vary widely. For example, *Macrobrachium rosenbergii* mate, ventral surfaces together, with the female normally underneath (Ismael & New 2000) but in sneak matings, the pair can be side by side (Karplus *et al.* 2000). This contrasts with redclaw crayfish (*Cherax quadricarinatus*) where the female likes to be on top (Merrick & Lambert 1991) while in *Penaeus monodon* the male curls himself orthogonally around the female to complete the mating act (Dall *et al.* 1990). Mating with hard-shelled females occurs in spiny lobsters and in penaeid shrimp that have 'open' thelyca (see Glossary). These crustaceans may rely on bristles or cement to hold the spermatophores in place externally. Clawed lobsters and certain species of penaeid shrimp have a 'closed' thelycum or pouch to retain the spermatophores until spawning occurs. In captive clawed lobsters (*Homarus americanus*) successful intermoult mating occurs regularly when suitably sized males and females are held together without shelters.

Spawning is the release of eggs either directly into the sea in the case of penaeid shrimp, or to the brood chamber beneath the abdomen in all other farmed groups. The eggs are fertilised as they are spawned but in species with 'internal' sperm storage this may occur several hours or even months after mating according to species. Sperm from one mating are sufficient to fertilise more than one batch of eggs in lobsters, spiny lobsters and penaeid shrimp with 'closed' thelyca. Artificial spermatophore extraction, impregnation of females and subsequent fertilisation has been accomplished (section 8.10.1.4) and

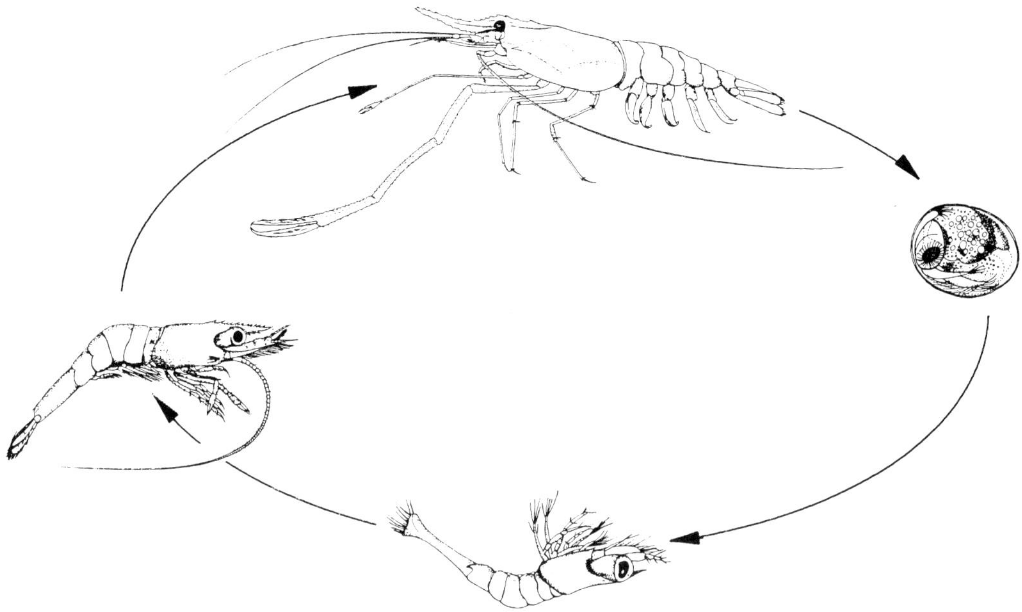

Fig. 2.3b Caridean prawns and lobsters (the illustration shows *Macrobrachium rosenbergii* and the adult is a blue claw male): egg (incubated by the female for 21–25 days); zoea (5–12 instars in 20–40 days); post-larva/juvenile to maturity (120–210 days).

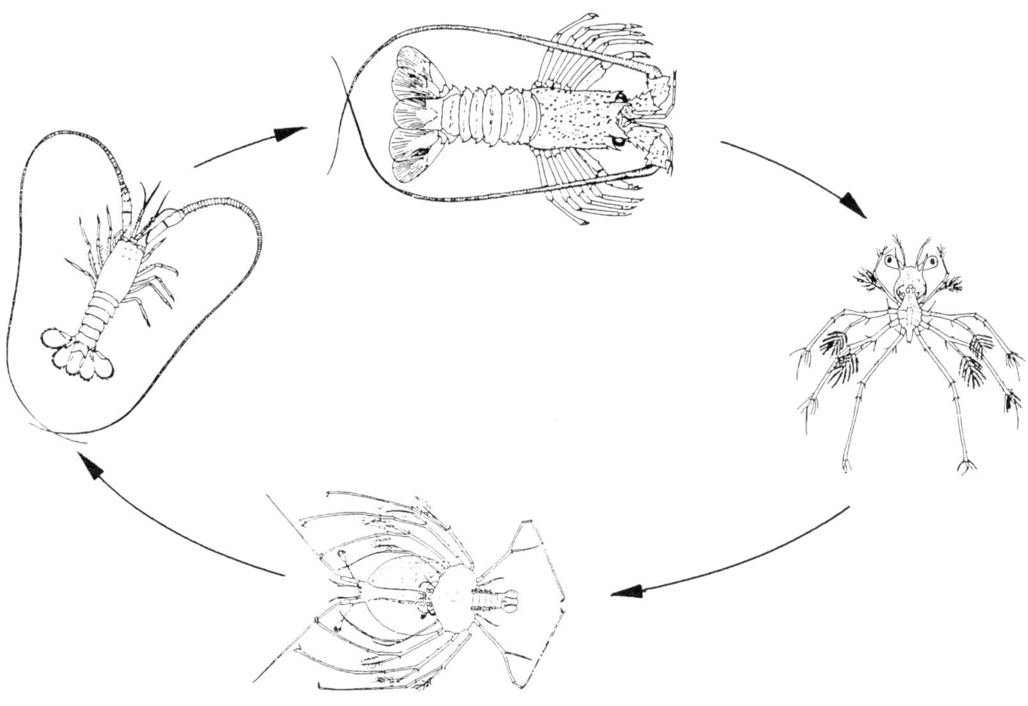

Fig. 2.3c Spiny lobster: egg (not shown) incubated by female for 7–180 days; early and late phyllosoma larvae (9–25 instars in 65–391 days); puerulus (1 stage in 7–56 days); juvenile (not shown) to maturity (730–1460 days).

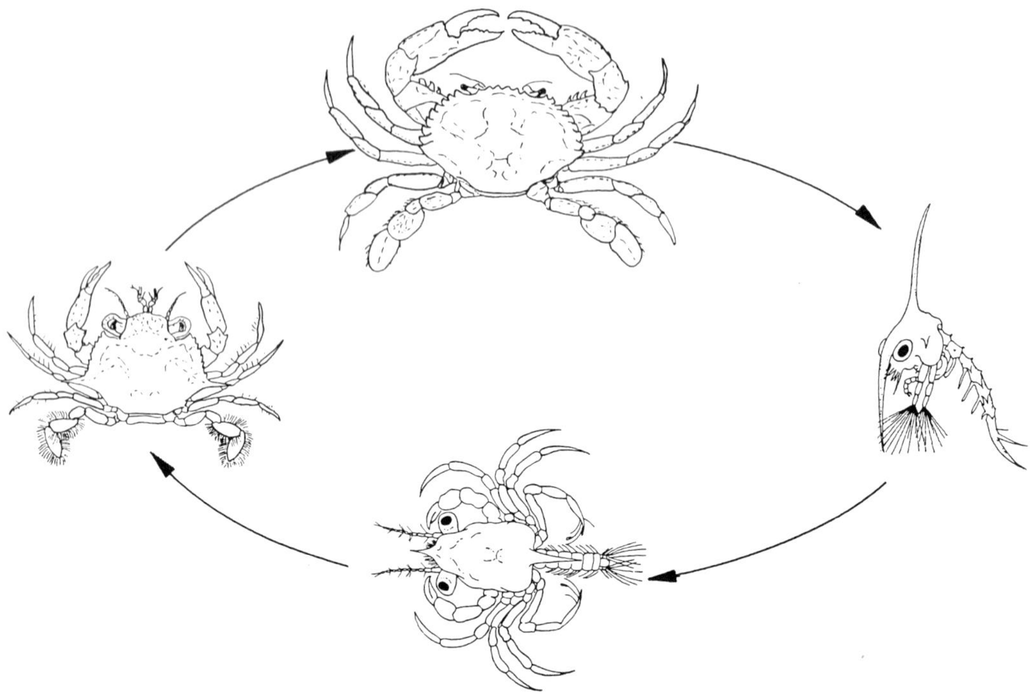

Fig. 2.3d Crab: egg (not shown) incubated by female for 6–25 days; zoea (4–7 instars in 12–24 days); megalopa (1 stage in 5–7 days); first crab stage to maturity (120–540 days).

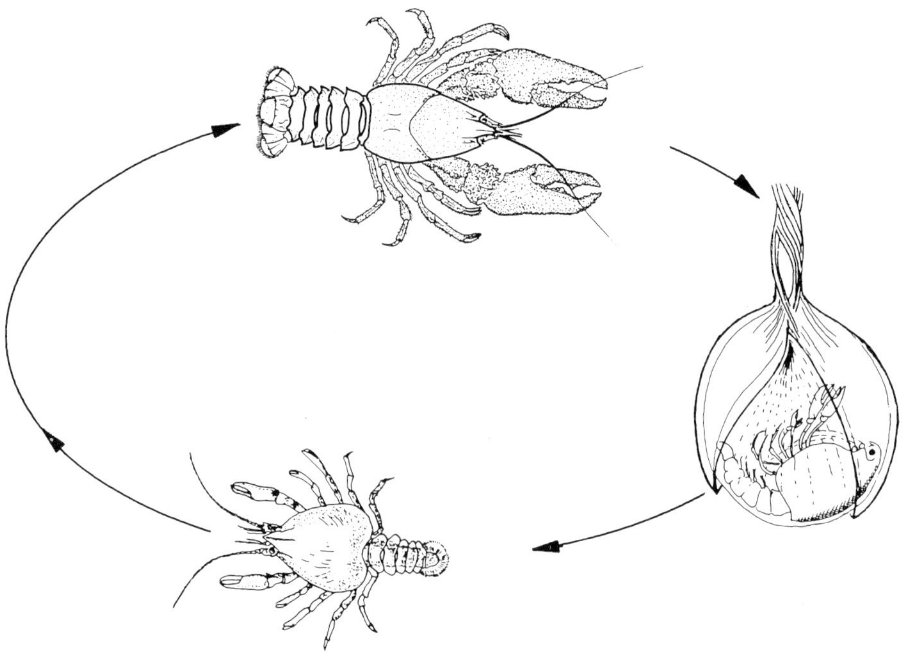

Fig. 2.3e Freshwater crayfish: egg (incubated by female for 7–180 days); juvenile (clings to female until second or third moult usually 7–30 days); juvenile to maturity (90–1095 days).

has permitted hybridisation of shrimp (Lester & Pante 1992); prawns (Karplus *et al.* 2000); lobsters (Talbot & Helluy 1995) and closely related spiny lobsters (MacDiarmid & Kittaka 2000) (section 2.6.3).

Penaeid eggs hatch a few hours after spawning and each larva is left to fend for itself as it develops through about 12 free-swimming planktonic instars through the nauplius, protozoea and mysis stages before metamorphosing into a post-larva (Fig. 2.3a). Egg incubation in the non-penaeid decapods lasts from a few weeks in prawns to as long as 4–9.5 months in lobsters. Throughout this time the female tends and ventilates the clutch until hatching occurs. Substantial egg losses sometimes occur in laboratory-held lobsters due to unsuitable temperature and salinity regimes (Wickins *et al.* 1995), nemertean infestations (Kuris 1991), disturbance during spawning and attachment or abnormal egg tending behaviour by the female (Talbot & Helluy 1995). During incubation the early nauplius and protozoea stages are often by-passed in the egg so that when hatching occurs the larvae are sufficiently advanced to be able to catch and feed on zooplankton almost immediately. The most extreme cases of abbreviated development occur among the crayfish where there is no free-living larval phase and post-larvae hatch directly (Fig. 2.3e). The post-larvae cling to their mother for the first one to three moults, nourished by internal yolk until they are able to begin foraging for food. All farmed crustaceans are cannibalistic and unless the young can escape the mother they risk being eaten.

The larvae of farmed crustaceans exhibit a wide range of feeding habits which often changes as they progress through each developmental stage. In penaeids, the nauplii feed on internal stores of yolk while the protozoea stages filter unicellular algae from the water. The mysis stages feed voraciously on zooplankton (rotifers, *Artemia* nauplii, copepods) and in this respect they are like the larvae of caridean prawns and lobsters (Lavalli & Factor 1995) (Fig. 2.3b), spiny lobsters (Fig. 2.3c) and crabs (Fig. 2.3d). There is some evidence that particulate and dissolved organic matter can supplement the diet of spiny lobster (*Panulirus japonicus*) phyllosoma larvae (Souza *et al.* 1999).

Immediate post-metamorphic juveniles (post-larvae) begin to develop the feeding habits they will need for adult existence. At this stage most species are omnivorous (though they may also be selective in what they eat), and crayfish, shrimp and juvenile lobsters in particular can exist for considerable periods browsing on detritus and microscopic organisms in the substrate. Newly settled clawed lobsters are also well equipped to filter small particles (1 mm or less) that become suspended in the water during their burrowing activities (suspension feeding) as well as to take larger organisms raptorially (Lawton & Lavalli 1995). Indeed, all farmed crustaceans seem equally able to browse on detritus and benthic micro-organisms, scavenge non-living material and become active (and selective) predators during the ongrowing phases. All species can also become cannibalistic when overcrowded, or underfed, and newly moulted individuals are particularly at risk. In nature, crayfish consume a higher proportion of vegetable material than lobsters, prawns and most shrimp.

Plate 2.1 Courtship behaviour in redclaw crayfish (*Cherax quadricarinatus*). (Photo courtesy Clive Jones, Department of Primary Industries, Queensland, Australia.)

2.3 Moulting, growth, maturation and excretion

The external shell (exoskeleton) of crustaceans is made up of a basement membrane, cellular epidermis and cuticle. The cuticle is composed of four layers, two of which are calcified matrices of chitin (section 2.4.3) and protein. The cuticle is capable only of limited expansion. Growth occurs through moulting (shedding the exoskeleton or ecdysis) at intervals throughout life. The rate of growth is a function of the frequency of moulting and the increase in size at each moult. Adverse nutritional or environmental conditions can decrease both functions. The main sequence of events in the cycle is:

(1) Accumulation of mineral and organic reserves;
(2) Removal of material from the old shell and formation of the new exoskeleton;
(3) Ecdysis (moulting) accompanied by an uptake of water;
(4) Molecular strengthening of the exoskeleton by rearrangement of organic matrices and deposition of inorganic salts;
(5) Replacement of fluid by tissue growth.

The frequency of moulting varies naturally between species, with size and with age. Young shrimp larvae moult two or three times in a day, juveniles every 3–25 days depending on temperature and species, while adult lobsters and crayfish may only moult once every one or two years. Crustaceans often eat cast shells, a convenient source of minerals that would otherwise be lost. Mineralisation of the new shell is affected by the availability of particular ions (calcium, bicarbonates and pH) in the surrounding waters, in the diet, and especially in freshwater animals, from materials stored in the body, e.g. gastroliths, prior to moulting (sections 2.4.5, 3.3.3.2 and 7.3.5.3). The changes that arise in water composition during intensive culture and particularly in densely stocked recirculation systems can have a major effect on the mineralisation process and on the animal's ability to control blood pH (sections 8.4.4 and 8.4.5).

Newly moulted individuals are particularly vulnerable to cannibalism especially under crowded culture conditions. Neither the presence of shelters nor the availability of adequate food eliminates cannibalism although their absence may increase it. Attempts to reduce cannibalism by synchronising moulting, ameliorating aggressive behaviour through claw removal or by giving drugs have met with little success. Apart from the specialised markets for soft-shelled crayfish and crabs, only hard-shelled crustaceans fetch worthwhile prices. It is therefore necessary to minimise the proportion of the population moulting at the time of harvesting. Species grown in outdoor ponds may tend to moult in phase with the lunar cycle or in response to a change of water. Such effects are taken into account when the decision to harvest is made (section 7.2.7).

The moulting cycle and sexual maturation are two vital physiological processes influenced by a complex of glands situated in crustacean eyestalks. Other organs exerting hormonal control over these processes include the thoracic ganglia, Y-organ, androgenic gland, the mandibular organ and parts of the brain. The ability to interfere with normal moulting has implications both for increasing growth rate and for the commercial production of soft-shelled individuals (sections 5.6, 7.5.8 and 12.8.6). Control over maturation is important for the production of domesticated strains (sections 2.6.2, 8.10.1 and 12.3) and for the culture of species outside their natural range (sections 4.2, 4.3.2, 11.3.2 and 12.4). For further reading on crustacean endocrinology the reviews of Quackenbush (1986) for crustaceans in general; Beltz (1995) and Waddy *et al.* (1995) for clawed lobsters; Chang (1992) and Huberman (2000) for penaeid shrimp and Vogt (2001) for crayfish are recommended.

Surgical removal of both sets of eyestalk glands and the resulting decrease in the moult-inhibiting hormone levels substantially increases growth in lobsters (Koshio *et al.* 1989) and spiny lobsters (Radhakrishnan & Vijayakumaran 1984), provided the diet is sufficient to support the extra growth. However there is evidence that ablation may also accelerate a change from gregarious to solitary behaviour in some species, e.g. *Panulirus homarus*, with obvious implications for their communal culture (Radhakrishnan & Vijayakumaran 1998). In the long term, however, bilaterally ablated lobsters are vulnerable to stress and do not survive well. Unilateral ablation gives increased growth without undue mortality in spiny lobsters (*Panulirus ornatus*) (Juinio-Meñez & Ruinata 1996) but does not enhance growth to the same extent in clawed lobsters (Waddy *et al.* 1995). In the crab *Portunus pelagicus*, unilateral ablation increased moulting frequency and dry meat yield (Germano 1994). The effect in *Macrobrachium rosenbergii* depends on an individual's rank within the social hierarchy of the population (section 7.3.7), increasing growth in 'laggards' but not in 'jumpers' (Karplus & Hulata 1995).

Removal of only one set of the glands from *Macrobrachium* spp. increases growth and fecundity (section 7.3.2) while in penaeid shrimp that do not readily mature in captivity unilateral ablation is sufficient to reduce the circulating gonad-inhibiting hormone to levels that permit rapid maturation in adults (section 7.2.2.5). The effect is similar though less reliable in lobsters (Waddy *et al.* 1995) and in redclaw crayfish may only work with females that have not previously spawned (Sagi *et al.* 1997a). Ablation may also affect hormones involved in the mobilisation of food reserves and lead to reduced egg quality. Other means of inducing maturation less likely to have unwanted side effects include control of light intensity, wavelength and photoperiod, usually in association with increases in temperature. Positive effects seem to occur more readily in *Fenneropenaeus chinensis* which migrates to spawn in response to changing day length and temperature, and in burrowing species like *Marsupenaeus japonicus*. In some other species (including *Penaeus monodon*) the effects of these triggers are minimal but are slightly enhanced when a penetrable substrate is present to reinforce cyclic behaviour patterns. Water quality also seems important and dissolved organic matter, low pH and calcium levels can inhibit maturation. Alternatively, placing *Marsupenaeus japonicus* in seawater previously exposed to ultraviolet irradiation can induce maturation.

Transplantation of the male androgenic gland into young females has been used to induce sex reversal (section 2.6.3) and along with eyestalk ablation, such surgical techniques provide researchers with useful tools in the search for factors that might ultimately be useful for the less traumatic control of crustacean growth and reproduction (sections 7.2.2.5 and 12.4). In the meantime, eyestalk ablation is widely used to achieve immediate commercial objectives.

The gills are a crustacean's major effectors of gaseous exchange (oxygen, carbon dioxide), while, additionally, the gut and antennal glands are involved in regulating body salts and water, acid–base balance and nitrogenous excretion. The gills are situated beneath the lateral portions of the carapace and are kept irrigated by the beating of blade-like scaphognathites. Oxygen is taken up from the circulating water across the gill surface and binds to the respiratory pigment haemocyanin for distribution throughout the body. Carbon dioxide and ammonia are eliminated largely, but not solely, across the gill membranes where selective transport processes also occur to ensure regulation of internal ionic content, water and acid–base balance. The ability to regulate varies with species and life cycle stage, for example, estuarine penaeids are markedly more tolerant of salinity fluctuations than lobsters (section 4.6.2). For further reading on this aspect of crustacean physiology the reviews of Dall *et al.* (1990) for shrimp, McMahon (1995) for clawed lobsters and McMahon (2001) for crayfish are recommended.

2.4 Nutrition

This section summarises selected features of crustacean nutrition that are considered relevant to understanding project resource needs. Although our understanding is not as advanced as it is for chicken and trout, it is now well known that the dietary requirements of crustaceans differ in several important ways from those of farmed fish, birds and mammals. Also, most farmed crustaceans manipulate and fragment their food with their claws and mouthparts outside the body prior to ingestion; a habit that is particularly wasteful of pelleted feeds since it aids leaching, and creates numerous small particles that are scattered by water currents coming from the gills (section 8.3.6.3). It is important therefore that pelleted rations are properly formulated, processed and bound (section 8.8.2). Crustaceans are, however, well equipped with digestive enzymes that allow a wide range of food organisms to be exploited. Good reviews of their digestive processes include those for crustaceans in general (Ceccaldi 1998); shrimp (Dall 1992); freshwater crayfish (Vogt 2001) and clawed lobsters (Conklin 1995).

In the 1970s and 1980s most key studies on nutritional requirements were made with *Marsupenaeus japonicus* (Kanazawa 1985). Much work was also done with *Macrobrachium rosenbergii* (Fox *et al.* 1994), *Homarus* spp. (Conklin 1995) and crayfish (D'Abramo & Robinson 1989) but now, increasingly, the requirements of other species, e.g. *Litopenaeus vannamei*, *Penaeus monodon*, *Procambarus clarkii* (Jover *et al.* 1999) and *Eriocheir sinensis* (Mu *et al.* 1998), are being investigated. Following the exhortations of New (1976), the adoption of standardised experimental approaches to nutrition studies in recent years has enhanced comparability between species and researchers, leading to greater efficiency in international research effort (D'Abramo & Castell 1997).

The paragraphs that follow give a brief outline of crustacean nutrition. For further reading and comprehensive academic reviews of the subject, useful publications have been prepared for crustaceans in general (D'Abramo *et al.* 1997; Stickney 1998); shrimp (Tacon 1990; Shiau 1998); freshwater prawns and shrimps (Fox *et al.* 1994; D'Abramo & Conklin 1995); freshwater prawns (D'Abramo & Sheen 1994; New 1995;

D'Abramo & New 2000); freshwater crayfish (Brown 1995; Nyström 2001); clawed lobsters (Conklin 1995); crustacean larvae (Jones *et al.* 1997a,b) and spiny lobsters (Kanazawa 2000). The special nutritional requirements of crustacean broodstock were reviewed by Harrison (1990, 1997).

2.4.1 Protein

The protein requirement of crustaceans tends to be higher than that of land animals, possibly because the natural aquatic environment combines an abundance of proteins and lipids with a relative scarcity of carbohydrates. Protein is not only used for tissue growth but can also be used by many species as an energy source. Because of this ability, increasing protein levels in a diet beyond that needed to satisfy basic requirements can improve growth. Protein is, however, expensive. Provided the diet contains sufficient quantities of the ten essential amino acids required by most decapods in a readily assimilable form, the gross protein content can be reduced by substituting an alternative source of energy (e.g. carbohydrate or lipid). A correct balance is essential since at moulting, a crustacean can lose 50–80% of its body protein, some of which it recovers, together with other valuable nutrients, by eating the cast shell. Previously reported optimum protein levels in diets ranged from about 30% to 60% but more recent re-evaluations made using diets containing more easily digested proteins, combined with alternative energy sources, indicate that many of these may be overestimates. For redclaw crayfish, for example, diets containing 20–30% protein and 5–10% lipid and based largely on plant rather than animal material are proving cost-effective. Low protein diets containing a high carbon to nitrogen ratio can be used in zero water exchange systems with some shrimp species (e.g. *Litopenaeus vannamei*) because they promote the development of bacterial flocs that provide the remainder of the protein requirements (sections 7.2.6.6 and 8.3.7). Protein digestibility is around 75–93% but in *Homarus gammarus*, *Palaemon serratus* and *Litopenaeus vannamei*, for example, inherent digestive enzyme activity has been found to adapt to diet composition (Guillaume 1997). The environment may also influence apparent protein requirements. For example, *Penaeus monodon* when reared under low salinity conditions (Shiau 1998) and *Litopenaeus vannamei* when grown in high salinity water (Robertson *et al.* 1993) benefited from an elevated level of dietary protein, possibly because they were using it as an energy source in osmoregulation. It appears from dietary and physiological studies (Jones *et al.* 1997a,b) that protein is by far the most important component in larval nutrition.

The ten essential amino acids are the same for most farmed crustaceans: arginine, histidine, isoleucine, leucine, lysine, methionine, phenylalanine, threonine, tryptophan and valine, although the ideal quantitative distribution for different species has yet to be determined. Tyrosine and cystine are semi-essential and can reduce the requirement for phenylalanine and methionine in diets, but some freshwater crayfish also require aspargine. Microencapsulation techniques are providing a useful tool for the elucidation of quantitative requirements for amino acids but it is still not clear if the reported differences in protein requirements between species result from different digestibilities of the protein and/or energy source, the extent to which protein is utilised as an energy source, or from a need for a high level of a particular amino acid(s) (Guillaume 1997). Some protein sources with good amino acid profiles, e.g. crab, are readily digestible, others (mussel, squid) contain factors that enhance crustacean growth to a greater extent than would be expected from their amino acid profile alone, while some plant protein sources may also contain anti-nutritional factors such as trypsin inhibitors and antivitamins (Fox *et al.* 1994; Chamberlain 1995) (section 8.8.2).

2.4.2 Lipids and sterols

In marine decapod crustaceans, the predominant lipids are neutral, storage lipids (triacylglyceride – TAG – reviewed by D'Abramo, 1997) and polar, membrane and transport lipids including phospholipids (Teshima 1997) and plasma lipoproteins (Yepiz-Plascencia *et al.* 2000). Research since 1980 has highlighted the 'essential' nature of particular lipid components and sterols for most species. These include phospholipids (often included at 2% of the diet) especially phosphatidylcholine, cholesterol – or phytosterols for some freshwater species – (0.5–1%), and certain long chain poly- and highly-unsaturated fatty acids (PUFAs and HUFAs) (see Glossary), which crustaceans have no, or only limited ability to synthesise *de novo*. Recent studies have highlighted the benefits of lipid components such as phosphatidylcholine in increasing resistance to stress (Coutteau *et al.* 1996). Phospholipid components other than phosphatidylcholine can enhance shrimp growth and survival, although the precise requirement for dietary phospholipid (if any) has yet to be determined for *Macrobrachium* and marron. Four fatty acids are particularly important dietary ingredients for crustaceans: linoleic (18:2 n-6), linolenic (18:3 n-3), eicosapentaenoic (20:5 n-3, often referred to as EPA) and docosahexaenoic (22:6 n-3, also called DHA). The latter two n-3 highly unsaturated fatty acids (HUFAs) are the most indispensable and can increase resistance to environmental stress and disease.

The distribution of lipid classes and fatty acid composition in crustaceans varies with habitat (marine or freshwater), temperature, tissue, life cycle and moult stage, and readily reflects the composition of the diet (Clarke & Wickins 1980). The freshwater prawn *Macrobrachium*, for example, contains more n-6 polyunsaturated fatty acids (especially linoleic) than marine shrimp where n-3 HUFAs (especially DHA) predominate. Lipid levels of 2–8% in compounded diets generally meet essential fatty acid and energy requirements but absolute requirements may vary according to species (Merican & Shim 1996) and the level and quality of carbohydrate and protein in the diet, for example, for crayfish (Ackefors *et al.* 1992), shrimps and prawns (Fox *et al.* 1994) and freshwater prawns (New 1995). Most crustaceans seem able to digest lipids efficiently (>94%) (D'Abramo 1998).

2.4.3 Carbohydrates, dietary fibre and chitin

No specific dietary requirement for carbohydrates has been established for crustaceans, yet carbohydrates are cheap by comparison with proteins and lipids and their inclusion in diets is justified because they can reduce the amount of protein utilised for energy (protein sparing; section 2.4.4). Carbohydrates provide a store of energy and also glucosamine, a precursor in the synthesis of nucleic acids and chitin. In nature, *Macrobrachium* utilises carbohydrates for energy to a greater extent than many penaeids where lipid and protein are often more important. Mono- and disaccharides (sugars) are less useful in formulated feeds than polysaccharides such as wheat flour, dextrin and alpha starch, possibly because their rapid absorption into the tissues may exceed the animal's capacity to metabolise them satisfactorily. Digestibility of carbohydrates is generally high, 80–90%, but can vary with the source and degree of gelatinisation after processing (Cousin *et al.* 1996; section 8.8.2.1). Crayfish, such as *Cherax quadricarinatus*, seem well equipped with polysaccharide hydrolases (enzymes which may be of endogenous or exogenous origin, i.e. from gut bacteria), which enable them to utilise plant material easily (Xue *et al.* 1999). There are also differences in specific carbohydrases between species (Omondi & Stark 1995) that could ultimately affect choices (and hence costs) of dietary ingredients. Levels of incorporation are often around 20–40% in compounded feeds and it is suggested that mixtures of complex carbohydrates from different sources could be manipulated to maximise protein sparing opportunities (D'Abramo & Conklin 1995).

Dietary fibre is similar in chemical structure to carbohydrate but differs in its physiological action, being generally resistant to digestive enzymes. The presence of fibre may slow the passage of food through the digestive system, allowing more nutrients to be absorbed, and although it is often used as a filler in compounded diets, little is known of the interactions between fibre and other dietary components, particularly carbohydrates. Knowledge of the digestibility of dietary components by crustaceans is essential if cost-effective diets are to be formulated, and our present awareness has been comprehensively reviewed by Lee and Lawrence (1997).

Chitin forms a major part of the exoskeleton and most crustaceans readily consume their cast shells after moulting. In view of this, the ability to utilise dietary chitin seems surprisingly limited but does vary between species. The enzyme chitinase has been purified from *Homarus americanus* and *Marsupenaeus japonicus* while moderate levels (about 30%) of chitin digestion (either by endogenous secretion of chitinases or by chitinases from gut bacteria) have been reported in *Litopenaeus vannamei, L. setiferus* and *Farfantepenaeus duorarum* (Clark *et al.* 1993). However in laboratory populations of *Penaeus monodon,* chitin digestibility was found to be very limited (Fox 1993). Crustacean processing wastes contain chitin and are often included in compounded diets at levels of 1–2%.

2.4.4 Protein : energy ratios

In general, decapod crustaceans feed to satisfy an energy need (Sedgwick 1979). The protein : energy (P : E) ratio and the actual metabolisable energy content of prepared diets (which, due to different component digestibilities, may be less than the total energy content) are recognised as critical to diet ingestion, utilisation and subsequent cost-effective growth performance (Cuzon & Guillaume 1997). Good results with shrimp and some crayfish have been obtained with dry pelleted diets containing 3–4 kcal g^{-1} (approximately 13–17 MJ gross energy kg^{-1} diet) and a P : E ratio of 0.07–0.10 at 25–35% protein content. Freshwater prawns can fulfil much of their energy needs from lipids and carbohydrates (ratio 1 : 4) and are thus efficient at sparing protein (D'Abramo 1998).

Attempts to spare more protein by increasing dietary energy further may be counter-productive since energy in excess of requirements may prevent the intake of sufficient protein and other nutrients required for growth. High levels of dietary lipid (over 10–12%) or carbohydrate (over 40%) or carbohydrate in the wrong form (e.g. as simple sugars) in the food, can be detrimental.

2.4.5 Vitamins and minerals

It is common practice for vitamin and mineral mixes to be added to diets. Early formulations were unsatisfactory since they were based on mixes used for poultry or mammals. They usually contained soluble vitamins, especially vitamin C (that quickly leached out and was lost, often within a few minutes of immersion), together with high levels of iron and magnesium. Imbalances in the calcium and phosphorus content were also evident. Risk of losses resulted in the practice of adding excessive amounts of micronutrients to compensate, until increasing cost-consciousness stimulated research into the incorporation of insoluble or encapsulated forms of these key nutritional components. Knowledge of the bioavailability or potency of each form is critical to the calculation of the amount to be incorporated in the diet but much remains to be elucidated.

Eleven water-soluble and four fat-soluble vitamins are required by most shrimp and probably many other crustaceans. The 'essential' water-soluble vitamins are thiamin, riboflavin, pyridoxine, pantothenic acid, niacin, biotin, inositol, choline, folic acid, cyanocobalamine, ascorbic acid; fat-soluble vitamins are vitamins A, D, E and possibly K. The role of these vitamins and symptoms of their deficiencies in Crustacea are summarised by Akiyama *et al.* (1992) and Conklin (1997).

Research throughout the 1990s has begun to reveal differences in individual vitamin requirements between penaeid species (Shiau 1998; Reddy *et al.* 1999). Such studies have certainly led to more critical formulations, though much remains to be done particularly in relation to aspects of resistance to stress, moulting and shell mineralisation. Diets without expensive vitamin and mineral supplements can give acceptable results in carefully managed extensive and some semi-intensive systems where requirements are probably met from ingestion of naturally occurring pond organisms.

Minerals are important components of the exoskeleton, enzymes and co-factors, some proteins and in osmoregulation and nerve activity. Unlike terrestrial animals, aquatic crustaceans can utilise minerals dissolved in the water, which makes it more difficult to elucidate their quantitative requirements. Nevertheless, dietary supplements of some minerals do seem important, even in ponds where live crustacean prey are abundant, and especially considering the repeated losses that occur at moulting. Heavily calcified crayfish and lobsters withdraw calcium from the exoskeleton for storage in paired gastroliths (literally, stomach stones – section 3.3.3.2) prior to moulting. Minerals commonly added to diets include calcium, phosphorus, potassium, magnesium, copper, iron, iodine, manganese, zinc, cobalt and selenium. The importance and quantitative requirements of many are uncertain and will vary between individual species and with environmental conditions. Freshwater prawns, for example, benefit from a calcium supplement when grown in soft water but not when alkalinity is high (Zimmermann *et al.* 1994). Marine shrimp do not appear to need supplemental calcium. Carbon dioxide content, alkalinity and pH of the culture water (individually and in their interactions) can all affect calcium uptake and metabolism with consequent effects on shell (exoskeleton) mineralisation in both marine and freshwater species (sections 8.4.4, 8.5 and 8.9). For example, high alkalinity in the water appears to reduce the tolerance of *Macrobrachium rosenbergii* to levels of total hardness over 150 mg L^{-1}. In the laboratory, growth and survival of juvenile prawns were best when total hardness levels were between 25 and 100 mg L^{-1} and alkalinity was below 100 mg L^{-1} (Table 8.3).

Interactions governing bioavailability of nutrients have been reported between minerals and vitamins, minerals and dietary plant material, minerals and commercial binding agents and between minerals themselves (Conklin 1997; Davis & Lawrence 1997). Added calcium can reduce phosphorus bioavailability in feeds and in the past, this has given rise to overestimates of dietary phosphorus requirements. Furthermore, it is important that phosphorus should be kept to a minimum in diets in order to reduce its levels in farm effluents (Davis & Arnold 1998; section 11.4.3). However, much of the phosphorus in grains (67%) is in the form of indigestible phytate phosphorus, and if environmentally acceptable diets are to be prepared using grains, this will need to be released by predigestion (with consequent reduction in added mineral phosphorus). Alternatively, the grain content of the diet must be kept to a minimum. Further, marine species do not seem to need iron or extra magnesium in compounded diets, especially when plant components are present. An interactive effect between magnesium and potassium has also been suggested (Davis &

Lawrence 1997). Copper is a key component of the crustacean respiratory pigment haemocyanin. Since demand is not fully meet by absorption from the water, a dietary source is necessary. However, copper dissolved from metal tank fittings can be toxic (section 8.1.2).

2.4.6 Other additives

A wide range of compounds may be added to compounded diets including: binders to improve pellet stability, marine animal solubles to attract and stimulate feeding, pigments to improve shell and flesh appearance, antibiotics to combat bacterial infections, enzymes and probiotics to aid digestion and compounds such as glucans to stimulate general immune responses (D'Abramo *et al.* 1997; sections 8.3.6.3 and 8.8.2). The pigment astaxanthin, for example, when added at 100 mg kg^{-1} of diet, improved growth and survival of redclaw crayfish juveniles and enhanced the colour of older animals (Rouse & Rash 1999). The effectiveness of additives is, however, often questioned and the risk of residues of some of these additives occurring in product flesh and in farm and hatchery effluents may give cause for concern (sections 3.2.1 and 9.6). Many do not have government approval (e.g. US Food and Drug Administration; Chamberlain 1995), which could increase the vulnerability of the producer or exporter to consignment rejection or embargoes by an importing nation (section 11.3.4).

2.4.7 Broodstock nutrition

Because of our incomplete knowledge of their requirements, most captive broodstock are fed at least a proportion of natural foods during gonad maturation to ensure production of the highest-quality eggs (Browdy 1998). Little opportunity exists for manipulating egg quality through diet when wild-caught, egg-bearing females are used, as is commonly the case with lobsters and crabs. Roughly half the dry weight of a mature egg is protein and lipid in approximately equal amounts. These nutrients, once drawn from reserves in the hepatopancreas during ovarian maturation (vitellogenesis), must be replaced from the diet prior to re-maturation. Following eyestalk ablation to enforce repeated re-maturation (section 7.2.2.5), a high-quality diet becomes of paramount importance, although even under the best feeding regimes, reproductive exhaustion usually occurs, with consequent adverse effects on larval viability (Palacios *et al.* 1999).

The sources and fatty acid profiles of dietary lipids are critical especially in broodstock diets (Harrison 1990, 1997) both for gonad maturation and for the production of eggs in shrimp (Marsden *et al.* 1992) and prawns (Cavalli *et al.* 1999). For example, increased levels of arachidonic acid (20:4 n-6) are beneficial for ovarian maturation. In eggs during incubation and in young during other periods of prolonged starvation, TAG lipid provides a major portion of metabolic energy needs. Hatching in *Homarus gammarus*, for example, results in a drop in the n-3 : n-6 fatty acid ratio in TAG lipids but an increase in the ratio in the phospholipid fraction. These changes highlight the demand for n-3 fatty acids and the conservation of PUFAs in the phospholipid fraction that aid membrane function and flexibility during that critical event. High levels of these components frequently occur in early developing individuals from a brood at egg, larvae and juvenile stages where they provide a good indication of potential for subsequent ongrowing. They are therefore extremely beneficial in diets. However, the excessive utilisation of these important lipids during stressful incubation (e.g. exposure to low salinity), or as a result of an inadequate diet during the larval phase, can greatly reduce survival and predictability of hatchery performance (Wickins *et al.* 1995).

Protein, cholesterol, vitamins, carotenoids and many other dietary components are also critical for broodstock and may be needed at higher levels than are used in feeds for ongrowing. Although the specific requirements for many nutrients have yet to be determined, commercial maturation diets are available. They seldom meet every nutritional need, however, and in practice they are always supplemented with natural diets whose very use poses risks from viral infections, toxins and anti-nutritional factors.

2.4.8 Larvae nutrition

Larval digestive systems are physiologically adapted to the feeding habits at each developmental stage, and knowledge of the corresponding changes that occur in the types and activities of the digestive enzymes are of value in designing artificial feeds (Jones *et al.* 1997a,b). Diet may be used to manipulate enzyme activity in the later larval stages of some species but the relative inflexibility of response in early instars dictates the use of live foods for best results. Indeed for the most part, all larvae grow and survive best on living foods and early *Macrobrachium rosenbergii* larvae are even thought to rely heavily on exogenous sources of digestive enzymes (Jones *et al.* 1993).

Newly hatched *Artemia* nauplii are traditionally used in hatcheries to feed nearly all crustacean larvae, other than the algae-feeding protozoeal instars of penaeids. However, their nutritional quality, availability and size at hatching varies enormously according to their origin. Other non-decapod crustaceans are also grown as live food and several techniques exist for the enrichment of these, and *Artemia*, with selected lipids (especially n-3 highly unsaturated fatty acids – HUFAs) and pigments that increase their nutritional value to larvae. Enriched or not, live foods are expensive and their culture incurs additional facilities and management costs (section 7.11).

Much research effort has been applied to the development of microparticulate and microencapsulated larval feeds during the past ten years and today a wide range of proprietary non-living larvae diets exist (Jones 1998; Teshima *et al.* 2000). While it is often claimed that many of these diets can completely replace living foods, experience indicates that the majority of hatchery operators use them as dietary supplements or as partial replacements, particularly at times when the quality of the cultured food organisms may be suspect (section 7.2.4). Sangha *et al.* (2000) demonstrated the particular value of giving a single dose of live microalgae for the first feed of penaeid protozoea larvae before subsequently feeding microparticulate diets. For the most part, the use of microparticulate diets is restricted to penaeid larvae and late stage carideans, e.g. *Macrobrachium* (Jones *et al.* 1993; Lavens *et al.* 2000). Suspensions of artificial diets in larvae cultures provide a good substrate for bacterial blooms, some of which may provide nutritional benefits while others can be detrimental. In this regard microencapsulated feeds can help to maintain better water quality in that they leach nutrients only slowly. The additional presence of living microalgae has been reported to stimulate larval enzyme secretion, thereby aiding digestion of microencapsulated diets (Kumlu & Jones 1995). The value of microencapsulated diets to penaeid larvae can be further enhanced, at least in the laboratory, through inoculation of cultures with algal exudates or a balanced bacterial population (Alabi *et al.* 1999). Centrifuged or flocculated microalgae concentrates, which can be stockpiled during periods of low hatchery demand, are also being investigated as diets for penaeid larvae, to reduce hatchery costs (D'Souza *et al.* 2000). During the culture of spiny lobster larvae, living microalgae and populations of non-pathogenic (probiotic) bacteria are considered to be important for the maintenance of good water quality (Igarashi & Kittaka 2000) and possibly as feed supplements to the larval diets of sandfish larvae (*Actoscopus japonicus*), *Artemia* and mussel (Kanazawa 2000). Inadequate nutrition or adverse environmental conditions can delay metamorphosis and the resulting post-larvae are often inferior in subsequent growth performance (Wickins *et al.* 1995; Gebauer *et al.* 1999).

2.5 Disease

Throughout the past 15–20 years the global shrimp farming industry has experienced repeated, catastrophic crop failures with consequent severe financial losses. Animals already stressed by high-density culture practices, environmental degradation or inadequate diet readily succumbed to outbreaks of infectious diseases. Most infections were caused by viruses or bacteria; a few had fungal and protozoan aetiologies. The most serious outbreaks were frequently associated with pathogens found over a previously narrow range, but which had been transferred with stock to new areas in which susceptible hosts existed. The fewer outbreaks reported from freshwater prawns, Australian and North American crayfish and clawed and spiny lobsters may in part be due to their culture in smaller quantities and at lower stocking densities than penaeid shrimp. Nonetheless, outbreaks among these other species can be serious: in the early 1990s Taiwanese production of *Macrobrachium* fell by 50%, largely because of disease (New 2000).

In general, disease control is most effective in hatcheries. Beyond the nursery phase, disease prevention rather than control is often the only course of action, and medication seldom appears as an item in published costs. However, considerable financial losses occur during the ongrowing phase as a result of disease and can often be associated with poor diet, water quality or pond bottom chemistry.

In the past, disease outbreaks have been just as devastating to European crayfish and occasionally to other cultured species, but the current huge scale of shrimp farming and international trade in live, fresh and frozen crustaceans has re-emphasised the international need for effective disease diagnosis, prevention and control (sections 8.9.2, 8.9.3 and 8.9.4). Current research focuses on understanding crustacean defence mechanisms, the effect of stressors on them and on developing strategies to promote disease resistance; including genetic selection and engineering. Causes of non-infectious diseases, for example dietary deficiencies or exposure to metabolic wastes and toxic chemicals, are considered briefly in section 8.9.

2.5.1 Defence against infection

A crustacean's first line of defence against microbial invasion is its cuticle. This is physically tough and has antimicrobial properties, for example it contains inhibitors against enzymic attack. If it is penetrated, there is an immediate recognition of the 'non-self' material by haemocytes and plasma proteins (Vargas-Albores & Yepiz-Plascencia 2000). Clotting agents then congeal the haemolymph and attempt to immobilise the invaders pending their destruction. Cells (haemocytes) circulating in the haemolymph can also encapsulate or otherwise trap larger alien particles in melanised nodules (Johansson *et al.* 2000). It is generally believed that crustaceans do not possess an acquired (highly specific, immunoglobulin-mediated) immunity system equivalent to that of vertebrates and hence cannot readily be 'vaccinated' against particular pathogens (but see sections 2.5.5, 8.9.4.1 and 12.2). Instead, their defence systems, while effective, tend to be more general and based on haemocytes that can mount phagocytic (see Glossary) cytotoxic and inflammatory responses to invading microbes. These cell-mediated defences are normally activated by specific protein and carbohydrate molecules (lipopolysaccharides, peptidoglycans, glycans and mannins) on the surface cells of bacterial, fungal and protozoan pathogens that are recognised by the haemocytes as genetically different, i.e. non-self (Thörnqvist & Söderhäll 1997). Their ability to recognise viruses, however, is limited because many viruses have surface molecules similar to those on the host's cells.

Two of the three types of circulating haemocytes are involved in the prophenoloxidase system (Sritunyalucksana & Söderhäll 2000), which is an important component of cellular defence reactions. Once set in motion, the system cascades a whole range of physiologically active proteins to destroy the foreign material and stimulate phagocytosis by the third type of cell (hyaline cells). Humoral (i.e. non-cellular) effectors such as secretions of non-specific intracellular enzymes, lytic and cytotoxic molecules, including superoxide anions (Muñoz *et al.* 2000) are also produced. The crustacean's innate, non-adaptive immune system is based on different types of circulating molecules including antimicrobial peptides (Roch 1999; Bachère *et al.* 2000). The antimicrobial peptides found include agglutinins, lysins, precipitins, cytokine-like molecules, cell adhesion molecules (opsonins) and clotting agents (Smith & Chisholm 1992; Holmblad & Söderhäll 1998). Horseshoe crab, blue crab and shore crab haemolymph or haemocytes have broad-spectrum antimicrobial activity capable of inhibiting Gram-positive and Gram-negative bacteria and some fungi (Roch 1999) while a new family of crustacean defence molecules (penaeidins) with antibacterial and antifungal activities has recently been reported from *Litopenaeus vannamei* (Destoumieux *et al.* 1997, 2000).

2.5.2 Tolerance to infection

The natural tolerance of the host organism can vary according to its life cycle stage, its moult stage, its genetic make-up and the degree of stress to which it has been exposed during culture. Examples of stressors known to suppress crustacean immune responses include environmental factors such as adverse temperature, salinity and oxygen levels and industrial pollutants including heavy metals, and pesticides (Le Moullac & Haffner 2000). Most crustaceans harbour a mix of potential pathogens whose presence does not necessarily cause a disease. Many pathogens live in harmony with their hosts until some factor upsets the balance; stressed individuals, for example, are always highly susceptible to infection (section 12.2). In addition, the virulence of a disease-causing agent may also vary, e.g. over time or with season, and is likely to be substantially different in different hosts. One example of the latter is the crayfish plague fungus that is endemic in and tolerated by North American species but which is lethal to European and Australasian crayfish. Another is infectious hypodermal and haematopoietic necrosis virus (IHHNV), which seems benign in farmed shrimp in Asia but which causes serious losses in the Americas (Lightner 1999). These are all important considerations for the diagnostician who may be confronted with a stock behaving abnormally or suffering unusually high mortalities (section 8.9.2).

2.5.3 Stimulation of the immune system

Viral diseases differ from bacterial, fungal and protozoan diseases in that they are not susceptible to therapy. At best, therapy may reduce the risks of secondary diseases attacking animals already weakened by viruses. While exposure to the types of foreign material normally used in vaccines (usually dead pathogens) can enhance crustacean defensive reactions for a short period, there seems no long-term specific memory like that seen in vertebrates. Nevertheless it seems possible to stimulate the defence system of some crustaceans by adding certain yeast, fungi or bacterial cell wall components (beta-glucans, lipopolysaccharides, peptidoglycan) to the feed (Raa 2000) (sections 8.9.4.3 and 12.2). These compounds are known as immunostimulants and several are produced commercially. In the late 1990s a purified (96%) form of one of them, β-1,3-D-glucan, was marketed. The poly-branched, triple helix configuration of this molecule has a greater number of receptor sites than other glucans and is claimed to be more potent in activating the haemocytes to synthesise and release their cascade of antimicrobial molecules.

Once activated, a defence system has an increased metabolic requirement for certain micronutrients, e.g. vitamin C, carotenoids and minerals. Depletion of tissue reserves and inadequate levels of any of these in diets could account for some of the variability seen in trials with immunostimulants. Combining an immunostimulant with some antibiotic drugs or with other, different immunostimulants may confer even greater effectiveness. The addition of probiotics (section 8.9.4.2) to the feed may also stimulate the immune system. Rengpipat *et al.* (2000) showed that by feeding a probiotic bacterium (*Bacillus* S11) to 6–7 g *Penaeus monodon* under laboratory conditions, resistance to pathogenic, luminescent bacteria (*Vibrio harveyi*) was enhanced after 90 days through stimulation of both cellular and, in older shrimp, humoral defence systems. However, the protective effects of such treatments are short by comparison with those of vaccination typically seen in vertebrates.

2.5.4 Viruses

Viruses can cause considerable mortality within hatchery, nursery and ongrowing facilities and are readily transported from the wild to hatcheries, from nurseries to farms and from country to country with shipments of stock, other live crustaceans and even frozen product (section 8.9.3). There are over 30 viruses already identified in cultured crustaceans with more being discovered annually. Twenty or more are reported in penaeids (Lightner & Redman 1998), six to ten in crayfish (Edgerton 1999), three or four in *Macrobrachium* (Johnson & Bueno 2000) and at least three in crabs but not all have caused serious diseases. As yet no viral diseases of lobsters have been reported (Evans *et al.* 2000). Several crustaceans commonly found in shrimp ponds (*Portunus, Scylla, Acetes*) can act as reservoirs of infection (Lo *et al.* 1996). Published descriptions of many viruses vary in detail and may be referring to similar, if not the same virus. It is sometimes appropriate therefore to consider these as groups or complexes.

Four viruses (or virus groups) in particular have caused major problems within the shrimp industry: these are the DNA parvovirus called white spot syndrome virus (WSSV), the DNA penaeid baculo-like virus, infectious hypodermal and haematopoietic necrosis virus (IHHNV), the RNA rhabdovirus yellow head virus (YHV) and the ssRNA virus Taura syndrome virus (TSV). Other important groups include *Penaeus monodon*-type baculoviruses (MBV), baculoviral midgut gland necrosis type viruses (BMN), hepatopancreatic parvovirus (HPV) and *Baculovirus penaei* type viruses (BP).

2.5.5 Bacteria

Possibly the only true primary bacterial diseases of farmed crustaceans are Gaffkaemia (*Aerococcus viridans*) which causes serious losses among clawed lobsters held in storage pounds (Martin & Hose 1995) and some strains of *Vibrio harveyi* and *V. penaeicida* isolated from moribund shrimp; although the virulence of the vibrios is not fully understood (Saulnier *et al.* 2000). Gaffkaemia enters only through breaks in the exoskeleton and crowding in captivity could therefore predispose to infection. It can be experimentally induced in spiny lobsters (*Panulirus interruptus*) and the prawn *Pandalus platyceros*. Gaffkaemia is currently the only example of a disease that a crustacean can be protected against by vaccination, although short-term resistance to vibriosis following vaccination has been reported in *Penaeus monodon* and *Marsupenaeus japonicus* (30–50 days) (Teunissen *et al.* 1998) and *Litopenaeus vannamei* (7 days) (Alabi *et al.* 2000). Most other bacterial diseases are of secondary aetiology with many species co-existing benignly with farmed crustaceans until conditions predispose them towards pathogenicity (Gomez-Gil *et al.* 1998). Such conditions might arise from factors favouring rapid bacterial growth such as elevated levels of dissolved and particulate organic materials in the water or those causing stress, like inadequate nutrition or poor water quality and husbandry practices. Rainy season outbreaks of luminous bacterial disease in South-east Asian shrimp hatcheries are one example. Increased stocking densities and reduced pond reconditioning periods between crops may constitute others. Similarly, the stresses of live transportation have been shown to reduce serum bactericidal activity in shrimp and crabs (Ueda *et al.* 1999). Some infections such as *Vibrio parahaemolyti-*

cus can be transmitted via the food. The susceptibility of a host species can also vary with age or life cycle stage. Important pathogens include *Vibrio, Beneckea, Pseudomonas* and *Aeromonas* species, although many others including rickettsias (see Glossary) have been reported (e.g. from *Macrobrachium* hatcheries in Puerto Rico; Johnson & Bueno 2000). Several have been reported present together in diseased shrimp (Lightner & Redman 1998), crayfish (Edgerton & Owens 1999) and spiny lobsters (Evans *et al.* 2000).

Infections usually take the form of localised erosions of the cuticle (bacterial shell disease), localised tissue infections or general septicaemias and can cause mass mortalities in intensively farmed shrimp and crayfish. Bacterial erosion of the exoskeleton can allow entry of other pathogens (e.g. *Vibrio alginolyticus* and *V. anguillarum*) but otherwise the unsightly appearance of infected animals can cause significant financial loss in live storage operations with lobsters and crabs (Evans *et al.* 2000) and during the production of soft-shelled crustaceans. The normally innocuous, filamentous bacterium (*Leucothrix* spp.) can become lethal by physically smothering shrimp gills without harming underlying tissues.

2.5.6 Fungi

The most serious of the fungal diseases is that caused in the wild by the crayfish plague fungus (*Aphanomyces astaci*), which is lethal to all native European and Australian crayfish. North American crayfish are generally resistant but can carry the disease and under stress will also succumb. Other fungal infections (e.g. *Lagenidium* sp., *Sirolpidium* sp.) are mostly serious during the hatchery and nursery phases of any species where they cause systemic non-inflammatory mycosis. Penaeid larvae seem unable to mount a worthwhile inflammatory response to these phycomycetes which can therefore become rapidly established. Juveniles and adults are at risk from *Fusarium* spp. which typically penetrate the cuticle through any lesion or abrasion. The host's response to these is inflammatory, resulting in dark melanised regions such as those characteristic of black gill disease. Some fungi may produce toxins that inhibit osmoregulation and increase risks of mortality at moult (Souheil *et al.* 1999). Fungus-like yeasts (e.g. *Candida* spp.) in haemolymph and tissues of *Macrobrachium rosenbergii* adults and juveniles (but not larvae), caused significant mortalities in Taiwan during winter (Lu *et al.* 1997). As with other micro-organisms, poor husbandry or water quality can bring opportunistic colonisers to cause problems, e.g. *Haliphthoros* sp. in laboratory populations of *Homarus* spp. (Aiken & Waddy 1995) and wild-caught *Jasus edwardsii* (Diggles 1999).

2.5.7 Protozoa

Infestations of epicommensal protozoa, like those of bacteria and blue-green algae, are frequently associated with high levels of dissolved organic matter accumulating in the water. Examples of protozoans commonly infesting the gills and exoskeletons of cultured crustaceans include *Zoothamnium* spp., *Epistylis* spp. and *Vorticella* spp. Death may be from hypoxia or from interference with moulting, locomotion or feeding. Eggs of lobsters, freshwater prawns and crayfish may also become seriously infested during incubation.

Parasitic protozoa include the microsporidans *Ameson (=Nosema)* spp., *Agamasoma (=Thelohania)* spp. and *Pleistophora* spp. In penaeids they cause cotton or milk shrimp disease, so named because the body musculature often becomes opaque. Microsporidans do not always cause significant mortalities, perhaps because of their slow growth relative to that of shrimp (Bachère *et al.* 1995). They do not require an intermediate host, some being transmitted transovarially, others by direct infection, e.g. by consumption of spores that may be attached to cast exoskeletons. After plague fungus, telohaniasis or porcelain disease is probably the most serious disease to affect crayfish worldwide. Gregarines, which use an intermediate host, are commonly found in the gut of penaeids but are usually benign. Infections with histophagous ciliates are also known to occur in a range of marine and freshwater crustaceans in captivity.

For further information on crustacean defence mechanisms and diseases the reader is referred to the works of Sindermann and Lightner (1988) on farmed crustaceans in general; Lightner (1996) and Flegel and MacRae (1997) on penaeid shrimp; Brock (1993) and Johnson and Bueno (2000) on *Macrobrachium*; Alderman and Polglase (1988), and Evans and Edgerton (2001) for freshwater crayfish; Martin and Hose (1995) for clawed lobsters; Evans *et al.* (2000) for spiny lobsters and Fisher (1986) for embryo infestations. A comprehensive, illustrated guide to the histology of healthy, disease-free tissues of penaeid shrimp was prepared by Bell and Lightner (1988) and provides a reference base against which subtle changes in cell structure caused by inadequate diets, exposure to toxic materials as well as infection can be compared. Similarly, Vogt (2001) describes in detail the anatomy, ultrastructure and physiology of healthy crayfish organ systems.

2.6 Genetics

Crustaceans develop and grow under the control of instructions from their genes and in response to the influences of their environment. Domestication aims to modify the collective gene pool of a cultured stock to obtain the best possible attribute performances, e.g. rapid growth, disease resistance (sections 8.10.1.2, 8.10.1.3 and 12.3). Unlike most domesticated birds and mammals, the majority of crustaceans farmed currently are taken from natural populations with large stores of genetic variation. Many also have high fecundity, and offer the prospect of labile sexual differentiation, all advantageous features in the search for genetic improvement. On the other hand, moulting, combined with territorial instincts and plasticity of growth rate, makes them theoretically one of the least amenable groups for cultivation and eventual domestication (Wickins 1984). The three common approaches to making improvements are through traditional selective breeding, hybridisation and the application of new gene transfer technologies (genetic engineering).

2.6.1 Genetic variation and heritability

As a first step, measurement of genetic variation (heterozygosity) within and between species provides useful information for the identification of founder broodstocks (Fetzner & Crandall 2001). Electrophoretic studies of tissue protein variation in shrimp made during the 1980s suggested that penaeid genetic diversity was amongst the lowest reported for any animal, making this group particularly vulnerable to selection pressure (Dall *et al.* 1990). While other authors have suggested that this may in part have been due to the electrophoretic techniques used, there is good evidence that further reduction of genetic variation occurs in captive stocks reared from small numbers of founder individuals (Lester & Pante 1992) while growth and reproductive performance in *Macrobrachium* (Kutty *et al.* 2000) may, or in penaeids (Bédier *et al.* 1999) may not, be compromised.

Phenotypic characteristics such as growth rate or tail weight, as expressed in populations, are invariably controlled both by the activities of a gene or group of genes (genotypes) and the influence of the environment; only the genetic component of the characteristic's variability being passed from parent to offspring. For selective breeding purposes it is important to know the heritability (i.e. the ratio of genotypic to phenotypic variation) since this expresses the reliability of the measured characteristic as a guide to the breeding value.

Several studies of heritable characteristics of crustaceans have been reported in the last two decades (Malecha 1983; Lester & Pante 1992). Estimates of growth rate heritability for lobster (*Homarus americanus*), freshwater prawn (*Macrobrachium rosenbergii*) (New 1995), red swamp crayfish (*Procambarus clarkii*) (Lutz & Wolters 1989), redclaw crayfish (*Cherax quadricarinatus*) (Gu *et al.* 1995) and shrimp (*Marsupenaeus japonicus*) (Hetzel *et al.* 2000) range widely, from 10% to 72%, indicating some scope for improvement of favourable traits by selective breeding (sections 8.10.1.2 and 12.3). A heritability estimate of, for example, $h^2 = 0.24$ for a particular trait indicates that 76% of the variation for that trait is due to environmental causes. Indeed recent data for some penaeid shrimp, including *Penaeus monodon* (Benzie *et al.* 1997), suggest they have the advantage of an additive effect of genetic variance for growth (Hetzel *et al.* 2000). In *Macrobrachium rosenbergii* heritability for growth is a sexually dimorphic trait (sections 4.6.1 and 7.3.7). Male size is determined almost entirely by behavioural interactions while female size is under significant genetic control ($h^2 = 0.35$), suggesting that selective breeding programmes aimed at females could improve growth (Karplus *et al.* 2000). Growth heritability values, however, tend to decrease with age, probably due to the increasing size variability (heterogeneous growth) typically found in captive populations (Gu *et al.* 1995) while estimates of heritability for size in young *Penaeus monodon* increased with age, possibly as a result of declining maternal effects (Benzie 1997).

2.6.2 Selective breeding

In the 1980s attempts were made to assess the potential for the development of selective breeding programmes to improve tail weight in penaeids (Lester 1983) and crayfish (Lutz & Wolters 1989), and, in lobsters, to gain faster growth (Fairfull & Haley 1981), improved tolerance to crowding (Finley & Haley 1983) or select distinctively coloured individuals – colourmorphs (Aiken & Waddy 1995). However the existence of a short-lived, water-borne substance capable of inhibiting growth in lobsters (Nelson *et al.* 1980), and the behaviourally induced, morphological changes occurring in dominant male *Macrobrachium* (section 7.3.7), were just two examples of the difficulties faced by researchers investigating the establishment of domestication programmes for crustaceans. Most of these studies concluded that, although selection would be possible, improvements would only be moderate and might take a long time to achieve.

Recently, however, the perceptions of low genetic variability (often based on neutral markers with no relevance to commercially interesting traits) and difficulties in domestication (of shrimp in particular) have been challenged (Pullin *et al.* 1998). Already, selection for improvements in survival in *Macrobrachium nipponense* (New 1995) and for penaeid growth and disease resistance (Fjalestad *et al.* 1999; Goyard *et al.* 1999) have been reported. Similarly, studies with *Marsupenaeus japonicus* concluded that although heritability of growth was moderate, the rate of response to selection could be high because of high levels of natural variation (Hetzel *et al.* 2000). Commercial trials in Australia, for example, have shown a worthwhile increase in mean harvest weight from first to third generation pond stocks (Preston *et al.* 1999).

On the other hand, families of shrimp derived from four 'high-health' populations in the USA (designed to supply and support the industry) (section 8.9.4.4) were found to vary considerably in growth performance and susceptibility to viral diseases. As a result, a selective breeding programme was started to improve performance (Gjedrem & Fimland 1995) and genetic diversity is being closely monitored (Wolfus *et al.* 1997). Non-endemic species of shrimp are being routinely reared through several generations, for example in USA, Tahiti, New Caledonia, Australia and Venezuela (Browdy 1998), but globally, there are few large-scale breeding programmes based on practices sufficiently sound for the development of commercial strains (section 8.10.1). Systems for physically tagging animals are important in these programmes to maintain pedigree records (section 8.10.1.1).

2.6.3 Hybridisation, sex reversal and manipulation of chromosome number

Studies in the 1980s suggested that interspecific hybridisation among clawed lobsters might offer the most immediate prospects for introducing variability into broodstock, but crosses between *Homarus americanus* and *H. gammarus* have not regularly yielded families showing markedly improved characteristics. Interspecific crosses between *Macrobrachium* species, between *Astacus* species, as well as between several penaeid and closely related spiny lobster species, have been successfully achieved but few significant improvements to culture seem to have resulted from hybrids. As far as we are aware, the only fertile, hybrid offspring reported came from very closely related species such as those from female *Orconectes rusticus* and male *O. propinquus* (Hamr 2001) although *Homarus gammarus* females × *H. americanus* males may produce fertile eggs if back-crossed with wild stock (Talbot & Helluy 1995).

One potential advantage observed from crossing *Penaeus monodon* females with *P. esculentus* males was the possibility of enhancing market value by matching different aspects of colours and markings without compromising growth rate; another was the bias towards male offspring in the hybrids (Benzie *et al.* 2001). Interestingly, a crossbreeding combination with yabbies (*Cherax albidus* females and *C. destructor* males) yielded all-male offspring (Lawrence *et al.* 1998; Austin & Meewan 1999) but survival of the hybrids was low. In another study, far greater numbers of all-male offspring consistently resulted from crossing female *Cherax rotundus* with *C. albidus* males, a result of practical significance for the industry (Lawrence *et al.* 2000). Considerable advantages could accrue if these hybrids prove to grow and survive as well as the yabbies currently under commercial production (section 12.8.3).

Sex determination in decapods may be chromosomal (although distinct sex chromosomes have been found in only a few species) and is likely to be complex, labile and subject to environmental influences (e.g. in shrimp (Pérez Farfante & Robertson 1992), *Macrobrachium* (Malecha *et al.* 1992) and crayfish (Curtis & Jones 1995)). The development of monosex populations is typically aimed at preventing overpopulation of ponds due to breeding and at producing populations containing only the faster or more uniformly growing sex. For example, a comparison of (hand selected) all-male, mixed and all-female populations of yabbies (*Cherax albidus*) resulted in a 70% overall increase in the market value of the all-male population (Lawrence *et al.* 1998). Hand sexing is, however, laborious and most applied research to date has been on sex reversal techniques using *Macrobrachium rosenbergii* on account of the marked heterogeneous growth of the males of this species.

The first successful decapod sex reversal was accomplished by implanting androgen gland tissue from male *Macrobrachium* into sexually undifferentiated females and resulted in reproductively competent, masculinised females (neomales) and predominantly female offspring (female to male ratios of 1 : 0, 6 : 1 or 3 : 1) (Malecha *et al.* 1992). Later (despite the earlier publication date), Sagi and Cohen (1990) removed the androgen glands from two males to create neofemales that mated successfully with normal males to produce predominantly male progeny. Little further progress appears to have occurred with *Macrobrachium* possibly because of the considerable advances made in understanding the behavioural causes of heterogeneous growth (section 7.3.7). However, one report suggests it may be possible to increase the percentage of males in a brood (e.g. to 75%) by feeding methyl testosterone to the larvae and post-larvae (Phillips & Lira 2000). Alternatively, Karplus *et al.* (2000) suggest that because male growth is more heterogeneous at high stocking densities, commercial yields might be improved by culturing all-female populations at high densities instead. Improved yields are reported from all-male populations of redclaw (*Cherax quadricarinatus*) in Israel (Sagi *et al.* 1997b) and it has been suggested that the apparent sexual lability of *Cherax* spp. makes them promising candidates for sex reversal studies (Fowler & Leonard 1999).

Artificially increasing the ploidy or number of chromosomes in each cell to the triploid and tetraploid state with a view to increasing growth rate at the expense of gonad development or inducing sterility in non-indigenous or genetically modified stocks has been achieved in the Chinese mitten crab (*Eriocheir sinensis*) (Chen *et al.* 1997) and shrimp (*Fenneropenaeus indicus* and *F. chinensis*) using chemical and temperature shock treatments (section 8.10.2). One population of *F. chinensis* tetraploids grew 20% faster than normal diploid controls but died after 6 months (Benzie 1998). So far in these preliminary studies, yields of polyploid individuals and their subsequent survival has been poor.

2.6.4 Gene transfer

Gene transfer or genetic transformations offer certain advantages over traditional breeding programmes. They allow useful genes or DNA constructs from one species to be used in another and provide opportunities for inserting them into the host's genome without chromosome disruption. Genes to produce, for example, antimicrobial peptides (Mialhe *et al.* 1995) and those that modulate immune responses (Bachère *et al.* 1997) are of topical interest. For the inserted sequences to be expressed they must include sequences that serve as promoters and terminators for their transcription. DNA sequences have been inserted into the pronucleus of individual shrimp embryos by various techniques such as microinjection (section 8.10.2). However, the constructed sequence may not become incorporated in the chromosomal DNA until at least one, perhaps several, cell divisions have occurred. Control over where and in which chromosome it will be incorporated is often critical if it is to function properly and not interfere with the expression of important host genes.

At present, considerable improvements in growth rate, meat yield and reproductive potential can be gained by control of the culture environment, surgical manipulations, and possibly sex reversal, rather than by true domestication, although encouraging results with penaeids and Australian crayfish are already being reported.

2.7 References

Ackefors H., Castell J.D., Boston L.D., Räty P. & Svensson M. (1992) Standard experimental diets for crustacean nutrition research. II. Growth and survival of juvenile crayfish *Astacus astacus* (Linné) fed diets containing various amounts of protein, carbohydrate and lipid. *Aquaculture*, **104** (3–4) 341–356.

Aiken D.E. & Waddy S.L. (1995) Aquaculture. In: *Biology of the Lobster* Homarus americanus (ed. J.R. Factor), pp. 153–175. Academic Press, New York.

Akiyama D.M., Dominy W.G. & Lawrence A.L. (1992) Penaeid shrimp nutrition. In: *Marine Shrimp Culture: principles and practices* (eds A.W. Fast & L.J. Lester), pp. 535–568. Elsvier Science, New York.

Alabi A.O., Cob Z.C., Jones D.A. & Latchford J.W. (1999) Influences of algal exudates and bacteria on growth and survival of white shrimp larvae fed entirely on microencapsulated diets. *Aquaculture International*, **7**, 137–158.

Alabi A.O., Latchford J.W. & Jones D.A. (2000) Demonstration of residual antibacterial activity in plasma of vaccinated *Penaeus vannamei*. *Aquaculture*, **187** (1–2) 15–34.

Alderman D.J. & Polglase J.L. (1988) Pathogens, parasites and commensals. In: *Freshwater Crayfish: biology, management and exploitation* (eds D.M. Holdich & R.S. Lowery), pp. 167–212. Croom Helm, London.

Austin C.M. & Meewan M. (1999) A preliminary study of primary sex ratios in the freshwater crayfish, *Cherax destructor* Clark. *Aquaculture*, **174** (1–2) 43–50.

Bachère E., Mialhe E., Noël D., Boulo V., Morvan A. & Rogriguez J. (1995) Knowledge and research prospects in marine mollusc and crustacean immunology. *Aquaculture*, **132** (1–2) 17–32.

Bachère E., Cedeno V., Rousseau C., *et al.* (1997) Transgenic crustaceans. *World Aquaculture*, **28** (4) 51–55.

Bachère E., Destoumieux D. & Bulet P. (2000) Penaeidins, antimicrobial peptides of shrimp: a comparison with other effectors of innate immunity. *Aquaculture*, **191** (1–3) 71–88.

Bauer R.T. & Holt G.J. (1998) Simultaneous hermaphroditism in the marine shrimp *Lysmata wurdemanni* (Caridea: Hippolytidae): an undescribed sexual system in the decapod Crus-

tacea. *Marine Biology*, **132**, 223–235.
Bédier E., Goyard E., Oatrois J., *et al.* (1999) Inbreeding based selection in penaeid shrimp. In: *Book of Abstracts, World Aquaculture '99*, 26 April–2 May 1999, Sydney, Australia, p. 62. World Aquaculture Society, Baton Rouge, LA, USA.
Bell T.A. & Lightner D.V. (1988) *A Handbook of Normal Penaeid Histology*, 114 pp. World Aquaculture Society, Baton Rouge, LA, USA.
Beltz B. (1995) Neurobiology and neuroendocrinology. In: *Biology of the Lobster* Homarus americanus (ed. J.R. Factor), pp. 267–89. Academic Press, New York.
Benzie J.A.H. (1997) A review of the effects of genetics and environment on the maturation and larval quality of the giant tiger prawn *Penaeus monodon*. *Aquaculture*, **155** (1–4) 69–85.
Benzie J.A.H. (1998) Penaeid genetics and biotechnology. *Aquaculture*, **164** (1–4) 23–47.
Benzie J.A.H., Kenway M. & Trott L. (1997) Estimates for the heritability of size in juvenile *Penaeus monodon* prawns from half-sib matings. *Aquaculture*, **152** (1–4) 49–53.
Benzie J.A.H., Kenway M. & Ballment E. (2001) Growth of *Penaeus monodon* × *Penaeus esculentus* tiger prawn hybrids relative to the parental species. *Aquaculture*, **193** (3–4) 227–237.
Bliss D.E. (ed.) (1980–1985) *The Biology of Crustacea.* Vol. 1–10, Academic Press, London.
Brock J.A. (1993) A synopsis of pathology, diseases, and production problems of cultured *Macrobrachium*, with an emphasis on experiences in Hawaiian prawn farming. In: *Handbook of Mariculture,* 2nd edn, *Vol. 1 Crustacean aquaculture* (ed. J.P. McVey), pp. 361–91. CRC Press, Boca Raton, FL, USA.
Browdy C.L. (1998) Recent developments in penaeid broodstock and seed production technologies: improving the outlook for superior captive stocks. *Aquaculture*, **164** (1–4) 3–21.
Brown P.B. (1995) A review of nutritional research with crayfish. *Journal of Shellfish Research*, **14** (2) 561–568.
Cavalli R.O., Lavens P. & Sorgeloos P. (1999) Performance of *Macrobrachium rosenbergii* broodstock fed diets with different fatty acid composition. *Aquaculture*, **179** (1–4) 387–402.
Ceccaldi H.J. (1998) A synopsis of the morphology and physiology of the digestive system of some crustacean species studied in France. *Reviews in Fisheries Science*, **6** (1–2) 13–39.
Chamberlain G.W. (1995) Frontiers in shrimp nutrition research. In: *Swimming Through Troubled Water. Proceedings of the special session on shrimp farming* (eds C.L. Browdy & J.S. Hopkins), pp. 108–117. Aquaculture '95. World Aquaculture Society, Baton Rouge, LA, USA.
Chang E. (1992) Endocrinology. In: *Marine Shrimp Culture: principles and practices* (eds A.W. Fast & L.J. Lester), pp. 53–91. Elsevier Science, New York.
Chen L.Q., Zhao Y.L., Wang Y.F., *et al.* (1997) Triploidy and tetraploidy induction in the Chinese mitten-handed crab *Eriocheir sinensis* by cytochalasin B. *Asian Fisheries Science*, **10** (2) 131–137.
Clark D.J., Lawrence A.L. & Swakon D.H.D. (1993) Apparent chitin digestability in penaeid shrimp. *Aquaculture*, **109** (1) 51–57.
Clarke A. & Wickins J.F. (1980) Lipid content and composition of cultured *Penaeus merguiensis* fed with animal food. *Aquaculture*, **20**, 17–27.
Cobb J.S. & Phillips B.F. (eds) (1980) *The Biology and Management of Lobsters, Vol. 1, Physiology and Behavior*, 463 pp., and *Vol. 2, Ecology and management*, 390 pp. Academic Press, London.
Conklin D.E. (1995) Digestive physiology and nutrition. In: *Biology of the Lobster* Homarus americanus (ed. J.R. Factor), pp. 441–463. Academic Press, New York.
Conklin D.E. (1997) Vitamins. In: *Crustacean Nutrition* (eds. L.R. D'Abramo, D.E. Conklin & D.M. Akiyama), pp. 123–149. Advances in World Aquaculture Vol. 6, World Aquaculture Society, Baton Rouge, LA, USA.
Cousin M., Cuzon G., Guillaume J. & AQUACOP. (1996) Digestability of starch in *Penaeus vannamei*: in vivo and in vitro study on eight samples of various origin. *Aquaculture*, **140** (4) 361–372.
Coutteau P., Camara M.R. & Sorgeloos P. (1996) The effect of different levels and sources of dietary phosphatidylcholine on the growth, survival, stress resistance, and fatty acid composition of postlarval *Penaeus vanammei*. *Aquaculture*, **147** (3–4) 261–73.
Curtis M.C. & Jones C.M (1995) Observations on monosex culture of redclaw crayfish *Cherax quadricarinatus* von Martins (Decapoda: Parastacidae) in earthen ponds. *Journal of the World Aquaculture Society*, **26** (2) 154–159.
Cuzon G. & Guillaume J. (1997) Energy and protein:energy ratio. In: *Crustacean Nutrition* (eds L.R. D'Abramo, D.E. Conklin & D.M. Akiyama), pp. 51–70. Advances in World Aquaculture Vol. 6, World Aquaculture Society, Baton Rouge, LA, USA.
D'Abramo L.R. (1997) Triacylglycerols and fatty acids. In: *Crustacean Nutrition* (eds L.R. D'Abramo, D.E. Conklin & D.M. Akiyama), pp. 71–107. Advances in World Aquaculture Vol. 6, World Aquaculture Society, Baton Rouge, LA, USA.
D'Abramo L.R. (1998) Nutritional requirements of the freshwater prawn *Macrobrachium rosenbergii*: comparisons with species of penaeid shrimp. *Reviews in Fisheries Science*, **6** (1–2) 153–163.
D'Abramo L.R. & Castell J.D. (1997) Research methodology. In: *Crustacean Nutrition* (eds L.R. D'Abramo, D.E. Conklin & D.M. Akiyama), pp. 3–25. Advances in World Aquaculture Vol. 6, World Aquaculture Society, Baton Rouge, LA, USA.
D'Abramo L.R. & Conklin D.E. (1995) New developments in the understanding of the nutrition of penaeid and caridean species of shrimp. In: *Swimming Through Troubled Water. Proceedings of the special session on shrimp farming* (eds C.L. Browdy & J.S. Hopkins), pp. 95–107. Aquaculture '95. World Aquaculture Society, Baton Rouge, LA, USA.
D'Abramo L.R. & New M.B. (2000) Nutrition, feeds and feeding. In: *Freshwater Prawn Culture: the farming of* Macrobrachium rosenbergii (eds M.B. New & W.C. Valenti), pp. 203–220. Blackwell Science, Oxford, UK.
D'Abramo L.R. & Robinson E.H. (1989) Nutrition of crayfish. *Reviews in Aquatic Sciences,* **1**, 711–728.
D'Abramo L.R. & Sheen S. (1994) Nutritional requirements,

feed formulation, and feeding practices for intensive culture of the freshwater prawn *Macrobrachium rosenbergii*. *Reviews in Fisheries Science*, **2**, 1–21.

D'Abramo L.R., Conklin D.E. & Akiyama D.M. (eds) (1997) *Crustacean Nutrition*, 587 pp. Advances in World Aquaculture, Vol. 6, World Aquaculture Society, Baton Rouge, LA, USA.

Dall W. (1992) Feeding, digestion and assimilation in Penaeidae. In: *Proceedings, Aquaculture nutrition workshop*, Salamander Bay, 15–17 April 1991 (eds G.L. Allen & W. Dall), pp. 57–63. NSW Fisheries, Brackish Water Fish Culture Research Station, Salamander Bay, Australia.

Dall W., Hill, B.J., Rothlisberg P.C. & Sharples D.J. (1990) The biology of the Penaeidae. In: *Advances in Marine Biology*, Vol. 27 (eds J.H.S. Blaxter & A.J. Southward) 489 pp. Academic Press, London.

Davis D.A. & Arnold C.R. (1998) Bioavailability of feed grade calcium phosphate incorporated into practical diets for *Penaeus vannamei*. *Aquaculture Nutrition*, **4**, 209–215.

Davis D.A. & Lawrence A.L. (1997) Minerals. In: *Crustacean Nutrition* (eds L.R. D'Abramo, D.E. Conklin & D.M. Akiyama), pp. 51–70. Advances in World Aquaculture Vol. 6, World Aquaculture Society, Baton Rouge, LA, USA.

Destoumieux D., Bulet P., Loew D., Van Dorsselaer A., Rodriguez J. & Bachère E. (1997) Penaeidins: a new family of antimicrobial peptides isolated from the *Penaeus vannamei* (Decapoda). *Journal of Biological Chemistry*, **272**, 28398–406.

Destoumieux D., Muñoz M., Rodriguez J., Bulet P., Comps M. & Bachère E. (2000) Penaeidins, a family of antimicrobial peptides with chitin-binding activity: characterization and role in shrimp immunity. In: *Abstracts, Aqua 2000, Responsible Aquaculture in the New Millennium* (compiled by R. Flos & L. Creswell), p. 177. European Aquaculture Society, Special Publication No. 28.

Diggles B. (1999) Mortalities of *Jasus edwardsii* maintained in captivity. *The Lobster Newsletter*, **12** (1) 11–12.

D'Souza F.M.L., Lecossois D., Heasman M.P., Diemar J.A., Jackson C.J. & Pendrey R.C. (2000) Evaluation of centrifuged microalgae concentrates as diets for *Penaeus monodon* Fabricius larvae. *Aquaculture Research*, **31** (8–9) 661–670.

Edgerton B.F. (1999) A review of freshwater crayfish viruses. In: *Freshwater Crayfish 12* (eds M. Keller, M.M. Keller, B. Oidtmann, R. Hoffmann & G. Vogt), pp. 261–278. Weltbild Verlag, Germany.

Edgerton B.F. & Owens L. (1999) Histopathological surveys of the redclaw freshwater crayfish, *Cherax quadricarinatus*, in Australia. *Aquaculture*, **180** (1–2) 23–40.

Evans L.H. & Edgerton B.F. (2001) Pathogens, parasites and commensals. In: *Biology of Freshwater Crayfish* (ed. D.M. Holdich), pp. 377–438. Blackwell Science, Oxford, UK.

Evans L.H., Jones J.B. & Brock J.A. (2000) Diseases of spiny lobsters. In: *Spiny Lobsters: fisheries and culture*, 2nd edn (eds B.F. Phillips & J. Kittaka), pp. 586–600. Fishing News Books, Oxford, UK.

Factor J.R. (Ed.) (1995) *Biology of the Lobster* Homarus americanus, 528 pp. Academic Press, New York.

Fairfull R.W. & Haley L.E. (1981) The early growth of artificially reared American lobsters. *Theoretical and Applied Genetics*, **60**, 269–273.

Fast A.W. & Lester L.H. (eds) (1992) Marine shrimp culture: principles and practices, 862 pp. *Developments in Aquaculture and Fisheries Science*, **23**.

Fetzner J.W. Jr. & Crandall K.A. (2001) Genetic variation. In: *Biology of Freshwater Crayfish* (ed. D.M. Holdich), pp. 291–326. Blackwell Science, Oxford, UK.

Finley L.M. & Haley L.E. (1983) The genetics of aggression in the juvenile American lobster, *Homarus americanus*. *Aquaculture*, **33** (1–4) 135–139.

Fisher W.S. (1986) Defences of brooding decapod embryos against aquatic bacteria and fungi. In: *Pathology in Marine Aquaculture* (eds C.P. Vivares, J.R. Bonami & E. Jaspers), pp. 357–363. European Aquaculture Society, Special Publication No. 9.

Fjalestad K.T., Carr W.H., Lotz J., Sweeney J.N. & Gjedrem T. (1999) Genetic variation and selection response in body weight and disease resistance in Pacific white shrimp (*Penaeus vannamei*). *Aquaculture*, **173** (1–4) 10, abstract only.

Flegel T.W. & MacRae I.H. (eds) (1997) *Diseases in Asian Aquaculture III*, 409 pp. Fish Health Section, Asian Fisheries Society, Manila, Philippines.

Fowler R.J. & Leonard B.V. (1999) The structure and function of the androgenic gland in *Cherax destructor* (Decapoda: Parastacidae). *Aquaculture*, **171** (1–2) 135–148.

Fox C.J. (1993) The effect of dietary chitin on the growth, survival and chitinase levels in the digestive gland of juvenile *Penaeus monodon* (Fab.). *Aquaculture*, **109** (1) 39–49.

Fox C., Brown J.B. & Briggs M. (1994) The nutrition of prawns and shrimps in aquaculture – a review of recent research. In: *Recent Advances in Aquaculture*, Vol. 5 (eds J.F. Muir & R.J. Roberts), pp. 131–206. Blackwell Science, Oxford, UK.

Gebauer P., Paschke K. & Anger K. (1999) Costs of delayed metamorphosis: reduced growth and survival in early juveniles of an estuarine grapsid crab, *Chasmagnathus granulata*. *Journal of Experimental Marine Biology and Ecology*, **238**, 271–281.

Germano B.P. (1994) Effects of unilateral eyestalk ablation on growth in juvenile blue crabs *Portunus pelagicus* (L.) (Crustacea: Decapoda: Portunidae). *Asian Fisheries Science*, **7** (1) 19–28.

Gjedrem T. & Fimland E. (1995) Potential benefits from high health and genetically improved shrimp stocks. In: *Swimming Through Troubled Water. Proceedings of the special session on shrimp farming* (eds C.L. Browdy & J.S. Hopkins), pp. 60–65. Aquaculture '95. World Aquaculture Society, Baton Rouge, LA, USA.

Gomez-Gil B., Tron-Mayén L., Roque A., Turnbull J.F., Inglis V. & Guerra-Flores A.L. (1998) Species of *Vibrio* isolated from hepatopancreas, haemolymph and digestive tract of a population of healthy juvenile *Penaeus vannamei*. *Aquaculture*, **163** (1–2) 1–9.

Goyard E., Patrois J., Peignon J-M., Vanaa V., Dufour R. & Bédier E. (1999) IFREMER's shrimp genetics program. *Global Aquaculture Advocate*, **2** (6) 26–28.

Gu H., Mather P.B. & Capra M.F. (1995) Juvenile growth performance among stocks and families of red claw crayfish, *Cherax quadricarinatus* (von Martens). *Aquaculture*, **134** (1–2) 29–36.

Guillaume J. (1997) Protein and amino acids. In: *Crustacean*

Nutrition (eds L.R. D'Abramo, D.E. Conklin & D.M. Akiyama), pp. 26–50. Advances in World Aquaculture Vol. 6, World Aquaculture Society, Baton Rouge, LA, USA.

Haefner P.A., Jr. (1985) The biology and exploitation of crabs. In: *The Biology of Crustacea, Economic Aspects: fisheries and culture.* Vol. 10 (ed. A.J. Provenzano Jr.), pp. 111–166. Academic Press, London.

Hamr P. (2001) *Orconectes*. In: *Biology of Freshwater Crayfish* (ed. D.M. Holdich), pp. 585–608. Blackwell Science, Oxford, UK.

Harrison K. (1990) The role of nutrition in maturation, reproduction and embryonic development of decapod crustaceans: a review. *Journal of Shellfish Research*, **9** (1) 1–28.

Harrison K. (1997) Broodstock nutrition and maturation diets. In: *Crustacean Nutrition* (eds L.R. D'Abramo, D.E. Conklin & D.M. Akiyama), pp. 390–408. Advances in World Aquaculture Vol. 6, World Aquaculture Society, Baton Rouge, LA, USA.

Hetzel D.J.S., Crocos P.J., Davis G.P., Moore S.S. & Preston N.C. (2000) Response to selection and heritability for growth in the Kuruma prawn, *Penaeus japonicus*. *Aquaculture*, **181** (3–4) 215–223.

Holdich D.M. (ed.) (2001) *Biology of Freshwater Crayfish*, 702 pp. Fishing News Books, Oxford, UK.

Holdich D.M. & Lowery R.S. (eds). (1988) *Freshwater Crayfish: biology, management and exploitation*, 498 pp. Croom Helm, London.

Holmblad T. & Söderhäll K. (1998) Cell adhesion molecules and antioxidative enzymes in a crustacean, possible role in immunity. *Aquaculture*, **172** (1–2) 111–123.

Huberman A. (2000) Shrimp endocrinology: a review. *Aquaculture*, **191** (1–3) 191–208.

Igarashi M.A. & Kittaka J. (2000) Water quality and microflora in the culture water of phyllosomas. In: *Spiny Lobsters: fisheries and culture,* 2nd edn (eds B.F. Phillips & J. Kittaka), pp. 533–555. Fishing News Books, Oxford, UK.

Ismael D. & New M.B. (2000) Biology. In: *Freshwater Prawn Culture: the farming of* Macrobrachium rosenbergii (eds M.B. New & W.C. Valenti), pp. 18–40. Blackwell Science, Oxford, UK.

Johansson M.W., Keyser P., Sritunyalucksana K. & Söderhäll K. (2000) Crustacean haemocytes and haematopoiesis. *Aquaculture*, **191** (1–3) 45–52.

Johnson S.K. & Bueno S.L.S. (2000) Health management. In: *Freshwater Prawn Culture: the farming of* Macrobrachium rosenbergii (eds M.B. New & W.C. Valenti), pp. 239–258. Blackwell Science, Oxford, UK.

Jones D.A. (1998) Crustacean larval microparticulate diets. *Reviews in Fisheries Science*, **6** (1–2) 41–54.

Jones D.A., Kamarudin M.S. & Le Vay L. (1993) The potential for replacement of live feeds in larval culture. *Journal of the World Aquaculture Society*, **24** (2) 199–210.

Jones D.A., Kumlu M., Le Vay L. & Fletcher D.J. (1997a) The digestive physiology of herbivorous, omnivorous and carnivorous crustacean larvae: a review. *Aquaculture*, **155** (1–4) 285–295.

Jones D.A., Yule A.B. & Holland D.L. (1997b) Larval nutrition. In: *Crustacean Nutrition* (eds D. Akiyama, D. Conklin & L. D'Abramo), pp. 353–389. Advances in World Aquaculture Vol. 6, World Aquaculture Society, Baton Rouge, LA, USA.

Jover M., Fernandez-Carmona J., Del Rio M.C. & Soler M. (1999) Effect of feeding cooked-extruded diets, containing different levels of protein, lipid and carbohydrate on growth of red swamp crayfish (*Procambarus clarkii*). *Aquaculture*, **178** (1–2) 127–137.

Juinio-Meñez M.A. & Ruinata J. (1996) Survival, growth and food conversion efficiency of *Panulirus ornatus* following eyestalk ablation. *Aquaculture*, **146** (3–4) 225–235.

Kanazawa A. (1985) Nutrition of penaeid prawns and shrimps. In: *Proceedings of the first international conference on culture of penaeid prawns/shrimps*, Iloilo City, Philippines, 1984, pp. 123–130. Southeast Asian Fisheries Development Center, Aquaculture Department, Philippines.

Kanazawa A. (2000) Nutrition and food. In: *Spiny Lobsters: fisheries and culture,* 2nd edn (eds B.F. Phillips & J. Kittaka), pp. 611–624. Fishing News Books, Oxford, UK.

Karplus I. & Hulata G. (1995) Social control of growth in *Macrobrachium rosenbergii*. V. The effect of unilateral eyestalk ablation on jumpers and laggards. *Aquaculture*, **138** (1–4) 181–190.

Karplus I., Malecha S.R. & Sagi A. (2000) The biology and management of size variation. In: *Freshwater Prawn Culture: the farming of* Macrobrachium rosenbergii (eds M.B. New & W.C. Valenti), pp. 259–289. Blackwell Science, Oxford, UK.

Keenan C.P., Davie P.J.F. & Mann D.L. (1999) A revision of the genus *Scylla* de Haan, 1893 (Crustacea: Decapoda: Brachyura: Portunidae). *The Raffles Bulletin of Zoology*, **46**, 217–245.

Koshio S., Haley L.E. & Castell J.D. (1989) The effect of two temperatures and salinities on growth and survival of bilaterally eyestalk ablated and intact juvenile American lobsters, *Homarus americanus*, fed brine shrimp nauplii. *Aquaculture*, **76** (3–4) 373–382.

Kumlu M. & Jones D.A. (1995) The effect of live and artificial diets on growth, survival, and trypsin activity in larvae of *Penaeus indicus*. *Journal of the World Aquaculture Society*, **26** (4) 406–415.

Kuris A.M. (1991) A review of patterns and causes of crustacean brood mortality. In: *Crustacean Egg Production* (eds A. Wenner & A Kuris), pp. 117–141. A.A. Balkema, Rotterdam, Netherlands.

Kutty M.N., Herman F. & Le Menn H. (2000) Culture of other prawn species. In: *Freshwater Prawn Culture: the farming of* Macrobrachium rosenbergii (eds M.B. New & W.C. Valenti), pp. 393–410. Blackwell Science, Oxford, UK.

Lavalli K.L. & Factor J.R. (1995) The feeding appendages. In: *Biology of the Lobster* Homarus americanus (ed. J.R. Factor), pp. 349–393. Academic Press, New York.

Lavens P., Thongrod S. & Sorgeloos P. (2000) Larval prawn feeds and the dietary importance of *Artemia*. In: *Freshwater Prawn Culture: the farming of* Macrobrachium rosenbergii (eds M.B. New & W.C. Valenti), pp. 91–111. Blackwell Science, Oxford, UK.

Lawrence C., Morrissy N., Bellanger J. & Cheng Y.W. (1998) Final report, FRDC Project 94/075: Enhancement of yabby production from Western Australian farm dams. *Fisheries Research Report, Fisheries, Western Australia*, **112**, 1–134.

Lawrence C.S., Morrissy N.M., Vercoe P.E. & Williams I.H. (2000) Hybridization in Australian freshwater crayfish –

production of all male progeny. *Journal of the World Aquaculture Society*, **31** (4) 651–658.

Lawton P. & Lavalli K.L. (1995) Postlarval, juvenile, adolescent, and adult ecology. In: *Biology of the Lobster* Homarus americanus (ed. J.R. Factor), pp. 47–88. Academic Press, New York.

Lee P.G. & Lawrence A.L. (1997) Digestibility. In: *Crustacean Nutrition* (eds L.R. D'Abramo, D.E. Conklin & D.M. Akiyama), pp. 194–260. Advances in World Aquaculture Vol. 6, World Aquaculture Society, Baton Rouge, LA, USA.

Le Moullac G. & Haffner P. (2000) Environmental factors affecting immune responses in Crustacea. *Aquaculture*, **191** (1–3) 121–131.

Lester L.J. (1983) Developing a selective breeding program for penaeid shrimp mariculture. *Aquaculture*, **33** (1–4) 41–50.

Lester L.J. & Pante Ma. J. R. (1992) Genetics of *Penaeus* species. In: *Marine Shrimp Culture: principles and practices* (eds A.W. Fast & L.J. Lester), pp. 29–52. Elsevier, Amsterdam, Netherlands.

Lightner D.V. (1996) *A handbook of shrimp pathology and diagnostic procedures for disease of cultured penaeid shrimp* (*Pag. Var.*). World Aquaculture Society, Baton Rouge, LA, USA.

Lightner D.V. (1999) The penaeid shrimp viruses TSV, IHHNV, WSSV, and YHV: current status in the Americas, available diagnostic methods, and management strategies. *Journal of Applied Aquaculture*, **9** (2) 27–52.

Lightner D.V. & Redman R.M (1998) Shrimp diseases and current diagnostic methods. *Aquaculture*, **164** (1–4) 201–220.

Lo C.F., Ho C.H., Peng S.E., *et al.* (1996) White spot syndrome baculovirus (WSBV) detected in cultured and captured shrimp, crabs and other arthropods. *Diseases of Aquatic Organisms*, **27** (3) 215–225.

Lu C.-C., Tang F.-J., Ueno Y., Kou G.-H. & Chen S.-N. (1997) Yeast infection in prawns (*Macrobrachium rosenbergii* de Man) in Taiwan. *Acta Zoologica Taiwanica*, **8** (1) 33–45.

Lutz C.G. & Wolters W.R. (1989) Estimation of heritabilities for growth, body size, and processing traits in red swamp crawfish, *Procambarus clarkii* (Girard). *Aquaculture*, **78** (1) 21–33.

MacDiarmid A.B. & Kittaka J. (2000) Breeding. In: *Spiny Lobsters: fisheries and culture,* 2nd edn (eds B.F. Phillips & J. Kittaka), pp. 485–507. Fishing News Books, Oxford, UK.

Malecha S.R. (1983) Crustacean genetics and inbreeding: an overview. *Aquaculture*, **33** (1–4) 395–413.

Malecha S.R., Nevin P.A., Ha P., *et al.* (1992) Sex-ratios and sex-determination in progeny from crosses of surgically sex-reversed freshwater prawns, *Macrobrachium rosenbergii*. *Aquaculture*, **105** (3–4) 201–218.

Marsden G., McGuren J.M., Sarac H.Z., Neill A.R., Brock I.J. & Palmer C.L. (1992) Nutritional composition of some natural marine feeds used in shrimp maturation. In: *Proceedings of the aquaculture nutrition workshop*, Salamander Bay, 15–17 April 1991 (eds G.L. Allen & W. Dall), pp. 82–86. NSW Fisheries, Salamander Bay, Australia.

Martin G.G. & Hose J.E. (1995) Circulation, the blood and disease. In: *Biology of the Lobster* Homarus americanus (ed. J.R. Factor), pp. 465–495. Academic Press, New York.

McMahon B.R. (1995) The physiology of gas exchange, circulation, ion regulation, and nitrogenous excretion: an integrative approach. In: *Biology of the Lobster* Homarus americanus (ed. J.R. Factor), pp. 497–517. Academic Press, New York.

McMahon B.R. (2001) Physiological adaptation to environment. In: *Biology of Freshwater Crayfish* (ed. D.M. Holdich), pp. 327–76. Blackwell Science, Oxford, UK.

Merican Z.O. & Shim K.F. (1996) Qualitative requirements of essential fatty acids for juvenile *Penaeus monodon*. *Aquaculture*, **147** (3–4) 275–291.

Merrick J.R. & Lambert C.N. (1991) *The Yabby, Marron and Red Claw: production and marketing*, 185 pp. J.R. Merrick Publications, Artarmon, Australia.

Mialhe E., Bachère E., Boulo V. & Cadoret J.P. (1995) Strategy for research and international cooperation in marine invertebrate pathology, immunology and genetics. *Aquaculture*, **132** (1–2) 33–41.

Mu Y.Y., Shim K.F. & Gou J.Y. (1998) Effects of protein in isocalorific diets on growth performance of the juvenile Chinese hairy crab, *Eriocheir sinensis*. *Aquaculture*, **165** (1–2) 139–148.

Muñoz M., Cedeño R., Rodríguez J., Van der Knaap W.P.W., Mialhe E. & Bachère E. (2000) Measurement of reactive oxygen intermediate production in haemocytes of the penaeid shrimp, *Penaeus vannamei*. *Aquaculture*, **191** (1–3) 89–107.

Nelson K., Hedgecock D., Borgeson W., Johnson E., Dagget R. & Aronstein D. (1980) Density dependent growth inhibition in lobsters, *Homarus* (Decapoda, Nephropidae). *Biological Bulletin*, **159**, 162–176.

New M.B. (1976) A review of dietary studies with shrimps and prawns. *Aquaculture*, **9**, 101–144.

New M.B. (1995) Status of freshwater prawn farming: a review. *Aquaculture Research*, **26** (1) 1–54.

New M.B. (2000) Commercial freshwater prawn farming around the world. In: *Freshwater Prawn Culture: the farming of* Macrobrachium rosenbergii (eds M.B. New & W.C. Valenti), pp. 290–325. Blackwell Science, Oxford, UK.

Nyström P. (2001) Ecology. In: *Biology of Freshwater Crayfish* (ed. D.M. Holdich), pp. 192–235. Blackwell Science, Oxford, UK.

Omondi J.G. & Stark J.R. (1995) Some digestive carbohydrates from the midgut gland of *Penaeus indicus* and *Penaeus vannamei* (Decapoda: Penaedae). *Aquaculture*, **134** (1–2) 121–135.

Palacios E., Perez-Rostro C.I., Ramirez J.L., Ibarra A.M. & Racotta I.S. (1999) Reproductive exhaustion in shrimp (*Penaeus vannamei*) reflected in larval biochemical composition, survival and growth. *Aquaculture*, **171** (3–4) 309–321.

Pérez Farfante I. & Kensley B. (1997) Penaeoid and Sergestoid shrimps and prawns of the world. *Memoires Musée National D'Histoire Naturelle*, **175**, 1–233.

Pérez Farfante I. & Robertson L. (1992) Hermaphroditism in the penaeid shrimp *Penaeus vannamei* (Crustacea: Decapoda: Penaeidae) *Aquaculture*, **103** (3–4) 367–376.

Phillips B.F. & Kittaka J. (eds) (2000) *Spiny Lobsters: fisheries and culture*, 2nd edn, 679 pp. Fishing News Books, Oxford, UK.

Phillips H. & Lira A. (2000) Sex reversal of *Macrobrachium rosenbergii*. In: *Abstracts, Aqua 2000, Responsible Aquaculture in the New Millennium* (compiled by R. Flos & L. Creswell), p. 562. European Aquaculture Society, Special Publication No. 28.

Preston N.P., Crocos P.J. & Moore S.M. (1999) Comparative growth of wild and domesticated *Penaeus japonicus* in commercial production ponds. In: *Book of Abstracts, World Aquaculture '99*, 26 April–2 May 1999, Sydney, Australia, p. 612. World Aquaculture Society, Baton Rouge, LA, USA.

Pullin R.S.V., Williams M.J. & Preston N. (1998) Domestication of crustaceans. *Asian Fisheries Science*, **11**, 71–80.

Quackenbush L.S. (1986) Crustacean endocrinology, a review. *Canadian Journal of Fisheries and Aquatic Science*, **43**, 2271–2282.

Raa J. (2000) Modern science and ancient wisdom combined for the benefit of aquaculture. *International Aquafeed*, (2) 32–34.

Radhakrishnan E.V. & Vijayakumaran M. (1984) Effect of eyestalk ablation in spiny lobster *Panulirus homarus* (Linnaeus): 1. On moulting and growth. *Indian Journal of Fisheries*, **31** (1) 130–147.

Radhakrishnan E.V. & Vijayakumaran M. (1998) Bilateral eyestalk ablation induces morphological and behavioral changes in spiny lobsters. *The Lobster Newsletter*, **11** (1) 9–11.

Reddy H.R.V., Rai A. & Annappaswamy T.S. (1999) Essentiality of vitamins for juvenile white shrimp *Penaeus indicus*. *Israeli Journal of Aquaculture – Bamidgeh*, **51** (3) 122–132.

Rengpipat S., Rukpratanporn S., Piyatiratitivorakul S. & Menasaveta P. (2000) Immunity enhancement in black tiger shrimp (*Penaeus monodon*) by a probiont bacterium (*Bacillus* S11). *Aquaculture*, **191** (4) 271–288.

Robertson L., Lawrence A.L. & Castille F. (1993) Interaction of salinity and feed protein level on growth of *Penaeus vannamei*. *Journal of Applied Aquaculture*, **2** (1) 43–54.

Roch P. (1999) Defense mechanisms and disease prevention in farmed marine invertebrates. *Aquaculture*, **172** (1–2) 125–145.

Rouse D.B. & Rash J. (1999) Influence of astaxanthin in the diets of juvenile and adult red claw crayfish, *Cherax quadricarinatus*. In: *Freshwater Crayfish 12* (eds M. Keller, M.M. Keller, B. Oidtmann, R. Hoffmann & G. Vogt), p. 944. Weltbild Verlag, Germany.

Sagi A. & Cohen D. (1990) Growth, maturation and progeny of sex reversed *Macrobrachium rosenbergii* males. *World Aquaculture*, **21** (4) 87–90.

Sagi A., Shoukrun R., Levy T., Barki A., Hulata G. & Karplus I. (1997a) Reproduction and molt in previously spawned and first-time spawning red-claw crayfish *Cherax quadricarinatus* females following eyestalk ablation during the winter reproductive-arrest period. *Aquaculture*, **156** (1–2) 101–111.

Sagi A., Milstein A., Eran Y., *et al.* (1997b) Culture of the Australian red-claw crayfish (*Cherax quadricarinatus*) in Israel. II. Second growout season of overwintered populations. *Israeli Journal of Aquaculture – Bamidgeh*, **49** (4) 222–229.

Sangha R.S., Puello Cruz A.C., Chavez-Sanchez M.C. & Jones D.A. (2000) Survival and growth of *Litopenaeus vannamei* (Boone) fed a single dose of live algae and artificial diets with supplements. *Aquaculture Research*, **31** (8–9) 683–689.

Saulnier D., Haffner P., Goarant C., Levy P. & Ansquer D. (2000) Experimental infection models for shrimp vibriosis studies: a review. *Aquaculture*, **191** (1–3) 133–144.

Sedgwick R.W. (1979) Influence of dietary protein and energy on growth, food consumption and food conversion efficiency in *Penaeus merguiensis* de Man. *Aquaculture*, **16** (1) 7–30.

Shiau S-Y. (1998) Nutrient requirements of penaeid shrimps. *Aquaculture*, **164** (1–4) 77–93.

Sindermann C.J. & Lightner D.V. (eds) (1988) Disease diagnosis and control in North American marine aquaculture. *Developments in Aquaculture and Fisheries Science*, **17**, 1–431.

Smith V.J. & Chisholm J.R.S. (1992) Non-cellular immunity in crustaceans. *Fish and Shellfish Immunology*, **2**, 1–31.

Souheil H., Vey A., Thuet P. & Trilles J-P. (1999) Pathogenic and toxic effects of *Fusarium oxysporum* (Schlecht.) on survival and osmoregulatory capacity of *Penaeus japonicus* (Bate). *Aquaculture*, **178** (3–4) 209–224.

Souza J.C.R., Strussmann C.A, Takashima F., Sekine S. & Shima Y. (1999) Absorption of dissolved and dispersed nutrients from sea-water by *Panulirus japonicus* phyllosoma larvae. *Aquaculture Nutrition*, **5** (1) 41–51.

Sritunyalucksana K. & Söderhäll K. (2000) The proPO and clotting system in crustaceans. *Aquaculture*, **191** (1–3) 53–69.

Stickney R.R. (ed.) (1998) Crustacean nutrition symposium, *Reviews in Fisheries Science*, **6** (1–2) 1–167.

Tacon A.G.J. (1990) *Standard methods for the nutrition and feeding of farmed fish and shrimp.* Argent Laboratories Press, Redmond, USA **1**: 1–117; **2**: 1–129; **3**: 1–208.

Talbot P. & Helluy S. (1995) Reproduction and embryonic development. In: *Biology of the Lobster* Homarus americanus (ed. J.R. Factor), pp. 177–216. Academic Press, New York.

Teshima S. (1997) Phospholipids and sterols. In: *Crustacean Nutrition* (eds L.R. D'Abramo, D.E. Conklin & D.M. Akiyama), pp. 85–107. Advances in World Aquaculture Vol. 6, World Aquaculture Society, Baton Rouge, LA, USA.

Teshima S., Ishikawa M. & Koshio S. (2000) Nutritional assessment and feed intake of microparticulate diets in crustaceans and fish. *Aquaculture Research*, **31** (8–9) 691–702.

Teunissen O.S.P., Faber R., Booms G.H.R., Latscha T. & Boon J.H. (1998) Influence of vaccination on vibriosis resistance of the giant black tiger shrimp *Penaeus monodon* (Fabricius). *Aquaculture*, **164** (1–4) 359–366.

Thörnqvist P-O. & Söderhäll K. (1997) Crustacean immune reactions, a short review. In: *Diseases in Asian Aquaculture III* (eds T.W. Flegel & I.H. MacRae), pp. 203–217. Asian Fisheries Society, Manila, Philippines.

Ueda R., Sugita H. & Deguchi Y. (1999) Effect of transportation on the serum bactericidal activity of *Penaeus japonicus* and *Ovalipes punctatus*. *Aquaculture*, **171** (3–4) 221–225.

Vargas-Albores F. & Yepiz-Plascencia G. (2000) Beta glucan binding protein and its role in shrimp immune response. *Aquaculture*, **191** (1–3) 13–21.

Vogt G. (2001) Functional anatomy. In: *Biology of Freshwater Crayfish* (ed. D.M. Holdich), pp.53–151. Blackwell Science, Oxford, UK.

Waddy S.L., Aiken D.E. & De Kleijn D.P.V. (1995) Control of growth and reproduction. In: *Biology of the Lobster* Homarus americanus (ed. J.R. Factor), pp. 217–266. Academic Press, New York.

Wickins J.F. (1976) Prawn biology and culture. *Oceanography and Marine Biology: an annual review* (ed. H. Barnes), **14**, 435–507.

Wickins J.F. (1984) Crustacea. In: *Evolution of Domesticated Animals* (ed. I.L. Mason), pp. 424–428. Longman, London.

Wickins J.F., Beard T.W. & Child A.R. (1995) Maximizing lobster, *Homarus gammarus* (L.), egg and larval viability. *Aquaculture Research*, **26**, 379–392.

Wolfus G.M., Garcia D.K. & Alcivar-Warren A. (1997) Application of the microsatellite technique for analysing genetic diversity in shrimp breeding programs. *Aquaculture*, **152** (1–4) 35–47.

Xue X.M., Anderson A.J., Richardson, N.A., Anderson A.J., Xue G.P. & Mather P.B. (1999) Characterisation of cellulase activity in the digestive system of redclaw crayfish (*Cherax quadricarinatus*). *Aquaculture*, **180** (3–4) 373–386.

Yepiz-Plascencia G., Vargas-Albores F. & Higuera-Ciapara I. (2000) Penaeid shrimp hemolymph lipoproteins. *Aquaculture*, **191** (1–3) 177–189.

Zimmermann S., Leboute E.M. & De Sousa S.M. (1994) Effects of two calcium levels in diets and three calcium levels in the culture water on the growth of the freshwater prawn, *Macrobrachium rosenbergii*. In: *Book of Abstracts, World Aquaculture '99*, 26 April–2 May 1999, Sydney, Australia, p. 196. World Aquaculture Society, Baton Rouge, LA, USA.

Chapter 3
Markets

3.1 Overview

Crustaceans are among the most highly valued of luxury foods and the high prices they obtain serve to stimulate interest in crustacean farming and to underpin its economic viability. The potential for aquaculture is all the more apparent because wild fisheries have been unable to react to market demand and their yields have been stable or have increased only gradually. However, although aquaculture production has responded to generally favourable market conditions, many farmers have discovered that they cannot take for granted the ability to sell their output at profitable prices. The usual patterns linking supply and demand apply in crustacean markets too, and high prices are simply a reflection of limited supplies. Some crustacean markets, most notably for frozen shrimp, have undergone structural change and prices fell in the late 1980s as farm output of small and medium sizes increased sharply. In addition, the status of crustaceans as luxury foods is a mixed blessing – when economies grow it is reflected in buoyant demand and high prices, but in times of recession, demand and prices can slump disproportionately. Consumers in shrinking economies are more cautious with their disposable income and consumption of luxury foods falls sharply, either directly when cheaper alternatives are selected in the supermarket or indirectly when outings to restaurants are minimised. Seafood consumption in Japan, for example, suffered in the 1990s due to economic setbacks and shrimp consumption dropped to such an extent that in 1997 the USA overtook Japan as the world's largest importer.

Shrimp farming now produces around 800 000 mt per year, representing 27% of world shrimp supplies, and an increase of 23% from levels a decade ago (section 1.3). Improved supplies from aquaculture have enabled shrimp to compete more effectively in the mass market for everyday foods. For example, in the USA, shrimp is becoming a favourite in casual as well as white-tablecloth restaurants and it has displaced staples such as steak on some menus (Lang 2000). However this trend will slow if shrimp farms continue to suffer production problems related to environmental degradation and disease.

Large tonnages of *Macrobrachium* are also being produced, particularly in China (62 000 mt) and Bangladesh (48 000 mt), but while Chinese prawns are consumed on national markets live or fresh on ice, most of the output from Bangladeshi farms enters international trade in the form of frozen tails or frozen whole prawns. Part of the production from other Asian sources such as Thailand and Vietnam is also entering international trade but most prawn sales, as in China, are dependent on local or national markets or on individually targeted export markets where demand has been identified. *Macrobrachium* farmers that do not operate in Bangladesh, China or Vietnam or in similar low-cost production environments have encountered major difficulties in selling their product at cost-effective prices, particularly in the face of stiff competition with marine shrimp (section 3.3.2). This has restricted the expansion of freshwater prawn farming in many areas of the world despite the fact that this species is widely cultivated. Small producers can often obtain premium prices by supplying nearby restaurants and hotels. Success on a larger scale has been limited to countries where freshwater prawns are well established as a desirable food item, for example in Thailand. However, even in this country initial success has been tempered by falling prices resulting from increased farm output.

The majority of crayfish farmed in the USA are consumed within the same southern states that produce them, with some exports directed at the high price,

Swedish market. Increasing quantities of crayfish are produced in China and exported to Europe and the USA, although in the latter market trade has been held back by the imposition of penalty duties on Chinese crayfish. Spain is the leading supplier within Europe with much production sold to consumers in France and Sweden as well as home markets. Australian crayfish farmers sell nearly all of their product at home but continue to explore export markets in Europe, the USA and Japan (section 3.3.3.4).

Demand for clawed and spiny lobsters on world markets continues to be very strong. However, any future success with the marketing of farmed lobsters may rely on the successful development of new markets for small animals which, for clawed lobsters at least, are probably the only economically viable size for production by land-based culture systems (sections 3.3.4 and 7.8.9).

Farmed crabs are sold live in traditional local markets primarily in Asia. International trade does occur in frozen crab products and processed crab meat though this relies almost exclusively on crabs from capture fisheries (section 3.3.5).

Although the majority of farmed crustaceans are consumed in the USA, Japan and Western Europe, the economies of several developing and newly industrialised countries are strengthening, particularly in Asia, and consumption of luxury crustacean foods in these countries is increasing. South-east Asia, for example, now takes 5% of Indian exports that would traditionally have gone to Europe or Japan (Rao & Prakash 1999). Although this trend suffered a reverse during the Asian financial crisis of 1997–98 it has been notable in countries like Taiwan and Thailand where high-value aquaculture products are increasingly destined for home consumption as well as export. The relatively wealthy areas of China such as Beijing and the southern states are also consuming greater quantities of national farm output. Currently shrimp consumption in major shrimp farming nations such as Indonesia and the Philippines is based on small, low-value wild species that are often converted into products such as shrimp paste and shrimp crackers. But as wealthy city dwellers consume more crustaceans, farmed product sells increasingly well within producer countries. Increasing tourism in many developing countries is also serving to boost demand in the hotel and restaurant trade.

Eating habits are among the most deeply rooted elements of a culture and people tend to eat mostly what they have been raised on as children. As a reflection of this, preferences for different types of crustaceans and different product forms remain very strong within countries or regions (Hottlet 1992). In fact, observers note that there has been an overall trend towards increasing segmentation within crustacean markets and that one of the best strategies for farmers and processors is to make customised products that target individual market segments. Through product diversification and development an increasing variety of value-added crustacean foods are already being produced, with many finding ready acceptance with consumers (Traesupap *et al.* 1999). Some products are able to offer greater convenience and ease of preparation to the supermarket shopper, while others such as soft-shell crustaceans begin to open up novel markets (sections 3.3.3.2 and 3.3.5). In all, they hold out considerable hope for generating increased overall sales volumes and enhanced revenues. In their analysis of aquaculture markets and market research, Kinnucan and Wessels (1997) see the industry following the pattern currently under way in American agriculture with tighter linkages between farm production and consumer demand, and increased control of the vertical food system by large agribusiness entities. Such entities are well equipped to undertake the necessary market development (section 3.2.5).

Another important trend has been the shift of value-added processing activity away from importing countries and towards developing countries where the raw materials originate. Countries like India have started moving into value-added processing rather than remaining as suppliers of raw product for overseas processors (Rao & Prakash 1999). This makes good economic sense and is a welcome development. It has been held back by worries about poor quality control and hygiene standards and by a lack of market knowledge. In addition, the EU has obstructed progress by imposing high tariffs on value-added products to protect its labour-intensive fish processing industry (Josupeit & de Franssu 1992).

International trade in fresh and frozen shrimp is valued at $8.5–9bn per year (Ferdouse 1999) and frozen shrimp is now a commodity like coffee and orange juice with futures contracts and options (see Glossary) traded on the Minneapolis Grain Exchange (section 3.3.1). Aquaculture has been instrumental in this development because farm output has helped to provide the year-round supply of an homogeneous product that provides the basis for a viable futures contract. Aquaculture production has also reduced the need for large cold-storage holdings. In the USA over the period 1979–86, a more consistent supply of shrimp due to farming enabled average cold-storage holdings to be reduced from two to one month's supply. Nowadays when cold-storage holdings of shrimp build up to high levels, this is a sign of oversupply rather than insecurity about the availability of the product. In the late 1980s inventories climbed to record levels in Japan as buyers stockpiled shrimp awaiting improvements in prices. Despite an overall improved price stability for shrimp, up-to-date marketing information can be very useful to producers and buyers alike. Regular publications with market analyses and forecasts include:

- *Infofish International*, Kuala Lumpur
- *Seafood International*, Quantum Publishing, Croydon, UK
- *Shrimp Market Report*, LMR Fisheries Research, Del Mar, California

Some specialised Internet sources provide further market information and analysis:

Aquaculture Outlook
http://jan.mannlib.cornell.edu/reports/erssor/livestock/ldp-aqs/
Infofish Trade News
http://www.jaring.my/infofish
Market Price and Index
http://www.shrimpcom.com
Market Report, Online Seafood Business
http://www.seafoodbusiness.com/market.html
Shrimp Notes
http://www.shrimpcom.com
Urner Barry Publications Inc
http://www.seafoodnet.com
Shrimp Market Reports
http://www.fis.com

Avault (2000a) provides a useful guide to the terminology of international seafood trade.

If the crustacean farming industry is to market its output successfully in the future, not only must it produce competitively priced products, it must also satisfy consumer concerns about product safety and product quality. In addition it must acknowledge that many consumers are seeking more detailed information about the origins and production methods of their food. Most safety and quality concerns can be effectively addressed through implementation of the management tool known as Hazard Analysis, Critical Control Points (HACCP) and through established measures of quality control and sanitation (sections 3.2.2 and 9.6). But if the label 'farm-raised' is to fulfil its true potential as a marketing asset, crustacean farmers will also need to address wider issues. These include the need to minimise negative environmental impacts and the importance of adherence to industry codes of practice governing the use of antibiotics (Sze 2000). There are also concerns regarding growth hormones, probiotics, immunostimulants and other additives. At the moment there is little market interest in genetically manipulated products because of negative consumer perceptions and the need to demonstrate benefits to the consumer fairly and convincingly. But many consumers respond favourably to products bearing such labels as 'organic', so if confusion over the definition of eco-labelling terms can be cleared up (Ducherne 2000; Anon. 2000a) this trend will present further opportunities for responsive producers to satisfy consumer concerns and benefit the environment (Clay 1997). Small steps are already being taken in the right direction: one large shrimp processing operation in Honduras, conscious of the need to limit environmental damage, chose ammonia as its refrigeration medium rather than freon-based products that have been implicated in the depletion of the ozone layer (Hansen 2000).

3.2 Marketing crustaceans

3.2.1 Importance of correct handling and quality control

In general, crustacean flesh is rich in lipids, protein and free amino acids and has a tendency to perish very quickly. The attractive colours of shell and flesh are due to dietary pigments (carotenoids and carotenoproteins) but can be marred by blackening during freezing and thawing (Konosu & Yamaguchi 2000). Thus, in order to ensure that products reach the consumer in good condition, attention to quality control and careful handling is essential right through all stages of harvesting, processing and marketing. Quality control measures can be begin at the farm even before harvesting commences – for example, by taking a sample of animals and checking for soft shells and any abnormalities. Soft-shelled shrimp and prawns are liable to break up during harvesting and processing. For all farmed crustaceans the aim is to harvest when the majority are at mid intermoult stage because at this point the water content and quality of the flesh are optimal.

The use of ice is vital when dealing with fresh product. Ice serves to prevent desiccation, retard bacterial growth, and slow the rate at which flesh will spoil. In spite of this many small-scale traders in developing countries lack knowledge in its proper use. Common mistakes are the use of chunks rather than crushed or flaked ice (these forms are softer and have a much better contact area), the addition of ice only to the top of baskets of fish and shellfish, and the use of dirty ice which has been in contact with the market floor. Advances in ice-making technology now enable the production of forms of ice with characteristics superior even to those of flaked ice. Flaked ice is thin (1–2 mm), has a low temperature (–5°C) and tends to freeze together unless stored in a cooled bunker. Chip ice, in comparison, is thicker (7–8 mm) and has a higher temperature (–0.5°C). It has the advantage of easier storage, in insulated rather than chilled bunkers, and being closer to melting point it steadily moistens the product and has better cooling characteristics (von Rohr 1995). The latest ice technology, currently being tested with fish, uses a pumpable mixture of ice crystals and water known as slurry or binary ice (Wang *et al.* 2000).

Ice application should begin as soon as possible after harvest to swiftly lower temperatures and maintain them during sorting, grading, weighing and any other processing steps prior to cooking or freezing. Figure 3.1 shows a time and temperature profile for shrimp from one integrated farm and processing plant in Indonesia. All temperature control before freezing is accomplished with chip ice or with a mixture of chip ice and water. As a rough guide, 5 kg of ice are needed for each kg of product, half to be used at the farm and the remainder during processing.

When dealing with live animals careful handling is also critical and can greatly affect market value. Shipping and storage facilities provided for exports of live Canadian lobster must result in survival rates above 95% if premium prices are to be obtained. Improvements to acclimation and packing techniques have reduced mortality rates among airfreighted spiny lobsters to as low as 1% (Stevens & Sykes 2000; Sugita & Deguchi 2000). High survival rates for shipments of live *Marsupenaeus japonicus* to Japan are also used as an indicator that the whole consignment is likely to be of premium quality (Ovenden 1994).

The impact of quality can be readily observed in international crustacean markets when products fetch different prices depending on their country of origin. These differences are based on established reputations for quality rather than the condition of each batch being handled. Thus although China is now building a strong reputation as a supplier of high-quality seafood (Traesupap *et al.* 1999), the average price paid for white shrimp from China has traditionally been lower than that for equivalent Ecuadorian product. In Bangladesh the prices paid for shrimp and prawns have been 7–16% lower than the Asian average because of the poor record of some processors (Cato & Lima dos Santos 1998). In 1997 the EU temporarily banned seafood imports from Bangladesh, India and Madagascar when inspectors found serious deficiencies in hygiene standards (Anon. 1997). Unfortunately reputations for quality are rapidly damaged and only slowly repaired. The seizure of a single batch of home-produced shrimp by Australian fisheries inspectors in 1990, and subsequent declaration as unfit for human consumption, was potentially damaging to the whole of Australia's shrimp farming business (Ruello 1990). In Malaysia, problems with maintaining a high-quality image for shrimp exports were at one point attributed to a lack of confidence of investors in purchasing state-of-the-art processing machinery (Low 1988). Even with the concerted effort of a majority of farmers to raise standards, it is difficult to improve a country's reputation for quality, and action at a national level becomes essential. The Ocean Garden Products Corporation coordinates the sale of Mexican seafood on the US market

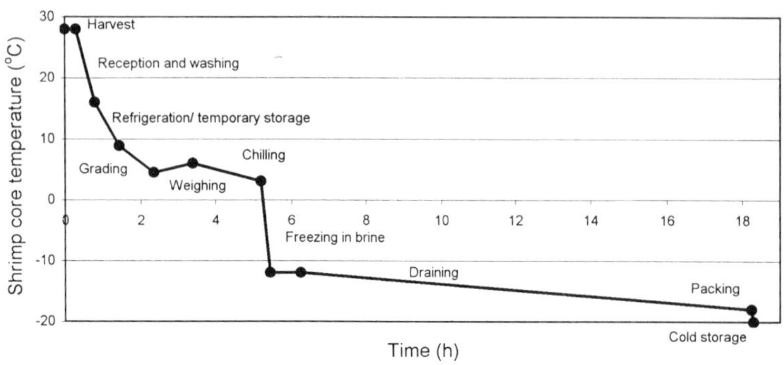

Fig. 3.1 Profile of shrimp core temperature over time following harvest and individual quick freezing (IQF) in brine at an integrated farm and processing plant. Between harvest and brine freezing, temperature is reduced with chip ice (D. Lee 1999, unpublished data).

Plate 3.1 Thai women processing farmed shrimp. Note that in this picture the face masks are not always covering the nose properly.

and by enforcing quality control standards it has been able to keep the prices of Mexican shrimp above those of many other exporting countries. There is some scope for the efforts of individual operations to have a positive impact on the marketing of their own products. For example, shrimp sold on Sydney Fish Market from one farm in New South Wales fetched 25% more than shrimp sold the same day from a Queensland farm, the distinction being based on the perceived difference in quality from the two sources (Ruello 1990).

Great care is needed to maintain the high-quality, health-food image of crustaceans (Nettleton 1992) and to assure they are safe to eat. Unfortunately shellfish (mainly molluscs) are responsible for serious outbreaks of food poisoning and their reputation is also vulnerable to fears about water pollution. Fourteen deaths due to *Shigella* in the Netherlands in 1984 were attributed to infected shrimp imported from Asia. Food safety concerns prompted precautionary measures in the USA in 1999 when the FDA banned interstate marketing of vacuum-packed fresh seafood including crayfish tails. Vacuum packing prolongs shelf life by removing air and by slowing the growth of spoilage bacteria, but the bacterium *Clostridium botulinum* can grow in anaerobic conditions at temperatures above 2°C and produces deadly toxins (Anon. 1999a). Such fears about the safety of crustacean products should however be kept in perspective. Shrimp, for example, continues to represent one of the safest forms of muscle protein consumed in the world and it is much safer than fish or chicken. Problems usually involve mishandling or cross contamination in retail or food service settings or in the home rather than defective product *per se* (Otwell & Flick 1995).

Further food safety concerns are prompted by the possible misuse of antibiotics on farms and the use of adulterated feed. The uncontrolled use of oxytetracycline and oxolinic acid in some shrimp ponds in Thailand has led to the detention of exports to the USA and Japan (Srisomboon & Poomchatra 1995). Penicillin residues in farmed shrimp triggered allergic reactions in consumers in Germany and prompted importers to seek guarantees that their shrimp supplies originated from farms that do not employ antibiotics. Delaying the harvest for an appropriate period following most drug treatments allows residues to be purged from the flesh (Mishra & Singh 1999). However, pesticides and heavy metal residues may also accidentally accumulate in farmed crustaceans from the water or from food. For example the contamination of animal feedstuffs by dioxins in Belgium in 1999 raised concerns about possible contamination of aquaculture products and highlighted the need for traceability in feeds and feed ingredients.

In most countries strict microbiological and quality specifications are laid down for imported foods by health or food authorities and shipments may be detained for inspection and destroyed if they are substandard. However, random sampling and detention of defective lots is an ineffective method of assuring food safety and it is being replaced by the improved approach known as HACCP. This can be characterised as a shift in emphasis away from quality control and towards quality assurance.

3.2.2 Food safety and HACCP

In response to consumer concerns about food safety, food regulatory bodies are requiring food producers and, in particular food processors, to adopt the HACCP system. This is a technique of quality assurance that focuses on food safety. It is designed to identify hazards, establish controls and set up a system for monitoring these controls and their effectiveness (Reilly & Käferstein 1997). In any particular operation HACCP may simply build upon established procedures for good sanitation and safe operation, but it is something of a revolution because it focuses very strongly on prevention rather than cure and because it aims to make seafood handlers, rather than government agencies, responsible for assuring food safety (Evans 1995). The system originated in the private sector as a means of producing defect-free food for astronauts and there are seven elements to its implementation, usually called the seven principles of HACCP:

(1) Analyse hazards and identify control methods
(2) Identify critical control points (CCPs)
(3) Establish critical limits for CCPs
(4) Establish monitoring and checking procedures
(5) Establish corrective action to be taken when critical limits are exceeded
(6) Establish a record keeping system
(7) Establish verification procedures

Under previous quality control systems, regulators or interested parties would inspect food processing plants and would only obtain a 'snapshot' of prevailing conditions. In contrast, once an HACCP approach has been adopted, a better insight into the workings of a plant can be obtained by inspecting the summary document known as the HACCP plan and by reviewing data from critical control points to ensure that either the critical limits are not exceeded or that appropriate corrective action has been taken. HACCP does not eliminate all risks but it brings safety and quality benefits to the food producer or processor since it reduces the chance that substandard product will be marketed. It has been endorsed worldwide by organisations such as the Codex Alimentarius Commission of the UN, and it is viewed as an efficient system because inspectors can more easily identify problematic operations and can thereby focus their limited resources with greater precision. Before the introduction of HACCP, there was an over-reliance on the testing of consignments of finished product and the rejection of any defective lots, for example, those containing noxious substances or pathogenic organisms. This 'curative' approach to food safety is potentially very costly when consignments are rejected or very dangerous when contaminated product proceeds undetected into the food supply chain.

Despite the clear advantages of HACCP, it can be problematic and initially it can be expensive (Lima dos Santos *et al.* 1994; Cato & Lima dos Santos 1999). Intensive staff training may be needed before the principles can be successfully implemented and significant investment may be necessary to bring processing plants up to the required sanitary standards. The required inputs are usually greatest in developing countries. To address this problem, aid programmes, such as the ASEAN–Canada Fisheries Post-Harvest Technology Project, have been initiated (Suwanrangsi *et al.* 1997; Wiryanti & Madakia 1997) and very positive results are being achieved. A major exercise to implement HACCP programmes in Bangladesh has also had a positive impact on quality. One important change was to incorporate all deheading and peeling within the processing line of the exporting plants to eliminate the risks associated with preprocessed product that may have been handled in unsanitary conditions (Cato & Lima dos Santos 1998). However, despite such progress there is a risk in developing nations that exporters view HACCP-based food safety regulations simply as unfair barriers to trade. This risk is particularly high when dealing with value-added seafood products for which more handling is required and for which the control measures need to be most stringent. Seafood producers need to be encouraged to produce more value-added products under effective HACCP plans rather than to be held back by apparent trade barriers.

HACCP is relevant to all stages of production from ongrowing through to final packing and distribution. On a crustacean farm the adoption of HACCP will focus attention on the risks of contamination from industrial and human waste and from agricultural and antibiotic residues. Post-harvest handling becomes the first point at which preventive measures can be taken to limit the effects of bacteria. These and other practical aspects of HACCP for crustacean farmers and processors are addressed in section 9.6. As long as buyers maintain their focus on food safety, producers will seek to comply with HACCP regulations to gain a commercial advantage over non-compliant competitors.

3.2.3 Importance of reliable supplies

In many situations, farmed crustaceans can provide a more reliable and consistent supply than wild sources. This can be a distinct advantage when it comes to satisfying the needs of a processing industry as well as the final market. Indeed, despite some initial resistance, farmed black tiger prawns (*Penaeus monodon*) have made a successful impact on world markets, precisely because they now have a reputation for consistent supply as well as high quality. Aquaculture output has also had beneficial implications for the US crayfish market (section 3.3.3.1). In contrast, the failure of pioneer freshwater prawn farms in the USA to provide the regular supplies needed to develop a new market for their products led to many business failures (section 3.3.2).

If availability is predictable and prices are steady, it is much more feasible for restaurants to include crustaceans on their menus. Reliable supplies of Canadian lobsters on the UK market have made this product available for virtually 365 days of the year and allowed it to be included on the menu for sporting events and banquets. The former British Crayfish Marketing Association, realising the importance of predictable supply, established a UK crayfish season lasting from January to April and set prices for signal crayfish at the beginning of each season to allow fixed menu pricing (Clarke 1989).

Some of the best prices for aquaculture products are obtained when seasonal demand cannot be met by wild catches. For example, demand for live kuruma shrimp (*Marsupenaeus japonicus*) in Japan peaks around New Year and during the flower viewing season in April, and shrimp farmers are able to time their harvests to coincide with these periods when wild catches are typically low (section 3.3.1.2).

3.2.4 Harvesting strategies

In most crustacean farms the overall approach to harvesting will be determined by market considerations. Shrimp and prawn ponds and some types of crayfish pond may be harvested either in a single complete operation or as a series of partial steps. Single complete harvests of large shrimp ponds are only feasible when facilities for handling and processing bulk quantities are available, for example when supplying frozen shrimp for an export market. Multiple partial harvesting however is suitable for the supply of smaller, sometimes local, markets, e.g. hotels, where limited batches are more acceptable than bulk quantities and can fetch higher prices. Live sales in particular rely on this latter approach.

For a large farm it can be beneficial to design and manage the operation so as to obtain a steady flow of product. This helps to maintain a constant market presence and can be particularly useful for farms that are part of an integrated concern that operates its own processing plant. In one shrimp farm study (Table 10.1) it was estimated that 78 harvests per year would be required to achieve the desired level of production continuity and that these harvests could be provided by 30 ponds, assuming 2.6 cycles could be obtained per pond per year. While this is attractive on paper, the reality of continuous production is rarely achieved in outdoor ponds, not least because of the vagaries of weather conditions and fluctuations in seedstock availability and water supply.

3.2.5 Market development

The need for aquaculture market development has been clearly identified by, among others, Filose (1988) who recommended that the emphasis in aquaculture must switch away from a 'pure production mentality', in which huge tonnage increases are 'thrown' at buyers in a haphazard manner, and move towards the formulation of considered objectives and strategies to create new sales. In a similar vein, Kinnucan and Wessels (1997) note that aquaculture is best served if marketing research precedes production-related research, and not vice versa.

The aim of market research is to develop an in-depth understanding of consumer perceptions, how they are constituted and how they might be influenced. Consumers base their choice of food on their established beliefs about a product and on a range of other factors including colour, cleanliness, size, apparent freshness, packaging, brand and price. Research, either through direct test marketing or questionnaires, aims to provide insights into this complex mix and it can help establish likely levels of acceptability for new products. The marketing of aquaculture products relies all the more on an understanding of consumer preferences because it involves getting consumers to recognise new or slightly differentiated products. Marketing concepts and market research methods are discussed in relation to aquaculture products by Kinnucan and Wessels (1997) and Avault (1999a) while Hanson *et al.* (1994) describe how the technique of consumer profiling can assist market research.

Market development aims to capitalise on the information gleaned by market research. The subject can best be approached by reference to a series of marketing concepts: product differentiation, market segmentation, demand function modification (see Glossary), market extension and segment development.

The goal of *product differentiation* is to create sufficient customer loyalty so that a product commands a premium price, a greater market share or both. Common methods of achieving this include pricing, branding, labelling and grading. An example would be to label crustaceans as 'Produce of Australia' to capitalise on the positive, pollution-free image of that country.

Market segmentation is technically defined as the situation in which similar products are perceived by the consumer to be unique with regard to at least one of their attributes. A particular market segment may be defined in terms of outlet point, e.g. supermarket or restaurant, or frequency of consumption. It may also be defined by demographic factors that relate to a specific product usage (e.g. income level, age, sex, family size, social class and occupation). Segments exist because consumer preferences are so heterogeneous and multifaceted. Some consumers will place more emphasis on nutritional qualities, others on taste, yet others on convenience. Accurate identification of market segments is critical to the development of effective marketing strategies. An ideal segment should be of a sufficiently large size, possess potential for growth, and not be over-occupied by competitors. Once a suitable segment has been identified promotional efforts can be targeted accordingly. For example, advertising can be delivered in the right place and at the right time once spatial and seasonal demand patterns are understood. One approach that may be especially suited to smaller organisations is to concentrate marketing efforts on just one particular segment of a market, rather than produce a wide range of different products.

Demand function modification is a managerial strategy that aims to influence the consumer's demand for a product by altering their assessment of its performance on a given attribute. To take an example from finfish marketing: catfish demand has been successfully boosted in the USA by generic advertising focusing on the farm-raised origins and 'tastiness' of the product. There is some potential in promoting these characteristics in farmed crustaceans since they are usually of equal or higher quality than equivalent products from the wild (most can usually be processed within an hour of harvest). However it may be difficult to replicate the success achieved with catfish because, if distinctions are made, consumer preference tends to favour wild sources as it already does in the case of salmon. It is often necessary to educate potential customers about the desirable features of crustacean products before a significant level of demand can be generated. Providing recipes, serving suggestions, cooking instructions, calorie counts and nutritional information on packages and during promotional campaigns can improve consumer uptake of food products. The value of cooking instructions is illustrated by the case of freshwater prawns that can be easily spoiled through overcooking if prepared in the same way as the more familiar marine shrimp.

Market extension involves expanding the geographical range of a product's consumption. For crustaceans this usually involves extension beyond the traditional areas in which they are eaten. Market extension for live and frozen products, however, requires that a network of suitable handling facilities be in place. Indeed, the rise of shrimp farming in regions like South-east Asia has only been possible through the evolution of infrastructure for cold storage and transportation, to facilitate access to valuable city and export markets. In an analysis of market channels for farmed shrimp in Hawaii, Macaulay *et al.* (1983) identified the need for a greater number of more widely dispersed selling points, and for increased speed of delivery to shorten the time between orders being placed and deliveries being received.

Segment development is a particular kind of demand function modification (see above) that aims to generate distinct groups or segments of consumers by creating within them a set of common perceptions about a product. In analytical terms, the demand functions of individuals are modified so that they become similar. In practice, pursuing the strategy of demand function modification often leads to the development of segments as by-products anyway. An example of segment development would be to promote the health-food benefits of crustaceans to create a market grouping of adult consumers with high blood pressure or cholesterol levels. Crustacean flesh in many product forms has gained widespread acceptance as a health food as a result of increasing consumer interest in lighter meals, balanced 'natural' diets and a general dislike for foods containing additives (Nettleton 1992). The low to moderate levels of cholesterol found in crustaceans such as shrimp do raise cholesterol levels compared to a low-cholesterol diet; however, they shift the balance between high-density lipoprotein (HDL, the so-called 'good' form of cholesterol) and low-density lipoprotein (LDL, the so-called 'bad' form of cholesterol) in favour of HDL and this can reduce susceptibility to heart disease (Jory 2000).

Avault (1999b,c, 2000b) describes the marketing of aquaculture products by reference to the four 'P's of the marketing mix: product, price, promotion and place (of distribution). Ten different pricing strategies are described including low, penetration pricing for a new product, and prestige pricing in which a high price is charged, in combination with different packaging and brand name, to appeal to those consumers who always purchase top-of-the-range products. Avault (1999b,c, 2000b) also considers product promotion through advertising, personal selling, general publicity and a sales promotion programme and stresses that advertising messages need to be clear, succinct and focused.

In many situations, people are only accustomed to eating shellfish in hotels and restaurants, so to encourage purchases for home consumption, products need to be presented in retail outlets in prepared or easy-to-cook forms. Good results have been achieved with shrimp and home consumption has risen to 50% in Japan and 30% in USA (Ferdouse 1994). This approach will, however, usually require investment in new processing plant and creation of a company market identity. Examples of alternative products to live crayfish include: frozen boiled whole crayfish; frozen peeled and unpeeled tails; and frozen or canned prepared products such as crayfish soup. Alternatives to live lobsters include cooked and frozen whole animals in brine, and blanched lobster sealed in a vacuum pouch. These product forms all have the benefit of an extended shelf life and cheaper freight rates than live animals.

Trade associations can play an important role in the co-ordination and execution of promotional efforts that may be beyond the means of individual producers or processors. They can also serve to lay down marketing standards. An example of one such organisation was the British Crayfish Marketing Association (BCMA) that had the objective of placing crayfish on the menu of leading UK hotels and restaurants. Before ceasing operations in 1991, it co-ordinated the marketing of 10 cm CL, hand-graded crayfish and established two grades of product: premium grade with undamaged claws at 15–19 kg^{-1} (approx. 50–70 g each) and standard quality with slight limb damage or smaller size (33 kg^{-1}). The marketing effort also involved encouraging regular buyers to install holding tanks (Richards 1988; Clarke 1989). The BCMA was established as a co-operative venture with the help of a government grant, and clearly governments can play a useful role in supporting marketing organisations (section 11.5.3.3). In the developing world, institutional support can be especially valuable to small-scale operators who as individuals do not possess adequate resources to efficiently market their products.

3.3 World crustacean markets

3.3.1 Shrimp

The world market for shrimp, around 3×10^6 mt, is dominated by the USA, Japan and Western Europe, which between them account for around 85% of the world trade. The remainder is sold in lesser markets such as Australia, Singapore and Hong Kong or consumed within producer countries. A significant section of the market (approximately 11–14%) is for cold-water species of which *Pandalus borealis* and *Acetes japonicus* are among the most important (Josupeit 1999). Their importance, however, varies greatly between the three main markets: around 50% of European shrimp consumption depends on cold-water species; the equivalent figures for the USA and Japan are close to 10% and 5% respectively. Warm-water species, of which *Fenneropenaeus indicus*, *F. chinensis*, *Penaeus monodon* and *Litopenaeus vannamei* are the most important, account for the rest of shrimp consumption in these three markets.

Shrimp farms rely on warm-water species and in 1999 two main species, *Penaeus monodon* and *Litopenaeus vannamei*, between them accounted for more than 70% of farmed output. In general, the great bulk of cultured shrimp flows from the developing tropical producer countries to the nearest of the three main markets in the developed world.

Frozen shrimp is now a traded commodity. The white shrimp futures and options contract was launched in 1993 and was followed by that for black tiger shrimp in 1994 (MGEX 2000). Futures and options enable shrimp buyers and sellers to lock in profit margins before the physical delivery of the product and provide insurance against sharp price fluctuations. They can thus provide the basis for more predictable cash flows. However the futures market may not always function efficiently (Anderson & Fong 1997). Often the relationship between shrimp buyers and sellers is secretive and there is a wish to avoid providing competitors with benchmark prices. Hence the disclosure of production and pricing data is limited and non-verifiable and this generates a poor statistical base on which to drive the futures market. Many shrimp traders view seafood futures simply as a speculator's ploy rather than as a valuable hedging mechanism (Jovellanos 1993).

The prospect of seafood trade via the Internet, as a means of cutting out middlemen and saving money, has not lived up to expectations. There are a dozen or more sites on the Internet that purport to trade seafood but transactions on these sites are almost non-existent (Vidali *et al.* 2000). International seafood trade relies on many intermediaries who carry out important functions such as providing technical help on terms and conditions, getting FDA clearance, providing finance, warehousing, breaking down shipments, and handling problems with defective lots. And these tasks are not always amenable to the usual efficiency gains achieved with e-commerce.

Despite the increasing importance of aquaculture, shrimp prices are still determined to an important degree by the ex-vessel prices of wild-caught shrimp. This means that fuel prices are very influential since diesel fuel constitutes around 70% of the cost of operating a shrimp trawler. By contrast, in a semi-intensive shrimp farm, expenditure on fuel usually represents less than 10% of operating costs and in consequence these operations are much less sensitive to fluctuations in fuel prices (section 10.6.1.6). Pricing is complicated by the fact that the size ranges of shrimp from fisheries and aquaculture do not coincide. While fisheries yield all sizes, aquaculture production centres on small and, especially, medium sized shrimp. The result is that, while supplies of large shrimp are limited by the wild catch and prices have remained high, supplies of small and medium sized shrimp have increased and competition is stiffer. Figure 3.2 provides an example of the different prices that can be obtained for various size categories which are usually based on 'counts' – the number of whole shrimp per kilogram or the number of tails per pound (Appendix 2). The supply of aquacultured shrimp grew so swiftly in the late 1980s that markets became highly vulnerable to oversupply and sharp falls in price. Problems with black tiger shrimp from Thailand were particularly acute in 1989 and they prompted trade associations and government departments to begin a co-ordinated promotional campaign to boost home shrimp consumption. The 1990s have also witnessed price volatility, with shocks coming from disease-related supply problems, notably the collapse of Chinese farm output in 1993 due to white spot virus and the impact of the same virus in Ecuador in 1999 and 2000. Economic depressions, particularly in Japan, have constrained demand and prices, and the Asian financial crisis prompted a slump in prices in 1998.

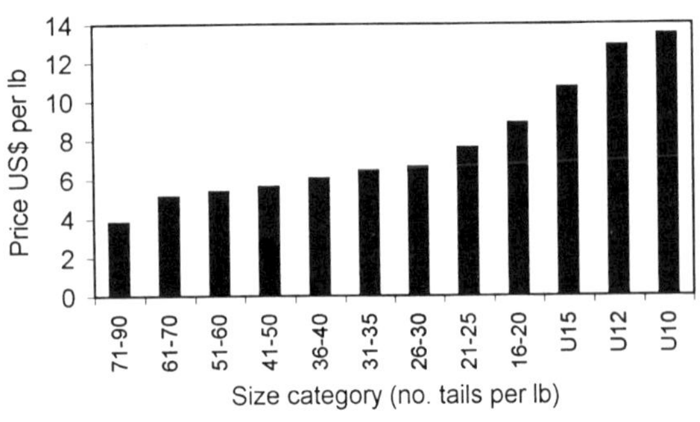

Fig. 3.2 Example of prices obtained for shrimp in different size categories (Appendix 2). U = under. US Gulf browns, ex-warehouse (New York, *Infofish Trade News*, April 2000).

To reduce reliance on any one particular market, diversification is advisable. For shrimp exporters, this may require selling product to Europe, the USA and Japan but it reduces the impact of unforeseen problems specific to any one of these markets. Apart from economic depression, such problems in the past have included uncertainties in the US and European market due to holding orders placed on shrimp imports and due to trade embargoes. Considerable confusion arose in 1996 following US proposals to ban imports of shrimp from nations that harvested wild shrimp with fishing technology harmful to sea turtles (sections 11.2.5 and 11.5.3.2).

Prices also vary depending on the country of origin and the shrimp type. Important categories on the US market include Ecuador white, Gulf brown (Gulf of Mexico) and Thai black tiger. On the Japanese market major categories include Indonesian black tiger, Indian black tiger, Thai black tiger and Indian white. In addition, shrimp are categorised according to product form, such as peeled and deveined (P&D), peeled un-deveined (PUD), headless, whole and individually quick frozen (IQF). Prices for categories with a relatively consistent supply, for example medium sized Ecuadorian white headless, are followed in trade journals (section 3.1) for use as benchmarks and as indicators of overall price trends.

At the level of the farm operator, a limited number of alternative strategies are available to improve the marketability of shrimp. One approach is to take advantage of the higher value of large shrimp and reduce competition with other aquaculture operations simply by growing larger animals. This is generally only possible through the use of lower stocking densities rather than prolonged ongrowing (Fig. 8.4). However, under pond conditions lower stocking densities usually result in reduced yields per hectare (Fig. 10.5), and this may offset any increase in the value of the crop due to the larger size of individual shrimp. The reality of the economics of this trade-off between yield and individual value is one of the main factors underlying the decision of most farmers to produce small and medium sized shrimp in the first place. If shrimp are harvested sooner rather than later there is certainly less risk of crop failure, and during the 1990s shrimp markets showed a remarkable ability to accept increased quantities of small farmed shrimp without suffering a fall in prices.

Diversifying production to include alternative shrimp species is another option for some farmers, assuming that the requirements of the new species can be met and its market characteristics are borne in mind. In the assessment of the culture potential of different shrimp species in Taiwan, aspects of market acceptability were given considerable priority (Liao & Chien 1990). More recently, trials with alternative species have been prompted by the desire to find stocks that are resistant to disease rather than attributes that improve marketing options. Both *Litopenaeus vannamei* and *L. stylirostris* have been introduced to Asia with this in mind (sections 2.6.2 and 8.10.1.3).

At the level of the processor, the output of more value-added products is helping to expand demand, particularly for the smaller sizes of shrimp. Such products are being produced increasingly in developing countries, and joint venture deals are often established with foreign importers. Accompanying this trend is a movement away from exporting traditional 5 lb and 2 kg blocks towards the production of smaller consumer packs for retail sale. Precooked and easy-to-cook products are increasingly favoured by consumers in place of fresh whole seafoods, both for their convenience and the absence of 'fishy' odours. Existing value-added products include:

- cooked peeled and deveined shrimp in 200 g trays
- breaded or battered shrimp
- shrimp on seafood skewers (brochettes) with cuttlefish and paprika
- oriental stir-fry dishes and ready-to-eat noodles
- shrimp in ready-to-eat consumer packs of sashimi
- small shrimp in tomato or cocktail sauce
- IQF raw headless shrimp with tail fan on
- raw peeled and deveined headless shrimp with tail fan on and splayed flesh (butterfly form)
- shrimp paste, soup and crackers
- shell-on easy-to-peel IQF

Warm-water shrimp can be peeled raw while cold-water shrimp are usually cooked or blanched prior to further processing. Polyphosphates are sometimes added to facilitate shell removal but can give a transparent appearance and slimy texture to the product if overdone (Henson & Kowalewski 1992). Shrimp is clearly an extremely flexible product that can be presented in a wide variety of market forms (Fig. 3.3). One seafood supplier (Ocean

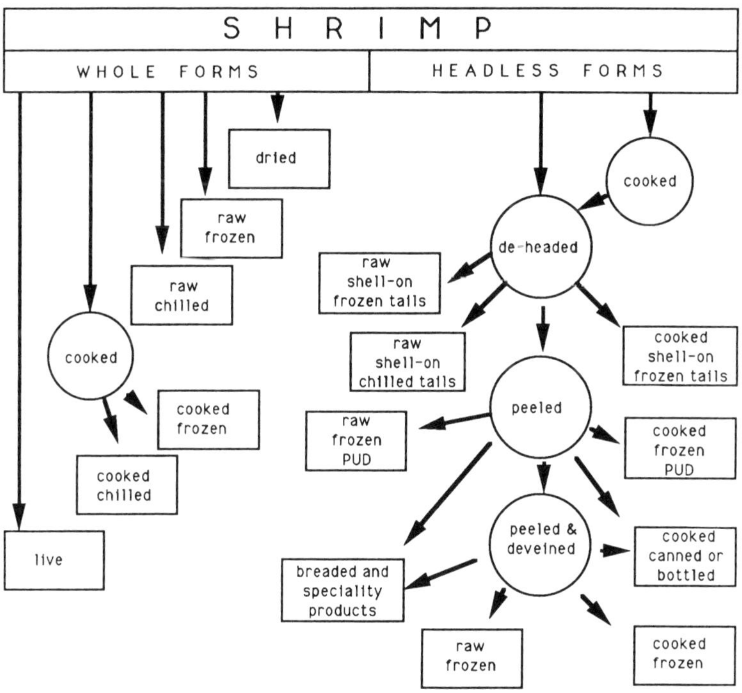

Fig. 3.3 Shrimp product forms. PUD = peeled un-deveined; see Glossary for unfamiliar terms.

Garden Products Inc.) offers 43 different shrimp products, differentiated by product form, species, colour, and freezing method.

3.3.1.1 USA

In 1997 the USA overtook Japan as the largest importer of shrimp (Shehadeh 1997). In 1998 it imported a total of 332 000 mt, approximately half of which were shell-on headless (tails), half were peeled, including cooked and peeled, and a very small amount were canned and breaded. Supplies come from as many as 70 different countries, of which Thailand and Ecuador were the leaders in 1998 (Chauvin 1999). Supply patterns are typically labile and have been heavily influenced by disease outbreaks. Chinese farms used to be a major source of white shrimp (*Fenneropenaeus chinensis*) and this product was readily accepted in the USA as a competitively priced alternative to Ecuadorian white shrimp (*Litopenaeus vannamei*). However, supplies fell off dramatically following the collapse of the Chinese farming industry in 1993 due to white spot virus. The same virus cut *L. vannamei* output from Ecuador and Central America in 1999 and 2000. Faced with increased risks of virus outbreaks, some *L. vannamei* farmers harvest shrimp earlier than usual and hence supply smaller sizes (Vidali *et al.* 2000). Some have resorted to harvesting small, so-called 'popcorn' shrimp at just 4 g.

Confronted with reduced supplies of farmed *L. vannamei* (white shrimp), US importers have substituted black tiger shrimp from Asia (Brown 2000). Japan remains the main destination for farmed black tiger shrimp but the USA steadily took more during the 1990s as the Japanese economy stalled. US demand has also been influenced by the state of its economy and during the recession of 1990–91 consumers became more careful with their discretionary income and restaurant consumption declined. As a result, wholesale prices of black tiger (21–25 tails per lb – Appendix 2) fell by one-third from $6.10 to $4.05 (Filose 1992). Back in the 1980s farmed black tiger made slow progress in US markets due to confusion over prices and quality, because of product arriving from many different sources, and suppliers who often by-passed traditional import channels. In addition, initial consumer unfamiliarity with the term 'black tiger' meant that it was necessary to present the product in a cooked form or as a cocktail shrimp (Filose 1988).

US shrimp prices, particularly the large sizes, are greatly influenced by the level of landings in the Gulf of Mexico. These vary seasonally and are usually at their greatest in the period from April to November. In a good season 10 000–20 000 mt can be landed in Texas and Louisiana in a 2 to 3-month period (Vidali *et al.* 2000). Patterns of shrimp purchasing in the USA can be divided between different shrimp sizes (Appendix 2) and regional consumer preferences:

(1) *Large* (21–25 tails per lb or fewer, equivalent to whole shrimp of 27 g or more). Consumed mostly in higher-quality restaurants. Most supplies are obtained from the fishing fleets of Mexico, Panama and the USA. Mexico is the preferred source because it has an established reputation for high quality and so is able to set prices according to its landings.
(2) *Medium* (26–50 tails per lb, equivalent to whole shrimp of 14–26 g). Shrimp in this size range fetch considerably lower prices than large shrimp and represent an economical choice for households and restaurant owners. These sizes are produced by aquaculture, and come from Ecuador and Central America (mostly *L. vannamei*) and from Asia (mostly *Penaeus monodon*).
(3) *Small* (more than 50 tails per lb, equivalent to whole shrimp of 13 g or less). As well as being consumed in restaurants and sold in supermarkets, these sizes are used for further processing. The smallest sizes (80 tails or more per lb) are peeled and sold as raw or cooked meats in a range of convenient products for home consumption that are distributed via supermarkets and other retail seafood stores.

Consumer preferences vary between regions and generally every major metropolitan region has its particular preference. Premium quality white shrimp (mostly *Litopenaeus setiferus*) are favoured on the west coast and in the north-east; brown shrimp (mostly *Farfantepenaeus aztecus*) from the Gulf of Mexico are preferred in the middle section of the country and around Baltimore, Philadelphia and Washington DC; pink varieties (mostly *F. duorarum* and *F. notialis*) are preferred in the south-east. A market niche in the USA has been reported for 'blue' tiger shrimp. These are farmed *Penaeus monodon* that take on a bluish appearance rather than the usual black colour. The unusual colour is thought to relate to the diet and although at first affected product was discarded as substandard it later appeared to be saleable (Anon. 1990).

US market channels have four basic levels: major importers, combined distributors/importers, pure distributors/wholesalers, and various types of end user. There has been consolidation on the distribution side but an increase in the number of importers. US-based individuals with relatives and friends involved in shrimp farming, set up their own sales offices to import shrimp and undercut the prices of established dealers. As Filose (1992) puts it, 'the cost of entry seems to be simply leasing a fax machine'. Shrimp farmers within the USA tend to adopt two different marketing strategies depending on the size of their operation. Small farms sell their product to coastal processors for packing and marketing, while larger farms contract a processor to pack under their own custom brand labels and then manage their own marketing. In South Carolina average ex-farm prices are generally influenced by regional wholesale prices, not just local supplies. Any differences in price between farms depend on the sizes of shrimp harvested and on success at targeting niche markets. The best prices are paid for live shrimp but supplying this market entails greater harvesting costs and complex shipping logistics (Rhodes *et al.* 1995).

The image of shrimp as health food is particularly strong in the USA and this is expected to boost demand. The consumer attitude towards shrimp is expected to remain very positive as long as sufficient emphasis is placed on quality at all levels of production, processing and marketing. The most significant development of late is the general acceptance of shrimp on restaurant menus across the board – from high-priced to low-priced operations, and including ethnic and fast-food specialists. Shrimp eating is not restricted to seafood restaurants and is establishing itself as part of the American culture (Vidali *et al.* 2000).

3.3.1.2 Japan

The Japanese are among the largest consumers of shrimp in the world, eating per head more than double the Americans and about four times as much as the Europeans. Annual per capita consumption appeared to be reaching a plateau of 3 kg in the mid-1990s until the seafood market suffered heavily as disposable income shrank in the economic recession (Ferdouse 1999). Shrimp, which had made progress in the market for everyday food during the 1980s, are once again treated as a luxury item and imports fell in 1998 for the fifth successive year to 239 000 mt (Chauvin 1999). Of this, 98% of shrimp were frozen, the remainder being live, fresh, chilled, dried, salted or brined, and Indonesia and India were the leading suppliers. The mix of frozen product forms is approximately, raw headless (60–70%), peeled (10–15%) and whole (10%).

Although shrimp consumption has not yet responded to the strengthening of the yen, import values have been improving because more value-added products are being taken. As confidence returns to the Japanese economy it will be able to absorb more shrimp, in the way it did in the 1980s. During this period shrimp consumption was fuelled by continuous growth in personal disposable income and the associated increasing popularity of commercial eating and drinking establishments such as tempura shops, noodle shops and sushi bars. In addition, family-style restaurants became more numerous and shrimp were typically included in business lunches. Young people were consuming more ready-to-eat food, and in response to people staying out late, food shops stayed open for extended periods.

In general, since 1980 shrimp prices have become more stable as a result of the development of the mass market, but they remain sensitive to shocks. This was highlighted during the illness of Emperor Hirohito that depressed Japanese shrimp consumption as celebrations were suspended. Also the Kobe earthquake led to the loss of major markets in the Kansai region of Japan as festivities were cancelled.

Overall, shrimp sales on retail markets have been encouraged by making product available in supermarkets in special consumer packs, prepared by defrosting large volume imports, repacking into smaller quantities, and refreezing. In the 1970s home consumption accounted for 30% of sales but more recent estimates place the proportion at close to 50% (Ferdouse 1994). This increase has also been helped by promotional campaigns successfully highlighting the health-food image of shrimp, and to lower retail prices resulting from the fact that the small and medium sizes generally consumed at home are plentiful as a result of aquaculture. Large shrimp (16–20 tails per lb or fewer), on the other hand, which are consumed in commercial eating houses, come primarily from wild sources and are in more limited supply.

Patterns of consumption of different types of shrimp vary between regions and are based partly on consumer preference for shrimp of particular colours. While Osaka and Kyoto have become important areas for black tiger shrimp (mostly farmed *Penaeus monodon*), southern Honshu shows a preference for economically priced Indian white shrimp, and Tokyo accepts all species.

Overall, the variety of imported species and product forms in Japan continues to increase. Cold-water shrimp, including species from the North Atlantic, have an important share of the market. The successful development of the Japanese market for black tiger shrimp has had much to do with the success of *P. monodon* farming. At first unfamiliar to the consumer because of its different colour and texture, its acceptance was slow, but reliability of supply, initially from Taiwan, competitive pricing and high quality overcame the initial reluctance of consumers and importers alike. Although *P. monodon* production has suffered setbacks, e.g. in Taiwan, this species dominates the output of shrimp aquaculture in Southeast Asia, and Japan remains its principal world market. At the same time, the high quality of farmed *P. monodon* is also helping to boost sales in the USA and Europe.

Many Japanese importers and major retail chains have become involved in establishing processing plants, often as a joint venture, in tropical nations such as Malaysia. Supermarkets have also set up their own buying houses to import products directly from producing countries and as a result import channels are becoming more direct (Ferdouse 1999). Processors outside Japan are increasing their sales of value-added shrimp and seafood products, which they are able to produce at much lower cost than Japanese-based operations. Home-based Japanese processors mostly buy peeled un-deveined shrimp (PUD) that are used for products such as shrimp croquettes or, in the case of small shrimp, as an ingredient in instant noodles. Strict adherence to quality guidelines is ensured by placing Japanese employees of importing companies on the production lines of foreign processing plants. This approach to quality control has also enabled products to be custom made to suit the needs of the Japanese market and has reduced the need to detain large amounts of shrimp in Japan for inspection by health officials. The emphasis on food safety control in Japan has not switched to HACCP as it has in the USA and the EU (section 3.2.2).

A small but highly priced section of the Japanese shrimp market is for live *Marsupenaeus japonicus* (kuruma). In fact, at \$25–60 kg^{-1}, these live shrimp are three or four times more expensive than frozen product and are the most costly shrimp for human food anywhere in the world. Colourful and with a delicious taste, they are served raw as sashimi in high-class restaurants, sold to gourmet customers in tempura shops, or sold live in gift packages in department stores. Prices fluctuate through the year and highest demand is centred on the New Year and the flower viewing season in April. Wild fisheries provide much of the supply but this premium market also supports aquaculture operations in the southern islands of Japan and in Taiwan. The warmer, more southerly, locations of Taiwan, Kagoshima, Amami and Okinawa are able to maintain culture operations through the winter and take advantage of demand from December to May when fishery supplies dwindle. Despite the demand, supplies of live and cultured kuruma prawns from Japan and Taiwan have declined and opened the way for supplies from Australian prawn farmers based in Queensland (Ovenden 1994). The annual market has been estimated at 8500 mt with some consumption also in Hong Kong and Korea. Great attention is given to careful handling because if the shrimp are dead on arrival at the market their value is halved. After harvesting or capture they are dropped into an aerated cooling tank and the temperature is reduced in stages to 12–13°C. After 20–30 minutes, dead and soft-shelled shrimp are removed and the remainder sorted by size and packed into cardboard boxes between layers of dry chilled sawdust. Packed in this way they are able to survive for 10–30 hours (Table 7.2). The market is complex and it tends to favour the most flexible suppliers. Factors that ultimately determine prices in central wholesale markets include the total quantity supplied, quantity by size category, physical quality (survival rate, colour, smell), the day of the week, and whether or not it is raining (fewer consumers dine out when it is raining). Size is important, with shrimp of 25–28 g fetching 4.5 times more than shrimp of 12–15 g. Unfortunately the colour of cultured *M. japonicus* is generally less intense than for wild specimens and there is a price difference, with the farmed product fetching 12–42% less (Liao & Chien 1990). Kuruma shrimp from Australia typically fetch 20% less than local Japanese product (O'Sullivan 1996), achieving values in the range of \$36–52 kg^{-1} (Pyper 2000). Australian prawns are harvested using baited mesh-covered traps at night when temperatures are 24°C or less, so that shrimp can withstand subsequent chilling to 13°C. Attention is paid to careful size grading.

3.3.1.3 Europe

Shrimp are a major import for the EU representing one-third of the value of all EU seafood imports (Anon. 1998). However, European shrimp consumption per head, at 0.5–0.8 kg per year, still lags far behind the estimated 1.3 kg and 2–3 kg consumed in the USA and Japan respectively. Although consumption increased markedly in the 1980s it has since levelled off, largely reflecting limits to wild and aquaculture supply.

The European market relies heavily on supplies of cold-water shrimp (and indeed their availability has a major influence on prices) but landings have either been stable or in steady decline. As a result, opportunities for the sale of warm-water shrimp have been steadily improving and countries such as India and Thailand have become major suppliers. No single country dominates European supplies and in fact more than 50 countries export shrimp to the EU, with developing nations accounting for two-thirds of the supply. While Japan and US markets are dominated by frozen headless product, Europe prefers whole or cooked and peeled product. As elsewhere in the world, shrimp in Europe benefit from a good, health-food image.

The European market is characterised as being very price-conscious, with importers readily altering their buying patterns to satisfy a price-sensitive retail market. Increasing receptiveness to new seafood products has enabled competitively priced, high-quality shrimp from aquaculture to gain acceptance. However, warm-water shrimp in general have to consistently outperform established products in terms of price and quality if they are to greatly increase their market share. This may in part be due to the conservatism of many wholesale buyers, particularly in the UK, who tend not to think in terms of promoting or extending new markets but rather deal only in products they are able to sell on quickly. Even so, traditional European market channels are being increasingly by-passed and the number of links in the distribution chains is being reduced. In the UK, for example, major retail organisations are buying direct from suppliers and eliminating the wholesale stage.

High-quality black tiger shrimp, particularly from Indonesia and Thailand, are making an impact, but the dark appearance of *Penaeus monodon* is limiting the acceptance of this product in countries like the Netherlands where light-coloured shrimp are preferred. Latin-American suppliers of farmed shrimp send about 30% of their production to Europe, particularly to Spain, Italy and France (Rosenberry 1999). By supplying Europe's market for whole shrimp rather than the US market for tails, they are able to obtain a 100% processing yield instead of only 57–68%.

Although the markets of countries in Western Europe are usually considered collectively as the European shrimp market, sharp distinctions in buying preferences exist between northern and Mediterranean regions and from one country to the next. In fact, an overall trend towards reinforcement of these distinctions and increased market segmentation has been observed. In countries bordering the Mediterranean the preferred market form is whole raw product that is cooked for dishes and meals requiring whole shrimp. Warm-water species are more important here than in northern Europe, and Cuba is a big supplier. Cold-water shrimp, e.g. red shrimp *Pleoticus muelleri* from Argentina, are also consumed in Italy (Josupeit & de Franssu 1992). In general, northern Europe prefers cooked and peeled shrimp mostly from cold-water sources.

Superimposed on this overall pattern are differences between countries. In Italy a large market exists for large pink or red shrimp and this remains buoyant despite the increasingly high prices of the large sizes. Spanish consumers on the other hand prefer white shrimp (Nierentz & Josupeit 1988). In both Spain and Italy, farmed *Marsupenaeus japonicus* is readily accepted because its banded appearance is very similar to the locally available *Melicertus kerathurus*. Production of *Marsupenaeus japonicus* within Europe however is no more than a few tonnes per year and shows little sign of increasing. Spain, France and Belgium have the largest per capita shrimp consumption in Europe. Although French consumers traditionally have preferred cold-water shrimp because of their superior flesh texture and taste, tropical shrimp now make up 70% of all imports, and the white shrimp *Litopenaeus vannamei* has become very important. Over 80% of shrimp marketed in France are head-on and they are primarily sold cooked and ready to eat, in contrast to Spain where frozen product dominates sales. Ecuador has become a key supplier of tropical shrimp to France. Frozen imports are thawed, cooked, refrozen, packaged and dispatched to sales outlets as ready-to-eat products with a sell-by or eat-by date usually 5 days post-processing. Processors are trying to extend this period with controlled atmosphere packaging. The French processing industry is experiencing fierce competition from large supermarket chains that account for 65% of national sales and import their products directly, and then have processing carried out by specialised service providers. At the time of writing, a shortage of Ecuadorian supplies and a strengthening of the US dollar have pushed up prices and resulted in a 30% drop in supermarket sales. This effect has been even more marked in Spain (Lucien-Brun 2000).

Many consumers in Belgium, the Netherlands and Germany prefer the mild taste of cold-water *Crangon* species, which is unlike that of tropical marine species. However the wild *Crangon* fishery has declined, opening the way for small-sized tropical shrimp. Germany and Belgium now consume large amounts of tropical prawns that have been repacked in the Netherlands. The shortage of the preferred *Crangon* species has prompted unscrupulous restaurateurs in Belgium to surreptitiously mix tropical cooked and peeled product in their dishes (Hottlet 1992). In Germany 70% of shrimp are consumed in the catering sector. Canteen services favour small and cheap cold-water and warm-water shrimp, while white-tablecloth restaurants demand the large sizes, mainly headless, there being less demand for medium sizes. The UK market has a distinct preference for cooked and peeled cold-water shrimp that are usually eaten cold in salad dishes. Nonetheless, increasing UK demand for shrimp in Indian and Chinese restaurants is favouring imports of warm-water species that have a tougher texture more suitable for oriental dishes. More than half UK sales are made to the catering trade and warm-water species now make up more than 50% of all shrimp consumption. To increase their acceptability, warm-water shrimp are packed in plastic pouches and presented in the same fashion as cold-water shrimp as cocktail and salad shrimp (Lyons Seafoods 1995). Around 5% of shrimp in the UK are destined for processing into value-added products.

3.3.1.4 Other markets

Other important shrimp markets are found in South-east Asia, Australia and Mexico. Imports into South-east Asia were hit hard by the Asian financial crisis and in 1998 declined to 110 000 mt, down from 130 000 mt the previous year. Hong Kong, Singapore and Taiwan between them account for more than half of these totals (Ferdouse 1999). The decline reflects, in particular, a decline in the restaurant trade. Shrimp is very much a luxury food so economic recovery will have a very positive effect on markets in the region.

Singapore has a high domestic per capita shrimp consumption at 2.8 kg per year. It acquires product from Indonesia, Malaysia and Thailand and re-exports part of it. Hong Kong imports large amounts of shrimp, mainly from China, consumes 60–70%, and re-exports the remainder back to China (Ferdouse 1999). Taiwan, following the crash of its domestic shrimp farming industry in the late 1980s, has become a net importer of shrimp. In China, strengthening domestic demand, particularly in the southern states and Beijing, takes an important part of the local farm production that has partially recovered from disease problems of the early 1990s. *Penaeus monodon* has now replaced *Fenneropenaeus chinensis* as the main farmed species.

In South-east Asia shrimp are largely consumed at the catering level in restaurants, hotels, ordinary eateries and in food stalls. Popular product forms are live, fresh or chilled whole, peeled and dried. Local wet markets sell mainly whole fresh or chilled or thawed product. Supermarkets also sell chilled, head-on, peeled, dried and fermented shrimp and shrimp crackers. Unlike in western markets, cooked and peeled shrimp are not very popular. Asian consumption rises in the fourth quarter of the year due to year-end festivities, school holidays, weddings and, most of all, Chinese New Year celebrations in January or February. As female employment grows in the newly industrialised economies of the region, housewives prefer more easy-to-cook convenience foods and supermarkets are responding with tray-packed products.

Australia tends to export high-quality shrimp from its own wild fisheries to the lucrative Japanese market, while importing cheaper and sometimes lower-quality product from Asia. Principally it buys cooked and peeled shrimp along with some raw products. Some *Litopenaeus stylirostris* farmed in New Caledonia has been flown to market in Sydney in cooked and uncooked forms and marketed under the name Paradise Prawns (Ruello 1990). Australian supplies also come from a small but developing home-based shrimp farming industry mostly producing *Penaeus monodon*. *Marsupenaeus japonicus* is also farmed and some farms have subsequently diversified into producing *Fenneropenaeus merguiensis* marketed as Crystal Bay prawns (O'Sullivan 1999).

Mexico produces around 100 000 mt of shrimp per year from fisheries and aquaculture and around half of this total is consumed in the country by tourists and locals (Vidali *et al.* 2000).

3.3.2 Freshwater prawns

After a slow start, farmed freshwater prawns are beginning to make an impact on international markets, with several countries in Asia producing frozen prawns at prices that compete effectively with the dominant rival product, marine shrimp. As an indication of progress on western markets, shell-on *Macrobrachium* tails from Bangladesh are becoming a common sight in the freezers of UK supermarkets, often under the label 'freshwater king prawns'. In 1998 farm output from China and Bangladesh was estimated at 62 000 mt and 48 000 mt respectively (New 2000a). Although China consumes virtually all of its own production, the great majority (70–98%) of Bangladeshi prawns are destined for export, and are sent to the EU, Hong Kong, Japan and the USA, mostly in the form of frozen whole product. Figures for Thai exports, although relatively low, shed further light on this trade: shipments of frozen tails in 1998 totalled 441 mt, of which 54% went to the UK and 25% to Belgium, and shipments of frozen whole prawns totalled 674 mt of which Italy took 38% and France 31% (New 2000b). Thailand imports some prawns from Bangladesh, Myanmar and Vietnam for processing and re-export (Philips & Lacroix 2000).

Important quantities of *Macrobrachium* are also produced in Taiwan and Vietnam, and the total world output of around 130 000 mt is valued at nearly $800 million (New 2000a). Despite the international demand for frozen shrimp and prawns, many farms are unable to make money supplying this market. For them, the marketing of *Macrobrachium* continues to represent a major challenge and acts as a constraint to expansion. Thus, although *Macrobrachium* farming has become a widespread activity, the scale of developments has generally not lived up to expectations and production has declined in some areas, particularly Hawaii and Israel.

The most reliable markets for *Macrobrachium* exist where freshwater prawns are traditionally eaten and where they have established an image as a desirable fishery product, for example in China, Thailand and Taiwan. Around 95% of Chinese prawns are marketed via wholesalers with the remainder sold at the farm gate. Live and fresh prawns are preferred to frozen product. Production in both Thailand and Taiwan is around 7000–8000 mt and has been quite stable, although in Thailand there has been some contraction as a result of increased prices for rice paddy and strong competition from shrimp farming. Although Thailand exports some frozen product, the majority of prawns are sold within the country, live or fresh and whole on ice. Thai consumers prefer male prawns with short orange claws rather than blue-claw males that give more claw waste. For live transport to nearby restaurants, prawns may need to be held in containers of aerated water as they have only a limited tolerance to air exposure. D'Abramo *et al.* (1995) note that prawns can be transported live on vertically stacked shelves in well-aerated water for at least 24 h at densities of 600 g L^{-1} and temperatures of 20–22°C.

Elsewhere in the world, where well-developed national markets do not exist, many small operations have centred their marketing efforts on the local hotel and restaurant trade and nearby retail outlets. This is probably the best near-term commercial strategy and it may offer significant opportunities for development programmes to stimulate *Macrobrachium* production at the level of artisanal farmers, especially where tourism is being encouraged (e.g. Cuba).

Some farming ventures that opted to produce freshwater prawns because they were seen as an easy species to culture, and which later encountered problems with marketing their product, were accused of following a production orientated approach to species selection instead of matching output to meet market needs (Chaston 1983). Some farmers mistakenly considered *Macrobrachium* to be an ideal substitute for marine shrimp, when in fact the species are substantially different. *Macrobrachium* has a tougher shell and is more difficult to peel; its flesh has a different taste and a more delicate consistency; when cooked in the same way as marine shrimp the flesh becomes unappetising; product texture is much more sensitive to storage time and temperature abuse, and specialised freezing technology is required. Even if equivalent prices can be obtained for headless product, the processing yield is 40% (or even less with large-clawed males) compared to 57–68% with marine shrimp. In addition, *Macrobrachium* cannot strictly use the desirable labels 'shrimp' or 'seafood', although, in the Caribbean region, producers disguise the freshwater origins of their product somewhat by using labels such as 'Langostinos del Caribe' and 'La crevette bleue des caraibes'.

New (1990) noted that the economics of *Macrobrachium* farming would improve if freshwater prawns could be sold on a market that did not differentiate between shrimps and prawns. One such market exists in Belgium, where freshwater prawns from Bangladesh are accepted because of their similarity in taste to the preferred cold-water species *Crangon* (Nierentz & Josupeit 1988). It has even been reported that headless freshwater prawns have to some extent become a substitute for langoustines and have been sold under the name 'scampi' – a term usually reserved for *Nephrops* (Hottlet 1992).

Examples of *Macrobrachium* market prices from around the world in 1999, compiled by New (2000b), show that farm gate prices tend to be highest in the USA, and in the French West Indies and French Polynesia, where they mostly fell in the range \$10–20 kg^{-1} (frozen, fresh and live) but often exceeded \$20 kg^{-1}. Some of the lowest prices were found in China at \$4–6 kg^{-1} (live), and in Thailand at \$1.3–6.1 kg^{-1} (live and fresh).

As a result of its different characteristics, successful market development for prawns has often relied on selling whole animals to a gourmet market, highlighting their differences to shrimp rather than their potential as a substitute product. The best approaches to promoting *Macrobrachium* sales in new markets were reviewed by Philips and Lacroix (2000). To boost restaurant sales they identify the need for chefs to be educated about the characteristics of freshwater prawns and for promotional exercises to be undertaken. Such exercises may include cooking contests, inviting chefs to farms, and 'two for the price of one' sales promotions. Consumer awareness about the product also needs to be raised and, if potentially lucrative farm sales are to be enhanced, a clean orderly farm is essential to give the product credibility. Selling other products at the farm, such as fish alongside the *Macrobrachium*, can help to create more sales traffic. A good sales network also needs to be established and farm output must be sufficient to maintain steady supplies but also be flexible enough to respond to increased demand during holiday periods. Even if all these measures are taken, consumer resistance can still arise when careless producers allow poorly stored product to reach the market and damage the product's reputation.

Transporting and storing prawns fresh on ice is feasible for short periods only, because the flesh rapidly turns mushy after about 4 days under these conditions. To extend these periods the 'kill chill' process can be employed in which prawns are dipped in iced water and then blanched (precooked) at 65°C for 15–20 s. Alternatively, flushing with carbon dioxide and storage at 4°C can also extend the life of fresh product. The critical subject of post-harvest handling and processing is described by Madrid and Philips (2000). For fresh or frozen product they stress the importance of minimising physical damage to the prawns during harvest; washing with clean chlorinated water to remove bacteria; and rapid killing by thermal shock in iced water (50 kg prawns with 50 L of water and 80 kg ice for 30 min). These measures help preserve flesh quality by limiting the impact of bacteria and proteolytic enzmes. The shelf life of fresh product can be also be extended by deheading, and if prawns are quick frozen they can be stored for up to 6 months at –35°C. However, the freezing process must be very rapid to avoid ice damage to the tissues, which is another cause of mushy texture. Small-scale operators who use domestic freezers to store their prawns risk producing a very low-quality product unless they pre-chill the prawns and only freeze small quantities at a time.

Despite estimates of a potential US market for some 4500 mt of whole prawns per year, attempts to develop and supply this market have failed through inability to provide a regular supply of prawns and weak acceptance of the product at cost-effective prices (New 1990). US companies such as Amfac and General Mills have retreated from prawn projects in Hawaii and Honduras respectively, and one Texas operation abandoned farming essentially because 'prawn proved impractical to market'. Unfortunately, the US consumer is not familiar with head-on shrimp and prawns and the general preference for headless shrimp results in consumer resistance to whole product. Test marketing of prawns in South Carolina showed good potential for sales but also recorded a preference for headless product (Liao & Smith 1981). More recently, some success has been achieved with large specimens supplied fresh, mostly heads-on, from Costa Rica, Dominican Republic, Mexico and Puerto Rico. The prawns are positioned between smaller shrimp and larger lobsters for consumption in white-tablecloth restaurants in major cities. Trials with breaded tails have been hampered by the fact that the shorter, wider tail of the freshwater prawn makes it appear smaller than equivalent penaeid tails (Chauvin 1992). Prawns produced in Hawaii are mostly sold on home markets live or freshly iced to ethnic Filipinos. The use of the label 'Hawaiian Prawns' has assisted the development of a premium speciality market but even so, prawn production through the 1980s in Hawaii gradually declined, primarily due to low pond yields and a limited domestic market. It has not shown signs of recovery in the 1990s. All the same, the farming of *Macrobrachium* within the USA presents the seafood retailer with some novel products: actual fresh prawns as opposed to supposedly 'fresh' marine shrimp that are actually frozen product recently thawed; live product for display in tanks; and very large sized prawns – 'jumbos' – for which there is no equivalent farmed shrimp product (Tidwell 2000).

Some prawns farmed in Israel have been sold in the EU, and a heads-on market is becoming established in Italy, France and Spain. Once again, irregularity of supply has been a problem in developing the potential of these markets, and the generally unfavourable economics of prawn farming have resulted in Israeli freshwater aquaculture being redirected towards tilapia production and trials with Australian redclaw crayfish (I. Karplus 2000, pers. comm.).

3.3.3 Crayfish

During the last decade international trade in crayfish has greatly increased, largely as a result of rising farm production and exports from China. Between 1992 and 1995, for example, Chinese exports to the USA increased eight-fold. Total Chinese production, based on the species *Procambarus clarkii* introduced from Japan in the 1930s, was put at 40 000 mt in the early 1990s (Pérez *et al.* 1997) and as high as 70 000 mt in 1999 (Huner

2001). In the USA, Chinese product is often priced much lower than local product (Caffey *et al.* 1996) and it has achieved rapid market penetration. This success has unfortunately prompted protectionist measures in the form of penalty duties averaging 123% and Chinese imports of tail meat, which totalled 2600 mt in 1998, are dropping as a result (Hempel 1999). The Louisiana Crawfish Farmers Association has responded more positively to the competition by sponsoring culinary shows, mounting advertising campaigns and publishing brochures to distinguish local US product from the imported one. Partly as a result of these efforts the wholesale prices of local product are now double those of imported product (Huner 2000). Chinese crayfish are mostly exported in the form of frozen peeled tails in size categories 80–100, 100–150 and 150–200 lb^{-1} (Robinson 2000). Other product forms, such as cooked and seasoned whole crayfish, are supplied along with some speciality products prepared for Scandinavian and US palates with dill sauce or dill brine and packed in specially shaped containers.

Despite the growing importance of Chinese production, most of the available market information for crayfish relates to the other three main regions where they are fished, farmed and consumed, i.e. the USA, Europe and Australia.

3.3.3.1 USA

Crayfish find a ready market in the southern states of the USA where they are obtained from both fisheries and extensive farm systems (section 7.5.4). The industry in Louisiana covers 50 000 ha and yields 16–20 000 mt per year (Caffey *et al.* 1996; Harvey 1999), accounting for 90% of US farmed crayfish. The bulk of Louisiana's production, around 75%, is consumed in the same state, where the appeal of crayfish rests largely on the popularity of Cajun cooking, part of the region's French heritage. Exports of US crayfish to Sweden account for about 7% of Louisiana production (Anon. 1999a), and include custom-made products such as graded whole crayfish packed in trays. Exports totalled around 4000 mt in 1994 but fell sharply in the face of strong foreign competition and totalled just 800 mt in 1998. The principal EU importers, Sweden and France, expressly forbid the movement and introduction of live *Procambarus* (sections 7.6.9, 11.3.2 and 11.3.3) and thus exclude US producers from the premium priced market for live crayfish. The US market for processed tail meat is facing stiff competition from cheaper Chinese product (around \$11 kg^{-1} wholesale) and some imports are now coming from Spain too. Ironically the demand for traditional Cajun cooking has come to depend on imported crayfish meat as well as local supplies (Huner 2000).

Between 1993 and 1998 the wild fishery in Louisiana yielded between 9000 and 31 000 mt per year (Harvey 1999). Production has always been highly seasonal and centres on March, April and May. Aquaculture has extended output to the period from November to May and this has assisted the trend towards expansion of sales in 'out-of-state' restaurant and retail markets. Recent drought conditions (1999–2000) however, have resulted in reduced supplies, and crayfish are both smaller and, at \$3 kg^{-1}, more expensive than usual (Huner 2000; Robinson 2000).

Traditionally most product has been sold live to local markets, but general over-reliance on such markets has placed stress on the marketing system and has only been alleviated by an increased emphasis on alternative product forms. While approximately half the crop is still sold alive, most of the remainder is now processed for tail meat. This enables gluts of production, which would otherwise have oversaturated the market, to be utilised profitably. Small amounts of small crayfish are sold for live bait at wholesale prices of \$5 kg^{-1} (Huner 2001).

Harvested live crayfish are washed in a bath or spray system and purged for 24–48 h to provide a better quality product. They are transported in small-mesh vegetable sacks, 16–23 kg per sack. In the early 1990s new price differentials based on size arose as a result of penetration into European markets and favoured the production of larger sizes of crayfish (>20 g). The industry adopted size grading as a standard practice: greater than 35 g destined for export; 20–34 g for restaurants/domestic live sales; and 13–20 g to be processed for tail meat. Normally only 10% of harvested animals are export grade (Caffey *et al.* 1996). Crayfish for tail meat are rinsed, blanched, cooled and, in the absence of a commercially successful machine for meat removal, peeled by hand. This results in high labour costs of \$3 kg^{-1} (Huner 2001). Also, the processing yield is a mere 15% (Table 4.6g) and although additional flesh is available in the claws, it is not considered economically viable to extract it.

There have been a number of other constraints regarding the processing and marketing of crayfish meat in the USA. These result from the seasonality and price instability of the supply of crayfish, rancidity in tails packed with the fat (hepatopancreas included), a lack of standard size categories and consumer ignorance about the product. All the same, significant improvements have been made in the areas of plant design; proper determination of cooking times; product stabilisation; freezing techniques; quality control programmes; and packaging (Roberts & Dellenbarger 1989).

In South Carolina most farmers are able to minimise their marketing costs by selling their product directly to consumers (farm gate sales) who do not require delivery, continuous supply, nor great attention to size grading. This method of direct selling probably represents the most profitable strategy for small-scale producers but it is only likely to be viable with the support of regular customers and a good passing trade. Larger crayfish farming operations in South Carolina also make deliveries to restaurants and seafood retail stores.

The different product forms in which crayfish can be marketed are illustrated in Fig. 3.4. For the restaurant trade the preference is for live product, while for domestic consumption both live product and processed meat are very important. Although overall consumer preference for live crayfish has been recorded in South Carolina, 40% of customers preferred to buy only cooked tail meat (Liao 1984). Negative impressions of crayfish were characterised by comments about the high price, and, for whole animals, the small meat yield and the difficulty of peeling. In order to compete more effectively with marine products, crayfish are often referred to incorrectly as seafood (Holdich 1990).

Small amounts of Australian redclaw crayfish are now being farmed in the USA. For marketing purposes they are positioned between the smaller red swamp crayfish and the larger clawed and spiny lobsters to avoid direct competition with these established products. Compared to native US species, redclaw not only grows larger, it also has a higher meat yield (Table 4.6g but see section 4.6.4) and the claws contain edible quantities of meat – all factors that enhance its market potential.

Pilot studies have been undertaken to investigate the fishery and market potential of crayfish found in northern USA and southern Canada. Swedish importers have shown interest in *Orconectes rusticus* larger than 9 cm (TL) and this species may have some potential for commercial fishery production or pond culture in Minnesota and Southern Ontario. An investigation into the possible exploitation of wild *O. virilis* in Saskatchewan revealed promising populations but not enough large animals to supply European markets. *Orconectes* gave favourable results in cooking and tasting trials but proved difficult to trap and had to be collected by hand. It was concluded that local markets could only be developed if reliable supplies could be secured (Hamr 2001).

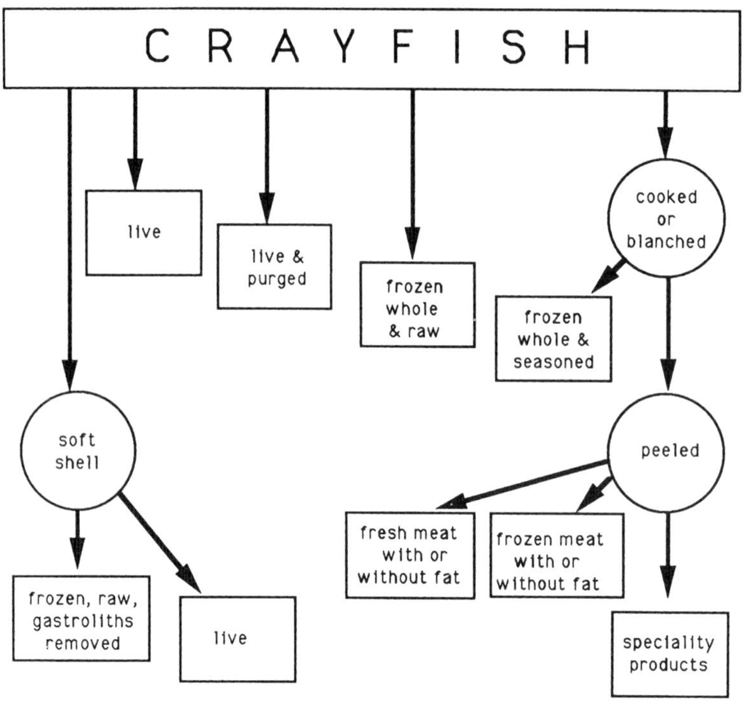

Fig. 3.4 Crayfish product forms. See Glossary for unfamiliar terms.

3.3.3.2 *Soft-shelled crayfish*

A specialised section of the US crayfish market is for soft-shell product. This is a very high-value food that once reached a price 15–20 times higher than hard-shelled crayfish (Clarke 1989). Following a boom in the late 1980s there were an estimated 150 producers of soft-shell crayfish, but since then there has been a considerable shake-up in this niche market and by 1994 there were fewer than a dozen operations remaining (Huner 1999). Problems arose because entrepreneurs simply could not generate demand at prices that could ensure profitability. They anticipated wholesale prices around \$17.60 kg^{-1} but realistic prices settled around \$13.20 kg^{-1}, at which level only the largest producers could realise economies of scale and remain viable. This decline occurred despite the efforts of seafood trade organisations such as the Louisiana Seafood Promotion and Marketing Board and the Louisiana Crawfish Promotion and Research Board that provided support through national and international trade show demonstrations and production of promotional materials including videos and recipe brochures.

Most (90%) soft-shell crayfish fall in the size range 12.5–20 g (50–80 crayfish kg^{-1}) because larger animals are not available in sufficient quantities and because smaller individuals are not efficient to produce (Culley & Duobinis-Gray 1989). Processing yield is excellent and there is no need for deveining or purging because crayfish do not eat in the period just before moulting. In fact processing yield would be 100% but for the need to remove the two gastroliths – a pair of calcareous secretions located behind the rostrum – which reduces the yield to 92% (sections 2.4.5 and 7.5.7).

Soft-shell crayfish are typically frozen in water (which serves to protect the delicate limbs) in 1 L bags containing 454–680 g of product. Packaging methods also include the use of vacuum shrink-wrap pouches, and some product is displayed on white plastic trays with six or twelve crayfish per pack. Only small amounts of soft-shell crayfish are sold live because of the likelihood of limb loss. Quality is a prime consideration. Buyers will not accept soft crayfish that have lost both claws, and will only take a limited number of one-claw animals. 'Paper-shelled' animals that have started to harden and turned leathery are also hard to sell, as are small specimens (>73 kg^{-1}). Ice is usually used to stop the process of shell hardening. Soft-shell crayfish are usually consumed batter-covered or fried, with or without a seafood stuffing, and a secondary product is broiled soft-shell crayfish.

3.3.3.3 *Europe*

Crayfish are successfully marketed in Europe in countries where there is a tradition for eating them, particularly in Sweden, Finland and France. The total consumption of the six main consumer countries has been put at 6300 mt of which Sweden accounts for 48%, Spain 38% and France 7%; Finland, Germany and Belgium between them account for the remaining 8% (McLeod 1998). European capture fisheries yield 2800 mt yr^{-1} (with 80% coming from Spain), while farms, mostly Swedish, produce a mere 70 mt yr^{-1} (Ackefors 1998). Other estimates of farm output indicate that Italy produces 30–40 mt yr^{-1} including the species *Cherax destructor* and *Cherax quadricarinatus* introduced from Australia (D'Agaro *et al.* 1999). Thus there is a considerable shortfall in supply and this is filled by frozen imports from the USA and China.

The Turkish fishery for *Astacus leptodactylus* used to be a major supplier but production declined markedly as a result of industrialisation and crayfish plague. Precise production figures are unavailable (possibly around 1000 mt in 2000) but at its peak Turkey may have produced 8000 mt per year. Spain became an important exporter following the introduction of the plague-resistant North American crayfish *Procambarus clarkii*. A wild fishery for this species was established and rapidly expanding output saturated domestic demand and provided material for export.

In Scandinavia it is traditional to eat crayfish during the summer, a custom upheld most strongly in Sweden and Finland. Crayfish can fetch as much as \$100 kg^{-1} during the first 2 weeks of the Swedish crayfish season in August. Gourmets prefer the noble crayfish (*Astacus astacus*) and there is a clear market preference for this species. All the same, Louisiana has become a significant exporter of *Procambarus* to Sweden following the decline in the Turkish *Astacus* fishery. Small amounts of rusty crayfish, *Orconectes rusticus*, produced in Wisconsin, USA, have also been exported to Sweden and because this species is similar in appearance to the noble crayfish it has obtained better prices than *Procambarus* spp.

The best prices for crayfish are obtained for live animals and in Europe there is little experience with products such as crayfish meat, soft-shelled crayfish and unpeeled tails. This situation is changing as more frozen product arrives from the USA and China. Turkey exports both live and processed crayfish. Live shipments leave by airfreight in lightweight wooden boxes containing

5 kg of product, and processors sell cooked whole crayfish (18–25 kg^{-1}) packed in brine (Koksal 1988). Russia and Georgia also send crayfish to Western Europe (Ackefors 1998).

Quality requirements for crayfish on European markets whether alive or frozen include:

- total length greater than 10 cm
- intact symmetrical appendages (especially chelae)
- clean flexible shell
- well filled-out body (both meat and hepatopancreas)
- uniform colour (orange/red preferred in cooked product)

Prices vary greatly depending on season, size, species and market place. Live *Procambarus clarkii* may fetch as little as \$1–5 kg^{-1} in Spain and Portugal while live *Astacus astacus* can fetch \$30–40 kg^{-1} in France and \$70–108 kg^{-1} in Sweden. Frozen *Procambarus clarkii* may sell for \$11 kg^{-1} in Sweden and \$3.6 kg^{-1} in France (Pérez *et al.* 1997).

The production of crayfish from fisheries and a few farms in Italy has been estimated at 30–40 mt yr^{-1} (D'Agaro *et al.* 1999). A variety of species are produced including *A. astacus* which sells for \$20–26 kg^{-1}, *Cherax spp.* for \$8–14 kg^{-1} and *A. leptodactylus* which sells for \$10–13 kg^{-1} for food and for \$0.5–0.7 each as individuals for restocking. The stocking of public waters has become a profitable business on the basis of good prices for juveniles.

There is a small crayfish farming industry in the UK but it is severely constrained by a lack of consumer interest in, or awareness of, a product that is not traditionally consumed. The industry tried to expand but between 1987 and 1993 prices fell by almost half and many operations were abandoned (Rogers & Holdich 1995). Some farms survive today by expending a lot of effort in marketing. Others make ends meet, by selling juveniles, advice and equipment to hobbyists and prospective new entrants. Such problems are perhaps typical for any entrepreneur trying to farm crustaceans on a small scale and at the same time develop new markets. The prospects for selling crayfish in supermarkets appear bleak since, in the view of large UK seafood retailers, crayfish are not a serious proposition for the expansion of existing product ranges. They foresee problems with obtaining consistent supplies at predictable prices, with competition from other established crustaceans, and with quality problems linked to receiving supplies from numerous small farms (Young 1996). Crayfish farmers are left to supply niche markets in restaurants, hotels, pubs, embassies, private parties and some local wholesalers. They dispatch live crayfish arranged three or four deep in boxes lined with damp paper and with watercress or long grass on top. Each box contains 4.5 kg of crayfish and ice is included on hot days. Polystyrene boxes are used for deliveries lasting more than 24 h and with care, transport for up to 48 h is feasible. Good presentation and grading are also important with the main categories being 'standard' 17–22 crayfish kg^{-1} and 'giants' 11–14 kg^{-1}. The latter may represent 10% of the catch and can fetch premium prices (Richards & Campbell 1996).

3.3.3.4 Australia

Crayfish, with their excellent flavour, are a much-prized food in Australia where they are often prepared on barbecues or served as entrées in restaurants. The main farmed species are the marron (*Cherax tenuimanus*), the redclaw (*C. quadricarinatus*) and the yabbies (*C. destructor* and *C. albidus*). Although wild crayfish catches are generally in decline, erratic output from some fisheries, for example for the yabby, has the potential to swamp markets and depress prices.

Australian crayfish are mostly sold live. Prior to sale, a processor must purge, or depurate, the crayfish in clean water to empty food from the digestive tract in the tail, to improve the flavour of the flesh and to prevent stress during transport caused by faecal wastes (WA Fisheries 1999a). Asian markets, including Japan, prefer live product and although crayfish in this form also get the best prices in Europe, European markets will accept cooked and frozen crayfish, which incur lower shipping and maintenance costs and reduced risk of losses. Live marron have been airfreighted for sale in Europe, retailing in some fish shops at promotional prices four times lower than clawed lobster. Yabbies have also reached fish counters in the UK.

Within Australia, market outlets for crayfish farmers include the catering trade, direct sales, retailers, wholesalers and auctions. Direct sales may be the best option for small operators who are located close to population centres or tourist routes because the requirements for consistent prices, regular sizes and supply are lower than for the catering trade. Wholesalers, retailers and auctioneers, on the other hand, are more able to deal with bulk quantities. Alternative specialised markets include the aquarium trade, bait crayfish and restocking (conservation) programmes.

Some of the more persistent problems facing the marketing of Australian crayfish include the fact that freshwater products are generally considered to have an inferior taste to their saltwater equivalents, even though in taste trials crayfish compare very favourably with marine lobster (Jones 1990). Also, if export markets are to be developed, particularly in Europe and the USA, Australian product needs to be made available in sufficiently large quantities to sustain the interest of buyers (Wingfield 2000) and it must be able to compete with crayfish produced at relatively lower costs in extensive US systems and in China (section 10.6.2.3). In anticipation of these problems O'Sullivan (1989) proposed an aggressive and co-ordinated marketing effort to provide regular supplies of high-quality, clean, intact and purged crayfish while emphasising the Australian, 'pollution-free' origin of the product.

Marron are the crayfish most highly prized by chefs and they attract much higher prices than either yabbies or redclaw. Their flesh has a subtle, sweet and nutty flavour with a fine grain and firm consistency. The hepatopancreas, commonly known as mustard, is also very popular because of its sweet flavour and agreeable consistency (Cupitt 1999). Commercial tail recovery at 42–43% (with shell on) compares favourably with spiny lobsters (WA Fisheries 1999b) and there is further meat in the claws. Total meat yield is around 31%. Marron are usually sold live following gill-washing in clean water and well-handled product can be successfully transported out of water in moist, cooled insulated boxes. They are most commonly sold at 100–200 g but the economics of farming favour the production of smaller 40 g animals that can be produced after only one year of ongrowing. Smaller animals are still considered suitable for restaurants and indeed animals below 100 g appear to have an enhanced flavour (Jones 1990).

The marketing attributes of redclaw are its attractive appearance, both live and cooked, and its resemblance to clawed lobster. They also compare favourably to lobsters in terms of taste, tail to body ratio and meat yield (Rubino 1992). Being smaller than lobsters and yet larger than shrimp they can potentially complement these products rather than compete directly. However Huner (2000) warns against over-optimism about putting redclaw of 100–200 g on the US market in close competition with small, clawed lobsters, noting that the latter are a well-established product that is readily available in the price range \$10–15 kg^{-1}. Queensland is the main source of redclaw and currently 20% is exported, 30% sold interstate and 50% consumed locally. For Australian sales there are usually three steps in the marketing chain: producer, wholesaler and restaurateur, but some production is sold at the farm gate. Keast (2000) reports a more consistent presence of redclaw on Sydney Fish Market that is helping to boost the confidence of buyers looking for regular supplies. There has also been a trend away from larger sizes down to a preference for animals in the 60–80 g and 40–60 g categories, which is a good production proposition for farmers because of shorter crop times. Less emphasis is being placed on exports while domestic demand increases. Several market size grades exist from 35 g to over 100 g (Lobegeiger 1999) with premium prices paid for the largest specimens that tend to feature in à-la-carte restaurants as entrées and main course dishes. Smaller grades are used in buffet-style presentations (Lawrence & Jones 2001). They can be held out of water with minimal stress for several days if kept cool and moist. The processing of redclaw is uncommon although there is interest in alternative products such as frozen uncooked and chilled cooked that would combine cheaper transport and good consumer acceptability. Some farmers present frozen, hand-picked and graded redclaw in 3 kg boxes.

Live yabbies are supplied within Australia to top restaurants and served as entrées under the name baby lobsters. Individual growth is very variable but animals of the minimum market size (30 g) can be produced in less than 6 months. Tail meat recovery from deheaded and shelled yabbies is 15–20%. Although yabbies fetch much lower prices than marron, they can be produced with very inexpensive technology (WA Fisheries 1999a). Nenke and Nenke (2000) bemoan a lack of market coordination and stress the need for a steady year-round flow of production rather than seasonal gluts that drive prices down.

For all types of Australian crayfish, market development will benefit from the production of alternative product forms, e.g. frozen boiled whole crayfish, frozen peeled and unpeeled tails, and frozen or canned prepared products such as crayfish soup (Clarke 1989). Studies carried out by Jones (1990) using a tasting panel are significant in this respect and address some basic questions relevant to processing and the development of recipes. The tests centred on redclaw and established that for frozen product there was no deterioration in quality over a 6-month period of storage. However, frozen cooked whole crayfish had a marginally less acceptable flavour than frozen raw whole or headless product that was cooked after defrosting. Average tail meat yields for cooked product were established at 26.2%, 22.8% and 12.0% for marron, redclaw and yabby respectively (section 4.6.4). In the redclaw an extra 4.5% yield was obtained by extracting claw meat from the larger (>100 g) specimens and although this yield was low, claw flesh was recognised as having a sweeter, more delicious flavour than tail meat. An optimal cooking method of 7 min boiling in fresh water was established for whole redclaw to retain the characteristic flavour and provide flesh with a tender, slightly resilient texture. Alternative methods of cooking, i.e. boiling in saltwater, steaming or microwave cooking, had no noticeable effects on the flavour although the latter two techniques resulted in inferior meat yields. Of particular significance to farmers are results that indicate that the flavour of redclaw can be enhanced by growing them in saline water (up to 2.4% saline) rather than fresh water, and that the same effect on flavour could be induced in animals grown in fresh water by conditioning them for 48 h in 3.0% saline water, immediately prior to harvest.

Small producers find it difficult to predict yields and to guarantee supplies each year so most of them recognise that they could smooth out fluctuations in supply and demand by co-operative efforts. They also realise the importance of minimum quality standards to ensure proper purging, handling and marketing, but despite this few people are actually committed to a group marketing strategy (Mosig 1999; Keast 2000). Some recent developments are however encouraging. A certain amount of marron is now marketed by co-operatives that combine the output of numerous small farms to fulfil larger orders, and despite general fragmentation in the redclaw industry localised marketing groups have formed to establish quality standards, brand names and combine promotional efforts (Lawrence & Jones 2001).

3.3.4 Clawed and spiny lobsters

The potential for profitable lobster aquaculture has always been associated with the existence of a well-established, high-value market for the final product. Certainly demand continues to be very strong and the word insatiable is often used to describe the world's appetite for lobster.

One of the main marketing problems facing the putative lobster farmer is the fact that lobster markets are geared to receiving animals of the sizes that are supplied from wild fisheries, and that the culture of animals to these sizes may not be economically viable. Size regulations are laid down by fishery authorities to protect wild stocks and enable lobsters to breed before they are captured. In the case of clawed lobsters, markets and fisheries usually deal with live animals of a minimum size 350–500 g, yet lobster farms for *Homarus americanus* or *H. gammarus*, if they are to have a reasonable chance of making profits, may have to grow animals to only half this size (section 7.8.9 and 10.6.3.6). So it would probably be necessary to create a new market for small size lobsters while retaining the high market value through promotion as a gourmet food item. Considerable market research would be needed to establish the acceptability of this or any other new crustacean product and some consumer resistance might develop against farmed product as it has done in the case of salmon. In some countries changes in legislation might also be required to exclude cultured lobsters from existing regulations. In Britain, however, it would be a defence to show that the undersized lobsters in question were reasonably believed to have been produced by farming (Howarth 1990). In the case of clawed lobster, a smaller product would probably find ready market acceptance and would complement rather than compete with larger wild specimens. Provided adequate supplies of the wild-caught juveniles can be obtained, the economic problems are less severe with spiny lobsters since they are grown in less costly communal systems and reach marketable sizes in a shorter time than their clawed counterparts (section 7.9.5). All the same, profitable spiny lobster culture will rely on greatly reducing the infrastructure and operating costs of land-based farming operations, as well as lowering feed and labour costs (section 10.6.4).

3.3.4.1 Clawed lobsters

The world supply of clawed lobster has been relatively stable since 1990, with some temporary shortages linked to stormy weather in the main fisheries. In consequence, prices have also been relatively stable, excepting the usual fluctuations linked to seasonal variations in catches and to increased demand during festive periods. Canada and the USA dominate production and between them they produce around 75 000 mt yr^{-1} of *Homarus americanus*. The European lobster fishery is small in comparison, yielding just 2500 mt yr^{-1} of the related species *H. gammarus* (Renard 1999). While the USA is the largest market for clawed lobster and consumes the majority of its own production, Canada exports most of its landings, including around 16 000 mt yr^{-1} to its neighbour (Smith 1995). It also sends large quantities to Asia and Europe. One large Canadian exporter sends 40–50% to Asia, 25–30% to the USA and 25–30% to Europe (Renard 1999). France is the main European destination for North American lobster followed by Italy and Spain. European sales have been depressed somewhat following a fall in value of the Euro with respect to the US dollar.

The farming of clawed lobsters would provide an ideal opportunity to capitalise on seasonal price fluctuations caused by uneven supply and demand. While most landings are made during the spring, summer and autumn, particularly during periods of calm weather, year-round demand causes prices to rise in the winter. Demand in Europe is particularly high around Christmas and the New Year. The lobster trade has responded by using holding pounds as a way of retaining fished lobsters for sale when market conditions are more favourable (section 7.8.9). The largest exporters in Canada operate land-based systems that can hold as much as 3000 mt of lobsters in individual cells in recirculating seawater. Temperatures are kept low at –2°C to minimise mortality and maximise storage time, and live product can be supplied year-round with mortality rates of the order of only 1%. The net impact of such holding systems has been to dampen price fluctuations.

North American clawed lobster has made steady progress in European markets because of consistency of supply and quality but there is still a preference and price differential in favour of European lobster (*H. gammarus*), based on a perceived difference in flavour (M. Esseen 2000, pers. comm.). For example, in March 1998 the price paid for European lobster on Billingsgate Market, London, was 80% higher than for American lobster. In France, to help differentiate locally caught product from its imported counterpart, fishermen attach tags to their lobsters indicating the port of origin and the name of the fishing vessel.

Lobsters for live export are packed in lined cardboard or polystyrene boxes with frozen gel-packs to reduce the temperature. They are mostly destined for the hotel and catering trades but are also found in retail outlets. New-shell lobsters, harvested from July to October, are preferred by some consumers because the meat is tender and easier to eat and no tools are required to crack open the shell. Canadian lobsters are a big draw for supermarkets in France and are used for in-store promotions leading up to Christmas and the New Year. Medium sizes of 400–600 g are most in demand and the consumer pays the equivalent of $12 kg^{-1} (Anon. 1999b). Frozen lobster products are mainly supplied to the retail sector and the USA takes most of the Canadian exports of frozen canned meat.

Although the best prices are obtained for live lobster, increased emphasis on alternative product forms may be needed in the future. This may be especially important if culture operations ever develop on a significant scale and larger volumes of product need to be marketed. The bulk of the output from the Canadian wild fishery is sold live, but already large quantities are marketed in other forms. Live sales rely on the larger lobsters, which are divided into eleven size categories between 454 g and 2270 g (Smith 1995), while alternative products tend to rely on smaller lobsters of 0.5–1 lb (227–454 g) or weak specimens not suited to live sales. Some zones within the Canadian fishery, for example those around Prince Edward Island, are specially managed to provide smaller lobsters for processing operations. Different product forms are detailed below.

Whole frozen forms:

- Cooked and sealed in plastic vacuum pouches with brine ('popsicle pack') and sold in six size grades.
- Blanched (cooked for 2 min) and vacuum-packed (cooking is to be completed by consumer). Blanching prevents the meat from sticking to the shell and produces a product similar to uncooked lobster.
- Cooked or blanched and sealed within special vacuum skin pack for prolonged shelf life (up to 24 months).

Frozen meat:

- Lobster is cooked, and the meat picked from shell and packed in metal cans or plastic containers, without further heat treatment, and then frozen. Tail and claw meat is the most valuable grade and broken meat, often from very large 'jumbo' lobsters, is the lowest grade.

Heat processed meat:

- Lobster is cooked and the meat picked from shell and packed in metal cans and then sealed and sterilised by heating.

Tails:

- Frozen raw, ranging in size from 85 g to 170 g each.

Speciality, value-added consumer products:

- Canned lobster paste made with lobster hepatopancreas and/or roe mixed with meat.
- Cooked and frozen lobster in the half shell.
- Lobster pâté – made from the same ingredients as lobster paste, plus flour and spices.
- Minced lobster loaf – 'deboned' lobster body meat packed in bags and frozen.

- Frozen cocktail claws that have been scored for easy snapping by the consumer.
- Tomalley (lobster hepatopancreas) – a greenish paste sold as a spread.

The occurrence of genetically determined red and blue colourmorphs of the American clawed lobster *Homarus americanus* may provide opportunities to develop the culture of animals with a unique appearance. This could enhance the marketability of the live product: red, for example, is considered by many consumers to be the 'natural' colour of lobsters (Aiken & Waddy 1995).

3.3.4.2 Spiny lobsters

In common with clawed lobsters, spiny lobsters are a highly prized seafood delicacy and limited supplies are reflected in high prices. Australia is the world's largest source and lands between 10 000 and 17 000 mt yr^{-1}. Other major producers are Cuba, Brazil, the Bahamas and New Zealand. While clawed lobsters have marketable meat in the claws and body as well as the tail, in the spiny lobster the tail is usually the only meat source. As a reflection of this, much of the international trade consists of frozen tails. However, premium prices are paid for live animals and more suppliers, particularly in Australia and New Zealand, are providing live product. As an illustration of this trend, between 1987 and 1999 New Zealand's exports in live form increased from 15% to 95%, the remainder being frozen and processed. A similar pattern prevails in Australia with live product now accounting for around 50% of exports, and rising (Riepen 1997; Stevens & Sykes 2000).

US imports of spiny lobster account for one-third of world output but demand in Asia and Europe is also strong and growing more quickly than in the USA. Japan prefers small plate-sized specimens for use as sashimi and Hong Kong and China require large lobsters for banquet settings. Glass tanks containing live seafood are a common sight in restaurants in Asian cities and Asian customers pay a premium for the added quality of live product that is seen as an absolute guarantee of freshness. In Asian markets lobsters fill a niche based on rich consumers and business entertainment and they are one of the most expensive live foods, wholesaling at \$34–37 kg^{-1} compared to \$22–34 kg^{-1} for live prawns and \$8–11 kg^{-1} for live crab (Riepen 1997). Spiny lobsters are also popular in Japan at traditional congratulatory occasions such as weddings and New Year's Day celebrations when their long antennae symbolise happiness and longevity (Tsuruta & Kittaka 2000). Demand has dropped somewhat since 1997 due to economic depression as dinners have been simplified and major celebrations have been modified to less formal ones.

In Europe, France has the largest consumption. When Australian exporters faced a drop in demand in Asia following the Asian financial crisis they turned their attention to supplying Europe with live and frozen whole lobsters and frozen tails. The potential of the EU market is constrained by a 15% tariff, but all the same Australian suppliers have become deft at exploiting the potential of live sales and respond quickly to the needs of festivals or other special occasions. For example, an order placed in Europe on a Monday morning can be despatched the next day and arrive after a total of just 48 h. Careful handling and low temperatures (<12°C) are essential to ensure product quality. Lobsters arrive in 5 kg boxes with the antennae intact and may be accompanied by a guarantee stating that they will remain alive in viviers for 9–10 days (Morineau 1999). The catch in Australia and New Zealand is sold to processors who generally process for export. Lobsters are always size graded but there is no standard system. In Australia some 24% of production is consumed nationally and in New Zealand the equivalent proportion is 6% (Stevens & Sykes 2000). New Zealand live lobsters are kept in controlled low-temperature conditions and the temperature is dropped further to 6°C just prior to export. They are packed between wood wool in polystyrene bins, 3–18 per bin and a small frozen container of saltwater or gel is placed inside to allow shipping with minimal mortality. Live lobster can withstand 30 h of transport, normally with less than 5% mortality. The main destinations for live exports from Australia and New Zealand are Japan, Taiwan, Hong Kong and China (via Hong Kong), and the USA is by far the biggest market for frozen tails. The processing for lobster tails involves chilling, deheading, vacuum extraction of the hindgut, washing, draining, grading, packing in 10 lb and 5 lb cartons, and finally freezing.

Small quantities of spiny lobster are produced in ongrowing operations that rely on supplies of wild-caught juveniles or subadults. In Taiwan such operations produced 400 000 lobsters in 1987, for domestic consumption. However, mass imports of Chinese spiny lobsters threatened to undermine prices and the profitability of ongrowing units. In Taiwan most if not all lobsters are sold live. They are transported for up to 24 h in cardboard boxes containing 16 kg of lobster in four layers with chilled sawdust. A valuable market exists for juvenile lobsters used to stock the ongrowing operations, and animals above 25 g are sold for \$42 kg^{-1}. There are no size restrictions on the capture of wild lobsters in Taiwan (Chen 1990). Newly collected pueruli (see Glossary) for ongrowing in New Zealand are estimated to cost \$0.20–0.70 each (section 10.6.4). There is a potential gourmet market for soft-shell spiny lobster of 40–60 g that could be produced from wild-caught pueruli.

3.3.5 Crabs

The crabs produced by aquaculture are sold live to markets that are based on traditional local consumption of the same species caught in the wild. International trade in crab is mostly reliant on frozen crabs and processed meat of fishery origin. When the need arises however there is little to impede transport of live crabs over long distances. Mud crab (*Scylla* spp.) for example can survive out of water for one week if sprinkled occasionally with water to keep them cool and moist (section 7.10.6). Crabs can also be transported in viviers (section 8.4.6) and maintained in the same tanks used for holding other seafood such as lobsters.

In the industrialised world crabs are regarded as a luxury food item and are mostly eaten in restaurants and hotels. Japan is by far the largest market and the largest importer, taking some 60% of world exports of crabs and crab products (Globefish 1995). China, the USA and Vietnam have the most productive crab fisheries.

Crab fishermen mostly employ baited traps or 'pots' but sometimes use lines, nets, or dredges and crabs are mostly supplied live to processors. The main market forms are: whole cooked (in boiling water) and either chilled or frozen; frozen crab parts such as claws or groups of legs (clusters); frozen 'snap and eat' legs in which the shell segments are scored to permit easy cracking by hand; or hand-picked crab meat. Freezing technique has a bearing on the quality and price of the final product, with US consumers preferring the taste and appearance of crab frozen in brine rather than blast-frozen crab (Globefish 1995). Crab meat is shipped fresh or frozen in cans or in vacuum packs and can vary in the proportions of claw and leg meat. Small mud crabs from Thailand are a major source of crab meat and Japan is the main buyer. Prices have been held down by increased competition from substitute products, including surimi, and because of recession in Japan. Nevertheless, there is strong competition in international trade in canned and cooked crab meat for which consumers have difficulty differentiating species and origins. In the USA a big surge in imports of blue crab meat (actually from swimming or sand crabs (*Portunus pelagicus*) rather than blue crabs (*Callinectes sapidus*)) from countries like Indonesia has prompted the US crab fishing and processing industries to lobby for protection measures (Robinson 1999).

Plate 3.2 Marketing live mud crabs (*Scylla* spp.). (Photo courtesy D.J. Macintosh, University of Aarhus, Denmark.)

Cultured crabs are highly valued in certain Asian markets. In Taiwan, female mud crabs that are filled with orange coloured roe are especially prized and fetch far more than males or immature females. After spawning, their value drops by 90%. Harvested mud crabs are bound with twine or thick rope interwoven with rice straw to immobilise claws. Sometimes they are transported in ventilated polystyrene containers at 16–20°C and 95% relative humidity and with a cushion of moist wood shavings they can survive for 7 days (Dagoon 1997). While roe-bearing females are sold by number, other mud crabs are sold by weight and the weight of the

Plate 3.3 A mature female green mud crab (*Scylla olivacea*). Crabs with ripe ovaries are called egg crabs and command high prices in many Asian markets. (Photo courtesy D.J. Macintosh, University of Aarhus, Denmark.)

rope is included in the price (Chen 1990). Female portunid crabs with full ripe ovaries are a luxury food item in Japan. Although they are not farmed, restocking programmes are aimed at enhancing the Japanese fishery for *Portunus trituberculatus*. Again price is very sensitive to condition: if the female is berried rather than roe-bearing her value is 60–70% less.

The viability of soft-shell crab production in shedding systems in the USA (section 7.10.9) relies on the high processing yield and the high market value of this gourmet seafood. More than 1400 mt of soft-shelled blue crab, *Callinectes sapidus*, are produced each year from Maryland to Georgia, for sale as a local delicacy or distribution to top-line restaurants throughout the eastern USA. Prices vary each year depending on the productivity of the fishery. Experiments have been carried out with the production of soft-shell mud crabs in Queensland, Australia (Beale 2000). Prior to consumption the gills and a few other minor body parts are trimmed off and there is no need for purging because crabs do not eat prior to moulting.

3.3.6 Analogue products

The sales of crustacean analogue products are increasing worldwide and have important implications for some traditional crustacean products. Surimi, the most common product, can be produced in various forms such as crab sticks and shrimp tails to take advantage of the high prices obtained for crustaceans. It is made from mechanically deboned, washed and stabilised white fish flesh, produced mainly in the USA and Japan usually using Alaska pollock as the raw material. The washing removes odorous substances and the whole process produces a bland white product that can be combined with flavourings and other additives such as polyphosphates and sugar (Putro 1991). Various shapes such as curled shrimp tails can then be easily produced by extrusion but surimi producers are becoming ever more ingenious in their attempts to imitate high-value crustacean products. Surimi is shaped to represent the cooked meat from the claws of crabs and lobsters; it is moulded and then given coloured bands to appear like cooked and peeled tiger shrimp and lobster tails; and it is even moulded around long thin claws from real crabs and breaded to produce a deceptive hybrid product. The advantages of surimi are its relatively low price, consistent quality and availability, and the fact that it can use the desirable label 'seafood'.

Surimi crab analogues in the form of sticks, flakes and chunks made a big impact on the US crab market where they were able to fill lower-priced market niches in which natural product was unable to compete (Vondruska 1986). Shrimp imitations were less successful (Holmes 1988) although surimi does compete effectively with breaded and battered shrimp. Shrimp analogues do not sell well in restaurants because the small savings that customers can make do not outweigh the preference for the genuine article rather than an imitation product. Another use for imitation shrimp products is in toppings for pizzas, but surimi cannot compete with the gourmet image of shrimp and because of this its impact will continue to be restricted.

3.3.7 By-products

The potential of various crustacean by-products has been investigated as a way of generating additional revenue from processing operations. Shrimp meal is derived from a mixture of shrimp heads and shells by drying and then pulverising in a hammer mill. The resulting product is rich in protein, calcium, chitin and pigments, and it can be included in livestock and aquaculture feeds. It is particularly useful in some fish diets because it contains chemo-attractants and carotenoids, the latter imparting an attractive reddish colour to the flesh of salmon or trout. Some animal feeds are formulated with as much as 30% shrimp meal. Shrimp waste can also be used for animal feed following transformation into shrimp silage. This involves hydrolysis of the proteins by enzymatic action enhanced by the addition of acid and, as a final step, neutralisation (Hall & de Silva 1994; Guillou *et al.* 1995). Hansen (2001) describes some industrial processing methods for the waste from shrimp and other small crustaceans.

Sometimes food products can be recovered from crustacean waste. Broken pieces of shrimp flesh, for example, can be incorporated in soups and chowders and if the quality is good enough they can be used for shrimp crackers. To obtain good quality product it is important to prevent spoilage of fresh waste by proteolytic enzymes and bacteria so chilling, freezing or prompt cooking are essential. Pasteurisation is a necessary safety step for any recovered meat (Schoemaker 1991). The process of extracting pulp from shrimp head waste involves cooking, pressing to remove excess moisture and then a meat–bone separator usually employed for fish. In a similar way meat–bone separators can be used to make shrimp paste and flavourings from shrimp heads. Meyers (1987) extracted the meat from shrimp heads and used it to make a novel product – shrimp sausages.

Ideally the waste heads and shells from shrimp should be kept separately because they have different compositions and are suited to different products. Crustacean shells can be deproteinised and demineralised to produce chitin. However there are only limited uses for this natural polymer so most chitin is converted to chitosan which has haemostatic and wound healing properties (No & Meyers 1995). Chitosan is also incorporated in healthcare products, many of which are becoming very popular because they are claimed to aid weight loss (Subasinghe 1999). Commercial shrimp waste contains 12–18% chitin and the markets for chitin and its derivatives have been reviewed by van Ornum (1992).

When US crayfish are processed for tail meat a massive 85% becomes residue. Traditionally this has been deposited in landfill sites but some is now used for extraction of carotenoid pigments such as astaxanthin for incorporation in fish diets. Also, dried crayfish waste meal is used in crayfish baits, shrimp feeds and fish attractants (Caffey *et al.* 1996). The chitin recovery rate from crayfish waste is 23.5% on a dry basis. Raw crayfish waste can also be used as fertiliser for agricultural crops and for high-quality compost (Huner 1994; 2001).

Spiny lobster heads from processing operations in Australia and New Zealand are not discarded but are graded and frozen in 10 kg cartons, and a paste from finely macerated heads is blast frozen in blocks (Stevens & Sykes 2000). In the USA crab shells are used for calcium and other vitamin supplements (Anon. 2000b).

3.4 Markets for aquaculture technology, products and services

As the crustacean farming industry expands, opportunities for the provision of goods and services also increase. Already demand for a large variety of supplies and various technical and support services has persuaded some companies to diversify into crustacean aquaculture and others to form and dedicate themselves exclusively to supplying the needs of the sector. In 1987 the UK Department of Trade and Industry reviewed the market opportunities for intensive marine aquaculture technologies and services (DTI 1988). The market was valued at approximately £850m annually, and a number of opportunities for UK companies were identified. Markets relevant to crustacean farming (* indicates technology and services already well developed in the UK, and therefore offering the best opportunities for export) were:

- feedstuffs*
- speciality food components, e.g. pigments*
- microencapsulated and spray-dried algal diets for larvae*
- feeding equipment
- harvesting pumps and graders
- security systems for farms
- predator scaring devices*
- pump technology for farms
- recirculation and biofiltration systems
- aeration equipment

- heat pumps
- water filtration and UV sterilisation equipment
- equipment for live transport
- water quality monitoring equipment
- consultancy for new farms, in turnkey projects, environmental monitoring
- insurance and site assessment*
- biotechnological approaches, e.g. transgenic manipulation and cryopreservation*
- lobster culture technology
- diagnostic aids based on poly- and monoclonal antibody systems*
- new vaccines*
- education, training and associated instructional aids and software*
- insurance services*
- venture capital for foreign based projects
- specialised processing plant for shellfish

If a new survey was undertaken now it might additionally include stock enhancement techniques, and artificial reef and habitat design technologies.

3.4.1 Supplies

Feed companies have been quick to respond to the growth in crustacean farming and have developed a range of specialised diets. Of these, shrimp diets are the most significant and, once the fastest growing sector of the global market for animal feeds, are now among the most important of all aquaculture feeds. The total quantity of shrimp feed consumed in 2000 has been estimated at 1489×10^3 mt and this amount is projected to rise to 2425×10^3 mt by the year 2010 (Barlow 2000). Shrimp diets are produced in, and exported from, Japan, the USA and Europe, but their output is becoming increasingly centred in the major shrimp farming nations themselves. In some of these countries the governments limit imports of feed to protect home-based production. Partly as a result of this and partly to circumvent trade barriers, foreign feed companies are often encouraged to set up overseas joint ventures.

Accompanying the production of crustacean feed is the demand for its various ingredients, notably fishmeal but also soybean meal, wheat and various additives, including binding agents, pigments and vitamin and mineral mixtures (sections 2.4 and 8.8.2). Consumption of fishmeal for shrimp diets in 2000 was estimated at 360 000 mt and this is expected to increase by 20–30% over the next 10 years (Barlow 2000). Total fishmeal requirements for all aquaculture species are projected to be 2.8×10^6 mt in the year 2010.

On a global scale, the fishmeal industry expects to be able to cope with the increasing requirements of aquaculture even though fishmeal production is unlikely to rise. Aquaculture currently accounts for around 40% of consumption, with the remainder destined for poultry and mammalian feeds. During the formulation of animal as opposed to shrimp feeds, greater reliance can be placed on terrestrial sources of protein (such as soybeans and wheat) and can release an increasing portion of the available fishmeal to aquaculture (I. Pike 1990, pers. comm.). Even so, increasing demand overall will push up the price of fishmeal and adversely effect the economic viability of some intensive and semi-intensive crustacean farms (section 11.5.1.4).

Other aquaculture supplies include specialised feeds for use in hatcheries. Live feeds such as *Artemia* are consumed in nearly all operations that rear crustacean larvae and early post-larvae. Hatcheries for fish and cephalopods may require copepods and mysids (section 7.11). Live algae for penaeid larvae are grown on the spot in hatcheries but there is a small market for sterile starter cultures. Specialised artificial diets have been developed to reduce dependence on live feeds and satisfy demand for more convenient forms of feed (section 8.8.1). They are typically highly priced (sometimes in excess of $\$100\,kg^{-1}$), and because of the relatively small volumes required they are easily shipped around the world. Several companies, based either in Europe, Japan or the USA, currently compete for a leading position in this expanding market. Most of the diets are size-graded compounded microparticles although innovations include microencapsulated diets, heterotrophically grown and spray-dried algae, nutrient-enriched dried yeast, and special nutrient mixtures for feeding to *Artemia* and rotifers to boost their nutritional value (section 7.11.2.1). Recent innovations include liquid suspensions containing nutrient droplets and probiotic bacteria.

A significant component of the cost of trapping crayfish in the USA is bait (section 7.5.5; Fig. 10.18). Trash fish is traditionally used, but recognising the need for an effective and more stable product with good storage properties, some 13 US companies market artificial baits.

Other supplies used by crustacean farmers include chemical products for ponds, such as: inorganic fertilisers; the fish eradicators rotenone and saponin; and various treatment and conditioning products such as bentonite, zeolite and lime (sections 8.3.3 and 8.3.6). In hatcheries smaller quantities of chemicals are used, including nutrients for algae culture, disinfectants and various therapeutants. In the future, markets for diagnostics and alternative therapeutants such as vaccines may expand and reduce dependence on antibiotics (sections 8.9.4 and 11.3.4). Diagnostic reagents can enable the rapid recognition of diseases and already there are at least 17 companies worldwide that are involved in aquaculture diagnostics. Medications can be either administered to the culture water or added to feeds, but manufacturers of new therapeutants for aquaculture need to conform with the drug and chemical registration processes (section 11.5.3.2). Although there is a great need for disease diagnosis and advice, usually only the services of veterinary surgeons or fish disease specialists are available. The number of crustacean disease specialists in the whole world is probably fewer than ten.

3.4.2 Equipment

A sizeable market exists for the equipment used to establish hatcheries and farms (Tables 10.8 and 10.11). There are several aquaculture supply companies that specialise in providing a wide range of equipment and some can offer complete start-up packages. Other companies provide specialised or customised equipment in addition to the regular items needed during production. Artificial substrate in the form of long arrays of felt-like buoyant ribbons is now commercially available for use in hatcheries, nurseries and super-intensive ongrowing systems.

A large part of the investment in a crustacean processing plant lies in its equipment. Most units require cold-storage facilities, refrigeration plant and ice machines and often additional specialised equipment depending on the type of processing to be carried out (section 10.6.6). Equipment needs are greatest for the production of value-added foods and these are becoming increasingly important in world crustacean markets (sections 3.1 and 3.3.1). The various processes involved in feed production (Fig. 8.7) also require large amounts of hardware and again offer opportunities to the suppliers of specialised equipment to sell and install their systems, and provide spare parts and back-up services.

3.4.3 Broodstock, nauplii and juveniles

Significant opportunities exist to satisfy the hatchery demand for broodstock, and in the case of shrimp, for nauplii. Demand for penaeid broodstock in South-east Asia has stimulated substantial international trade, some of which is carried out in contravention of legal restrictions. As disease problems have intensified the market for specific pathogen-free (SPF) and specific pathogen-resistant (SPR) broodstock, post-larvae, and nauplii has opened up. SPF and SPR *Litopenaeus vannamei*, *L. stylirostris* and more recently *Penaeus monodon* are produced in carefully quarantined facilities in Hawaii and Taiwan for sale worldwide. The market for SPF *Litopenaeus vannamei* has been estimated at 29.6×10^9 nauplii worth \$13.5m; 20.8×10^9 post-larvae worth \$68.4m; and 120 000 broodstock worth \$3m (Kuljis & Brown 1992). The possibility of supplying disease-free shrimp to four Asian countries (Philippines, Indonesia, Thailand and Taiwan) from Australia was considered by O'Sullivan (1997). He estimated that the potential market for SPF *Penaeus monodon* was some \$85m for post-larvae and another \$5m for nauplii and broodstock (1992 data). Some nauplii are traded internationally (section 7.2.3) and the potential of world markets has encouraged experimentation with nauplii cryopreservation techniques to facilitate storage and long-distance transport (section 12.4).

Markets also exist for juvenile and broodstock crayfish. In Europe, for example, more than one million crayfish juveniles are sold each year either as second or third instar hatchlings or as one-summer-old juveniles (Huner *et al.* 1987; section 10.6.2.1). Once the problems of rearing the larvae of spiny lobsters have been solved, there is a potential market for juveniles to stock ongrowing operations, and if clawed lobsters can be successfully ranched, there could also be a sizeable market for their juveniles.

3.4.4 Services

Basic services required by shrimp farms can include pond cleaning and harvesting, for which extra manpower and special equipment may be needed. Professional services required by the crustacean farming business as a whole include market analysis, financial planning and appraisal, accounting, legal services and insurance (sections 9.3.6 and 11.5.2).

In addition, many different forms of technical service are in demand. Complete turnkey systems usually comprise comprehensive engineering, design and management packages (section 9.3.8). The services of consultants or consulting companies may be hired for a range of different subjects, including the preparation of prefeasibility and feasibility studies, project supervision, training and technical assistance (section 9.3.6).

The specialised aquaculture expertise of highly trained technical advisors and consultants is easier to sell when advanced and intensive technology is required. At the operational level most extensive culture systems necessitate only basic management skills (MacPherson & Mackay 1987). This is one reason why Westerners, and technology sellers in general, tend to encourage the development and installation of intensive systems since these are most likely to require their input. This should not however be allowed to distort the choice of culture systems in situations where extensive and semi-intensive methods are clearly the most appropriate (section 11.2).

Demand for up-to-date information on developments in research, culture technology, markets and general industry news supports the publication of a whole range of books, newsletters and periodicals. Some of these are dedicated to crustaceans while most are concerned with aquaculture in general or with aquaculture in a particular country.

Aquaculture information sources on the Internet have been compiled by the Institute of Aquaculture, Stirling, Scotland: http://www.stir.ac.uk/Departments/NaturalSciences/Aquaculture/; and by the Aquaculture Network Information Centre in the USA: http://ag.ansc.purdue.edu/aquanic/.

Many managers now make use of computers running specialised commercial software to improve the production efficiency of hatcheries and farms. A listing and descriptions of aquaculture software have been compiled at Oregon State University: http://biosys.bre.orst.edu/aquacult/aquasoft.htm.

Once software systems become established, future revenue for software companies will rely on the development of updated and improved packages. Using computers to monitor and control pond or hatchery water conditions, however, can easily lead to problems when technicians accept the data returned as accurate without question. The system that can distinguish between a fouled probe and deteriorating water quality has yet to be invented.

Finally, there are many opportunities to provide or sell educational programmes or technical courses, for students and other interested parties such as businessmen and entrepreneurs. Popular short courses on shrimp aquaculture are run in the USA, especially for people who are new to crustacean aquaculture. They combine hands-on experience with a basic introduction and training in techniques and provide background information on subjects such as economics and marketing. Aquaculture education is reviewed in *World Aquaculture* (1999, Vol. 30, No. 2) and lists of practical and academic courses relating to aquaculture can be obtained from the World Aquaculture Society: http://www.was.org.

3.5 References

Ackefors H. (1998) The culture and capture crayfish fisheries in Europe. *World Aquaculture*, **29** (2) 18–24; 64–67.

Aiken D.E. & Waddy S.L. (1995) Aquaculture. In: *Biology of the Lobster* Homarus americanus (ed. J.R. Factor), pp. 153–175. Academic Press, New York.

Anderson J.L. & Fong Q.S.W. (1997) Aquaculture and international trade. *Aquaculture Economics and Management*, **1** (1) 29–44.

Anon. (1990) Black and blue tigers. *Seafood International*, June 1990 p. 11.

Anon. (1997) Globefish market summary. *Infofish International*, (5) 33.

Anon. (1998) Overseas trade, a structural defecit. *Produits de la Mer*, (48) 134.

Anon. (1999a) FDA Ban on crawfish. *Aquaculture Magazine*, **25** (3) 14–16.

Anon. (1999b) The French market for live Canadian lobsters is beginning to look up. *Produits de la Mer*, (52) 24.

Anon. (2000a) GAA technical committee meets at Aqua 2000. *Global Aquaculture Advocate*, **3** (3) 3–4.

Anon. (2000b) Vitamins from shellfish waste. *Infofish International*, (1) 63.

Avault J.W. (1999a) Research report. Researching your market. *Aquaculture Magazine*, **25** (6) 49–53.

Avault J.W. (1999b) Research report. The marketing mix, part 1. *Aquaculture Magazine*, **25** (4) 71–76.

Avault J.W. (1999c) Research report. The marketing mix, part 2. *Aquaculture Magazine*, **25** (5) 80–82.

Avault J.W. (2000a) Research report. International marketing and sales terms. *Aquaculture Magazine*, **26** (2) 72–74.

Avault J.W. (2000b) Research report. Your market position. *Aquaculture Magazine*, **26** (1) 81–84.

Barlow S. (2000) Fishmeal and fish oil: sustainable ingredients for aquafeeds. *Global Aquaculture Advocate*, **3** (2) 85–88.

Beale R. (2000) Soft shell, hard sell. *The Bulletin*, **118** (6247) 44–45.

Brown P. Jr. (2000) The USA shrimp market. *Global Aquaculture Advocate*, **3** (2) 51–52.

Caffey R.H., Romaire R.P. & Avault J.W. Jr. (1996) Sustainable aquaculture: crawfish farming. In: *Freshwater Crayfish 11* (ed. W.T. Momot), pp. 587–598. Louisiana State University, USA.

Cato J.C. & Lima dos Santos C.A. (1998) Bangladesh: shrimp processors invest in plant upgrades and HACCP to produce safe products. *Infofish International*, (5) 51–56.

Cato J.C. & Lima dos Santos C.A. (1999) Bangladesh shrimp processing sector: economic cost of implementing HACCP. *Infofish International*, (3) 49–57.

Chaston I. (1983) *Marketing in Fisheries and Aquaculture*, 144 pp. Fishing News Books, Blackwell Scientific Publications, Oxford, UK.

Chauvin W.D. (1992) Freshwater prawns in the USA – a market

in transition. *Infofish International*, (4) 17–19.

Chauvin W.D. (1999) US market situation. *Shrimp Notes*, **22** (3) 1. http://www.shrimpcom/9903/htm.

Chen L-C. (1990) *Aquaculture in Taiwan*, 273 pp. Fishing News Books, Blackwell Scientific Publications, Oxford, UK.

Clarke S. (1989) Freshwater crustacean farming: the world scene. South Australian Dept. Fisheries, *SAFISH,* **13** (4) 10–12.

Clay J.W. (1997) Toward sustainable shrimp aquaculture. *World Aquaculture*, **28** (3) 32–37.

Culley D.D. & Duobinis-Gray L. (1989) Soft-shell crawfish production technology. *Journal of Shellfish Research,* **8** (1) 287–291.

Cupitt W. (1999) *Marron information.* Marron Marketing Australia Pty Ltd. http://www.midwest.com.au/~bringo/

D'Abramo L.R., Daniels W.H., Fondren M.W. & Brunson M.W. (1995) Management practices for culture of freshwater prawn (*Macrobrachium rosenbergii*) in temperate climates. *Mississippi Agriculture and Forestry Research Station Technical Bulletin 1030*. Mississippi State University, MS, USA.

D'Agaro E., De Luise G. & Lanari D. (1999) The current status of crayfish farming in Italy. In: *Freshwater Crayfish 12* (eds M. Keller, M.M. Keller, B. Oidtmann, R. Hoffman & G. Vogt), pp. 506–517. International Association of Astacology. Weltbild Verlag, Germany.

Dagoon N.J. (1997) Post-harvest, processing. *Asian Aquaculture*, **14** (3) 23.

DTI (1988) Feasibility study on the technology of mariculture, Vol. 2. Review of technologies and services, 259 pp. Department of Trade and Industry, *Resources from the Sea Programme, report 87/27*, Aberdeen University Marine Studies Ltd, Aberdeen, UK.

Ducherne L. (2000) Food fight shaping up over 'organic' wild seafood. *Online Seafood Business*, June 2000. http://www.seafoodbusiness.com/issue_feat.html

Evans T. (1995) Seafood safety – what exporters must know about HACCP. *Infofish International*, (3) 48–52.

Ferdouse F. (1994) The international market for cultured shrimp. *Infofish International*, (6) 16–22.

Ferdouse F. (1999) Japanese and other Asian markets for shrimp – an overview. *Infofish International*, (6) 23–28.

Filose J. (1988) The North American market perspective. In: *Shrimp '88, Conference Proceedings*, pp. 20–25, Bangkok, Thailand, 26–28 January 1988, Infofish, Kuala Lumpur, Malaysia.

Filose J. (1992) Surviving the recession in the US shrimp market. *Infofish International*, (6) 18–24.

Globefish (1995). The world market for crab. *FAO/Globefish Research Programme*, **37**, 1–59.

Guillou A., Khalil M. & Adambounou L. (1995) Effects of silage preservation on astaxanthin forms and fatty acid profiles of processed shrimp (*Pandalus borealis*) waste. *Aquaculture,* **130** (4) 351–360.

Hall G.M. & de Silva S. (1994) Shrimp waste ensilation. *Infofish International*, (2) 27–30.

Hamr P. (2001) *Orconectes*. In: *Biology of Freshwater Crayfish* (ed. D.M. Holdich), pp. 585–608. Blackwell Science, Oxford, UK.

Hansen L. (2000) Design considerations for a modern shrimp-cooking facility. *Global Aquaculture Advocate*, **3** (4) 65–68.

Hansen P. (2001) Separation of meat and shell components from crustaceans. *Infofish International*, (1) 56–59.

Hanson G.D., Rauniyar G.P. & Hermann R.O. (1994) Using consumer profiles to increase U.S. market for seafood: implications for aquaculture. *Aquaculture*, **127** (4) 303–316.

Harvey D. (1999) *Aquaculture Outlook*. Economic Research Service, US Department of Agriculture, 5 March; 4 October. http://usda.mannlib.cornell.edu/reports/erssor/livestock/ldp-aqs/

Hempel E. (1999) The impact of the Asian financial crisis on the shrimp industry. *Infofish International*, (3) 19–24.

Henson L.S. & Kowalewski K.M. (1992) Use of phosphates in seafood. *Infofish International*, (5) 52–54.

Holdich D. (1990) "Crawfish state" Louisiana plays host to astacologists. *Fish Farmer,* **13** (4) 32–33.

Holmes K. (1988) Impact of analogue products. In: *Shrimp '88, Conference proceedings*, Bangkok, Thailand, 26–28 January 1988, pp. 44–49. Infofish, Kuala Lumpur, Malaysia.

Hottlet O. (1992) The European market for frozen shrimp. *Infofish International*, (5) 16–20.

Howarth W. (1990) *The Law of Aquaculture*, 271 pp. Fishing News Books, Blackwell Scientific Publications, Oxford, UK.

Huner J.V. (1994) *Freshwater Crayfish Aquaculture in North America, Europe and Australia. Section II Freshwater crayfish processing,* pp. 91–115. Haworth Press, Binghamton, New York.

Huner J.V. (1999) The fate of the Louisiana soft-shell crawfish. *Aquaculture Magazine*, **25** (3) 46–51.

Huner J.V. (2000) Crawfish the big picture. In: *Australian Crayfish Aquaculture Workshop Proceedings* (eds C. Lawrence & G. Whisson), pp. 1–4. International Association of Astacology, Curtin University of Technology, Perth, Australia.

Huner J.V. (2001) *Procambarus*. In: *Biology of Freshwater Crayfish* (ed. D.M. Holdich), pp. 541–84. Blackwell Science, Oxford, UK.

Huner J., Gydemo R., Haug J., Jarvenpåa, T. & Taugbøl T. (1987) Trade, marketing and economics. In: *Crayfish Culture in Europe*. Report from the workshop on crayfish culture, 16–19 November 1987, Trondheim, Norway, pp. 54–62.

Jones C.M. (1990) *The Biology and Aquaculture Potential of the Tropical Freshwater Crayfish* Cherax quadricarinatus, 109 pp. Information series QI 90028, Queensland Department of Primary Industry, Australia.

Jory D.E. (2000) Shrimp and cholesterol. *Global Aquaculture Advocate*, **3** (4) 83.

Josupeit H. (1999) An overview on the world shrimp market. http://www.globefish.org/presentations/worldshrimpproduction/index.htm

Josupeit H. & de Franssu L. (1992) EEC markets for value-added shrimp from developing countries. *Infofish International*, (2) 16–20.

Jovellanos C. (1993) Managing financial risk in international seafood trading. *Infofish International*, (3) 13–19.

Keast W. (2000) The redclaw industry in Queensland: production and marketing. In: *Australian Crayfish Aquaculture*

Workshop Proceedings (eds C. Lawrence & G. Whisson), pp. 21–25. International Association of Astacology, Curtin University of Technology, Perth, Australia.
Kinnucan H.W. & Wessels C.R. (1997) Marketing research paradigms for aquaculture. *Aquaculture Economics and Management*, **1** (1) 73–86.
Koksal G. (1988) *Astacus leptodactylus* in Europe. In: *Freshwater Crayfish: biology and exploitation* (eds D.M Holdich & R.S. Lowery), pp. 365–400. Croom Helm, London.
Konosu S. & Yamaguchi K. (2000) Colour and taste. In: *Spiny Lobsters: fisheries and culture,* 2nd edn (eds B.F. Phillips & J. Kittaka), pp. 625–632. Fishing News Books, Oxford, UK.
Kuljis A.M & Brown C.L. (1992) *A market study for specific pathogen free shrimp*, 54 pp. Publication No. 112. Centre for Tropical and Subtropical Aquaculture, Honolulu, HI, USA.
Lang J.M. (2000) Seafood takes over – restaurants swap shrimp for steak. *Online Seafood Business*, March 2000. http://www.seafoodbusiness.com/00mar/issue/html.
Lawrence C. & Jones C. (2001) *Cherax*. In: *Biology of Freshwater Crayfish* (ed. D.M. Holdich), pp. 635–69. Blackwell Science, Oxford, UK.
Liao D.S. (1984) Market analysis for crawfish aquaculture in South Carolina. *Journal of the World Mariculture Society,* **15**, 106–107.
Liao D.S. & Smith T.I.J. (1981) Test marketing of freshwater shrimp *Macrobrachium rosenbergii* in South Carolina. *Aquaculture*, **23** (1–4) 373–379.
Liao I-C. & Chien Y-H. (1990) Evaluation and comparison of culture practices for *Penaeus japonicus*, *P. penicillatus*, and *P. chinensis* in Taiwan. In: *The Culture of Cold-tolerant Shrimp* (eds K.L. Main & W. Fulks), pp. 49–63. The Oceanic Institute, Honolulu, HI, USA.
Lima dos Santos C., Josupeit H. & Chimisso dos Santos D. (1994) EEC quality and health requirements and seafood exports from developing countries. *Infofish International*, (1) 21–26.
Lobegeiger R. (1999) Redclaw crayfish. *DPI Note*, Department of Primary Industries, Queensland. http://www.dpi.qld.gov.au/dpinotes/animals/aquaculture/f99014.html
Low J. (1988) Market outlook for marine shrimps. In: *Proceedings of Seminar on Marine Prawn Farming in Malaysia*, pp. 85–91. Malaysian Fisheries Society, Serdang, Malaysia.
Lucien-Brun H. (2000) The shrimp market in France. *Global Aquaculture Advocate*, **3** (4) 86–88.
Lyons Seafoods (1995) Coldwater shrimp. *Seafood International*, March 1995, 23–25.
Macaulay P.J., Samples K.C. & Shang Y.C. (1983) Assessing the performance of market distribution channels for aquacultured products: the case of Hawaii-produced marine shrimp. In: *Proceedings of the First International Conference on Warm Water Aquaculture – Crustacea* (eds G.L. Rogers, R. Day & A. Lim), pp. 34–42. 9–11 February 1983, Brigham Young University, Hawaii, USA.
MacPherson N. & MacKay T. (1987) *Worldwide market opportunities for mariculture technology and services*, 48 pp. Phase 1 report for Aberdeen University Marine Studies, Mackay Consultants, Inverness, Scotland.
Madrid R.M.M. & Phillips H. (2000) Post-harvest handling and processing. In: *Freshwater Prawn Culture: the farming of* Macrobrachium rosenbergii (eds M.B. New & W.C. Valenti), pp. 326–344. Blackwell Science, Oxford, UK.
McLeod D.A. (1998) Crustacean farming in Europe: a commercial prospect? *Scottish Fish Farmer*, (117) 7.
Meyers S.P. (1987) Sausages from shrimp wastes. *Infofish International*, (5) 36.
MGEX (2000) White shrimp. http://www.mgex.com/whiteshrimp.htm
Mishra A.K. & Singh M. (1999) Drug, vaccine and pesticide use in aquaculture. *Infofish International*, (5) 25–30.
Morineau D. (1999) L'Australie séduit l'Europe. *Produits de la Mer*, (54) 125 [in French].
Mosig J. (1999) Grower questions the economics of yabbies. *Austasia Aquaculture Magazine,* **13** (2) 37–39.
Nenke M. & Nenke M. (2000) The Western Australian yabby industry. In: *Australian Crayfish Aquaculture Workshop Proceedings* (eds C. Lawrence & G. Whisson), pp. 14–19. International Association of Astacology, Curtin University of Technology, Perth, Australia.
Nettleton J.A. (1992) Greening your diet means seafood. *Infofish International*, (2) 49–53.
New M.B. (1990) Freshwater prawn culture: a review. *Aquaculture*, **88** (2) 99–143.
New M.B. (2000a) History and global status of freshwater prawn farming. In: *Freshwater Prawn Culture: the farming of* Macrobrachium rosenbergii (eds M.B. New & W.C. Valenti), pp. 1–11. Blackwell Science, Oxford, UK.
New M.B. (2000b) Commercial freshwater prawn farming around the world. In: *Freshwater Prawn Culture: the farming of* Macrobrachium rosenbergii (eds M.B. New & W.C. Valenti), pp. 290–325. Blackwell Science, Oxford, UK.
Nierentz J.H. & Josupeit H. (1988) The European shrimp market. *Infofish International*, (5) 14–18.
No H.K. & Meyers S.P. (1995) Preparation and characterization of chitin and chitosan. *Journal of Aquatic Food Product Technology*, **4** (2) 27–52.
van Ornum J. (1992) Shrimp waste – must it be wasted? *Infofish International*, (6) 48–52.
O'Sullivan D. (1989) *Marketing freshwater crayfish*, 14 pp. (mimeo). Presented at: Symposium on Tropical Aquaculture, 5–6 August 1989. N. T. University, Australia.
O'Sullivan D. (1996) Asian competition for Australian wild-caught and farmed prawns. *Austasia Aquaculture Magazine,* **10** (2) 5–8.
O'Sullivan D. (1997) Opportunities for disease-free prawn PLs. *Austasia Aquaculture Magazine,* **11** (3) 28–32.
O'Sullivan D. (1999) Seafarm has new prawn on the market. *Austasia Aquaculture Magazine,* **13** (6) 21–25.
Otwell W.S. & Flick G.J. Jr. (1995) A HACCP program for raw, cultured penaeid shrimp. In: *Proceedings of the Special Session on Shrimp Farming* (eds C.L. Browdy & J.S. Hopkins), pp. 214–226. World Aquaculture Society, Baton Rouge, LA, USA.
Ovenden C. (1994) *Japan's kuruma prawn market*, 21 pp. Information Series QI 94041, Department of Primary Industries, Queensland, Australia.
Pérez J.R., Carral J.M., Celada J.D., Sáez-Royuela M., Muñoz C. & Sierra A. (1997) Current status of astaciculture production and commercial situation of crayfish in Europe. *Aquaculture Europe,* **22** (1) 6–13.

Philips H. & Lacroix D. (2000) Marketing and preparation for consumption of *Macrobrachium rosenbergii*. In: *Freshwater Prawn Culture: the farming of* Macrobrachium rosenbergii (eds M.B. New & W.C. Valenti), pp. 345–368. Blackwell Science, Oxford, UK.
Putro S. (1991) Processing of surimi and fish jelly products, 23 pp. *Infofish Technical Handbook Series*, No. 2. Infofish, Kuala Lumpur.
Pyper W. (2000) Making a splash on the market. *Ecos*, (104) 13–23.
Rao D.R.M. & Prakash D.V. (1999) Indian seafood exports. *Infofish International*, (1) 19–24.
Reilly A. & Käferstein F. (1997) Control of food safety hazards in aquaculture. *Aquaculture Research*, **28**, 735–752.
Renard A.C. (1999) Incertitude sur l'Amérique du nord. *Produits de la Mer*, (57) 71–72 [in French].
Rhodes, R.J., Browdy C.L. & Stokes A.L. (1995) Made in the USA: what have we learned about marketing US farmed shrimp? Abstract in *Proceedings of the Special Session on Shrimp Farming* (eds C.L. Browdy & J.S. Hopkins), pp. 246–247. World Aquaculture Society, Baton Rouge, LA, USA.
Richards M. (1988) Co-operative marketing of signal crayfish in the United Kingdom. In: *Proceedings of First Australian Shellfish Aquaculture Conference*, Perth, 1988 (eds L.H. Evans & D. O'Sullivan), pp. 326–330. Curtin University of Technology, Australia.
Richards K. & Campbell C. (1996) *Marketing signal crayfish in the UK*, 11 pp. (mimeo). NRA Crayfish Workshop, 1 March 1996, University of Nottingham, UK.
Riepen M. (1997) The Asian market for live seafood. *Infofish International*, (1) 21–25.
Roberts K.J. & Dellenbarger L. (1989) Louisiana crawfish product markets and marketing. *Journal of Shellfish Research*, **8** (1) 303–307.
Robinson F. (1999) Species focus: blue crab. *Online Seafood Business*, Oct. 1999. http://www.seafoodbusiness.com/99oct/issue/html
Robinson S. (2000) Focus: look east for crawfish. *Online Seafood Business*, April 2000. http://www.seafoodbusiness.com/market
Rogers W.D. & Holdich D.M. (1995) Crayfish production in Britain. In: *Freshwater Crayfish 10* (eds M.C. Geddes, D.R. Fielder & A.M.M. Richardson), pp. 583–595. Baton Rouge, Louisiana State University, LA, USA.
von Rohr R. (1995) Ice in fisheries. *Infofish International*, (1) 47–50.
Rosenberry R. (1999) *World Shrimp Farming 1999*, 320 pp. Rosenberry, San Diego, USA.
Rubino M.C. (1992) Economics of red claw (*Cherax quadricarinatus* (von Martens, 1868)) aquaculture. *Journal of Shellfish Research*, **11** (1) 157–162.
Ruello N. (1990) Prawn market update. *Austasia Aquaculture Magazine*, **4** (10) 20.
Schoemaker R. (1991) Shrimp waste utilisation. *Infofish Technical Handbook Series*, No. 4, 20 pp. Infofish, Kuala Lumpur.
Shehadeh Z. (1997) Review of the state of world aquaculture. *FAO Fisheries Circular* No. 886 FIRI/C886. http://www.fao.org/fi/publ/circular/c886.1
Smith G.G. (1995) The world market for lobster. *FAO/Globefish Research Programme*, **36**, 1–91.
Srisomboon P. & Poomchatra A. (1995) Antibiotic residues in farmed shrimp and consumer health. *Infofish International*, (4) 48–52.
Stevens R.N. & Sykes D. (2000) Export marketing of Australian and New Zealand spiny lobsters. In: *Spiny Lobsters: fisheries and culture,* 2nd edn (eds B.F. Phillips & J. Kittaka), pp. 641–653. Fishing News Books, Oxford, UK.
Subasinghe S. (1999) Chitin from shellfish waste – health benefits overshadowing industrial uses. *Infofish International*, (3) 58–65.
Sugita H. & Deguchi Y. (2000) Shipping. In: *Spiny Lobsters: fisheries and culture,* 2nd edn (ed. B.F. Phillips & J. Kittaka), pp. 633–640. Fishing News Books, Oxford, UK.
Suwanrangsi S., Sophonphong K., Masae S., *et al.* (1997) *Quality Management for Aquacultured Shrimp*, 129 pp. ASEAN–Canada Fisheries Post-harvest Technology Project Phase II, South East Asian Fisheries Development Centre, Singapore.
Sze C.P. (2000) Antibiotic use in aquaculture: the Malaysian perspective. *Infofish International*, (2) 24–28.
Tidwell J. (2000) Culture of *Macrobrachium rosenbergii* in temperate climates. In: *Abstracts, Aqua 2000, Responsible Aquaculture in the New Millennium* (compiled by R. Flos & L. Creswell), p. 708. European Aquaculture Society, Special Publication No. 28.
Traesupap S., Matsuda Y. & Shima H. (1999) Diversification of shrimp products in the Japanese markets during the 1990s. *Aquaculture Economics and Management*, **3** (2) 167–175.
Tsuruta M. & Kittaka J. (2000) Marketing and distribution in Japan. In: *Spiny Lobsters: fisheries and culture,* 2nd edn (eds B.F. Phillips & J. Kittaka), pp. 655–663. Fishing News Books, Oxford, UK.
Vidali C., Filose J. & Blake D. (2000) *Ocean Garden Products*. http://members.aol.com/brosenberr/OceanGardenProducts.html
Vondruska J. (1986) *Blue crab markets and analog products*, 19 pp. Prepared for National Blue Crab Industry Association Annual Meeting. 26–28 February 1986, New Orleans, LA. National Marine Fisheries Service, USA.
WA Fisheries (1999a) *Farming yabbies*. http://www.wa.gov.au/westfish/aqua/broc/aqwa/yabby/index.html
WA Fisheries (1999b) *Farming marron*. http://www.wa.gov.au/westfish/aqua/broc/aqwa/marron/index.html
Wang M.J., Hansen T.M., Kauffield M., Christensen K.G. & Goldstein V. (2000) Slurry ice in fish preservation. *Infofish International*, (2) 42–46.
Wingfield M. (2000) An overview of the Australian freshwater crayfish farming industry. In: *Australian Crayfish Aquaculture Workshop Proceedings* (eds C. Lawrence & G. Whisson), pp. 5–13. International Association of Astacology, Curtin University of Technology, Perth, Australia.
Wiryanti J. & Madakia H. (1997) *Improved Quality Control for Fresh and Frozen Shrimp*, 110 pp. ASEAN–Canada Fisheries Post-harvest Technology Project Phase II, South East Asian Fisheries Development Centre, Singapore.
Young N.W. (1996) *Marketing crustaceans*, 12 pp. (mimeo). National Rivers Authority, Crayfish Workshop, 1 March 1996. University of Nottingham, UK.

Chapter 4
Candidates for Cultivation

4.1 Introduction

The primary goal of most (but not all) crustacean aquaculture enterprises is to generate profits as quickly as possible. In the simplest terms an 'ideal' species must therefore command a high price in an established market, it must grow rapidly to a marketable size on inexpensive, readily available diets, be resistant to disease while not having stringent environmental or biological requirements, and be readily available for culture from wild-caught or hatchery-reared stock. The only species to possess the majority of these attributes are to be found among the penaeid shrimp, which is of course why they form the bulk of farmed crustaceans (Table 1.1). There are no known freshwater species possessing so many favourable characteristics although the Australian redclaw crayfish (*Cherax quadricarinatus*) is rapidly growing in popularity among aquaculturists. However, even if the present rate of progress continues it will be many years before production levels could even begin to approach those of *Macrobrachium rosenbergii.*

It is now technically possible to culture for the table a large number of commercially valuable crustaceans through all or at least a major part of their life cycle, in many cases at a considerable distance from their home range. However, it is not yet feasible to culture many of these animals profitably because of low market value, slow growth rate or because of the need to create an expensive culture environment. Such considerations have ruled out several superficially attractive species for aquaculture, at least for the foreseeable future. Among these are: *Nephrops norvegicus* (scampi) unsuitable, along with several other nephropid lobsters, because of a pronounced burrowing habit or a preference for offshore environments; most caridean prawns (other than three or four *Macrobrachium*, and perhaps one *Pandalus* species) on the basis of small size, slow growth and low fecundity; and the majority of crabs, on account of low value and sometimes technical difficulties associated with lack of data on their culture requirements.

Recent advances in the laboratory culture of spiny lobster larvae in Japan, Australia and New Zealand are exciting (Kittaka 2000) and survivals of up to 10% have been achieved with *Jasus verreauxi* (Kittaka 1997); however, commercially viable mass rearing techniques have yet to be developed. Despite these erstwhile drawbacks, some of the larger crustaceans, notably *Homarus*, and some crabs and spiny lobsters, may be suitable for sea ranching provided problems concerning production costs and ownership can be resolved (sections 7.8.11, 7.10.8, 10.6.3 and 11.5.3.1). Hatchery culture and release of juvenile shrimp and crabs for the enhancement of natural stocks is usually publicly funded but opportunities do exist for commercial involvement (sections 7.2.9, 8.11.1 and 12.7).

While not strictly farming, the culture of several small crustaceans (e.g. *Artemia*, *Moina*, *Daphnia*, *Amphiascoides*, *Tisbe*, *Acartia*, *Eurytemora*, *Mysidopsis*) as live food for the young stages of other species has expanded following the culture of novel shellfish and fish (section 7.11). Similarly, the culture of valuable ornamental crustaceans, e.g. *Lysmata*, *Stenopus*, is being investigated in response to the demands of a lucrative display aquarium trade and at the same time to reduce serious over-exploitation of natural stocks and their habitats (section 7.4.4.1). Also significant quantities of shrimp, notably (*Litopenaeus setiferus* and *Farfantepenaeus duorarum*) and crayfish (*Orconectes* spp.) are reared in the USA as live bait for anglers (sections 7.2.5 and 7.5.9).

For some species or locations several quite different approaches to culture methodology are worth considering; in other situations a species may be grown profit-

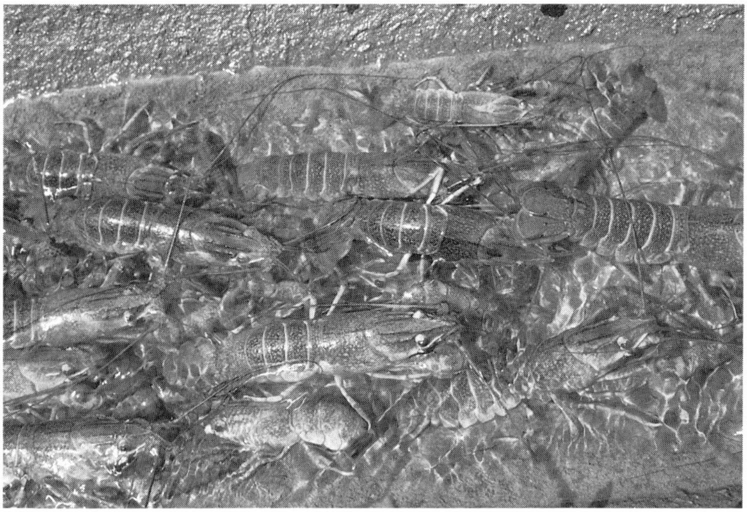

Plate 4.1 Australian redclaw crayfish (*Cherax quadricarinatus*) moving upstream towards a flow-trap (section 7.7.8). (Photo courtesy Clive Jones, Department of Primary Industries, Queensland, Australia.)

ably using methods modified only to take account of increasing intensification. This chapter examines the factors governing the selection of species to be cultured for the table in relation to environment, biology, product marketability and other aquatic organisms (polyculture). At the same time possible candidates for 'catch crops' or off-season cultures that might be employed to improve cash flows or increase marketing opportunities are considered, e.g. for shrimp such as *Penaeus esculentus* (section 7.2.9), *Fenneropenaeus penicillatus* (Liao 1988) or crabs (*Scylla* spp.) (Shelley & Field 1999).

4.2 Location

Compatibility with the local environment (including endemic, putative pathogens) and the availability of brood- or seedstock (post-larvae or juveniles) are perhaps the most important reasons for choosing to culture a species within its home range. Sometimes however, the only native species available is, for one reason or another, perceived to perform less well in culture than exotic species imported from similar climatic regions elsewhere. An example is *Litopenaeus schmitti* where respectable yields have only been obtained periodically in Cuba and Colombia (Currie 1998). The inconsistency in performance was due initially to lack of experience and knowledge of the species and subsequently to inadequate feed quality (Cuba) or a disease outbreak (Colombia). The potential for improved performance through better husbandry and by selective breeding programmes has now been recognised. If implemented successfully, farming the endemic species would promote a valuable resource, especially in the Cuban context, and be demonstrating sound ecological practices. Table 4.1a shows the broad environmental requirements of important cultivable crustaceans while Table 4.1b indicates their natural geographic distribution and gives examples of areas into which they have been transplanted for the purposes of cultivation.

When a species is cultured outside or near the limits of its normal distribution but within a comparable climatic zone, the farmer must establish reliable supplies of juveniles or broodstock and/or maintain broodstock production facilities locally. Trials with cryopreserved penaeid eggs and nauplii (Diwan & Kandasami 1997) have not, as yet, led to the development of useful, cost-effective techniques (sections 3.4.3 and 12.4) for buffering fluctuations in availability or quality of seed (section 11.3.1.1). Reliable importation of juveniles is seldom satisfactory in the long term and becomes particularly difficult and expensive as scale increases. A species that reproduces readily in captivity in the new environment is therefore more suited to transplantation than one which does not. The early development of *Macrobrachium* farms in Hawaii, the Caribbean Islands and Israel, of *Marsupenaeus japonicus* farms in Brazil, *Procambarus* and *Pacifastacus* culture in Europe and of *Cherax* in Ecuador provide good examples of this. However, transportation of crustaceans to new locations carries with it the risk of transferring diseases, and ecological problems may arise when the introduced species escapes and becomes established outside the farm. Increasingly legislation is being

Table 4.1a The environmental habitats of the main species and groups of cultivable crustaceans.

	Marine/brackish	Freshwater
Tropical	*Penaeus monodon* *Fenneropenaeus indicus* *F. merguiensis* *Litopenaeus stylirostris* *L. vannamei* *Metapenaeus* spp. *Panulirus* spp. *Thenus* spp. *Scylla* spp.	*Macrobrachium* spp. *Cherax quadricarinatus*
Subtropical/warm temperate	*Fenneropenaeus chinensis* *F. penicillatus* *Marsupenaeus japonicus* *Jasus* spp. *Panulirus* spp *Portunus* spp.	*C. destructor, C. albidus* *C. tenuimanus* *Procambarus clarkii* *P. zonangulus* *Orconectes* spp. *Eriocheir sinensis*
Temperate	*Pandalus platyceros* *Homarus americanus* *H. gammarus*	*Astacus astacus* *A. leptodactylus* *Pacifastacus leniusculus* *Orconectes* spp. *Paranephrops* spp.

invoked or adapted to control such movements in an attempt to protect the industry and local fisheries (sections 8.9.4, 11.3.2, 11.3.3 and 12.2).

Under some circumstances (for example to escape disease or where proximity to lucrative markets is a prime advantage) it may be thought desirable to culture a species in an environment or climatic region to which it is not well adapted. For example *Litopenaeus vannamei* is being raised inland in low-salinity groundwater in Arizona (section 7.2.6.5) while tropical freshwater prawns are grown commercially in geothermal waters in New Zealand (Table 5.5). Similarly, attempts have also been made to increase growth rate of temperate water species such as the shrimps *Pleoticus muelleri*, *Artemesia longinaris* and *Sclerocrangon boreas* (Miglavs 1992a) by raising the temperature of the water, either through direct or indirect use of heated effluent or geothermal waters (Wickins 1982). Where temperatures are only suitable for maximum growth during part of the year, sheltered nursery systems may be required to overwinter stocks or 'head-start' juveniles prior to ongrowing in outdoor ponds, for example, redclaw crayfish (Sagi *et al.* 1997); yabbies (Mosig 1999) and shrimp (Moya *et al.* 1999). In addition to the problems of animal supply, such approaches often require the creation and maintenance of some kind of heat exchange or controlled-environment system.

Currently, reasons for considering controlled-environment culture in closed recirculation systems include reducing the risks from imported diseases and compliance with regulations governing farm effluents (Davis & Arnold 1998). In the past, many commercial attempts were made to farm crustaceans intensively in recirculation systems, particularly shrimp (section 7.2.6.6) and clawed lobsters (section 7.8.9), in an attempt to emulate highly controlled, production-line industries. The tolerance of several species to controlled-environment systems was evaluated on a laboratory or pilot scale as a necessary prerequisite. Examples included shrimp and prawns (Forster & Beard 1974; Hanson & Goodwin 1977; Ogle & Lotz 1992), freshwater crayfish (section 7.7.6.4) and clawed lobsters (section 7.8.9). At least one such shrimp facility has survived, albeit now producing 2–4 g 'popcorn' shrimp (*Litopenaeus vannamei*) for local Hawaiian markets (section 3.3.1.1). The concept of total control over production remains attractive and new enterprises have recently been established in Ecuador (Kagel 1998) and China (section 7.2.6.6). However, few controlled-environment ongrowing systems have proved to be commercially viable over the long term. Increasingly, however, broodstock facilities are maintained either in isolation to provide 'high-health' seed to farmers or in 'closed-cycle' shrimp farms to protect hatchery and farm ponds from imported diseases (Clifford 1997) (sections 7.2.2.3 and 8.9.4.4).

Table 4.1b The geographic distribution of the major species and groups of cultivable crustaceans with examples of known transplantations.

Species	Natural range	Transplantations
Penaeus monodon	Indo-Pacific, Australia	Middle East, West Africa, Middle America, China, Italy
Fenneropenaeus chinensis	Korea, China, Gulf of Bohai, Northwest Pacific	New Zealand, USA, France
Litopenaeus stylirostris	Eastern Central Pacific	USA, Caribbean Islands, Tahiti, Hawaii, Taiwan, Brunei
L. vannamei	Eastern Central Pacific	USA, Caribbean Islands, Tahiti, Hawaii, Taiwan, China, West Africa
Marsupenaeus japonicus	Japan, Korea, Western Central Pacific	Brazil, Mediterranean, France
Macrobrachium rosenbergii	Indo-Pacific, Australia	Caribbean Islands, USA, Hawaii, USSR, Southern Africa, New Zealand, Israel, Brazil, China, Mauritius, Mexico, UK, Taiwan, Uruguay, Argentina, Paraguay, Surinam
Pandalus platyceros	Eastern Central Pacific	UK (experimental)
Astacus astacus	Europe (except Iberian Peninsula)	Morocco, Siberia
Astacus leptodactylus	Turkey, Greece, Hungary	UK, Italy, France, Switzerland, Austria, Belgium, Czech Republic, Denmark, Finland, Germany, Latvia, Lithuania, Netherlands, Poland, Spain
Cherax albidus	South-eastern Australia	Western Australia
C. albertsii	Papua New Guinea	—
C. destructor	South-eastern Australia	South Africa
C. tenuimanus	Western Australia	USA, Caribbean Islands, Chile, Panama, South Africa, Taiwan, China, Malaysia, Japan, New Zealand, France
C. quadricarinatus	Northern Australia (Queensland)	USA, Caribbean Islands, Ecuador, China, Southern Africa (Swaziland, Zambia), Italy, Israel, New Caledonia
Orconectes spp.	North America	France, Germany, Mexico, Austria, Belgium, Russia, Sweden, Italy
Pacifastacus leniusculus	North-western USA	Scandinavia, UK, Italy, Spain, Denmark, France, Netherlands, East Europe, Russia, Greece, Japan
Procambarus clarkii	Southern USA, NE Mexico	South Africa, East Africa, France, Italy, Spain, Middle East, Caribbean Islands, Central and South America, Japan, Taiwan, Malaysia, China, Thailand, Philippines, Singapore
Homarus americanus	Canada, Atlantic USA	Japan, Pacific USA, Italy
H. gammarus	Europe, Mediterranean	—
Jasus spp.	Australia, New Zealand, South Africa	Japan, France
Panulirus spp.	Japan, Australia, India, Africa, Caribbean, Eastern Central Pacific	Japan
Eriocheir spp.	China, Korea	Europe, USA
Portunus spp.	Japan, Korea	—
Scylla spp.	Indonesia, Thailand, Australia	—

4.3 Broodstock

4.3.1 Seasonal availability

Security and predictability of broodstock supply are prerequisites for any successful crustacean hatchery enterprise. Where there is an established fishery for the species to be cultured, or where natural breeding populations are abundant, ripe impregnated or egg-bearing females may be caught for at least part of the year. Special dispensation to fish for breeding animals may be required in some areas at certain times of the year, while in other areas a complete ban may be enforced. When assessing a new or expanding project it is wise to consider potential competition for broodstock supplies, particularly in areas where aquaculture developments are being

. Alternative sources of supply should also be investigated.

4.3.2 Ease of establishing and maintaining a broodstock

A number of species mature, mate and spawn readily in production ponds, e.g. *Macrobrachium*, while others must experience specific seasonal photoperiod and temperature regimes or be induced to breed through surgical or dietary manipulations that act on their hormone systems. Indeed, some crustaceans (e.g. *Cherax* spp.) breed so readily in ponds that overcrowding and stunting occur. To combat this the potential for producing monosex or sterile male-only populations of *C. destructor*, *C. albidus* and *C. quadricarinatus* are being investigated (Lawrence & Jones 2001) (section 2.6.3). In Table 4.2 important cultivable species are categorised according to the ease with which breeding populations may be maintained. Interestingly, although most wild-caught *Litopenaeus vannamei* require ablation to mature and spawn reliably, stocks held in maturation systems in Colombia for just one or two generations and in Venezuela for over 14 generations are reported to spawn predictably without ablation (A. Guillaumin 2000, pers. comm.). Daniels *et al.* (2000) suggest that the development of methods to synchronise spawning in *Macrobrachium* held in temperate climates would be useful since it is typical for only 5–10% of broodstock females to hatch eggs in a one- to three-day period of time. Details of the conditions for breeding each species group in captivity are described in Chapter 7.

A very different problem arises with *Pandalus platyceros* whose aquaculture potential in temperate regions has been investigated (C. Campbell & Y. Alabi 2001, pers. comm.; sections 7.4.2 and 7.4.3). These prawns are protandrous hermaphrodites. They mature and function first as males but 1.5–2.5 years later mature and function as females. This habit combined with a low fecundity (Table 4.6f) poses formidable difficulties in the management of captive broodstock unless they are farmed where wild stocks are available.

Table 4.2 Broodstock categories.

Species that mature, mate and spawn readily in captivity	
Shrimps	*Fenneropenaeus chinensis, Fenneropenaeus indicus, F. merguiensis, Marsupenaeus japonicus,* (captive stocks of *Litopenaeus vannamei* – see text, section 4.3.2)
Prawns	*Macrobrachium rosenbergii, Macrobrachium* spp.
Crayfish	*Astacus leptodactylus, Cherax albidus, C. destructor, C. quadricarinatus, C. tenuimanus, Pacifastacus leniusculus, Procambarus clarkii, Paranephrops planifrons, P. zealandicus.*
Crabs	*Eriocheir sinensis, Scylla* spp.
Species requiring environmental or hormonal manipulation to breed reliably	
Shrimps	*Penaeus monodon, Litopenaeus stylirostris* (wild stocks of *Litopenaeus vannamei* – see text, section 4.3.2)
Prawns	*Pandalus platyceros*
Clawed lobsters	*Homarus americanus, H. gammarus*
Spiny lobsters	*Jasus* spp., *Panulirus* spp. ?

4.4 Larvae

4.4.1 Duration and complexity of larval life

The duration and complexity of the larval phase as well as the fecundity of cultivable crustaceans have a major impact on hatchery design, running costs and the technical skills required to maintain a predictable output of post-larvae. However, a lengthy larval phase is not always associated with greater complexity although abbreviated, less complex larval development is generally associated with reduced fecundity (Fig. 4.1). For example, penaeid larvae pass through three quite distinct morphological stages, each with different modes of nutrition, in half the time *Macrobrachium rosenbergii* larvae take to reach their first significant metamorphosis. Fecundity is inevitably low (generally less than 1000 eggs) in freshwater crayfish and in the large cold-water shrimp *Sclerocrangon boreas*, only reaches about 340 eggs (Miglavs 1992b). This is because sufficient yolk must be deposited in each egg to sustain complete development so that juveniles hatch directly from the eggs. In contrast, spiny lobsters incubate large numbers of small eggs with very little yolk but have an enormously protracted larval life, e.g. from 64 days (*Palinurus elephas*) to 391 days (*Panulirus japonicus*) (section 7.9.3).

4.4.2 Resistance to disease

All larvae seem susceptible to disease and infestations and it is widely recognised that the penaeid hatchery industry is particularly vulnerable to introduced viral and bacterial diseases, partly due to the worldwide trade in broodstock and nauplii, and partly to the widespread

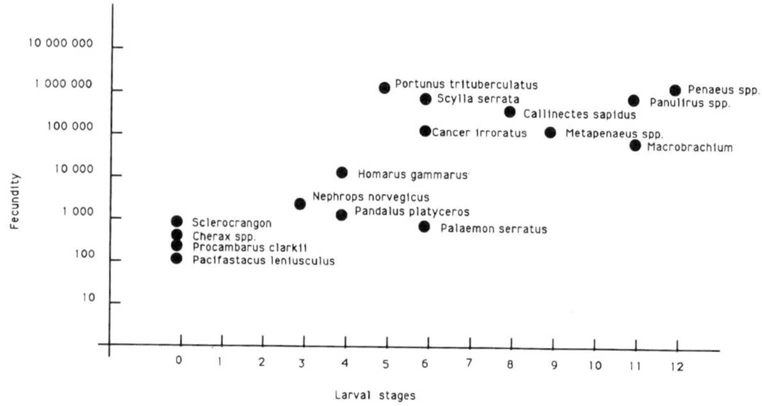

Fig. 4.1 The relationship between fecundity and the number of free-swimming larval instars in selected crustaceans.

culture of penaeid species outside their natural range (section 8.9.3). In species other than penaeids, however, most mortalities are believed to be the result of, or exacerbated by, stresses arising from poor water, water management or food; the longer the larval life the greater the difficulty of maintaining good culture conditions.

4.5 Post-larvae and juveniles

4.5.1 Availability from the wild

Many thriving extensive culture operations (for example penaeid farms in South-east Asia and Middle America) are based on supplies of naturally occurring post-larvae or juveniles. With intensification and expansion, such industries become increasingly vulnerable to natural fluctuations in wild seed supply and the price changes that follow. In the past 10 years, hatcheries have certainly increased their contribution to overcoming deficits in supply (Table 4.3a), and although the numbers of shrimp hatcheries have increased considerably (Table 4.3b), they seldom completely fulfil the demand. Nurseries stocked with either wild or hatchery produce can go some way to maximising the use of the available seed, and a high tolerance of a species to crowding, compounded diets and, in some cases, recirculated water is a distinct advantage.

In the absence of a proven, commercial-scale hatchery technology, attempts to initiate spiny lobster ongrowing and fattening operations based on the collection of wild pueruli or juveniles have often been frustrated by the difficulty of locating reliable supplies or by the costs involved in collection. Recent advances in the understanding of juvenile ecology, in collector design, and in obtaining special dispensation for the capture of juveniles (sections 7.9.4 and 11.5.3.3), however, have rekindled interest in the prospects for ongrowing and ranching operations (sections 7.9.5, 7.9.8 and 10.6.4).

Significant effort is now also being applied to the development of hatchery technologies to support mud crab cultivation in response to concerns about overfishing of both juveniles for fattening and adults (sections 7.10.1 and 11.3.1.2).

4.5.2 Nursery

Following metamorphosis or capture, crustacean post-larvae benefit from special attention (weaning and acclimation) as they pass through a period of transition from epipelagic to benthic life. Fundamental physiological and behavioural changes occur at this time as they are weaned away from the diets they enjoyed as larvae to compounded diets, and acclimated to the new environmental conditions of nursery ponds, sometimes (for example, during *Macrobrachium* and *Eriocheir* culture) with a concurrent change of salinity. Interestingly, at least one strain of the oriental river prawn, *Macrobrachium nipponense*, has been found that can be reared throughout its life in fresh water, and is increasingly being cultured in China (Kutty *et al.* 2000). The rate at which weaning can be successfully completed and the juveniles grown to a size suitable for stocking in ongrowing ponds varies between species from a few days to a month or two. A short period is generally advantageous although some ongrowing strategies are dependent on indoor nursing throughout a winter period (e.g. *Macrobrachium* and *Cherax* culture in the southern USA and Israel) so that ponds can be stocked with juven-

Table 4.3a Hatchery capacity and the demand for post-larvae.

Country	Number of hatcheries	Hatchery output (post-larvae × 10^6 yr^{-1})	Anticipated demand (wild + hatchery) (post-larvae × 10^6 yr^{-1})	Year
1980s (Shrimps, freshwater prawns and crabs)				
China				
Penaeus	10 govt. + backyard	72 443	72 443	1987
	340	42 000	—	1989
Ecuador				
Penaeus	70	3000	6000	1986
	85 (+ 30 not in production)	6000	11 000	1989
	40 producing	4320	15 000	1990
Indonesia				
Penaeus	38 + 85 planned	34–110	1750	1987
	93	1460	7300	1988
Indonesia				
Macrobrachium	5	3.6	—	1987
Japan				
Portunus	17	16	—	1982
Korea				
Penaeus	3 govt. + backyard	78.3	—	1989
Malaysia				
Penaeus	32	539–600	—	1987
Philippines				
Penaeus	56	100	—	1987
Taiwan				
Penaeus	1200	4500	4500+	1987
Taiwan				
Macrobrachium	20	25	—	1987
Thailand				
Penaeus	5 major + >1000 backyard	41	152	1984
		18.8	180 (*P. monodon*)	1985
		6000	6000+	1989
Thailand				
Macrobrachium	44	80	—	1987
USA				
Penaeus	6	138 (not yet achieved)	800–1000	1989
1990s (Shrimps)				
Australia (Queensland)				
Penaeus monodon	5–9	155–175	—	1996–98
Marsupenaeus japonicus	5–7	33–43	—	1996–98
Costa Rica	1	480	—	1997
Ecuador	200–300	—	30 000	1999
Honduras	13–15	3924	6540	1997
India	90–150	1500	6500	1997–98
Iran	17	370	—	1999
Mexico	30–33	6 000–7000	8200	1997–98
New Caledonia	5	168 (capacity)	—	1997
Nicaragua	3	360	900	1997
Panama	11	1744	1200	1997
Singapore	1	25	—	1997
Thailand	800–1000	15 000	—	1995
(Freshwater prawns)				
Bangladesh	11–40	—	>200	2000
Brazil	18	5600	—	2000
China	20–250	10–800	6000–10 000	1999
India	16–22	250	—	1999
Taiwan	50	500	1300	1998
Thailand	100–300	5000	—	1999

Table 4.3b Estimated (range) of penaeid shrimp hatchery numbers and distribution of sizes in the 1990s. Source: Rosenberry (1998, 1999).

Country	Total number of hatcheries	Distribution of hatchery sizes (%) Small	Medium	Large	Predominant species
Australia	8–16	0	80–90	10–20	*P. monodon, M. japonicus*
China	2000	80	10	10	*F. chinensis*
India	150–225	20	60	20	*P. monodon, F. indicus*
Indonesia	300	85	10	5	*P. monodon, F. merguiensis*
Iran	17	10	80	10	*F. indicus*
Malaysia	100	20–50	50–70	0–10	*P. monodon, F. indicus, F. merguiensis*
New Caledonia	5	100	0	0	*L. stylirostris*
Philippines	90–120	5	95	0	*P. monodon*
Sri Lanka	66	50	30	20	*P. monodon*
Taiwan	220	80	20	0	*P. monodon, M. japonicus, L. vannamei*
Thailand	800–1000	80	15	5	*P. monodon, F. indicus, F. merguiensis*
Vietnam	1000	90	10	0	*P. monodon, F. merguiensis, F. indicus*
Other (Bangladesh, Myanmar, etc.)	2000	—	—	—	spp. var.
Belize	1	0	0	100	*L. vannamei*
Brazil	18–20	10	40	50	*L. vannamei*
Colombia	11–12	10	80	10	*L. vannamei, L. stylirostris*
Ecuador	200–350	60	30	10	*L. vannamei, L. stylirostris*
Honduras	13–15	60	30	10	*L. vannamei, L. stylirostris*
Mexico	30–33	30	60	10	*L. stylirostris, L. vannamei*
Nicaragua	3–5	40	40	20	*L. vannamei, L. stylirostris*
Panama	11	20	40	40	*L. vannamei, L. stylirostris*
Peru	3	0	100	0	*P. vannamei*
USA(inc. Hawaii)	8–10	30–50	30–40	20–30	*L. vannamei, L. stylirostris, L. setiferus*
Venezuela	6–7	0	0	100	*L. vannamei*
Other (Argentina, Brazil, Costa Rica)	20	—	—	—	spp. var.

iles as soon as it is warm enough for growth outdoors. Indoor and outdoor *Macrobrachium* nurseries are also used in tropical regions to maximise the efficient use of ongrowing ponds and food utilisation (section 7.3.4). Examples of the survival and growth rate of selected species during the nursery phase of culture are given in Table 4.4.

Some species (*Homarus*) are so territorial and cannibalistic that severe losses and mutilation occur if they are not placed in individual rearing containers soon after metamorphosis. Various attempts to rear juvenile homarids communally for 6–12 months have been made (section 7.8.8) and some workers have suggested it might be worth while to periodically remove or immobilise the claws to reduce aggression and fighting during this stage (Aiken & Waddy 1995). Similar studies have been made with *Macrobrachium* (Karplus *et al.* 1989). Such treatments, however, can increase the risk of disease, incur postoperative losses and would not be considered humane in some countries.

4.6 Ongrowing

4.6.1 Growth rate and size distribution

Rapid growth rate is one of the most important attributes of a candidate for aquaculture in that it maximises cash flow, minimises the period in which crop loss would be financially most damaging, and minimises time taken to recover from crop failure. The farmer, however, is interested not only in average growth rate but also in the proportion of the crop that fetches the highest price. This is usually that which contains the largest animals (Table 4.5), but in some crab and crayfish operations it may be ovigerous or soft-shelled animals (sections 7.5.8, 7.10.4 and 7.10.9). Both sex and stocking density can affect the final distribution of sizes in the crop and here again *Macrobrachium* presents particular problems in that dominant (blue claw) males grow substantially larger than females and hierarchies develop that encourage the further spread of sizes. The expression of male growth and the development of the characteristic

Table 4.4 Growth rate and survival of crustaceans during the nursery phase of culture.

Species	No. m^{-2}	Period (days)	Survival (%)	Growth (g) from	Growth (g) to	Reference
Shrimp						
F. chinensis	25	14	100	0.15	0.97	Forster & Beard 1974
	166	28	92	0.13	2.21	Forster & Beard 1974
	150–300	45	86–100	7 mm TL	50 mm TL	Main & Fulks 1990
	135–234	18–23	33–46	6–7 mm TL	0.06–0.21	Main & Fulks 1990
F. indicus	25	14	93	0.26	0.61	Forster & Beard 1974
	166	28	68	0.23	1.08	Forster & Beard 1974
	50–200	32–80	>60	PL_5	1	AQUACOP 1984
M. japonicus	25	14	100	0.2	1.13	Forster & Beard 1974
	166	28	99	0.2	1.62	Forster & Beard 1974
P. monodon	25	14	100	0.15	1.27	Forster & Beard 1974
	166	28	100	0.15	1.87	Forster & Beard 1974
	50–200	35–85	<50	PL_5	1	AQUACOP 1984
	50–150	45–50	47–85	PL_{10-15}	0.5–1.8	Tacon A. 1986 pers.comm.
	125–1000	35	83–98	0.013	0.25–0.87	Briggs M. 1990 pers.comm.
L. stylirostris	50–200	30–45	<50	PL_5	1	AQUACOP 1984
	150–250	45–50	55–85	PL_{4-6}	0.8	Tacon A. 1986 pers.comm.
L. vannamei	150	21–28	65–75	PL_{7-10}	0.6–0.8	Villalon 1993
	50–200	45–50	>60	PL_5	1	AQUACOP 1984
	200–300	45–50	69–95	PL_{4-6}	0.8	Tacon A. 1986 pers.comm.
	4000–7800	35	95–100	0.002	0.08	Samocha *et al.* 1993
	100	30–40	?	PL_{10-12}	1.0	Peterson & Griffith 1999
Prawns						
M. rosenbergii	25	14	100	0.13	0.33	Forster & Beard 1974
	166	28	98	0.15	1.04	Forster & Beard 1974
	688	42–63	80–90	0.009–0.016	0.5	Heinen & Menzi 1991
	32–800	30–105	32–95	0.01–0.7	0.2–2.1	Alston & Sampaio 2000
(in cages)	2–4 L^{-1}	20	88–95	0.01	0.051–0.061	Marques *et al.* 2000
	100–200	60	84–88	0.053	0.65–0.75	Marques *et al.* 2000
Pandalus platyceros	—	60	98	0.63	2.8	Oesterling & Provenzano 1985

Crayfish						
Astacus astacus	800	112	49 (85**)	0.07	0.63	Keller 1988
A. leptodactylus	20–30	120	85–90	0.03	3.34	Tcherkashina 1977
	130	90	51	0.04	0.84	Koksal 1988
Cherax destructor	100–300	42	90	0.10	1.0	Geddes *et al.* 1995
Cherax quadricarinatus	1278	36	52	0.02	0.24	Jones 1990
	250	60	64	—	1.13	Rouse 1995
Pacifastacus leniusculus	100	120	67	0.03	2–3	Lewis 2001
				instar 3	30–40 mm CL	
Procambarus clarkii	200–300	30	80–90	0.02	0.50	Huner & Barr 1984
Lobsters						
Homarus americanus	—	60	95	0.44	1.48	Norman-Boudreau & Conklin 1983
H. gammarus	400*	30	80	Instar 4	0.16	Beard *et al.* 1985
H. gammarus	400*	77–88	95–100	5–6 mm CL	11–12 mm CL	Ali & Wickins 1994
Spiny lobsters						
Panulirus argus	115*	70	93	0.2	0.99	Lellis 1992
Panulirus japonicus	8 per 800 L tank	120	89	8 mm CL	20 mm CL	Norman *et al.* 1994
Jasus edwardsii	90	112	98	2.3	10.1	Crear *et al.* 2000
Jasus edwardsii	100	365	85–95	Puerulus	20–30	Jeffs & Hooker 2000
Crabs						
Eriocheir sinensis	50–120	120	—	6.25 mg	5–25	Li 1998
Mithrax spinosissimus	181	56	67	20 mg	710 mg	Lellis 1992
Portunus trituberculatus	3090	12	58	Megalopa	6–12 mm CW	Chu *et al.* 1996
Scylla spp.	ND	28	35–53	Megalopa	3 g	Quinitio & Parado-Estepa 2000

*held individually. **M Keller, 2001 pers. comm.

Table 4.5 Examples of the proportion (%) of harvested crustaceans reaching marketable sizes.

Penaeus monodon, Thailand, examples of seasonal variations

Size (g)	<25	26	29	32	36	>38
January	1.3	0	27	5.5	32.6	25.4
February	5.7	0.8	31.4	10.8	23.6	22.6
March	7.4	38.1	14.7	6.1	20	10.8
April	9.5	15.6	19.5	0.2	16	29.9
May	45	2.7	22.5	5.9	9	6.9
June	50.5	1.5	22.7	3.1	6.4	5.3
July	32.6	0.7	22.5	9.4	7.8	13.8

P. monodon, Philippines

Size (g)	<22.1	24.2	32.1	52.6	>52.7
%	5.5	5.6	54.1	34.3	0.2

P. monodon, India

Size (g)	9	12	15	19	24	29	34	41
%	20.3	25.3	7.0	7.9	8.3	10.4	12.4	8.3

L. stylirostris, Belize

Size (g)	<11	11	13	16	19	22	26
%	5.8	2.9	11.6	39.1	26.1	11.6	2.9

L. vannamei, USA, recirculation system

Headless count/lb	61–70	31–35	26–30
%	9	71	20

Macrobrachium rosenbergii, Israel

Size (g)	<20	20–25	25–30	30–35	35–40	40–45	45–90
%	9.9	6.6	8.5	14.5	18.4	35.5	6.6

Macrobrachium rosenbergii, Mississippi, USA

Tails kg^{-1}	>110	110–100	99–89	88–78	77–67	66–56	55–45	45–35	<35
%	5.3	2.8	3.6	10.5	20	31.6	22.6	4.2	0.5

Macrobrachium rosenbergii, S. Carolina, USA

Tails kg^{-1}	>154	111–154	78–110	56–77	<56
Pond					
Alcolu 1	16	26	40	14	4
Alcolu 3	5	0	25	60	10
Cameron	6	4	27	32	31
Moncks C.	14	22	35	24	5
Murrells	5	9	30	33	23
Ridgeland	23	37	27	10	3
Sumter	8	27	44	18	3
Walterboro	21	43	21	11	4

Macrobrachium rosenbergii, Jamaica							
Count lb^{-1}	21–25	16–20	13–15	8–12			
%	30	35	15	20			
Macrobrachium rosenbergii, Egypt							
Count kg^{-1}	>61	41–61	21–41	1–21			
% (stocked @ 0.3 g)							
2 m^{-2}	2	6	79	13			
5 m^{-2}	9	28	45	18			
% (stocked @ 2 g)							
2 m^{-2}	2	9	48	41			
5 m^{-2}	7	3	57	33			
Cherax	*Cherax destructor*, Australia		*C. tenuimanus*, Australia				
Size (g) (after 1 yr)	<40	>40	>40				
%	56–64	36–44	70				
Cherax quadricarinatus, pond trials, south-east USA							
Mean wt (g)	<8	8–14	15–24	25–49	50–79	>80	
%	0	0.2	3.7	57.2	32.0	6.9	
Cherax quadricarinatus, demonstration farm, N. Queensland, Australia							
Size group (g)	<30	30–50	51–70	71–90	91–110	>110	berried females
Approx. weight harvested (kg)	2.5	65	75	10	6	4	22
Homarus gammarus, UK (change in percentage reaching 85 mm CL with time)							
Years	<2	2.25	2.5	2.75	3	3.25	3.5
%	3	18	38	22	12	6	1

morphological types of males (blue claw, strong and weak orange claw, and small) are now known to be regulated by social interactions based on the relative size ranking of very early juveniles rather than by directly heritable genetic factors (section 2.6.1). Early hatching and early metamorphosis may also play a part in conferring an initial size advantage (Karplus *et al.* 2000). Similarly, growth and survival advantages may be expressed in early hatching and early metamorphosing clawed lobster larvae, but in *Homarus* are probably related to physiological rather than behavioural factors (sections 7.8.6 and 7.8.7). Specific stocking and harvesting strategies are employed by *Macrobrachium* farmers to minimise these effects (section 7.3.7) although promising research on the production and usefulness of monosex populations in both *Macrobrachium* and *Cherax* spp. has yet to achieve its full potential (sections 2.6.3, 7.3.7 and 12.8.3). Even when lobsters are held individually, the range of sizes that develops may mean that only a proportion of the crop reaches a legally saleable size at any one time (Table 4.5). In the absence of behavioural interaction the variation may be due to both genetic and husbandry (including environmental) components (section 2.6.1).

4.6.2 Tolerance to water quality changes

Species that survive and grow well under culture conditions are generally those that are adapted to a shallow water, naturally changeable environment (Fig. 4.2). Nowhere is this more apparent than in the cases of inshore and estuarine penaeids, crabs, *Macrobrachium*, the

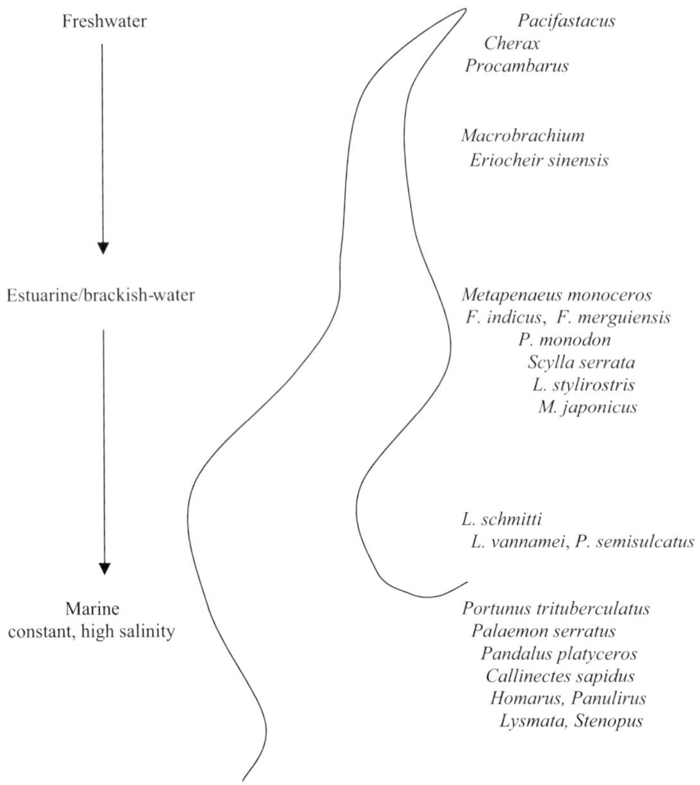

Fig. 4.2 Habitat preferences of selected crustaceans.

swamp crayfish (*Procambarus clarkii*) and perhaps the yabby (*Cherax destructor*). Species from more stable or open oceanic environments (e.g. spiny lobster larvae; clawed lobsters) do not usually perform so satisfactorily and often require more stringently controlled culture conditions.

4.6.3 Resistance to disease

A high degree of resistance to disease is a feature of many successfully cultured species but three features are worthy of note in the context of assessing a culture project. Firstly, no species is immune to disease and serious losses can occur if the animals or their environment are overstressed. The near collapse of the Taiwanese shrimp culture industry in 1988 following the unprecedented demands made by the rapidly expanding industry (sections 1.1 and 11.3.3) heralded a series of failures similarly linked to environmental degradation throughout the tropics. Secondly, the use of seed from outside the farm carries with it the risk of importing disease, particularly if the exporting hatchery itself imports broodstock from elsewhere. In this context, specific pathogen-free (SPF) or specific pathogen-resistant (SPR) stocks of shrimp and some crayfish have been developed (sections 8.9.4.4 and 8.10.1.3 respectively). Thirdly, European and Australian crayfish are all susceptible to the crayfish fungus plague likely to be present wherever North American species are found (section 11.3.3). Substantial investment in cultivating these groups outdoors in such areas may be unwise unless it can be unequivocally established, for example in the case of *Cherax quadricarinatus*, that a tropical temperature regime militates against infection (section 4.7).

4.6.4 Other factors

Among other factors that may influence the choice of a culture species or an assessment of a project are the requirement for a penetrable substrate in which to bury (e.g. for *Marsupenaeus japonicus* – section 7.2.6.6), the ability of an animal to escape from ponds and the ease

Plate 4.2 A feast of cooked male Koura (*Paranephrops planifrons*) culled from a pilot farm in New Zealand. The crayfish are males of about 110 mm TL, or 85 g. (Photo courtesy David Smythe and Peter Wilhelmus, New Zealand Clearwater Crayfish (Koura) Ltd.)

and efficiency with which harvesting may be accomplished. *Pacifastacus leniusculus* and some species of *Cherax*, *Paranephrops*, *Macrobrachium* and crab can travel easily over damp terrain, and anti-escape fences may be necessary. Burrow-dwelling species of crayfish, for example *Procambarus clarkii*, have to be harvested by trapping rather than pond draining, with the result that an unknown number may remain unharvested. Some species of yabby, e.g. *Cherax destructor*, may burrow more extensively and consequently damage levees and dam walls more than others such as *C. albidus*. The distance of the farm from the market and the ability of an animal to withstand the rigours of live transportation and storage are of importance if valuable speciality live sales are sought (sections 3.2.1 and 8.4.6). Lobsters, crayfish, crabs and spiny lobsters are traditionally sold live, but most shrimp and prawns are sold frozen or on ice.

Despite the acknowledged disadvantages, the farming of *Macrobrachium* and *Procambarus* arguably offers greater prospects of long-term sustainability than some shrimp farming operations due to the lower culture densities employed, the reduced or nil requirement for dietary marine fishmeal and protein, its suitability for rural family businesses away from expensive coastal sites and the minimal risk of soil salinisation. Concerns exist about effluent discharges but a major consideration is likely to be the effects of transplantations for example it is reported that an Indian species (*Macrobrachium malcolmsonii)* is being examined for possible introduction to China (Valenti & New 2000).

Two morphological features of decapod Crustacea are of interest when choosing a species to culture. These are the presence of large chelae, typically indicative of a territorial or predatory species, and the proportion of the body that is eaten, usually the tail and claw muscle. *Homarus*, *Nephrops*, most *Macrobrachium* species and probably most northern crayfish are naturally territorial or solitary, and when crowded together are likely to become unduly aggressive. The density at which fighting and cannibalism becomes a problem varies between species, with crayfish being far more tolerant of crowding than lobsters. Limb loss, shell blemishes and increased size variation are the chief manifestations of overcrowding, and all of them seriously affect the market value of the crop.

Although both head and tail are eaten in some parts of South-east Asia and also during the consumption of soft-shelled crabs and crayfish, most markets are based on the edible muscle of tail and claws. The processing yield, i.e. the weight of the tail (peeled or unpeeled) plus claw meat if present, per unit weight of animal is greater in penaeids (50–68%) than in caridean prawns (35–48%), crayfish (10–40%) and lobsters (30–35%) (Tables 4.6e–i). Caution should however be exercised when considering meat yields expressed as a percentage of whole animal weight. In crayfish, for example, values can vary with the degree of mineralisation of the exoskeleton, season, sex and age. Additionally, quoted figures do not always specify whether the claw meat and/or hepatopancreas were included in the yield (the latter being common practice in the USA), or if the meat was fresh or cooked (Harlioğlu & Holdich 2001). It is also worth noting that it is conventional in Australia to include the shell when determining meat yield. Thus the figures of around 40% often found in the literature really equate to about 25% meat yield if the shell is excluded (Morrissy *et al.* 1990).

The appearance and coloration of the farmed product becomes increasingly important as competition increases between farms and between farmed and fished crustaceans. The ability to influence coloration and flesh composition during culture – through, for example, tank background colour, light regimes and diet – is available for a range of species and could be especially useful for those that would be grown under a reasonable degree of environmental control (sections 2.4.6, 2.6.3 and 3.2.1). The flavour of freshwater prawns (*Macrobrachium*) can be enhanced by acclimation to seawater for 18 h although some would argue against the loss of the prawn's natural, subtle flavour (Madrid & Phillips 2000).

4.7 Comparison of species

The most significant biological and husbandry aspects that affect the main operational phases of crustacean farming are presented for broodstock (Table 4.6a), larvae culture (Table 4.6b), nursery (Table 4.6c) and ongrowing (Table 4.6d). These tables show the broad differences and similarities between the major groups of interest and give an indication of the range in survival rates or yields that might be anticipated. Wherever possible, data judged to represent performances obtainable under commercial conditions were selected. In other cases, ranges of values from pilot or laboratory trials were used. The average values can be used to compare general performance attributes of the groups, but it is emphasised that the following examples should not be used in financial calculations; they are shown here simply to give a sense of perspective when considering which species to culture.

As an example, rough estimates of the numbers of harvestable penaeids that can be reared from one female can be obtained by multiplying the mean fecundity by broodstock productivity (600 000 × 0.55, Table 4.6a). This gives 330 000 nauplii which, when multiplied by the estimated overall survival (0.36, Table 4.6d) gives 118 800 harvestable shrimp. Estimates for the remaining groups are:

M. rosenbergii	7020
temperate crayfish	37
tropical crayfish	237
clawed lobsters	3187
spiny lobsters*	1377 animals

(* assuming broodstock productivity is 30%)

No figure is calculated for crabs because of the widely different estimates of broodstock performance between species.

Table 4.6a Generalised data for captive broodstock.

Species/group	Maturation frequency (yr^{-1})	Fecundity range	Source	Facility	Productivity (%)*
Penaeids	3–6	200 000–1 000 000	Wild, ponds or tanks	Indoor spawning tanks 0.1–40 mt	30–80
Macrobrachium	3–6	20 000–80 000	Ponds	Covered incubation/ hatching tanks 1–10 mt with hides	50–80
Crayfish					
Temperate	1–2	60–260	Ponds	Covered incubation/ hatching tanks or cages, hides, mesh floor	50–80
Tropical	3–5	140–1000	Wild or ponds	As above	80
Clawed lobsters	0.5–1.0	5000–80 000	Wild	Covered incubation tanks, claws banded	30
Spiny lobsters					
Warm temperate	1–2	80 000–400 000	Wild	Experimental, covered incubation tanks	30E
Tropical	1–2	50 000–1 000 000	Wild	As above	
Crabs	3–4	50 000–3 000 000	Wild or ponds	Covered tanks, sand bottom	4–80

*After allowing for female mortality, egg and hatching losses. E = Estimated.

Similar, very approximate, comparative estimates may be calculated for the area of ongrowing pond or tank bottom required to produce 1 tonne of market sized crustaceans. For example, survival multiplied by average initial stocking density (0.7 × 15, Table 4.6d) yields 10.5 shrimp of 30 g each or 0.315 kg m^{-2}; thus 1000/0.315 = 3174 m^2 of pond bottom are needed to produce 1 tonne of shrimp. This figure equates roughly to a yield of 3.15 mt ha^{-1} per harvest; typical of well-managed semi-intensive farms (Table 7.6; section 10.6.1.5). Comparative values for the other species/groups, in m^2, again assuming a good water supply and competent husbandry, are:

M. rosenbergii (single batch harvest)	8888
Temperate crayfish	9524
Tropical crayfish	2778
Lobsters (single layer battery)	351
Spiny lobsters	287
Crabs	2538

The low figures for clawed and spiny lobsters are due in part to their large size at harvest (about 350 g); the high value for crabs harvested at a similar size is a reflection of poorer survivals during nursery and ongrowing phases.

A further simple calculation (yield per unit area multiplied by number of crops per year) serves to illustrate the potential annual productivity of the land. It must, however, be emphasised that seasonally variable factors such as temperature, rainfall, seed or broodstock availability may severely curtail the useful growing season. In fact, few open-air farms are able to realise year-round pro-

Table 4.6b Generalised data for larvae culture.

Species/group	Duration of larval life (days)	Density (no. L^{-1})	Feed	Rearing facility	Expected survival (%)
Penaeids	12–14	30–200	Algae, rotifers, *Artemia*, flakes, microparticulates, microcapsules	Static water tanks 0.1–200 mt	50–80
Macrobrachium	20–40	30–200*	*Artemia*, compounded diets, sieved fish/invertebrates	Static or recycled brackish water, 1–20 mt tanks	30–60
Crayfish					
Temperate	0	—	—	—	—
Tropical	0	—	—	—	—
Clawed lobsters	9–14	25–50	Fresh/frozen shrimp, mollusc flesh, adult *Artemia*, mysids	40–100 L kreisels (section 7.8.7)	20–60
Spiny lobsters Warm temperate and tropical	64–391	5–15	*Artemia*	Experimental	up to 10%** i.e. 1.7%
Crabs	15–30	6–50	Algae, rotifers, *Artemia*, sieved bivalve flesh	Static 10–200 mt tanks	4–70

*Two-phase culture (section 7.3.3). **To puerulus then 17% to juvenile.

Table 4.6c Generalised data for nursery culture.

Species/group	Duration of nursery phase (days)	Stocking density (no. m^{-2})	Rearing facility	Expected survival (%)
Penaeids	10–60	50–100	Small (0.04–1 ha) ponds, open and covered tanks/raceways	70–90
Macrobrachium	7–90	100–5000	Static or recycled covered tanks, small ponds	70–90
Crayfish, temperate	70–120	2–10	Shallow troughs, numerous hides	40–80
Crayfish, tropical	30–90	2–15	as above, shaded	60–80
Clawed lobsters	20–30	350–400	Individual cells	70–80
Spiny lobsters	90–120*	200E	Experimental,	>60
Warm temperate	90–100**		Shaded tanks with hides	85–95**
Crabs	14–28	2000–3000 max	Shaded tanks, hanging net shelters	20–70

E = Estimated. *Includes puerulus stage of 7–56 days. ***Jasus edwardsii* (Jeffs & Hooker 2000).

Table 4.6d Generalised data for ongrowing.

Species/group	Duration of ongrowing phase (months)	Stocking density (no. m^{-2})	Basic facility	Size at harvest (g)	Expected survival (%)	Estimated mean overall survival (%) (hatch to harvest)
Penaeids	3–4	5–25	Earth ponds, 0.1–100 ha, 1 m deep	15–45	60–80	36
Macrobrachium	3–5 or continuous	5–10	Earth ponds, 0.1–1.0 ha, 1 m deep	15–45	40–60	18
Crayfish, temperate	24–36	3–4	Earth channels, 10–60 m long, 5–12 m wide, 1.5 m deep	40–60	30–90	36
Crayfish, tropical	12–18	3–5	As above, and ponds 0.1–0.5 ha	40–200	60–90	52
Clawed lobsters	24–30	90–100*	Individual cells, transfer to larger cells periodically	345–400	80–90	25
Spiny lobsters	8–30**	5–25**	Shaded tanks with hides, ponds	250–350	75–80	0.012 70***
Crabs	3–8	0.5–4	Shaded tanks, ponds	200–500	30–70	8

*9 m^{-2} at harvest. **Depends on size at stocking. ***From first juvenile stage (Jeffs & Hooker 2000).

duction. However, to highlight further the differences between the groups of cultivable crustaceans, the hypothetical annual yields calculated from averaged values in Table 4.6d, in mt ha^{-1} are:

Penaeids 11.0
M. rosenbergii 3.6
Crayfish: temperate 0.4
Crayfish: tropical 3.0
Clawed lobsters 12.8
Spiny lobsters 22.0
Crabs 8.9

The superficially attractive value for clawed and spiny lobsters belies the difficulties of high rearing facility costs (clawed lobsters) and as yet unsolved mass larvae rearing techniques (spiny lobsters). The high value for crabs results from the fact that so many farms stock moderately large, wild-caught juveniles for fattening. This practice is also undertaken with spiny lobsters where estimated yields can be up to 25 kg m^{-2} (Booth & Kittaka 2000). Jeffs and Hooker (2000) developed a hypothetical, indoor, intensive farm for *Jasus edwardsii* based on results from pilot laboratory studies. The model anticipates yields of 5 tonnes of 300 g lobsters from 2000 m^2 of stacked trays per year from the fourth year of operation. A final density of around 20–25 lobsters m^{-2} or 6–7.5 kg m^{-2} of tray (their fig. 1) in a stack three trays high gives lobster yields of around 18–22.5 kg m^{-2} of farm floor.

Additional species-specific information is presented for penaeid shrimps (Table 4.6e), caridean prawns (Table 4.6f), freshwater crayfish (Table 4.6g) and clawed lobsters, spiny lobsters and crabs (Table 4.6h). In Table 4.6i the most important features are compared in order to assist further in the choice of species for the various culture options described in Chapter 5. Penaeid shrimp clearly have the best biological attributes for culture although increases in the availability of *Macrobrachium* and its widespread acceptance in European markets during the past decade show that it now seems set to claim second place from freshwater crayfish. Commercial production of warm-water crayfish, notably the Australian redclaw (*Cherax quadricarinatus*), is growing slowly in several countries. Redclaw seems to have three major advantages over *Macrobrachium*: no independent larval existence, no requirement for saline water, and a possible greater tolerance to crowded conditions. Disadvantages are poorer fecundity and the potential risk from plague fungus disease, although this has not stopped its culture in Ecuador, USA and elsewhere. It is extremely difficult to compare objectively the disease susceptibility of different crustaceans, since all may succumb to infection or infestation under stressful culture conditions. Reports suggesting that tropical temperature regimes inactivate the fungus need to be substantiated

Table 4.6e A comparison of attributes to be considered when choosing a species to culture: marine and brackish-water shrimp.

	Fenneropenaeus chinensis	*Fenneropenaeus indicus*	*Marsupenaeus japonicus*	*Fenneropenaeus merguiensis*	*Penaeus monodon*	*Fenneropenaeus penicillatus*	*Penaeus semisulcatus*	*Litopenaeus stylirostris*	*Litopenaeus vannamei*
Age at first maturity (months)	9(v)	7(v)	9(a)	6–7(y)	10–12(v)	—	—	10(v)	6–12(g,v)
Size at first maturity (g)	M 20–30(a,v)	M 6(v)	M 15–20(a,v)	M 6–15	M 30–40 (m,v)	M 22–40(a)	—	M 35–40(v)	M 30–40(v)
	F 30–60(a,v)	F 8(v)	F 20–25(a,v)	F 6–15(z)	F 60–70 (m,v)	F 35–60(a)	—	F 35–55(v)	F 35–45(v)
Control of breeding	With and without ablation(v)	Ablation not necessary(v)	With and without ablation(v)	With and without ablation(y)	With ablation(h)	With and without ablation(v)	With and without ablation(s)	With ablation(v)	With ablation*(v)
Interval between spawning (days)	5–30(c,v)	1–9(y)	60(v)	—	4–20(l,v)	4–7(f)	12–15(s)	3–19(y)	3–40(g,v)
Fecundity, captive	100 000–300 000 (a,v)	23 000–75 000	200 000(v)	10 000–90 000(y)	150 000–300 000(v)	50 000–200 000(f)	58 000–284 000(s,ad)	70 000–400 000	80 000–250 000(g,v)
Fecundity, wild	400 000–500 000 (e)	—	up to 1 000 000	—	400 000–1 000 000 (h)	100 000–500 000(f)	—	—	100 000–500 000(g)
Incubation period (h)	—	—	10–14(a)	18(y)	13(m)	—	—	12–18(g)	12–18(g)
No. of larval stages	12(m)	11–12	12(ac)	11–12	12(m)	11	12(ac)	11	11
Duration of larval phase (days)	17(a)	—	12–13(a)	10–12(z)	8–10	9–14	11–14(d)	8–12(g)	8–12(g)
Age stocked for ongrowing (days post-metamorphosis)	10–21(e)	3>40(q,t)	50(m,n)	42(z)	25–55 (i,j,m)	5–10(m)	43–51(r)	45–60(ab)	33–75(g,t)

Table 4.6e *(continued).*

	Fenneropenaeus chinensis	*Fenneropenaeus indicus*	*Marsupenaeus japonicus*	*Fenneropenaeus merguiensis*	*Penaeus monodon*	*Fenneropenaeus penicillatus*	*Penaeus semisulcatus*	*Litopenaeus stylirostris*	*Litopenaeus vannamei*
Size stocked for ongrowing (g)	25–30 mm TL>PL_{25}(e)	<0.01–1.48(q,t)	0.5–0.8(m)	0.2(z)	0.5(l)	0.7–1.0 cm(m)	0.17–2.4(r)	0.5–1.0(ab)	0.6–2.0(t) PL_{10} (x)
Density stocked (no. m^{-2})	7.5–15(e)	2.5–515(q,t)	1–400(n,v)	4–162(o,z)	2–66(i,af)	20–330(a,m)	3–45(p,r)	6.5–93(t,u)	3–122(t,u,x)
Ongrowing period (days)	126–190(b)	70–319(q,t)	122–183(m,n)	76–112(z,aa)	80–225 (i,l)	95–141(a)	49–162(r)	252(t)	67–164(t,x)
Size at harvest (g)	9.4–43.5(b)	4–20(t,ae)	10–25(v)	7–12.5(z,aa)	21–40 (i,af)	9–21(a)	7–21(r)	15–28(t,u)	7–23(t,u,x)
Survival (%)	25–55(b)	32–91(q)	40–70(m)	47–73(z)	30–93 (i,af)	45–90(a)	44–87(r)	5–70(t,u)	40–90(u,x)
Yield (kg ha^{-1} $crop^{-1}$)	314–2308(b)	231–15 000(p,t)	300–30 000 (n,v)	200–5850(o,z)	500–14 500(i,l)	3400–12 300(a)	135–2740 (p,r)	2 500–50 000(t,u)	490–20 000(t,u,x)
Meat yield (%)	57(k)	—	—	65(k)	49 PUD 59 shell on headless (g)	more than other penaeids(a,m)	—	—	63–68(k)
Crops yr^{-1}	1	1–2+(q,t)	1–2(m)	—	1–3(i,l)	1–2	1(r)	1(t)	1–3(w)

(a) Liao & Chien 1990; (b) Zhang *et al.* 1983; (c) Wang & Ma 1990; (d) Samocha & Lewinsohn 1977; (e) Qingyin 1992; (f) Hu 1990; (g) D.O'C Lee, unpubl. data; (h) Hansford & Marsden 1995; (i) Chen *et al.* 1989; (j) IFC 1987; (k) Rosenberry 1989; (l) Chiang & Liao 1985; (m) Chen 1990; (n) Spotts 1984; (o) Maguire 1979; (p) Nandakumar 1982; (q) Gopalan *et al.* 1982; (r) Issar *et al.* 1988; (s) Browdy & Samocha 1985; (t) AQUACOP 1984; (u) McIntosh 1999; (v) AQUACOP undated report; (w) Chamberlain 1989; (x) Aragon-Noriega & Calderon-Aguilera 1997; (y) J.F. Wickins unpubl. data; (z) Wickins & Beard 1978; (aa) Briggs 1988; (ab) Pretto 1983; (ac) Samocha *et al.* 1989; (ad) Samocha 1980; (ae) Al-Thobaiti & James 1998; (af) Lin 1995.

*See Table 4.2.

Table 4.6f A comparison of attributes to be considered when choosing a species to culture: marine and freshwater prawns.

	Macrobrachium rosenbergii(a)	*M. acanthurus*	*M. amazonicum*(j)	*M. birmanicum*	*M. carcinus*	*M. malcolmsonii*	*M. nipponense*	*Pandalus platyceros*(b,c)	*Palaemon serratus*(b)
Age at first maturity	4–7 mo	4–7 mo	168 d (in lab.)	4–7 moE	4–7 moE	4–7 moE	—	24–36 mo(*)	8 mo
Size at first maturity	20–30 g	—	33–59 mm TL	—	55 g(j)	42–83 mm TL(j)	~50 mm TL	25–30 g	3–3.5 g
Control of breeding	yes, not widespread	not practised	not practised	not practised	not practised	yes, not widespread	not practised	no	not practised
Interval between spawning	3–4 mo	—	1–2 mo	—	1–2 mo	1–2 mo in captivity	—	12 mo	4 mo
Fecundity	80 000–96 000	2000–13 400(d)	2200	2100–42 000 (h)	10 000–80 000(d)	3500–94 000(j)	500–5000	2500–4500	1–2000
Incubation period (days)	19–21	14–18	12–24	—	19(j)	13(j)	20	4–5 mo	30–40
No. of larval stages	11	10(e)	8–11	5(g)	12(i)	11(k)	9	4	6
Duration of larval phase (days)	20–40	32–40(e)	23–33	20–25(g)	32–65(j)	28(l)	18–25	15–24	18–35
Age stocked for ongrowing (days post-metamorphosis)	30–60	30–60E	—	—	30–60E	30–60E	20	1 mo	2 mo
Size stocked for ongrowing	0.5–1.0 g	15–20 mm TL(f)	0.9 g	—	13-30 mm TL(f)	20-50 mm TL(m,n)	0.2–0.5 g	0.5 g	0.5 g
Density stocked (no. m^{-2})	5–10	7–11(f)	8	—	1–6(f)	3–6(m,n,o)	60–70	50–100	50–100
Ongrowing period (days)	3–5 mo, or continuous	125–188 or continuous(f)	3–6 mo	—	166–180 or continuous (f)	120–150 or continuous(m,n); 6 mo(o)	6–8 mo	18 mo	18–24 mo
Size at harvest	25–45 g	74–121 mm TL(f)	70 mm TL	—	28–80 mm TL(f)	124–180 mm TL(m,n); 17–200 g(o)	4–6 g	10–18 g	5–8 g
Survival (%)	40–60	8–15(f)	—	—	5(f)	30–70(m,n,o)	—	10–85	15–75
Yield (kg ha^{-1} $crop^{-1}$)	1–2500 3000–4000(p)	1100 yr^{-1}(j)	245–900	—	1200(j)	534–925(m,n) 440–565(o)	390–1875	492 g m^{-2}*	315 g m^{-2}*
Meat yield (%)	37, peeled, raw 48(q)	—	—	—	—	—	—	35E	35E
Crops yr^{-1}	1–3 or continuous	2	—	—	—	2 or continuous(j)	continuous in season	0.5–0.7	0.5–0.7

(a) New 1990; (b) Wickins 1972; (c) Oesterling & Provenzano 1985; (d) Coelho *et al.* 1982; (e) Choudhury 1970; (f) Dobkin *et al.* 1974; (g) Khan *et al.* 1984; (h) Prakash & Agarwal 1985; (i) Choudhury 1971; (j) Kutty *et al.* 2000; (k) Qureshi *et al.* 1993; (l) Sankolli & Shenoy 1980; (m) Rao *et al.* 1986; (n) Rajyalakshmi *et al.* 1980; (o) Kanaujia *et al.* 1997; (p) Valenti & New 2000; (q) Madrid & Phillips 2000.
*Laboratory tanks (Wickins & Beard 1978). E = estimated.

Table 4.6g A comparison of attributes to be considered when choosing a species to culture: freshwater crayfish.

	Astacus astacus	*Astacus leptodactylus*	*Pacifastacus leniusculus*	*Procambarus clarkii*	*Orconectes* spp. (ae)	*Cherax destructor/ albidus*	*Cherax tenuimanus*	*Cherax quadricarinatus*	*Paranephrops planifrons* (c,r)
Age at first maturity	3–5 yr	2–3 yr(g)	1–3 yr(ac)	3–9 mo	1–1.5 yr	<12 mo(o)	2–3 yr(p)	6 mo(e)	15 mo 24 mo***
Size at first maturity (mm)	62–85 TL (ab)	82 TL(g) 75–85 TL(ab)	25–42 TL(ac) (15 g)	45–125 TL	17–23 CL	35–45 CL 20 g(af)	30–50 CL(b)	40 CL(h)	50–60 TL 35–40 CL***
Control of breeding	Yes, not widespread	Yes, not widespread	Yes, not widespread	Yes, not widespread	Possible not practised	Yes(q)	Possible	Yes(s)	Yes; ?***
Interval between spawning (mo)	12–24	12	12(b)	2–5(h)	12	3–6(o,p); 2–12(af)	12(b)	1–2(e)	12; 24***
Fecundity	100–250	200–400(g)	70–300(ac)	50–600(ad)	55–575	124–960(h,n); 300–500(u)	200–800(b); 100–400(u)	150–250(e); 200–800(n)	150–400; ?***
Incubation period	5–8 mo	5–6 mo(g); 1–2 mo @ 16–18°C(g)	4–9 mo(ac)	1–3 weeks	1 mo	20–30 days(h)	4–10 weeks(b) 12–16 weeks (f)	30–71 days(i,p)	6–7mo; ?***
Brooding period (days)	21–28(ab)	14–25(ab,l)	14(l)	14–21(ad)	15–20	12–23	15–30	7–33	42–56; 21***
Age stocked for ongrowing (from hatching)	3–4 mo	3–4 mo	3–4 mo	mature adult*	adults*	28 days	30–60 days	50–60 days(i)	10 weeks; 3–4 weeks***
Size stocked for ongrowing (g)	0.5–1 (15–50 mm TL)	0.5(g)	0.5	mature adult*	adults or ovigerous females*	0.1–0.5	0.2–1	0.5–1(i); 5–10(af)	10–15 mm TL; ?***
Density stocked (no. m^{-2})	0.2–25	5–10	2–10	25–100 kg ha^{-1}; 1–16(z)	0.25 (or adults*)	3–30(z)	5–15(z)	2–10(i); 5–10(af)	25; 1–10***
Ongrowing period	2–4 yr	1–2 yr	1–2 yr	1 yr	12–14 mo	4–12 mo	12–24 mo	6–13 mo(t)	2 yr; 2 yr***
Size at harvest (g)	30–80	30–80	30–80	17–80	1–5 (bait); 25–29 (table)	50–100	40–120	40–100+(d)	45–50;
30–40***									
Survival (%)	—	60(g)	30–40	47–88(j)	—	18–36(j)	60	49–94(k)	?; 60–70***
Yield (kg ha^{-1} crop^{-1})	60–430(ab)	500–1000(g)	500–1000(b)	200–3000(ad)	323–807	300–1500(z)	1000–4000(z)	1000–2300(i,y); 3000(af)	3300–4400P; 5250P***
Meat yield (%)	18–24	14–23(g,v)	25**(v)	10–26(a)	—	25(h)	22–30(i)	14–40	23
cooked	15(aa)	9–13(aa)	11–15(aa)	11–25(m)	—	7.7–17.4(i)	—	mean 22(i,w) max 34(w)	23*** 31 shell on***
Crops yr^{-1}	0.25–0.5	0.5	0.5	1	1	1–2.5	0.5–1	0.5–2(t)	0.5; 0.5***

(a) Moody 1989; (b) Wickins 1982; (c) Smythe 1998; (d) Jones 1988; (e) Sammy 1988; (f) Holker 1988; (g) Koksal 1988; (h) Sokol 1988; (i) Jones 1990; (j) Mills & McCloud 1983; (k) O'Sullivan 1988; (l) D. Holdich 2001, pers. comm.; (m) Huner 1993; (n) Austin 1998; (o) Merrick & Lambert 1991; (p) Rouse 1995; (q) Mills *et al.* 1994; (r) D. Smythe & P. Wilhelmus, 2001 pers. comm.; (s) Barki *et al.* 1997; (t) Romero 1997; (u) Semple *et al.* 1995; (v) Mackevičienė 1999; (w) Gu *et al.* 1994; (x) J. Hollows, 2001 pers. comm.; (y) O'Sullivan 1995; (z) Evans & Jussila 1997; (aa) Harlioğlu & Holdich 2001; (ab) Skurdal & Taugbøl 2001; (ac) Lewis 2001; (ad) Huner 2001; (ae) Hamr 2001; (af) Lawrence & Jones 2001; (ag) Diver 1998.

*Initial stocking of self-perpetuating, extensive cultures. **Fresh males, including claw meats. ****Paranephrops zealandicus* (x,ag). P = Projected yield (pre-commercial).

Table 4.6h A comparison of attributes to be considered when choosing a species to culture: lobsters and crabs.

	Homarus americanus	*Homarus gammarus*	*Panulirus, Jasus* spp.	*Scylla* spp.	*Portunus trituberculatus*	*Eriocheir sinensis*
Age at first maturity	3–4 yr(a)	3–6 yr(a)	3–5 yr(a)	4–6 mo	12 mo	12–18 mo(f)
Size at first maturity	200–700 g	400 g	70 mm CL	10 cm CW; 200 g(h)	10 cm CW	55 mm CW(f)
Control of breeding	Yes, experimental	Partial	Yes, experimental	Yes, not widespread	Possible not widespread	Not needed
Interval between spawning	10–24 mo	12–24 mo	2–12 mo(m)	2–3 mo(g)	3–6 wks(b)	12 mo(q)
Fecundity	5000–50 000	5000–10 000	0.5–1 million(c)	0.8–2.9 million(e,n)	1–3 million(b)	0.25–1 million(f)
Incubation period	4–7.5 mo	4–9.5 mo	1–9 wks(c,m); 1–6 mo(k, p)	6–17 d(e,l)	20–25 d(b)	15 d(q)
No. of larval stages	4–5	4	9–25(j)	5–6(e)	5(e)	5–6(f)
Duration of larval phase (days)	9–14	10–14	65–132† 189–391‡	15–24(e,l)	23	<30(f) 30–40(g)
Age stocked for ongrowing (post-metamorphosis)	2 wks	2 wks	6–12 mo	2–4 wks	1–4 wks	1–4 mo(g)
Size stocked for ongrowing	5–6 mm CL	5–6 mm CL	2–10 g	1.5–3 cm CW(h)	1–2 cm CW	5–25 g(g); 50–100**(q)
Density stocked (no. m^{-2})	90–100*	90–100*	5–25***	0.5–1(e)	Release at sea(b)	0.4–0.8(g); 1.5–4.5(r); 0.1–0.2**(q)
Ongrowing period	1.5–2.5 yr	2–3 yr	1–3 yr(c,o)	3–6 mo(e)	1–2 yr	6–9 mo(g); 4–5 mo**(q)
Size at harvest (g)	345–400	345–400	250–350 (i)	8–9 cm CW; 200–250 g(h)	10–15cm CW	125 g(g); 150–200 for export(q,r); 250**(q)
Survival (%)	80–90	80–90	75–80	40(e); 50–70 (h)	20–40(b)	60(g); 40(r)
Yield (kg ha^{-1} $crop^{-1}$)	28 000	28 000	35 000, 60 000**	55–1800(h), 8000 crabs ha^{-1}(e)	3–12% recaptures	450–1500(r); 300–500(q)
Meat yield (%)(cooked)	30–40	30–40	40–45(d)	25–30(d)	25–30(d)	20(q)
Crops yr^{-1}	0.5	0.5	0.3–0.75	1–2(h)	0.5–1	1(g)

(a) Wickins 1982; (b) Cowan 1983a; (c) Oesterling & Provenzano 1985; (d) Jones 1990; (e) Cowan 1983b; (f) Veldhuizen & Stanish 1999; (g) Li 1998; (h) Macintosh 1982; (i) Rahman & Srikrishnadas 1994; (j) Kittaka 1994; (k) Tong *et al.* 2000; (l) Quinitio *et al.* 2000; (m) MacDiarmid & Kittaka 2000; (n) Surtida 1997; (o) Booth & Kittaka 2000; (p) Chubb 2000; (q) culture with rice, K.M. Li, 2001 pers. comm.; (r) monoculture, Liu Fengqi, 2001 pers. comm.

*Moved to larger compartments 9 m^{-2} at harvest. **Fattening from 100 to 250 g. ***Depends on size for fattening operations (see text). †*Palinurus elephas*. ‡Australian species.

Table 4.6i Culture attributes of selected crustacean groups. Best = *****; worst = *.

Species/group	Control of breeding	High fecundity	Short larval life	Rapid growth	Tolerant of crowding	High meat yield	Simple technology	Compounded diets available	Commercial viability	Score (total *)
Penaeids	****	****	*****	*****	*****	*****	*****	*****	*****	43
Macrobrachium	*****	***	***	****	**	****	****	*****	****	34
Crayfish, temperate	***	*	*****	***	***	**	*****	**	****	28
Crayfish, tropical	*****	*	*****	****	***	***	****	***	***	31
Lobsters, clawed	****	**	*****	**	*	****	*	****	*	24
Lobsters, spiny	**	****	*	**	***	****	*	**	*	20
Crab	***	*****	***(a)	***(a)	**	***	***	**	**	26

(a)Varies with species.

(O'Sullivan 1996). Marketing a novel product can also be a problem and this was reported to be a contributory factor in the decline of redclaw crayfish production in Ecuador which started in 1994–95, peaked at 200–300 mt in 1998, but is expected to fall to around 80–100 mt in 2000–2001 (Lawrence & Jones 2001). A further potential advantage of crayfish is their ability to feed low down in the food chain. However, juveniles and subadults in particular grow best on diets containing high-quality protein and attempts to replicate detrital diets for crayfish in indoor controlled environment systems may not be the best approach.

Cultivable crabs and temperate water crayfish rank similarly, according to the criteria used in Table 4.6i (total number of stars) although market considerations and difficulties of rearing some crab larvae probably account for the differences in the extent to which the two groups are farmed.

The two main biological features that militate against the farming of clawed and spiny lobsters are the prolonged larval life of spiny lobsters and the need to rear clawed lobsters in individual compartments to avoid cannibalism. As a direct result of these features it has not yet been possible to develop a mass culture method for larval spiny lobsters, nor yet an economic battery system for rearing clawed lobsters to a marketable size (sections 7.8.9 and 10.6.3.6).

4.8 References

Aiken D.E. & Waddy S.L. (1995) Aquaculture. In: *Biology of the Lobster* Homarus americanus (ed. J.R. Factor), pp. 153–175. Academic Press, New York.

Ali Y.O. & Wickins J.F. (1994) The use of fresh food supplements to ameliorate moulting difficulties in lobsters, *Homarus gammarus* (L.), destined for release to the sea. *Aquaculture and Fisheries Management*, **25**, 483–496.

Alston D.E. & Sampaio C.M.S. (2000) Nursery systems and management. In: *Freshwater Prawn Culture: the farming of* Macrobrachium rosenbergii (eds M.B. New & W.C. Valenti), pp. 112–125. Blackwell Science, Oxford, UK.

Al-Thobaiti S. & James C.M. (1998) Saudi Arabian shrimp success in hypersaline waters. *Fish Farmer*, **21** (4) 40–41.

Anon. (1990) Seed shortage faces US shrimp farms. *Fish Farming International*, **17** (4) 25.

AQUACOP (1984) Review of ten years of experimental penaeid shrimp culture in Tahiti and New Caledonia (South Pacific). *Journal of the World Mariculture Society*, **15**, 73–91.

AQUACOP (undated) *Commercial shrimp culture: selection of species and rearing techniques*, 25 pp. Centre Oceanologique du Pacifique, BP 7004, Taravao, Tahiti.

Aragon-Noriega E.A. & Calderon-Aguilera L.E. (1997) Feasibility of intensive shrimp culture in Sinaloa, Mexico. *World Aquaculture*, **28** (1) 64–65.

Austin C.M. (1998) A comparison of clutch and brood size in the red claw, *Cherax quadricarinatus* (von Martens) and the yabby, *C. destructor* Clark (Decapoda: Parastacidae). *Aquaculture*, **167** (1–2) 135–145.

Barki A., Levi T., Hulata G. & Karplus I. (1997) Annual cycle of spawning and moulting in the red-claw crayfish, *Cherax quadricarinatus*, under laboratory conditions. *Aquaculture*, **157** (3–4) 239–249.

Beard T.W. & Wickins J.F. (1992) Techniques for the production of juvenile lobsters, 22 pp. *Fisheries Research Technical Report* (92). MAFF Directorate Fisheries Research, Lowestoft, UK.

Beard T.W., Richards P.R. & Wickins J.F. (1985) The techniques and practicability of year-round production of lobsters, *Homarus gammarus* (L.) in laboratory recirculation systems, 22 pp. *Fisheries Research Technical Report* (79). MAFF Directorate Fisheries Research, Lowestoft, UK.

Booth J.D. & Kittaka J. (2000) Spiny lobster growout. In: *Spiny Lobsters: fisheries and culture,* 2nd edn (eds B.F. Phillips & J. Kittaka), pp. 556–585. Fishing News Books, Oxford, UK.

Briggs M. (1988) *Techniques and constraints of shrimp farming in Thailand*, 43 pp. Report on Shrimp Team Visit. 2–21 May 1988. Institute of Aquaculture, Stirling, Scotland.

Browdy C.L. & Samocha T.M. (1985) Maturation and spawning of ablated and nonablated *Penaeus semisulcatus* de Haan (1844). *Journal of the World Mariculture Society*, **16**, 236–249.

Chamberlain G.W. (1989) *Status of shrimp farming in Texas*, 16 pp. Presented at 20th Meeting of the World Aquaculture Society, Los Angeles, 12–16 February 1989.

Chen L.C. (1990) *Aquaculture in Taiwan*, 273 pp. Fishing News Books, Blackwell Scientific Publications, Oxford, UK.

Chen J.C., Liu P.C. & Lin Y.T. (1989) Culture of *Penaeus monodon* in an intensified system in Taiwan. *Aquaculture*, **77** (4) 319–328.

Chiang P. & Liao I.C. (1985) The practice of grass prawn (*Penaeus monodon*) culture in Taiwan from 1968 to 1984. *Journal of the World Mariculture Society*, **16**, 297–315.

Choudhury P.C. (1970) Complete larval development of the palaemonid shrimp *Macrobrachium acanthurus* (Weigmann, 1836) reared in the laboratory. *Crustaceana*, **18**, 113–132.

Choudhury P.C. (1971) Complete larval development of the palaemonid shrimp *Macrobrachium carcinus* (L.) reared in the laboratory (*Decapoda, Palaemonidae*). *Crustaceana*, **20**, 51–69.

Chu H., Zhang Y., Yi J., Zhang X. & Sun S. (1996) On technique for medium rearing blue swimming crab with net cages. *Shandong Fish. Qilu Yuye*, **13** (5) 24–26 [in Chinese, English abstract seen].

Chubb C.F. (2000) Reproductive biology: issues for management. In: *Spiny Lobsters: fisheries and culture,* 2nd edn (eds B.F. Phillips & J. Kittaka), pp. 245–275. Fishing News Books, Oxford, UK.

Clifford H.C. III (1997) Shrimp farming in Venezuela. *World Aquaculture*, **28** (1) 60–61.

Coelho A.P., Porto M.R. & Soares C.M.A. (1982) Biologia e cultivo de camarones del agua doce. *Serie Aquicultura*, (1) 1–55. Recife, Pernambuco.

Cowan L. (1983a) Hatchery production of crabs in Japan. In: *Proceedings of the First International Conference on Warm Water Aquaculture – Crustacea*, 9–11 February 1983 (eds G.L. Rogers, R. Day & A. Lim), pp. 215–220. Brigham Young University, HI, USA.

Cowan L. (1983b) *Crab farming in Japan, Taiwan and the Philippines*, 85 pp. Information Series QI 84009, Queensland Department of Primary Industries, Queensland, Australia.

Crear B.J., Thomas C.W., Hart P.R. & Carter C.G. (2000) Growth of juvenile southern rock lobsters, *Jasus edwardsii*, is influenced by diet and temperature, whilst survival is influenced by diet and tank environment. *Aquaculture*, **190** (1–2) 169–182.

Currie D. (1998) Colombian shrimp farmers prepare for the 21st century. *World Aquaculture*, **29** (3) 18–21.

Daniels W.H., Cavalli R.O. & Smullen R.P. (2000) Broodstock management. In: *Freshwater Prawn Culture: the farming of* Macrobrachium rosenbergii (eds M.B. New & W.C. Valenti), pp. 41–51. Blackwell Science, Oxford, UK.

Davis D.A & Arnold C.R. (1998) The design, management and production of a recirculating raceway system for the production of marine shrimp. *Aquacultural Engineering*, **17**, 193–211.

Diver P. (1998) *Sweet Koura Enterprises*, 10 pp. (mimeo). Paper presented to the Maori and the Business Aquaculture Conference, 1998, Auckland, New Zealand.

Diwan S.D. & Kandasami K. (1997) Freezing of viable embryos and larvae of marine shrimp, *Penaeus semisulcatus* de Haan. *Aquaculture Research*, **28** (12) 947–950.

Dobkin S., Azzinaro W.P. & van Montfrans J. (1974) Culture of *Macrobrachium acanthurus* and *M. carcinus* with note on the selective breeding and hybridization of these shrimps. *Proceedings of the Fifth Annual Meeting of the World Mariculture Society*, pp. 51–62. Charleston, SC, USA.

Evans L.H. & Jussila J. (1997) Freshwater crayfish growth under culture conditions: proposition for a standard reporting approach. *Journal of the World Aquaculture Society*, **28** (1) 11–19.

Forster J.R.M. & Beard T.W. (1974) Experiments to assess the suitability of nine species of prawns for intensive cultivation. *Aquaculture*, **3** (4) 355–386.

Geddes M.C., Smallridge M. & Clarke S. (1995) Potential for a nursery phase in the commercial production of the Australian freshwater crayfish, the yabbie *Cherax destructor.* In: *Freshwater Crayfish 8* (ed. R.P. Romaire), pp. 504–517. Louisiana State University, LA, USA.

Gopalan U.K., Purushan K.S., Santhakumari V. & Meenakshikunjamma P.P. (1982) Experimental studies on high density, short term farming of shrimp *Penaeus indicus* in a 'pokkali' field in Vypeen Island, Kerala. *Proceedings of Symposium on Coastal Aquaculture 1982*, (1) 151–159. Marine Biological Association of India, Cochin, India.

Gu H., Mather P.B. & Capra M.F. (1994) The relative growth of chelipeds and abdomen and muscle production in male and female redclaw crayfish, *Cherax quadricarinatus* von Martens. *Aquaculture*, **123** (3–4) 249–257.

Hamr P. (2001) *Orconectes*. In: *Biology of Freshwater Crayfish* (ed. D.M. Holdich), pp. 585–608. Blackwell Science, Oxford, UK.

Hansford S.W. & Marsden G.E. (1995) Temporal variation in egg and larval productivity of eyestalk ablated spawners of the prawn *Penaeus monodon* from Cook Bay, Australia. *Journal of the World Aquaculture Society*, **26** (4) 396–405.

Hanson J. & Goodwin H. (eds) (1977) *Shrimp and Prawn Farming in the Western Hemisphere*, 220 pp. Dowden, Hutchinson & Ross Inc., Stroudsberg, PA, USA.

Harlioğlu M.M. & Holdich D.M. (2001) Meat yields in the introduced freshwater crayfish, *Pacifastacus leniusculus* and *Astacus leptodactylus*, from British waters. *Aquaculture Research*, **32**, 411–417.

Heinen J.M. & Mensi M.J. Jr. (1991) Feeds and feeding schedules for indoor nursery culture of postlarval freshwater prawns *Macrobrachium rosenbergii*. *Journal of the World Aquaculture Society*, **22** (2) 118–127.

Holker D.S. (1988) Marron – *Cherax tenuimanus*, 16 pp. Unpublished report.

Hu Q. (1990) On the culture of *Penaeus penicillatus* and *P. chinensis* in southern China. In: *The Culture of Cold-tolerant Shrimp* (eds K.L. Main & W. Fulks), pp. 77–91. The Oceanic Institute, Honolulu, HI, USA.

Huner J.V. (1993) Recovery of edible products from some common North American orconectid and procambarid crayfish (Cambaridae) with emphasis on *Procambarus clarkii* (Girard) and *Procambarus zonangulus* Hobbs & Hobbs. In: *Freshwater Crayfish 9* (eds D.M. Holdich & G.F. Warner), pp. 28–37. Louisiana State University, USA.

Huner J.V. (2001) *Procambarus*. In: *Biology of Freshwater Crayfish* (ed. D.M. Holdich), pp. 541–84. Blackwell Science, Oxford, UK.

Huner J.V. & Barr J.E. (1984) *Red Swamp Crawfish: biology and exploitation*, 148 pp. Sea Grant Publication No. LSU-T-80–001. Louisiana Sea Grant College Program, Louisiana State University, LA, USA.

IFC (1987) *Marine Shrimp Farming: a guide to feasibility study preparation*, 90 pp. Aquafood Business Associates, International Finance Corporation, P.O. Box 16190, Charleston, SC 29412, USA.

Issar G., Seidman E.R. & Samocha Z. (1988) Preliminary results of nursery and pond culture of *Penaeus semisulcatus* in Israel. *Fisheries and Fishbreeding in Israel*, **21** (4) 27–37.

Jeffs A. & Hooker S. (2000) Economic feasibility of aquaculture of spiny lobsters *Jasus edwardsii*, in temperate waters. *Journal of the World Aquaculture Society*, **31** (1) 30–41.

Jones C.M. (1988) Aquaculture potential of *Cherax quadricarinatus*: research objectives and preliminary studies. In: *Proceedings of the First Australian Shellfish Aquaculture Conference*, Perth, 1988 (eds L.H. Evans & D. O'Sullivan), pp. 73–78. Curtin University of Technology, Perth, Australia.

Jones C.M. (1990) *The Biology and Aquaculture Potential of the Tropical Freshwater Crayfish* Cherax quadricarinatus, 109 pp. Information Series QI 90028, Queensland Department of Primary Industries, Queensland, Australia.

Kagel K. (1998) Achieving sustainable high yields in a closed shrimp mariculture system. In: *Aquaculture and Water: fish culture, shellfish culture and water usage* (compiled by H. Grizel & P. Kestemont), pp. 126–127. Abstracts presented at Aquaculture Europe '98, Bordeaux, France, 7–10 October 1998.

Kanaujia D.R., Mohanty A.N. & Tripathi S.D. (1997) Growth

and production of Indian river prawn *Macrobrachium malcolmsonii* (H. Milne Edwards) under pond conditions. *Aquaculture*, **154** (1) 79–85.

Karplus I., Malecha S.R. & Sagi A. (2000) The biology and management of size variation. In: *Freshwater Prawn Culture: the farming of* Macrobrachium rosenbergii (eds M.B. New & W.C. Valenti), pp. 259–289. Blackwell Science, Oxford, UK.

Karplus I., Samsonov E., Hulata G. & Milstein A. (1989) Social control of growth in *Macrobrachium rosenbergii*. 1. The effect of claw ablation on survival and growth of communally raised prawns. *Aquaculture*, **80** (3–4) 325–335.

Keller M. (1988) Finding a profitable population density in rearing summerlings of European crayfish *Astacus astacus* L. In: *Freshwater crayfish 7* (ed. P. Goeldlin de Tiefenau), pp. 259–266. Musée Zoologique Cantonal, Lausanne, Switzerland.

Khan S., Khanam S.F & Shahadet A. (1984) Development of early larval stages of *Macrobrachium birmanicum* (Schenkel, 1902) (Crustacea, Decapoda, Palaemonidae). *Bangladesh Journal of Zoology*, **12** 79–89.

Kittaka J. (1994) Culture of phyllosomas of spiny lobster and its application to studies of larval recruitment and aquaculture. *Crustaceana*, **66** (3) 258–270.

Kittaka J. (1997) Application of ecosystem culture method for complete development of phyllosomas of spiny lobster. *Aquaculture*, **155** (1–4) 319–331.

Kittaka J. (2000) Culture of larval spiny lobsters. In: *Spiny Lobsters: fisheries and culture,* 2nd edn (eds B.F. Phillips & J. Kittaka), pp. 508–532. Fishing News Books, Oxford, UK.

Koksal G. (1988) *Astacus leptodactylus* in Europe. In: *Freshwater Crayfish: biology, management and exploitation* (eds D.M. Holdich & R.S Lowery), pp. 365–400. Croom Helm, London.

Kutty M.N., Herman F. & Le Menn H. (2000) Culture of other prawn species. In: *Freshwater Prawn Culture: the farming of* Macrobrachium rosenbergii (eds M.B. New & W.C. Valenti), pp. 393–410. Blackwell Science, Oxford, UK.

Lawrence C. & Jones C. (2001) *Cherax*. In: *Biology of Freshwater Crayfish* (ed. D.M. Holdich), pp. 635–69. Blackwell Science, Oxford, UK.

Lellis W.A. (1992) A standard reference diet for crustacean nutrition research VI. Responses of postlarval stages of the Caribbean king crab *Mithrax spinosissmus* and the spiny lobster *Panulirus argus*. *Journal of the World Aquaculture Society,* **23** (1) 1–7.

Lewis S.D. (2001) *Pacifastacus*. In: *Biology of Freshwater Crayfish* (ed. D.M. Holdich) pp. 511–40. Blackwell Science, Oxford, UK.

Li K. (1998) Rice aquaculture systems in China: a case of rice-fish farming from protein crops to cash crops, 15 pp. In: *Integrated Bio-systems in Zero Emissions Applications* (eds E-L. Foo & T. Della Senta). Proceedings of the Internet Conference on Integrated Bio-Systems 1998. http://www.ias.unu.edu/proceedings/icibs

Liao I.C. (1988) Feasibility study for alternative culture species in Taiwan – *Penaeus penicillatus*. *Journal of the World Aquaculture Society*, **19** (4) 227–236.

Liao I.C. & Chien Y.H. (1990) Evaluation and comparison of culture practices for *Penaeus japonicus*, *P. penicillatus*, and *P. chinensis* in Taiwan. In: *The Culture of Cold-tolerant Shrimp* (eds K.L. Main & W. Fulks), pp. 49–63. The Oceanic Institute, Honolulu, HI, USA.

Lin K.C. (1995) Progression of intensive marine shrimp culture in Thailand. In: *Swimming Through Troubled Water. Proceedings of the Special Session on Shrimp Farming* (eds C.L. Browdy & J.S. Hopkins), pp. 13–23. Aquaculture '95. World Aquaculture Society, Baton Rouge, LA, USA.

MacDiarmid A.B. & Kittaka J. (2000) Breeding. In: *Spiny Lobsters: fisheries and culture,* 2nd edn (eds B.F. Phillips & J. Kittaka), pp. 485–507. Fishing News Books, Oxford, UK.

Macintosh D.J. (1982) Fisheries and aquaculture significance of mangrove swamps, with special reference to the Indo-West Pacific region. In: *Recent Advances in Aquaculture* (eds J.F. Muir & R.J. Roberts), pp. 3–85. Croom Helm, London.

Mackevičienė G. (1999) A comparative study of physiological and biochemical indices of native European and alien species of crayfish in Lithuania. In: *Freshwater Crayfish 12* (eds M. Keller, M.M. Keller, B. Oidtmann, R. Hoffmann & G. Vogt), pp. 205–220. Weltbild Verlag, Germany.

Madrid R.M.M. & Phillips H. (2000) Post-harvest handling and processing. In: *Freshwater Prawn Culture: the farming of* Macrobrachium rosenbergii (eds M.B. New & W.C. Valenti), pp. 326–344. Blackwell Science, Oxford, UK.

Maguire G.B. (1979) *A report on the prawn farming industries of Japan, the Philippines and Thailand*, 110 pp. Brackish Water Fish Culture Research Station, C/-P.O., Salamander Bay, NSW., 2301, Australia.

Main K.L. & Fulks W. (eds) (1990) *The Culture of Cold-tolerant Shrimp*, 215 pp. The Oceanic Institute, Honolulu, HI, USA.

Marques H.L. de A., Lombardi J.V. & Boock M.V. (2000) Stocking densities for nursery phase culture of the freshwater prawn *Macrobrachium rosenbergii* in cages. *Aquaculture*, **187** (1–2) 127–132.

McIntosh R.P. (1999) Changing paradigms in shrimp farming: 1. General description. *Global Aquaculture Advocate*, **2** (4–5) 42–47.

Merrick J.R. & Lambert C.N. (1991) *The Yabby, Marron and Red Claw: production and marketing*, 185 pp. J.R. Merrick Publications, Artarmon, Australia.

Miglavs I.J. (1992a) Effects of temperature and density on *Artemia*-fed postlarvae of the sculptured shrimp *Sclerocrangon boreas*. In: *First European Crustacean Conference, Abstracts*, 31 August–5 September 1992, pp. 100–101. Paris, France.

Miglavs I.J. (1992b) *Sclerocrangon boreas*, the sculptured shrimp: a new species for mariculture. Some primary results of cultivation. In: *First European Crustacean Conference, Abstracts*, 31 August–5 September 1992, p. 100. Paris, France.

Mills B.J. & McCloud P.I. (1983) Effects of stocking and feeding rates on experimental pond production of the crayfish *Cherax destructor* Clark (Decapoda: Parastacidae). *Aquaculture*, **34** (1–2) 51–72.

Mills B.J., Morrissy N.M. & Huner J.V. (1994) Cultivation of freshwater crayfishes in Australia. In: *Freshwater Crayfish Aquaculture in North America, Europe, and Australia* (ed. J.V. Huner), pp. 217–289. Food Products Press. An imprint

of The Haworth Press, Inc. New York.
Moody M.W. (1989) Processing of freshwater crawfish: a review. *Journal of Shellfish Research*, **8** (1) 293–301.
Morrissy N.M., Evans L.E. & Huner J.V. (1990) Australian freshwater crayfish: aquaculture species. *World Aquaculture*, **21** (2) 113–122.
Mosig J. (1999) Hothouse yabbies keep growing all winter. *Austasia Aquaculture Magazine*, **13** (5) 16–18.
Moya M.E., Lawrence A.L., Collins C.A. & Samocha T.M. (1999) Acclimation of *Penaeus vannamei* postlarvae to 2 ppt ground saline water in Sonora desert, Arizona. In: *Book of Abstracts, World Aquaculture '99*, 26 April–2 May 1999, Sydney, Australia, p. 424. World Aquaculture Society, Baton Rouge, LA, USA.
Nandakumar G. (1982) Experimental prawn culture in coastal ponds at Mandapam Camp. *Proceedings of Symposium on Coastal Aquaculture 1982*, (1) 103–111. Marine Biological Association of India, Cochin, India.
New M.B. (1990) Freshwater prawn culture: a review. *Aquaculture*, **88** (2) 1–44.
Norman C.P., Yamakawa H. & Yoshimura T. (1994) Habitat selection, growth rate and density of juvenile *Panulirus japonicus* (Von Siebold, 1824) (Decapoda, Palinuridae) at Banda, Chiba Prefecture, Japan. *Crustaceana*, **66** (3) 365–383.
Norman-Boudreau K. & Conklin D.E. (1983) Purified protein in artificial diets for juvenile lobster. In: *Proceedings of First International Conference on Warm Water Aquaculture – Crustacea*, 9–11 February 1983 (eds G.L. Rogers, R. Day & A. Lim), pp. 343–351. Brigham Young University, Hawaii, USA.
Oesterling M.J. & Provenzano A.J. (1985) Other crustacean species. In: *Crustacean and Mollusk Aquaculture in the United States* (eds J.V. Huner & E. Evan Brown), pp. 203–234. AVI Inc. Westport, CT, USA.
Ogle J.T. & Lotz J.M. (1992) Closed system culture of *Penaeus vannamei*. *Gulf Research Reports*, **8** (4) 401–413.
O'Sullivan D. (1988) Marron farmers anticipate a bright future. *FINS*, **21** (1) 21–25. Fishing Industry News Service, Perth, Western Australia, Australia.
O'Sullivan D. (1995) Techniques for semi-intensive culture of freshwater crayfish in Australia. In: *Freshwater Crayfish 10* (eds M.C. Geddes, D.R. Fielder & A.M.M. Richardson), pp. 569–582. Louisiana State University, Baton Rouge, LA, USA.
O'Sullivan D. (1996) Ecuador red claw tie-up with Australian industry. *Austasia Aquaculture Magazine*, **10** (1) 10–13.
O'Sullivan D. (1997) Scientists meet on mud crab culture. *Austasia Aquaculture Magazine*, **11** (4) 47–51.
Peterson J.J. & Griffith D.R.W. (1999) Intensive nursery systems. *Global Aquaculture Advocate*, **2** (6) 30–31.
Prakash S. & Agarwal G.P. (1985) Studies on the effect of non-pollutional interference on freshwater prawn (*Macrobrachium birmanicum*) choprai fishery of middle stretch of River Ganga. *Uttar Pradesh Journal of Zoology*, **5**, 42–48.
Pretto R. (1983) *Penaeus* shrimp pond grow-out in Panama. In: *Handbook of Mariculture, Vol. 1 crustacean aquaculture* (ed. J.P. McVey), pp. 169–178. CRC Press, Boca Raton, FL, USA.
Qingyin W. (1992) Shrimp culture in China. *Infofish International*, (4) 50–53.
Quinitio E.T. & Parado-Estepa F. (2000) Transport of *Scylla serrata* megalopae at various densities and durations. *Aquaculture*, **185** (1–2) 63–71.
Quinitio E.T., Parado-Estepa F., Rodriguez E. & Millamena O.M. (2000) Seed production of mud crab *Scylla* spp. In: *Abstracts, Aqua 2000, Responsible Aquaculture in the New Millennium* (compiled by R. Flos & L. Creswell), p. 580. European Aquaculture Society, Special Publication No. 28.
Qureshi T.A., Basha S.M. & Biswas R. (1993) Larval rearing of Indian river prawn, *Macrobrachium malcolmsonii* through synthetic sea water. In: *World Aquaculture '93. From discovery to commercialization* (eds M. Carrillo, L. Dahle, J. Morales, P. Sorgeloos, N. Svennevig & J. Wyban), p. 158. European Aquaculture Society, Special Publication No. 19.
Rahman M.K. & Srikashnadhas B. (1994) The potential for spiny lobster culture in India. *Infofish International*, (1) 51–53.
Rajyalakshmi T., Reddy O.R., Rao A. & Ramakrishna R. (1980) Growth and production of riverine prawn *Macrobrachium malcolmsonii* (H. Milne Edwards) in pond culture, Andhra Pradesh, India. *Proceedings of First National Symposium on Shrimp Farming*, August 1978, p. 101. Bombay, India.
Rao K.G., Reddy O.R., Rao P.V.A.N.R. & Ramakrishna R. (1986) Monoculture of Indian freshwater prawn, *Macrobrachium malcolmsonii* (Milne Edwards). *Aquaculture*, **53** (1) 67–73.
Romero X.M. (1997) Production of redclaw crayfish in Ecuador. *World Aquaculture*, **28** (2) 5–10.
Rosenberry R. (1989) *World shrimp farming 1989*, 28 pp. Aquaculture Digest, San Diego, USA.
Rosenberry R. (1999) *World shrimp farming 1999*, 320 pp. Rosenberry, Shrimp News International, **12**. San Diego, USA.
Rouse D.B. (1995) Australian crayfish culture in the Americas. *Journal of Shellfish Research*, **14** (2) 569–572.
Sagi A., Milstein A., Eran Y., *et al.* (1997) Culture of the Australian red-claw crayfish (*Cherax quadricarinatus*) in Israel. II. Second growout season of overwintered populations. *Israeli Journal of Aquaculture – Bamidgeh*, **49** (4) 222–229.
Sammy N. (1988) Breeding biology of *Cherax quadricarinatus* in the Northern Territory. In: *Proceedings of First Australian Shellfish Aquaculture Conference*, Perth, 1988 (eds L.H. Evans & D. O'Sullivan), pp. 79–88. Curtin University of Technology, Perth, Australia.
Samocha T.M. (1980) *Developmental aspects of* Penaeus semisulcatus *de Haan, 1844 (Crustacea Decapoda) under laboratory and pond conditions*, 14 pp. Extract from PhD Thesis, Tel-Aviv University, Israel.
Samocha T. & Lewinsohn C. (1977) A preliminary report on rearing penaeid shrimps in Israel. *Aquaculture*, **10** (3) 291–292.
Samocha T.M., Uziel N. & Browdy C.L. (1989) The effect of feeding two prey organisms, nauplii of *Artemia* and rotifers, *Brachionus plicatilis* (Muller), upon the growth and survival of larval marine shrimp, *Penaeus semisulcatus* (de Haan). *Aquaculture*, **77** (1) 11–20.
Samocha T.M., Castille F.L., Lawrence A.L. & Talley S.E. (1993) Early spring growth trial of *Penaeus vannamei* at high stocking densities in raceways. In: *Techniques for Mod-*

ern Aquaculture (ed. J.K. Wang), pp. 230–240. American Society of Aquacultural Engineers, St. Joseph, USA.

Sankolli K.N. & Shenoy S. (1980) *Macrobrachium malcolmsonii* a prospective competitor for the jumbo prawn *M. rosenbergii*. *Proceedings of First National Symposium on Shrimp Farming*, August 1978, pp. 151–153. Bombay, India.

Semple G.P., Rouse D.B. & McLain K.R. (1995) *Cherax destructor*, *C. tenuimanus* and *C. quadricarinatus* (Decapoda: Parastacidae): a comparative review of biological traits relating to aquaculture potential. In: *Freshwater Crayfish 8* (ed. R.P Romaire), pp. 495–503. Louisiana State University, LA, USA.

Shelley C. & Field D. (1999) The culture of portunid crabs. *Infofish International*, (4), 46–49.

Skurdal J. & Taugbøl T. (2001) *Astacus*. In: *Biology of Freshwater Crayfish* (ed. D.M. Holdich), pp. 467–510. Blackwell Science, Oxford, UK.

Smythe D.R. (1998) *New Zealand Clearwater Crayfish Ltd.*, 7 pp. (mimeo). Paper presented at the New Zealand Marine Farming Association Inc. Annual conference, 27–30 August 1998. New species forum: freshwater crayfish (koura). Nelson, New Zealand.

Sokol A. (1988) The Australian Yabby. In: *Freshwater Crayfish: biology, management and exploitation* (ed. D.M. Holdich & R.S. Lowery), pp. 401–425. Croom Helm, London.

Spotts D. (1984) The development of Shigueno-style shrimp culture in southern Japan. *Aquaculture Magazine*, **10** (4) 26–31.

Surtida M. (1997) AQD's mudcrab R&D. *Asian Aquaculture*, **19** (3) 17–20.

Tcherkashina N.Y. (1977) Survival, growth and feeding dynamics of juvenile crayfish (*Astacus leptodactylus*) in ponds and the river Don. In: *Freshwater Crayfish 3* (ed. O.V. Lindqvist), pp. 95–100. University of Kuopio, Kuopio, Finland.

Tong L.J., Moss G.A., Pickering T.D. & Paewai M.P. (2000) Temperature effects on embryo and early larval development of the spiny lobster *Jasus edwardsii*, and description of a method to predict larval hatch times. *Marine and Freshwater Research*, **51** (3) 243–248.

Valenti W.C. & New M.B. (2000) Grow-out systems – monoculture. In: *Freshwater Prawn Culture: the farming of* Macrobrachium rosenbergii (eds M.B. New & W.C. Valenti), pp. 157–176. Blackwell Science, Oxford, UK.

Veldhuizen T.C. & Stanish S. (1999) Overview of the life history, distribution, abundance, and impacts of the Chinese mitten crab, *Eriocheir sinensis*, 14 pp. (mimeo). California Department of Water Resources, Sacramento, USA.

Villalon J.R. (1993) Commercial semi-intensive penaeid grow-out techniques in Ecuador. In: *Handbook of Mariculture*, 2nd edn, *Vol. 1 Crustacean aquaculture* (ed. J.P. McVey), pp. 237–273. CRC Press, Boca Raton, FL, USA.

Wang K. & Ma S. (1990) Advances in larval rearing techniques for *Penaeus chinensis* in China. In: *The Culture of Cold-tolerant Shrimp* (eds K.L. Main & W. Fulks), pp. 42–48. The Oceanic Institute, Honolulu, HI, USA.

Wickins J.F. (1972) Experiments on the culture of the spot prawn *Pandalus platyceros* Brandt and the giant freshwater prawn *Macrobrachium rosenbergii* (de Man). *Fish. Invest.*, London, Ser. 2, **27** (5) 1–23.

Wickins J.F. (1982) Opportunities for farming crustaceans in western temperate regions. In: *Recent Advances in Aquaculture* (eds J.F. Muir & R.J. Roberts), pp. 87–177. Croom Helm, London.

Wickins J.F. & Beard T.W. (1978) *Ministry of Agriculture, Fisheries and Food. Prawn culture research*, 41 pp. Lab. Leafl., (42), MAFF Directorate Fisheries Research, Lowestoft, UK.

Zhang N., Lin R.,Cao D., Zhang W., Gao H. & Liang X. (1983) Preliminary observations on the relationship of body weight, length with the daily food requirements of Chinese shrimp, *Penaeus orientalis* Kishinouye. *Oceanologia et Limnologia Sinica*, **12** (5) 482–487.

Chapter 5
Ongrowing Options

5.1 Introduction

Once a crustacean develops beyond the hatchery and nursery phases of culture, a range of options exists for the longer and thus more expensive operation of ongrowing. For convenience these options are grouped in relation to the climatic zone that provides the most suitable temperatures for growth; operations in some regions, however, may be seasonally constrained by variations in broodstock availability, seed supply or water quality. Controlled environment farming operations are included in the zones in which they evolved despite their potential to be independent of climate.

The process termed 'fattening' refers to holding, and often feeding, wild-caught subadult or adult crayfish, lobsters and crabs for a short period of time, perhaps 2–12 weeks, to gain extra revenue from an increase in size or seasonal market value. In Atlantic North America the process is called pounding and is also designed to improve meat condition of clawed lobsters after moulting (sections 3.3.4.1 and 7.8.9). Additionally, the value of crabs to gourmet markets may be increased by a change in their reproductive condition (sections 3.3.5 and 7.10.4). In the literature the term 'fattening' may occasionally be used more loosely to describe the ongrowing of wild-caught spiny lobster and crab juveniles that are not easily reared in a hatchery (sections 7.9.4 and 7.10.1) or of hatchery-reared juveniles that have been overwintered to make best use of a short ongrowing season (sections 7.3.2, 7.3.4, 7.7.5 and 7.10.3). The production of soft-shelled crustaceans and the options for stock enhancement, sea ranching or other forms of hatchery-based fisheries are discussed independently of geographical location towards the end of this chapter.

5.2 Tropical climates

In the tropics, where temperatures are generally suitable for year-round growth, methods for the ongrowing of juveniles may be usefully categorised under four headings, based broadly on expected yields and husbandry practices (e.g. for shrimp, Table 7.6). It is necessary to emphasise, however, that these categories cannot be defined precisely, and caution should be exercised if they are used in business negotiations.

5.2.1 Extensive

Typically, penaeid shrimp, crabs, some crayfish and *Macrobrachium* species are farmed extensively and produce yields of up to 1000 kg ha^{-1} when harvested, with up to three partial or sometimes full harvests per year (Table 5.1). Much reliance is placed on natural food production within the pond, which may be enhanced by fertilisation with animal manure or chemical fertilisers. The ponds are stocked with naturally occurring juveniles but with little prospect of controlling density. There is usually some water exchange of up to 5–10% per day, although in marine and brackish-water ponds this may be restricted to spring tide periods. Pond sizes range from about 0.5 to 100 ha.

5.2.2 Semi-intensive and intensive

The crustaceans farmed at higher densities are predominantly penaeid shrimp (Table 7.6), *Macrobrachium rosenbergii* and increasingly, crabs and Australian crayfish (*Cherax* spp.). Cultures yield from around 500 to 15 000 kg ha^{-1} each year, frequently from 2 to 2.5 crops

Table 5.1 Examples of tropical extensive aquaculture: shrimp, prawns, crayfish, crabs.

Species	Location	Annual yield range (kg ha^{-1})	Crops yr^{-1}	Live weight (g)	Reference
Penaeus monodon	India	200–1000	2–3	21–26	Chakraborty *et al.* 1986
Litopenaeus vannamei, *L. stylirostris*	Ecuador	240–1200	2–3	19–22	CPC 1989
Macrobrachium rosenbergii	Vietnam, China	<500	1–2	Variable	Valenti & New 2000
	India	500–1500	1–2	50–120	New 2000
M. nipponense	China	150	1–2 E in rice paddy	Variable	Kutty *et al.* 2000
Cherax quadricarinatus	USA	1000–1400	1–2 experimental	38–48 (from 3 g in 158 days)	Pinto & Rouse 1996
Scylla spp.	Andamans, India	675–918	1–1.5	250+	Dorairaj & Roy 1995

E = Estimated.

per year (Tables 5.2a and b). Mainly compounded foods are used, occasionally in conjunction with fertilisation to enhance natural production. Performance depends on controlled stocking with wild and/or hatchery-reared post-larvae. Before the 1990s, good yields were usually only obtained with carefully managed tidal or pumped water exchange of up to 30–40% per day, and the provision of supplementary aeration towards the end of the culture period. More recently however, equivalent yields are being obtained from carefully managed ponds to which little or no new water is added beyond that needed to compensate for evaporation and leakage (section 8.3.7). Also some farms are beginning to recycle and treat their water to reduce the risk of disease or in response to legislation governing effluent quality (section 8.3.6.8). Pond sizes range from around 0.2 to 3 ha and modern farms may have square, rectangular, or occasionally round, purpose-built earthen ponds with separate inlet and outlet channels to assist water management. In some localities the ponds are lined with clay, concrete or butyl rubber linings (sections 8.1.4, 8.3.9 and 10.6.1.5). Productivity of many semi-intensive ponds is limited to a maximum of about 3000 kg ha^{-1} per harvest by low and fluctuating oxygen levels, poor phytoplankton management and irregular water exchange.

Culture of crustaceans to marketable size in cages developed in the 1980s and typically involves shrimp, spiny lobsters and crabs. Some experimental culture of freshwater prawns in cages and tanks is also reported from Brazil, China and Thailand with yields equivalent to around 10 000 kg ha^{-1} yr^{-1} (Table 5.2c). Cages are also employed in fattening operations that add value to wild-caught spiny lobsters and crabs (Table 5.2c and sections 7.9.5 and 7.10.4). Cage and pen culture may be employed when land is scarce or to avoid some of the difficulties caused by poor water exchange. However, when unrestricted development occurs, the number and proximity of the cages can seriously impede natural water circulation within and between the cages (section 11.2.5).

5.2.3 Super-intensive

This category of cultivation is dominated by penaeid shrimp which can provide yields equivalent to over 10 000–50 000 kg ha^{-1}, with two to four crops per year. Super-intensive farming demands almost total reliance on compounded feeds, precisely controlled feeding rates, and the use of hatchery-reared, nursed juveniles (see Glossary) stocked at specific densities. Continuous exchange of new water, generally over 50–200% per day, as well as aeration, is typical (section 7.2.6.6). However, an increasing number of farms either operate on a minimal water exchange or recycle a substantial proportion of the culture water (McNeil 2000). Shrimp are reared in concrete, plastic or fibreglass tanks and raceways ranging in size from 0.03 to 0.1 ha. In Japan, the high prices paid for *Marsupenaeus japonicus* permits its culture in circular concrete tanks with a sand substrate on a false floor (Shigueno tanks – Table 5.3 and section 7.2.6.6).

Indoor 'battery' farming of lobsters and some large Australian crayfish, while not yet commercially viable, has the potential to provide similar yields (Table 5.3 and section 7.8.9) and, in theory at least, could be independent of climate. One company experimented with the production of individually confined lobsters (presumably *Homarus americanus*) in Hawaii. Interestingly, they obtained water at 22°C from an experimental plant that utilised the thermal difference between warm surface seawater and cold seawater pumped from the depths to generate electricity (Rosenberry 1990).

Table 5.2a Examples of tropical semi-intensive aquaculture: shrimps, freshwater prawns, crayfish, crabs.

Species	Location	Annual yield range (kg ha^{-1})	Crops yr^{-1}	Live weight (g)	Reference
Penaeus monodon	Sri Lanka (artisanal farms)	1200–1700	2	31	De Silva & Jayasinghe 1993
P. monodon	Australia (zero exchange trial)	7500	1.5 A	30	O'Sullivan & Kiley 1997
P. monodon	South-east Asia	2044	2–6	28	IFC 1987
P. monodon	Indonesia	3776 (in 20 weeks)	2.1	32	Lee *et al.* 2000
Litopenaeus stylirostris	New Caledonia	3000	1 rotated with *P. monodon*	18–25	AQUACOP 1984
L. stylirostris, L. vannamei	Ecuador	2270	2 A	15–20 A	Kagel 1998
L. vannamei	Peru	2500	2.5	15	Viacava 1995
Macrobrachium rosenbergii	Hawaii	1396–2206	Continuous	>30	Malecha 1983
M. rosenbergii	Saudi Arabia	1350–2610 (in 168 days)	1, experimental (in 3×2×1 m tanks)	17–30	Siddiqui *et al.* 1997
M. rosenbergii	Taiwan	2000–3000	Continuous	30	New 2000
Cherax quadricarinatus	Australia	2000–4000	0.5–1	40–70 (1 yr) 100–200 (2 yr)	Jones 1990
C. quadricarinatus	Australia (demonstration farm)	2700–3360	1	45–55*	Jones & Grady 1997
C. quadricarinatus	Australia (demonstration farm)	4600–8460	2	50–70**	Jones & Grady 1997
C. quadricarinatus	Central America	1500–2000	2	80–120	Rouse 1995
Scylla spp.	Taiwan	up to 1800	2–3	200–300	Chen 1990

A = Assumed. *Stocked with 5 g juveniles. **Stocked with 16–22 g juveniles.

Table 5.2b Examples of tropical intensive pond aquaculture: shrimps, spiny lobster.

Species	Location	Annual yield range (kg ha^{-1})	Crops yr^{-1}	Live weight (g)	Reference
Penaeus monodon	Thailand	4000–12 000	2	33	Briggs 1989
P. monodon	Thailand (inland)	6000–8000	2 (low salinity)	20–25	Shivappa & Hambrey 1997
P. monodon	Taiwan	8700–13 700	1, but 2 crops possible	23–33	Chen *et al.* 1989
Fenneropenaeus penicillatus	Taiwan	4298–12 286	1, generally rotated with *P. monodon*	9.5–12.3	Liao & Chien 1990
Litopenaeus vannamei	USA	9000	1, experimental	13	Bratvold *et al.* 1999
L. vannamei	USA	5807–8163	1, experimental (zero water exchange)	16–21	Hopkins *et al.* 1995
L. vannamei	Venezuela	3600–5600	2 A	18.5	Clifford 1994
Panulirus homarus	Taiwan	up to 150 000 lobsters/1.3 ha	0.5–1(fattening of wild 5–250 g juveniles)	300–800	Chen 1990

A = Assumed.

Table 5.2c Examples of tropical cage aquaculture: shrimps, prawns, spiny lobster, crabs.

Species	Location	Annual yield range (equivalent kg ha^{-1})	Crops yr^{-1}	Live weight (g)	Reference
Fenneropenaeus indicus	India	1975 kg ha^{-1} in 90 days	1 experimental	13	Shanmugam *et al.* 1995
F. merguiensis	Singapore	20 000–30 000	2	12	Lovatelli 1990
Penaeus monodon	Thailand	32 000–48 000	2 experimental	25 E	Lovatelli 1990
Litopenaeus vannamei	Brazil	16 000–20 000	2–2.5	15–18	Paquotte *et al.* 1998
Macrobrachium rosenbergii	Thailand	10 000	1 experimental	10	Valenti & New 2000
M. rosenbergii	China	10–15 000	? (in 15–20 m^2 cages*)	—	Valenti & New 2000
M. nipponense	China	5000 (equivalent)	2 (in 10×3×1.3 m cages)	—	Kutty *et al.* 2000
Panulirus argus	Brazil	28 000	3-mo experiment (in sea pens)	Fattening from 283 to 338 g	Assad 1998
P. polyphagus	Singapore	up to 45 kg m^{-3}	2–3	Fattening from 100 to 300 g	Lovatelli 1990
P. homarus, P. versicolor	India (floating cages)	up to 1800	1–1.5	220–310	Kuthalingam 1990
Scylla spp.	Singapore	9–10 kg m^{-3}	10–20 day fattening	200–300	Lovatelli 1990
Scylla spp.	Philippines	600	2–3	250	Tabigoon 1998

E = Estimated. *Used four-stage transfer system with progressively larger mesh sizes.

Table 5.3 Examples of super-intensive aquaculture: shrimps, tropical crayfish, lobsters.

Species	Location	Annual yield range (kg ha^{-1})	Crops yr^{-1}	Live weight (g)	Reference
Marsupenaeus japonicus	Japan	20 000–30 000	1	20	Yano 1993
Litopenaeus vannamei	Tahiti	17 000–22 000	1–2	14–20	AQUACOP 1989
L. vannamei	Belize (pilot commercial)	22 466	2 A	15–18	McIntosh 1999
L. vannamei	Hawaii	up to 45 000	3	15.7	Wyban & Sweeney 1993
L. vannamei	Texas, USA	up to 114 000	Potentially continuous	11–14	Reid & Arnold 1992
L. vannamei	Chicago, USA (stacked trays)	1.07–1.51 kg m^{-2}	Continuous (2–3)	20	McCoy 1986
L. vannamei, L. stylirostris	Hawaii	50 000–100 000	2	13–20	Moore & Brand 1993
L. stylirostris	Belize (pilot commercial)	14 900	2 A	15	McIntosh 1999
*Cherax quadricarinatus**	Australia (pilot battery)	32 kg per module of 4 trays	Potentially continuous	70 g	O'Sullivan 1990
Homarus americanus	PEI Canada (battery)	6–7 kg m^{-2}	Continuous (0.3–0.5)	350–450	Waddy 1988

A = Assumed. *Pilot system operated with *Cherax tenuimanus* but also applicable to *C. quadricarinatus*. Tray dimensions not reported.

5.3 Warm temperate and Mediterranean climates

Warm temperate or Mediterranean climates have well-defined growth seasons. Most farmers must therefore obtain and stock juveniles as soon as the water temperature is high enough for growth. Alternatively they must provide additional cover, heating or a controlled indoor environment to sustain stock, and in some cases growth, throughout the cool season. Cultures are typically extensive or semi-intensive and utilise shrimp, prawns and crayfish (Table 5.4). Very often, juveniles only begin to occur naturally in sufficient numbers for ongrowing after the temperature has risen beyond the minimum needed for growth to begin. In these circumstances and where only hatchery-reared juveniles are available, it may be advantageous to produce young from broodstocks maturing at the end of the summer, nurse the young through the winter in an indoor controlled-environment system and stock them outside as soon as temperatures rise (section 7.3.4). An added advantage occurs if the juveniles have increased in size during winter. Hatcheries and overwintering nurseries are seldom used in the farming of the red swamp crayfish *Procambarus clarkii* since the majority of populations are self-sustaining. Increasing attention is, however, being paid to their use as the farming of *Cherax* species intensifies (Mosig 1999) and spreads beyond Australia for example to Italy (D'Agaro *et al.* 1999), America and Israel (section 7.7.5). However the approach is not always successful, with problems

Table 5.4 Examples of crustacean aquaculture in warm temperate and Mediterranean type climates: shrimps, freshwater prawns, crayfish.

Species	Location and culture type	Annual yield range (kg ha^{-1})	Crops yr^{-1}	Live weight (g)	Reference
Fenneropenaeus chinensis	N. China	1123–2308	1	19.9–43.5	Zhang 1990
Marsupenaeus japonicus	Japan	4200	1	15–25	New 1988a
M. japonicus	France	210–590	1	15–25 E	New 1988a
M. japonicus	Portugal	300–1200	1	13	Arrobas *et al.* 1993
M. japonicus	Spain	150–250	1	15–20	Cañavate & Sanchez 1989
M. japonicus	Italy	189–525	1	25–30	Lumare *et al.* 1989
P. monodon	Italy (+ nursery ponds)	1208	1	17.9	Lumare *et al.* 1993
P. semisulcatus	Israel	739–2740	1	7–21	Issar *et al.* 1987
L. vannamei	Arizona, USA	1000–4200 (best)	1	15–20	Fitzsimmons 1999, Samocha *et al.* 1999
L. vannamei	S. Carolina, USA	7187	1	17.1	Sandifer *et al.* 1993
L. setiferus	S. Carolina, USA	7995	1	15.2	Sandifer *et al.* 1993
Macrobrachium rosenbergii	Israel	up to 2800	1	20	New 1988b
M. rosenbergii	S. USA	1000–1800	1	30–35	Tidwell & D'Abramo 2000
M. rosenbergii	China	2250–3000	1	—	New 2000
Orconectes virilis	USA	700–800	1	14–18	Brown *et al.* 1990
Procambarus clarkii	Spain	350	1	40–80 E	Lorena 1986
Cherax albidus	W. Australia farm dams	340	7–8 partial harvests	20–100	Lawrence *et al.* 1998
C. destructor	Australia (simulation)	2000–3800	1	55	Staniford 1989
		1200–1500	1	20–80	Mills *et al.* 1994
C. tenuimanus	Australia	1866–5093	0.5–1	48–217	Morrissy *et al.* 1995
Eriocheir sinensis	China (monoculture)	450–1500	0.5–1	125–250	Liu Fengqi, 2001 pers.comm.

E = Estimated.

arising from post-handling mortalities, regulatory constraints, high temperatures and prolific breeding in production ponds.

It may be possible to offset nursery heating costs in the winter through the use of industrial thermal effluents or geothermal waters. Where the quality, temperature or quantity of the heated water is suboptimal it is technically possible to transfer the heat to the culture water using heat exchangers or heat pumps (section 8.4.4). Recently, at least two farms in Arizona began experimenting with growing marine shrimp inland (*Litopenaeus vannamei*) in low salinity, thermal groundwaters (Fitzsimmons 1999; Samocha *et al.* 1999) and produced yields of 1–4 kg m^{-2} (sections 7.2.2.3 and 7.2.6.5).

The farming of Crustacea in conjunction with other valuable species (for example, molluscs like abalone) either in polyculture or during alternate seasons, represents another potential option for warm temperate environments (Wickins 1982). Yet another is the seasonal rotation of species, such as *Macrobrachium* with *Procambarus*, to provide prawns for the American soft-shell trade when crayfish are not available (section 5.6, Table 5.6). Since, however, the necessary level of investment is likely to be higher in these regions than it is for polyculture in the tropics, it is usually best to ensure that the culture of each species is economically viable in its own right rather than relying on obtaining optimal performance from both to ensure success.

5.4 Temperate climates

There are far fewer culture options and species available for crustacean farmers in cool climates than there are for mollusc and fish farmers (Table 5.5). This is because cultivated molluscs are lower in the food chain (they filter or graze natural algae) and require neither additional feed nor daily attention. The majority of fish successfully farmed in cool waters are far more tolerant of crowding than are crustaceans and many species can utilise the whole water column rather than just the floor area. They are thus able to make more effective use of the culture facility than crustaceans. The slow growth of temperate water crustaceans, coupled with their need for daily attention and feeding, and the comparatively low yields obtainable per unit floor area, effectively reduce the choices available to extensive or semi-intensive crayfish farming and, potentially, to indoor controlled environment or 'battery' farming for the table (Wickins 1982) or to supplying juveniles for restocking and ranching (section 5.7.2).

Even so, it is worth mentioning that several attempts to rear clawed lobsters in coastal waters have been reported (Van Olst *et al.* 1977; Wickins 1982) although none achieved commercial viability. The systems included groups of floating individual containers, seabed cages serviced by divers, and a fixed seabed-to-surface cage system serviced at the surface by means of an

Table 5.5 Examples of crustacean aquaculture in temperate climates: shrimps, freshwater prawns, crayfish.

Species	Location and culture type	Yield range	Crops yr^{-1}	Live weight (g)	Reference
Marsupenaeus japonicus	UK (power stn. effluent ponds)	Experimental	2 E	10–15	Wickins 1982
M. japonicus	Italy (power stn. effluent ponds)	Experimental	Winter catch crop	11–15	Palmegiano & Saroglia 1981
Macrobrachium rosenbergii	USSR (power stn. effluent ponds)	up to 1000 kg ha^{-1} yr^{-1}	1	6.5–60	Khmeleva *et al.* 1989
M. rosenbergii	New Zealand (geothermal water)	Commercial, 2500–3000 kg ha^{-1} yr^{-1}	2	30–35	New 2000
M. nipponense	China, semi-intensive ponds, cages, or with rice	150–1875 kg ha^{-1} yr^{-1}	1–2	—	Kutty *et al.* 2000
Pacifastacus leniusculus	Europe, UK extensive and semi-intensive lakes, ponds	3 crayfish/ metre of bank	0.5	60–100	Alderman & Wickins 1996
Paranephrops planifrons	New Zealand extensive and semi-intensive ponds	Commercial trials, 500–1000 kg ha^{-1} yr^{-1} E	0.5	50–60	Smythe 1998

E = Estimated.

access pier. In some cases hatchery-reared lobsters were to be reared to market size, while in others, partly grown, wild-caught lobsters were held until they grew large enough to command a higher price (Bombace *et al.* 2000). Similarly, approximately 10% of wild-caught, soft-shelled lobsters in the State of Maine are routinely impounded and fed for several months to increase their value (sections 3.3.4.1 and 7.8.9). In 1998 an experimental Canadian hatchery produced 330 000 *Pandalus platyceros* post-larvae from wild-caught broodstock, with a view to ongrowing the juveniles in net pens and floating tanks (C. Campbell 2000, pers. comm.; section 7.4.2). The principal advantages of offshore systems were thought to be lower capital, water treatment and pumping costs. The disadvantages were lack of environmental control, difficulty of inspection, feeding and costly maintenance and predator control.

The use of heated industrial effluent waters for ongrowing, either directly or following heat exchange, was considered from both biological and financial standpoints on several occasions during the 1970s and 1980s, e.g. for clawed lobsters (Johnston & Botsford 1981; Sakurai *et al.* 1993; section 7.8.9), crayfish (Aston 1981), marine shrimp and prawns (Palmegiano & Saroglia 1981; Wickins 1982). The conclusions led to a number of experimental and pilot scale trials with crustaceans as well as fish, which included attempts to grow juveniles both outdoors and indoors in raceways, ponds or tanks. To the best of our knowledge, however, most of these did not maintain the reliability in flow and quality of supply necessary to justify significant further investment in crustacean farm facilities. Many of the problems seemed to arise from the disparity in size and in the operational priorities of the heat-generating industry and the aquaculture unit (section 11.5.1.1). One farm in New Zealand, however, makes use of warm water effluent from a geothermal power station to produce two crops of *Macrobrachium* per year. Details of the farm's production and methodology are given by New (2000) (section 7.3.5.3).

Options for battery farming and intensive, controlled environment culture in temperate regions have also been investigated for crustaceans in general (Wickins 1982; McCoy 1986), crayfish (section 7.7.6.4), clawed lobsters (section 7.8.9) and shrimp (section 7.2.6.6). Although such systems are not necessarily confined to temperate climates, they allow optimum temperatures for growth to be maintained throughout the year. Animals are stocked and harvested from the unit at regular intervals regardless of season. Implicit in the concept are complete control over broodstock and juvenile supply and the ability to maintain water quality, typically by recirculation through suitable treatment plant (section 8.4.4).

Even though it is a high-risk operation, controlled environment culture has attracted a number of entrepreneurs because of its similarity to a manufacturing process. Control over production processes allows a consistent product quality and the ability to alter the appearance of the product (through diet or illumination) to suit the customer. The ability to produce continuously permits cost-effective processing, distribution and marketing operations. There seems little doubt that it is technically feasible to grow lobsters, some shrimp and crayfish in such systems but the high capital and labour costs involved seem likely to render the operation unduly sensitive to market forces or possibly even uneconomic. Nevertheless, competition for coastal sites and water, the costs of complying with increasingly stringent effluent discharge regulations and the risks of water-borne disease all conspire to maintain interest in such approaches (Davis & Arnold 1998).

5.5 Polyculture

In a polyculture system, penaeid shrimp, *Macrobrachium*, crayfish, crabs or (less commonly) marine caridean prawns, are grown as an addition to a crop of seaweed, e.g. *Gracilaria* (Wang 1994), rice (Li 1998), molluscs (*apud* Wang *et al.* 1998), or fish such as milkfish (de la Cruz 1995), carp (Hoq *et al.* 1996) or tilapia (Sadek & Moreau 1998) which are grown for local sale (Table 5.6). For further reading on these approaches, Rhodes (2000) reviews the economics of polyculture with special reference to freshwater prawns (section 10.6.1.8) while Zimmermann and New (2000) and Kutty *et al.* (2000) provide comprehensive accounts of prawn polyculture and systems integrated with crops (section 7.3.5.2).

About 70% of crayfish in Louisiana are farmed in combination with rice. The cultivated crustacean provides a useful supplementary income and may even be the most valuable crop. Occasionally, because of juvenile scarcity, unseasonable temperatures or some incompatibility of the animals chosen, each species may have to be grown consecutively in a form of 'crop rotation' For example *Macrobrachium* and *Procambarus* (Avault & Granados 1995), *Pandalus platyceros* and mussels or abalone (Hunt *et al.* 1995) and *Litopenaeus vannamei* with *Farfantepenaeus californiensis* (Martinez-Cordova *et al.* 1999). Additionally, *Macrobrachium* is being grown in rotation with trout and ex-

Table 5.6 Examples of polyculture and crop rotation: shrimps, caridean prawns, crayfish, crabs.

Species	Location	Annual yield range (kg ha^{-1}) (crustaceans)	Crops yr^{-1}	Live weight (g)	Reference
Fenneropenaeus chinensis with tilapia hybrids	China	500	1, experimental (in enclosures)	10	Wang *et al.* 1998
F. merguiensis with *Artemia* and salt. (pilot projects on 150 ha)	Vietnam	164 kg shrimp 84 kg *Artemia* cysts	1	—	Binh & Lin 1995
Metapenaeus dobsoni, *M. monoceros*, *Fenneropenaeus indicus*, *Penaeus monodon* with fish	India	200 (shrimp and fish)	1 (rotation with rice)	0.8–12.5	Nagaraj & Neelakantan 1982
Litopenaeus vannamei with *Farfantepenaeus californiensis*	NW Mexico	2208–2462; 1200–1500	2, experimental (species rotation)	—	Martinez-Cordova *et al.* 1999
Penaeids with clams	Hawaii	2000 expected	1–2	15–25 E	York 1983
M. rosenbergii with carp	Bangladesh	428	1	34	Hoq *et al.* 1996
M. rosenbergii with tilapia & carp	Egypt	400–600	1	35–40	Sadek & Moreau 1996
M. rosenbergii with tilapia & carp	Various	400–1300	1–2 or continuous	20–40	Zimmermann & New 2000
M. rosenbergii with bait fish	USA	533	1	17–29	Scott *et al.* 1988
M. nipponense	China	1500	1	—	Kutty *et al.* 2000
M. malcolmsonii	India	300–400	2 or continuous	20–30 E	Kutty *et al.* 2000
Cryphiops caementarius with mullet	Chile	—	—	—	Zuniga-Romero *et al.* 1987
Pandalus platyceros in coho salmon cages	USA	74 g m^{-2} of cage net surface	1	8.6	Rensel & Prentice 1979
Procambarus clarkii with *Macrobrachium rosenbergii*	USA	1444; 855	0.5; 1, experimental (species rotation)	—	D'Abramo & Daniels 1992
Procambarus clarkii with fish	USA	400–1000	0.5–1	20–30 E	Huner 2001
Cherax quadricarinatus with Nile tilapia	USA	590	1–2 experimental	53	Brummett & Alon 1994
Eriocheir sinensis with rice	China	300–450	1	125	Li 1998
Eriocheir sinensis fattening	China	450–750	1	250	Li 1998
Scylla spp. with shrimp or milkfish	Philippines	340–500	1–2	200–230	Macintosh 1982

E = Estimated.

periments with yellow perch held in cages in prawn ponds are under way in the USA (Tidwell & D'Abramo 2000). However, experiments to culture *Cherax* species with fish such as Nile tilapia have resulted in less than satisfactory growth of either the crayfish (Kotha & Rouse 1997) or the tilapia (Brummett & Alon 1994) despite similar stocking densities: an argument for caution in the interpretation of results of polyculture trials. The theoretical possibility that pheromones from, for example, mating fish may influence crustaceans or vice versa during polyculture was raised by Zimmermann and New (2000).

Other polyculture prospects include crayfish with ducks or horticultural products, the latter two being either for sport, food or display (Smolowitz & Cimetti 1998). A recent conservation initiative in Vietnam has been the integration of shrimp with mangrove silviculture (section 7.2.6.2). Elsewhere, for example in Indonesia, Thailand and the Philippines, mangrove silviculture is practised with crab, shrimp and fish either in ponds or in cages (sections 7.10.4 and 11.4.1). Attempts to utilise nutrient-rich waste waters from crustacean farms to nourish filter-feeding bivalves, while promising, have seldom proved commercially viable (section 8.3.6.8).

Plate 5.1 Small-scale, integrated mangrove forest–crab pen for *Scylla* spp. in Sarawak, Malaysia. (Photo courtesy D.J. Macintosh, University of Aarhus, Denmark.)

Recent developments in the USA and Mexico include experiments in which effluents from shrimp ponds are used to produce microalgae (e.g. *Chaetoceros* spp.) that are then fed to American cup oysters held in upwelling (or fluidised bed) systems (Wang 1998) or to *Artemia* in attempts to improve effluent quality (Rodriguez-Canché *et al.* 2000). The keys to the oyster operation are control over the species of algae that predominate, the balance between production and consumption by the oysters and the management of other suspended particles. Herbivorous fish, such as grass carp, are sometimes added to ponds containing the crustaceans to control plant growth (they may be called 'sanitary' fish) but this is not considered to be true polyculture even though the fish may have value when harvested.

5.6 Production of soft-shelled crustaceans

Most if not all of the large edible crustaceans (40 g or over) are potentially marketable as a gourmet food item immediately after moulting while the shell is still soft. At present, however, only crabs and crayfish are marketed in significant quantities in this form (Table 5.7) but prices can be high and there would seem to be considerable potential for the development of markets for soft-shelled *Macrobrachium*, Australasian crayfish and possibly sub-adult spiny lobsters and clawed lobsters. In the northern USA, 1–5 g soft-shell crayfish (*Orconectes* spp.) are used as bait by anglers (section 7.5.9).

Soft-shelled crustaceans (sections 3.3.3.2 and 3.3.5) must be eaten or processed within the few hours between casting the shell and attaining what is known as the 'paper shell' stage, when water enters the tissues and spoils the texture. Crabs and crayfish about to moult are identified from catches and placed in shallow trays of clean, running water for up to about 7–10 days. Many soft-shell crab producers now use recirculation systems because of the greater control over water quality (Webster 1998). Inspections may be made as often as every 15 minutes, so harvesting and any initial processing will be continuous during the season (sections 7.5.7 and 7.10.9). One Louisiana producer is reported to alternate soft-shell crayfish production with that of soft-shell blue crabs after the crayfish season ends. The system water is, of course, made brackish for the crabs (Huner 1999).

Soft-shelled crustaceans may be produced from wild or cultured stock, although in practice selection of pre-moult animals from ponds without repeatedly disturbing the stock would be difficult. It is not yet practicable to induce synchronous moulting within a pond-reared population, although it would be relatively easy to harvest newly moulted individuals from a 'battery' farming system. To increase yields and predictability of moulting, attempts have been made to control levels of moult-inhibiting hormones (section 12.8.6), for example through eyestalk ablation, although consumer acceptance of ablated product has yet to be determined (sections 7.5.8 and 12.1).

Table 5.7 Examples of soft-shelled crustacean production operations: freshwater prawns, crayfish, crabs.

Species	Location	Holding system	Wild or cultured	Product features	Animal size	Reference
Macrobrachium rosenbergii	USA	Crayfish shedding trays	Cultured	Experimental study	25–35 g (assumed)	Lutz *et al.* 1990
Procambarus clarkii	S. USA	Wooden trays	Wild and cultured	Vacuum packed or frozen in trays or bags	12–20 g	Huner 1999
Orconectes spp.	USA	Wooden trays	Wild and cultured	Angler's bait	1–5 g	Huner 1997
Callinectes sapidus	USA	Wooden trays floating boxes	Wild	Dressed, IQF	80–140 mm CW	Oesterling & Provenzano 1985
Procambarus clarkii with *Macrobrachium rosenbergii*	USA	25 kg of prawns from 43 m^2 of crayfish shedding trays in 50 days	Wild and cultured	Experimental (species rotation)	—	Caffey *et al.* 1993

5.7 Hatchery supported fisheries, ranching and habitat modification

At present, there are few natural stocks of fished crustaceans that are not fully or even over-exploited and some are additionally threatened by environmental degradation or habitat loss. Restocking and habitat restoration programmes have been periodically established in several regions in an attempt to maintain both supplies and employment opportunities in coastal communities (Welcomme & Bartley 1998). Although the concept of enhancement is not new (lobster releases have been conducted since the late nineteenth century – section 7.8.11) few, if any of the early attempts were evaluated sufficiently critically to confirm success or failure. Since 1980, however, more rigorous approaches have been adopted and have produced tangible evidence of technical success. Nevertheless, serious questions still remain concerning the financial and socio-economic framework in which enhancement programmes can be justified. Concerns are also expressed about the impact of releasing animals that are genetically different from indigenous populations, the risks of spreading disease and the effect of modified habitat on the local ecology and sediment distribution patterns (sections 11.3.1.3 and 11.3.2).

5.7.1 Restocking and stock supplementation

A number of hatcheries in the Far East, North America and Europe are employed to produce juvenile crustaceans to restock inland fresh or coastal marine waters as part of national programmes aimed at restoring, stabilising or enhancing depleted fisheries. Such operations are known by a variety of names, the most common being restocking, stock enhancement or supplementation, reseeding and hatchery supported fishery. Some may be set up in mitigation for overfishing or stock losses following intentional or accidental changes to the local environment. One of the most successful marine restocking operations reported was with *Fenneropenaeus chinensis* in the Bohai Sea in northern China, where catches were stabilised at a level five times higher than previously recorded (Rothlisberg *et al.* 1999). Well over 20 different crustacean species have been reared for restocking worldwide, the greatest variety being in Japan (Munro & Bell 1997). Shrimp and crabs are the most popular in the tropics while crayfish and lobsters predominate in more temperate regions (Table 5.8 and Chapter 7). Positive effects from restocking *Macrobrachium* in Indian lakes and rivers, such as increased average daily catches by artisanal fisherfolk, have been reported but with little or no substantive scientific monitoring (New *et al.* 2000). In at least one country (Taiwan) effort is also directed towards increasing the availability of wild penaeid broodstocks following intensive overfishing for hatchery use (Su & Liao 1999).

Fluctuations in catches caused by varying natural recruitment, catchability, environmental conditions and weather patterns as well as indeterminate survival of released juveniles, can mask the effects of stock enhancement programmes in open waters (Van der Meeren *et al.* 1998). These factors make it difficult to judge the returns accurately (Xu *et al.* 1997) and an element of faith is usually necessary when justifying large national restocking programmes or even the funding for the basic research needed to underpin the approach and conduct pilot release and monitoring projects (Hallenstvedt 1999).

Plate 5.2 Woven mesh nursery cages (*hapas*) suspended in a lagoon in Sri Lanka for the rearing of shrimp post-larvae prior to their release in a stock-enhancement project. The cage area is protected from birds by discarded fishing nets and from catfish by submerged fencing previously used for local barrier traps (*kraals*).

The development of techniques enabling the hatchery-reared or transplanted animals to be identified in catches, several months (e.g. penaeids, section 7.2.9) or perhaps years after release (e.g. lobsters, section 7.8.11.2) has provided the essential keys for effective evaluation (sections 10.6.3.3 and 12.7). When a restocking project has been judged successful (section 8.11), the use of tagged animals in further releases and the associated specialised monitoring programmes may not be necessary, thereby reducing costs considerably.

Once a restocking scheme is started, public opinion may make it difficult to stop even though there may be little or no scientific evidence that enhancement of the stock or fishery has occurred (J. Hughes, 1980 pers. comm.; Lorec 1999). The potential benefits of restocking are, however, unlikely to be realised without the concurrent implementation of effective fisheries management procedures (Munro & Bell 1997) and it seems likely that, in many fisheries, greater productivity would result from improvements to fisheries management, environment or habitat than from releases of hatchery-reared juveniles.

5.7.2 Ranching and habitat modification

Some types of extensive crayfish farm, for example the wooded and marsh pond farms in the southern USA, the 'stock and forget' farms in UK and some farm dams in Australia (sections 7.5.4.1, 7.6.6.1 and 7.7.6.1), more closely resemble private fisheries or ranches than conventional aquaculture operations (Table 5.8). Broodstock or juveniles are stocked once, or over just 3–5 years and become self-sustaining provided they are not overfished. Very little management is involved but ownership is clearly defined and any investment can be protected.

Attempts to ranch marine or brackish-water crustaceans are only likely to be of commercial interest in areas where legislation permits leasing or ownership and the stock can be maintained within a defined boundary or indisputably distinguished from wild stocks (section 11.5.3.1). Examples include the use of net fences to confine hatchery-reared juveniles (embayment) or the introduction of an easily identifiable exotic species such as *Homarus americanus* to Japanese waters (section 7.8.11.2).

A variation on the ranching theme involves investment in the creation or modification of seabed habitat in areas where adults occur naturally, so that the stock becomes more concentrated and fishing becomes more efficient (section 7.9.8). This approach generally features spiny and clawed lobsters and crabs, concentrating them in a convenient locality for fishing. In the Caribbean the use of artificial shelters in the fishery for *Panulirus argus* has been so successful that the concept is being introduced in Australia, Kenya, Sri Lanka and the Seychelles. In Japan artificial reefs of concrete blocks, rocks or stones are set in seaweed beds (*Gelidium amansii*) and have become good fishing grounds for *Panulirus japonicus*. The deployment of similar reefs to modify currents in such a way as to enhance the settlement of the puerulus stages is also being examined (section 8.11.2). To protect such investment, access must be regulated by suitable

Table 5.8 Examples of crustacean stock supplementation and ranching projects: shrimps, freshwater prawns, crayfish, crabs, clawed and spiny lobsters.

Species	Location	Culture system	Releases	Returns/results	Reference
Fenneropenaeus chinensis	China	Hatchery	4×10^9	4.6–8.2% E; 4800 mt, worth $24m	Shang 1989
F. chinensis	Korea	Hatchery/nursery	8×10^6–11×10^6 yr^{-1}	Unknown	Park 1990
Marsupenaeus japonicus	Japan	Hatchery/nursery	300×10^6 yr^{-1}	5–8% recapture at market size	New 1988a
Penaeus monodon	Taiwan	Ponds	Tagged subadults, 30–50 g	0.17–16% returns of broodstocks	Su & Liao 1999
P. monodon	Sri Lanka	Commercial hatchery	Out of season releases into wild fishery	3.5% recaptured enhancing annual catch by 1400%	Davenport *et al.* 1999
Macrobrachium rosenbergii	Thailand	Hatchery/nursery	3×10^6 over 3 years	2% recaptured	NACA 1986
M. rosenbergii	Thailand	Hatchery/nursery	300×10^6 yr^{-1}	Catch increases reported	New *et al.* 2000
M. rosenbergii	Malaysia	Hatchery/nursery	1–4×10^6 yr^{-1}	Aim to improve fished stocks	New *et al.* 2000
M. rosenbergii	India, Kerala	Wild caught	1.9×10^6 (1999)	Some improvement in production	New *et al.* 2000
M. malcolmsonii	India, Andhra Pradesh	Wild caught	9×10^6 (1994–99)	Not reported	New *et al.* 2000
Astacus astacus	Europe	Adults, hatchery, artificial incubation	40 000–170 000 yr^{-1}, 1–30 inds. m^{-1}of bank	45–60% survival	Pérez *et al.* 1997; Ackefors 2000
P. leniusculus	Europe	Adults, hatchery	20 000–500 000 yr^{-1}	Increasing catches	Lewis 2001
Procambarus clarkii	USA, China, Spain	Adults	25–100 kg ha^{-1}	Self-sustaining populations	Huner 2001
Homarus americanus	USA	Hatchery	500 000 yr^{-1}	Unknown, additions made to existing fishery	Syslo 1986
Homarus americanus	USA	Hatchery	175 000 yr^{-1} for 5–6 yr	Unknown, additions made to existing fishery	Beal *et al.* 1998
Homarus americanus	Japan	Hatchery	<5000	Experimental studies	Kittaka *et al.* 1983
H. gammarus	UK	3 hatchery/nursery projects	10 000 yr^{-1} for 5 yr	1–5.5% returns of tagged lobsters sampled from existing fishery	Bannister & Addison 1998
H. gammarus	France	3 hatchery projects	250 000 yr^{-1} for 10 yr	Unknown, additions made to existing fishery	Latrouite 1998
H. gammarus	Norway	Hatchery	150 500 over 5 yr	6–7% returns from depleted fishery	Van der Meeren *et al.* 1998
Panulirus argus	Caribbean	Habitat modification	Some juveniles transplanted	Unknown, projects within existing fisheries	Butler & Herrnkind 2000
Portunus trituberculatus	Japan	Hatchery	10×10^6–50×10^6 yr^{-1}	3–12% estimated	Muroga *et al.* 1994
Eriocheir japonica	S. China	Wild seed from Changjiang estuary	Unknown numbers into Zhujiang estuary	'Good fishery established within a few years'	Li *et al.* 1993

E = Estimated.

legislation, possibly including the establishment of sanctuary areas for broodstock (Briones-Fourzán *et al.* 2000; Cruz & Phillips 2000).

Unlike young clawed lobsters, spiny lobsters (*Panulirus* spp.) do not appear to construct burrows or modify existing crevices. This means that in years when spiny lobsters become particularly abundant, the carrying capacity of a ground for these species may become limited by the number of available crevices (Briones-Fourzán *et al.* 2000). In his review of habitat modification methods employed to enhance fisheries for spiny lobsters (*Panulirus* spp.), Miller (1983) suggested that a real increase in survival, and therefore natural production, might be induced in areas where food was abundant by providing additional habitat (crevices) for the natural population. Taking the concept a step further, survival of juveniles of some species (*Panulirus argus*, *P. marginatus*) but not necessarily others (*P. cygnus*) (Butler & Herrnkind 2000) might also be enhanced if species-specific habitats were designed and deployed in natural nursery areas. This advance has now been made and investigators in Cuba, New Zealand and Australia are trying to determine the effectiveness of transplanting the juvenile spiny lobsters caught in the artificial nursery habitats to seabed areas that are either naturally more favourable for ongrowing or are made so by the deployment of artificial habitat (section 7.9.8). However, results of experiments with *P. argus* involving predator and shelter abundance warn against optimistic extrapolation from small-scale experiments to large-scale projects (Butler & Herrnkind 2000). Over the past two decades there has been considerable discussion as to whether artificial habitats can actually increase natural productivity rather than merely increase mortality of the stock due to fishing. However, the relevance of this 'attraction–production' debate has now been declared redundant in favour of more practical issues (Lindberg 1997).

It may also be possible to ranch marine crustaceans with low migratory instincts (e.g. *Homarus gammarus*) by creating or modifying seabed habitat so that it becomes suitable for occupation by hatchery-reared juveniles (Jensen *et al.* 2000) (sections 7.8.12 and 8.11.2). If the area created does not receive natural settlement, and if currents carry away any larvae eventually produced by the occupants, the population will not become self-sustaining in the long term and will require regular replenishment from a hatchery (Wickins 1997). Ownership or lease arrangements would allow protection of commercial investment (Pickering 2000) but such a scheme might also be of interest to organisations concerned with the environmentally acceptable utilisation of surplus non-toxic, solid industrial materials (Collins & Jensen 1997). Suitably stabilised and deposited as a reef, such material could, with maturity, provide the ecosystem needed to support new, hatchery-based fisheries (Wickins 1995; Jensen *et al.* 1998). Indeed, the maintenance of a hatchery and stocking programme might be imposed as a condition when granting permission to deploy the reef. In similar vein, a reef has been constructed in Rhode Island, USA, in mitigation for environmental damage following the wreck of a tanker. The reef will provide opportunities for experiments on lobster (*Homarus americanus*) recruitment to the reef together with a study of stocking with hatchery-reared juveniles (Cobb *et al.* 1998). In another example, clusters of mixed cobble and large granite boulders placed in Lake Tahoe in mitigation for habitat disturbance were found to be quickly inhabited by crayfish (Shaffer & Reiner 1994).

Studies conducted in Britain on artificial reefs of cement-stabilised, pulverised fuel ash blocks (Jensen & Collins 1997) and in Israel with car tyre reefs (Spanier 2000) revealed rapid natural immigration and colonisation from the surrounding area by clawed (*Homarus gammarus*) and slipper lobsters (*Scyllarides latus*) respectively. These observations lend support to the hypothesis that, if deployed in suitable areas, reefs made from such 'materials of opportunity' may have the potential to revitalise impoverished seabed areas (section 8.11.2). Experiments will be required to determine if, and at what scale of operation, a worthwhile level of fishing can be sustained and whether repeated releases of hatchery-reared juveniles will be necessary.

5.8 References

Ackefors H.E.G. (2000) Freshwater crayfish farming technology in the 1990s: a European and global perspective. *Fish and Fisheries*, **1**, 337–359.

Alderman D.J. & Wickins J.F. (1996) *Crayfish Culture*, 22 pp. Lab. Leafl., (76). MAFF Directorate Fisheries Research, Lowestoft, UK.

AQUACOP (1984) Review of ten years of experimental penaeid shrimp culture in Tahiti and New Caledonia (South Pacific). *Journal of the World Mariculture Society*, **15**, 73–91.

AQUACOP (1989) Production results and operating costs of the first super-intensive shrimp farm in Tahiti. *Aquaculture '89 Abstracts*, p. 56, from World Aquaculture Society Meeting 1989, Los Angeles, USA.

Arrobas I., Zhang L.S. & Song X.H. (1993) Experimental semi-intensive culture of shrimp *Penaeus japonicus* in Portugal. In: *World Aquaculture '93 From discovery to commercialization* (eds M. Carrillo, L. Dahle, J. Morales, P. Sorgeloos, N. Svennevig & J. Wyban), p. 108. European Aquaculture

Society, Special Publication No. 19.

Assad L.T. (1998) Growout of juvenile *Panulirus argus* in cages. *The Lobster Newsletter*, **11** (1) 11–13.

Aston R.J. (1981) The availability and quality of power station cooling water for aquaculture. In: *Aquaculture in Heated Effluents and Recirculation Systems* (ed. K. Tiews), **1**, 39–58. Heenemann Verlagsgesellschaft, Berlin.

Avault J.W. Jr. & Granados A.E. (1995) Economic implications of pond rotation of the prawn *Macrobrachium rosenbergii* with the redswamp crawfish *Procambarus clarkii*. In: *Freshwater Crayfish 8* (ed. R.P. Romaire), p. 703. Louisiana State University, LA, USA.

Bannister R.C.A. & Addison J.T. (1998) Enhancing lobster stocks: a review of recent European methods, results, and future prospects. *Bulletin of Marine Science*, **62** (2) 369–387.

Beal B.F., Chapman S.R., Irvine C. & Bayer R.C. (1998) Lobster (*Homarus americanus*) culture in Maine: a community-based, fishermen-sponsored public stock enhancement program. In: *Proceedings of a workshop on lobster stock enhancement held in the Magdalen Islands (Quebec)*. 29–31 October, 1997 (ed. L. Gendron), Canadian Industry Report of Fisheries and Aquatic Sciences, **244**, 47–54.

Binh C.T. & Lin C.K. (1995) Shrimp culture in Vietnam. *World Aquaculture*, **26** (4) 27–33.

Bombace G., Fabi G. & Fiorentini L. (2000) Artificial reefs in the Adriatic Sea. In: *Artificial Reefs in European Seas* (eds A.C. Jensen, K.J. Collins & A.P.M. Lockwood), pp. 31–63. Kluwer Academic, Netherlands.

Bratvold D., Lu J. & Browdy C.L. (1999) Disinfection, microbial community establishment and shrimp production in a prototype biosecure pond. *Journal of the World Aquaculture Society*, **30** (4) 422–432.

Briggs M. (1989) Shrimp farming developments in Thailand. *Aquaculture News*, (7) 12–13. Institute of Aquaculture, University of Stirling, Scotland.

Briones-Fourzán P., Lozano-Álvarez E. & Eggleston D.B. (2000) The use of artificial shelters (casitas) in research and harvesting of Caribbean spiny lobsters in Mexico. In: *Spiny Lobsters: fisheries and culture,* 2nd edn (eds B.F. Phillips & J. Kittaka), pp. 420–446. Fishing News Books, Oxford, UK.

Brown P.B., Hooe M.L. & Blythe W.G. (1990) Preliminary examination of production strategies and forages for *Orconectes virilis*, the northern or fan-tail crayfish. *Journal of the World Aquaculture Society*, **21** (1) 53–58.

Brummett R.E. & Alon N.C. (1994) Polyculture of Nile tilapia (*Oreochromis niloticus*) and Australian red claw crayfish (*Cherax quadricarinatus*) in earthen ponds. *Aquaculture*, **122** (1) 47–54.

Butler M.J. IV & Herrnkind W.F. (2000) Puerulus and juvenile ecology. In: *Spiny Lobsters: fisheries and culture,* 2nd edn (eds B.F. Phillips & J. Kittaka), pp. 276–301. Fishing News Books, Oxford, UK.

Caffey R.H., Lutz C.G., Avault J.W. Jr. & Higgins K.K. (1993) Observations on production of soft-shell prawns, *Macrobrachium rosenbergii*, in commercial crawfish shedding facilities. *Journal of Applied Aquaculture*, **2** (2) 93–102.

Cañavate J.P. & Sanchez M.P. (1989) *Effect of nursery stocking density on further growing of* Penaeus japonicus *cultured in earthen ponds.* European Aquaculture Society, Special Publication No. 10, 51–52.

Chakraborti R.K., Halder D.D., Das N.K., Mandal S.K. & Bhowmik M.L. (1986) Growth of *Penaeus monodon* Fabricius under different environmental conditions. *Aquaculture*, **51** (3–4) 189–194.

Chen J-C., Liu P-C. & Lin Y.T. (1989) Culture of *Penaeus monodon* in an intensified system in Taiwan. *Aquaculture*, **77** (4) 319–328.

Chen L-C. (1990) *Aquaculture in Taiwan*, 273 pp. Fishing News Books, Blackwell Scientific Publications, Oxford, UK.

Clifford H.C. III (1994) Semi-intensive sensation. *World Aquaculture*, **25** (3) 6 12 & 98–104.

Cobb J.S., Castro K., Wahle R.A. & Catena J. (1998) An artificial reef for lobsters in Rhode Island, USA. In: *Proceedings of a workshop on lobster stock enhancement held in the Magdalen Islands (Quebec)*. 29–31 October 1997 (ed. L. Gendron), Canadian Industry Report of Fisheries and Aquatic Sciences, **244**, 75–78.

Collins K. & Jensen A. (1997) Acceptable use of waste materials. In: *Proceedings of the First Conference of the European artificial reef research network*, Ancona, Italy, 26–30 March 1996 (ed. A.C. Jensen), pp. 377–390. Southampton Oceanography Centre, Southampton, UK.

CPC (1989) *Libro blanco del camarón*, 79 pp. Cámara de Productores de Camarón, May 1989, Guayaquil, Ecuador.

Cruz R. & Phillips B.F. (2000) The artificial shelters (pesqueros) used for the spiny lobster (*Panulirus argus*) fisheries in Cuba. In: *Spiny Lobsters: fisheries and culture,* 2nd edn (eds B.F. Phillips & J. Kittaka), pp. 400–419. Fishing News Books, Oxford, UK.

de la Cruz C.R. (1995) Brackishwater integrated farming systems in Southeast Asia. In: *Towards Sustainable Aquaculture in Southeast Asia and Japan* (eds T.U. Bagarinao & E.E.C. Flores), pp. 23–36. SEAFDEC Aquaculture Department, Iloilo, Philippines.

D'Abramo L.R. & Daniels W.H. (1992) Joint production of the red swamp crayfish *Procambarus clarkii* and the freshwater prawn *Macrobrachium rosenbergii* as part of a crop rotation system. In: *Aquaculture '92: growing toward the 21st-century.* 21–25 May 1992, Orlando, Florida, USA, p. 74. Mississippi State University, MS, USA.

D'Agaro E., De Luise G. & Lanari, D. (1999) The current status of crayfish farming in Italy. In: *Freshwater Crayfish 12* (eds M. Keller, M.M. Keller, B. Oidtmann, R. Hoffmann & G. Vogt), pp. 506–517. Weltbild Verlag, Germany.

Davenport J., Ekaratne S.U.K., Walgama S.A., Lee D. & Hills J.M. (1999) Successful stock enhancement of a lagoon prawn fishery at Rekawa, Sri Lanka using cultured postlarvae of penaeid shrimp. *Aquaculture*, **180** (1–2) 65–78.

Davis D.A & Arnold C.R. (1998) The design, management and production of a recirculating raceway system for the production of marine shrimp. *Aquacultural Engineering*, **17** 193–211.

De Silva J.A. & Jayasinghe J.M.P.K. (1993) The technology and economics of small-scale commercial shrimp farms in the west coast of Sri Lanka. *Journal of Aquaculture in the Tropics*, **8** (2) 141–149.

Dorairaj K. & Roy S.D. (1995) Mudcrab culture in the Andamans. In: *Proceedings of the Seminar on Fisheries – a multibillion dollar industry* (eds B. Krishnamoorthi, K.N. Krishnamoorthy, P.T. Meenakshisundaram & K.N. Nayar), p. 136. Aquaculture Foundation of India, Madras, India.

Fitzsimmons K. (1999) Shrimp farming in saline groundwater in Arizona, USA. In: *Book of abstracts, World Aquaculture '99*, 26 April–2 May 1999, Sydney, Australia, p. 263. World Aquaculture Society, Baton Rouge, LA, USA.

Hallenstvedt A. (1999) Management of enhanced stocks: social, economic, and access issues. In: *Stock Enhancement and Sea Ranching* (eds B.R. Howell, E. Moksness & T. Svåsand), pp. 493–508. Fishing News Books, Oxford, UK.

Hopkins J.S., Sandifer P.A. & Browdy C.L. (1995) Effect of two feed protein levels and feed rate combinations on water quality and production of intensive shrimp ponds operated without water exchange. *Journal of the World Aquaculture Society*, **26** (1) 93–97.

Hoq M.E., Islam M.M. & Hossain M.M. (1996) Polyculture of freshwater prawn (*Macrobrachium rosenbergii*) with Chinese and Indian carps in farmer's pond. *Journal of Aquaculture in the Tropics,* **11**, 135–141.

Huner J.V. (1997) The capture and culture fisheries for North American crawfish. *World Aquaculture*, **28** (4) 44–50.

Huner J.V. (1999) The fate of the Louisiana soft-shell crawfish. *Aquaculture Magazine*, **25** (3) 46–51.

Huner J.V. (2001) *Procambarus*. In: *Biology of Freshwater Crayfish* (ed. D.M. Holdich) pp. 541–84. Blackwell Science, Oxford, UK.

Hunt J.W., Foster M.S., Nybakken J.W., Larson R.J. & Ebert E.E. (1995) Interactive effects of polyculture, feeding rate, and stocking density on growth of juvenile shellfish. *Journal of Shellfish Research*, **14** (1) 191–197.

IFC (1987) *Marine shrimp farming: A guide to feasibility study preparation*, 90 pp. Aquafood Business Associates, International Finance Corporation, P.O. Box 16190, Charleston, SC 29412, USA.

Issar G., Seidman E.R. & Samocha Z. (1987) Preliminary results of nursery and pond culture of *Penaeus semisulcatus* in Israel. *Israeli Journal of Aquaculture – Bamidgeh*, **39** (3) 63–74.

Jensen A. & Collins K. (1997) The use of artificial reefs in crustacean fisheries enhancement. In: *Proceedings of the First Conference of the European artificial reef research network*, Ancona, Italy, 26–30 March 1996 (ed. A.C. Jensen), pp. 115–121. Southampton Oceanography Centre, Southampton, UK.

Jensen A.C., Hamer B.A. & Wickins J.F. (1998) Ecological implications of developing coastal protection structures. In: *Coastlines, Structures and Breakwaters* (ed. W. Allsop), pp. 70–81. Thomas Telford, London.

Jensen A., Wickins J. & Bannister C. (2000) The potential use of artificial reefs to enhance lobster habitat. In: *Artificial Reefs in European Seas* (eds A.C. Jensen, K.J. Collins & A.P.M. Lockwood), pp. 379–401. Kluwer Academic, Netherlands.

Johnston W.E. & Botsford L.W. (1981) Systems analysis for lobster aquaculture. In: *Aquaculture in Heated Effluents and Recirculation Systems*, Vol. 2 (ed. K. Tiews), pp. 455–464. Heenemann Verlagsgesellschaft, Berlin.

Jones C.M. (1990) The biology and aquaculture potential of the tropical freshwater crayfish *Cherax quadricarinatus*, 109 pp. Information Series QI 90028. Queensland Department of Primary Industries, Queensland, Australia.

Jones C.M. & Grady J. (1997) Growout of redclaw at two demonstration sites in North Queensland. *Austasia Aquaculture Magazine*, **11** (2) 52–56.

Kagel K. (1998) Achieving sustainable high yields in a closed shrimp mariculture system. In: *Aquaculture and Water: fish culture, shellfish culture and water usage* (compiled by H. Grizel & P. Kestemont), pp. 126–127. European Aquaculture Society, Special Publication No. 26.

Khmeleva N.N., Kulesh V.F. & Guiguiniak Y.G. (1989) Growth potentialities of the giant tropical prawn, *Macrobrachium rosenbergii* (de Man), in waste-heat discharge waters of a thermoelectric power station. *Aquaculture*, **81** (2) 111–117.

Kittaka J., Henocque Y., Yamada K. & Tabata N. (1983) Experimental release of juvenile lobster at Koshiki Islands in south Japan. *Bulletin of the Japanese Society of Scientific Fisheries*, **49** (9) 1347–1354.

Kotha S.R. & Rouse D.B. (1997) Polyculture of red claw crayfish (*Cherax quadricarinatus*) with Nile tilapia (*Sarotherodon niloticus*) *Journal of Shellfish Research*, **16** (1) 318–319.

Kuthalingam M.D.K. (1990) Studies on the culture of spiny lobsters, *Panulirus homarus* and *P. versicolor* in marine floating cages with an account of lobster fishery resources of India. *Abstract from II congreso de ciencias del mar*, 18–22 Junio 1990, p. 170. La Habana, Cuba.

Kutty M.N., Herman F. & Le Menn H. (2000) Culture of other prawn species. In: *Freshwater Prawn Culture: the farming of* Macrobrachium rosenbergii (eds M.B. New & W.C. Valenti), pp. 393–410. Blackwell Science, Oxford, UK.

Latrouite D. (1998) The French experience with enhancement of European lobster *Homarus gammarus*. In: *Proceedings of a workshop on lobster stock enhancement held in the Magdalen Islands (Quebec)*. 29–31 October 1997 (ed. L. Gendron), Canadian Industry Report of Fisheries and Aquatic Sciences, **244**, 55–58.

Lawrence C., Morrissy N., Bellanger J. & Cheng Y.W. (1998) Final report, FRDC Project 94/075: Enhancement of yabby production from Western Australian farm dams. *Fisheries Research Report*, **112**, 1–134. Fisheries, Western Australia.

Lee D.O'C., Hillion B. & Letessier L. (2000) An analysis of factors influencing pond yields in a semi-intensive shrimp farm. In: *Abstracts, Aqua 2000, Responsible aquaculture in the new millennium* (compiled by R. Flos & L. Creswell), p. 379. European Aquaculture Society, Special Publication No. 28.

Lewis S.D. (2001) *Pacifastacus*. In: *Biology of Freshwater Crayfish* (ed. D.M. Holdich), pp. 511–40. Blackwell Science, Oxford, UK.

Li G., Shen Q. & Xu Z. (1993) Morphometric and biochemical genetic variation of the mitten crab, *Eriocheir*, in southern China. *Aquaculture*, **111** (1–4) 103–115.

Li K. (1998) Rice aquaculture systems in China: a case of rice-fish farming from protein crops to cash crops, 15 pp. In: *Integrated bio-systems in zero emissions applications* (eds E-L. Foo & T. Della Senta). Proceedings of the Internet Conference on Integrated Bio-Systems 1998. http://www.ias.unu.edu/proceedings/icibs

Liao I.C. & Chien Y.H. (1990) Evaluation and comparison

of culture practices for *Penaeus japonicus, P. penicillatus* and *P. chinensis* in Taiwan. In: *The Culture of Cold-tolerant Shrimp* (eds K.L. Main & W. Fulks), pp. 49–63. Oceanic Institute, Honolulu, HI, USA.

Lindberg W.J. (1997) Can science resolve the attraction-production issue? *Fisheries*, **22** (4) 10–13.

Lorec A. (1999) Can the conditions for a successful enhancement or sea ranching be defined? In: *Stock Enhancement and Sea Ranching* (eds B.R. Howell, E. Moksness & T. Svåsand), pp. 1–5. Fishing News Books, Oxford, UK.

Lorena A.S.H. (1986) The status of the *Procambarus clarkii* population in Spain. In: *Freshwater Crayfish 6* (ed. P. Brinck), pp. 131–133. International Association Astacology, Lund, Sweden.

Lovatelli A. (1990) *Regional Seafarming Resources Atlas*, 83 pp. FAO/UNDP Regional seafarming development and demonstration project, RAS/86/024. January 1990, FAO, Rome, Italy.

Lumare F., Amerio M., Arata P., *et al.* (1989) Semi-intensive culture of the kuruma shrimp *Penaeus japonicus* by fertilizer and feed applications in Italy. In: *Aquaculture – a biotechnology in progress* (eds N. De Pauw, E. Jaspers, H. Ackefors & N. Wilkins), pp. 401–407. European Aquaculture Society, Bredene, Belgium.

Lumare F., Di-Muro P., Tenderini L. & Zupo V. (1993) Experimental intensive culture of *Penaeus monodon* in the cold-temperate climate of the north-east coast of Italy (a fishery 'valle' of the River Po delta). *Aquaculture*, **113** (3) 231–241.

Lutz C.G., Caffey R.H. & Avault J.W. Jr. (1990) Production of soft-shell prawns *Macrobrachium rosenbergii* in commercial crawfish shedding facilities. *Abstracts from World Aquaculture 90*, p. 85. 10–14 June 1990, Halifax, Nova Scotia, Canada.

Macintosh D.J. (1982) Fisheries and aquaculture significance of mangrove swamps, with special reference to the Indo-West Pacific region. In: *Recent Advances in Aquaculture* (eds J.F. Muir & R.J. Roberts), pp. 3–85. Croom Helm, London.

Malecha S.R. (1983) Commercial pond production of the freshwater prawn *Macrobrachium rosenbergii* in Hawaii. In: *Handbook of Mariculture, Vol. 1 Crustacean aquaculture* (ed. J.P. McVey), pp. 231–260. CRC Press, Boca Raton, FL, USA.

Martinez-Cordova L.R., Porchas-Cornejo M.A., Villarreal-Colmenares H., Campaña-Torres A., Perez-Alvidrez L.A. & Naranjo-Paramo J. (1999) Alternate culture of white shrimp, *Litopenaeus vannamei*, and yellowleg shrimp, *Farfantepenaeus californiensis*, during the summer–fall and fall–winter season in Sonora, Mexico. *Journal of Applied Aquaculture*, **9** (3) 67–73.

McCoy H.D. II. (1986) Intensive culture systems past, present and future. Part 1, *Aquaculture Magazine*, **12** (6) 32–35.

McIntosh R.P. (1999) Changing paradigms in shrimp farming: 1. General description. *Global Aquaculture Advocate*, **2** (4–5) 42–47.

McNeil R. (2000) Zero exchange, aerobic, heterotrophic systems: key considerations. *Global Aquaculture Advocate*, **3** (3) 72–76.

Miller D.L. (1983) Shallow water mariculture of spiny lobster (*Panulirus argus*) in the Western Atlantic. In: *Proceedings of First International Conference on Warm Water Aquaculture – Crustacea.* 9–11 February 1983 (eds G.L. Rogers, R. Day & A. Lim), pp. 238–245. Brigham Young University, HI, USA.

Mills B.J., Morrissy N.M. & Huner J.V. (1994) Cultivation of freshwater crayfishes in Australia. In: *Freshwater Crayfish Aquaculture in North America, Europe, and Australia* (ed. J.V. Huner), pp. 217–289. Food Products Press. An imprint of The Haworth Press, Inc. New York, USA.

Moore D.W. & Brand C.W. (1993) The culture of marine shrimp in controlled environment superintensive systems. In: *Handbook of Mariculture, 2nd edn, Vol. 1 Crustacean aquaculture* (ed. J.P. McVey), pp. 315–348. CRC Press, Boca Raton, FL, USA.

Morrissy N.M., Bird C. & Cassells G. (1995) Density-dependent growth of cultured marron, *Cherax tenuimanus* (Smith 1912). In: *Freshwater Crayfish 10* (eds M.C. Geddes, D.R. Fielder & A.M.M. Richardson), pp. 560–568. Baton Rouge, Louisiana State University, USA.

Mosig J. (1999) Hothouse yabbies keep growing all winter. *Austasia Aquaculture Magazine*, **13** (5) 16–18.

Munro J.L. & Bell J.D. (1997) Enhancement of marine fisheries resources. *Reviews in Fisheries Science*, **5** (2) 185–222.

Muroga K., Suzuki K., Ishamaru K. & Nogami K. (1994) Vibriosis of swimming crab *Portunus trituberculatus*. *Journal of the World Aquaculture Society*, **25** (1) 50–54.

NACA (1986) *Giant freshwater prawn breeding and farming in Thailand: an introduction.* NACA TV Video production, FAO network of Aquaculture Centres in Asia.

Nagaraj M. & Neelakantan B. (1982) Prawn culture in the paddy fields of Uttara Kannada District, Karnataka. *Proceedings of Symposium on Coastal Aquaculture*, (1) 392–393. Marine Biological Association of India, Cochin, India.

New M.B. (1988a) Shrimp farming in other areas. In: *Shrimp '88, Conference proceedings*, 26–28 January 1988, Bangkok, Thailand, pp. 102–122. Infofish, Kuala Lumpur, Malaysia.

New M.B. (1988b) *Freshwater Prawns: status of global aquaculture, 1987.* NACA Technical Manual No. 6, 1–58. Network of Aquaculture Centres in Asia.

New M.B. (2000) Commercial freshwater prawn farming around the world. In: *Freshwater Prawn Culture: the farming of* Macrobrachium rosenbergii (eds M.B. New & W.C. Valenti), pp. 290–325. Blackwell Science, Oxford, UK.

New M.B., Singholka S. & Kutty M.N. (2000) Prawn capture fisheries and enhancement. In: *Freshwater Prawn Culture: the farming of* Macrobrachium rosenbergii (eds M.B. New & W.C. Valenti), pp. 411–428. Blackwell Science, Oxford, UK.

Oesterling M.J. & Provenzano A.J. (1985) Other crustacean species. In: *Crustacean and Mollusk Aquaculture in the United States* (eds J.V. Huner & E. Evan Brown), pp. 203–234. AVI Inc. Westport, CT, USA.

O'Sullivan D. (1990) Intensive freshwater crayfish system tested. *Austasia Aquaculture Magazine*, **5** (4) 3–5.

O'Sullivan D. & Kiley T. (1997) Prawn ponds operate with zero water exchange. *Austasia Aquaculture Magazine*, **11** (2) 4–6.

Palmegiano G. & Saroglia M.G. (1981) Winter shrimp culture

in thermal effluents. In: *Aquaculture in Heated Effluents and Recirculation Systems* (ed. K. Tiews), **2**, 297–302. Heenemann Verlagsgesellschaft, Berlin.
Paquotte P., Chim L., Martin J.-L.M., Lemos E., Stern M. & Tosta G. (1998) Intensive culture of shrimp *Penaeus vannamei* in floating cages: zootechnical, economic and environmental aspects. *Aquaculture*, **164** (1–4) 151–166.
Park B-H. (1990) The status of shrimp culture in Korea. In: *The Culture of Cold-tolerant Shrimp* (eds K.L. Main & W. Fulks), pp. 3–15. Oceanic Institute, Honolulu, HI, USA.
Pérez J.R., Carral J.M., Celada J.D., Sáez-Royuela M., Muñoz C. & Sierra A. (1997) Current status of astaciculture production and commercial situation of crayfish in Europe. *Aquaculture Europe*, **22** (1) 6–13.
Pickering H. (2000) Legal framework governing artificial reefs in the European Union. In: *Artificial Reefs in European Seas* (eds A.C. Jensen, K.J. Collins & A.P.M. Lockwood), pp. 469–487. Kluwer Academic, Netherlands.
Pinto G.F. & Rouse D.B. (1996) Growth and survival of the Australian red claw crayfish *Cherax quadricarinatus* at three densities in earthen ponds. *Journal of the World Aquaculture Society*, **27** (2) 187–193.
Reid B. & Arnold C.R. (1992) The intensive culture of the penaeid shrimp *Penaeus vannamei* Boone in a recirculating raceway system. *Journal of the World Aquaculture Society*, **23** (2) 146–153.
Rensel J.E. & Prentice E.F. (1979) Growth of juvenile spot prawn, *Pandalus platyceros*, in the laboratory and in net pens using different diets. *US Fish Wildlife Service Fisheries Bulletin*, **76** (4) 886–890.
Rhodes R.J. (2000) Economics and business management. In: *Freshwater Prawn Culture: the farming of* Macrobrachium rosenbergii (eds M.B. New & W.C. Valenti), pp. 369–392. Blackwell Science, Oxford, UK.
Rodriguez-Canché L.G., Maldonado-Montiel T.J. & Rosas C. (2000) Preliminary evaluation of *Artemia franciscana* as a biological filter of the wastewater of a shrimp culture. In: *Abstracts, Aqua 2000, Responsible aquaculture in the new millennium* (compiled by R. Flos & L. Creswell), p. 606. European Aquaculture Society, Special Publication No. 28.
Rosenberry R. (1990) Shrimp farming in the United States. *World Shrimp Farming*, **15** (11) 11.
Rothlisberg P.C., Preston N.P., Loneragan N.R., Die D.J. & Poiner I.R. (1999) Approaches to reseeding penaeid prawns. In: *Stock Enhancement and Sea Ranching* (eds B.R. Howell, E. Moksness & T. Svåsand), pp. 365–378. Fishing News Books, Oxford, UK.
Rouse D.B. (1995) Australian crayfish culture in the Americas. *Journal of Shellfish Research*, **14** (2) 569–572.
Sadek S. & Moreau J. (1996) Prawn (*Macrobrachium rosenbergii)* culture in earthen ponds in the Nile Delta, Egypt: culture parameters and cost benefits. *Israeli Journal of Aquaculture – Bamidgeh*, **48** (4) 201–218.
Sadek S. & Moreau J. (1998) Culture of *Macrobrachium rosenbergii* in monoculture and polyculture with *Oreochromis niloticus* in paddies in Egypt. *Israeli Journal of Aquaculture – Bamidgeh*, **50** (1) 33–42.
Sakurai N., Hagiwara K., Sugihara T. & Kittaka J. (1993) Experimental culture of the American lobster (*Homarus americanus*) utilizing warm waste water from a power station in Japan. In: *Programme and Abstracts, Fourth International Workshop on Lobster Biology and Management*, 25–31 July 1993, p. 52. Sanriku, Iwate, Japan.
Samocha T.M., Lawrence A.L., Bray W.A., *et al.* (1999) Production of marketable *Penaeus vannamei* in greenhouse-enclosed raceways in the Arizona desert using ground saline water. In: *Book of abstracts, World Aquaculture '99*, 26 April–2 May 1999, Sydney, Australia, p. 669. World Aquaculture Society, Baton Rouge, LA, USA.
Sandifer P.A., Hopkins J.S., Stokes A.D. & Browdy C. (1993) Preliminary comparisons of the native *Penaeus setiferus* and Pacific *P. vannamei* white shrimp for pond culture in South Carolina, USA. *Journal of the World Aquaculture Society*, **24** (3) 295–303.
Scott G., Perry W.G. Jr. & Avault J.W. Jr. (1988) Polyculture of the giant Malaysian prawn and the golden shiner in southwestern Louisiana. *Journal of the World Aquaculture Society*, **19** (3) 118–126.
Shaffer J.E. & Reiner J. (1994) Reefs as habitat restoration in an ultra-oligotrophic lake. *Bulletin of Marine Science*, **55** (2–3) 1352.
Shang Y.C. (1989) Marine shrimp farming in P.R. China. *Infofish International*, (2) 16–17.
Shanmugam A., Rajamanickam S. & Kannupandi T. (1995) Cage culture of *Penaeus indicus* in Vellar Estuary. *Journal of the Marine Biological Association of India*, **37** (1–2) 166–170.
Shivappa R.B. & Hambrey J.B. (1997) Tiger shrimp culture in freshwater? *Infofish International*, (4) 32–36.
Siddiqui A.Q., Al-Hafedh Y.S., Al-Harbi A.H. & Ali S.A. (1997) Effects of stocking density and monosex culture of freshwater prawn *Macrobrachium rosenbergii* on growth and production in concrete tanks in Saudi Arabia. *Journal of the World Aquaculture Society*, **28** (1) 106–112.
Smolowitz R. & Cimetti M. (1998) Fresh water polyculture of crayfish and salad greens. *Journal of Shellfish Research*, **17** (1) 362.
Smythe D.R. (1998) *New Zealand Clearwater Crayfish Ltd.* Paper presented at the New Zealand Marine Farming Association Inc. Annual Conference 1998. New species forum: freshwater crayfish (koura). 7 pp (mimeo).
Spanier E. (2000) Artificial reefs of the Mediterranean coast of Israel. In: *Artificial Reefs in European Seas* (eds A.C. Jensen, K.J. Collins & A.P.M. Lockwood), pp. 1–19. Kluwer Academic, Netherlands.
Staniford A.J. (1989) The effect of yield and price variability on the economic feasibility of freshwater crayfish *Cherax destructor* Clark (Decapoda: Parastacidae) production in Australia. *Aquaculture*, **81** (3–4) 225–235.
Su M-S. & Liao I.C. (1999) Research and development of prawn stock enhancement in Taiwan. In: *Stock Enhancement and Sea Ranching* (eds B.R. Howell, E. Moksness & T. Svåsand), pp. 379–392. Fishing News Books, Oxford, UK.
Syslo M. (1986) Getting the "bugs" out. *Massachusetts Wildlife*, **36** (3) 4–10.
Tabigoon L. Jr. (1998) Should you use cages and pens instead of ponds? *Asian Aquaculture*, **20** (5) 21–30.
Tidwell J.H. & D'Abramo L.R. (2000) Grow-out systems – culture in temperate zones. In: *Freshwater Prawn Culture: the farming of* Macrobrachium rosenbergii (eds M.B. New

& W.C. Valenti), pp. 177–186. Blackwell Science, Oxford, UK.

Valenti W.C. & New M.B. (2000) Grow-out systems – monoculture. In: *Freshwater Prawn Culture: the farming of* Macrobrachium rosenbergii (eds M.B. New & W.C. Valenti), pp. 157–176. Blackwell Science, Oxford, UK.

Van der Meeren G.I., Agnalt A-L. & Jørstad K.E. (1998) Lobster (*Homarus gammarus*) stock enhancement in Norway, experiences from a large scale release project, from 1990 to 1997. In *Proceedings of a workshop on lobster stock enhancement held in the Magdalen Islands (Quebec).* 29–31 October 1997 (ed. L. Gendron), Canadian Industry Report of Fisheries and Aquatic Sciences, **244**, 63–68.

Van Olst J.C., Carlberg J.M. & Ford R.F. (1977) A description of intensive culture systems for the American lobster (*Homarus gammarus*) and other cannibalistic crustaceans. *Proceedings of the Eighth Annual Workshop of the World Mariculture Society*, 9–13 January 1977, pp. 271–292. San Jose, Costa Rica.

Viacava M. (1995) Feeder trays for commercial shrimp farming in Peru. *World Aquaculture*, **26** (2) 11–17.

Waddy S.L. (1988) Farming the Homarid lobsters: state of the art. *World Aquaculture*, **19** (4) 63–71.

Wang H. (1994) Research on polycultivating seaweeds together with shrimps. 1. Experiments of polycultivating *Gracilaria* together with *Metapenaeus* and *Scylla* in fishpond. *Trans. Oceanol. Limnol. Haiyang Huzhao Tongbao*, (3) 52–59 [in Chinese, English abstract seen.]

Wang J-K. (1998) An integrated recirculating system for the production of oysters and shrimp. In: *Proceedings of the Second International Conference on Recirculating Aquaculture*, 16–19 July 1998, pp. 385–386. Virginia Tech, VA, USA.

Wang J-Q., Li D., Dong S., Wang K. & Tian X. (1998) Experimental studies on polyculture in closed shrimp ponds. I. Intensive polyculture of Chinese shrimp (*Penaeus chinensis*) with tilapia hybrids. *Aquaculture*, **163** (1–2) 11–27.

Webster D. (1998) Soft crabs and recirculating systems. *Aquaculture Magazine*, **24** (2) 23–24.

Welcomme R.L. & Bartley D.M. (1998) Current approaches to the enhancement of fisheries. *Fisheries Management and Ecology*, **5**, 351–382.

Wickins J.F. (1982) Opportunities for farming crustaceans in western temperate regions. In: *Recent Advances in Aquaculture* (eds J.F. Muir & R.J. Roberts), pp. 87–177. Croom Helm, London.

Wickins J.F. (1995) Experimental studies of lobster habitat requirements in reef structures. In: *Ecoset '95 International conference on ecological system enhancement technology for aquatic environments. The Sixth International Conference on aquatic habitat enhancement.* 29 October–2 November 1995, **2**: 652–657. Tokyo, Japan.

Wickins J.F. (1997) Strategies for lobster cultivation. *CEFAS Shellfish News*, (4) 6–10.

Wyban J.A. & Sweeney J.N. (1993) Intensive shrimp production in round ponds. In: *Handbook of Mariculture,* 2nd edn, *Vol. 1 Crustacean aquaculture* (ed. J.P. McVey), pp. 275–287. CRC Press, Boca Raton, FL, USA.

Xu J., Xia M., Ning X. & Mathews C.P. (1997) Stocking, enhancement, and mariculture of *Penaeus orientalis* and other species in Shanghai and Zhejiang Provinces, China. *Marine Fisheries Review*, **59** (4) 8–14.

Yano I. (1993) Ultraintensive culture and maturation in captivity of penaeid shrimp. In: *Handbook of Mariculture,* 2nd edn, *Vol. 1 Crustacean aquaculture* (ed. J.P. McVey), pp. 289–313. CRC Press, Boca Raton, FL, USA.

York R.H. (1983) Economics of pond polyculture of clams (*Mercenaria mercenaria*) and shrimp (*Penaeus* spp.) in Hawaii. In: *Proceedings of First International Conference on warm water aquaculture – Crustacea* (eds G.L. Rogers, R. Day & A. Lim), pp. 43–45. Brigham Young University, HI, USA.

Zhang N. (1990) On the growth of cultured *Penaeus chinensis* (Osbeck). In: *The Culture of Cold-tolerant Shrimp* (eds K.L. Main & W. Fulks), pp. 97–102. Oceanic Institute, Honolulu, HI, USA.

Zimmermann S. & New M.B. (2000) Grow-out systems – polyculture and integrated culture. In: *Freshwater Prawn Culture: the farming of* Macrobrachium rosenbergii (eds M.B. New & W.C. Valenti), pp. 187–202. Blackwell Science, Oxford, UK.

Zuñiga-Romero O., Ramos-Diaz R. & Reveulta-Duran J. (1987) Policultivo de camaron de rio (*Cryphiops caementarius*) y lisa (*Mugil cephalus*) en estanques. *Estud. Oceanol. Inst. Invest. Oceanol.* Univ. Antofagasta, (6) 67–77.

Chapter 6
Site Selection

6.1 Introduction

Siting decisions have a critical influence on the success of a crustacean farming project because they play a large part in determining potential yield levels and greatly affect the costs of construction and operation. Evidence of their importance is provided by the many abandoned projects in which poor siting has been identified as either the principal reason or a major contributory factor for failure. Of course no site can be optimal in all respects, but it is important to realise that some siting requirements are fundamental to success, and serious deficiencies with respect to these cannot be compensated by any amount of other favourable site characteristics (Muir & Kapetsky 1988). Muir and Lombardi (2000) describe a practical approach to the selection of a site for a freshwater prawn farm and note that it usually boils down to a process of balancing favourable and less favourable characteristics and deciding if the overall mix is suitable.

In this chapter the factors that govern site selection have been arranged, firstly into those that operate at the level of countries or regions, and secondly into those that have much more to do with the specific locality and proposed ground site for the operation. Though both types of factor are equally critical to successful site selection, this distinction is useful because if a country or region is unsuitable it can be omitted from the selection process without the need to identify possible sites within it.

Ideally, in the pre-investment stages of a crustacean project, the process of site selection will follow an appraisal of which species to produce, which techniques and phases of culture to perform, and what represents an economically viable scale of operation (Chapter 9). Although this is the most logical sequence for planning purposes, in some cases interest in the potential of crustacean aquaculture arises from possession of a particular piece of land and the wish to use it more productively. In these cases no site selection is performed, rather a species and the approach to its culture are chosen to fit the site's attributes. Crustacean production may be just one of several possibilities under consideration that, apart from other forms of aquaculture, may include projects involving agriculture or leisure amenities. In any circumstance it is certainly worth considering what value a site may have for alternative uses should crustacean farming prove unprofitable.

Some of the most critical aspects of site selection relate to climate, water supply, soil type, the availability of broodstock and seedstock, and the proximity of markets. In this chapter these are considered along with a whole range of other factors that, depending on the project being planned, are likely to have an impact on overall profitability. For any particular species, investigating the disease status of a country or region is becoming increasingly important (section 6.3.7). Countries are applying stricter quarantine measures and new regulations regarding the movement of live animals to limit the spread of diseases, but in the shrimp industry only a few parts of the world have escaped viral epidemics. Similarly it is also advisable to ensure adequate controls over key environmental factors (e.g. water abstraction and effluent discharge regulations) are in place to avoid having the new project constrained or polluted by existing enterprises.

6.2 Country or region

Site selection factors that can be conveniently taken into account at the level of countries and regions include: seasonal temperature range, the availability of essential inputs (broodstock, seedstock, labour, materials and services), the size and type of market to be exploited, and various political, institutional and legal considerations.

6.2.1 Climate

Knowledge of the usual weather patterns of a country and the possible diurnal, seasonal and annual extremes is essential to the site selection process. Armed with accurate and detailed information, it is possible to assess whether conditions are suitable for continuous year-round operations, whether some phases of culture will have to be restricted to certain months of the year, or if some environmental control will be required. In the light of present climate perturbations, the vulnerability of the site and access routes to floods and storms should also be assessed (section 6.2.1.6).

The climate will not only influence production but also the timing and duration of construction work, since at some sites earthworks can only be performed during the dry season. Any delays in generating revenue from production can strongly influence the financial viability of a project (section 10.4.1.1).

6.2.1.1 Temperature

Temperature is a fundamental determinant of crustacean growth rates and can only be cost-effectively managed through site selection. If it falls outside optimum ranges (Table 8.3; Appendix 1) for significant periods, production potential will be impaired.

Both mean temperatures and likely extremes must be considered. For example, although the mean annual air temperature in Kuwait is around 25.7°C, within the optimal range for the culture of most penaeid shrimp, it varies between a mean of 37.4°C in July and 12.7°C in January, both well outside ideal limits. In addition, the difference between night and day temperatures averages 14°C, greatly reducing the potential viability of outdoor culture operations (Farmer 1981). As a general rule, inland regions and those with little maritime influence on their climate will experience the greatest seasonal and diurnal fluctuations in temperatures. Water does not heat up or cool down immediately in response to changes in air temperatures and this provides some protection against daily extremes and short-lived spells of unusually hot or cold weather. This buffering effect is only significant however in embayments, deep ponds, lakes and some lagoons.

6.2.1.2 Rainfall

The quantity and seasonal patterns of rainfall should be studied for proposed aquaculture sites. Heavy rainfall, for example during monsoons, can cause severe flooding and embankment erosion and hinder farm access by turning earthen tracks to mud and by inundating waterways. In 1988 in southern Thailand, 1000 shrimp farms suffered stock losses and severe damage when five days of heavy rain caused serious flooding. Dry seasons can also have an impact on farming operations. In Central America and along the Caribbean coast of Colombia the dry season is associated with depressed growth in shrimp farms. This effect is not fully understood but occurs despite near-optimal salinities and temperatures in ponds. One explanation links low growth to reduced natural productivity in the water (Scura 1995).

The salinity levels in marine and brackish water ponds will also be strongly influenced by rainfall, or a lack of it (section 6.3.1.4).

6.2.1.3 Wind

Air movement over outdoor ponds promotes gaseous exchange between the atmosphere and the water and generates surface currents that destratify the water column. Although this effect is desirable, wind, particularly in combination with low humidity and high temperature, can cause significant evaporation in ponds that raises the salinity of brackish water and thereby increases requirements for water renewal. On the other hand, calm periods accompanied by high temperatures or heavy rainfall can induce thermal or salinity stratification of the water, and prevent oxygen from reaching the pond depths. If these periods are frequent in a particular region, then requirements for aeration, water circulation or water exchange will be increased. At coastal sites on sunny days, sea breezes are usually generated as a result of the differential heating and cooling rates of land and water. These breezes can ameliorate the sun's heating effects. In regions with very strong winds, wave damage to pond embankments can be severe unless sheltered sites can be located (section 6.3.2.3).

Wind-blown dust can also present problems for aquaculture in some desert regions. Outdoor tanks in Kuwait trapped on average 11.6 g of dust m^{-2} per month and as much as 2.43 g m^{-2} per day during a particularly severe storm (Farmer 1981). Even low levels of dust (0.02 g L^{-1}) suspended in larvae cultures have been found to cause heavy mortalities (Al-Hajj *et al.* 1981).

6.2.1.4 Evaporation and humidity

Evaporation is most significant when low-humidity levels are combined with wind and elevated temperatures. In arid regions at exposed sites mean evaporation rates can exceed 15 mm per day.

The water supply for an aquaculture operation must replace evaporative losses, and for marine and brackish water permit some control over salinity levels. For the latter, particular problems may arise during hot, dry seasons when the salinity levels rise both in ponds and in the incoming water supply, hampering attempts at salinity control through water renewal (section 6.3.1.4).

Very humid conditions result in reduced pond evaporation but make indoor work environments very unpleasant. They may necessitate air-conditioning in offices, laboratories and staff accommodation, and lead to increased energy costs.

6.2.1.5 Insolation (sunshine)

Insolation levels are determined largely by cloud cover and latitude. At high latitudes intensities will be less than at tropical sites, but total insolation will greatly increase in spring and summer with the longer days.

Insolation is a very important factor in the calculation of energy budgets in controlled-environment and greenhouse systems. In unheated structures it should be sufficient to raise temperatures and promote growth in early spring. However, in all greenhouses in the summer insolation can cause over-heating unless adequate ventilation and shading is provided.

For outdoor algae cultures made in support of hatcheries, very strong sunlight may inhibit growth and raise temperatures excessively, necessitating the use of shade cloth. Shade cloth is also required in some nursery operations and for transit/holding tanks, where it prevents algal growth as well as reducing temperatures.

6.2.1.6 Climate change

It is worthwhile considering the likely impacts of climate change on an aquaculture venture (sections 1.4 and 11.4.4). When sea levels rise, coastal and estuarine sites will be exposed to increased risk of wave damage and flooding, particularly on extreme tides associated with storm surges. Also, where rainfall increases, it will cause more flooding of rivers and surrounding areas and thereby lead to increased embankment erosion and greater levels of suspended solids in water. Silt that accumulates in ponds and canals demands periodic removal. Conversely, absence of rain at crucial times can upset behaviour patterns and affect productivity. For example, rainfall during September to December is essential if Louisiana crayfish are to emerge from their burrows with their young; drought at this time seriously impairs recruitment of young to the ponds (J.V. Huner, 2001 pers. comm.).

6.2.2 Availability and costs of essential inputs

The costs and availability of essential inputs will have a direct bearing on the economics and ease of operating a crustacean farm or hatchery. Although it is possible to transport men and materials almost anywhere in the world, it makes economic sense to set up operations where essential resources are readily available and inexpensive, or at least competitively priced.

6.2.2.1 Broodstock and seedstock

Broodstock and/or seedstock are essential to crustacean aquaculture and much attention should be paid to locating reliable supplies. This can be done either by choosing a site where wild stocks are available or alternatively by working with species or strains that will breed readily in captivity and give some independence from wild sources (Chapter 4).

Although techniques exist for shipping live material over long distances, there are advantages to having a local supply. Not least of these is the simple fact that local supplies are usually cheaper than imports and, because broodstock and seedstock costs are often significant components of operating budgets, even modest savings can influence profitability. Although natural stocks may only be available for certain periods each year, when they do occur they may be abundant and therefore cheap. Production schedules can often be arranged to capitalise on patterns of seasonal abundance.

A conveniently located wild stock can support a culture industry at more than one level of operation by providing both wild-caught juveniles for farms and broodstock for hatcheries. The bulk of the Ecuadorian shrimp farming industry has relied upon supplies of wild *Litopenaeus vannamei* seed. When these supplies became limiting to production in the mid-1980s, local supplies of broodstock were utilised for the production of postlarvae in hatcheries. Once hatcheries become established they can underpin a whole shrimp farming industry. For example, Thailand, the largest producer of farmed shrimp, relies on hatchery output of around

Plate 6.1 Semilleros of Ecuador using scissor push nets to catch wild post-larval shrimp.

15×10^9 *Penaeus monodon* post-larvae per year (Kongkeo 1995). The convenience of this situation can be contrasted with that of farms in Gambia (Skabo 1988) and Guinea, which, faced with the absence of a locally available fast growing penaeid species, were forced to go to the expense of importing *P. monodon* broodstock from Asia and *Litopenaeus vannamei* broodstock from Venezuela.

The situation for freshwater prawns and crayfish is somewhat different from that for the most important penaeid species: when they become established in farms, ample supplies of egg-bearing broodstock can usually be obtained from production or holding ponds. This provides independence from wild stocks and has given great impetus to the spread of *Macrobrachium rosenbergii* farming throughout the world; it has also led to the establishment of populations of several crayfish species outside their natural range.

Shrimp production in countries that rely on introduced species has grown slowly in comparison with production from countries where suitable cultivable species occur in the wild. Output in Brazil, for example, originally relied on the slow-growing indigenous shrimp *Litopenaeus schmitti* and *Farfantepenaeus brasiliensis,* but the desire to obtain better yields led to the introduction of *Litopenaeus vannamei*, *Penaeus monodon* and *Marsupenaeus japonicus*. Despite these measures and the presence of good sites and a favourable climate, shrimp production in Brazil initially increased only slowly. At the end of the 1980s it was producing around 2000 mt per year compared to 73 000 mt per year in Ecuador.

The comparison between Brazil and Ecuador in the 1980s illustrates the advantages of convenient wild stocks but it should be recognised that the situation has changed greatly during the 1990s. While Ecuador has continued to rely on wild stocks, countries like Brazil and Venezuela (Clifford 1997) have (of necessity) made progress with the domestication of *Litopenaeus vannamei* and *L. stylirostris* and are starting to reap benefits through the selection of favourable traits governing growth, survival and reproductive performance under farm and hatchery conditions. Thus while Ecuador's shrimp output has largely stagnated, Brazil's has risen more than ten-fold to 25 000 mt per year (Rosenberry 2000) and looks set to continue expanding. The absence of native stocks, which was once seen as an impediment to the industry, is proving indirectly to be a source of advantage. Captive stocks also give significant benefits in terms of health management (section 9.7.2). Animals can be selected for disease resistance traits (Persyn 1999) or kept in isolated sites and screened to provide supplies of SPF broodstock, nauplii and post-larvae. Currently Brazil's captive shrimp stocks are free of white spot syndrome virus but the same disease is widespread in Ecuador.

If a particular crustacean farming operation is to rely on the shipping of broodstock or seedstock over long distances, several disadvantages should be borne in mind. Firstly, live transport necessitates crowding for extended periods, and stress and mortality of the animals is likely to result. Secondly, obtaining supplies from distant sources often allows fewer opportunities for the buyer to inspect the quality of the purchase. This can be especially problematical if prepayment is demanded and a shipment subsequently arriving in poor condition cannot be returned to the supplier. Thirdly, imported stocks can have a significant ecological impact (sections 11.3.2 and 11.3.3) and may be subject to complex regulations regarding movements of live animals (causing delays at airports). Sometimes shrimp broodstock and seedstock are smuggled out of countries where legislation exists specifically to prevent the export of resources needed for domestic farming activities. Such supplies are usually sold at inflated prices and would become vulnerable if the relevant anti-contraband laws were strictly enforced. Distance is certainly a disadvantage in the face of competition for declining stocks.

Any factor likely to jeopardise sources of broodstock or seedstock should be investigated wherever large-scale projects are proposed. Sometimes fishing regulations specify periods each year when wild stocks cannot be fished, and sometimes complete bans operate to protect wild stocks (e.g. white-clawed crayfish in Great Britain and koura in New Zealand; sections 7.6.1 and 7.7.11). Capture of some species of spiny lobster for ongrowing may be allowed, with conditions, under licence (section 7.9.4). In the case of shrimp farming, if a fishery for wild post-larvae does not exist it can take time to establish one. It may require the training of seed fishermen in capture and handling techniques, and some research to establish which species are available and what is the pattern of seasonal abundance (section 7.2.6.1). Ideally, governments that wish to promote shrimp farming should support investigations into the availability of wild seed. However, resistance to the fishing of both post-larvae and broodstock may come from established fishing interests concerned about real or perceived detrimental impacts on wild stocks and competition from aquaculture (section 11.3.1). Indeed, confrontations occurred in Mexico following the capture of increasing numbers of shrimp post-larvae for aquaculture (P. Wood, 1989 pers. comm.).

When the best places to catch mature adults of the preferred crustacean species do not coincide with the most productive fishing grounds, special and often costly arrangements may be necessary. For example, where fishermen are not already engaged in supplying a live trade or broodstock for shrimp hatcheries, it may be necessary to go to the expense of periodically chartering a trawler.

6.2.2.2 Feeds and feed raw materials

Large quantities of feed are needed for semi-intensive and intensive culture operations, and they represent a major production expense. Feed of suitable quality must be readily available at a price that does not significantly narrow profit margins. Specialised crustacean diets are manufactured in countries where significant farming operations are established – there are for example 26 feed mills in Ecuador (Griffith & Schwartz 1999) – but outside these locations suitable diets must be imported. In West Africa, for example, the absence of a shrimp feed manufacturer led one farming operation to seek alternative sources in Europe, the USA and Asia (Skabo 1988). Extra costs for shipping and storage can be expected with all bulk imports of feed.

A feed mill to manufacture diets, at or close to the farm, may be set up either to support the requirements of the particular project or as a larger, independent commercial operation serving many customers. In either case, raw materials such as fishmeal, wheat and soybeans will be required, probably together with binding agents and vitamin and mineral mixes. Some or all of these materials may be unavailable locally, creating dependence on imports once more. By-products from animal processing operations can be used, sometimes including shrimp head meal, and if trash fish (especially marine species) is available this can be a valuable ingredient in simply prepared moist diets. Such diets however cannot be stored easily, requiring new batches to be prepared every few days, may only be suitable for small farms and may even harbour pathogenic viruses especially if crustacean offal is a component. For freshwater prawns it is possible to use diets made for pigs or poultry, though these are not designed to be stable in water and they disintegrate rapidly (section 8.8.2).

Diets for use in hatcheries are mostly produced in Japan, the USA and Europe but because they are used in relatively small volumes they can be easily shipped to other parts of the world at little extra expense (section 8.8.1). Some companies even offer their products at a fixed basic price all over the world. *Artemia* cysts, mostly produced or processed and packed in the USA, are also easy to transport and have long storage lives (up to 5 years). Many countries harvest *Artemia* but care should be taken with cysts from unrecognised sources because hatch rates, nutritional value and levels of contamination with pesticides vary greatly between batches and sources (Dhont *et al.* 1993).

6.2.2.3 Fertiliser and lime

Supplies and prices of inorganic fertilisers (section 8.3.6.2) should be easy to find because these products are widely used in agriculture. Organic fertilisers, which are mostly by-products of poultry and livestock production, are bulky, and if sources are distant transport costs may be prohibitive. Liming materials are often widely available because of their use in farming and the manufacture of cement (section 8.3.3).

6.2.2.4 Energy

Many important factors relating to energy supplies are dependent on the locality chosen (section 6.3.5). At the level of a country or region it may be important to investigate the overall national energy policy. Sometimes electricity prices for industry are subsidised to promote regional development. Some oil exporting countries provide fuel at subsidised prices that can make the installation and operation of generators a competitive alternative to reliance on mains electricity. Countries with a well-developed electrical network can provide a more reliable supply and, of course, the national voltage must be considered when electrical equipment is ordered.

6.2.2.5 Staff

Total labour costs are greater than simply the sum of wages and salaries. In most countries detailed legislation covers subjects such as overtime pay, national holidays, social benefits, health-care benefits, seasonal bonuses and company liabilities if permanent staff are dismissed. The full impact of these regulations on labour costs should be investigated and understood. Labour rates for construction workers will have an impact on the overall cost of setting up a project.

Customs relating to work and the differing roles of men and women may be significant. For instance, in Moslem and Christian communities the workload may need to be reduced during the periods of Ramadan or Christmas. In processing plants women often make up large part of the workforce, but this may not be considered acceptable in certain societies. Some other aspects of labour in relation to locality are discussed in section 6.3.6.

Suitably qualified technical staff can usually be more easily obtained in regions where crustacean aquaculture is established. People without suitable 'hands-on' experience, but with some technical background and an education in science, can be found in most developed and some developing countries and can be trained accordingly. Requirements for well-trained and experienced technical staff are at their greatest in more intensive projects.

People with good managerial skills, biological knowledge and experience immediately relevant to crustacean farming are often needed to run farms and hatcheries, but are in short supply worldwide. Although this problem is somewhat independent of location, it is worth considering that very remote regions can result in demands for higher salaries and deter the long-term involvement of expatriate staff (section 11.2.6).

6.2.2.6 Construction materials and engineering services

The availability of essential building materials should be investigated and unit costs calculated to include allowances for delivery. Materials needed may include timber, cement, sand, stones, aggregate, bricks, building blocks, steel reinforcement bars, steel roof supports, and roofing materials such as transparent corrugated sheets. For materials such as sand and aggregate, the main part of the cost is usually transport, so full costs cannot be established until possible sites are located.

Earth-moving is the main expense in the construction of ponds, so the likely cost per cubic metre should be investigated. Unless ponds are to be built by hand, bulldozers, graders, compactors and excavators may be needed. These can be easily located in most countries because of their use in road building.

A considerable quantity of plumbing materials, including plastic valves and PVC piping will be needed for most hatcheries. Decisions on what to use for making hatchery tanks should be influenced by the availability of local materials and the possibility of using prefabricated tanks of plastic or GRP. Some plastics may need to be imported and can be very expensive. The suitability of different materials is discussed in section 8.1. In one alternative building technique, inexpensive tanks can be created using concrete reinforced with chicken wire.

The availability of professional services, for example for quantity surveying, site engineering and contracting, well drilling and electrical installation, will probably need to be checked (section 9.3.6).

6.2.2.7 Equipment

Reliance on imported equipment can be high in some

developing nations, especially if a 'high-tech' turnkey package is acquired. This can create dependence on outside supplies of spare parts, which may result in high maintenance costs and extended periods during which equipment is under repair. It is therefore often worth investigating locally made alternatives before resorting to imports. Often development aid agencies specify equipment sources to ensure quality, even though this results in increased dependence on imports. Examples of the kind of equipment needed to establish a shrimp hatchery and a shrimp or prawn farm are given in Tables 10.5 and 10.8 respectively.

In some countries imports of various chemicals and therapeutants may be essential, particularly in order to maintain a hatchery operation. If algae are to be grown from stock cultures a series of high-grade nutrients will be required, and sometimes compressed carbon dioxide. For live transport compressed oxygen may be needed, though this is generally widely available because of its use in hospitals and for welding torches.

6.2.2.8 Technical services and support

In addition to those professional and engineering services mentioned above, others that may be needed include legal assistance, specialist plumbing, and pond cleaning and harvesting services. Obviously the requirement for and availability of each of these will vary from one location to the next. Technical support is especially difficult to obtain in some countries but it can be very important when problems arise during production (section 9.3.6). It is often provided by extension services, disease specialists and laboratory services which can perform water quality and nutritional analyses, albeit at a cost. Some feed companies take on the role of extension agents in return for purchase of their feeds (section 11.5.2).

6.2.3 Markets

The identification of the market to be exploited (Chapter 3) is very important to site selection. Although much crustacean farming relies on distant, bulk-frozen markets, in which proximity is not essential, when live and fresh crustaceans are aimed at local specialist markets there are usually cost and quality benefits associated with basing production near to the consumer.

Frozen crustacean products like whole and headless shrimp are traded worldwide and the best locations to farm them are not simply determined by the fact that most consumption is centred in Japan, the USA and Europe. For example, some shrimp farming businesses that are targeting the North American market originally started their operations within the USA but have now relocated to Central or South America to take advantage of more favourable climates and cheaper production costs. The savings easily outweigh any extra shipping costs. Apart from penaeids, other farmed crustaceans that are shipped over long distances include freshwater prawns from China and Bangladesh and crayfish from China, sold in Europe and the USA.

If live or fresh products are to reach the consumer before their condition and value deteriorates, the duration of transport should be just a matter of hours. Proximity to the market, or at least very efficient transport links, is essential. When limited small and local markets are targeted – for example for live crabs in Asia, for live shrimps in Japan, or for sales of *Macrobrachium* and crayfish either at the farm gate or direct to the catering trade – attention should be given to finding sites within reach of population centres or tourist areas. In southern Spain many crayfish farmers have been able to take advantage of the large numbers of holidaymakers arriving each summer.

Knowledge of regional and local market conditions can also be significant to the site selection process. Often consumption rates vary greatly with season and sometimes religious taboos are significant and can severely limit local marketing opportunities. In Israel, shrimps and prawns are not considered 'kosher' and are either exported or consumed by tourists.

6.2.4 Processing facilities

Processing requirements may be minimal if live or fresh products are sold on local markets, but processing facilities for washing, deheading, freezing, packing or canning will usually be required for bulk and export markets.

Farms should be located so that only short journeys are required for perishable products to reach the processor, and the availability of processors and the quality of their installations should be investigated to ensure they meet acceptable standards of hygiene and quality assurance. Reputations for quality vary between plants and countries and will influence the prices that can be obtained. Not all processing facilities are capable of producing the range of high-quality and value-added products increasingly demanded by consumers, and this shortcoming has been identified as a likely constraint for some proposed crustacean farming operations.

Processing plants often exist to handle wild-caught products and they can easily deal with additional yields from aquaculture. As part of their service some supply trucks and ice during a harvest, provided that the farm is not too distant and has good road access. In other situations a source of ice may be needed. Ice suppliers are often easy to find in coastal areas where fishing is an important activity, but they may be rare inland.

6.2.5 Political, institutional and legal factors

6.2.5.1 Civil stability

Some countries or regions are troubled by guerilla or terrorist activity, military coups, separatist conflicts and internal security problems related to drug production, drug smuggling and resulting violent crime. Such problems have an impact on a country's overall economic and investment climate. Some conflicts have impinged directly on crustacean farming enterprises. For example, a massacre of Tamil workers by military personnel took place on one Sri Lankan shrimp farm and at least one shrimp farm in Mindanao, Philippines, has relied on access by helicopter because of guerilla activity in surrounding areas. Remote farms producing valuable crops in countries where security cannot be guaranteed are certainly easy targets for extortion or payroll robberies by armed gangs.

While potential dangers should not be underestimated, knowledge of the actual conditions in a country cannot be gained simply from the coverage of events in international news media. These often give impressions of widespread turbulence in countries that are for the most part stable and peaceful except for occasional or localised violent incidents that attract disproportionate attention. A more complete picture of the security situation of a country can be obtained by speaking to knowledgeable residents or experienced travellers who can make judgements based on first-hand experience.

Also of great importance to any proposed business is a measure of political and economic stability. Commenting on the situation in some African states, Skabo (1988) noted that high inflation rates, depreciation of the local currencies, coupled with monetary and political instability make long-term investments unattractive, in spite of the projected returns. Problems arose for shrimp farms in Panama during the political crisis that culminated in the US invasion of 1990 (Anon. 1999). For any project that operates in an unpredictable economic environment, financial planning can be a very complicated process.

6.2.5.2 Taxes and duties

Tax regulations differ widely between countries and always have an important impact on profitability. Of particular interest to investors will be levels of corporation tax, the availability of tax holidays and the possibility of tax concessions or tax write-offs on research and development projects.

Levels of import duties need to be known in order to calculate the true costs of imports. In many countries the relevant legislation is complex, with differing levels of duty applying to materials destined for different purposes. For instance, while medical supplies may be duty-free, some items of equipment destined for industry may attract duties of up to 50% or more. Sometimes very high duty is applied to discourage the importation of goods such as motor vehicles that are assembled or produced by a home-based industry. Often high rates extend to materials for which no real equivalent product is locally available. Nevertheless, when a strong case is made for the value of crustacean farming to a national economy, special rates of import duty may be introduced. In Ecuador, when shrimp farms were faced with acute shortages of wild shrimp post-larvae, all import duties on equipment destined for the construction of shrimp hatcheries were waived. The importance of positive government attitudes in general is discussed in section 11.5.3.3.

6.2.5.3 Exchange controls

In an attempt to regulate the flow of money in and out of some countries, currency exchange rates are maintained at artificial levels. These can delay and complicate the processes of importing and exporting and moving capital in and out of a country. Sometimes special regulations exist specifically to limit the repatriation of profits. One result of such measures is that two rates of exchange exist, one for free (or in some cases 'black') market exchange and another based on control by a central bank. Unpredictable movements in exchange rates can greatly confuse financial planning, especially for an export-based or foreign-funded operation.

6.2.5.4 Land costs and concessions

Many questions may need to be answered regarding the availability and cost of land. For example: can low-cost leases be obtained from government or regional authorities, or will land need to be purchased? In some countries the ability to buy land is restricted to nationals. Sometimes a government will provide land for a project and take a portion of the equity in return. Obtaining large tracts of contiguous land can be a particular problem if areas are divided into many small parcels, each with a different owner (section 11.5.3.1).

Often regulations exist which specify that certain zones be used preferentially for industry or tourism, or be reserved for wildlife. Even if sites are specifically designated as suitable for aquaculture development, the full site selection process should not be by-passed because 'suitable for aquaculture' can often be translated merely as meaning 'unsuitable for anything else so why not try aquaculture?' (section 11.4.1).

6.2.5.5 Availability of loans and grants

The sources and nature of finance need to be investigated, together with the availability of investment incentives (section 10.2). Projects likely to meet identified goals for economic development may qualify for grants or loans on especially favourable terms. The form of financial assistance can vary with respect to interest rates, the duration of grace periods before repayments are due, the proportion of total project cost that can be met by the loan or grant, and whether money is available to cover all project expenses or just capital investment in facilities and infrastructure. There may be a minimum amount of capital assistance that can be provided, and unfortunately when this level is set too high it renders the credit unsuitable for small-scale operations and only benefits large schemes. For smaller operations credit extension schemes operating at an appropriate scale may need to be located (section 11.2).

The interest rates payable on bank overdrafts should be taken into account, particularly for small operations in which cash may be needed at short notice to cover pre-operating and operating expenses.

6.2.5.6 Traditions

It is more problematic to start an aquaculture operation in regions like Africa, where the activity is largely unknown, than it is in areas such as South-east Asia where aquaculture has been practised for centuries. A lack of tradition and technology in some countries can lead to strong feelings of isolation for aquaculture pioneers. When crustacean farming does become established, trade organisations are often formed which can promote the exchange of information and serve as a lobbying force, for example in promoting responsible farming practices (section 11.1). In Thailand these organisations have included the Surat Thani Shrimp Farmers Association and the Thai Marine Shrimp Farmers Association, both of which have had a positive influence on the shaping of the aquaculture industry (Polioudakis 2000).

Local customs can affect aquaculture projects in many ways and need to be investigated during the site selection process. Strict vegetarian communities, anti-aquaculture lobbies and traditional multiple land ownership can all exert significant influence (section 11.2.5).

6.2.5.7 Legal requirements

Legal requirements affecting aquaculture are often complex and stringent and in some regions the bureaucratic 'red tape' they generate stifles interest in new projects. Many laws are simply inappropriate because they were originally drafted for industry, fisheries or agriculture (section 11.5.3). Areas where legal requirements are likely to affect crustacean farming projects include:

Land ownership: Laws may differ for foreigners and for land with sea frontage;
Water usage: Laws are often particularly strict with respect to groundwater abstraction and permits may be needed;
Water discharge: In some countries farm effluents must comply to set standards;
Introduction of new species: See transplantations (section 11.3.2);
Joint ventures: Equity limits may operate for foreigners (section 10.2.3);
Construction: Permits will be needed and many planning regulations usually apply (section 11.5.3.2);
Visa requirements: For foreign investors and technical staff and the need for foreign experts to train local staff;
Import/export: Restrictions relating to materials, broodstock, seedstock, and product.

McCoy (1996a,b,c, 1997) provides useful accounts of laws of concern to aquaculture in the USA – there are over 120 federal statutes with a direct or indirect impact and in one state as many as 26 approvals are required before a coastal shrimp farm can be operated. Before approvals are granted, a detailed plan of the operation including an analysis of water and flow rates needs to be drawn up along with a management plan indicating operating procedures. The burden of compliance has increased markedly in Texas as anti-aquaculture lobbies of conservationists, competing industries and local residents have added their weight to debates on coastal development. The resulting state of affairs has prompted one commentator (Mattei 1995) to note that the political temperature has as much impact as the water temperature in determining whether a shrimp farm succeeds! In Queensland, Australia, the need to comply with environmental legislation is also starting to have an influence on decisions about where to locate new aquaculture projects (section 11.5.3.2).

Particular problems can arise when a farm is set up under one regulatory regime and is then subject to modified requirements. In parts of the USA shrimp farms that were established when no restrictions were in place are now required to obtain discharge permits because aquaculture effluent has been reclassified with industrial wastewater and municipal sewage output. To avoid such complications it may be safer to select sites where coastal zone management plans have already been devised and already incorporate an aquaculture component. In this regard the impacts of coastal zone management plans and state aquaculture plans in the USA have been surveyed by Nelson *et al.* (1999).

The effects of legislation aimed at safeguarding aquaculture can also influence siting decisions (section 11.5.3.3). The Venezuelan shrimp industry is a small industry that has been regulated more effectively than in other Latin American countries and partly as a result of this it has suffered much less severely from disease epidemics. There has been no uncontrolled proliferation of farms, nearest neighbour distances of at least 50 km are enforced and imports of live shrimp are discouraged (Clifford 1997). For certain types of projects involving stock-enhancement, sea ranching or the construction of artificial reefs for lobsters, it would not be realistic to proceed without the necessary legal framework in place. In the UK the prospects for sea ranching lobsters have improved because of legal changes affecting the ownership rights of released animals (Pickering 2000) (section 11.5.3.3).

6.3 Locality

The most obvious requirement for a specific site is that it should provide enough land and water of sufficient quality for the proposed project. Soil characteristics are especially important if earthen ponds are to be constructed and attention must also be paid to the suitability of the local infrastructure, communications and labour force, as well as to a number of social and environmental factors.

The use of geographical information systems and remote sensing in aquaculture site selection are described by Kapetsky (1993). The techniques can provide information on vegetation types and surrounding land use and may be particularly useful in the absence of up-to-date maps. Satellite imagery and aircraft have some potential for providing useful water quality information but the operational use of remote sensing in aquaculture remains largely experimental (Egna 1994). To select the location for a shrimp farm in Saudi Arabia, Falaise and Boël (1999) set up study camps on proposed sites to make 24 h observations. They collected weather and current data, took soundings and mapped areas of seabed and lagoons.

6.3.1 Water

6.3.1.1 Quantity

Water at aquaculture sites may be obtained from surface waters or from underground reserves. Outdoor ongrowing operations have the greatest requirements (section 8.3.6.4). If surface water is to be taken from estuaries or rivers, seasonal variations in flow rates can be expected. During the dry season in some tropical river estuaries, it is possible for negative flow to occur with seawater moving inland to replace freshwater lost by evaporation. Under these conditions salinity levels can rise sharply.

Ideally streams for freshwater farms should not be allowed to flow directly through ponds because of the risk of siltation and flooding. Ponds should be built to one side of the stream or the stream diverted to one side of the ponds. For simplicity, however, some farm dams in Australia that have been stocked with crayfish consist of a straightforward transverse embankment across the path of a stream. Sites are usually selected so that water runoff each year will at least be sufficient to fill the dam.

Groundwater is more consistent year-round than surface water but it is still important to know how much is available and from what depth it needs to be pumped. The latter will influence pumping costs. Where underground freshwater is already being tapped for agriculture or to provide a supply of municipal water, the reserve may already be known. If not, a series of test boreholes will be needed. For all sites requiring freshwater, the needs of any nearby irrigation schemes should be reviewed in relation to the amount of freshwater available (section 11.4.2). If a water shortage ever arises, priority will usually be given first to domestic needs and then to the irrigation of agricultural crops rather than crustacean farms. In streams and rivers, traditional usage for such things as washing clothes and watering livestock should be investigated.

Ideally, hatcheries should be located where an unlimited supply of high-quality water is available. Marine seawater hatcheries located adjacent to an ocean, and most freshwater hatcheries with a reliable borehole supply, are more likely to have good quality water than estuarine or lagoon-side hatcheries. Borehole water and seawater taken from beach wells contains only very small amounts of suspended solids, thereby greatly reducing or even eliminating the need for further filtration (section 8.4.1). An ideal site for a *Macrobrachium* hatchery is one where wells sunk at different depths provide both fresh and saltwater that can then be mixed to achieve the optimum salinity of 12‰ (New & Singholka 1985). However certain groundwaters contain elevated concentrations of iron and manganese that can kill prawn larvae and if the problem cannot be remedied through the application of chelating agents (section 8.4.3), such waters will remain unsuitable for hatchery use (Correia *et al.* 2000).

A reliable supply of freshwater is also needed for washing operations in processing plants. If no mains supply exists then a well may need to be installed or surface water pumped and treated.

6.3.1.2 Distance from source

Water supplies should be as close as possible to a farm or hatchery to avoid the expense, ownership problems and maintenance costs of long supply channels or piping installations. A notable exception to this general rule is found in some *Macrobrachium* hatcheries and shrimp farms in Thailand, which are located at inland sites and rely on road tankers for a supply of concentrated brine from salt pans. The brine is diluted with freshwater for larvae culture, or added to freshwater ponds for ongrowing.

6.3.1.3 Tides

If tides are to be used for water exchange in brackishwater ponds, the tidal range must be sufficient for flushing but not be so great so as to require massive embankments and embankment protection. Tide tables should be studied to establish tidal ranges and variations between spring and neap tides. Spring tide ranges of around 2 m are about the minimum necessary, while anything greater than about 4 m will require excessive embankment protection (ASEAN 1978). Little or no water exchange will be possible during neap tides. Embankments should be able to cope with unusually high tides caused by storm surges. Where land is scarce in Japan, some farmers have built tidal impoundments in which the embankments become submerged at high tides (section 7.2.6.4).

Plate 6.2 An inland backyard hatchery for freshwater prawns in Thailand. The owner and her family are seen discussing alternative seawater supply prospects with an extension worker from the Department of Fisheries.

6.3.1.4 Quality

Details of water quality and likely fluctuations at a site can never be known with precision. An estimate has to be made of the risk of water quality falling at times to unacceptable levels. An initial site survey will usually only be done on one or two days and even a full survey will usually draw heavily on previously gathered information which may not be completely suitable. Data may have been gathered by workers or prospectors employed for purposes unrelated to aquaculture.

Aspects of water quality that are of particular significance at the site selection stage are covered here while the water quality tolerance of crustaceans is discussed later (section 8.5).

Although inland groundwaters can be saline, salinity levels are usually only of relevance to marine and brackish-water farms, where seasonal fluctuations in pond salinities can be expected as a result of precipitation and evaporation. Water at some estuarine sites can become virtually fresh during rainy seasons and super-saline following hot dry periods. During the latter, an alternative source of abundant fresh or low salinity water is at a premium.

Suboptimal levels for more than one water quality factor can have deleterious synergistic interactions and, although crustaceans can sometimes display remarkable resilience to suboptimal conditions, site selection should pay close attention to meeting water quality requirements if reliable survival and growth rates are to be obtained. Having said this, the range of salinities over which penaeid shrimp are successfully cultured is now quite wide and, while assumptions about optimum ranges are probably still valid, tolerance of suboptimal conditions is a feature of many systems. Reports of this include:

- *Penaeus monodon* culture at salinities of up to 50‰ following monsoon failure in Sri Lanka (Chamberlain 1988).
- Culture of *P. monodon* in very low-salinity water in inland areas of Thailand following the addition of hypersaline water to freshwater ponds (sections 7.2.6.5 and 8.3.7).
- Successful production of nauplii and larvae of *P. monodon* using beach well water of 25–27‰ in Madagascar (Pironet *et al.* 1999).
- No variation in growth rates of *Litopenaeus vannamei* cultured at 2, 4 and 8‰ (Samocha *et al.* 1998).
- Normal growth and survival obtained with *L. vannamei* in areas previously used for solar salt production in north-east Brazil using a water supply at 42‰ and pond salinities of up to 50‰ (Henig 2000).
- Few problems with the culture of *Fenneropenaeus indicus* at salinities of 42–50‰ in Iran (Shakouri 2000).
- No problems with *F. indicus* culture at 7‰ and no adverse effects detected at 3‰ (Evans 2000).
- Culture of *F. indicus*, particularly the Red Sea strain, at up to 50‰ in Saudi Arabia (Al-Thobaiti & James 1998).

The temperature of a water supply can differ significantly from ambient air temperatures. In some temperate crayfish farms well water may be particularly cool, in which case shallow ponds would usually be employed to allow this water to warm up before use. On the other hand, excessively hot water from geothermal sources may need to be left to cool before use, blended with cooler water, or used indirectly as a heat source for controlled-environment cultures (sections 5.3, 5.4 and 8.4.4) or even for pond supplies (section 7.3.5.3). Sometimes geothermal water may need to be 'degassed' by exposure to the atmosphere, aeration and agitation to help eliminate hydrogen sulphide and ammonia and remove iron that may otherwise precipitate on the gills of crayfish. Many geothermal waters are acceptable for use in culture operations but a few may contain unsuitable ionic compositions, for example excessive hardness.

The concentrations of phytoplankton and zooplankton in surface water can give an indication of potential productivity. For outdoor ongrowing operations, clear waters may require extra fertiliser or higher-quality feeds to compensate for their lack of nutrients and natural productivity. On the other hand, an excess of nutrients, particularly nitrates and phosphates, can lead to eutrophication (see Glossary). Water contaminated with agricultural fertilisers or sewage should be avoided. Of particular significance to site selection are the likely seasonal changes in the levels of chemical nutrients and the risk of toxic algae blooms occurring. Neither of these can be assessed simply with spot checks. Seasonal storms can increase the sediment load in some coastal waters and some regions are known to have waters with different characteristic levels of nutrients. For example, on the Caribbean coast of Mexico waters are known to contain less nitrogen and phosphorus than on the Pacific coast. These nutrients have also been found to be deficient in oxygen-rich surface waters off Oman.

Pollution can arise from many different sources and adversely affect crustacean farms and hatcheries. A thorough assessment must be made of the proximity of the proposed site to agricultural land where fertilisers and pesticides are used, to any industrial activity where chemicals are used or discharged, and to any other human activity resulting in effluent or run-off that could contaminate water supplies (e.g. sewage or mine tailings). Cruz (1993) gives some examples of poor siting decisions in Negros, Philippines, where sugar cane farmers turned to shrimp culture but clustered their farms along rivers into which twelve sugar plants and alcohol plants discharged their effluents. Avault (1994) advised that ponds should not be constructed on land formerly used to grow cotton, unless soil tests reveal the absence of DDT and other persistent pesticides.

At many aquaculture sites the most significant sources of pollution are existing aquaculture operations. The best-known examples of this can be found in southern Taiwan and in the Bay of Bangkok. To avoid the worst effects, prevailing current directions and seasonal changes therein should be considered. Hatchery discharges can contain pathogens, including bacteria with resistance to antibiotics that can severely limit the impact of antibiotic therapy (sections 11.3.4 and 11.4.3).

6.3.2 Topography

The principal factors to be considered in selecting a farm site with suitable topography concern elevation, gradient and exposure to winds.

6.3.2.1 Elevation

Poernomo (1990) compared the suitability of intertidal and supratidal areas for coastal shrimp ponds. The elevation of marine and brackish-water ponds is a compromise between good tidal exchange and (where pumped) low pumping costs, and the need to drain completely. Since tidally flushed ponds require land with intertidal elevations this often involves occupying areas covered by mangroves, salt marshes or mud flats. Tidal ponds typically permit water exchange only during spring tides and allow complete drainage only at low water during these periods. Suitable areas for tidal ponds thus need to be selected based on knowledge of tidal levels and ranges.

Pumped ponds can be built above the level of high water (spring tides) to allow for free drainage at nearly all times and to enable effective pond bottom maintenance between crops. The use of supratidal areas for shrimp ponds in Thailand has been identified as one of the factors that gives Thai shrimp farmers an advantage over their counterparts in Indonesia and Philippines that use intertidal land (Kongkeo 1997).

Ground with an elevation of up to 5 m is often used for large shrimp farms, with pumping costs closely related to the height that the water must be raised. In some very large operations the water supply for high-level ponds (>5 m) is provided by two separate sets of pumps via an intermediate reservoir. When marine or brackish-water ponds are operated with little or no water exchange, pumping costs are minimised and relatively high ground can be occupied. Zero exchange ponds in Belize have been built 6 m above sea level, thereby reducing the threat of hurricane damage.

6.3.2.2 Gradient

Gradients have important implications for the construction and drainage of ponds. Sites with steep slopes necessitate small ponds, deep water or excessive soil move-

Plate 6.3 A disgruntled and mud-covered Thai farm worker displaying a meagre catch of exhausted and dying shrimp taken from an undrainable pond.

ment during construction and are thus inappropriate for large extensive and semi-intensive ponds. Land gradients for these should be between 0.1% and 0.5% whereas for ponds of less than 1 ha gradients of 0.5–5% are suitable. Tidwell and D'Abramo (2000) note that some hilly sites that would normally be considered unsuitable for pond construction could be suitable for *Macrobrachium* culture in deep water. Their viewpoint is based on the assumption that artificial substrates will be provided within ponds to encourage prawns to make better use of the available three-dimensional space.

For freshwater ponds relying on streams or rivers, valleys with a V-shaped cross-section may be acceptable for small and raceway-type ponds, if the slopes are not too acute.

In the design of some hatcheries a raised reservoir is included to allow for gravity flow down to hatchery level. If this approach to water supply is planned then a raised area of land in the vicinity of the hatchery is needed to avoid the cost of building a large base for the reservoir. Some hatcheries make do with a metal or fibrocement header tank.

6.3.2.3 Exposure

Although steady light breezes are beneficial for ponds because they promote gaseous exchange, high winds at very exposed sites can result in severe wave damage to embankments and excessive evaporation, unless measures are taken to protect downwind embankments and install wind breaks. In one shrimp farm built at an exposed site on a windy section of the Spanish coast, severe wave damage necessitated the remodelling of ponds. At considerable expense, long, narrow ponds were constructed with the long embankments at right angles to the prevailing wind.

The level of exposure and amount of vegetation at a site can influence evaporation rates. For example, while an evaporation rate of 6016 mm yr^{-1} has been recorded at Kuwait International Airport, in the same country another site with plant and tree cover, the equivalent rate was only 1598 mm yr^{-1} (Farmer 1981). Vegetation serves to cut evaporation from ponds by reducing temperatures, raising humidity levels and reducing wind speeds.

6.3.3 Soil

The suitability of a site for pond construction is greatly influenced by the physical and chemical characteristics of its soil. Physical properties must provide for reliable and sufficiently impermeable embankments and pond beds, and soil chemistry should ideally contribute to, or at least not detract from, the fertility of the pond water.

An initial reconnaissance survey can be carried out on a proposed site in order to make a simple map and to collect soil samples. Later on, when the most suitable areas of a site have been located, a more detailed survey is performed by collecting around 0.5–2 soil core samples per hectare. Soil profiles are examined in their natural state by excavating an open pit (0.8 × 1.5 × 2 m deep, or down to parent rock or water table). Soil samples can be taken from within the pits or collected using an auger and should be bagged and labelled in preparation for analysis. A thin-walled steel tube (30–60 × 4–7 cm in diameter) is inserted into the ground to collect the undisturbed sample cores needed for certain laboratory tests. Soil quality usually needs to be investigated to a depth of at least 1 m below the intended base of ponds.

To provide a detailed soil chemical analysis, samples weighing around 1 kg each can be sent to a soil testing laboratory. However one factor of particular significance to aquaculture – pH – is best measured in the field with a pH meter (test one part of soil mixed with two parts of distilled water). Soils with pH values of between 5.5 and 9.5 are preferable and the range pH 6.5–8.5 is optimal for pond productivity. To help in the interpretation of the results of soil chemical analysis, Boyd *et al.* (1994) provide reference data from soils taken from 358 freshwater and 346 brackish-water ponds at sites all around the world. The data are arranged into concentration categories to facilitate comparisons and establish whether a particular element is found in unusually high or low concentrations. The ponds in Boyd's analysis cover a very wide range of soil chemical properties but all were producing crops of fish or shrimp at the time of analysis. To further aid the site selection process for aquaculture ponds, Hajek and Boyd (1994) have provided a rating system to indicate suitable soil and water quality characteristics. The system is based on a similar approach used for evaluation of sites for agriculture, road construction, waste disposal and residential developments.

For pond construction the most important physical properties of soil are texture, consistency, structure, permeability and colour.

6.3.3.1 Texture

Texture is largely determined by the proportions of different sized soil particles. To establish the proportions with precision it is necessary to dry and sieve a sample.

Fine particles (<2 mm) are classified as sand, silt and clay in decreasing order of size. For low porosity in embankments a minimum clay content of around 25% is recommended, although as little as 10% may be acceptable for well-compacted soils. For pond construction in general a minimum of 50% silt and clay particles is recommended (Coche 1985). However, soils with a clay content in excess of 60% are not recommended because they are subject to excessive cracking when ponds are dried. Very heavy clay soils also make earthworking hard and therefore expensive.

Detailed particle analysis of soils can be performed in a specialised laboratory but a number of simple tests can be performed in the field that may suffice for small projects. A simple indication of texture can be gained by taking a handful of moist soil, squeezing it into a ball, and throwing it 50 cm into the air and catching it. If the ball falls apart it is probably too sandy and unsuitable for ponds, whereas if it sticks together the clay content is probably adequate.

Further simple and quick field tests to indicate basic soil texture are detailed by Coche (1985). The bottle test involves putting 5 cm of soil in a glass bottle, filling the bottle with water, suspending the soil by vigorous mixing and then leaving the sample for an hour. The proportions of sand, silt and clay in the sediment can be estimated from the layers that settle in order of size in the bottom of the bottle.

Another method that gives a better indication of soil texture is the manipulative test, which involves making a ball of moist soil in the hand and rolling it into a thin sausage 15–16 cm long. If the sausage cannot be formed the soil is probably too sandy for ponds. If it can be formed but cannot be bent into a semicircle the soil is a loam, which is a fair material for building embankments. If the semicircle can be formed and then bent into a ring without cracks forming, then the soil is clay, that is excellent for embankment construction. If cracks do form, the soil is a light clay that is also suitable for embankment building.

Another simple test of the suitability of soil for embankments involves kneading moist soil to make several 10 cm diameter balls and leaving them submerged in water for a day. Good material will remain intact while balls of unsuitable soil will disintegrate within a few hours. One of the factors favouring the success of shrimp farming in Thailand has been the use of well-constructed intensive ponds using soils with a high clay content that prevent water seepage (Kongkeo 1995).

6.3.3.2 Consistency

Consistency is a measure of the strength with which soil is held together when dry, moist or wet and has important implications for pond construction. It can be gauged roughly in the field by using simple tests, or assessed with precision in the laboratory by using standard methods to determine the Atterberg limits. An Atterberg limit corresponds to the moisture content at which a soil sample changes from one consistency to another. Of particular interest to pond construction are the liquid limit and the plastic limit. They represent the percentage moisture content at which a soil changes (with decreasing wetness) from liquid to plastic consistency and from plastic to semi-solid consistency, respectively. For normal pond embankments best compaction results are obtained with a liquid limit of 35%. Clay which is to be used for an impervious core within an embankment, or as a layer lining an embankment (section 8.2.2), should have a liquid limit below 60% and a plastic limit below 20%.

The range of moisture contents at which soil remains plastic is measured by the difference between the plastic and liquid limits. The resulting value is known as the plasticity index (PI). In general it depends only on the amount of clay present and indicates the fineness of a soil sample and its capacity to change shape without altering its volume. For normal pond embankments PI values between 8% and 20% are acceptable, with 16% ideal for compaction. For clay used for impervious cores the PI should be greater than 30%.

6.3.3.3 Permeability

Soil permeability has obvious significance for aquaculture ponds and depends largely on the number and size of pores present in the soil – two factors which are principally determined by soil texture and structure (the way individual soil particles are assembled into larger particles known as aggregates). Permeability can be measured as a seepage rate in cm per day. Ponds with daily seepage losses of more than 1–2 cm can be difficult to manage. Compaction can be used to alter soil structure and reduce pore size and permeability.

One procedure for testing permeability is: dig a hole about 0.8 m deep, line the walls with clay (to prevent horizontal seepage), fill with water in the morning, check the level in the evening, then top up and cover the hole with vegetation. If most of the water is still present the next morning, the soil permeability is acceptable for pond construction.

To test the permeability of soil material for embankments the bottleneck test can be performed: cut the bottom off a bottle and fill the neck with a slightly compacted soil sample. Fix the bottle upside down and fill with water. The soil will be suitable for dams if no water has leaked out in 24 h.

Precise measurements of permeability can be made in a laboratory using an 'undisturbed' soil sample carefully collected with a thin-walled tube. Normal embankments require soil with a coefficient of permeability of less than 1×10^{-4} m s^{-1}. Pond bottoms require less than 5×10^{-6} m s^{-1}.

If the soil proves to be light and porous, it may be necessary for clay to be brought in from an outside source, or for pond sealants or pond liners to be used to provide impermeability (section 8.2.2). Such measures will increase construction costs and it will be necessary to investigate the availability of the relevant materials. The porosity of earthen ponds can be a particular problem in desert areas (Farmer 1981). Perhaps the only advantage of light soils is that they are easy to move and can thus reduce the costs of earthworking. Ponds should not be established on seepage areas with gravel or sand seams or aquifers, and should not be constructed below the water table. It is obviously difficult to make ponds on a rock substrate but it is possible if the topography is suitable and clay embankments are used on impervious, fracture-free rock.

A site close to a beach with deep sand is ideal for a coastal hatchery because it allows a sub-sand filter to be installed for the water supply. If, however, a hatchery is planned with broodstock and nursery ponds, then the need for impermeable soils or pond liners must also be taken into account. Some sites that have been previously used as salinas (salt pans) may present problems for aquaculture because of brine upwelling.

6.3.3.4 Colour

Soil colour may be a useful indicator of organic content and drainage conditions. Dark shades can indicate high organic content and/or poor soil drainage. Abundant pale yellow mottles coupled with a low pH characterise acid sulphate soils (see next section). Although layers of topsoil rich in organic matter can and must be removed, a site with deep peaty soil is unsuitable for pond construction.

6.3.3.5 Acid sulphate soils

One particular type of soil that needs to be identified during site selection is characterised by its potential to become highly acidic when drained and exposed to air. Known as potential acid sulphate soil, it can reach pH 4 or less following earthworking and in general should be avoided for pond building because it necessitates special construction and management procedures that can delay the start-up of production and greatly add to overall project costs.

Potential acid sulphate soils are found in saline areas such as coastal mangroves and in freshwater areas such as river plains, where sediments with a high organic content accumulate. They contain iron sulphides precipitated as a result of sulphide excretion by sulphur-reducing bacteria that decompose organic matter under anaerobic waterlogged conditions. While these soils remain waterlogged and anaerobic they undergo little change, but in contact with air bacterial oxidation produces sulphuric acid and precipitates of ferric hydroxide. The latter often imparts a red colour to the soil surface and can accumulate in the gills of crustaceans (section 8.9), but more critically the sulphuric acid can increase acidity by as much as 3 pH units. Measuring pH before and after exposure to air can identify potential acid sulphate soils. The test procedure involves taking a handful of soil (moisten if dry), working it into a cake 1 cm thick and placing it in a thin plastic bag. The bag is sealed to retain moisture and encourage bacterial activity. If after 1 month the pH measures 4 or less, the soil is acid sulphate.

Actual acid sulphate soils (potential acid sulphate soils that have already been oxidised) are relatively rare but are easily recognised by their low pH values and often by mottles of the pale yellow mineral jarosite.

When ponds constructed with acid sulphate soils are filled, sulphuric acid leaches into the pond water. Embankments generally produce more acid than pond bottoms because, for the most part, the latter remain waterlogged and are less exposed to the air. Particularly heavy leaching of acid can occur from embankments following rainstorms. Acidic conditions are detrimental to pond productivity, adversely affecting bacteria and phytoplankton as well as the crop. Acid embankments usually restrict the growth of vegetation and are vulnerable to erosion.

Where crustacean farms have been constructed at sites with acid sulphate soils, a range of measures can be taken to limit pond acidity both during pond preparation and operation (sections 8.2.2.5 and 8.3.8). Useful accounts of ways of coping with acid sulphate soils include those given by Webber and Webber (1978), Simpson *et al.* (1983) and Boyd (1995).

6.3.4 Vegetation

A site with dense, deep-rooted vegetation will result in greatly increased land clearance costs. Former agricultural land has good soils but care must be taken to check for residual pesticides and other contaminants. Some tea plantations, for example, are sprayed with zinc.

6.3.5 Communications and infrastructure

Good approach routes either via land or water are essential to gain access to an operation all year round. For sites without access from an existing road network, the cost of building a road capable of withstanding the usual weather conditions must be taken into account.

The proximity of a site to a source of ice and to processing facilities will need to be considered. One possibility for remote sites is to construct a processing plant alongside an ongrowing operation. However, even if this strategy is followed, good transport links will still be needed to reach markets or market channels. For exports, international air- or seaports will be needed with appropriate facilities for storage and handling cargo and containers. Freight rates will need to be established for economic evaluation.

Hatcheries are most conveniently placed at the site of the farm, but for shrimps and prawns the different water quality requirements of the larvae rearing and ongrowing phases of culture often make this an impossibility. Nevertheless, this does not often represent a major constraint since post-larvae can be transported quite easily (sections 7.2.4 and 7.3.2).

Good telecommunications links are very useful for the efficient running of any business. In the absence of these, two-way radios can be used provided that the relevant authorities will allocate transmission frequencies for private use.

If mains electricity is to be used the extent of the existing electricity network should be taken into account, since the cost of extending power-lines can be prohibitive. The quality of the supply with regard to interruptions, overall capacity (some limited networks can only cope with extra demand on a domestic scale), voltage consistency and the number of phases available, should be investigated. This information is needed to properly compare the different options and costs for energy. Inadequate or non-existent electricity supplies at many sites necessitate the use of diesel powered generators either as a back-up or to provide for all electrical needs.

The location of other utility networks for freshwater and gas sometimes plays a part in the site selection process. If there is no mains supply or underground source of freshwater, a supply from water tankers may be needed and this may have significant cost implications. Municipal freshwater is likely to be chlorinated so it is not suitable for direct use in culture operations.

6.3.6 Labour force

The remoteness of a site can have implications for the availability and well-being of a project's workforce. An isolated site away from towns or villages may reduce the risks of poaching but can necessitate considerable expenditure on transport or the construction of housing. Distant community facilities, such as schools and health centres, can present problems for employees (section 11.2.3).

The availability of a casual and seasonal labour force should be investigated if it will be needed for construction work or for harvesting and processing operations. Sometimes there are other seasonal activities that will compete for a local workforce. The limited availability of experienced construction workers in some rural areas can mean bringing in labourers from urban centres. Other considerations concern the reliability of the available labour, local attitudes to work, and possible liaison difficulties.

6.3.7 Social, environmental and ecological factors

The social, ecological and environmental impacts of crustacean farming should be taken into account from the early planning stages of all projects (Chong 1990) (see also Chapter 11). Good site selection can help to minimise the negative impacts of crustacean farming while maximising the potential benefits. Dutrieux and Guélorget (1988) stress the importance of an ecological approach to the planning of aquaculture and note that very few sites on the French Mediterranean were chosen simply for their productivity potential. Indeed among 15 studied, 9 were chosen for socio-economic reasons such as availability of the plot or planning permission from the authorities. The kinds of issues that should be given attention include possible conflicts of land use with rights of way, recreation, and existing grazing or agriculture, and possible conflicting use of water resources including irrigation and navigation. Local fishermen may be antagonistic if the project intends to use wild broodstock or seedstock that form the basis of their livelihood (Avault 1994).

It is advisable, though not widely practised, to survey for disease before undertaking an aquaculture venture. Hudson and Lester (1994) for example, surveyed wild Australian mud crab populations to determine the types and occurrences of organisms that could jeopardise the emergent farming industry. They found four potential pathogens and a blood parasite similar to one that in Alaskan crabs causes a bitter aftertaste. Edgerton and Owens (1999) surveyed Australian crayfish farms not only to identify organisms that could constrain production but also to provide the basis for quantitative assessments of the risks inherent in stock and product movements. Without information on farmed species and their pathogens, effective quarantine measures cannot be implemented, which may have implications for both productivity and product exports. If a serious disease is present in a particular area, the only viable approach may be to work in closed systems and use SPF stocks.

6.4 Modifications to an existing facility

The first and most important task when acquiring a used facility is to establish why it is for sale or why it has been abandoned. The previous owners should be closely questioned, possibly with the guidance of an independent expert. Sites and facilities should be carefully inspected. Reasons for sale may include specific problems such as disease and poor water quality or wider problems relating to poor siting, inadequate management, lack of finance or changes in market conditions.

The roots of all general and site-specific problems must be exposed. Management problems can be related to the remoteness of a site and its lack of appeal to highly qualified staff. Disease may be due to increasing pollution from a whole range of sources including neighbouring aquaculture operations. Failure to operate profitably can sometimes relate to the use of extensive techniques that do not provide sufficient yields per hectare. Equally, failure may be attributed to the use of high-cost intensive or super-intensive methods that cannot achieve cost-effective production levels. Some shrimp hatcheries have been forced out of business because of low prices obtained for post-larvae; in Ecuador this has been associated with gluts of wild-caught post-larvae, and in Taiwan with the changing fortunes of the shrimp farming business as a whole. New owners who acquire an existing facility may be able, initially at least, to improve financial and managerial inputs, but they may be powerless to rectify many underlying problems.

6.4.1 Hatchery

When acquiring a used hatchery, special attention should be given to the quality of its water supply since larval crustaceans are more sensitive to water conditions than adults or juveniles. If the site appears to be favourable then the hatchery can be inspected to assess the quality and value of its installations. Abandoned hatcheries may seem to be complete from a visual inspection, but the process of bringing them back into production can be lengthy and may reveal many hidden system problems and design deficiencies. The purchase of a working hatchery may be preferable, although this is no guarantee that smooth running can be maintained with the existing facilities. Hatchery designs usually closely match the intended operating methods so if new techniques are to be applied, remodelling may be required (e.g. to permit more intensive management, the size of larval rearing tanks may need to be reduced).

6.4.2 Farm

In the acquisition of existing farms, the quality and value of the installations should be assessed, together with the attributes of the site. Investment in renovation will only be worthwhile if the site meets all the important site selection criteria. If extensive farms are to be upgraded, the kinds of modifications that can be considered include deepening ponds and increasing the bottom slope, strengthening bunds, separating inlet and outlet gates and incorporating nursery ponds.

6.5 References

Al-Hajj A.B., Farmer A.S.D., Al-Hassan K.E. & Saif M.A. (1981) *The effect of airborne dust and suspended marine sediments on the survival and larval stages of* Penaeus semisulcatus *de Haan.* KISR **201**, 1–22. Mariculture and Fisheries Department, Kuwait Institute for Scientific Research, Kuwait.

Al-Thobaiti S. & James C.M. (1998) Saudi Arabian shrimp success in hypersaline waters. *Fish Farmer*, **21** (4) 40–41.

Anon. (1999) Technology, experience and perseverance: success in Panama. *Aquaculture Magazine*, **25** (5) 42–54.

ASEAN (1978) *Manual on Pond Culture of Penaeid Shrimp*, 131 pp. ASEAN National Coordinating Agency of the Philippines, Ministry of Foreign Affairs, Manila, Philippines.

Avault J.W. Jr. (1994) Where to locate your aquaculture operation. *Aquaculture Magazine*, **20** (1) 76–80.

Boyd C.E. (1995) *Bottom Soils, Sediment, and Pond Aquaculture*, 348 pp. Chapman & Hall, New York.

Boyd C.E., Tanner M.E., Madkour M. & Masuda K. (1994) Chemical characteristics of bottom soils and freshwater and brackishwater aquaculture ponds. *Journal of the World Aquaculture Society*, **25** (4) 517–534.

Chamberlain G.W. (1988) Shrimp culture news from around the world. *Coastal Aquaculture*, **5** (1) 13–15.

Chong K.C. (1990) Economic and social considerations for aquaculture site selection. In: *Technical and Economic Aspects of Shrimp Farming*, Proceedings of the Aquatech '90 Conference, Kuala Lumpur, 11–14 June 1990 (eds M.B. New, H. de Saram & T. Singh), pp. 24–35, Infofish, Kuala Lumpur, Malaysia.

Clifford H.C. (1997) Shrimp farming in Venezuela. *World Aquaculture*, **28** (1) 60–61.

Coche A.G. (1985) Soil and freshwater fish culture. *Simple Methods for Aquaculture, FAO training series,* (6) 1–165. FAO, UN, Rome.

Correia E.S., Suwannatous S. & New M.B. (2000) Flow-through hatchery systems and management. In: *Freshwater Prawn Culture: the farming of* Macrobrachium rosenbergii (eds M.B. New & W.C. Valenti), pp. 52–68. Blackwell Science, Oxford, UK.

Cruz P.S. (1993) *Shrimp farming in the Philippines: culture practices and problems*, 30 pp. (mimeo). Kabukiran Enterprises, Davao City, Philippines.

Dhont J., Lavens P. & Sorgeloos P. (1993) Preparation and use of *Artemia* as food for shrimp and prawn larvae. In: *Handbook of Mariculture,* 2nd edn, *Vol. 1 Crustacean aquaculture* (ed. J.P. McVey), pp. 61–93. CRC Press, Boca Raton, FL, USA.

Dutrieux E. & Guélorget O. (1988) Ecological planning: a possible method for choice of aquacultural sites. *Ocean and Shoreline Management*, **11** (6) 427–447.

Edgerton B.F. & Owens L. (1999) Histopathological surveys of the redclaw freshwater crayfish, *Cherax quadricarinatus*, in Australia. *Aquaculture,* **180** (1–2) 23–40.

Egna H.S. (1994) Monitoring water quality for tropical freshwater fisheries and aquaculture: a review of aircraft and satellite imagery applications. *Fisheries Management and Ecology*, (1) 165–178.

Evans L. (2000) *Shrimp: high salinity/low salinity*. Correspondence, 6/12/2000. shrimp@egroups.com

Falaise F. & Boël L. (1999) A new technology for sustainable shrimp farming. *Infofish International*, (3) 33–99.

Farmer A.S.D. (1981) Prospects for penaeid shrimp culture in arid lands. In: *Advances in Food Producing Systems for Arid and Semi-arid Lands*, pp. 859–897. Academic Press Inc., London.

Griffith D.W. & Schwarz L. (1999) Shrimp farming in Ecuador: development of an industry. *Aquaculture Magazine*, **25** (1) 46–50.

Hajek B.F. & Boyd C.E. (1994) Rating soil and water information for aquaculture. *Aquacultural Engineering*, **13** (2) 115–128.

Henig O. (2000) *Shrimp: high salinity*. Correspondence, 27/11/2000. shrimp@egroups.com

Hudson D.A. & Lester R.J.G. (1994) Parasites and symbionts of wild mud crabs *Scylla serrata* (Forskal) of potential significance in aquaculture. *Aquaculture*, **120** (3–4) 183–199.

Kapetsky J. Mc D. (1993) Aquaculture and geographical information systems. *Infofish International*, (4) 40–3.

Kongkeo H. (1995) How Thailand made it to the top. *Infofish International*, (1) 25–31.

Kongkeo H. (1997) Comparison of intensive shrimp farming systems in Indonesia, Philippines, Taiwan and Thailand. *Aquaculture Research*, **28** 789–796.

Mattei. E. (1995) Shrimp farming in a regulatory climate: South Texas focus. *Aquaculture Magazine*, **21** (3) 54–63.

McCoy H.D. II (1996a) Aquaculture and the law: a primer. *Aquaculture Magazine*, **22** (1) 38–45.

McCoy H.D. II (1996b) Aquaculture and the law: a primer. Article two of a series. *Aquaculture Magazine*, **22** (2) 68–80.

McCoy H.D. II (1996c) Aquaculture and the law: a primer. Article three of a series. *Aquaculture Magazine*, **22** (3) 72–79.

McCoy H.D. II (1997) Aquaculture and the law: a primer. Article four of a series. *Aquaculture Magazine*, **23** (1) 60–66.

Muir J.F. & Kapetsky J.M. (1988) Site selection decisions and project cost: the case of brackish water pond systems. In: *Aquaculture Engineering Technologies for the Future.* Institution of Chemical Engineers Symposium series No. 111, pp. 45–63. EFCE Publication series No. 66, Hemisphere, London.

Muir J.F. & Lombardi J.V. (2000) Grow-out systems – site selection and pond construction. In: *Freshwater Prawn Culture: the farming of* Macrobrachium rosenbergii (eds M.B. New & W.C. Valenti), pp. 126–156. Blackwell Science, Oxford, UK.

Nelson G.R., DeVoe M.R. & Jensen G.L. (1999) Status, experiences, and impacts of state aquaculture plans and coastal zone management plans on aquaculture in the United States. *Journal of Applied Aquaculture*, **9** (1) 1–21.

New M.B. & Singholka S. (1985) *Freshwater Prawn Farming: a manual for the culture of* Macrobrachium rosenbergii, 118 pp. FAO Fisheries Technical Paper 225, Rev. 1.

Pickering H. (2000) Legal framework governing artificial reefs in the European Union. In: *Artificial Reefs in European Seas* (eds A.C. Jensen, K.J. Collins & A.P.M. Lockwood), pp. 469–487. Kluwer Academic, Netherlands.

Pironet F., Bosc F., Delaune J.M. & Lucien-Brun H. (1999) Production management in a *Penaeus monodon* hatchery in Madagascar under low salinity conditions. In: *Book of Abstracts, World Aquaculture '99*, 26 April–2 May 1999, Sydney Australia, p. 470. World Aquaculture Society, Baton Rouge, LA, USA.

Persyn H. (1999) *The use of a closed breeding cycle and selection for disease resistance as a possible means of combating white-spot syndrome*, 16 pp. (mimeo). Presented at Conferencia Regional de Camaronicultura, 7–8 July 1999, Hotel Plaza Paitilla Inn, Panama, Grupo FCE, Panama.

Poernomo A. (1990) Site selection for coastal shrimp ponds. In: *Technical and Economic Aspects of Shrimp Farming*, Proceedings of the Aquatech '90 Conference, Kuala Lumpur, 11–14 June 1990 (eds M.B. New, H. de Saram & T. Singh), pp. 3–23, Infofish, Kuala Lumpur, Malaysia.

Polioudakis E. (2000) *Synopsis of results of research on southern Thai shrimp farming; some recommendations.* http://www.shrimpaquaculture.com/Thai%20article.htm

Rosenberry R. (2000) Shrimp news = Brazil. Correspondence, 5/10/2000. shrimp@egroups.com

Samocha T.M., Lawrence A.L. & Pooser D. (1998) Growth and survival of juvenile *Penaeus vannamei* in low salinity water in a semi-closed recirculating system. *Israeli Journal of Aquaculture – Bamidgeh,* **50** (2) 55–59.

Scura E.D. (1995) Dry season production problems on shrimp farms in Central America and the Caribbean Basin. In: *Proceedings of the Special Session on Shrimp Farming* (eds Browdy C.L. & Hopkins J.S.), pp. 200–213. World Aquaculture Society, Baton Rouge, LA, USA.

Shakouri M. (2000) *Shrimp: high salinity*. Correspondence, 28/11/2000. shrimp@egroups.com

Simpson H.J., Ducklow H.W., Deck B. & Cook H.L. (1983) Brackish-water aquaculture in pyrite-bearing tropical soils. *Aquaculture*, **34** (3/4) 333–350.

Skabo H. (1988) Shrimp farming developments in West Africa. In: *Shrimp '88, Conference proceedings*, Bangkok, Thailand, 26–28 January 1988, pp. 95–102. Infofish, Kuala Lumpur, Malaysia.

Tidwell J.H. & D'Abramo L.R. (2000) Grow-out systems – culture in temperate zones. In: *Freshwater Prawn Culture: the farming of* Macrobrachium rosenbergii (eds M.B. New & W.C. Valenti), pp. 177–186. Blackwell Science, Oxford, UK.

Webber R.J. & Webber H.H. (1978) Management of acidity in mangrove sited aquaculture. *Revista de Biologia Tropical*, **26,** (Suppl. 1), 45–51.

Chapter 7
Techniques: Species/groups

7.1 Introduction

To assess an aquaculture proposal, a reasonable knowledge is needed of the techniques and management practices appropriate for each stage of the operation. It is equally important to know, at least in broad terms, the water quality and food requirements and the environmental tolerances of the animals being farmed. In this chapter the main farming methodologies are outlined for each species or group in accordance with the scheme shown in Table 7.1. However, to avoid undue repetition, fine differences in methods for each species within a group have been omitted. References to the literature cited for each species/group have been deliberately kept separate and are listed at the end of each section in order to facilitate the selection of material for further study.

While the majority of farming techniques are specific to species, some aspects relate to nearly all the cultivable species, for example, materials toxicity, food and feeding, disease diagnosis and control, genetic improvement, ranching, artificial habitats and the general principles of animal husbandry. These are discussed in Chapter 8.

7.2 Penaeid shrimp

7.2.1 Species of interest

Giant tiger shrimp (*Penaeus monodon*); whiteleg shrimp (*Litopenaeus vannamei*); fleshy shrimp (*Fenneropenaeus chinensis*); banana shrimp (*Fenneropenaeus merguiensis*); Indian white shrimp (*Fenneropenaeus indicus*); blue shrimp (*Litopenaeus stylirostris*); kuruma shrimp (*Marsupenaeus japonicus*); redtail shrimp (*Fenneropenaeus penicillatus*); green tiger shrimp (*Penaeus semisulcatus*); southern white shrimp (*Litopenaeus schmitti*); red spotted shrimp (*Farfantepenaeus brasiliensis*); southern pink shrimp (*Farfantepenaeus notialis*); brown tiger shrimp (*Penaeus esculentus*); greasyback shrimp (*Metapenaeus ensis*); speckled shrimp (*Metapenaeus monoceros*). Note that trade names (section 3.3.1) may vary from the above, often including the country of origin, for example, black tiger shrimp (*Penaeus monodon*), Chinese white shrimp (*Fenneropenaeus chinensis*) and Ecuadorian white shrimp (*Litopenaeus vannamei*).

The techniques for farming penaeid shrimp vary greatly in complexity during the ongrowing phase, and in dependence on hatcheries for the supply of juveniles for stocking. Traditional extensive methods, for example, simply trap and grow wild juveniles in shallow tidal impoundments (section 5.2.1) while modern super-intensive systems rely on a regular supply of hatchery-reared juveniles for ongrowing at high densities in specially designed concrete tanks or covered raceways (section 5.2.3). This section attempts to cover the wide range of practices commonly applied both in the Far East, where aquaculture has a long tradition, and in the Americas, where shrimp farming is a relatively recent activity. For more detailed accounts of shrimp farming techniques the reader is referred to the various authoritative manuals and technical publications produced since 1990: Chavez (1990); Villalon (1991); Wyban and Sweeney (1991); Fast and Lester (1992); Wyban (1992); McVey (1993).

7.2.2 Broodstock

7.2.2.1 Acquisition

Despite the increasing importance of shrimp stocks reproduced and maintained in captivity, most penaeid broodstock are still obtained from the wild. Capture techniques are basically the same as for normal shrimp fishing, although when trawling is employed the dura-

Table 7.1 Steps in the farming of crustaceans.

	Phase	Inputs	Facilities/Operation
	Acquisition of broodstock		
	Maturation	Natural feeds	Ponds/tanks/pens
	Spawning	Minimal disturbance	
Penaeid shrimp			
	Incubation	Minimal disturbance	Tanks/bins/kreisels
	Hatching	Clean, filtered water	
Crayfish			
	Larvae rearing	Hatching and culture of live food	Tanks/bins/kreisels
	Nursery	Hatchery reared or wild-caught juveniles; live and compounded feeds	Ponds/tanks/raceways/trays
	Ongrowing	Natural or compounded feeds	Ponds/tanks/raceways/trays/cages/ artificial reefs/protected fishery areas
	Harvesting	Bait, ice, seasonal labour	Seine/electro-netting/draining/ trapping/direct collection
Live			
	Processing	Ice, preservatives seasonal labour	Chilling/washing/sorting/value added treatment/freezing
	Marketing	Packaging	Distribution/sales

tion of the tows must be reduced to avoid crushing and stress which can lead to ovary resorption. Trawlers in Australia, for example, tow for 30 min at a speed of 3 knots and their nets incorporate devices to exclude large fish and turtles that would otherwise damage the contents of the cod end (section 11.2.5). To prevent interrupting the trawling operation it is possible to send a small boat in the wake of the trawler to collect the shrimp soon after capture. This involves packing the shrimp 5–15 at a time in plastic bags containing oxygen and 10 L of water, and depositing the bags overboard to be retrieved by the smaller vessel (Evans 2000). Other ways of catching broodstock apart from trawling include the use of tangle nets and traps. These methods are often used by artisanal fishermen and are particularly suited to catching live shrimp since they generally provide animals in better condition than trawls.

Most hatcheries obtain their broodstock from middlemen or directly from fishermen. Sometimes, however, supplies can be interrupted by close seasons and special dispensation may be needed to catch shrimp during these periods. In Japan, the market for live *Marsupenaeus japonicus* for human consumption provides a ready source of broodstock. But for *Penaeus monodon*, shortages and seasonal availability of wild broodstock constrain the development of the shrimp farming industry in many areas (section 12.4).

7.2.2.2 ***Transport***

Three methods are commonly used for the transport of broodstock from the point of capture to the hatchery:

(1) *En masse* in tanks or bins;
(2) In plastic bags with water and oxygen;
(3) In chilled sawdust.

Table 7.2 includes further details and some results obtained with these methods. The use of tanks or bins is usually suited only to short journeys since it involves transporting a large mass of water. To reduce physical damage shrimp can be placed within individual perforated plastic cylinders capped at each end with pieces of netting, or have protective tubes or bands located on sharp rostrums and telsons. These precautions are also necessary to prevent puncturing when shrimp are transported in plastic bags.

The lightweight transport method involving chilled sawdust is particularly suited to airfreight and is regularly performed with *Marsupenaeus japonicus* destined for live sales in Japan (section 3.3.1.2). By lowering the temperature to 4–10°C the metabolic rate of the shrimp is reduced and the period of survival extended to around 14 h (Schoemaker 1991). This transport technique, however, is not suitable for all penaeids: *Penaeus semisulcatus* for example, die rapidly out of water, and *P. monodon* cannot withstand temperatures below about 12°C.

Table 7.2 Examples of methods and results for transporting penaeid broodstock.

Species	Duration (h)	Survival (%)	Size	Temp. (°C)	Method and density	Reference
F. chinensis	3	95		18	*En masse* in oxygenated tank, 78 shrimp m^{-3}	Zhang *et al.* 1980
	3.5	51		18	*En masse* in oxygenated tank, 400 shrimp m^{-3}	Zhang *et al.* 1980
L. setiferus	6	100*		18–20	20 h fasted shrimp in 8 L of water in double plastic bags topped-up with oxygen. Placed in 40 L styrofoam shipping boxes with ice pack located on underside of lid. Density 59–68 g L^{-1}	Robertson *et al.* 1987
M. japonicus	10		30–50 g	11–17	Packed in cardboard boxes (1 kg per box) between layers of chilled dry sawdust. 5 mm thick bag of frozen wet sawdust included on warm days	Yamaha 1989; Chen 1990
M. japonicus	>14			4–10	Packed in insulated boxes between layers of chilled untreated low-resin sawdust	Richards-Rajadurai 1989
L. vannamei	24	100	13–16 cm (37 g)		Shrimp packed in individual PVC tubes (20 × 3.75 cm) in 30 L plastic bags with minimal water (0.6–1.2 L) topped-up with oxygen and placed within styrofoam boxes. Ice placed in box and separated from bag with 60 mm layer of plywood. 10 shrimp per bag	Johnson *et al.* 1984
F. indicus	42	99			Packed with coconut mesocarp dust within perforated PVC tubes suspended on lines within an aerated, water-filled tank, chilled by an outer jacket of ice and salt	Babu & Marian 1998

*Suffered delayed mortality of 34–46% in the 5 days following shipment.

For all methods involving water, attention must be paid to its quality, particularly for extended journeys. Most important are the levels of oxygen, temperature, ammonia and pH. Oxygen is essential for transport using polythene bags, and a supply of oxygen or compressed air is often needed for tanks and bins. Temperature can be controlled using chill packs and insulated boxes, and in some situations the build-up of ammonia can be delayed by starving the shrimp for 12–24 h prior to shipping (section 8.4.6). Although various products including pH buffers and ammonia control chemicals can be added to improve water quality their overall benefits may be negligible (Robertson *et al.* 1987). For air shipments, packaging needs to conform to IATA regulations, and labels indicating 'keep cool' and 'consignee inspect on arrival' can be useful.

If broodstock are to be retained in the hatchery, a period of quarantine is advisable and the use of a dilute formalin treatment dip (200–400 ppm for 1–2 h, then rinse) can help prevent the introduction of epifauna. Animals heavily infected with the virus *Baculovirus penaei* (BP) can sometimes be identified by observing polyhedral inclusion bodies (PIBs) in samples of faeces, but completely effective screening in this way is impossible since not all infected animals shed PIBs. Elimination of BP and other viruses may be impractical in many instances, and the only approach with any likelihood of success is to acquire certified high-health stock and keep them as far as possible stress-free during culture (sections 8.9.4.4 and 12.2).

The tagging of broodstock at an early stage is useful to identify individual animals for selective breeding purposes and for the elimination of specimens that give positive results when tested for viral infections. Small, numbered and coloured plastic bands, designed for canaries, can be placed on eyestalks for this purpose. An alternative, which can give better tag retention, is to insert specially designed pieces of fluorescent pigmented elastomer into the tail muscle, a technique that is most effective for light coloured shrimp like *Litopenaeus vannamei* rather than the darker *Penaeus monodon* (section 8.10.1.1).

7.2.2.3 Production of broodstock in captivity

The techniques for rearing broodstock in captivity are essentially the same as those used for ongrowing, but with modifications to promote growth beyond normal market sizes. The principal modification is a reduction in the stocking density. In semi-intensive farming *Penaeus monodon* are normally reared at a density of 5–20 shrimp m^{-2} but equivalent ponds managed for broodstock production may be stocked with just 0.7–1 shrimp m^{-2} (Simon 1982). AQUACOP and Patrois (1990) describe the rearing of *P. monodon, Fenneropenaeus indicus* and *Litopenaeus stylirostris* broodstock in outdoor earthen ponds (400–700 m^2), at densities declining from 3 m^{-2} to 0.5 m^{-2} as the animals increase in size. They also describe a more intensive system using smaller, aerated ponds (30–100 m^2) with firm bottoms of compacted coral, fibreglass or cement. This system is stocked at densities declining from 100 m^{-2} to 20 m^{-2} and is suitable for *L. vannamei* and *Fenneropenaeus indicus*, which are tolerant to crowding. The rearing process is divided into two or three stages and at the end of each stage the largest animals are selected. *Litopenaeus vannamei* are reared for 10–24 months but best results for *L. stylirostris* are obtained with animals grown for 7–12 months. *Penaeus monodon*, which are particularly sensitive to crowding, are used after 12 months of rearing, and *Fenneropenaeus indicus* after 6–24 months. Shrimp receive a high-quality formulated diet supplemented by squid and the natural productivity of the ponds. Moore and Brand (1993) recommend broodstock production in independent quarantine facilities and stress the importance of careful management to avoid brother–sister matings. Tanks are initially stocked with a 50% excess at 500 m^{-2} and slow growing shrimp are discarded. Density is reduced to 30 m^{-2} after the shrimp reach 2 g and later reduced again to 20 m^{-2}, to produce *Litopenaeus stylirostris* of 30–35 g and *L. vannamei* of 45–55 g. The process is repeated every 6 weeks for a continuous supply of broodstock.

For many species the ideal salinity for broodstock ponds is close to full strength seawater (34‰) so that shrimp do not require re-acclimatisation prior to introduction to maturation systems that operate best at 30–34‰. However some deviation appears feasible because *Penaeus monodon* reared in ponds at 38–50‰ have successfully spawned in maturation systems (Bray & Lawrence 1998). At the other end of the scale *Litopenaeus vannamei* can reproduce in very low salinities of 2.3‰ (Bray *et al.* 1999). To provide animals in prime condition and reduce stress, broodstock production ponds should be designed for easy harvesting and should be located near to the hatchery.

Hatchery operators have traditionally favoured the use of wild broodstock rather than captive-reared stock because of concerns about the quality of the eggs and nauplii produced by the latter. But it is now clear that the broodstock performance of most species can be improved rapidly in captive generations by the simple process of selecting the most fecund animals (Wyban & Sweeney 1991). *Litopenaeus vannamei* has been successfully reared for more than 17 generations in Venezuela and, significantly, the females no longer require ablation to induce spawning. Selective breeding with captive stocks has other advantages, clearly demonstrated by the case of *L. stylirostris* for which a strain has been developed that is resistant to IHHN virus (sections 2.5.4 and 8.9.4.4), a pathogen that commonly limits the productivity of this species in pond culture. The strain, which has been bred for 20–30 generations, is marketed in the Americas and Asia as Super Shrimp™ (Rosenberry 2000a) (section 8.10.1.3). In contrast, progress with captive stocks of *Penaeus monodon*, the most important farmed penaeid, has been much slower. Research work with captive spawners in the Philippines was halted because hatchery operators refused to use them (Walker 1994). The Australian Institute of Marine Science now has animals ready for open pond trials in partnership with commercial operators, but this has come as the result of a 12-year breeding programme (sections 1.4, 2.6.2 and 12.3).

7.2.2.4 Overwintering

It is possible to rear broodstock in subtropical or warm temperate regions (section 5.3), where the cultured species are likely to be non-native or at the limit of their natural distribution. The procedure usually involves an overwintering phase in which shrimp are retained from one season's harvests to become broodstock for the next production cycle. Shrimp are maintained in holding tanks or in ponds that are sometimes covered with transparent greenhouse structures. These structures serve to raise water temperatures and advance the onset of maturity in the stock.

7.2.2.5 Maturation in captivity

Some hatcheries are able to rely on supplies of gravid females from the wild that are transferred directly to tanks in which they spawn. But the absence of such supplies, or periodic shortages and variations in quality (Hansford & Marsden 1995), induce many hatcheries to opt for the alternative of inducing gonad development in captive shrimp.

The facility with which broodstock will mature, copulate and spawn in captivity varies greatly between different species (Table 4.2). Even in a carefully controlled environment some important species will not mature with regularity, so, in the absence of a proven alternative, the technique of unilateral eyestalk ablation is usually applied (section 2.3). Each eyestalk contains a complex of glands that functions to inhibit gonad development but their effect can be countered by the simple surgical removal either of one of the two eyestalks (extirpation), or of the eyestalk contents (enucleation). As many as nine different techniques have been described (Liao & Chen 1983). Unfortunately broodstock quality gradually declines after ablation and the energy reserves in eggs and nauplii can drop to the point where survival rates to the protozoea stages are compromised (Palacios *et al.* 1998). Reliance on ablation, however, is declining as progress is made with domestication (sections 4.3.2 and 12.4).

In the most simple approach to maturation, many hatcheries in South-east Asia obtain wild gravid females (usually *Penaeus monodon*) and, after the initial spawning, induce four or five additional spawns by eyestalk ablation. Fertilisation in such cases relies upon sperm retained by the female throughout the intermoult period (section 2.2).

Indoor facilities are usually favoured for maturation systems since they permit greater control of temperature, light intensity and photoperiod, all of which influence ovarian development. Nevertheless, outdoor tanks, pens or ponds can also be used. In one such outdoor system in the Philippines, male and ablated female *P. monodon* (sex ratio 1 : 1) were stocked at a density of 0.8–1.6 shrimp m^{-2} in rectangular net pens measuring $250\,m^2 \times 3.5\,m$ deep (Maguire 1979). The shrimp received a diet of fresh mussels, and ripe females were selected weekly by lifting the net. However, the successful application of this type of maturation system depends upon the availability of suitably sheltered seawater sites.

In most indoor maturation systems circular tanks (3–5 m in diameter) made of fibreglass, plastic or cement, are stocked with shrimp at 6–10 m^{-2}. The best results are usually obtained with a steady water temperature around 28°C and a salinity close to normal seawater (30–35‰), although *P. monodon* has also been successfully matured on a commercial scale in borehole water with a salinity of 25–27‰ (Pironet *et al.* 1999). Facilities located near an unpolluted water source that meets the basic water quality requirements usually operate with a flow-through of water (200–300% of tank volume per day), while in other circumstances much reliance may be placed on recirculation and water treatment. Recirculation can also serve to reduce the quantities of water and heat that are required in maturation facilities situated outside the tropics. Broodstock shrimp generally seem to adapt well to recirculation systems (Millamena *et al.* 1991).

The way different maturation systems are operated varies in detail according to: type of substrate; photoperiod and light wavelength and intensity; diet; stocking rate; sex ratio; and the method of fishing and selecting ripe females. A maturation tank must provide the shrimp with sufficient area and depth of water for successful courtship and mating behaviour. Ogle (1992) found that a depth of 60 cm resulted in twice as many spawns with viable nauplii as a water depth of 20 cm. Most commercial systems operate at a water depth of 0.3–1 m. Wyban and Sweeney (1991) provide details of a *Litopenaeus vannamei* maturation system used in Hawaii and AQUACOP (1983) describe another system employing tanks with a sand substrate that was successfully used with *Penaeus monodon, Litopenaeus vannamei* and other penaeids. In all cases special attention is given to husbandry – feeding rate, cleaning routines and minimising disturbance.

If females develop ripe ovaries but are not impregnated, it is possible to resort to artificial insemination to produce fertilised eggs. The basic technique, however, differs between penaeid species depending on whether the females possess a closed or open thelycum. With the former, for example with *Penaeus monodon*, females must be inseminated just after they moult and before the new shell hardens. Whole spermatophores, manually extracted from the males (section 8.10.1.4), are transferred to the still flexible thelycum where they will be retained during the intermoult period (Lin & Ting 1986). Females of open-thelycum species, such as *Litopenaeus vannamei* and *L. stylirostris*, require insemination prior to each spawn, and it is usual to apply only the sperm mass to the thelycum after squeezing it out from extracted spermatophores. Although artificial insemination has clearly demonstrated its viability and is routinely used in some *L. vannamei* hatcheries, most maturation units continue to rely on natural impregnation occurring within the maturation tanks since less labour is involved, mortality rates are lower, and more consistent fertilisation rates are obtained.

The nutritional requirements for maturation and production of high-quality eggs from shrimp broodstock are specialised (section 2.4.7), and despite considerable research effort into diet formulations, best results are only achieved with natural, fresh or frozen foods incorporating items such as mussels, squid, marine polychaete worms, krill and adult *Artemia.* Marine polychaete worms are highly favoured. In a survey of 21 maturation units for *Litopenaeus vannamei* and *L. stylirostris* Kawahigashi (1992) found that all used fresh or frozen worms. Adult *Artemia*, however, are also readily accepted by broodstock and they can reduce reliance on polychaetes without impairing reproductive performance (Naessens *et al.* 1997). Shrimp heads were once a popular fresh feed but, despite their good nutritional qualities, they are rarely used nowadays because of fears of virus transmission. Commercial 'maturation' diet formulations are available but are normally used only as a supplement to fresh feeds (Sangpradub *et al.* 1994). While most may be nutritionally adequate for egg yolk production, the claims by some manufacturers that their diets can actually induce maturation should be verified with the species to be reared.

7.2.3 Spawning and hatching

Healthy gravid female shrimp, whether taken from the wild or matured in captivity, will usually spawn viable eggs providing they have been successfully impregnated and are handled with care. Penaeid eggs, unlike those of most other farmed crustaceans, are released directly into the water rather than held beneath the abdomen for a period of incubation (section 2.2). Hence after spawning is completed, females can be discarded, or alternatively if a maturation system is in operation, they can be retained for rematuration and further spawning. Larvae hatch out 12–18 hours after the eggs are spawned (Fig. 2.3a).

In the most straightforward hatchery systems, originally developed in Japan, gravid females release their eggs directly within the larvae rearing tank. Tanks as large as 200 mt (water capacity) may receive between 100 and 200 gravid *Marsupenaeus japonicus,* placed in the tanks within nets for easy removal. The large tank approach has also been employed in China with *Fenneropenaeus chinensis*. Alternative systems separate the spawning and hatching phases. For example, in the system described by AQUACOP (1983), females are placed into individual spawning tanks, eggs are transferred to incubators, and nauplii hatch to be separated, counted and sometimes dipped in dilute iodine-based disinfectant, before being stocked into larvae rearing tanks. Some operators prefer disinfectants containing chloramine T for dipping nauplii (100 mg L^{-1} for 30 s for nauplii at or beyond instar 4). This kind of system allows the performance of individual females to be monitored and may help prevent the passage of diseases to the larvae culture tanks. Removing the female shrimp from spawning tanks shortly after the eggs have been released can have a beneficial effect on water quality and on subsequent egg hatching rates (Ogle & Beaugez 1997).

A less common approach, used in conjunction with maturation systems, leaves females to spawn undisturbed in the same tanks where they mature and copulate. Eggs are either collected in the outflowing water (Simon 1982) or left to hatch, with the nauplii being siphoned off later from near the water surface. Less handling is involved but there is a greater risk of nauplii picking up contamination from within the maturation tank.

Hatcheries that do not handle spawning females but instead acquire batches of nauplii from outside sources, are more correctly termed larvae culture or rearing facilities. Such facilities obtain supplies of nauplii either from specialist nauplii producers, who spawn wild ovigerous females, or from other hatcheries that operate maturation units. In Ecuador, nauplii are produced by numerous small spawning stations, many of which are based in Esmeraldas province and take advantage of seasonal supplies of wild gravid *Litopenaeus vannamei*. In Southeast Asia, particularly in Thailand and Taiwan, specialist nauplii producers working mostly with *Penaeus monodon* spawn wild gravid females and also achieve repeated spawnings through eyestalk ablation.

Nauplii are sometimes traded internationally and are relatively easy to transport (section 3.4.3). As many as 300 000 can be held in a 30 L sealed plastic bag with 15 L of seawater and 15 L of oxygen for up to 24 h at 18–24°C. One very large maturation facility in Panama, produces up to 6×10^9–7×10^9 nauplii per year and exports part of its output to six countries in South and Central America (Anon. 1999a).

7.2.4 Larvae culture

Although the biological features of penaeid larvae culture are the same in all hatcheries, different operations can be broadly characterised depending on three factors: overall size of the operation, the size of the larval rearing tanks, and the use of western or oriental culture techniques.

Hatcheries range in size from large operations producing more than 100×10^6 post-larvae (PL) yr^{-1}, with a large often highly trained workforce, down to small 'backyard' family concerns producing less than 10×10^6 post-larvae per year, staffed only by a handful of workers or family members. While large operations represent large investments and need to produce either year-round or at least on a regular seasonal basis, small hatcheries tend to keep investments to a bare minimum and have very flexible production schedules (section 10.6.1.1). The shrimp farming industry in Thailand relies heavily on around 1500 backyard hatcheries.

Although the distinction between western and oriental-style larvae rearing techniques is becoming increasingly blurred, the use of these terms has some historical significance and serves to distinguish between the more traditional methods originally developed in Japan and Taiwan (Liao 1992), and the more intensive approaches subsequently developed in the USA, particularly Galveston, Texas (Smith *et al.* 1992), and in French Polynesia. Oriental methods typically rear larvae at lower densities (30–100 L^{-1}) and the bulk of algal feed is produced by encouraging blooms within the larvae rearing tank. For this reason, water exchange rates are kept comparatively low (5–100% per day) to avoid flushing away all the algae. However in the systems developed in Taiwan, tanks are kept in the dark to promote a microbial community dominated by bacteria rather than the microalgae and fresh diatoms, traditionally including *Skeletonema*, that were grown in outdoor tanks to be added each day (Fegan 1992; Liao 1992). The western practice is to stock 50–200 larvae L^{-1} and use water exchange rates of 50–200% per day to maintain stricter control over water quality. However, the resulting higher requirement for algae leads to particularly heavy reliance on independently cultured algae stocks (section 8.8.1).

Tank size and design also tend to differ between the oriental and western approaches. Whereas tanks of any size up to 200 mt can be successfully operated using oriental culture methods, only small tanks of 0.1–15 mt are suited to the more rigorous water management required in western-style hatcheries. In addition, while oriental-style tanks typically have flat, gently sloping bottoms, western techniques benefit from the use of cylindro-conical tanks or tanks with 'V', 'U' or parabolic cross-sections which, combined with strong aeration, can help to prevent the accumulation of organic detritus. Flat-bottomed tanks tend to develop 'dead spots' where detritus accumulates and this may necessitate routine removal by siphoning. Although the western approach typically requires greater management inputs and is sometimes labelled 'high-tech', healthy larvae can be produced by either system and regular success depends on the skill and experience of the technical staff rather than the adoption of one style of culture or another.

The use of large rearing tanks presupposes that sufficient quantities of broodstock can be obtained to fill them with nauplii. Whereas this is often the case for *Marsupenaeus japonicus* in Japan, *Fenneropenaeus chinensis* in China, and in Taiwan, where large numbers of *Penaeus monodon* broodstock may be obtained from specialist importers; in many other situations limited supplies of broodstock favour the use of smaller tanks.

Hatcheries rear shrimp through three larval sub-stages (nauplius, protozoea and mysis) to produce post-larvae. The whole rearing process may take place in a single tank or be split into two separate rearing operations: nauplius through to mysis, and mysis onwards. Such two-stage operations use a smaller elevated tank (5–10 mt) from which mysis larvae are transferred by gravity to a larger (10–30 mt) tank for culture to post-larvae.

Water quality is more critical during larvae culture than at any other stage in the life cycle. It can often be improved marginally by the addition of EDTA, a chelating agent. EDTA may also have a beneficial antibacterial action (Fegan 1992). The need to control populations of pathogenic bacteria is an enduring feature of most rearing systems and antibiotics are commonly applied even though there are several problems associated with their routine use (section 11.3.4). The alternative probiotic approach has had mixed results. When specially selected non-pathogenic bacteria are used to inoculate larvae cultures to pre-empt and displace pathogenic bacteria, it is very difficult to obtain reproducible results and demonstrate scientifically that a change in bacterial flora has resulted in better larvae survival rates (N. Simões, 2000 pers. comm.). The results from one series of trials are perhaps typical (Anon. 1998): larvae cultures were inoculated with the harmless bacteria *Vibrio alginolyticus* to reduce problems with pathogenic luminescent strains of *V. harveyi* and as a result a 2-day delay in the development of *V. harveyi* populations was recorded. However when *Artemia* nauplii were introduced as a live feed the levels of *V. harveyi* rose very sharply and the benefits from the probiotic were quickly negated. In the end the highest larvae survival rates were achieved in parallel trials employing antibiotics. Nevertheless, bacteria-based systems are used in Taiwan (see above) and the positive results obtained in some hatcheries suggest continued research in this area would be worthwhile (sections 8.9.4.2 and 12.2).

While a hatchery is in operation the levels and detrimental effects of pathogens, especially bacteria, increase to the point where larvae survival rates become unacceptably low or at least highly unpredictable. In the face of this, most operators resort to the use of sanitary 'dry-out' periods in which production is suspended while all tanks, reservoirs, apparatus and water networks are disinfected and left to dry. Most hatcheries perform 'dry-outs' at the end of each production cycle for a period of 7–30 days (Wilkenfield 1992). One advantage of small hatcheries is that they can be more straightforward to disinfect than larger units, and because of their smaller capacities, they can be more rapidly restocked when production is resumed.

During the natural planktonic existence of penaeid larvae, live phytoplankton and zooplankton are the most important components of the diet (sections 2.2 and 2.4.8). Correspondingly, in culture, best results are obtained with live feeds such as microalgae and *Artemia* nauplii (section 7.11.2.1). A variety of compounded microparticulate and microencapsulated diets, and novel liquid suspensions containing nutrient droplets (section 8.8.1) and probiotic bacteria, have been developed to reduce dependence on live feeds. Although some are claimed to be suitable replacements, their role in commercial units has been mostly limited to partial replacement of live feeds, or as supplements. Autrand and Vidal (1995) have described one system that has succeeded in replacing microalgae (but not *Artemia*) with artificial particulate diets. To achieve a relatively stable community of heterotrophic and autotrophic bacteria no water was exchanged in the rearing tanks until the larvae had completed the mysis stage. The tanks were only filled to 60% of their capacity when stocked with nauplii and were then progressively topped up each day. Pathogenic bacteria were held in check with the antibiotic furazolidone and fungi were controlled with a trifluraline-based fungicide. Pironet *et al.* (1999) employed a rearing system that retained this reduced water exchange regime but still relied on microalgae – *Chaetoceros* sp.

Manuals and other relevant publications covering penaeid larvae culture in more detail include: Chavez (1990); Treece and Yates (1990); Wyban and Sweeney (1991); Liao (1992); Smith *et al.* (1992); McVey (1993); Treece and Fox (1993).

After metamorphosis food consumption, particularly of *Artemia*, increases substantially. In an attempt to reduce reliance on such expensive feeds, some hatcheries install long arrays of felt-like buoyant ribbons in pre-nursery tanks to provide a kind of artificial seagrass substrate. Prior to use, the ribbons are immersed for 7–10 days in outdoor conditioning tanks inoculated with algae such as *Navicula* and *Amphora* to develop an attached periphyton community that additionally includes nitrifying bacteria. When the newly metamorphosed post-larvae are transferred from the larvae culture to these pre-nursery tanks, the ribbons encourage their dispersal throughout the water volume, thereby reducing crowding. The ribbons may also have a beneficial impact on water quality by acting as an in situ biofilter (Peterson 2000). Such ribbons are commercially available and have been deployed successfully in some intensive nursery and ongrowing systems. The key to their success in hatcheries would appear to be establishing a dense community of potential food organisms on the surface of the ribbons in the pre-nursery tanks. Providing extra substrate alone does not appear to have significant benefit for post-larvae (Samocha *et al.* 1993a).

The stage at which post-larvae leave a hatchery for transfer to the nursery or ongrowing site varies considerably between different species and countries. In Taiwan, *Penaeus monodon* hatcheries usually produce PL_{10-15} (subscripts denote days from hatching), and in Ecuador and Panama, *Litopenaeus vannamei* hatcheries typically rear post-larvae to PL_{4-15}. Some hatcheries in the Americas and Asia, however, move very young PL_{3-4} from larvae rearing tanks into larger on-site post-larvae (or pre-nursery) tanks (see above) to produce PL_{10-20}. In Chinese hatcheries, *Fenneropenaeus chinensis* post-larvae are harvested between PL_7 and PL_{25}. Table 7.3 shows a summary of methods and results for transporting penaeid post-larvae and these data have been combined with those collected by Olin and Fast (1992) to generate Fig. 7.1 which illustrates safe shipping densities for post-larvae as a function of both post-larvae age and duration of journey.

The process of accurately counting penaeid post-larvae can be problematic, particularly with post-larvae older than PL_4, but it needs to be performed to the satisfaction of both the seller and the buyer. The most common approach is to concentrate post-larvae in a tank of known volume (300–500 L, ideally with the depth about the same as the diameter), agitate the water vigorously by hand (avoiding creating circular currents) and swiftly take multiple samples with small vessels of 0.1–1 L. Counts are made by slowly pouring the post-larvae over the lip of the sampling vessel or after pouring the post-larvae onto the surface of a flat sieve. The mean density,

Table 7.3 Methods and results for shipment of penaeid post-larvae.

Duration (h)	Post-larvae size (TL) or age*	Density (no. L^{-1})	Species	Temp. (°C)	Survival (%)	Transport method	Reference/source
unspecified	PL_{20}	500	*M. japonicus*	—	>90	2000 L tank + O_2	Maguire 1979
unspecified	5–25 mm	260–1320	mixed***	24–25	—	Tank + O_2	Pretto 1983
unspecified	PL_{20-25}	500–1250	*P. monodon*	21–24	—	8–10 L seawater in polythene bag + O_2	Primavera 1983
3	PL_{12-15}	2000	*L. vannamei*	25–27	>95	14 L water in 30–35 L plastic bag + O_2, ice optional	Ecuador, unpubl. data
3–6	PL_{12-15}	2000	*L. vannamei*	25–27	>95	14 L water in 30–35 L plastic bag + O_2, ice optional	Ecuador, unpubl. data
6	10–20 mm	70	mixed**	27.2	100	5 L water in 20 L polythene bags + O_2	Franklin *et al.* 1982
6	10–20 mm	100+	mixed**	27.2	>95	5 L water in 20 L polythene bags + O_2	Franklin *et al.* 1982
6–8	PL_2	500–1000	*P. monodon*	—	>80	20 L bags + O_2	Wickins, unpubl. data
6–8	PL_{10-15}	1500	*P. monodon*	—	>80	20 L bags + O_2	Wickins, unpubl. data
6–9	PL_{12-15}	1500	*L. vannamei*	25–27	>95	14 L water in 30–35 L plastic bag + O_2, ice optional	Ecuador, unpubl. data
8			*L. vannamei*	22	99	with wood chips or seaweed	Sanjuan 2000
< 12	PL_{20-25}	375–833	*P. monodon*	24	—	6–8 L plastic bags in styrofoam box + ice & sawdust	NACA 1986
<12	<PL_{15}	2000–3000		20–22	—	packed	Clifford 1992
12	unspecified	700	*M. japonicus*	—	—	1000 L tank with aeration	Kurian 1982
12	PL_5	833	*F. merguiensis*	28.5	—		FAO 1979
12	10–20 mm	70	mixed**	27.2	>95	20 L polythene bags 5 L water + O_2	Franklin *et al.* 1982
>12	>PL_{15}	1000		18–20	—	packed	Clifford 1992
18	10–20 mm	40	mixed**	27.2	100	20 L polythene bags 5 L water + O_2	Franklin *et al.* 1982
18	10–20 mm	50	mixed**	27.2	>95	20 L polythene bags 5 L water + O_2	Franklin *et al.* 1982
18	17 mm	190	*L. vannamei*	18	>99	Double plastic bags, 12 L water + O_2, styrofoam box	Smith & Ribelin 1984
24	10 mm	2500	mixed**	—	—	6 L water in double polythene bags + O_2 & ice	ASEAN 1978
24	10–20 mm	40	mixed**	27.2	97–98	20 L polythene bags 5 L water + O_2	Franklin *et al.* 1982
24	10–20 mm	40–50	mixed**	27.2	>95	20 L polythene bags 5 L water + O_2	Franklin *et al.* 1982
24	17–18 mm	830	mixed**	—	—	6 L water in double polythene bags + O_2 & ice	ASEAN 1978
24	20–24 mm	500	mixed**	—	—	6 L water in double polythene bags + O_2 & ice	ASEAN 1978
< 26	11–14 mm	500	*P. monodon*	28.5	good	18 L plastic bags + O_2	Singh *et al.* 1982
27–30	11–14 mm	375	*P. monodon*	28.5	good	18 L plastic bags + O_2	Singh *et al.* 1982
85	PL_{6-8}	250	*F. penicillatus*		60–70	10 L water in plastic bag + O_2	Liao 1992
85	PL_{15}	250	*P. monodon*		20–30	10 L water in plastic bag + O_2	Liao 1992

*Days past metamorphosis. ***F. indicus*, *P. monodon*, *P. semisulcatus*. ****L. vannamei*, *L. stylirostris*.

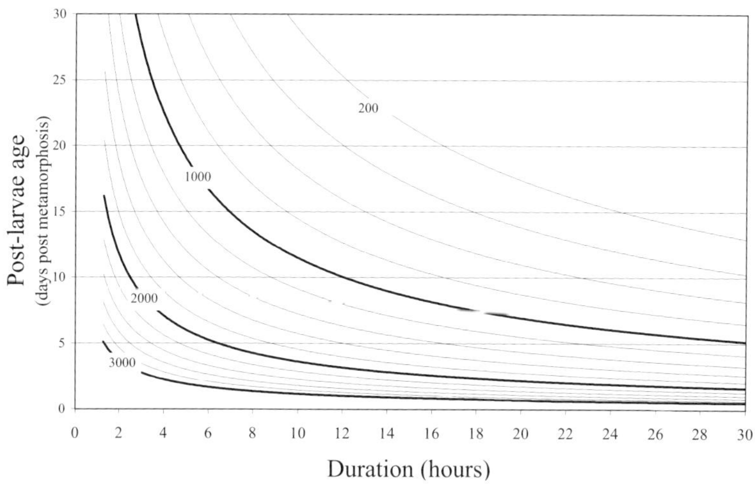

Fig. 7.1 Safe densities (number of post-larvae per litre) for the shipment of penaeid post-larvae as a function of age (days post-metamorphosis) and duration. Packed in plastic bags containing 5–15 litres of water and inflated with oxygen; temperature maintained between 20 and 25°C. (Based on data in Table 7.3 and in Olin & Fast 1992.)

based on the samples, is then multiplied by the volume of the tank to achieve a population estimate. Another counting method, which is popular with small-scale operators in Indonesia, is to concentrate all the post-larvae from a particular rearing tank in a vessel of 100–500 L. Dozens of circular white plastic bowls (40 cm diameter) are half filled with water and laid out on the floor of the hatchery or in an adjacent shaded area. The contents of the tank are then mixed vigorously by hand and 0.5–1 L of concentrated post-larvae transferred to a plastic bowl using a beaker. This process is repeated until the tank is empty and all the post-larvae are in the bowls. The buyer then selects any two bowls and the contents are counted by progressively scooping out the post-larvae with a small white dish into an empty vessel. The average of the two counts is assumed to be the number of post-larvae in each bowl. This process is very visual, giving buyers confidence in their purchase and enabling the condition and behaviour of the post-larvae to be inspected before they are bagged for shipment.

Variations in the quality of post-larvae can have an influence on subsequent performance in nursery or ongrowing ponds. Castille *et al.* (1993) for example found that growth and size variability in *Litopenaeus vannamei* post-larvae were useful indicators of overall quality because they could presage problems with IHHN virus, a pathogen that results in poor growth and survival during ongrowing. Hatchery operators routinely assess a range of post-larvae characteristics to try to evaluate the quality of production batches. Over the past decade stress tests have also become a popular method (Tables 7.4 and 7.5). Despite the appeal of stress tests, there is little or no quantitative evidence confirming the link between performance in a particular test and subsequent performance in a nursery or ongrowing pond (Fegan 1992; Griffith 2000) (sections 8.5 and 12.2). Indeed, the ability of post-larvae to withstand a particular test may be more closely related to their age than to variations in quality. For example, tolerance to salinity fluctuations improves as post-larvae grow and this is related to gill development and improving osmoregulatory capacity. Some studies have looked in detail at the link between measures of post-larvae quality and pond results only to conclude that other factors (feeding, water quality, seasonal effects) seem to have a dominant role (Rigolet *et al.* 1999). Although there is little consensus about which characteristics are most useful in determining overall post-larvae quality, some quality control measures are of value in ensuring that a hatchery is producing a consistent product. A combination of gross inspection, microscopic examination and at least one stress test can form the basis of a useful database and help in the elimination of substandard batches which could give rise to concerns about performance at the farm.

Before post-larvae are transferred to outdoor ponds they are usually acclimatised to their new environment. Post-larvae older than PL_{10} are more resilient to water quality changes, and salinity can be safely reduced at a rate of 3–5‰ h^{-1} down to 15‰, and more slowly thereafter (Jory 1998). It is usually most convenient to perform salinity adjustments within hatchery tanks before post-larvae are sent to the farm.

7.2.5 Nursery

The inclusion of a nursery phase in shrimp culture

Table 7.4 Assessment of post-larvae quality.

Characteristic	Acceptable	Not acceptable	Reference/source
Gross inspection			
Behaviour	Active swimming, occasional tail flicks, clinging to side of sample vessel	Zig-zagging or spiralling	Olin & Fast 1992
Behaviour	Benthic	Planktonic	Fegan 1992
Length heterogeneity	CV <15%	CV >15%	Clifford 1992
Fouling	Clean cuticle and appendages	Fouled/fuzzy cuticle and appendages	Fegan 1992
Average weight	(Temperature, age, species specific)	(Temperature, age, species specific)	D. Lee, 1999 unpublished
Growth rate	(Temperature, age, species specific)	(Temperature, age, species specific)	Samocha & Lawrence 1992
Size	(Temperature, age, species specific)	(Temperature, age, species specific)	Wilkenfield 1992
Morphological deformities (%)	Not specified	Not specified	Clifford 1992
Colour	Not specified	Reddish or pink	Wilkenfield 1992
Microscopic examination			
6th abdominal segment	Shorter than carapace length	Longer than carapace length	Fegan 1992
Rostral spines (*P. monodon*)	6–7 dorsal, 2–3 ventral	1–5 dorsal, 1–2 ventral	Fegan 1992
Gill development	Branched	Unbranched	Fegan 1992
Fouling	Clean cuticle and appendages	Excessive fouling with *Zoothamnium*, *Epistylis*, *Vorticella*, or *Leucothrix*	Fegan 1992
Chromatophores on ventral side of 6th segment (*P. monodon*)	Individual spots with slight spreading of pigment	Coalesced chromatophores forming wide band which dominates tail segment and gives PL red appearance on gross examination	Alday de Graindorge & Flegel 1999
Chromatophores in tail	Developed	Not yet developed	Fegan 1992
Muscle : gut ratio in 6th abdominal segment	4:1 or greater	Less than 4:1	Fegan 1992
Tail or 'back' muscle appearance	Translucent, clear	Opaque, cloudy, white, discoloured, shrunken or striated	Olin & Fast 1992
Hepatopancreas	Full, dark, well developed; lipid vacuoles visible	Empty, shrunken and pale; no lipid vacuoles visible	Olin & Fast 1992
Appendages	Intact, no deformities, no necrosis	Damaged, deformed	Olin & Fast 1992
Anterior gut	Full, well developed	Not full, not well developed	Olin & Fast 1992
Monodon bacculovirus occlusion bodies (*P. monodon*)	Absent	Present	Alday de Graindorge & Flegel 1999

provides farmers with hardy juveniles that have been fully acclimatised to the environment they are likely to encounter during ongrowing. Post-larvae mortality is often most significant among hatchery reared animals produced in very artificial conditions. Nursed juveniles are usually transferred to ongrowing ponds when they have reached 0.1–2 g, by which time most weak animals will have died and a realistic estimate can be made of the number stocked. Although during the ongrowing phase survival rates in a pond can never be determined with precision, a good initial count of the number of juveniles stocked improves control over stocking densities and greatly assists in the calculation of feeding levels (sections 8.3.4 and 8.3.5.1). Other advantages of nursery ponds are the relative ease with which predators can be eliminated from small ponds rather than larger ongrowing ponds, and the fact that they can be used for 'stockpiling' reserves of juveniles for periods as long as 6 months.

The principal problem with nurseries arises when juveniles are transferred to ongrowing units. The process usually involves draining into a netted sump and be-

Table 7.5 Examples of stress tests for assessing the quality of post-larvae.

Stress test	Acceptable	Not acceptable	Species, age	Reference/source
Temperature: 20°C for 1 h	>80% survival	<80% survival	*L. vannamei*	Villalon 1993
Salinity: drop by 15‰	100% survival	<100% survival	*P. monodon*	Bauman & Jamandre 1990
Salinity: 0‰ for 0.5 h then return to normal hatchery salinity	>70% survival	<70% survival	*P. monodon* PL_{17}	D. Lee, 1999 unpubl.
Salinity: 5‰ for 1 h	>80% survival	<80% survival	*L. vannamei*	Villalon 1993
Salinity: 3‰ for 2 h	>40% survival	<40% survival	*L. vannamei* PL_7	Samocha *et al.* 1998
Salinity: drop from 30 to 5‰	>85% survival	<85% survival	*L. vannamei* PL_{10}	Olin & Fast 1992
Salinity: drop from 30 to 10‰	>85% survival	<85% survival	*P. monodon* PL_{20}	Olin & Fast 1992
Combined salinity and temperature: 20‰ and 10°C for 4 h	>60% survival	<60% survival		Clifford 1992
Combined salinity and temperature: drop by 20‰ and 10°C, hold for 4 h	>70% survival	<70% survival		Clifford 1994
Combined salinity and temperature: drop to 20°C and 7.5‰ or 10‰, or raise to 45‰			*P. monodon* PL_{15}	Briggs 1992
Formalin exposure: 600 ppm for 2 h	>40% survival	<40% survival	*L. vannamei* PL_7	Samocha *et al.* 1998
Formalin exposure: 100 ppm for 2 h	100% survival	<100% survival	*P. monodon*	Bauman & Jamandre 1990

cause it involves much handling of the shrimp, substantial mortalities can occur unless great care is taken. Sensible precautions include working during cool periods of the day and avoiding prolonged overcrowding and any marked deterioration in water quality. To simplify the whole procedure some farms incorporate nurseries adjacent to ongrowing ponds, which enables transfer to be accomplished by gravity flow. Yet although this method avoids handling mortalities, it does not permit an accurate population estimate to be made. Fish pumps can be successfully employed to transfer shrimp juveniles from ponds to collecting tanks. This pumping process results in minimal stress when pond levels are dropped slowly, gradually concentrating the juveniles at a rate at which they can be successfully handled by the transfer crew.

In modern semi-intensive farms, particularly in the Americas, nursery ponds are usually incorporated in the design of the whole farm. They may represent between 6% and 15% of the total culture area. They are usually made with earthen embankments, and, at sizes of 0.04–1 ha, are much smaller than the ongrowing ponds. Stocking densities are typically 100–200 juveniles m^{-2}. In South-east Asia, nursery facilities also take the form of concrete tanks, concrete walled ponds with sand bottoms, staked net pens and floating cages. These are either managed as independent operations by specialist nursery growers or are included as an integral part of a farm or hatchery. Stocking densities are typically 50–100 post-larvae m^{-2}. Fixed or floating cages (sometimes known as 'hapas') are made from fine mesh netting (0.5 mm) attached to a wooden frame and may contain 30 000 PL_5 per cage (3.7×2.7×1.3 m deep). The shrimp are fed a paste of finely ground trash fish placed on feeding trays, but later mussel meat may be provided, suspended from an array of hooks. The juveniles are transferred for ongrowing after 15–25 days.

In extensive farms which rely on trapping wild juveniles, nursery ponds can play an additional role since young fish predators entering alongside shrimp seed can be eradicated with rotenone or saponin (section 8.3.6.1) before shrimp are transferred for ongrowing. Some extensive operations have created nursery ponds especially for this purpose by dividing existing large ponds into smaller units. In China, post-larvae were traditionally stocked directly into ongrowing ponds, thus by-passing the nursery phase altogether, but it has become popular to use small nursery ponds for 2–3 weeks before ongrowing (Qingyin 1992). This approach is more efficient because ongrowing ponds that used to be stocked with post-larvae at densities of around 50 m^{-2}, in anticipation of low survival rates, can produce equivalent yields when stocked with hardier, nursed juveniles at just 7.5–15 m^{-2}.

Nursery management, particularly within integrated projects, requires good organisation and is especially critical with the tight production schedules of intensive farms. Attention must be given to the duration of ongrowing, the timing of output, and the quantity and size of juveniles at harvest. Table 4.4 summarises the results of nursery rearing trials with various crustaceans.

More intensive nursery systems, sometimes within greenhouse structures, have been developed, particularly in the USA, as a means of saving space, improving biosecurity (see Glossary), rearing bait shrimp, and producing juveniles ready for early stocking in a limited ongrowing season (Samocha *et al.* 1993b). Systems in Hawaii have made use of 6 m diameter circular tanks with fibreglass walls and concrete bottoms, stocked at densities of 800–1200 shrimp m^{-2} in which *Litopenaeus vannamei* PL_{5-10} reached 0.3–2 g after 30–50 days (Wyban & Sweeney 1991). Higher densities, up to 7800 m^{-2}, may also be feasible with careful management and attention to waste removal (Samocha & Lawrence 1992). The need for biosecurity has stimulated renewed interest in enclosed tank or raceway systems as an alternative to the use of open ponds. When aeration and artificial substrates are employed, it may be possible to stock these systems with virus-screened ('high-health' – section 8.9.4.4) PL_{10-12} at densities as high as 20 000 m^{-2} (Peterson & Griffith 1999). Greenhouse enclosed raceways have also shown potential for producing bait shrimp, *Farfantepenaeus duorarum*, at the 7–12 g size favoured by sport fishermen in the Gulf of Mexico. Techniques are being commercially developed in Florida to produce harvests after 60–70 days of rearing (Dixon 1999).

7.2.6 Ongrowing

The different techniques used for ongrowing penaeids are conveniently categorised into four groups, primarily on the basis of the expected yield at harvest and stocking density (section 5.2). Table 7.6 shows typical features of the categories: extensive, semi-intensive, intensive and super-intensive. Considerable overlap can occur, however, between the different terms in published literature since some definitions attempt to include levels of feed and water management. The economic implications of growing shrimp at different levels of intensity are discussed in Chapter 10 (section 10.5) while some of the merits of the various options are also discussed in Chapter 5.

Table 7.6 Typical features of penaeid ongrowing systems.

	Extensive	Semi-intensive	Intensive	Super-intensive
General features	Low-density culture often with fish & crabs, sometimes in rotation with salt or rice	Moderate density culture of mixed or single shrimp species	High-density monoculture	Very high-density or controlled environment monoculture
Water system	Tidal flushing via single inlet/outlet mainly on spring tides	Daily tidal &/or pumped water exchange often through separate inlet/outlet channels	Reduced or zero water exchange relying instead on water treatment and reuse	Continuous, high rate of exchange with new or treated water; none in zero exchange systems
Source of juveniles	Wild juveniles trapped in pond during filling or collected from wild	Hatchery-reared and wild juveniles. Often separate nursery ponds	Hatchery- and nursery-reared juveniles	Hatchery- and nursery-reared juveniles
Fertilisation	None or organic prior to filling	Organic or inorganic on filling and as required	Organic or inorganic as required	None
Feed	Shrimp rely wholely or mostly on natural productivity	Natural productivity + supplementary feeds of low-price pellets, molluscs or trash fish	Artificial formulated diets + occasional fresh fish/mollusc flesh	Artificial formulated diets
Aeration	None	Paddlewheels or propeller-aspirator pumps used when needed	Paddlewheels or propeller-aspirator pumps used regularly	Continuous aeration, sometimes compressed oxygen
Pond sizes (ha)	0.5–100 0.3–1 m deep	0.2–3 0.8–1.5 m deep	0.1–1.0 1.2–1.5 m deep	0.03–0.1 2 m deep (Japan; lined ponds); 0.6 m deep (raceways)
Stocking density (no. m^{-2})	0.2–5.0	5.0–20	15–50	50–250
Shrimp size (g)	10–50	15–40	15–40	15–25
Yield $crop^{-1}$ (mt ha^{-1})	<1	0.5–5.0	5–15	10–50

7.2.6.1 Extensive

The simplest forms of penaeid shrimp farming are the traditional extensive methods of southern and Southeast Asia, which involve the trapping and growing of wild shrimp juveniles in shallow earthen impoundments. Farms are built on low-lying tidal flats and marshlands along coastal and estuarine margins, and stocking relies upon the influx of naturally occurring shrimp juveniles when the impoundments are filled. In general however, the trapping of shrimp seed in this way allows little control over the quantity and species stocked and permits the entry of predatory and competitive crabs and fish. Although rotenone or saponin can be applied to kill the fish (section 8.3.6.1), many extensive operations now make use of wild shrimp juveniles netted or trapped in nearby waters. Total stocking densities in extensive systems, however, remain low – between 0.2 and 5 shrimp m^{-2}.

Catches of wild shrimp juveniles vary with latitude and seasonally both in quantity and component species, and effective management of an extensive farm requires knowledge of the usual patterns of seed availability. In the backwaters and estuaries of India, George and Suseelan (1982) found that *Fenneropenaeus indicus* and *Metapenaeus dobsoni* were the most abundant species and that the highest concentrations of *Penaeus monodon* juveniles were found on the middle and northern regions of the east coast. Catches are often greatest around spring tides. Also, in Vietnam, the peak abundance of penaeid seed occurs from April to June and the equinoctial tide in late April is a favoured period for trapping (Quynh 1990).

Significant fisheries for wild shrimp juveniles have developed. In Ecuador a seasonal workforce of 32 000 'semilleros' provides the shrimp industry with around $5–13 \times 10^9$ *Litopenaeus vannamei* juveniles per year. In Vietnam a typical family engaged in fishing shrimp seed may be able to collect 20 000 *Fenneropenaeus merguiensis* juveniles in a day (Quynh 1989). The capture methods employed depend largely upon the nature of the collection site and include:

- boat and raft-mounted scissor nets and fine mesh trawl nets;
- one-man operated scissor nets (push nets);
- beach seines;
- dip-nets used in conjunction with lures made from grass and twigs;
- nets set at high tide on shallow tidal banks parallel to the shore to catch post-larvae as the tide recedes.

Unwanted larvae and juveniles caught along with the shrimp are sometimes picked out by hand, and fish can be eliminated by the use of selective poisons. Fortunately, juvenile shrimp are generally more resistant than fish to the stresses of capture and transport and they tend to survive in greater numbers. Unfortunately, however, the whole process is very wasteful and it has been estimated that for every single shrimp juvenile grown in a pond, up to 100 shrimp and fish juveniles are killed (section 11.3.1). Improvements in handling, such as keeping containers in the shade and providing oxygen and/or

Plate 7.1 Artisanal collection of prawn juveniles. Inevitably a variety of other species are also trapped. (Photo courtesy John Dallimore, JD & Associates, Hamburg, Germany.)

aeration during prolonged holding or transport, would reduce the wastage although some buyers like to view their seed in outdoor, unshaded tanks, believing that this treatment will toughen the seed.

Healthy wild juveniles can usually be identified by observing their level of activity. Conflict between the buyers and sellers of seed arises over the quantity involved (reliable counts are difficult and laborious to obtain) and the relative abundance of preferred species. In Ecuador buyers only pay for the juveniles of *Litopenaeus vannamei* and *L. stylirostris*; other shrimp, notably *Farfantepenaeus californiensis* and *Litopenaeus occidentalis*, are considered 'weed' species because they suffer high mortality in ponds and individuals do not attain more than a few grams in weight. Identifying the different species, however, is another laborious task, which can rarely be performed with precision in the field and often relies on examining the juveniles under a low-power binocular microscope. Guides to the identification of post-larval shrimp have been prepared for various areas including: Australia (Heales *et al.* 1985), Ecuador (Yoong & Reinoso 1983), the Philippines (Motoh & Buri 1981) and the Indian Ocean (Paulinose 1982).

Extensive farming systems use little or no supplementary feed, relying instead upon the natural productivity of a pond to provide the shrimp with food in the form of small benthic invertebrates, algae and organic detritus. This productivity is sometimes boosted by the addition of organic fertilisers such as poultry and cow dung. Table 5.1 shows some examples of extensive cultures and the resulting yields and harvest sizes. In the Philippines extensive farming is practised at densities of 0.2–1 shrimp m^{-2}, and trials with *Penaeus monodon* to assess the productivity of ponds receiving a mixture of chicken manure and inorganic fertiliser alone yielded 100–200 kg ha^{-1} per crop of 28–35 g shrimp after 86 days (Subosa & Bautista 1991). Extensive farming, as practised in the Mekong Delta of Vietnam, involves the use of intertidal ponds which are small (1–3 ha) in the case of family concerns but may be larger (100–300 ha) in the case of state farms. Ponds are shallow, rarely deeper than 0.5 m and are stocked with wild post-larvae of up to six different penaeids: *Fenneropenaeus merguiensis, F. indicus, Penaeus semisulcatus, P. monodon, Metapenaeus ensis* and *M. affinis*. But unreliable recruitment and acid sulphate soils limit productivity to 100–400 kg ha^{-1} yr^{-1} (Binh & Lin 1995).

Extensive ponds are typically shallow (0.3–1.0 m) and often irregular in size and shape. They usually require routine maintenance of embankments and inlet/outlet sluices, and because of their low-lying positions often need repairs after storm and flood damage. Typically the pond bottoms contain undrainable pools which are difficult or impossible to dry after harvest and may have to be netted or treated with fish poisons before the pond is restocked. In some cases extensive ponds are constructed by raising the bunds of rice paddies or salt pans and sometimes, in Bangladesh and Vietnam, salt or rice is produced in seasonal rotation with shrimp.

7.2.6.2 Polyculture

In extensive systems, shrimp are usually harvested along with an incidental mixture of fish and crabs. Alternatively, in places such as Taiwan, the Philippines and Indonesia, shrimp ponds may be deliberately stocked with the fry of selected species such as milkfish (*Chanos chanos*) and mud crab (*Scylla* spp.) (section 7.10.4). In Taiwan, *Penaeus monodon* stocked at 150 000–500 000 ha^{-1} may be combined with milkfish at 200–300 ha^{-1} and mud crabs at up to 1000 ha^{-1} (Chen 1990). Clams are also polycultured with shrimp in Taiwan (Liao 1992).

In more intensive systems in Thailand, filter feeding fish such as tilapia and mullet, and mussels, cockles, polychaete worms and seaweeds have all been proposed for production with shrimp. However, to avoid problems with competition for feed, space and oxygen, they are typically put in reservoirs or inlet canals rather than into ongrowing ponds. Results are not always encouraging, with slow growth of fish even at low stocking densities; also tilapia can actually become a pest because they breed so rapidly and because their nest-building behaviour can disturb sediments. If pond effluent is to be reused, effective sedimentation processes must not be disturbed. Chanratchakool *et al.* (1998) conclude that the possible financial benefits and extra work involved in polyculture must be carefully assessed before it is attempted.

In Vietnam shrimp yields of 300–400 kg ha^{-1} yr^{-1} are feasible in rotation with rice (4–5 mt ha^{-1} yr^{-1}) in those rice paddies that are affected by seawater intrusion during the dry season. Surprisingly the overall impact of flooding dry rice paddies with seawater may actually be positive because it waterlogs the soil and prevents acid sulphate problems (section 8.3.8). In the rainy season most residual salt is flushed out and salt-tolerant rice varieties can be grown. Another farming system in Vietnam involves rotation of shrimp and salt. Salt is produced in the dry season and is the main product, with a value ten times greater than that of the shrimp. Shrimp production is carried out in the wet season and is limited to 100–150 kg ha^{-1} yr^{-1} because ponds are particularly shallow (20–40 cm).

Some salt operations in Vietnam produce *Artemia* in their evaporation ponds during the dry season while simultaneously extracting salt in adjoining ponds. However in the rainy season the salinity falls too low for *Artemia* cyst production and shrimp are stocked instead, with adult *Artemia* making a useful contribution to the shrimp's diet. On an annual average, yields per hectare may be 84 kg cysts (wet weight), 164 kg of shrimp and 374 mt of salt. The system has not been widely adopted because it is still experimental and it requires additional inputs of management time, capital and technical expertise, all of which are in short supply (Binh & Lin 1995) (Table 5.6 and section 7.11.2.1).

Vietnam has also pioneered work with mixed shrimp–mangrove forestry (silviculture) farms. Each operation covers 2–17 ha and, in theory, 70% of the surface area is dedicated to mangrove. Shrimp are reared in long, shallow (40–60 cm), narrow (3–4 m) channels between raised areas of soil that may or may not be covered with mangrove (*Rhizophora apiculata* at 20 000 ha^{-1}). Farmers are required to manage their allotted mangroves as part of their lease agreement and the mangrove provides timber, thatching material, fuel wood and charcoal and a natural habitat for, among other things, juvenile shrimp. The channels are stocked with wild shrimp post-larvae on high tides but problems with acid sulphate soils and severe disease outbreaks have limited shrimp yields and undermined the financial viability of the system. The difficulty of enforcing the rules of the scheme has also resulted in farmers eliminating too much mangrove, to the detriment of the local ecology (Johnston *et al.* 1999). Shrimp–mangrove systems can potentially yield 100–600 kg of shrimp ha^{-1} yr^{-1} (Binh & Lin 1995).

The crop rotation, or polyculture, of shrimp with tilapia has been tried as a means of ameliorating the impact of white spot syndrome virus (WSSV) by breaking the cycle of infection. Tilapia may indeed help by consuming dead shrimp and other crustacean hosts of the virus (Frese 2000). It is certainly appealing to stock a crop of fish in shrimp ponds that would otherwise lay empty because of disease, but there may be problems for newcomers to tilapia farming who try to sell fish on markets that are prone to rapid saturation (Chamorro 2000).

7.2.6.3 Semi-intensive

Semi-intensive farming generates greater yields than extensive methods through the stocking of selected shrimp species at higher densities, the provision of supplementary feed, and the use of improved water exchange rates. Reliance is often placed on supplies of hatchery-reared seed, and fast-growing species such as *Penaeus monodon*, *Litopenaeus vannamei*, *L. stylirostris* and *Fenneropenaeus chinensis* are usually preferred. Often an intermediate phase of nursery rearing is performed (section 7.2.5) and ongrowing ponds are typically stocked at densities between 5 and 20 shrimp m^{-2}. Table 5.2a gives examples of yields from a range of different semi-intensive crustacean farming operations.

Natural productivity in ponds is often boosted by the application of organic or inorganic fertilisers (section 8.3.6.2), but at the shrimp densities employed in semi-intensive farming this productivity alone is insufficient to provide for rapid shrimp growth, and supplementary feeding becomes essential (section 8.3.6.3). Although compound diets are mostly used, these are sometimes combined with fresh foods such as trash fish, and in China, a diet of crushed whole mussels is sometimes provided. In Ecuador an average food conversion ratio of 2.46 : 1 has been recorded at stocking densities around 5.2 m^{-2}. Food conversion ratios can be improved by the use of feeding trays as an alternative to broadcasting feed over the whole pond. This feed management technique is becoming popular and is described in section 8.3.6.3. In northern China as much as 30 mt of crushed whole mussels may be used to produce a tonne of shrimp, and although at first this may appear to be excessive, for mussels of 40–45 mm this equates to a food conversion ratio of around 1.5–2 : 1 (dry weight mussel meat : live weight shrimp harvested).

Farms originally built for extensive farming can be transformed for semi-intensive operation by a process which usually involves deepening ponds, strengthening embankments, building independent inlet and outlet sluices, creating nursery ponds, and providing a pumped water supply (section 8.2). This process requires capital investment, but if combined with controlled stocking and supplementary feeding it can greatly boost yields.

Semi-intensive shrimp ponds are usually 0.5–1.5 m deep, rectangular in shape, and are equipped with separate inlet and outlet sluices in opposing embankments. Within one farm, ponds are often built with uniform sizes because this can simplify the routine management of stocking, feeding and water exchange rates. Water exchange may rely on tides or pumps, but if ponds are built above the level of high tide, as is often the case in Ecuador, pumping becomes essential. The advantage of

Plate 7.2 Aerial view of a large Ecuadorian shrimp farm showing the pumping station (bottom right) and supply canal (centre) bounded by nursery ponds. In the background are a number of irregularly shaped ongrowing ponds of 4–25 ha.

raised ponds is that they can be constructed without the destruction of large areas of intertidal habitat and, because they can be drained at any state of the tide, harvesting and pond bottom preparation are simplified. In some farms an enlarged supply canal serves as a reservoir and is sometimes used to hold broodstock at low densities. Paddlewheel or other aeration devices are often installed in ponds and operated at night to avoid critically low oxygen tensions, especially in the early hours of the morning (section 8.3.6.6). In most semi-intensive systems water quality is maintained by water exchange alone.

In Central and South America semi-intensive farms may be large. In Ecuador, for example, the average farm size is in excess of 100 ha and 500 ha is not unusual. Individual ponds measure 10–20 ha and water is usually supplied by diesel powered pumps via a central reservoir. The national average productivity is 700 kg ha^{-1} yr^{-1} (Griffith & Schwarz 1999) based largely on production of *Litopenaeus vannamei*, but individual ponds and farms can generate greater yields if they are constructed and managed well. Clifford (1994) made a case study of the methods used to achieve good productivity with *L. vannamei* in semi-intensive farms in Venezuela. Ponds of 10 ha were stocked with PL_{12} at a density of 18–22 m^{-2} and 100 days later crops of 1800–2800 kg ha^{-1} were harvested. This level of output is probably close to the limit for large semi-intensive ponds that do not utilise aerators. The shrimp used were derived from captive pond-reared stock held for eight generations. Before the ponds were filled with water, the dry pond beds were tilled and fertilised with chicken manure at 1000 kg ha^{-1}. Incoming water was filtered through screens of 700 μm and additional fertilisers – urea (60 kg ha^{-1}) and triple superphosphate (3 kg ha^{-1}) – were added. Two weeks later postlarvae were stocked. Food was distributed by boat four times per day at 07:00, 13:00, 18:00 and 23:00 in the proportions 20 : 10 : 40 : 30 respectively. Food conversion ratios of 1.1 : 1 were achieved using a pelleted diet containing 35% protein through careful adjustments of daily rations based on checking consumption rates of small amounts of feed placed on feeding trays. Algae densities were managed by additions of urea (23 kg ha^{-1}) and triple superphosphate (2.3 kg ha^{-1}) to maintain turbidity around 45 cm (measured by Secchi disc). Water was exchanged at a rate of 2–12% of pond volume per day to maintain water quality.

Unfortunately reliance on water exchange has turned out to be a weak point in semi-intensive systems because it can compromise biosecurity and lead to the rapid spread of pathogens, particularly viruses (section 9.7.2). The effluents generated by a typical semi-intensive farm may also have potentially negative environmental consequences (section 11.4.3) so for the development of new systems more attention is now being focused on methods with reduced and zero-water exchange (sections 7.2.6.5 and 7.2.6.6).

7.2.6.4 Cage and pen culture

Pens and cages are relatively rare for shrimp farming, although they can be used with minimal investment and as a part-time activity, if suitable sites are available. Pen culture is practised in the Philippines, Singapore and India, and in Thailand it has demonstrated high productivity and has attracted some small-scale operators. Cages are often employed as shrimp nurseries (section 7.2.5). A significant drawback of all pens and cages, however, is their vulnerability to adverse weather conditions. Suitable sites must be relatively pollution-free, sheltered from waves and excessively strong currents, and provide a water depth of around 1–2 m on lowest tides.

Pens normally consist of plastic mesh nets supported by rectangular bamboo or wood frames, and are fastened by the corners to bamboo poles. The bases of the nets are in contact with the mud. The mesh, which should be as coarse as possible without permitting the shrimp to escape (1.4 mm for juveniles, then 5 mm after 1 month), allows water exchange and also provides a surface for the attachment of microalgae on which the shrimp can graze. In Thailand, *Penaeus monodon* are stocked at densities of 40–110 m^{-2}, and either fresh or pelleted feed or a combination of both is provided three or four times per day. The food is lowered into the pens on mesh feeding trays (0.5 × 0.5 m). Feeding rates for pellets are initially 10–20% body wt per day and fall steadily to 3.0% by the end of the 3–4-month culture period. The successful elimination of predators usually results in survival rates around 70%, and 100 kg of 30 g shrimp might be expected from a pen measuring 6 × 6 × 6 m (Tookwinas 1990). In the cage culture of *Fenneropenaeus merguiensis* in Singapore, annual yields of shrimp (average live weight 12 g) equivalent to 20 000–30 000 kg ha^{-1} have been obtained (Table 5.2c).

Although in theory pens should allow efficient exchange of water and thereby avoid problems with oxygen depletion, seaweed and other fouling organisms can block the mesh. In such cases shrimp may require transfer to clean pens and aeration may become necessary towards the second month of the culture period. When large numbers of pens become concentrated in suitable sites (section 11.2.5) water quality problems often arise.

Pens offer the advantage of relatively straightforward harvesting using scoop nets or cast nets, and some can simply be lifted in their entirety. After the harvest, netting can be cleaned or replaced prior to restocking, and, if adequate supplies of nursery-reared juveniles are available, three crops per year may be possible.

One particular type of pen or impoundment, often referred to as the Amakusa-style after the island in Japan where they are operated, comprises an intertidal area bounded by low intertidal concrete walls about 1 m high. The walls are topped with a mesh barrier supported by a wooden frame, to prevent the escape of the crop when the pens are flooded. Pens range in size from 0.2 to 1 ha, and when stocked with *Marsupenaeus japonicus* at 20–30 m^{-2} can produce 3–4 mt ha^{-1} annually (Kungvankij & Kongkeo 1988).

7.2.6.5 Intensive

Intensive shrimp farming is performed in small ponds of 0.1–1.0 ha and results in yields of between 5 and 15 mt ha^{-1} per crop. Nursed juveniles of hatchery origin are stocked at high densities of between 15 and 50 shrimp m^{-2}, and are fed on compounded diets. With dependable supplies of juveniles, intensive farms in the tropics can produce up to three crops per year. Pond conditions are carefully controlled, sometimes through water exchange, but aeration is usually provided on a continuous or regular basis and indeed aeration is becoming the preferred tool of water quality management in intensive systems. This reflects a move towards intensive farming with little or no water renewal at all, so-called zero exchange systems, that hold out potential for production with greatly reduced disease problems and environmental impacts.

Intensive ponds are usually well-made structures with concrete walls or plastic lined embankments. They are typically 1–1.5 m deep, usually rectangular or circular in shape, and can be efficiently drained between crops to allow for the removal of accumulated organic sediment. Pond bottoms are usually natural earth or an artificial layer of 10 cm of sand or gravel. Ponds completely lined in plastic are generally deeper (up to 2 m water depth). After harvest, earthen pond beds may be conditioned (section 8.3.3) by sun-drying, ploughing and dressing with lime (200–300 kg ha^{-1}). In Taiwan, ponds are sometimes treated with saponin (2.5–10 ppm), added in the form of tea seed cake in a shallow layer of water, to eliminate pests.

Natural productivity is usually encouraged by the use of fertilisers: sometimes manure (1 mt ha^{-1} annually) but more usually with daily or weekly additions of inorganic fertiliser (section 8.3.6.2). For *Penaeus monodon*, an example of recommended feeding rates using compound diets containing 35–39% protein is:

Daily ration (% body wt per day)	Shrimp size (g)
25–15%	<2
15–7%	2–5
10–7%	6–15
5–3%	>15

These rates, however, require daily adjustment in accordance with observations of food remains on nets placed on the pond bottoms. Frequent feeding, four to six times per day, improves the utilisation of feed and conversion ratios of less than 2 : 1 are reported in Taiwan (Chen 1990).

Water is supplied to the ponds by pumping, and paddlewheels or propeller-aspirator pumps are arranged to impart a circular motion to the water. This action provides a more even distribution of dissolved oxygen, and in centrally drained ponds sweeps waste towards the centre where it can be voided. Long-shaft paddlewheels, with 10–20 wheels per shaft, are a becoming a popular tool for aeration and water circulation. Water quality is carefully monitored. If an algae bloom becomes excessively dense, denoted by a low Secchi disc reading (section 8.3.5.2), it can be treated with an algaecide such as simazine or cutrine or diluted by increasing the water exchange rate (section 8.3.6.4).

The intensive farming systems for *Penaeus monodon*, originally developed in Taiwan in the early 1980s, placed great reliance on high water exchange rates that increased during the production cycle from around 5% per day in the first month to around 30% per day in the fourth month. However this rate of water usage would now be considered excessive. The world leaders in shrimp production are, at the time of writing (2000), the Thais and they have adapted intensive systems to operate with reduced or zero water exchange. The success of their intensive farms is illustrated by the fact that some 80% of all Thai production, some 160 000 mt yr^{-1}, comes from just 27 000 ha of intensive ponds. Most farms are owned by small-scale operators, comprise just one or two square or round ponds of 0.16–1.6 ha, and are built on good quality soils previously used for rice farming. They are small, flexible operations that can respond swiftly to changes of season, disease outbreaks and market price fluctuations.

Of necessity in the early 1990s farms cut their reliance on water exchange because of environmental and disease problems caused by the over-concentration of farms. They started to set aside ponds for use as reservoirs and for the sedimentation of waste to enable water reuse. Clean water is brought into farms once per cycle only on good spring tides, is held in a reservoir and progressively transferred to ongrowing ponds until it is used up. Water is then pumped from the ongrowing ponds back into the reservoir and held for sedimentation. Later on it is reused in ongrowing ponds after passing down a canal equipped with paddlewheel aerators. The pond salinity may rise to 40‰ but if new water is needed it is first disinfected by chlorination (Kongkeo 1995).

In another approach to avoiding seawater-borne diseases, in areas with salty soils, ponds may actually be filled with freshwater. The freshwater leaches salt from the soil and may attain a salinity of 5‰ after 1–2 weeks. Surprisingly 30% of Thai production actually comes from freshwater areas, sometimes as much as 200 km from the sea (Kongkeo 1997). Hypersaline water, brought in by road tankers, is mixed with freshwater prior to stocking the shrimp. No water exchange is performed except for some topping up with freshwater. This approach has however potential negative environmental impacts if precautions are not taken to prevent saline water affecting surrounding land (section 11.4.3). Farms located next to the Andaman Sea make use of high-quality full-strength seawater, also in an attempt to avoid diseases. Pond salinity may reach 45‰ in such cases, but the shrimp tolerate these conditions.

The variety of different water sources used for shrimp culture continues to expand. In Arizona saline groundwater is used for intensive and semi-intensive farming of *Litopenaeus vannamei*. The water tends to favour blooms of blue-green algae rather than the more usual diatoms, but yields of 4200 kg ha^{-1} per crop have been achieved and it has even proved feasible to reuse the waste water for irrigating olive groves and fields of sorghum (Fitzsimmons 1999).

The zero exchange approach to shrimp farming has been taken one step further in experimental trials with partitioned aquaculture systems. Such systems separate the culture functions and the wastewater treatment into different parts of the same body of water to facilitate the management of both production and waste treatment. The cultured species is reared at high density in a small portion of the total water volume in a pen and the water is continually circulated between the two parts of the system (Brune *et al.* 2000).

Small plastic-lined ponds (0.1–0.36 ha × 1.5–2.0 m deep) have demonstrated their suitability for intensive shrimp production. Durable liners, often made of high-density polyethylene (HDPE), can prevent embankment erosion, simplify harvesting and facilitate pond cleaning and disinfection between crops. In Oman, ponds of 0.1 ha have provided harvests averaging 5900 kg ha^{-1} using imported *Penaeus monodon* (P. Fuke, 1990 pers. comm.), and trials in Texas with *Litopenaeus vannamei* have demonstrated potential for high survival rates (>90%) and productivity equalling or exceeding that of ponds with conventional soil or sand substrates (Anon. 1990). Although more expensive (section 10.6.1.5), lined ponds have the advantage of preventing salt intrusion into groundwater and have been successfully used in sites where sandy permeable or highly acid soils would normally preclude pond construction. They can also save time because there is no need to condition or till soil between crops. It has also been found that lined ponds can improve feed conversion ratios and reduce problems with anoxic sludge (Singh 1993).

The need for efficient water circulation and waste removal in intensive ponds has focused attention on the advantages of circular ponds. The Japanese originally developed concrete round ponds for the super-intensive farming of *Marsupenaeus japonicus* (section 7.2.6.6). Researchers in Hawaii developed cheaper round ponds made of compacted earth with ferrocement berms, or with plastic liners. Ponds of 0.2 ha were considered an ideal size and with four paddlewheel aerators of 1 hp and a central 6 m diameter sump area, good circulation patterns and elimination of settled waste were achieved (Wyban & Sweeney 1991). Earthen round ponds have been adopted on a fully commercial scale by, among others, an intensive farm on the Red Sea coast of Saudi Arabia (Falaise & Boël 1999). This particular project combines the attributes of a well-engineered farm, round ponds and the use of large pre-treatment reservoirs to 'green' incoming full-strength seawater with algae prior to use in ongrowing ponds. Around 100 ha of circular ponds are supplied by 100 ha of reservoirs and effluent passes into a large settlement pond before rejoining the Red Sea.

7.2.6.6 Super-intensive

Harvests in excess of 2 kg m^{-2} are possible in super-intensive systems by using very high stocking densities (50–250 shrimp m^{-2}), relying almost totally on compounded feeds, and by very careful manipulation of the culture conditions. By providing optimal conditions, assisted by water treatment and high levels of aeration, good survival rates can be obtained. Sometimes controlled, enclosed environments or 'battery systems' have been used (see below).

Several distinct super-intensive systems have been developed to a commercial scale. In Japan, the shortage of land suitable for building large shrimp ponds, and the existence of an exceptionally high-value market for live *Marsupenaeus japonicus* (section 3.3.1.2), stimulated the development of increasingly intensive farming systems. Circular concrete tanks of 1000–2000 m^2 and 2 m deep, designed by and named after Kunihiko Shigueno, are operated commercially in Kagoshima at sites with high-quality seawater. Post-larvae (PL_{30}) may be stocked initially at as many as 300–400 m^{-2} and thinned out later to 140–150 m^{-2} when they reach 5 g. Sometimes, however, lower densities of 70–80 m^{-2} are employed. High-protein formulated diets (minimum 55% protein) are provided in four daily feedings initially at 12–18% body wt day^{-1} but reduced to 10% at 1 g and 3–5% at 5 g. Market size animals around 20 g can be obtained after 5–6 months (Table 5.3), and with good quality feeds and careful adjustment of feeding levels (following routine observation of food remains), food conversion ratios almost as low as 1 : 1 have been recorded.

Water is pumped into Shigueno tanks from a pipe arranged across the diameter, and angled jets generate a circular flow pattern. This helps concentrate food remains, faecal waste and cast shells in the centre of the tank from which point they can be voided through a central drain. *Marsupenaeus japonicus* only grows well if provided with a substrate into which it can burrow, so tanks are given a 10–15 cm layer of sand on a false bottom. Water normally exits downwards through the sand, although the direction of water flow can be reversed to help clean the substrate. A high exchange rate (250–400% per day) is provided, sometimes in combination with aeration, and serves to maintain oxygen levels throughout the tank. Since water in the tank is usually quite clear, seaweed is able to grow on the bottom and this requires routine removal. To combat this problem some tanks were subsequently built to allow a greater depth of water (2.2–2.5 m).

In Hawaii another round pond system has been developed which, unlike the Shigueno system, employs a simple compacted earthen substrate. It has been operated with *Litopenaeus vannamei,* a species which does not require sand in which to burrow and which can be reared on pellets with lower protein content than those required for super-intensive *Marsupenaeus japonicus* culture. Trials on both an experimental scale (0.034 ha ponds) and a commercial scale (0.2 ha ponds) have been performed. The greatest productivity of 1.91 kg m^{-2} of shrimp averaging 15.1 g was achieved at a density of 150 shrimp m^{-2}, but a more valuable crop of 1.71 kg m^{-2} of larger (26.1 g) shrimp was achieved in a trial with a lower stocking density of 75 m^{-2}. High survival rates around 85% were normal, and feed conversion ratios averaging 2.0–2.5 : 1 were recorded using a 42.2% protein diet manufactured in Taiwan for the culture of *Fenneropenaeus penicillatus* (Wyban & Sweeney 1993). A commercial operation stocked with 100 shrimp m^{-2} achieved three crops per year, each crop yielding 1.5 kg m^{-2} of shrimp averaging 15–17 g. However, although these ponds are cheaper to build and operate than Shigueno tanks, there is unfortunately no high price market for *Litopenaeus vannamei* that is equivalent to that in Japan for live *Marsupenaeus japonicus*. Interestingly, the approach to water management in this Hawaiian system differs significantly from that in Shigueno tanks. Lower water exchange rates (10–100% per day, average 60% per day) are employed and blooms of diatoms are maintained since these have been observed to promote growth and survival in the crop. In addition, greater reliance is placed on paddlewheels (20 hp ha^{-1} is recommended) to oxygenate the water and maintain circulation in the tank.

Although super-intensive culture trials often yield excellent results, it should be remembered that they are usually performed in a research and development environment with well-motivated management and the back-up of laboratory facilities. Under these favourable conditions the prospects of success are greatly increased.

Another super-intensive system, developed jointly by the Universities of Arizona, USA and Sonora, Mexico, consisted of PVC-lined raceways (3.4 × 6 × 0.3 or 0.6 m deep) beneath inflated plastic greenhouses ('aquacells') (Salser *et al.* 1977). High-density rearing trials were performed with several penaeid species and *Litopenaeus stylirostris* gave the most promising results. Water was pumped from seawater wells and sprayed onto the surface of the raceways to give turnover rates of as much as 700% per day. Only compounded diets were used and waste was routinely siphoned from the tank bottoms. Refinements of this system led to the establishment of a commercially operated 2-ha *L. stylirostris* farm in Hawaii, Marine Culture Enterprises (MCE), which comprised 52 raceways and an integral hatchery. When the system worked well shrimp grew from 1.9 g to 24.1 g in 18 weeks and from 2.7 g to 18.2 g in 14 weeks, and target densities of almost 5.0 kg m^{-2} were achieved. Unique features of the system included continuous cropping and a transfer system that enabled shrimp to be moved from raceway to raceway to maintain optimum biomass densities. Transfers were performed by herding shrimp to one end of the raceway with a divider and then by using gravity flow or a fish pump. Unfortunately the facility, which represented an investment of $25m, was dogged by disease problems, most notably IHHN virus, and eventually closed. However, much useful information was generated and many lessons learned:

- Diet quality is of critical importance in super-intensive systems.
- Isolating effluents from groundwater is important because seepage from the drainage canal can contaminate well water.
- In highly loaded systems any power cut or pump failure can result in a rapid drop in oxygen tension and shrimp losses.
- Well-maintained back-up systems are essential.
- At densities of 7.9 kg m^{-2} it is very hard to maintain oxygen levels even when pure oxygen is used.
- Raceways of 200 m^2 give better results than raceways of 500 m^2.
- At high densities startled shrimp jump in self-propagating waves and barriers are needed to prevent losses.
- *Litopenaeus stylirostris* can give better results than *L. vannamei* but the former is highly susceptible to IHHN virus.
- Effective quarantine procedures are of critical importance to limit problems with pathogens.
- Routine treatments with formalin or copper compounds are needed to prevent gill fouling.
- Too much light results in shrimp becoming covered with epiphytes so white plastic covering was chosen for the raceways to permit only 40–60% light penetration.

The supply of well water for the raceways went into a vertical downwelling tube into which oxygen was bubbled to induce supersaturation. The resulting water was delivered along the length of each raceway in multiple small tubes that discharged at the raceway bed. Aeration and circulation were achieved with airstones and surface water spray jets (Moore & Brand 1993).

Research work with the MCE system suggested that 400% exchange per day was needed to support a shrimp density of 5 kg m^{-2}. Indeed, MCE and most other early super-intensive systems have been highly complex indoor systems that have relied very heavily on water exchange to maintain water quality. Few of them have been successful. An alternative approach to managing water quality (but one which is not suited to shallow raceway systems) is not to flush away uneaten feed and faecal waste but to keep it in suspension in the culture tanks and maintain aerobic decomposition. In this way shrimp can, in effect, get an opportunity to eat their feed more than once and nitrogen assimilation efficiency can be improved. As a result the level of protein in feeds can be cut back, perhaps to as little as 20%, to greatly reduce production costs (Chamberlain 1998) (sections 2.4.1, 8.3.6.3 and 12.5). This approach can be feasible in very dynamic systems powered by aerators and circulators that generate water velocities adequate to continually scour the bottom of ponds or tanks.

A good example of how this kind of system can work in practice is provided by a 'zero exchange' farm in Belize that has been designed partly on the basis of research carried out at the Waddell Mariculture Centre in South Carolina (McIntosh 1999). The ponds are constructed in sandy soil on relatively high ground, 4.5–6 m above sea level, have plastic (30 mm HDPE) liners and are square and deep (1.4–2.3 m). Incoming water is initially filtered to 250 μm and blooms of algae are initiated before use. Once on the farm, however, the water is reused from crop to crop after passing through a sedimentation basin. Production ponds are very strongly aerated using aerators to the equivalent of 30 hp ha^{-1}. Shrimp are stocked at densities of 125 m^{-2} to yield on average a little over 1.1 kg m^{-2} at harvest. The aerators are positioned to try to keep sediment in suspension and the energy consumption totals 3–4.5 kW h per kg of shrimp produced. For the culture of *Litopenaeus vannamei* the protein level in the feeds has been reduced from 30% to 22–24%. No inorganic fertilisers are added but a cheap feed containing only plant proteins is used to fertilise the pond water and stimulate heterotrophic bacteria. Both types of feed are fed simultaneously but the proportion of plant protein feed is steadily reduced to about 20% of the total by the end of the cycle.

The management of zero exchange ponds is a new and rapidly evolving field (section 12.5). In a system in Australia, using 20 hp ha^{-1} of aeration in ponds of 2.5 ha, ammonia levels are routinely monitored and, if they exceed 0.5 ppm and are rising with no sign of an algal bloom, a soluble carbon source is applied in the form of molasses (10 ppm) to promote heterotrophic activity and convert the ammonia to biomass (Body 2000). In Belize, after 7 weeks in production, sludge that has accumulated in the centre of ponds is pumped to drainage canals and thence into settling basins. This process is repeated every 2 weeks thereafter. The main ponds are 0.4 ha in size but smaller ponds of 700 m^2 are also used and have yielded as much as 2.7 kg m^{-2} when operated with increased aeration (equivalent to 50 hp ha^{-1}) plus artificial seagrass substrate in the form of arrays of felt-like, buoyant ribbons. Other super-intensive experiments have also made use of artificial seagrass and promising results have been obtained with *L. vannamei* at densities of 130 m^{-2} (Browdy & Bratvold 1999). Plankton densities were generally lower than in normal ponds but this was more than compensated for by rich periphyton on the artificial substrate and by an overall improvement in water quality with higher pH and lower ammonia levels. Such systems frequently make use of SPF post-larvae to limit disease problems.

Environmental considerations have also prompted experimental trials with the culture of *L. vannamei* in freshwater because the effluent waste can be productively used for crop irrigation without fears of a build-up of salt. Trials with hard water (0.4–0.5‰) in Florida have been successful (Scarpa 1998). Recirculating freshwater culture has also been tested in the same state. Densities of up to 150 shrimp m^{-2} have been used in three-phase systems using progressively larger tanks. Feeds with elevated calcium, phosphorus, potassium and vitamin C were employed to achieve yields of 1.14 kg m^{-2}. In theory, the three-phase system allows for six crops per year from the main tanks (Van Wyk *et al.* 2000).

A 'battery' operation, King James Shrimp Ltd of Illinois, USA, produced shrimp in indoor stacked raceways within a heated 'closed' recirculating system, but went bankrupt in 1982. The system employed artificial seawater, of which 10–15% was exchanged each day. Yields of 1–1.5 kg m^{-2} were obtained (McCoy 1986). Another super-intensive system with stacked raceways, nine units high, has been set up more recently in Michigan (Rosenberry 2000b). Raceways are made of reinforced concrete, are compartmentalised for two-phase ongrowing and are sloped for self-cleaning. Water cascades between levels to aid oxygenation and the whole system is operated in the dark.

In Europe, highly intensive systems for fish farming have been redesigned to provide a habitat suited to high-density shrimp farming, and commercial trials are being performed in Japan and China (P. Wood, 2001 pers. comm.) By using artificial seaweed to encourage shrimp to occupy the whole volume of water, yields as high as 12.8 kg m^{-2} per crop are anticipated. Apart from the artificial seaweed, which is negatively buoyant and is strung along the surface of the tanks, other innovations include:

- A reverse flow system for water circulation in which new water is added at the surface in pulsed waves at one end of the raceway to displace old water that returns in the reverse direction along the bed of the raceway.

- An endlss conveyor belt running along the V-shaped bottom of the raceway to remove faeces, sludge and moults.
- A self-propelled feeding robot running on a steel rail above the raceways (Anon. 1999b).

7.2.7 Harvesting

Shrimp can be harvested from ponds by a variety of methods including draining, seining, cast-netting, trapping and electro-fishing.

The most commonly applied technique for completely harvesting a pond is by draining. In this process the shrimp are first concentrated within the pond by lowering the water level and keeping screens in position in the sluices. Then the screens are removed and the shrimp are swept into a collection net. The net may be in the form of a long (3–4 m) tapering bag, a staked enclosure or an open-topped rigid cage. As the pond is drained it is important to maintain a steady outflow of water because if the flow is repeatedly stopped and started shrimp tend to resettle in the mud and many may become stranded as draining proceeds. To handle a large amount of shrimp with minimal damage, the harvest can be spread over several consecutive nights, refilling the pond each time. Most importantly for an efficient harvest, however, a pond must drain well. Any remaining pools in channels or bottom depressions require laborious seining and hand-picking, during which shrimp deteriorate in condition and may accumulate mud beneath their carapaces.

The efficiency of a drain harvest can be improved by taking into account shrimp behaviour. Penaeids in general spend much of their time buried in the substrate, with just their eyes protruding, and in this position they are impossible to harvest effectively. Activity, however, does increase at night, especially during periods of new and full moon, and swimming behaviour can be encouraged by water movement; so by performing harvests during the hours of darkness and sharply lowering and raising the pond level in advance, better catches can usually be made. In addition, during a night harvest, lamps can be placed alongside the collection area to attract the shrimp. Behaviour, however, shows some variation between species; *Penaeus monodon*, for example, is reported to be more reluctant than other penaeids to swim out of a pond during draining.

The methods of handling shrimp during a harvest are influenced by the type of market to be supplied. If shrimp are to be deheaded they can be scooped out of the collection net or sump, mixed with an equal weight of flake or chip ice and placed in plastic boxes in an insulated truck. However, if the shrimp are destined to be sold whole, then the collection net must be emptied frequently, at least each 15–20 minutes, so that all shrimp are live when transferred to an ice slurry containing sodium metabisulphite. This chemical is an antioxidant and it retards the process of melanisation in which shrimp blacken through the action of enzymes, particularly in the head region. The metabisulphite concentration needs to be maintained at around 10% and fresh ice added during harvesting (Villalon 1993).

Traps rely on shrimp moving around in a pond and so are most efficient when the shrimp are active. A leader made of netting or bamboo slats is arranged at right angles to the embankment and guides shrimp into bamboo or net catching chambers. Frequent collections may be necessary to avoid overcrowding and keep the shrimp alive. An advantage of traps is that slat spacing or mesh size can be set to harvest only the largest shrimp.

Electro-fishing makes use of an electric current to stimulate shrimp to jump. Electrodes are located either on the base of a drag-net, which is drawn across the pond bottom, or at the ends of hand-held poles. In the latter system the cathode forms the ring of a scoop net and the operator, equipped with an accumulator, wades through the pond collecting the shrimp as they jump. Electro-fishing is conveniently performed in small ponds or tanks and is often chosen for intensive and super-intensive operations. It enables shrimp to be collected in excellent condition and has proved to be popular with professional harvesting teams in Taiwan.

Since the value of the crop is reduced by the presence of newly moulted soft-shelled shrimp (because of their susceptibility to physical damage during collection and processing), harvesting should be timed to avoid the peak occurrence of moulting, which may coincide with full moons. The period 2–3 days after new moon has been recommended for harvesting, although in practice cast-net samples are usually taken to check the condition of shrimp before any decision is taken to commence a full-scale harvest. Sometimes flushing a pond with new water when harvesting (to wash shrimp towards the outlet) has been found to precipitate moulting.

Harvesting may be performed in a single complete operation or as a series of partial harvests spread over a period of several weeks or months. The harvesting strategy adopted by any particular farm will be determined by processing capabilities and the market requirement for either bulk or small quantities (section 3.2.4).

In some countries it can be important to minimise labour costs and this can be achieved through mechanised harvesting. There are three main mechanical systems: Archimedes screw, reciprocating vacuum/pressure pumps, and recessed impeller pumps. Some people favour the screw system because it is easily repaired and because only a small motor is needed since only the shrimp are lifted. All the same, pumping shrimp with water can also be efficient. As a pond drains, not all the water enters the pump and most escapes through the sides of a net or cage. The water flow carries shrimp towards a stainless steel collection cone (which is specially designed to fit on the harvest box at the pond outlet) where a submersible pump or suction hose is located. Pumps are designed for moving delicate products such as tomatoes or live trout and the impeller acts on the water alone and does not actually touch the shrimp. Submersible pumps have the advantage that they do not need priming. Up on the pond bank the excess water drains through a grating and the shrimp tumble over the same grating with a minimal amount of water, to be collected for chilling in an ice slurry (Hodgson *et al.* 2000).

7.2.8 Processing

Large amounts of shrimp are sold as peeled or unpeeled tails after washing, grading, and deheading. Some are sold whole, either chilled, individually quick frozen (IQF), or bulk frozen in blocks with water. Small amounts are sold live. *Marsupenaeus japonicus,* for example, destined for specialist Japanese markets (section 3.3.1.2), are packed in insulated boxes between layers of pre-chilled sawdust after washing, grading and chilling. The many market forms of shrimp are summarised in Chapter 3 (Fig. 3.3), and a range of value-added and speciality shrimp products is listed in section 3.3.1.

7.2.9 Hatchery supported fisheries, ranching

Shrimp release programmes have been undertaken in several countries to enhance fishery stocks, although in Taiwan releases are made more with a view to improving broodstock availability.

In Japan around $200–300 \times 10^6$ young shrimp are released annually (Imamura 1999) and recaptures of 5–8% are reported (Table 5.8). The shrimp are liberated either as post-larvae (PL_{20}) or as nursery-reared juveniles (3 cm TL). Nursery rearing is performed primarily to reduce the predation losses associated with the direct release of small hatchery-reared post-larvae. Three types of nursery are employed: fenced enclosures, man-made lagoons and artificial tidelands.

Fenced enclosures are supported by sticks, floats and anchors but a major problem remains their vulnerability to damage from wind, waves and tidal currents. Prior to stocking, fish predators within the enclosure are either netted or poisoned. After 2–3 weeks of rearing, juveniles are released simply by removing the fence.

Lagoons are created by constructing temporary embankments along the shoreline, and so can only be sited where suitable embankment-building material is available. Stocked juveniles are released when they reach 3 cm in length, by rupturing the embankment. In both fenced enclosures and lagoons, the juveniles are very vulnerable to adverse weather, especially heavy rains and typhoons. Red tides (see Glossary) can also be problematic and can contribute to the general unpredictability of survival rates (range 0–70%).

Artificial tidelands consist of a series of low walls, which retain water to a depth of about 5 cm when the tide recedes. They are protected by a breakwater and a seaward fence to exclude fish, and are specially designed to provide conditions which are favourable to shrimp juveniles but not to their predators. For example, a comparatively hard substrate is provided which is unsuitable for the burrowing habit of gobies. Juveniles are stocked at a maximum density of $100\,m^{-2}$ and they rely on natural productivity for their sustenance. Although the substrate sometimes becomes completely exposed between tides, the shrimp juveniles can survive for a few hours in just a few millimetres of water. A pumped water supply may be provided during low tide. Survival rates are very variable and particular problems may arise with the excessive growth of seaweed.

Figures for shrimp releases in China indicate that very significant efforts have been made to enhance *Fenneropenaeus chinensis* stocks in and around the Yellow Sea and Gulf of Bohai. In 1986 an estimated total of 4×10^9 hatchery-reared shrimp were liberated (Table 5.8). In the semi-enclosed Jiaozhou Bay, the release of over 350×10^6 juveniles during 1984–86 appears to have had a dramatic effect on the depleted local fishery. Stocks increased by factors of between 4.7 and 7.3 during the 3 year release period, but declined immediately after restocking ceased. Estimated average survival was a very impressive 32% (Liu 1990). In another programme, up to 224×10^6 3–4 cm TL juvenile *F. chinensis* were released per year into Xiang Shan Bay. Prior to restocking, there was almost no fisheries catch, but afterwards catches (of smallish shrimp at 10–12 cm TL) increased and there was a strong correlation between the number released and total fishery landings. The recapture rate was estimated at 6.8–13.3% but the state funding of releases was eventually stopped because there was no practical way to recover the costs of the operation. The programme failed to establish a breeding population and catches swiftly returned to zero (Xu *et al.* 1997).

Over the period 1972–79, 120×10^6 post-larvae, mostly *Penaeus semisulcatus*, were released into Kuwaiti waters, and although recapture rates were unknown, Farmer (1981) estimated that the operation would break even at a recapture rate of 2% (section 10.6.1.9). In Taiwan, sea ranching is considered a promising method of increasing supplies of wild gravid females and recapture rates as high as 15% have been estimated (Chiang & Liao 1985). Pond-reared subadults of disease-free stocks were released after tagging and return rates of between 0.17% and 15% were achieved (Table 5.8).

One ranching operation, set up in the 1970s as an enclosed, stocked fishery by Marifarms Inc. in Florida, USA, incorporated a netted embayment of 1000 ha and two lagoons of 120 ha each. A total of 19 km of netting was required, which was subject to frequent damage from storms and tides. Within the embayment, hatchery-reared seed were stocked into a two-stage nursery which consisted of staked net enclosures. Juveniles were released and fed with trash fish and pellets, but no attempt was made to eliminate predators and competitors. Harvesting was performed with a trawler and in the first year yielded a total of 95 mt of shrimp (after processing). However, the operation eventually proved to be uneconomic, in part due to legislative constraints that required access by leisure craft, and an undertaking to restock the fishery from the hatchery (Hanson & Goodwin 1977).

Investigations are under way for the enhancement of *Penaeus esculentus* stocks in Exmouth Gulf, Western Australia, and this could become a model for the best approach to shrimp stock enhancement, largely because the licence to fish this area is held by a single company. As a result there should be few of the problems of ownership or dissipation of benefits associated with the enhancement of an open access fishery. Because the fishing company can capture any benefits, it has the incentive to fund a hatchery and nursery without relying on the state for funding. Preliminary research so far has concentrated on developing economical raceway systems to nursery rear the post-larvae prior to release and on the selection of the most suitable release sites. The fishery usually yields 200–680 mt yr^{-1} and the objective of the exercise is to smooth out annual catch fluctuations that may be linked to recruitment levels. A bioeconomic model estimated that it would be necessary to release 7×10^6 juveniles to increase catches by 100 mt. This would assume a recapture rate of around 20–30%, which is optimistic, particularly for an open water fishery, but it could be feasible because the intention is to nursery rear the juveniles prior to release. It is hoped that the released animals can be genetically 'fingerprinted' by using microsatellite techniques (see Glossary and section 8.10.1.3) so that they can subsequently be detected in the commercial catches (Die *et al.* 1999; Loneragan *et al.* 2000).

The ability to distinguish the hatchery-reared stocks from their wild counterparts is essential if the impact of a release programme is to be evaluated with any confidence. This has been achieved with lobsters by using micro-wire tags that can later be identified with a metal detector (section 7.8.11.2). With shrimp, Davenport *et al.* (1999) were able to identify the impact of stock enhancement by releasing *Penaeus monodon* juveniles out of phase with the normal recruitment patterns to detect a spike in fishery yield 3 months later. The exercise was successfully repeated two years running and juveniles released at 2 cm (TL) were recaptured at a rate of 3%. The value of the extra landings compensated for the costs of buying and releasing the juveniles (section 10.6.1.9).

7.2.10 References

Alday de Graindorge V. & Flegel T.W. (1999) *Diagnosis of shrimp diseases with emphasis on the black tiger shrimp* (Penaeus monodon). (CD Rom) Multimedia Asia, Bangkok Thailand.

Anon. (1990) Plastic liner trial shows encouraging results: initial tests showed liner-use potential. *Newsline*, **3** (3) 4–5. Oceanic Institute, Honolulu, HI, USA.

Anon. (1998) Guidelines for controlling luminous vibriosis in *Penaeus monodon* hatcheries. *Austasia Aquaculture Magazine*, **11** (5) 59–62.

Anon. (1999a) Technology, experience and perseverance: success in Panama. *Aquaculture Magazine*, **25** (5) 42–54.

Anon. (1999b) Shrimp progress. Mega Fisch unveils reverse-flow system. *Fish Farming International*, **26** (10) 9–10.

AQUACOP (1983) Constitution of broodstock, maturation, spawning, and hatching systems for penaeid shrimps in the Centre Oceanologique du Pacifique. In: *Handbook of Mariculture, Vol. 1 Crustacean aquaculture* (ed. J.P. McVey), pp. 105–127. CRC Press, Boca Raton, FL, USA.

AQUACOP & Patrois J. (1990) *Élevage des géniteurs pénéides en captivité au C.O.P. Etat des techniques 1989–1990*, 18 pp. (mimeo). IFREMER, Tahiti [in French].

ASEAN (1978) *Manual on pond culture of penaeid shrimp*,

131 pp. ASEAN National Coordinating Agency of the Philippines, Ministry of Foreign Affairs, Manila, Philippines.
Autrand M. & Vidal F. (1995) An alternative technology for shrimp larviculture. *Infofish International*, (5) 37–41.
Babu M.M. & Marian M.P. (1998) Live transport of gravid *Penaeus indicus* using coconut mesocarp dust. *Aquacultural Engineering*, (18) 149–155.
Bauman H.R. & Jamandre D.R. (1990) A practical method for determining quality of *Penaeus monodon* (Fabricius) fry for stocking in grow-out ponds. In: *Technical and Economic Aspects of Shrimp Farming*, Proceedings of the Aquatech '90 Conference, Kuala Lumpur, 11–14 June 1990 (eds M.B. New, H. de Saram & T. Singh), pp. 124–137, Infofish, Kuala Lumpur, Malaysia.
Binh C.T. & Lin C.K. (1995) Shrimp culture in Vietnam. *World Aquaculture*, **26** (4) 27–33.
Body A. (2000) *Sugar/molasses, C:N ratios*. Correspondence, 23/09/00. shrimp@egroups.com
Bray W.A. & Lawrence A.L. (1998) Successful reproduction of *Penaeus monodon* following hypersaline culture. *Aquaculture*, **159** (3–4) 275–282.
Bray W.A., Moya M.E., Lawrence A.L. & Samocha T.M. (1999) Broodstock culture of *Litopenaeus vannamei* in low salinity desert groundwater of 2.3 ppt: summary of growth and sperm development. In: *Book of Abstracts, World Aquaculture '99,* 26 April–2 May 1999, Sydney, Australia, p. 101. World Aquaculture Society, Baton Rouge, LA, USA.
Briggs M.R.P. (1992) A stress test for determining vigour of post-larval *Penaeus monodon* Fabricius. *Aquaculture and Fisheries Management*, **23**, 633–637.
Browdy C.L. & Bratvold D. (1999) Effects of substrate enhancement and water disinfection on microbial dynamics and shrimp production. In: *Book of Abstracts, World Aquaculture '99,* 26 April–2 May 1999, Sydney, Australia, p. 112. World Aquaculture Society, Baton Rouge, LA, USA.
Brune D.E., Schwartz G., Reed S.M., Collier J.A., Schwedler T.E. & Eversole A.G. (2000) Designing the partitioned aquaculture system for marine shrimp culture. In: *Abstracts, Aqua 2000, Responsible Aquaculture in the New Millennium* (compiled by R. Flos & L. Creswell), p. 97. European Aquaculture Society, Special Publication No. 28.
Castille F.L., Samocha T.M., Lawrence A.L., He H., Frelier P. & Jaenike F. (1993) Variability in growth and survival of early postlarval shrimp (*Penaeus vannamei* Boone 1931). *Aquaculture*, **113** (1–2) 65–81.
Chamberlain G. (1998) The feeds industry assessment of biosecure systems. In: *Proceedings of the US marine shrimp farming program biosecure workshop*, 14 February 1998 (ed. S.M. Moss), pp. 81–84. The Oceanic Institute, Hawaii.
Chamorro R. (2000) *Re: crop rotation with tilapia*. Correspondence. 06/07/00. shrimp@egroups.com
Chanratchakool P., Turnbull J.F., Funge-Smith S.J., MacRae I. H. & Limsuwan C. (1998) *Health management in shrimp ponds*, 152 pp. Aquatic Animal Health Research Institute, Bangkok, Thailand.
Chavez C. (ed.) (1990) *The Aquaculture of Shrimp, Prawn, and Crayfish in the World: basics and technologies*, 400 pp. Haworth Press, Food Products Press Inc., Binghamton, New York.
Chen L.C. (1990) *Aquaculture in Taiwan*, 273 pp. Fishing News Books, Blackwell Scientific Publications, Oxford.
Chiang P. & Liao I.C. (1985) The practice of grass prawn (*Penaeus monodon*) culture in Taiwan from 1968 to 1984. *Journal of the World Mariculture Society,* **16**, 297–315.
Clifford H.C. III (1992) Marine shrimp pond management: a review. In: *Proceedings of the Special Session on Shrimp Farming* (ed. J. Wyban), pp. 110–137. World Aquaculture Society, Baton Rouge, LA, USA.
Clifford H.C. III (1994) Semi-intensive sensation. *World Aquaculture*, **25** (3) 6–12; 98–104.
Davenport J., Ekaratne S.U.K., Walgama S.A. Lee D. & Hills J.M. (1999) Successful stock enhancement of a lagoon prawn fishery at Rekawa, Sri Lanka using cultured postlarvae of penaeid shrimp. *Aquaculture*, **180** (1–2) 65–78.
Die D.J., Watson R., Loneragan N. & Kailis G. (1999) Feasibility of enhancing a trawl fishery for a tropical penaeid: *Penaeus esculentus* in Exmouth Gulf, Western Australia. In: *Book of Abstracts, World Aquaculture '99,* 26 April–2 May 1999, Sydney, Australia, p. 215. World Aquaculture Society, Baton Rouge, LA, USA.
Dixon H.M. (1999) Bait shrimp production in Florida. *Global Aquaculture Advocate*, **2** (6) 20.
Evans L. (2000) *Harvesting and shipping live shrimp*. Correspondence. 13/09/00. shrimp@egroups.com
Falaise F. & Boël L. (1999) A new technology for sustainable shrimp farming. *Infofish International*, (3) 33–99.
FAO (1979) *Brackishwater shrimp and milkfish culture, applied research and training project, Indonesia.* 34 pp. Food and Agriculture Organisation, Rome.
Farmer A.S.D. (1981) Prospects for penaeid shrimp culture in arid lands. In: *Advances in Food Producing Systems for Arid and Semi-arid Lands*, pp. 859–897. Academic Press, New York.
Fast A.W. & Lester L.J. (eds) (1992) Marine shrimp culture: principles and practices. *Developments in Aquaculture and Fisheries Science* (23), 862 pp. Elsevier, Amsterdam.
Fegan D.F. (1992) Recent developments and issues in the shrimp hatchery industry. In: *Proceedings of the Special Session on Shrimp Farming* (ed. J. Wyban), pp. 55–70. World Aquaculture Society, Baton Rouge, LA, USA.
Fitzsimmons K. (1999) Shrimp farming in saline groundwater in Arizona USA. In: *Book of Abstracts, World Aquaculture '99.* 26 April–2 May 1999, Sydney, Australia, p. 263. World Aquaculture Society, Baton Rouge, LA, USA.
Franklin T., Sondara Raj R. & Prabhavathy G. (1982) Transport of prawn seed. In: *Proceedings of Symposium on Coastal Aquaculture 1982*, Pt 1, pp. 395–396. Marine Biological Association of India, Cochin, India.
Frese T. (2000) *Crop rotation and/or polyculture with Tilapia.* Correspondence. 04/07/00. shrimp@egroups.com.
George M.J. & Suseelan C. (1982) Distribution of species of prawns in the backwaters and estuaries of India with reference to coastal aquaculture. In: *Proceedings of Symposium on Coastal Aquaculture 1982*, Pt 1, pp. 273–284. Marine Biological Association of India, Cochin, India.
Griffith D. (2000) *Nauplii and PL quality*. Correspondence. 23/08/00, shrimp@egroups.com.
Griffith D.W. & Schwarz L. (1999) Shrimp farming in Ecuador: development of an industry. *Aquaculture Magazine*, **25** (1) 46–50.

Hansford S.W. & Marsden G.E. (1995) Temporal variation in egg and larval productivity of eyestalk ablated spawners of the prawn *Penaeus monodon* from Cook Bay, Australia. *Journal of the World Aquaculture Society*, **26** (4) 396–405.

Hanson J.A. & Goodwin H L. (Eds.) (1977) *Shrimp and prawn farming in the western hemisphere*, 439 pp. Dowden, Hutchinson & Ross Inc., Stroudsburg, PA, USA.

Heales D.S., Polzin H.G. & Staples D.J. (1985) Identification of the commercially important *Penaeus* species in Australia. In: *Second Australian National Prawn Seminar* (eds P.C. Rothlisberg, B.C. Hill & D.J. Staples), pp. 41–46. Cleveland, Australia.

Hodgson L., Gregg K. & McIntosh R. (2000) Mechanized shrimp harvesting. *Global Aquaculture Advocate*, **3** (4) 57–61.

Imamura K. (1999) The organization and development of sea farming in Japan. In: *Stock Enhancement and Sea Ranching* (eds B.R. Howell, E. Moksness & T. Svåsand), pp. 91–102. Blackwell Science, Oxford, UK.

Johnson S.K., Cichra C.E. & Chamberlain G.W. (1984) *Wet transport of mature marine shrimp in sealed containers*, 4 pp. FDDL-S14, Fish Disease Diagnostic Lab., Texas Agricultural Extension Service, Texas A&M University, College Sta., TX, USA.

Johnston D., Clough B. Xuan T.T. & Phillips M. (1999) Mixed shrimp–mangrove forestry farming systems in Ca Mau Province, Vietnam. *Aquaculture Asia*, (2) 6–12.

Jory D.E. (1998) Postlarvae acclimation before stocking. *Aquaculture Magazine*, **24** (2) 82–91.

Kawahigashi D. (1992) A survey of commercial maturation technology in the western hemisphere. In: *Proceedings of the Special Session on Shrimp Farming* (ed. J. Wyban), pp. 52–54. World Aquaculture Society, Baton Rouge, LA, USA.

Kongkeo H. (1995) How Thailand made it to the top. *Infofish International*, (1) 25–31.

Kongkeo H. (1997) Comparison of intensive shrimp farming systems in Indonesia, Philippines, Taiwan and Thailand. *Aquaculture Research*, **28,** 789–796.

Kungvankij P. & Kongkeo H. (1988) Culture system selection. In: *Shrimp '88, Conference Proceedings*, 26–28 January 1988, pp. 123–136. Bangkok, Thailand. Infofish, Kuala Lumpur, Malaysia.

Kurian C.V (1982) Some constraints in prawn culure. In: *Proceedings of Symposium Coastal Aquaculture 1982*, (1), 98–102. Marine Biological Association of India, Cochin, India.

Liao I.C. (1992) Penaeid larviculture: Taiwanese method. In: *Marine Shrimp Culture: principles and practices* (eds A.W. Fast & J.L. Lester), pp. 193–215. Elsevier, Amsterdam.

Liao I.C. & Chen Y.P. (1983) Maturation and spawning of penaeid prawns in Tungkang Marine Laboratory, Taiwan. In: *Handbook of Mariculture, Vol.1 Crustacean aquaculture* (ed. J.P. McVey), pp. 155–160. CRC Press, Boca Raton, FL, USA.

Lin M.N. & Ting Y.Y. (1986) Spermatophore transplantation and artificial fertilization in grass shrimp. *Bulletin of the Japanese Society of Scientific Fisheries*, **52** (4) 585–589.

Liu J.Y. (1990) Resource enhancement of Chinese shrimp, *Penaeus orientalis*. *Bulletin of Marine Science*, **47** (1) 124–133.

Loneragan N., Preston N., Crocos P. & McCulloch R. (2000) Developing techniques for the stock-enhancement of prawns in Australia: brown tiger prawns (*Penaeus esculentus*) in Exmouth Gulf, Western Australia. In: *Abstracts, Aqua 2000, Responsible aquaculture in the new millennium* (compiled by R. Flos & L. Creswell), p. 411. European Aquaculture Society, Special Publication No. 28.

Macintosh D.J. (1987) *Aquaculture Production and Products Handling in ASEAN*, 365 pp. ASEAN Food Handling Bureau, Kuala Lumpur, Malaysia.

Maguire G.B. (1979) *A Report on the Prawn Farming Industries of Japan, the Philippines and Thailand*, 110 pp. Brackish Water Fish Culture Research Station, Salamander Bay, Australia.

McCoy H.D. II. (1986) Intensive culture systems past, present and future: parts I, II and III. *Aquaculture Magazine*, **12** (6) 32–35; **13** (1) 36–40; **13** (2) 24–29.

McIntosh R.P. (1999) Changing paradigms in shrimp farming: 1. General description. *Global Aquaculture Advocate*, **2** (4–5) 42–47.

McVey J.P. (ed.) (1993) *Handbook of Mariculture*, 2nd edn, *Vol. 1 Crustacean aquaculture*, 442 pp. CRC Press, Boca Raton, FL, USA.

Millamena O.M., Casalmir C.M. & Subosa P.F. (1991) Performance of recirculating systems for prawn hatchery and broodstock maturation tanks. *Aquacultural Engineering*, **10**, 161–171.

Moore D.W. & Brand C.W. (1993) The culture of marine shrimp in controlled environment superintensive systems. In: *Handbook of Mariculture*, 2nd edn, *Vol. 1 Crustacean aquaculture* (ed. J.P. McVey), pp. 315–348. CRC Press, Boca Raton, FL, USA.

Motoh H. & Buri P. (1981) Identification of postlarvae of the genus *Penaeus* appearing in shore waters. *Researches on Crustacea*, **11** 3–11.

NACA (1986) *A prototype warm water hatchery*, 32 pp. Technology series, selected publication No. 4. NACA, Bangkok, Thailand.

Naessens E., Lavens P., Gomez L., *et al.* (1997) Maturation performance of *Penaeus vannamei* co-fed *Artemia* biomass preparations. *Aquaculture*, **155** (1–4) 87–101.

Ogle J.T. (1992) *Design and operation of the gulf coast research laboratory penaeid shrimp maturation facility I* Penaeus vannamei, 41 pp. Gulf Coast Research Laboratory, Ocean Springs, MS, USA.

Ogle J.T. & Beaugez K.A. (1997) A study of the factors influencing the hatch rate of *Penaeus vannamei* eggs. III. Presence of the female after spawning. *Gulf Research Reports*, **9** (4) 365–366.

Olin P.G. & Fast A.W. (1992) Penaeid PL harvest, transport, acclimation and stocking. In: *Marine Shrimp Culture: principles and practices* (eds A.W. Fast & J.L. Lester), pp. 301–320. Elsevier, Amsterdam, Netherlands.

Palacios E., Ibarra A.M., Ramirez J.L., Portillo G. & Racotta I.S. (1998) Biochemical composition of eggs and nauplii in white Pacific shrimp, *Penaeus vannamei* (Boone), in relation to the physiological condition of spawners in a commercial hatchery. *Aquaculture Research*, **29,** 183–189.

Paulinose V.T. (1982) Key to the identification of larvae and postlarvae of the penaeid prawns (Decapoda: Penaeidea) of the Indian Ocean. *Mahasagar Bulletin of the National Institute of Oceanography,* **15,** 223–229.

Peterson J.J. (2000) Use of substrate technology in shrimp larviculture. *Global Aquaculture Advocate*, **3** (1) 68–69.

Peterson J.J. & Griffith D.R.W. (1999) Intensive nursery systems. *Global Aquaculture Advocate*, **2** (6) 30–31.
Pironet F., Bosc F., Delaune J.M. & Lucien-Brun H. (1999) Production management in a *Penaeus monodon* hatchery in Madagascar under low salinity conditions. In: *Book of Abstracts, World Aquaculture '99*, 26 April–2 May 1999, Sydney Australia, p. 470. World Aquaculture Society, Baton Rouge, LA, USA.
Pretto R. (1983) *Penaeus* shrimp pond grow-out in Panama. In: *Handbook of Mariculture, Vol.1 Crustacean aquaculture* (ed. J.P. McVey), pp. 169–178. CRC Press, Boca Raton, FL, USA.
Primavera, J.H. (1983) Prawn hatcheries in the Philippines. *Asian Aquaculture*, **5** (3) 5–8.
Qingyin W. (1992) Shrimp culture in China. *Infofish International*, (4) 50–53.
Quynh V.D. (1989) Viet Nam, coastal aquaculture in the southern provinces. *World Aquaculture*, **20** (2) 22–28.
Quynh V.D. (1990) Shrimp larviculture in Viet Nam. *Larviculture and Artemia Newsletter*, (16) 22–26. State University of Ghent, Belgium.
Richards-Rajadurai P.N. (1989) Live storage and distribution of fish and shellfish. *Infofish International*, (3) 23–27.
Rigolet V., Bador R. & Ducharne F. (1999) A step towards the definition of objective criteria to evaluate quality of *P. monodon* and *L. vannamei* postlarvae. In: *Book of Abstracts, World Aquaculture '99*, 26 April–2 May 1999, Sydney, Australia, p. 39. World Aquaculture Society, Baton Rouge, LA, USA.
Robertson L., Bray, W.A. & Lawrence, A.L. (1987) Shipping of penaeid broodstock: water quality limitations and control during 24 hr shipments. *Journal of the World Aquaculture Society*, **18** (2) 45–56.
Rosenberry R. (2000a) *Super shrimp, new animals, new feeds, new systems*. http://www.shrimpnews.com/SuperShrimp.html
Rosenberry R. (2000b) *Designed, built and managed by Russ Allen*. http://members.aol.com/brosenberr/AllenRuss.html
Salser B., Mahler L., Lightner D., *et al.* (1977) *Controlled Environment Aquaculture of Penaeids*, 19 pp. Environmental Research Laboratory, Tucson, AZ, USA.
Samocha T.M. & Lawrence A.L. (1992) Shrimp nursery systems and management. In: *Proceedings of the Special Session on Shrimp Farming* (ed. J. Wyban), pp. 87–105. World Aquaculture Society, Baton Rouge, LA, USA.
Samocha T.M., Lawrence A.L. & Beidenbach J.M. (1993a) The effect of vertical netting and water circulation pattern on growth and survival of *Penaeus vannamei* postlarvae in an intensive raceway system. *Journal of Applied Aquaculture*, **2** (1) 55–64.
Samocha T.M., Lawrence A.L. & Bray W.A. (1993b) Design and operation of an intensive nursery raceway system for penaeid shrimp. In: *CRC Handbook of Mariculture*, 2nd edn, *Vol. 1 Crustacean Aquaculture* (ed. J.P. McVey), pp. 173–210. CRC Press, Boca Raton, FL, USA.
Samocha T.M., Guajardo H., Lawrence A.L., *et al.* (1998) A simple stress test for *Penaeus vannamei* postlarvae. *Aquaculture*, **165** (3–4) 233–242.
Sangpradub S., Fast A.W., Piyatiratitivorakul S. & Menasveta P. (1994) Effect of different feeding regimes on ovarian maturation and spawning of pond-reared giant tiger prawn in Thailand. *Aquaculture International*, **2** (1) 49–58.
Sanjuan H. (2000) *Shrimp larvae and sawdust*. Correspondence. 24/10/2000, shrimp@egroups.com
Scarpa J. (1998) Freshwater recirculating systems in Florida. In: *Proceedings of the US marine shrimp farming program biosecure workshop*, 14 February 1998 (ed. S.M. Moss), pp. 67–70. The Oceanic Institute, Honolulu, HI, USA.
Schoemaker R. (1991) *Transportation of live and processed seafood*, 29 pp. Infofish Technical Handbook (3). Infofish, Kuala Lumpur, Malaysia.
Simon C.M. (1982) Large-scale, commercial application of penaeid shrimp maturation technology. *Journal of the World Mariculture Society*, **13**, 301–312.
Singh H., Chowdhury A.R. & Pakrasi B.B. (1982) Experiments on the transport of postlarvae of tiger prawn *Penaeus monodon* Fabricius. In: *Proceedings of Symposium on Coastal Aquaculture* 1982, Part 1, pp. 232–235. Marine Biological Association of India, Cochin, India.
Singh T. (1993) The use of pond liners in aquaculture. *Infofish International*, (5) 50–53.
Smith L., Beidenbach J.M. & Lawrence A.L. (1992) Penaeid larviculture: Galveston method. In: *Marine Shrimp Culture: principles and practices* (eds A.W. Fast & J.L. Lester), pp. 171–191. Elsevier, Amsterdam, Netherlands.
Smith T.I.J. & Ribelin, B. (1984) Stocking density effects on survival of shipped post-larval shrimp. *Progressive Fish-Culturist*, **46** (1) 47–50.
Subosa P.F. & Bautista M.N. (1991) Influence of stocking density and fertilization regime on growth, survival and gross production of *Penaeus monodon* Fabricius in brackishwater ponds. *Israeli Journal of Aquaculture – Bamidgeh*, **43** (2) 69–76.
Tookwinas S. (1990) Pen culture techniques of marine shrimp in Thailand. *Infofish International*, (2) 38–40.
Treece G.D. & Fox J.M. (1993). *Design, operation and training manual for an intensive culture shrimp hatchery*, 187 pp. Marine Advisory Service, Texas A&M Sea Grant College Program, Galveston, TX, USA.
Treece G.D. & Yates M.E. (1990) *Laboratory manual for the culture of penaeid shrimp larvae*, 95 pp. Marine Advisory Service, Sea Grant College Program, Texas A & M University, TX, USA.
Van Wyk P., Davis M., Laramore R., Main K., Mountain J. & Scarpa J. (2000) *Farming Marine Shrimp in Recirculating Freshwater Systems*, 320 pp. Harbour Branch Oceanographic Institution, Fort Pierce, FL, USA.
Villalon J.R. (1991) *Practical Manual for Semi-intensive Commercial Production of Marine Shrimp*, 100 pp. (approx.). TAMU-SG-91-501. Texas Sea Grant College Program, Galveston, TX, USA.
Villalon J.R. (1993) Commercial semi-intensive penaeid grow-out techniques in Ecuador. In: *Handbook of Mariculture*, 2nd edn *Vol. 1 Crustacean aquaculture* (ed. J.P. McVey), pp. 237–273. CRC Press, Boca Raton, FL, USA.
Walker T. (1994) Improving the return from prawn spawners: research shows the way. *Austasia Aquaculture Magazine*, **8** (4) 7–9.
Wilkenfield J.S. (1992) Commercial hatchery status report: an industry panel viewpoint. In: *Proceedings of the Special Session on Shrimp Farming* (ed. J. Wyban), pp. 71–86. World Aquaculture Society, Baton Rouge, LA, USA.

Wyban J. (ed.) (1992) *Proceedings of the Special Session on Shrimp Farming*, 301 pp. World Aquaculture Society, Baton Rouge, LA, USA.
Wyban J.A & Sweeney J.N. (1991) *The Oceanic Institute Shrimp Manual*, 158 pp. The Oceanic Institute, Honolulu, HI, USA.
Wyban J.A. & Sweeney J.N. (1993) Intensive shrimp production in round ponds. In: *Handbook of Mariculture,* 2nd edn, *Vol. 1 Crustacean aquaculture* (ed. J.P. McVey), pp. 275–287. CRC Press, Boca Raton, FL, USA.
Xu J., Xia M., Ning X. & Mathews C.P. (1997) Stocking, enhancement, and mariculture of *Penaeus orientalis* and other species in Shanghai and Zhejiang Provinces, China. *Marine Fisheries Review*, **59** (4) 8–14.
Yamaha (1989) Prawn culture. *Yamaha Fishery Journal* (30), 8 pp. Yamaha Motor Co., Shizuoka-ken, Japan.
Yoong F. & Reinoso B. (1983) *Manual practico para la identificacion de post-larvas y juveniles de cuatro especies de camarones marinos*, 41 pp. Boletín Científico y Técnico VI (II), Instituto Nacional de Pesca, Guayaquil, Ecuador [in Spanish].
Zhang W., Dengong C., Rujie L., Naiyu Z., Hengzu G. & Xuanyuan L. (1980) Notes on factors affecting the successful artificial rearing of the Chinese prawn (*Penaeus orientalis* Kishinouye) from eggs to post-larvae [in Chinese, translated by Chau-ling Chan, Lancaster Polytechnic] *Transactions Oceanology and Lim. Sinica*, (1) 39–44.

7.3 *Macrobrachium*

7.3.1 Species of interest

Giant river prawn (*Macrobrachium rosenbergii*); oriental river prawn (*M. nipponense*); monsoon river prawn (*M. malcolmsonii*); painted river prawn (*M. carcinus*); Amazon river prawn (*M. amazonicum*); African river prawn (*M. vollenhovenii*); cinnamon river prawn (*M. acanthurus*).

The culture techniques described in this section almost all relate to *Macrobrachium rosenbergii,* the species that dominates commercial freshwater prawn farming. All the same, in the past decade the culture potential of other species has been demonstrated and significant quantities are being produced – mainly of *M. nipponense* in China but also of *M. malcolmsonii* in India. Since the culture techniques for these and for the other *Macrobrachium* species differ only slightly from those for *M. rosenbergii*, the information supplied here can be applied generally. The accounts by Malecha (1983), Corbin *et al.* (1983), New and Singholka (1985) and New (1995) provide details of the operating techniques for prawn hatcheries, nurseries and farms, as does the more recent, comprehensive book by New and Valenti (2000). The latter also includes a review of the culture status of freshwater prawn species other than *M. rosenbergii* (Kutty *et al.* 2000). Further information on farming techniques, with an emphasis on production in subtropical and temperate regions, is provided on the Internet by D'Abramo *et al.* (1998a, b).

7.3.2 Broodstock, incubation and hatching

Where *M. rosenbergii* occurs naturally, local fisheries can provide a source of broodstock females but because this species will readily mature, copulate and spawn in ongrowing ponds, they are more commonly and conveniently selected from farm production ponds, or in some cases from broodstock holding ponds. Maturation of *M. rosenbergii* in captivity often occurs at a small size (<25 g) and while this minimises the effort and facilities required for basic broodstock management, care should be taken to avoid unwittingly selecting for low fecundity and slow growth (see below and section 8.10.1.2). Hatcheries do not tend to maintain their own captive stocks. If temperatures are adequate (over 25°C), ponds can provide a year-round supply of egg-carrying or 'berried' females (but see section 12.8.2). These females, instantly recognisable by the clutch of eggs held in the brood chamber on the underside of the abdomen, are selected for the hatchery on the basis of large size, healthy appearance and egg colour. The eggs are bright orange when spawned, gradually darken to a brownish orange and eventually turn dark grey 2–3 days prior to hatching. Usually only females with eggs at the late stage of development (dark grey) are chosen. When the eggs have hatched, the females can be sold for food or returned to broodstock or ongrowing ponds.

The fecundity of female *M. rosenbergii* is influenced by size. Large, fully mature females produce well over 100 000 eggs per brood whereas smaller animals taken from production ponds may produce between 10 000 and 30 000 eggs. The number of eggs produced has been estimated to average 870 per gram of female body weight (Sureshkumar & Kurup 1998) and usually falls in the range 500–1000. Care should be taken in the choice of female broodstock to ensure that favourable traits are retained and enhanced. Fast-growing young females are preferable to females that are simply large since the former can lead to improvements in growth rates (Daniels *et al.* 2000). Although small females may produce fewer eggs per clutch than larger females, overall fecundity is boosted because they produce more clutches in a given period. It is best to select broodstock prawns at the intermoult stage. Other desirable characteristics are

Plate 7.3 Pond-reared broodstock *Macrobrachium rosenbergii* in a Caribbean hatchery. Plastic pipe sections are placed in the tank to provide shelter.

good pigmentation, intact appendages and the absence of any injuries, external parasites or epicommensal infestations (section 2.5.7).

Careful handling of berried females is essential to minimise egg loss. Shipping in aerated tanks or bins is convenient for short journeys of 1 h or less. For longer periods, transport in plastic bags with water and oxygen gives good results. To protect bags from punctures, sheathing sharp rostrums and telsons with rubber tubing is preferable to blunting them with scissors. However, immobilising prawns by wrapping them individually with mesh or webbing is not advisable since it can significantly reduce survival rates (Smith & Wannamaker 1983). Lowering the temperature to between 18 and 23°C using crushed ice applied externally to the bags can reduce the effects of stress and a 24 h fasting period prior to shipment may also be beneficial. Details of shipping methods and some results for shipping small adults (17 g) are included in Table 7.7. After arrival at the hatchery, the use of a disinfectant dip containing 0.2–0.5 mg L^{-1} copper or 15–20 ppm formalin for 30 min is advisable to reduce the likelihood of introducing fouling organisms. The use of higher concentrations of formalin, 50–100 ppm for 1 h, has also been reported for this purpose (Daniels *et al.* 2000).

Berried females are often kept in water at a salinity of 12–15‰ so that newly emerged larvae do not require subsequent acclimatisation to the brackish water used for rearing. But some hatcheries keep their broodstock in water of just 3–5‰, or even freshwater, and then raise the salinity once the larvae have hatched. The optimal temperature range for the female prawns is 27–32°C.

The use of tanks specifically for the hatching process is rare but, in Taiwan, hatching tanks equipped with numerous shelters may be stocked with 20–40 g berried females at a rate of about 13 kg m^{-2} (Chen 1990). When the aeration supply is interrupted, newly hatched larvae can be siphoned from the water surface into separate rearing tanks. The more common practice is for females to be held in the same tanks that will be used for larvae rearing. A stocking rate of three prawns (measuring 10–12 cm TL) per cubic metre of water has been recommended in order to achieve a density of 30–50 larvae per litre (New & Singholka 1985). In a typical hatchery in Thailand, 40–50 berried females may produce 700 000 larvae in a 15 m^3 capacity larvae rearing tank filled to a depth of 0.3 m with water of 3–5‰ salinity. When the females are removed the water level is increased to 0.8–0.9 m and the salinity is raised to 12–15‰ (Correia *et al.* 2000). Females held just for the purpose of releasing larvae are usually not fed but broodstock held for extended periods may receive a pelleted diet supplemented with squid and beef liver.

The management of broodstock in temperate or subtropical regions is more complicated than in the tropics because wild gravid females may only be available in the warmer months of the year. The best approach is usually to take females from autumn harvests and overwinter them indoors to provide broodstock for next year's production. Usually the largest females are selected, in order to achieve a large number of eggs per spawn, along

Table 7.7 Methods and results for shipping post-larval, juvenile and adult *Macrobrachium rosenbergii*.

Stage and size	Duration (h)	Density (no. L^{-1} or g L^{-1})	Temp. (°C)	Survival (%)	Transport method	Reference/source
PL	short	750			40 L container	Macintosh 1987
PL	<1	750			Aerated tank	New 1990
PL	<12	300			Plastic bags with water + O_2	New 1990
PL	<24	833			40 L bags in cartons, 6 L water	SICA 1988
PL <20 mg	<24	200	18–23	90	24-h starved PLs*	Smith & Wannamaker 1983
PL	24–36	667			Plastic bags (30 × 40 cm), 3 L water + O_2	Correia *et al.* 2000
PL	24–36	100			Plastic bags with water + O_2	New 1990
PL <20 mg	24–48	150	18–23	90	24-h starved PLs*	Smith & Wannamaker 1983
Juvenile 6 g	24	18 g	19–20	>90	24-h starved juveniles, 40 per box*	Smith & Wannamaker 1983
Juvenile 6 g	48	9–11 g	19–20	>90	24-h starved juveniles, 20–25 per box*	Smith & Wannamaker 1983
Adult 17 g	24	12–15 g	19–20	100	24-h starved adults, 10–12 per box*	Smith & Wannamaker 1983
Adult 17 g	48	12–15 g	19–20	75	24-h starved adults, 10–12 per box*	Smith & Wannamaker 1983

*Styrofoam box (38 × 38 × 20 cm deep) containing double plastic bags with 13.6 L water and oxygen.

with males of an equal or larger size. Synchronisation of hatching can, however, be difficult (section 12.4). Over-wintering ponds or tanks can be stocked at around one individual per 20–60 L of water and require a minimum temperature of 25°C. Artificial substrates can reduce aggressive interactions but mortality rates of 40–50% for females and 60–70% for males can still be anticipated. For long-term holding, a sex ratio of one or two blue claw (BC) males (section 4.6.1) per 20 females is recommended. For production of larvae in winter and early spring an additional two to three orange claw (OC) males per 20 females should be included (Daniels *et al.* 2000).

Unilateral eyestalk ablation can be employed to accelerate the onset of spawning and the rate of spawning. Increased broodstock productivity and faster growth have been demonstrated in India with *M. malcolmsonii* using this technique. Ablated males increased growth by 36%, females by 41% while moult frequency increased by 24% and the number of eggs per brood rose 31%. A higher percentage hatch was also claimed (Murugadass & Marian 1989). Similar results were obtained with *M. nobilii* (Kumari & Pandian 1987) and extrapolation from these results indicates that scope for the enhancement of reproductive performance through ablation may also exist in *M. rosenbergii* although the technique is not used commercially (New 1990).

7.3.3 Larvae culture

Techniques for large-scale culture of *Macrobrachium* larvae were successfully developed by Takuji Fujimura in Hawaii in the mid-1960s. His method is known as the green-water technique because a population of algae, usually a green coloured *Chlorella* spp., is maintained in the culture water. It differs from the so-called clear-water technique, subsequently developed in mainland USA and French Polynesia, which as the name suggests uses water without added algae that is frequently changed or recycled. The algae in green-water cultures do not represent a source of nutrition for the prawn larvae and are only ingested incidentally with other items of prey or food. They do however, provide feed for *Artemia*, rotifers or other living prey, and help condition the culture water through the removal of toxic substances, particularly ammonia. The disadvantages of the green-water technique are that the algae often grow erratically and can cause the pH to fluctuate widely. In order to maintain 'green-water' conditions, water renewal must not flush away more algae than can be replaced by reproduction within the tank or by topping-up from separate algae stocks.

The clear-water system avoids these drawbacks. It allows greater exchange rates, better control over live and compounded food additions and helps stabilise water conditions. If a water quality or disease problem does occur in any particular tank, water exchange can be increased in response. These advantages have led to a widespread preference for clear-water techniques and green-water hatcheries are now rare.

Larvae rearing tanks can range in size from a few hundred to 30 000 L, but most are at the lower end of this range. Tanks with U-shaped bottoms are usually the most efficient but despite this many hatcheries use rectangular concrete tanks with gently sloping flat bottoms that are easier and cheaper to construct. Alternatives include fibreglass cylindro-conical and hyperbolic shaped vessels; modified, circular, concrete drainage pipe sections; fibrocement (asbestos) tanks; and plastic-lined tanks with frames of wood or bamboo. Operating water depths are usually between 0.4 and 1 m. Shading is provided to protect larvae from the harmful effect of direct sunlight known as 'sun-cancer'. Some light is needed ($6–75\,\mu E\,m^{-2}\,s^{-1}$ = 500–6000 lux) because prawn larvae are partly visual predators. There is some evidence that dark coloured tanks produce better batches of larvae than light coloured tanks (Correia *et al.* 2000).

Rearing densities usually lie between 30 and 100 larvae L^{-1}, but may be initially raised to 200 L^{-1} in systems in which the larvae are later to be divided between two or more culture vessels. Some Brazilian hatcheries use separate tanks for larvae up to instar 5 or 6 stocked at densities as high as 300–500 L^{-1}. After instar 6 the larvae are transferred to larger tanks and reared at densities of 60–100 L^{-1}. Other hatcheries initially concentrate larvae in the bottom of the rearing tank and progressively raise the water level to reduce density. The average salinity may range between 8‰ and 20‰, but 12 ± 2‰ is generally satisfactory (section 12.8.2). Some strains of *M. nipponense* complete larval development entirely in freshwater (section 12.8.2). Optimal temperatures and pH ranges are 26–31°C and pH 7.0–8.5, respectively. Prior to use, brackish water is often treated for 24 h by chlorination with 20–50 mg L^{-1} sodium hypochlorite. Any residue is removed by strong aeration or neutralised with sodium thiosulphate (Correia *et al.* 2000).

Larvae receive a combination of live and prepared feeds. Newly hatched *Artemia* are provided (to maintain a concentration of 1–5 nauplii mL^{-1} or to provide between 5 and 50 nauplii per prawn larva) along with egg custard and minced flesh of fish such as tuna. Mussel and fish flesh are sometimes incorporated within egg custards and soybean curd is also used. In all cases, prepared feeds are sieved through a mesh prior to feeding to ensure that appropriately small particles are created that can be easily maintained in suspension in the rearing tank. As a rough guide, for each production cycle, a total of 1.2–1.6 kg of prepared feed may be needed per cubic metre of culture (stocked at 30–50 larvae L^{-1}) (New & Singholka 1985). Daily feeding levels are adjusted to satisfy the demand of the larvae yet avoid water pollution through overfeeding. Commercially produced diets for crustacean larvae are available but their use has not become routine in all *Macrobrachium* hatcheries.

Aeration is used to maintain dissolved oxygen concentrations, and if sufficiently strong helps keep particles of food and detritus suspended in the water column. This enables much particulate waste to be flushed away with outgoing water during a water exchange and slows down the accumulation of organic material within the larvae rearing tank. Nevertheless, in most systems it is necessary to siphon out settled waste at least once every two days once non-live feeds are included in the diet. Daily siphoning initially leads to a water exchange of 20–30% of tank volume per day, increasing to 50–60% per day after metamorphosis to post-larvae.

Larvae survival rates usually average between 30 and 60% and result in yields of between 10 and 35 post-larvae L^{-1}. Production of 90 post-larvae L^{-1} and survival rates above 90% have been achieved in small intensively managed cylindro-conical rearing vessels (of 0.8 and 2 m^3 capacity) (AQUACOP 1982). However, for most hatcheries high survival rates are not essential to maintain a viable production level because supplies of berried females are usually abundant. Effective procedures for water treatment and sanitation are essential to avoid problems with *Zoothamnium, Epistylis* and hydroids (section 2.5.7). The use of antibiotics to control outbreaks of bacterial disease was widespread in the 1980s (New 1990) and probably remains so today (section 11.3.4).

Before metamorphosis into post-larvae, *M. rosenbergii* larvae pass through 11 successive moult stages, usually over a period of 18–23 days. Sometimes, however, adverse nutrition or water quality can prolong larval life and lengthen the time during which a population is metamorphosing. This increases the opportunity for cannibalism and can greatly reduce survival. Post-larvae have the appearance of miniature adult prawns and switch from a planktonic to a largely bottom-dwelling existence. They also acquire the euryhaline capability characteristic of the species, and can be safely exposed to a drop in salinity in preparation for the freshwater conditions they will encounter during nursery rearing and ongrowing. In Brazilian hatcheries, once 90% of larvae have metamorphosed the salinity is lowered over a period of 2 days, but if necessary the acclimatisation process can be carried out in just 2–8 h without stressing the post-larvae (Correia *et al.* 2000). Prawn post-larvae are usually harvested from tanks with scoop nets and any remainder collected by draining. Methods and results for shipping prawn post-larvae are included in Table 7.7.

Recirculating water systems incorporating rotating biodiscs or other types of biological filters have been developed to operate where supplies of good quality fresh or seawater are limited or where heat needs to be conserved (AQUACOP 1983; Cange *et al.* 1987; Valenti & Daniels 2000). They also have the advantage of providing relatively stable water conditions and they can reduce the amounts of effluent. However they are not favoured in all situations. In Thailand, for example, they have generally been abandoned in favour of simpler systems employing daily water exchange (Correia *et al.* 2000). The biological filters used in recirculating systems are living systems that must be 'activated', fed and maintained; they therefore require careful management (section 8.4.5). Efficient systems require less than 10% new water per day and water can be reused for two production cycles. In their review of prawn hatcheries using recirculation systems, Valenti and Daniels (2000) recommend high recirculation rates of 10–24 times the larvae rearing volume per day to minimise the size of the required biofilters. Many systems function however with lower exchange rates of 1.4–5 volumes per day. Biofilters are best kept in darkness because sunlight encourages algae growth. System management involves careful adjustment of larvae feeding rates to suit consumption rates while good aeration is important to maintain solid waste in suspension so that the filters can process it. Disease management relies heavily on cleanliness and minimising the chance of introducing disease organisms, with emphasis on chemical footbaths, washing of hands, and clean food, supplies and equipment (Valenti & Daniels 2000).

Healthy prawn larvae are active, display a red-brown pigmentation, and in the absence of aeration congregate at the water surface. In contrast, when they are not healthy they feed poorly, resort to cannibalism, develop a blue pigmentation and swim irregularly. Ideally substandard batches should be identified as early as possible and eliminated. To help in this process and provide a systematic basis for the comparison of batch quality and subsequent culture performance, a condition index can be employed. The quality index of Tayamen and Brown (1999) is derived from a series of scores based on whether or not larvae have a full gut and high lipid content; well dispersed chromatophores; amber body coloration; a straight, intact rostrum; high muscle to gut ratio in sixth abdominal segment; a clear/transparent appearance of abdominal muscle; absence of black spots and fouling organisms; and swim actively with a fast, jumpy motion. Larvae that score well in this system have shown better growth and survival (as far as 45 days post-metamorphosis) and have reached metamorphosis more quickly.

7.3.4 Nursery

Nurseries produce hardy prawn juveniles that are better equipped than young post-larvae to withstand the conditions of ongrowing ponds and thus they lead to better survival rates and farm productivity. In subtropical and temperate climates nursery ponds have the additional role of making best use of a limited ongrowing season (section 5.3). Nursery rearing can also be important in certain polycultures with large or carnivorous fish to avoid losses due to predation that would otherwise result from stocking small post-larvae straight from hatchery tanks (section 7.3.5.2). The process of size-screening nursed juveniles provides a valuable opportunity to counteract the heterogeneous individual growth that typically impairs the farming potential of *Macrobrachium* (section 7.3.7).

Nursery rearing is performed either indoors or outdoors in concrete tanks, or in ponds that may have concrete or brick walls or an entirely earthen construction. Cages have also been employed but mostly on an experimental basis (Alston & Sampaio 2000). To accommodate the aggressive and territorial behaviour of juveniles and to reduce losses through cannibalism, additional substrate material may be provided. This may take the form of draped netting or rows of netting arranged lengthways. In some indoor systems, mesh is fixed horizontally or vertically on frames of wood, aluminium or PVC. Some farmers provide shelters for their juveniles using cheaper alternatives such as aquatic plants, palm leaves, branches, pebbles or shells, but these types of substrate are less convenient for cleaning and reuse than suspended plastic netting.

Outdoor nurseries usually comprise ponds of 300–2000 m^2 that are stocked either with hatchery post-larvae or with juveniles from indoor nurseries. Stocking densities range between 75 and 1500 post-larvae m^{-2} and it may take 4–10 weeks to reach an average size of 0.8–1.5 g. To

Plate 7.4 Handpicking the last few juvenile freshwater prawns from a nursery pond in Mauritius. Netting material, previously suspended within the pond to provide additional substrate, has been removed prior to harvesting and can be seen piled on the far embankment.

eliminate dragonfly nymphs and other predatory aquatic insects, incoming water is usually filtered, but some farmers pretreat their pond water with a short-lived insecticide such as trichlorfon to achieve the same ends. Daily water exchange may be 15–30% of pond volume and at high stocking densities aerators may be deployed. Juveniles are fed at a rate of around 15–20% of biomass per day and are harvested with a 5–6 mm mesh seine, or by draining (Alston & Sampaio 2000).

In some nurseries in Thailand, post-larvae are stocked in concrete tanks at densities between 1000 and 5000 m^{-2} and are held for periods of 7–28 days (New & Singholka 1985). Earthen nursery ponds in the same country may be stocked at lower densities of 20–25 post-larvae m^{-2} and harvested after longer periods of 75–90 days (New 1988). In West Malaysia post-larvae are stocked at 100 m^{-2} for between 28 and 42 days (Lee 1982). Higher densities of 800 post-larvae m^{-2} have been reported in Taiwan for rearing cycles lasting 30–50 days in ponds of around 400 m^2 (Chen 1990).

Prawn post-larvae readily accept the same type of prepared feeds that are used for ongrowing, including diets designed for other species like shrimp or catfish. They will consume particles of shrimp, squid, fish, fish roe, corn, soybean, wheat, cooked eggs and spinach. For feeding young post-larvae in concrete tanks, the use of floating diets, such as those used for catfish, can help to visually assess food consumption rates and adjust feeding levels to demand.

In Mediterranean and warm temperate climates, the use of enclosed nurseries enables the production of juveniles in spring as soon as outdoor pond temperatures are adequate for ongrowing. Systems typically rely on greenhouses to capture and retain solar heat, but thermostatically controlled heaters may also be needed. Water is usually biologically filtered and recycled (3–24 tank volumes per day). Tanks may be made of concrete, fibreglass or asbestos and have a surface area of 10–50 m^2 and a depth of 1 m. Stocking densities range from less than 200 post-larvae m^{-2} up to 6000 post-larvae m^{-2}, but higher mortality is usually associated with densities at the top of this range. Once the juveniles reach 0.02–0.2 g they are transferred either to outdoor nursery ponds or directly to ongrowing ponds (Alston & Sampaio 2000).

In early enclosed system trials in South Carolina, USA, cylindrical fibreglass and rectangular concrete or aluminium tanks (3.8–9.5 m^3 water capacity) equipped with strip layer habitats of plastic netting or fibreglass screen were used. These habitats permitted densities of over 6000 post-larvae m^{-2} (of tank bottom), but the most consistent survival rates (60–90%) were generally obtained in the range 1200–5400 post-larvae m^{-2}. At mean temperatures of 25.2–28.6°C, post-larvae stocked at a size of 0.01 g reached 0.2–0.44 g after 77–92 days. Feed was supplied three times per day *ad libitum*, and consisted of a compounded diet, chopped fish, squid, spinach and egg. Water plants (*Egeria* sp. and *Lemna* sp.) provided supplemental food and substrate, and assisted in water purification (Smith *et al.* 1983). Both greenhouses and geothermal heat were employed to raise water temperatures in Israeli prawn nurseries (Cohen & Barnes 1982).

7.3.5 Ongrowing

7.3.5.1 Extensive

The operation of extensive prawn ponds that rely entirely on natural productivity is not widely reported. Such ponds are stocked at densities of 1–4 post-larvae m^{-2} and are capable of yielding around 200–300 kg ha^{-1} annually; more when fertilisers are applied. In Vietnam extensive culture in ponds or within bamboo pens relies on stocking wild-caught juveniles (Valenti & New 2000).

7.3.5.2 Polyculture

Although commercial polyculture involving freshwater prawns was quite rare in the 1980s, *Macrobrachium rosenbergii* has repeatedly shown potential in mixed species cultures both as the main crop or as a valuable secondary crop in fish ponds. The advantage of mixing prawns and other species relies on the occupation of complementary ecological niches and the absence of direct competition for food or space. The most commonly used species are fish such as tilapias; common, silver, grass and bighead carps; and catfish. Fish may improve the water quality by filtering out excess plankton and thereby decrease the need for water exchange. The disadvantage of polyculture is the fact that pond management becomes considerably more complicated and that some compatible fish species have relatively low market value. Carp, for example, are typically bony and this can limit their market appeal (Zimmerman & New 2000).

Most feeding regimes in prawn-fish polyculture are designed to suit the needs of the fish, in recognition of the fact that the fish will usually consume the feed before it reaches the prawns (Zimmerman & New 2000). Culture trials by García-Pérez *et al.* (2000) confirm that when there is competition for food the fish will fare better than the prawns. The comparison of the polyculture of prawns (7 m^{-2}) and *Oreochromis niloticus* (1 m^{-2}) with monocultures at the same densities produced the same fish yields with or without prawns. In contrast, the prawns fared much better without the fish and growth in monoculture was twice as fast as in polyculture.

Early Taiwanese research demonstrated that polyculture with milkfish (*Chanos chanos*) and grey mullet (*Mugil cephalus*) leads to efficient utilisation of pond resources (Liao & Chao 1982). Similarly, Israeli studies showed how *Macrobrachium* integrated well with tilapia, carps (common, silver and grass) and mullet. Stocking densities of 0.5–1.5 prawns m^{-2} combined with about 1.2 fish m^{-2} (carp and tilapia) resulted in yields of 6700–10 100 kg of fish ha^{-1} per crop and 220–780 kg ha^{-1} per crop of large prawns averaging 45–90 g. Food conversion ratios for the whole system were between 0.87 : 1 and 1.7 : 1, and conveniently, prawns and fish could be harvested and processed with the same equipment (Cohen 1984). In extensive polyculture trials, relying only on natural productivity enhanced by swine manure, Malecha *et al.* (1981) stocked four carp species at a combined density of 0.55 fish m^{-2} with freshwater prawns at 7.9 m^{-2}. Resulting prawn yields averaged 322 kg ha^{-1}, although mean size was only 12.7 g. Polyculture of *M. rosenbergii* (1.2–1.8 m^{-2}) with five species of Indian and Chinese carp (total density 0.85 m^{-2}) produced prawn yields of 162–428 kg ha^{-1} and fish yields of 4604–5821 kg ha^{-1} after 10 months (Hoq *et al.* 1996) (Table 5.6).

Size at stocking is critical in polyculture systems because small prawns are vulnerable to fish predation and, conversely, because big prawns can eat juvenile fish. The farming of prawns with catfish, for example, has demonstrated technical feasibility as long as prawn juveniles are stocked large enough to avoid predation by the catfish (Cohen 1984; Lamon & Avault 1987). In prawn–catfish polyculture, ponds can be stocked at densities of 0.7–1.2 prawns m^{-2} before the introduction of the catfish fry but managing the harvesting of the two crops can be complicated. Difficulties arise if the prawns are due to be harvested at the onset of cooler weather but the catfish have an off-flavour and are not ready for harvesting (D'Abramo *et al.* 1998b).

Other alternatives to prawn monoculture include prawn production with rice and prawn production in rotation with various other crops. Low yields (13–16 kg ha^{-1}) of *M. lanchesteri* have been obtained with rice in paddies in the Philippines (Guerrero *et al.* 1982) and in Vietnam, rice–prawn polyculture using *M. rosenbergii* typically produces prawn yields of 150–180 kg ha^{-1} per crop. Further data from Vietnam, from a farm where prawns were grown in rice paddies and in trenches, indicate how prawn farming can be integrated with established agriculture. Rice yields were 8.5 mt ha^{-1} yr^{-1} in two crops and prawn yields were 354 kg ha^{-1} yr^{-1} in a single crop. Rice was destined for local consumption but prawns were exported and contributed nearly one-third of net farm income. Prawns were fed cassava, coconut waste, broken rice and rice bran, and sludge from the trenches was used as a fertiliser for fruit trees (Hien *et al.* 1998).

In temperate regions of the USA, farming possibilities include crop rotation of prawns with rainbow trout and the overlapping of crops of prawns and crayfish (*Procambarus clarkii*) (Tidwell & D'Abramo 2000). In the latter case juvenile prawns can be stocked (4 m^{-2}) in late May and harvested between August and early October, and adult mature crayfish can be stocked (0.9 m^{-2}) in late June or early July. In late February the seine harvesting of crayfish begins and continues to late June before new crayfish are stocked. The new juvenile prawns are small enough to pass through the mesh of the seine (Granados *et al.* 1991; D'Abramo *et al.* 1998b).

7.3.5.3 Semi-intensive

The majority of prawn farming operations can be categorised as semi-intensive. Typically annual yields of between 0.5 and 5 mt ha^{-1} are achieved by stocking prawns at densities in the range 5–20 m^{-2} and by applying supplemental feeds.

Despite the theoretical efficiency of keeping ponds continually in production, under a regime of multiple stocking and harvesting, this approach to pond management is now rarely used and batch culture predominates with ponds being drained and restocked each year, although harvests of large individuals are made periodically (section 7.3.7). Under batch culture, for example in Thailand, ponds stocked with 4–10 post-larvae m^{-2} can yield 500–1500 kg ha^{-1} yr^{-1} through the application of low-cost feeds and with minimal water management. If nursery-reared juveniles are stocked at similar densities, yields can rise to 1500–2500 kg ha^{-1} yr^{-1} as long as more balanced, water-stable feeds are used. Yet higher annual production, 2500–4500 kg ha^{-1}, is feasible if three crops per year can be produced in areas with a year-round ongrowing season. This involves the stocking of juveniles from a two-stage nursery system at densities of 12–20 m^{-2} and the use of ponds that benefit from aeration and more intensive management of feeding and water quality (Valenti & New 2000).

Ongrowing ponds are usually of earthen construction and range in size from a few hundred square metres up to 2–3 ha. Larger units are difficult to harvest efficiently (section 7.3.6). Various materials such as bamboo, bricks and pipes have been used in production ponds to provide shelters for prawns and reduce aggressive interaction. The benefit gained, however, is usually outweighed by the practical disadvantages of increased labour and interference with harvesting. Most operators therefore favour uncluttered pond beds or, in a few cases, adopt the approach used in some nurseries, of using suspended netting which can be easily removed before harvesting takes place.

Water is obtained from rivers, wells, irrigation channels or reservoirs and pumped or gravity fed to ponds either individually or sequentially. Some ponds, for example in Hawaii, operate with flow-through systems, but in Thailand and elsewhere water addition rarely represents much more than just topping-up to replace evaporative or seepage losses. Optimal stocking densities are partly determined by water exchange rate. A density of 10 prawns m^{-2} was considered suitable for ponds in Hawaii with a more or less constant flow-through of water (Fujimura 1982) while 3–5 m^{-2} were more appropriate for many ponds that contained primarily static and hence warmer water. Stocking strategies must also be closely tied to the harvesting regime (section 7.3.7).

In experimental ongrowing trials *Macrobrachium* has been shown to grow and survive well in water up to a mean salinity of 16‰ (Smith *et al.* 1982), and some farmers take advantage of its euryhaline nature to utilise tidal water with a fluctuating but low salinity. Nevertheless, *M. rosenbergii* is predominantly cultured in freshwater, and a reliable supply is normally a prerequisite for prawn farming. To achieve optimum growth and survival, water temperatures in the range 26–31°C are recommended. In temperate zones, short growing seasons can only support single annual crops unless some source of thermal effluent or geothermal water can be exploited. Such heat sources have demonstrated potential to support prawn farming (section 5.4), though density limitations due to aggression preclude intensive cultures. Optimum water hardness levels appear to lie between 30 and 100 mg $CaCO_3$ L^{-1} (Table 8.3). Some very hard well waters (registering 305–638 mg $CaCO_3$ L^{-1}) have been found not only to depress growth but also to lead to prawns becoming encrusted with bryozoans and protozoans (Cripps & Nakamura 1979). Elsewhere, Bartlett and Enkerlin (1983) found that growth was not adversely affected by hardness levels between 940–1060 mg $CaCO_3$ L^{-1}, but the water they used had relatively low alkalinity. It is now known that the adverse effects of water hardness on prawn growth are exacerbated by high levels of alkalinity (section 8.5 and Table 8.3).

Effective predator control usually increases survival rates to 50% or more during ongrowing. In addition to basic precautions (section 8.3.6.1), in South-east Asia small netting fences of 60 cm height may need to be fixed around ponds to keep out invading catfish and snakehead fish. In addition, jars sunk into pond banks can serve as crab traps. Prior to stocking, air-breathing insects can be eliminated with a 2 : 1 mixture of motor oil and diesel fuel (9–19 L ha^{-1}) applied to the pond surface (D'Abramo *et al.* 1998b). Stocking nursery-reared juveniles can reduce early losses to predation and dogs can be used to keep predatory birds away. If ponds are operated on a continuous basis and rarely drained, the eradication of fish predators cannot be performed as part of pond preparation (section 8.3.3) and must be done during seine harvests.

The growth of vegetation is usually promoted on embankments to reduce erosion, even though this can reduce the efficiency of seine harvesting. Plant growth within the ponds has the same drawback, but can be prevented by filling ponds to a depth of about 1 m and by maintaining a bloom of algae sufficiently dense to shade the pond bottom. In semi-intensive *Macrobrachium* ponds, fertiliser is rarely needed to maintain algae growth since nutrient release through the breakdown of supplemental feeds is usually more than adequate for this purpose. All the same, in temperate climates a bloom of algae can be encouraged 1–2 weeks before stocking with juveniles by the addition of liquid fertiliser (NPK = 10 : 34 : 0 or 13 : 38 : 0) at rates of 4.7–9.4 L ha^{-1} (D'Abramo *et al.* 1998b).

Aeration equipment is sometimes used on prawn farms to permit increased stocking densities and productivity, and it can be valuable as a temporary response to critically low oxygen concentrations when the alternative of rapid water renewal is unavailable. One farm in the Dominican Republic practised polyculture with *M. rosenbergii* (10 m^{-2}) and tilapia (3 m^{-2}) in ponds of 0.5–0.8 ha and each pond was equipped with one or two aerators (1 hp each) (D. Lee, 1993 unpublished data). Aspects of water quality management in crustacean ponds are considered in section 8.3.

A range of different fresh foods, prepared diets and commercially produced pelleted feeds are used for prawn ongrowing. The latter include pellets produced for chickens, pigs and shrimp as well as prawns, though non-aquatic animal feeds usually display poor stability in water. Various recipes exist for the preparation of feeds at the farm, which take advantage of locally available foodstuffs such as rice bran, cooked broken rice, chopped trash fish, fishmeal, cornmeal, soybean meal or cake, meat, bonemeal, alfalfa and brewery waste. D'Abramo and New (2000) give various recipes for producing moist pellets, some of which incorporate binding agents such as agar or guar gum to provide water stability. After grinding, mixing and simple extrusion, these diets are sun-dried to a moisture content of less than 10% in order to extend storage life (section 8.8.2). The protein and lipid requirements of *M. rosenbergii* seem less stringent than for some penaeid shrimp and other marine crustaceans (D'Abramo 1998) and diets often incorporate relatively inexpensive ingredients. In Bangladesh the freshwater snail, *Pila globosa*, has proved to be an economical prawn feed and the meat is extracted and fed at rates of 60–74 kg ha^{-1} to yield prawn harvests of 490–620 kg ha^{-1} (Ahmed 1998). However, the direct application of fresh foods such as chopped trash fish, chicken carcasses, animal bones with flesh remains, soft-shelled snails and mussels is usually reported along with cautionary notes about the dangers of severe pollution associated with overfeeding these items.

Prawns are usually fed once or twice per day. In many artisanal operations, feeding levels are determined solely in response to water quality changes and observations of the quantity of uneaten food remaining in shallow pond margins; ideally they should also be based on estimates of the crop biomass (section 8.3.5.1). If sufficient care is taken to adjust feeding rates to suit demand, the worst effects of pollution are avoided, but if dissolved oxygen concentrations become critically low (best measured just before dawn – section 8.3.2), feeding may need to be suspended until conditions improve. Food conversion ratios between 2 : 1 and 4 : 1 for dry diets and 7 : 1 and 9 : 1 for wet feeds such as trash fish are reported (Wulff 1982a; New & Singholka 1985). Feeding trays (section 8.3.6.3) have greatly helped feed management in penaeid shrimp ponds but for freshwater prawns the aggressive territorial behaviour of the species may render them ineffective.

The farming of *Macrobrachium* in subtropical and temperate climates poses particular problems because of the abbreviated outdoor ongrowing season. In central USA, for example, only 120–140 growing days are available each year. Following the analysis of Tidwell and D'Abramo (2000), for prawn farming to achieve commercial viability in such regions, particular attention must be paid to three aspects of methodology: firstly ponds should be stocked with advanced juveniles; secondly the juveniles should be size-graded; and thirdly, artificial substrate should be provided. Efficient broodstock and hatchery management are also necessary to produce post-larvae within a critical 3–4 week 'window of opportunity' at the start of each season (section 12.4). Advanced juveniles, weighing 0.3–0.5 g, can be produced after 45–60 days of nursery culture and size-grading can be performed using bar graders that are typically used for finfish. If the small and large juveniles are then cultured independently this can result in better overall yields than if all the different sizes are cultured together. This is because, in the absence of size grading, the larger juveniles suppress the growth potential of the smaller juveniles (section 7.3.7). The benefits of artificial substrates have been demonstrated in one set of trials by an 18% increase in total production, a 13% increase in mean size, and a 17% improvement in feed efficiency (Tidwell & D'Abramo 2000).

For prawn farms in temperate regions ongrowing ponds cannot be stocked with more than 6–10 juveniles m^{-2} because growth rates are density dependent and at higher densities the prawns would not reach market size within one season. However, under such constraints and by employing a commercially available sinking catfish feed, Tidwell and D'Abramo (2000) note that it is possible to consistently produce prawns averaging above 40 g and to achieve yields of 2500 kg ha^{-1} yr^{-1}. When it comes to harvesting the pond, seining without draining is insufficient to extract the whole crop and can leave as many as 25% of prawns behind. The use of a harvesting sump, or catch basin (which needs to be aerated) within a pond at the drain end can simplify the process of drain harvesting, since it concentrates the prawns as the water level is dropped. Additionally, in a pond that is drain harvested rather than seined, the artificial substrates can be left in place between crops.

One *Macrobrachium* farm in the North Island of New Zealand has managed to overcome the constraints of a temperate climate by using an industrial heat source to raise pond temperatures. Direct use of the geothermal waters used by the Wairakei power station is impossible because of high levels of sulphur, lithium and arsenic so the hot water must be passed through a heat exchanger to transfer heat to the river water feeding the ponds. The farm includes a hatchery and nursery, and 6 ha of ongrowing ponds that are stocked at densities of 10 juveniles m^{-2}. The ponds are drain harvested twice per year to produce annual yields of 2.5–3 mt ha^{-1}. The operation also capitalises on its value as a tourist attraction and receives 25 000 visitors per year (New 2000).

New (1988) noted that pens are sometimes used for commercial prawn farming in Thailand, but no indication of yields was given. Cages are used in lakes in China and yield up to 1.5 kg m^{-2} stocked at 15–20 prawns m^{-2} (Valenti & New 2000).

Although the optimal temperature range for farming *M. rosenbergii* is usually 26–31°C, rapid growth and good results have been obtained in the USA at temperatures averaging 25°C. Interestingly culture at these 'sub-optimal' temperatures appears to have a favourable impact on the prawn population structure and results in fewer stunted small males and more of the faster growing orange-clawed males (section 7.3.7). Female maturation also appears to be delayed resulting in better somatic growth (section 12.8.2).

Species of freshwater prawn, other than *M. rosenbergii*, that are demonstrating culture potential include *M. nipponense* and *M. malcolmsonii*. The former is the second most important species reared in China and yields of 390–1875 kg ha^{-1} are reported (Wang & Qianhong 1999). The latter is cultured in India in areas where seed prawns and a natural fishery exist, and where *M. rosenbergii* is unavailable. The feasibility of year-round hatchery production has been demonstrated (Kanaujia *et al.* 1999) and ponds stocked with 3–5 juveniles m^{-2} have produced crops of 440–565 kg ha^{-1} after 6 months (Kanaujia *et al.* 1997).

7.3.5.4 Intensive

Although *M. rosenbergii* is not among species suited to intensive or super-intensive farming, largely because of its aggressive territorial behaviour, some attempts have been made to assess its performance in high-density trials.

In Brazil, concrete tanks have yielded the equivalent of 6.21 mt ha^{-1} yr^{-1}, and yields from controlled environment (indoor) tanks are now being investigated (Valenti & New 2000). In Great Britain, very small (0.62 m^2) indoor tanks stocked with 162 juveniles m^{-2} yielded 486 g m^{-2} (equivalent to 4860 kg ha^{-1}) after 112 days of culture. Prawns averaged only 7 g at harvest and the survival rate was 45% (Wickins & Beard 1978). Although these results were comparable with those obtained under the same conditions with some species of penaeid shrimp, they were notably inferior to the productivity of 1908 g m^{-2} and survival rate of 91% achieved with *Penaeus monodon*. Similar yields were obtained from a much larger outdoor concrete tank of 173 m^2 in South Carolina, USA, stocked with nursed juveniles (mean size 1.0 g) at 83 m^{-2}. This produced a crop equivalent to 4700 kg ha^{-1} after 110 days, but the mean harvest weight of prawns again was low at 8.5 g. In a parallel trial, larger animals averaging 16.5 g were produced using a lower stocking density of 32 individuals m^{-2}

and a culture duration of 138 days. However, total yield under this regime was reduced to 3828 kg ha^{-1}. In both trials, survival rates were high at 66.5 and 73.2% and the tanks were well aerated and contained artificial substrate in the form of draped netting. Using a compounded diet containing 25% protein, feed conversion ratios of 2.3 : 1 and 1.4 : 1 were obtained in the higher and lower density trials respectively (Sandifer *et al.* 1982).

7.3.6 Harvesting

Harvesting techniques rely on draining or seining or a combination of both. Seining employs a net (about 2 m high when stretched vertically) extended across the pond and drawn along the banks usually by a team of workers. Wulff (1982b) described a method using two such nets positioned across the centre of the pond and drawn independently towards opposite ends. This process shortens the duration of each haul and reduces stress and physical damage to the catch. Narrow pond design (max. 30–50 m) facilitates the seining procedure and, as a result, ponds to be harvested in this way rarely exceed 1 ha in size.

Unfortunately seines are cumbersome to operate and ineffective on uneven pond banks and beds and as a means of selective harvesting their efficiency is only rated at 50–75% (Karplus *et al.* 2000). A mechanised seine harvest system incorporating a boom mounted on a tractor was described by Losordo *et al.* (1986). In trials in heavily silted ponds a tractor operator and three workers were able to collect 63.5% of the marketable prawns (sizes not given). All the same, mechanised harvesting is hardly ever applied in commercial prawn farms (Valenti & New 2000).

During drain harvesting by gravity, prawns are carried towards the pond exit gate where they collect in a sump, net bag or net enclosure. While the prawns are being concentrated they can be manually scooped up or transferred by pump to containers on the pond bank, but great care must be taken to avoid damage. Large ponds (>2–3 ha) do not usually drain sufficiently well for efficient drain harvests. Prawns remaining in the pond require laborious netting or hand-picking that must be performed quickly if the animals are to be collected in marketable condition. Unfortunately *Macrobrachium* have a tendency to swim upstream against any flow of water that is supplied to sweep them to the collection point.

7.3.7 Stocking and harvesting regimes and the management of size variation

The management of stocking and harvesting in freshwater prawn ponds is greatly complicated by the heterogeneous growth rates characteristic of cultured populations. Within a pond stocked with a single batch of postlarvae some animals have been observed to grow 15 times faster than others and in one typical example, after 6 months of ongrowing, prawns reached an average size of 48 g yet ranged from 10 g to 110 g (Menasveta & Piyatiratitivokul 1982) (Table 4.5).

The majority of the size heterogeneity is found among the male prawns and has been shown to arise from behavioural interactions within a population rather than from differences in genetic growth potential (sections 2.6.1 and 4.6.1). Early maturing males develop large blue claws (BC), take dominant positions in a hierarchy and at the same time undergo a reduction in growth rate. Meanwhile other males, characterised by smaller orange claws (OC), grow more rapidly and eventually transform at a single moult to become even larger, newly dominant BC males. In addition, a significant proportion of small males (SM) experience stunted growth and remain at the bottom of the hierarchy. Although these possess the potential to become OC and eventually BC males, they remain as runts unless the density of the larger and more aggressive BC and OC males is reduced. The practical result of this is that the growth potential of much of the prawn population in a pond can only be released by repeatedly harvesting the largest animals.

In straightforward batch culture, a single stocking operation is followed by a single total harvest at the end of the ongrowing period and then the pond is drained and prepared for the next cycle. This approach, however, produces a wide size range among the prawns and complicates the processes of size grading and marketing. As a result the alternative strategy of multiple harvesting is more often employed.

The multiple harvest approach involves a single stocking followed about 5 months later by a series of partial seine harvests at monthly or fortnightly intervals. By the use of a seine with an appropriate mesh size (3.8–5 cm), the larger prawns (>25–40 g) are selected while the smaller ones remain to continue growing at a reduced density. After a period of about 8 months the pond is drained and all remaining prawns are harvested. This approach is efficient and popular in Thailand where it results in the production of manageable quantities of prawns at regular intervals and maximises the productivity of a culture cycle that is often limited by the onset of the dry season. In situations where the growing season is not limited, after the harvesting is completed the pond can be immediately prepared for treatment, refilling and restocking.

Yet another method, continuous culture, can also reduce size heterogeneity but it is now rarely practised. This method involves multiple stocking as well as multiple harvesting, and was considered an effective way to take advantage of a year-round water supply and growing season. In theory it has the potential to maximise annual output and proponents of the approach have recorded annual yields of up to 4000 kg ha^{-1}. Taiwanese operations have been reported to produce even higher yields with the application of good pond management (Liao & Chao 1982). Hawaiian ponds managed for continuous culture were stocked with 16–22 prawns m^{-2} and could produce around 2500 kg ha^{-1} annually (Shang & Mark 1982). The problems with the method are that in ponds (but not in indoor controlled environment systems) populations of predators and competitors can steadily build up and that incomplete harvesting leaves behind some large dominant prawns that can depress the growth rates of newly stocked juveniles. As a result, pond productivity declines over time and undermines the viability of the farm. Seasonal water shortages in places like Thailand have also discouraged continuous culture (Valenti & New 2000).

Apart from modifying stocking and harvesting regimes, the most practical ways of reducing size heterogeneity at harvest are by providing additional artificial substrates or shelters and by size-grading juveniles. Ponds equipped with such shelters tend to have fewer stunted males, because aggressive interactions are less frequent. Substrates and shelters thus have the same effect as reducing prawn density and can increase survival rates by providing newly moulted prawns a refuge from cannibalism. To be commercially successful, prawn shelters need to be attractive to prawns; to provide a large surface area; to be easy to install, remove and store; and to be cheap (Karplus *et al.* 2000).

The size-grading of juveniles and the separate culture of the larger and smaller animals is a fruitful approach to raising prawn yields because the developmental pathway in the male population is determined early in their development. Large morphotype determination is irreversible and the upper fraction of size-graded juveniles produces a greater proportion of faster growing BC and OC males than non-graded populations stocked at the same density. In contrast the stunted condition of small males is not irreversible and once larger dominant males are removed they can undergo rapid compensatory growth. The potential of the small fraction is best realised by providing an ongrowing period of at least 130–165 days (Karplus *et al.* 2000).

Research in the late 1980s revealed good prospects for reducing size heterogeneity at harvest and increasing overall yields by culturing monosex populations (section 2.6.3). Pond culture trials, however, indicated that all-male populations provided higher yields than all-female populations (Cohen *et al.* 1988). Thus the usual objective was to seek all-male populations. However it is known that females are more tolerant of crowding than males and one commercial strategy would be to rear all-female populations at high densities. An alternative might be to enhance female growth by delaying their sexual maturation (section 12.8.2). Manual selection of monosex populations requires skilled workers, is highly labour intensive and would probably not be economic. The production of single sex populations through surgical and genetic manipulation is possible but the techniques have yet to be refined for application on a commercial scale. Creation of all-male populations may also be possible through the application of an androgenic hormone (section 2.6.3).

7.3.8 Processing

Freshwater prawns are usually sold to local markets either live or whole on ice, but much farmed product is now being exported in the form of frozen tails. Unfortunately the processing yield of headless prawns, at around 40%, is inferior to that of marine shrimp (57–68%). Large claw (BC) males yield 5–8% less meat than OC males.

To avoid mushiness in fresh product, harvested prawns need to be immediately washed in clean water, killed in a mixture of ice and water at 0°C and then washed in lightly chlorinated water (5 mg L^{-1}) (Valenti & New 2000). With whole marine shrimp the blackening of the head that occurs even when ice is applied is commonly controlled with the use of sodium metabisulphite. However, with *Macrobrachium* the blackening process is much slower so there is little to be gained from using this chemical (Griessinger *et al.* 1991). More information on the post-harvest handling, processing and marketing of prawns is provided in section 3.3.2.

7.3.9 Hatchery supported fisheries, ranching

Many wild stocks of *M. rosenbergii* and other species of freshwater prawn have been depleted through pollution, overfishing and the interruption of migration routes with irrigation schemes. Augmentation of fisheries through the release of hatchery-reared juveniles would therefore seem worth investigating. New and Singholka (1985) note some success with this in Thailand in providing additional food and income for local fishermen. In one stocking programme lasting 3 years, 3×10^6 juveniles were released in a lake measuring 410 km$^2 \times$ 15 m deep. Four thousand families fished the lake and recapture rates were estimated at 2% (NACA 1986).

More recently in Thailand, government hatcheries that were originally established to promote freshwater prawn farming now produce juveniles for stocking lakes, canals, rivers and reservoirs. More than 300×10^6 juveniles were released in 1999. Fishermen provide positive feedback but a shortage of hard data persists on the overall scale of impacts. *M. rosenbergii* juveniles (PL_{20}) have been stocked into Vembanad lake and some small reservoirs and rivers in Kerala, India, as part of a social fisheries programme. In 1999, for example, 1.9×10^6 were released. The juveniles are believed to have contributed to an improvement in production (Table 5.8). A programme to stock *M. malcolmsonii* in the upper reaches of the Godavari river, in Andhra Pradesh, has been run for several years. In this case juveniles did not come from hatcheries but were captured in the lower reaches of the same river. In Malaysia, just under 12×10^6 *M. rosenbergii* juveniles were stocked into rivers between 1988 and 1990 to rehabilitate depleted fisheries. Improved catches were reported and releases continue (New *et al.* 2000).

7.3.10 References

Ahmed N. (1998) Freshwater snail, *Pila globosa* play an important role for prawn culture in Bangladesh. *CEFAS Shellfish News*, (5) 35–36.

Alston D.E. & Sampaio C.M.S. (2000) Nursery systems and management. In: *Freshwater Prawn Culture: the farming of* Macrobrachium rosenbergii (eds M.B. New & W.C. Valenti), pp. 112–122. Blackwell Science, Oxford, UK.

AQUACOP (1982) Mass production of juveniles of freshwater prawn *Macrobrachium rosenbergii* in French Polynesia: Predevelopment phase results. In: *Proceedings of Symposium on Coastal Aquaculture 1982*, Part 1, pp. 71–75. Marine Biological Association of India, Cochin, India.

AQUACOP (1983) Intensive larval rearing in clear water of *Macrobrachium rosenbergii* (de Man, Anuenue stock) at the Centre Oceanologique du Pacifique. In: *Handbook of Mariculture, Vol. 1, Crustacean aquaculture* (ed. J.P. McVey), pp. 179–188. CRC Press, Boca Raton, FL, USA.

Bartlett P. & Enkerlin E. (1983) Some results of rearing giant prawn, *Macrobrachium rosenbergii* in asbestos asphalt ponds in hard water and on a low protein diet. *Aquaculture*, **30** (1–4) 353–356.

Cange S.W., Pavel D.L., Lamon L.S. & Avault J.W. Jr. (1987) Development of larval rearing systems for the Malaysian prawn *Macrobrachium rosenbergii* in southern Louisiana. *NOAA Technical Report, 47*, pp. 43–49.

Chen L.C. (1990) *Aquaculture in Taiwan,* 273 pp. Fishing News Books, Blackwell Scientific Publications, Oxford, UK.

Cohen D. (1984) Prawn production in catfish ponds: proposed strategy and test trials. *Aquaculture Magazine*, **10** (2) 14–20.

Cohen D. & Barnes A. (1982) The *Macrobrachium* programme of the Hebrew University, Jerusalem. In: Giant prawn farming (ed. M.B. New). *Developments in Aquaculture and Fisheries Science*, **10**, 381–385.

Cohen D., Sagi A., Ra'anan Z. & Zohar G. (1988) The production of *Macrobrachium rosenbergii* in monosex populations. III. Yield characteristics under intensive monoculture conditions in earthen ponds. *Israeli Journal of Aquaculture – Bamidgeh*, **40** (2) 57–63.

Corbin J.S., Fujimoto M.M. & Iwai T.Y. Jr. (1983) Feeding practices and nutritional considerations for *Macrobrachium rosenbergii* culture in Hawaii. In: *Handbook of Mariculture, Vol. 1, Crustacean aquaculture* (ed. J.P. McVey), pp. 391–412. CRC Press, Boca Raton, FL, USA.

Correia E.S., Suwannatous S. & New M.B. (2000) Flow-through hatchery systems and management. In: *Freshwater Prawn Culture: the farming of* Macrobrachium rosenbergii (eds M.B. New & W.C. Valenti), pp. 52–68. Blackwell Science, Oxford, UK.

Cripps M.C. & Nakamura R.M. (1979) Inhibition of growth of *Macrobrachium rosenbergii* by calcium carbonate water hardness. *Proceedings of the World Mariculture Society*, **10**, 575–580.

D'Abramo L.R. (1998) Nutritional requirements of the freshwater prawn *Macrobrachium rosenbergii*: comparisons with species of penaeid shrimp. *Reviews in Fisheries Science*, **6** (1–2) 153–163.

D'Abramo L.R. & New M.B. (2000) Nutrition feeds and feeding. In: *Freshwater Prawn Culture: the farming of* Macrobrachium rosenbergii (eds M.B. New & W.C. Valenti), pp. 203–220. Blackwell Science, Oxford, UK.

D'Abramo L.R., Brunson M.W., Daniels W.H. & Fondren M.E. (1998a) *Freshwater prawns hatchery and nursery management*. Mississippi State Extension Service. http://ext.msstate.edu/pubs/pub2002.htm

D'Abramo L.R., Brunson M.W. & Daniels W.H. (1998b) *Freshwater prawns pond production and grow-out*. Mississippi State Extension Service. http://ext.msstate.edu/pubs/pub2003.htm

Daniels W.H., Cavalli R.O. & Smullen R.P. (2000) Broodstock management. In: *Freshwater Prawn Culture: the farming of* Macrobrachium rosenbergii (eds M.B. New & W.C. Valenti), pp. 41–51. Blackwell Science, Oxford, UK.

Fujimura T. (1982) Discussion session – practical farming. In: Giant prawn farming (ed. M.B. New). *Developments in Aquaculture and Fisheries Science*, **10**, 469–498.

García-Pérez A., Alston D.E. & Cortés-Maldonado R. (2000) Growth, survival, yield and size distribution of freshwater prawn *Macrobrachium rosenbergii* and tilapia *Oreochromis niloticus* in polyculture and monoculture systems in Puerto Rico. *Journal of the World Aquaculture Society*, **31** (3) 446–451.

Granados A.E., Avault J.W. & Cange S.W. (1991) Double-cropping prawns, *Macrobrachium rosenbergii*, and red swamp crawfish, *Procambarus clarkii*. *Journal of Applied Aquaculture*, **1** (1) 65 77.

Griessinger J.M., Lacroix D. & Gondouin P. (1991) *L'élevage de la crevette tropicale d'eau douce*, 160 pp. IFREMER, Plouzané, France.

Guerrero L.A., Circa A.V. & Guerrero R.D. III (1982) A preliminary study on the culture of *Macrobrachium lanchesteri* (de Man) in paddy fields with and without rice. In: Giant prawn farming (ed. M.B. New). *Developments in Aquaculture and Fisheries Science*, **10**, 203–206.

Hien T.T.T., Minh T.H., Phuong N.T. & Wilder M.N. (1998) Current status of freshwater prawn farming in the Mekong Delta of Vietnam. *JIRCAS Journal*, **6**, 89–100.

Hoq M.E., Islam M.M. & Hossain M.M. (1996) Polyculture of freshwater prawn (*Macrobrachium rosenbergii*) with Chinese and Indian carps in farmer's pond. *Journal of Aquaculture in the Tropics*, **11**, 135–141.

Kanaujia D.R., Mohanty A.N. & Tripathi S.D. (1997) Growth and production of Indian river prawn *Macrobrachium malcolmsonii* (H. Milne Edwards) under pond conditions. *Aquaculture*, **154** (1–2) 79–85.

Kanaujia D.R., Mohanty A.N. & Tripathi S.D. (1999) Year-round breeding and seed production of Indian river prawn *Macrobrachium rosenbergii* (H. Milne Edwards) under controlled conditions. *Journal of Aquaculture in the Tropics*, **14** (1) 27–36.

Karplus I., Malecha S.R. & Sagi A. (2000) The biology and management of size variation. In: *Freshwater Prawn Culture: the farming of* Macrobrachium rosenbergii (eds M.B. New & W.C. Valenti), pp. 259–289. Blackwell Science, Oxford, UK.

Kumari S.S. & Pandian T.J. (1987) Effects of unilateral eyestalk ablation on moulting, growth, reproduction and energy budget of *Marcobrachium nobilli*. *Asian Fisheries Science*, **1**, 1–17.

Kutty M.N., Herman F. & Le Menn H. (2000) Culture of other prawn species. In: *Freshwater Prawn Culture: the farming of* Macrobrachium rosenbergii (eds M.B. New & W.C. Valenti), pp. 393–410. Blackwell Science, Oxford, UK.

Lamon M.S. & Avault Jr., J.W. (1987) Polyculture stocking strategies for channel catfish, *Ictalurus punctatus*, and the prawn, *Macrobrachium rosenbergii*, using one catfish density and three prawn densities with two prawn sizes. Abstract in: *Journal of the World Aquaculture Society*, **18** (1) 23A.

Lee C.L. (1982) Discussion session – practical farming. In: Giant prawn farming (ed. M.B. New). *Developments in Aquaculture and Fisheries Science*, **10**, 469–498.

Liao I.C. & Chao N.H. (1982) Progress of *Macrobrachium* farming and its extension in Taiwan. In: Giant prawn farming (ed. M.B. New). *Developments in Aquaculture and Fisheries Science*, **10**, 357–359.

Losordo T.M., Wang J-K., Mark J.B. & Lam C.Y. (1986) A mechanised seine harvest system for freshwater prawns. *Aquacultural Engineering*, **5** (1) 1–16.

Macintosh D.J. (1987) *Aquaculture production and products handling in ASEAN*, 365 pp. ASEAN Food Handling Bureau, Kuala Lumpur, Malaysia.

Malecha S.R. (1983) Commercial pond production of the freshwater prawn, *Macrobrachium rosenbergii*, in Hawaii. In: *Handbook of Mariculture, Vol. 1, Crustacean aquaculture* (ed. J.P. McVey), pp. 231–260. CRC Press, Boca Raton, FL, USA.

Malecha S.R., Buck D.H., Baur R.J. & Onizuka D.R. (1981) Polyculture of the freshwater prawn *Macrobrachium rosenbergii*, Chinese & common carps in ponds enriched with swine manure. 1. Initial trials. *Aquaculture*, **25** (2–3) 101–116.

Menasveta P. & Piyatiratitivokul S. (1982) Effects of different culture systems on growth, survival and production of the giant freshwater prawn (*Macrobrachium rosenbergii* de Man). In: Giant prawn farming (ed. M.B. New). *Developments in Aquaculture and Fisheries Science*, **10**, 175–189.

Murugadass S. & Marian M.P. (1989) Maximization of seed production by eyestalk ablation technique in *Macrobrachium malcolmsonii*. *Aquaculture 89 Abstracts*, p. 103, World Aquaculture Society, Los Angeles, USA.

NACA (1986) *Giant freshwater prawn breeding and farming in Thailand: an introduction*. FAO Network of Aquaculture Centers in Asia, NACA TV Video production.

New M. B. (1988) Freshwater prawns: status of global aquaculture, 1987. *NACA Technical Manual 6*, 58 pp. World Food Day 1988. Publication of the Network of Aquaculture Centres in Asia, Bankok, Thailand.

New M.B. (1990) Freshwater prawn culture: a review. *Aquaculture*, **88** (2) 99–143.

New M.B. (1995) Status of freshwater prawn farming: a review. *Aquaculture Research*, **26** (1) 1–54.

New M.B. (2000) Commercial freshwater prawn farming around the world. In: *Freshwater Prawn Culture: the farming of* Macrobrachium rosenbergii (eds M.B. New & W.C. Valenti), pp. 290–325. Blackwell Science, Oxford, UK.

New M.B. & Singholka S. (1985) Freshwater Prawn Farming: a manual for the culture of *Macrobrachium rosenbergii*, 118 pp. *FAO Fisheries Techical Paper 225, Rev. 1*.

New M.B. & Valenti W.C. (eds) (2000) *Freshwater Prawn Culture: the farming of* Macrobrachium rosenbergii, 443 pp. Blackwell Science, Oxford, UK.

New M.B., Singholka S. & Kutty M.N. (2000) Prawn capture fisheries and enhancement. In: *Freshwater Prawn Culture: the farming of* Macrobrachium rosenbergii (eds M.B. New & W.C. Valenti), pp. 411–428. Blackwell Science, Oxford, UK.

Sandifer P.A., Smith T.I.J., Stokes A.D. & Jenkins W.E. (1982) Semi-intensive grow-out of prawns (*Macrobrachium rosenbergii*): preliminary results and prospects. In: Giant prawn farming (ed. M.B. New). *Developments in Aquaculture and Fisheries Science*, **10**, 161–172.

Shang Y.C. & Mark C.R. (1982) The current state-of-the-art of freshwater prawn farming in Hawaii. In: Giant prawn farming (ed. M.B. New). *Developments in Aquaculture and Fisheries Science*, **10**, 351–356.

SICA (1988) *General information about our hatchery*, 2 pp. Trade brochure, Sica Guadeloupéene D'Aquaculture, Les Plaines, Pointe Noire, Guadeloupe, FWI.

Smith T.I.J. & Wannamaker A.J. (1983) Shipping studies with juvenile and adult Malaysian prawns *Macrobrachium rosenbergii* (de Man). *Aquacultural Engineering*, **2** (4) 287–300.
Smith T.I.J., Sandifer P.A. & Jenkins W.E. (1982) Growth and survival of prawns, *Macrobrachium rosenbergii*, pond reared at different salinities. In: Giant prawn farming (ed. M.B. New). *Developments in Aquaculture and Fisheries Science*, **10**, 191–202.
Smith T.I.J., Jenkins E.W. & Sandifer P.A. (1983) Enclosed prawn nursery systems and effects of stocking juvenile *Macrobrachium rosenbergii* in ponds. *Journal of the World Mariculture Society*, **14**, 111–125.
Sureshkumar S. & Kurup B.M. (1998) Fecundity indices of giant freshwater prawn, *Macrobrachium rosenbergii* (de Man). *Journal of Aquaculture in the Tropics*, **13** (3) 181–188.
Tayamen M. & Brown J.H. (1999) A condition index for evaluating larval quality of *Macrobrachium rosenbergii* (De Man, 1879). *Aquaculture Research*, **30**, 917–922.
Tidwell J.H. & D'Abramo L.R. (2000) Grow-out systems – culture in temperate zones. In: *Freshwater Prawn Culture: the farming of* Macrobrachium rosenbergii. (eds M.B. New & W.C. Valenti), pp. 177–186. Blackwell Science, Oxford, UK.
Valenti W.C. & Daniels W.H. (2000) Recirculation hatchery systems and management. In: *Freshwater Prawn Culture: the farming of* Macrobrachium rosenbergii (eds M.B. New & W.C. Valenti), pp. 69–90. Blackwell Science, Oxford, UK.
Valenti W.C. & New M.B. (2000) Grow-out systems – monoculture. In: *Freshwater Prawn Culture: the farming of* Macrobrachium rosenbergii (eds M.B. New & W.C. Valenti), pp. 157–176. Blackwell Science, Oxford, UK.
Wang G. & Qianhong S. (1999) Culture of freshwater prawns in China. *Aquaculture Asia*, **4** (2) 14–17.
Wickins J.F. & Beard T.W. (1978) *Prawn culture research*, 41 pp. Lab. Leafl. (42). MAFF Directorate Fisheries Research, Lowestoft, UK.
Wulff R.E. (1982a) The experience of a freshwater prawn farm in Honduras, Central America. In: Giant prawn farming (ed. M.B. New). *Developments in Aquaculture and Fisheries Science*, **10**, 445–448.
Wulff R.E. (1982b) Practical farming discussion session. In: Giant prawn farming (ed. M.B. New). *Developments in Aquaculture and Fisheries Science*, **10**, 469–498.
Zimmermann S. & New M.B. (2000) Grow-out systems – polyculture and integrated culture. In: *Freshwater Prawn Culture: the farming of* Macrobrachium rosenbergii (eds M.B. New & W.C. Valenti), pp. 187–202. Blackwell Science, Oxford, UK.

7.4 Other caridean shrimps and prawns

7.4.1 Species of interest

Spot prawn (*Pandalus platyceros*); common prawn (*Palaemon serratus*); sculptured shrimp (*Sclerocrangon boreas*); freshwater prawns and shrimps (*Atya* spp., *Cryphiops caementarius*); ornamental carideans (*Lysmata* spp., *Stenopus* spp.).

7.4.2 Broodstock, larvae culture and nursery

The species that has attracted the most attention in this group is *Pandalus platyceros* but it is a temperate water species and cannot compete with the penaeids, *Macrobrachium rosenbergii* and many crayfish in terms of growth rate or with penaeids in terms of meat yield (Kelly *et al.* 1977; Wickins 1982). One particular difficulty with *Pandalus platyceros* would be the need to rely on a supply of wild-caught ovigerous females or to maintain large broodstock and incubation facilities in support of a farming enterprise. The same constraint would apply to many of the other large carideans such as *Sclerocrangon boreas* (which has been bred in captivity; sections 4.2 and 4.4.1), all of whom have low fecundity. Techniques for *Pandalus platyceros* broodstock maintenance, larvae culture and nursery are similar to those described for *Macrobrachium*, albeit conducted at lower temperatures, and are described further by Wickins (1972) and Wickins and Beard (1978).

In commercial trials in Canada wild-caught, egg-bearing females were held in darkened 3000 L cylindrical tanks supplied with a constant flow of ambient seawater (8–14°C). Newly hatched larvae swam towards a light placed at the outflow and into a collection vessel. They were reared at a density of 100 L^{-1} in 500–2000 L cylindrical vessels in 5 μm filtered seawater that was changed completely each time the majority moulted to a new instar. The feed was *Artemia* (10 nauplii mL^{-1}) and hanging artificial seaweed was provided as a settlement substrate at the fourth instar. Survival after 13–22 days averaged 25% (range 13–90%) giving 330 000 post-larvae in the 1998 season (Y. Alabi, 2001 pers. comm.).

Post-larvae were transported to the nursery, a 1–2 h journey, in aerated seawater and grown on in experimental tanks (6 × 1 × 1 m deep) floating in a salmon net-pen station. Although some prawns grew to 15 g in 15 months, initial survival rates were too variable for commercial evaluation (C. Campbell, 2001 pers. comm.).

7.4.3 Ongrowing

Attempts to grow *Pandalus platyceros* in raceways in power station effluents were made in Britain in the early 1970s (Wickins 1982). The best results suggested that only one crop per year of 6–8 g prawns could be obtained and that culture was unlikely to be profitable. It has been suggested that the species has promise for polyculture, perhaps with abalone (Kelly *et al.* 1977) although Hunt *et al.* (1995) found growth in monoculture significantly better than in polyculture with either abalone or mussels and proposed that the prawns would be best cultured sequentially with the molluscs in any future pilot commercial studies. Attempts have also been made to culture spot prawns in salmon cages (Table 5.6) but a major problem with the latter scheme was the need to stock prawns large enough to be retained by the mesh used for the salmon (Oesterling & Provenzano 1985). Cultivation of the river prawn (*Cryphiops caementarius*) in ponds and in polyculture with mullet (Zuniga-Romero *et al.* 1987) in Chile is constrained by the scarcity of wild juveniles, but the recent development of hatchery techniques could enhance the availability of post-larvae for ongrowing (Merino 1998).

7.4.4 Other prospects

Smaller species that have been considered for cultivation because of their value as gourmet food items, bait for sport anglers, or as laboratory bioassay animals include *Palaemon serratus* (Wickins 1982), *Atya lanipes* and *A. scabra* (Cruz-Soltero & Alston 1992), and *Palaemonetes kadiakensis*, *P. pugio*, *P. paludosus* (Oesterling & Provenzano 1985). Although generally small in size, atyids differ from other cultivable shrimps and prawns in that the post-larval stages are able to filter particulate material from the water column. The largest atyid shrimp are *Atya gabonensis* of west Africa and *A. innocous* from the West Indies, which may attain 90–124 mm total length.

7.4.4.1 Ornamental shrimp

Ornamental marine shrimp are highly valued because of their attractive coloration and behaviour; the aquarium trade prizes over 18 species. Their widespread collection from the wild throughout the tropical and subtropical regions causes concern not only because they play a key role in ecosystems but also because of the destructive methods used in their capture (section 11.3.1.3). Techniques used for their culture are similar to those for *Macrobrachium* (section 7.3) but generally are on a much smaller scale (Lin *et al.* 1999).

Several species have been reared in the laboratory but three in particular (*Lysmata wurdemanni, L. debelius* and *L. amboinensis*) have been more thoroughly studied with a view to defining conditions for larger scale, commercial culture (Zhang *et al.* 1998; Somões & Jones 1999a). Other species such as *L. grabhami* and *Stenopus hispidus* have potential, but larval survival is often poor (Lin *et al.* 1999). In general the larval life of *Lysmata* spp. in laboratory cultures seems to be very variable (30–140 days) possibly as a result of insufficient knowledge of their culture and nutritional requirements (D.A. Jones, 2000 pers. comm.). The duration of larval life in *Stenopus hispidus* is long (119–210 days) compared to that of *S. scutellatus* (43–77 days) for which larvae survival rates of 50% are reported (Zhang *et al.* 1997). Although the techniques used for the culture of ornamental shrimp are similar to those used for other carideans, there are important differences. These include the prolonged larval life, the risk of precipitating lethally aggressive behaviour between adults by inappropriate handling (Somões & Jones 1999b) and incomplete knowledge of nutritional requirements.

Prices can be as high as $20 per shrimp (Lin *et al.* 1999) and demand continues to exceed supply. The potential market may be substantial but an increase in supply may depress prices. Possible reasons why no commercial culture is performed include the high capital cost of a hatchery and the high labour costs of larval rearing and food culture when compared to the ease of catching and transporting specimens through the established channels of the ornamental fish trade. As far as we are aware no successful large-scale culture for these colourful animals has yet been reported. Nevertheless, the situation may change as research improves understanding of larvae culture requirements and as natural supplies of shrimp come under increasing pressure from overexploitation and habitat destruction.

7.4.5 References

Cruz-Soltero S. & Alston D.E. (1992) Larval rearing of two decapod freshwater shrimp, *Atya scabra* (Leach) and *A. lanipes* (Holthuis) in the laboratory. *Journal of Shellfish Research,* **11** (1) 193.

Hunt J.W., Foster M.S., Nybakken J.W., Larson R.J. & Ebert E.E. (1995) Interactive effects of polyculture, feeding rate, and stocking density on growth of juvenile shellfish. *Journal of Shellfish Research*, **14** (1) 191–197.

Kelly R.O., Haseltine A.W. & Ebert E.E. (1977) Mariculture potential of *Pandalus platyceros* Brandt. *Aquaculture,* **10** (1) 1–16.

Lin J., Zhang D. & Creswell R.L. (1999) Marine ornamental shrimp: status and prospects. *Aquaculture Magazine,* **25** (3) 52–55.

Merino G.E. (1998) The Chilean aquaculture industry and the role played by the Universidad Catolica Del Norte in its development. *Proceedings of Second International Conference on Recirculating Aquaculture,* 16–19 July 1998, pp. 399–402, Virginia Tech, VA, USA.

Oesterling M.J. & Provenzano A.J. (1985) Other crustacean species. In: *Crustacean and Mollusk Aquaculture in the United States* (eds J.V. Huner & E. Evan Brown), pp. 203–234. AVI Inc., Westport, CT, USA.
Somões F. & Jones D.A. (1999a) The entrainment of marine cleaner shrimps *Lysmata debelius* and *L. amboinensis* (Crustacea, Caridea) to feed on inert maturation diets. In: *Book of Abstracts, World Aquaculture '99*, 26 April–2 May 1999, Sydney, Australia, p. 701. World Aquaculture Society, Baton Rouge, LA, USA.
Somões F. & Jones D.A. (1999b) Pair formation in the tropical marine cleaner shrimp *Lysmata debelius* (Crustacea, Caridea). In *Book of Abstracts, World Aquaculture '99*, 26 April–2 May 1999, Sydney, Australia, p. 700. World Aquaculture Society, Baton Rouge, LA, USA.
Wickins J.F. (1972) Experiments on the culture of the spot prawn *Pandalus platyceros* Brandt and the giant freshwater prawn *Macrobrachium rosenbergii* (de Man). *Fisheries Investigations, London, Series 2*, **27** (5) 1–23.
Wickins J.F. (1982) Opportunities for farming crustaceans in western temperate regions. In: *Recent Advances in Aquaculture* (eds J.F. Muir & R.J. Roberts), pp. 87–177. Croom Helm, London.
Wickins J.F. & Beard T.W. (1978) Ministry of Agriculture, Fisheries and Food. *Prawn culture research*, 41 pp. Lab. Leafl. (42). MAFF Directorate Fisheries Research, Lowestoft, UK.
Zhang D., Lin J. & Creswell R.L. (1997) Larviculture and effect of food on larval survival and development in golden coral shrimp *Stenopus scutellatus*. *Journal of Shellfish Research,* **16**, 367–369.
Zhang D., Lin J. & Creswell R.L. (1998) Effects of food and temperature on survival and development in the peppermint shrimp *Lysmata wurdemanni*. *Journal of the World Aquaculture Society,* **29** (4) 471–476.
Zuniga-Romero O., Ramos-Diaz R. & Reveulta-Duran J. (1987) Policultivo de camaron de rio (*Cryphiops caementarius*) y lisa (*Mugil cephalus*) en estanques. *Estud. Oceanol. Inst. Invest. Oceanol.* Univ. Antofagasta, (6) 67–77.

7.5 Crayfish: USA

7.5.1 Species of interest

Red swamp crayfish (*Procambarus clarkii*); white river crayfish (*P. zonangulus* formerly *P. acutus*); crayfish reared mainly for bait (*Orconectes* spp.). The signal crayfish (*Pacifastacus leniusculus*) is also cultured to a small extent but is considered more fully in section 7.6 (Europe). Of the two crayfish raised for the table, *Procambarus clarkii* makes up 90% of production in the USA where, in many respects, the 'farms' more closely resemble well managed private fisheries than aquaculture operations. Nevertheless, production of *P. clarkii* in North America is regarded by some as a legitimate example of sustainable aquaculture (Caffey *et al.* 1997). Considerable production now also arises from China and has brought about changes in western markets (section 3.3.3.1). In the west, both wild and farmed crayfish (*Orconectes* and *Procambarus*) may also be used as bait by sport anglers (Anon. 1997). Two species of *Cambarus* are also believed to have commercial potential (Guiaşu 2001) (section 12.8.3).

This section includes information from trade articles and technical papers. It may be extended through reference to Avault and Huner (1985); Momot (1988); Brunson (1989); Roberts and Dellenbarger (1989); McClain and Romaire (1995); Guiaşu (2001); Hamr (2001); Huner (2001).

7.5.2 Broodstock

Red swamp crayfish broodstock are obtained from natural fisheries or from managed ponds. Most farmers only stock once as self-maintaining populations provide the basis for most production. The adults are transported in sacks or plastic trays by road, and stocking of broodstock into new ponds occurs from April to June at around 20–65 kg ha^{-1}, animals being evenly distributed all along levees at a sex ratio of about 1 : 1. The animals burrow well below water level and after 2 weeks (longer if no old burrows are present) the ponds are drained, leaving the crayfish submerged in their burrows. Often white river crayfish become established in the ponds and can, on occasions, become the dominant population (Lutz & Wolters 1999). The well-established burrows typically associated with pond perimeters and internal levees are the main source of the next season's stock since crayfish do not burrow successfully while the pond is being drained (Huner 1999a).

7.5.3 Hatchery and nursery

Procambarus clarkii breeds year round at the high temperatures in the deep south (Louisiana) but less frequently further north. *P. zonangulus* has lower fecundity and a later, more restricted spawning period. Hatchery production or stocking with known numbers of juveniles is neither practised nor required on a commercial scale. It is likely that photoperiod control could extend the breeding season in *Orconectes* spp.

7.5.4 Ongrowing

7.5.4.1 Natural/extensive

The crayfish in the southern United States are produced primarily in Louisiana where extensive ongrowing may be grouped into four categories:

(1) Marsh and swamp ponds (up to 100 ha in area) that exist by the coast and are filled and drained by pumping. Typically they are on soils with a high peat content and contain slightly brackish water. The ponds are filled in autumn and drained by the following June. Oxygen depletion is a major problem and circulation lanes are mown through the vegetation prior to flooding, to assist water movement and harvesting. The ponds are flooded at intervals to gradually increasing depths to discourage decomposition of plant material. Flooding is however delayed in hot weather.
(2) Wooded ponds are similar to the above but conditions are poorer as the water circulation is reduced by the trees and shrubs. Annual yields from marsh and wooded ponds are about 200–600 kg ha^{-1}.
(3) Rice fields where the banks are raised from about 0.1 m to 0.3–0.5 m. These may be operated by double cropping rice and crayfish. Rice is planted in March/April but care must be taken if fungicides and herbicides are applied to protect the rice since many are toxic to crayfish; it is likely that all insecticides are toxic to crayfish. The rice is harvested after 100–120 days and water that is on the rice field to reduce weed growth is drained and the field reflooded in mid-September to mid-October. The crayfish emerge and release young into the rice stubble that should be left standing. Trapping begins in November and is continued until April when the ponds are drained and rice planted again. Alternatively, trapping may continue until June when soybeans are planted instead of rice. Annual yields range from 1000–2000 kg ha^{-1} with this method.
(4) Open ponds of 8–20 ha are typically built in low-lying areas of marginal agricultural quality and may look as if they are full of grass. However, they often have boat or trapping lanes that aid water circulation. The soil is typically heavy clay and ponds with smooth bottoms can be readily constructed. It may be necessary to place anti-seep collars of metal, fibreglass or plastic around drainpipes to reduce the risk of leaks from some ponds. Feeding is effective but adds cost and is not much practised (Huner 1997a). Instead, volunteer or intentionally cultivated cover crops (grasses, rice) are grown in the dry summer season while crayfish are quiescent or spawning in their burrows, to provide the basis for detritus in the food chain. When the vegetation decomposes in autumn the ponds are flushed with water. As a forage, rice and a sorghum–Sudan grass hybrid are better than many natural grasses, particularly those varieties that produce a lot of foliage. This provides nutrition through decomposition and importantly, a supply of nutritious, detritus-feeding invertebrates (worms, snails, insects and small crustaceans) (McClain 1994). These may, however, become depleted by March when compounded feeds (cattle range pellets) or hay (rice, alfalfa or wheat straw) may be added (Brown 1995). Many of the Louisiana farmers do not plant forage although planting does provide a complex, shelter-rich environment advantageous to the juveniles and allows them access to the surface when oxygen levels fall (Figler *et al.* 1999). Yields may be around 500–1500 kg ha^{-1}. About 65–70% of the crayfish production area in Louisiana falls in this category.

A typical production schedule (de la Bretonne & Romaire 1989) is:

May: Stock 25–50 kg adults per hectare (for 'new' ponds);
May–June: Slowly drain pond over 1–2 weeks;
July–August: Plant rice or other vegetation for forage;
September–October: Reflood the pond gradually to avoid severe oxygen depletion;
November–May/June: Continuous harvesting;
May: Drain pond and repeat cycle without restocking crayfish.

When crayfish are rotated with rice the schedule (Huner 1997b) is:

March–April: Plant rice;
June: Stock 25–50 kg adults per hectare (for 'new' ponds);
August: Drain and harvest rice;
October: Re-flood the pond gradually to avoid severe oxygen depletion;
November–April: Continuous harvesting;
March–April: Plant rice and repeat cycle without restocking crayfish.

One of the most important problems in swamp crayfish ponds is the depletion of the forage substrate necessary for good growth during winter and early spring.

The depletion is exacerbated by underfishing and often results in stunting of the remaining stock. The supply of forage cannot be guaranteed when too much reliance is placed on volunteer plants. Selected varieties of rice and other crops produce better results but there is still a need to identify other semi-aquatic plants which will provide forage throughout the harvesting season, and to extend studies on food webs within the pond ecosystem. Many growers would be interested in compounded feeds but the costs are unlikely to be justified without intensifying production methods, and the crayfish burrowing habit together with the need to harvest by trapping precludes marked intensification of production.

Uncontrolled reproduction of crayfish within ponds often leads to an overabundance of juveniles with consequent reduction in growth rates (Jarboe & Romaire 1995). This is most likely to occur in smaller ponds, i.e. those with a large perimeter relative to pond area, because the established burrows which provide most of the offspring are confined to the perimeter. In small ponds, production can be high but mean size and hence crop value can be lower than in large ponds; ponds of 2–4 ha seem to provide the best compromise (Huner 1999a). Relaying the small individuals, preferably into rice fields early in the season (McClain & Bollich 1997), is sometimes practised but may not always be cost effective (Huner 1997a).

Predator control is not usually a problem provided the crayfish have been able to burrow effectively; screens are sometimes used against fish, birds and mammals, while rotenone may be used against fish larvae entering with the water (Huner 2001). In contrast to the European situation, no major disease or parasite problems have been reported, possibly because culture densities are low and because when stressful conditions (e.g. anaerobic water) predispose crayfish to infection and disease, they die unnoticed (Huner 1997a).

The prospects for double-cropping crayfish with prawns (*M. rosenbergii*) in Louisiana was investigated with the aim of producing 15–17 g prawns for the soft-shell market during mid-May to mid-October when crayfish ponds are not in use (Tables 5.6 and 5.7). Polyculture with fish is not widely practised although opportunities for polyculture of *Procambarus* spp. with ornamental goldfish and koi carp in Louisiana have been identified (Huner 2001). In South Carolina crayfish and waterfowl are sometimes combined (Eversole & Pomeroy 1989). The shooting season is autumn which delays trapping and results in lower annual yields of crayfish. In Louisiana non-destructive ecotourism activities such as bird-watching are also being explored (Huner 2001) but increasing populations of predatory birds are causing problems (section 11.2.5).

7.5.4.2 Intensive

No more than two or three commercial US firms are believed to have grown *Procambarus clarkii* in controlled environment conditions (Avault & Huner 1985) either for stocking or to adult size. Prices are too low to justify the costs of such systems although intensive indoor systems (section 7.5.8) for soft-shell crayfish production seem profitable.

7.5.5 Harvesting

Most harvesting is done by traps, with two types commonly used. The first is a 'stand-up' funnel or pyramid placed at the pond bottom, with an open top above the surface. These are about 90 cm tall × 45 cm diameter and made of 20 mm poultry netting coated in plastic and fitted with an anti-escape collar. The 'pillow' type of trap can also be laid on the pond bottom and has one or more funnel entrances. If pond oxygen levels fall, crayfish cannot climb to the surface and may die when this kind of trap falls over or is deployed on its side. Daily catches are variable and affected by environmental conditions, e.g. temperature and lunar phase. Often 25 traps ha^{-1} are inspected 3–5 days per week for 60–150 days but pyramid traps set at 50–60 ha^{-1} and fished for 3–4 days per week are likely to be the most profitable (Romaire 1995). Nowadays, 150–300 traps can be emptied and reset per hour using flat-bottomed, motorised boats. Even so, trap designs could still be improved. Unfortunately the prospects for using water flows to induce *P. clarkii* to concentrate in areas where they can be easily netted do not seem promising (Romaire & Lawson 1990) without further development of water circulation systems and gear (Romaire 1995). Several mechanical harvesting devices have been proposed, mainly using bow-mounted trawls, but none has yet been widely used commercially (Huner 1999b). Trapping costs contribute 60–80% of production costs (approximately 33–40% of total operating costs) (section 10.6.2.3) but in some years not enough of the population is caught and forage becomes depleted. This loss of food results in stunting or reduced growth. Useful reviews of harvesting techniques may be found in Romaire (1995) and Huner (2001).

The fishery and culture industries together use some 15 000–30 000 mt of bait annually. Baits typically include gizzard shad, striped mullet, pollack and menhaden but the development of an effective, pelleted, artificial bait has saved the industry significant time and cost in the buying, preparation and storage of highly perishable fish-based baits (Huner 2001). In warm waters (>20°C) both pellets and fish are used in the same trap. About 150 g of bait per trap seems optimal. Harvesting often begins in late November but traps may also be set periodically to check the condition of the stock (for example, to see if they are emaciated or ovigerous). Drop nets are also useful when checking the stock.

7.5.6 Transportation

Clean crayfish placed in 16–20 kg mesh sacks or plastic trays and held in a high humidity, cool container (4–6°C), may be transported alive for journeys lasting several days. Good air circulation around the sacks and minimal vibration will enhance survival rates. Washing for 1 h in aerated vats followed by purging provides a clean product although some mortality is inevitable.

7.5.7 Processing

Most crayfish are sold alive for boiling in seasoned water, although 30% of the Louisiana crop is processed, i.e. either cooked whole and frozen or cooked beheaded, peeled, deveined (the gut removed) then packed fresh-frozen or vacuum packed (Huner 1994). In the late 1990s the industry received a setback when the US Food and Drug Administration banned interstate marketing of vacuum packed fresh sea food which included crayfish (Anon. 1999; section 3.2.1). Increasingly, the live crop is purged (to evacuate unsightly gut contents) either by immersion in purging vats (Moody 1989) which were traditionally through-flow but increasingly use recycled, treated water, or by holding in shallow trays under a mist or constant spray of water for 12–48 h (McClain 2000). In the case of soft-shelled crayfish, the New Orleans and Washington markets take whole, newly moulted crayfish with only the gastroliths removed; this represents about 92% edible product. If the internal organs are also removed about 72–82% remains. The soft-shelled crayfish are sold frozen in bags or in tray packs (Huner 1999b). In the early 1990s new price differentials based on size arose as a result of selling into European markets (section 3.3.3.1) and size grading became a standard practice (Rollason & McClain 1996). Modified stock management and harvesting strategies (including pond rotation, culling, feeding, intermittent trapping and relaying) were adopted to maximise yields of the largest animals (McClain & Romaire 1995).

Processing wastes from Louisiana operations alone amount to 40 000 mt annually and the extraction of useful compounds such as chitin, pigments and flavours has been investigated (section 3.3.7). Processing wastes are potentially valuable additions in crustacean diets but at present the resource seems under-utilised.

7.5.8 Soft-shelled crayfish

Soft-shelled crayfish are prized as food (*Procambarus*) and also as bait (*Procambarus* and *Orconectes*). To keep the former in perspective, 50 mt were marketed by 150 producers in 1988 in comparison with 60 000 mt of hard-shelled crayfish. Huner (1988a) predicted that the potential total soft-shell food market in the USA was about 1500 mt. Indeed growth was rapid up to the end of the 1980s when there were about 300 producers. Since then the industry declined (section 3.3.3.2) and only 100 mt were produced in 1996–97 (Huner 1997a) by less than 12 producers (Huner 1999b).

Soft-shell crayfish do not normally enter traps as they do not feed; prototype electro-trawls and other mechanical trawl devices have been developed to catch these and paper-shell crayfish from ponds (Huner 1999b). Those caught up to 48 h after moulting are suitable for some markets. In essence, the selected crayfish are held captive in shallow 'shedding' trays until they moult. They are then removed, processed and sold. In the majority of cases pre-moult and immature intermoult animals (which moult every 10–30 days) are selected and placed in the trays at densities of 250 to 500, 10–20 g crayfish m^{-2} (about 5 kg m^{-2}). Typical tray systems described by Huner (1988a; 1999b) and Culley and Duobinis-Gray (1989) comprise a series of trays (0.91 × 2.74 × 0.15 m deep; water depth 5 cm) which receive water at 0.8–2.1 L water kg^{-1} min^{-1}. Control over water quality may be improved through the incorporation of a recirculation system (Malone & Burden 1988) although many of these as well as some through-flow systems are no longer economically viable (Huner 1999b). High protein feeds may be given at 1–3% of body weight per day in two to three feeds per day.

Two labour-saving intensive soft-shell systems have been proposed and were expected to be in operation in 1988–89 season (see US patents of Bodker 1984 and Malone & Culley 1988). Competent identification and careful handling of sexually immature, intermoult crayfish of 70 mm total length or over are necessary prerequisites of soft-shell crayfish production (Huner 1988b; Culley & Duobinis-Gray 1989).

The soft crayfish are identified by a change of colour and 3–4% of the population moult each day for the 4–7-month season. It may become possible to predict moulting using machine vision techniques (Timmermans *et al.* 1995). It is also possible to hold selected animals overnight in deionised or chilled water which prevents the shell from hardening and reduces the need for all-night collection. A prototype automated technique for separating soft- from hard-shelled crayfish involves water currents which flush the weaker, soft-shelled crayfish along a long, narrow, spiralling raceway to a collection box. An electric potential 'inhibition gate' discourages hard-shelled individuals from entering the box (Chen *et al.* 1995). Market forces have so far prevented commercial production of such devices (Huner 1999b). Research that would contribute most significantly to the production of soft-shell crayfish (and indeed all other crustaceans for which such a market exists or can be created) is centred on the prediction of when moulting will occur, how moulting can be induced and how hardening rates can be controlled (section 12.8.6). Attempts to control predictability and frequency of moulting include adjustment of temperature, eyestalk ablation and the interactions between the two (Chen *et al.* 1995) but it is not certain how consumers will react to the ablated product.

7.5.9 *Orconectes* spp.

These are mainly small species cultured primarily for bait and are raised incidentally with fish in some areas. A number of studies have shown that the larger orconectids (e.g. *Orconectes rusticus* >90 mm TL) have commercial potential as food (Hamr 2001; section 12.8.3). This species is, however, an effective competitor, displacing other crayfish species and readily destroying aquatic vegetation. It should not therefore be transplanted without exceptional justification. There are obvious risks associated with its use as live bait.

Annual production of *Orconectes immunis* from north-eastern and north-central USA is estimated at 175 mt and a further 400–500 mt are thought to come from wild catches (Huner 1997b). The adults are stocked into mud-bottomed ponds in autumn at 1500–2500 ha^{-1} or as berried females at 750–1200 ha^{-1}. The young hatch in spring and reach bait size (1–5 g) by July when seining begins. Those in a soft-shelled condition are most sought after by anglers. Moulting occurs within 1–4 days after the crayfish are placed in shallow trays. The ponds may be fertilised with agricultural fertiliser (NPK ratio of 6 : 12 : 6 or 0 : 12 : 0 at 220 kg ha^{-1}) at 3-week intervals or as required. Some supplemental feeding is practised, for example with cracked corn and potatoes. Yields of 56 000–156 000 animals ha^{-1} and 323–807 kg ha^{-1} have been recorded from research ponds with *O. virilis* although, in the wild, this species is not easily harvested by trapping. A spreadsheet model indicated that production of soft-shelled orconectid crayfish in trays, enhanced by eyestalk ablation, could also be profitable (Gunderson *et al.* 1997).

7.5.10 References

Anon. (1997) Crawfish for bait in Louisiana. *Aquaculture Magazine*, **23** (1) 18–21.

Anon. (1999) FDA ban on crawfish. *Aquaculture Magazine*, **25** (3) 14–16.

Avault J.W. Jr. & Huner J.V. (1985) Crawfish culture in the United States. In: *Crustacean and Mollusk Aquaculture in the United States* (eds J.V. Huner & E. Evan Brown), pp. 1–61. AVI Inc. Westport, CT, USA.

Bodker J.E. Jr. (1984) *Method and apparatus for raising soft-shell crawfish.* U.S. Patent No. 4,475,480, Washington, DC, USA.

Brown P.B. (1995) A review of nutritional research with crayfish. *Journal of Shellfish Research*, **14** (2) 561–568.

Brunson M.W. (1989) Forage and feeding systems for commercial crawfish culture. *Journal of Shellfish Research*, **8** (1) 277–280.

Caffey R.H., Romaire R.P. & Avault J.W. Jr. (1997) Sustainable aquaculture: crawfish farming. In: *Freshwater Crayfish 11* (ed. W.T. Momot), pp. 587–598. Louisiana State University, LA, USA.

Chen S., Drennan D.G. II & Malone R.F. (1995) Impact of friction, flow rate and loading density on automated soft-shell crawfish separation. *Aquacultural Engineering*, **14**, 1–14.

Culley D.D. & Duobinis-Gray L. (1989) Soft-shell crawfish production technology. *Journal of Shellfish Research*, **8** (1) 287–291.

de la Bretonne L.W. Jr. & Romaire R.P. (1989) Commercial crawfish cultivation practices: a review. *Journal of Shellfish Research*, **8** (1) 267–275.

Eversole A.G. & Pomeroy R.S. (1989) Crawfish culture in South Carolina: an emerging aquaculture industry. *Journal of Shellfish Research*, **8** (1) 309–313.

Figler M.H., Cheverton H.M. & Blank G.S. (1999) Shelter competition in juvenile red swamp crayfish (*Procambarus clarkii*): the influences of sex differences, relative size, and prior residence. *Aquaculture*, **178** (1–2) 63–75.

Guiaşu R.C. (2001) *Cambarus*. In: *Biology of Freshwater Crayfish* (ed. D.M. Holdich), pp. 609–34. Blackwell Science, Oxford, UK.

Gunderson J.L., Richards C. & McDonald M. (1997) Soft crayfish production by eyestalk ablation: can it be profitable? In: *Freshwater Crayfish 11* (ed. W.T. Momot), pp. 567–576. Louisiana State University, LA, USA.

Hamr P. (2001) *Orconectes*. In: *Biology of Freshwater Crayfish* (ed. D.M. Holdich), pp. 585–608. Blackwell Science, Oxford, UK.
Huner J.V. (1988a) Soft shell crawfish industry. In: *Proceedings of First Australian Shellfish Aquaculture Conference*, Perth, 1988 (eds L.H. Evans & D. O'Sullivan), pp. 28–42. Curtin University of Technology, Perth, Australia.
Huner J.V. (1988b) *Procambarus* in North America and elsewhere. In: *Freshwater Crayfish, Biology, Management and Exploitation* (eds D.M. Holdich & R.S. Lowery), pp. 239–261. Croom Helm, London.
Huner J.V. (1994) Section II Freshwater crayfish processing. In: *Freshwater Crayfish Aquaculture in North America, Europe, and Australia* (ed. J.V. Huner), pp. 91–115. Food Products Press, An imprint of the Haworth Press, Inc. New York.
Huner J.V. (1997a) How crawfish ponds work – some thoughts. *Aquaculture Magazine*, **23** (5) 61–69.
Huner J.V. (1997b) The capture and culture fisheries for North American crawfish. *World Aquaculture,* **28** (4) 44–50.
Huner J.V. (1999a) The relationship between pond size and crayfish (*Procambarus* spp.) production. In: *Freshwater Crayfish 12* (eds M. Keller, M.M. Keller, B. Oidtmann, R. Hoffmann & G. Vogt), pp. 573–583. Weltbild Verlag, Germany.
Huner J.V. (1999b) The fate of the Louisiana soft-shell crawfish. *Aquaculture Magazine,* **25** (3) 46–51.
Huner J.V. (2001) *Procambarus*. In: *Biology of Freshwater Crayfish* (ed. D.M. Holdich), pp. 541–84. Blackwell Science, Oxford, UK.
Jarboe H.H. & Romaire R.P. (1995) Effects of density reduction and supplemental feeding on stunted crayfish *Procambarus clarkii* populations in earthen ponds. *Journal of the World Aquaculture Society,* **26** (1) 29–37.
Lutz C.G. & Wolters W.R. (1999) Growth and yield of red swamp crawfish *Procambarus clarkii* stocked separately and in combination with white river crawfish *Procambarus zonangulus*. *Journal of the World Aquaculture Society,* **30** (3) 394–397.
Malone R.F. & Burden D. (1988) *Design Manual for Intensive Soft-shell Crawfish Production.* Louisiana Sea Grant program, Louisiana State University, Baton Rouge, LA, USA, not seen, cited in Huner (1988a).
Malone R.F. & Culley D.D. (1988) *Method and apparatus for farming soft-shell aquatic crustaceans.* U.S. Patent No. 4,726,321, Washington, DC, USA.
McClain W.R. (1994) Dispelling some misconceptions about crawfish forages. *Aquaculture Magazine,* **20** (6) 41–45.
McClain W.R. (2000) Assessment of depuration system and duration on gut evacuation rate and mortality of red swamp crawfish. *Aquaculture,* **186** (3–4) 267–278.
McClain W.R. & Bollich P.K. (1997) Relaying crayfish: summary of an intercropping approach. In: *Freshwater Crayfish 11* (ed. W.T. Momot), pp. 533–549. Louisiana State University, LA, USA.
McClain W.R. & Romaire R.P. (1995) Management considerations for the production of large procambarid crawfish. *Journal of Shellfish Research,* **14** (2) 553–559.
Momot W.T. (1988) *Orconectes* in North America and elsewhere. In: *Freshwater Crayfish, Biology, Management and Exploitation* (eds D.M. Holdich & R.S. Lowery), pp. 262–282. Croom Helm, London.
Moody M.W. (1989) Processing of freshwater crawfish: a review. *Journal of Shellfish Research,* **8** (1) 293–301.
Roberts K.J. & Dellenbarger L. (1989) Louisiana crawfish product markets and marketing. *Journal of Shellfish Research,* **8** (1) 303–307.
Rollason S.H. & McClain W.R. (1996) Development of a water-based grading apparatus for live crayfish. In: *Freshwater Crayfish 10* (eds M.C. Geddes, D.R. Fielder & A.M.M. Richardson), pp. 596–604. Baton Rouge, Louisiana State University, LA, USA.
Romaire R.P. (1995) Harvesting methods and strategies used in commercial procambarid crawfish aquaculture. *Journal of Shellfish Research*, **14** (2) 545–551.
Romaire R. & Lawson T. (1990) Evaluation of water circulation to improve crayfish (*Procambarus* spp.) harvest efficiency. *Abstract from World Aquaculture 90,* p. 92, 10–14 June 1990, Halifax, Nova Scotia, Canada.
Timmermans T., Sistler F.E. & Lawson T.B. (1995) Predicting crawfish molting with machine vision. *Journal of the World Aquaculture Society,* **26** (3) 234–239.

7.6 Crayfish: Europe

7.6.1 Species of interest

Signal crayfish (*Pacifastacus leniusculus*); Turkish or narrow-clawed crayfish (*Astacus leptodactylus*); noble crayfish (*Astacus astacus*); thick-clawed crayfish (*Astacus pachypus*); white-clawed crayfish (*Austropotamobius pallipes*). Most young are produced for restocking, particularly of the latter two species, but some *Pacifastacus leniusculus* are reared for bait or, especially in Britain, directly for ongrowing for the table. Preliminary culture experiments are being made with *Astacus pachypus*. Overall, only about 4% of crayfish come from farming (Ackefors 1998). The red swamp crayfish (*Procambarus clarkii*) is also cultured in southern Europe but the methods are similar to those used in the USA (section 7.5). Otherwise the methods currently used for rearing crayfish are similar throughout Europe (Arrignon 1981; Huner 1995; Rogers & Holdich 1995; Ackefors 1997; Mackevičienė *et al.* 1997; Ackefors 2000; Lewis 2001; Skurdal & Taugbøl 2001). A list of researchers and crayfish related programmes in Europe has been published (Westman & Mannine 1996).

7.6.2 Broodstock

Broodstock are obtained either from cultured stocks or from wild stock brought into the hatchery. They are fed a variety of natural foods including small crustaceans and water plants. Detritus seems to form a large part of the diet in nature. The number of eggs spawned as well as the number carried through to hatching varies considerably, even within species, which makes planning difficult for hatchery managers and has stimulated interest in artificial incubation where the eggs are physically removed from the mother to be incubated separately (section 7.6.4.1).

7.6.3 Mating and spawning

Dimensions of outdoor broodstock units seem to be based on individual preferences or existing facilities, for example they may be tanks (2–10 m × 0.5–1 m × 0.5–1 m deep) or ponds (100–500 m^2 and about 1 m deep). They are stocked with berried females or mature wild-caught broodstock taken in autumn. Density is typically 4 m^{-2} with a sex ratio of one male to three females; hides are considered essential (Koksal 1988). Good results have been reported with *Astacus astacus* in hatchery tanks at 50 m^{-2} and a sex ratio of 1 : 1. Mating frequency was improved when males were larger than the females (Mackevičienė *et al.* 1999). After mating, males are removed from tank systems and sold, but in ponds the sexes are not so readily separated. Alternative systems, used for example in France, include cages (floor to roof height 30–40 cm for adults and 10 cm for juveniles). These are floated in etangs (see Glossary) and are also used for holding juveniles prior to restocking.

7.6.4 Incubation and hatching

Hatcheries may be simple or complex depending on whether or not eggs are incubated artificially and temperature is controlled. They may house broodstock, mating and incubation tanks that are often rectangular (10 m × 2 m × 0.4 m deep) but can also be circular (0.8–1 m diameter and 0.8 m deep). They are often fitted with hides and may additionally be used to hold young prior to stocking.

Ponds or tanks containing ovigerous females are drained in April or early May and the females transferred to individual mesh cages or held communally in boxes (40 × 20 × 10 cm) fitted with a 1 cm mesh floor. Alternatively, shallow nursery ponds or tanks (approximately 3 × 0.5 × 0.6 m deep) fitted with perforated floors and containing one or more shelters per female are employed. The perforations allow the juveniles to escape from the females as soon as they become independent (section 2.2).

Hatching occurs in spring or early summer when temperatures rise to 14°C and above. Towards the end of the incubation period hatching may be encouraged if the temperature is raised to 18–24°C. In some hatcheries the water exchange rate is set at once per 24 h (Alderman & Wickins 1996), in others at 15 L min^{-1} for tanks containing up to 6–9 females m^{-2} (Koksal 1988). Closed, heated recirculation systems may also be used to accelerate development of *Astacus leptodactylus* prior to stocking in cold regions (Golubev & Bakulin 1998).

7.6.4.1 Artificial incubation

Astacid crayfish have low fecundity, one breeding period each year and/or a prolonged incubation period. Egg loss and embryo mortality during maternal incubation in captivity are frequently reported (Perez *et al.* 1999). Removal of eggs from brooding females for artificial incubation is being studied to achieve economies of space, greater control over environmental conditions, the ability to store embryos, and to reduce risks of disease (Gonzàlez *et al.* 1993). Eggs removed at an advanced stage of development (when thoracic limb buds can be seen) can survive as well as maternally incubated eggs (50–60%) but most mortalities occur at hatching or at the next one to two moults (Perez *et al.* 1999). Correct environmental conditions seem more important than the time when the eggs are taken from the female (Carral *et al.* 1992). Celada *et al.* (2000) found that signal crayfish eggs could be stored for periods of up to 4 weeks under suitable conditions (<12°C) with subsequent survival to instar 2 juveniles exceeding 63%. The techniques are not, however, widely used by farmers although acceptable results using commercial salmon egg incubators are reported (Gonzàlez *et al.* 1993). The water supplied to a hatchery where eggs are incubated artificially must be clean and of high quality. Temperature control is critical and recirculation systems employing biological filters are sometimes installed to conserve heated water (Ackefors & Lindqvist 1994). Eggs that have reached the eyed stage are carefully removed from the female by siphon or stripped using blunt forceps and stocked into conical jars or flat, mesh-bottomed vessels (approximately 2–10 L capacity) in which a gentle upwelling current of water keeps the eggs in suspension. Gently rocking, water-filled trays are also used (Nylund & Westman 1992). Survival is improved if temperature is held at about 5°C for 3 weeks following stripping and then raised to 15°C (Carral *et al.* 1992). The flow is about 1 L min^{-1} but is regulated at hatching so that the young

Plate 7.5 An *Astacus astacus* crayfish hatchery and nursery in Bavaria. Juveniles are reared mainly for restocking natural waters and the survival rate of hatchlings from June to October is around 85%. (Photo courtesy Max Keller, Erste Bayerische Satzkrebszucht, Germany.)

are maintained 1–2 cm from the bottom. Pieces of sponge may be added to which the young attach. As soon as hatching begins it is advisable to separate the juveniles from the unhatched eggs and put them into another incubator at a density of 6000–10 000 per vessel, with a flow 0.7 L min^{-1} so that they can cling onto one another in a ball 1–2 cm from the bottom (Arrignon 1981). When they moult to instar 2, the flow is reduced to 0.4–0.5 L min^{-1}. At this stage the density must be reduced to below 6000 per vessel and feeding can begin.

7.6.5 Nursery

Keller (1988) recommends lifting broodstock females every 10 days and shaking those with attached juveniles in a bucket of water to facilitate separation of the young. By this means he found that up to 10 000 juveniles per hour could be collected. Normally juveniles detach or leave the mother after 2–21 days, depending on species, and escape predation by going through the mesh floor. Nursery facilities are frequently shallow troughs or long rectangular tanks fitted with hides, especially around the margins. The juveniles are fed a range of foods including detritus, *Cladophera* (a green filamentous alga), and live and frozen zooplankton, especially crustaceans.

Widely different stocking densities have been reported for rearing juveniles to a size at which they may be on-grown; generally 2–3 cm TL for noble crayfish, but smaller for signal crayfish (Ackefors 1997). Figures range from 50 m^{-2} (Westman 1973) to 2500 m^{-2} (Arrignon 1981) although Keller (1988) claims a density of 400 m^{-2} was the most profitable (70–80% survival) for *Astacus astacus* juveniles (3.5 months old, 2–3 cm long). During this time they may be thinned to 100 m^{-2} and fed for 3–4 months until they are big enough to be stocked in ponds. Juveniles for restocking could also usefully be produced in salmonid farm facilities during their off-season. If for example salmonid rearing occurred from December to June, crayfish would be reared from July to October.

7.6.6 Ongrowing

Three levels of culture may conveniently be defined: natural or extensive, semi-intensive and intensive or controlled environment culture. Some examples of extensive and semi-intensive pond layouts are shown in Fig. 7.2.

7.6.6.1 Natural/extensive

In the European context, this category clearly embraces the concepts of both restocking (hatchery supported fisheries) and ranching as described in sections 5.7.1 and 5.7.2, as well as the more traditionally defined extensive culture methods. About 96–98% of crayfish consumed in Europe come from wild harvests or extensive systems (Holdich 1993) and most of the restocking is in response to the depredations of crayfish plague fungus and environmental degradation (including acidification of natural lakes; Momot 1992). Fisheries in rivers and streams, as well as those in lakes and etangs (often covering 200 ha or more), may be stocked with hatchery-reared young. While very little management other than conventional fisheries regulation may be practised, slightly more management influence may be applied to crayfish grown in smaller lakes (10 ha × 0.3–5 m deep) where vegetation or hides such as stone outcrops or islands may be provided to create refuges (Rogers *et al.* 1999) (Figs 7.2a,b). Lakes may be stocked with 200–1000 crayfish ha^{-1} for 3–5 years using groups of 50 crayfish placed at intervals along the bank. This type of 'stock and forget' culture in natural or semi-natural enclosed waters, such as irrigation ponds or gravel pits, can produce annual yields of 60–500 kg ha^{-1} in Britain, and 900–2400 kg ha^{-1} in France (Arrignon 1993). Trial conversion of a German carp pond to *Astacus astacus* culture suggested profitable yields of around 220 kg ha^{-1} might be feasible (Piwernetz & Balg 1999). In response to reduced growth caused by unwanted reproduction in ponds, the fattening of an all-male population of *A. astacus* in Bavaria was examined by Keller and Keller (1995). Selected males (mean weight 51 g) were stocked in a 0.4 ha pond at 1 m^{-2} and fed fish, corn and an artificial pellet during the summer growing season. After 1 year, the yield was a respectable 303 kg ha^{-1} of prime-sized crayfish (74–91 g). In some shallow waters (e.g. ponds and lakes 1.5–5 m deep) fertilisation with agricultural fertilisers can increase natural production while the increased turbidity associated with it can provide protection from birds and sunlight, albeit sometimes at the cost of increased oxygen consumption at night.

7.6.6.2 Semi-intensive

In this category crayfish are introduced into prepared ponds, at densities up to 6–10 individuals per habitable square metre of underwater surface. In Sweden, farmers growing *Astacus astacus* stock 8–25 individuals of 15–50 mm TL per square metre of pond bottom; *Pacifastacus leniusculus* are generally smaller (instar 2–3) and stocked less densely (3 m^{-2}) (Ackefors 1997). Management is minimal although fertilisation (up to 50 kg N ha^{-1}) may be practised. Fertilisation and water exchange need to be carefully controlled to prevent excessive growth of filamentous algae. Feeds, which may include potatoes, apples or other vegetable matter, are sometimes given (section 7.6.7) but care is taken not to cause fouling and subsequent deoxygenation. Mesh enclosure of parts of lakes or etangs has been tried, but, in some areas of France for example, cages seem to be preferred as they allow better control of stock. Existing ponds and lakes can also be modified to increase the underwater 'bank' area by constructing reefs of rock and stones or by placing tubes, pipes or plant holders in the ponds. As a guide, stream-dwelling *Austropotamobius pallipes* prefer to shelter under stones 3.7 times larger and 2.3 times wider than their carapace length while *Pasifastacus leniusculus* offered a range of tubes as refuges, chose those with a diameter two to three times their carapace width (Foster 1993). Careful planning is necessary to ensure good water circulation can be maintained (Figs 7.2a,b). Rectangular earth ponds (10–15 × 3 × 1–2 m deep), situated 2 m apart, are also employed as are square or round ponds around 2500 m^2 in area. These are now more popular in Sweden, being cheaper to construct and easier to

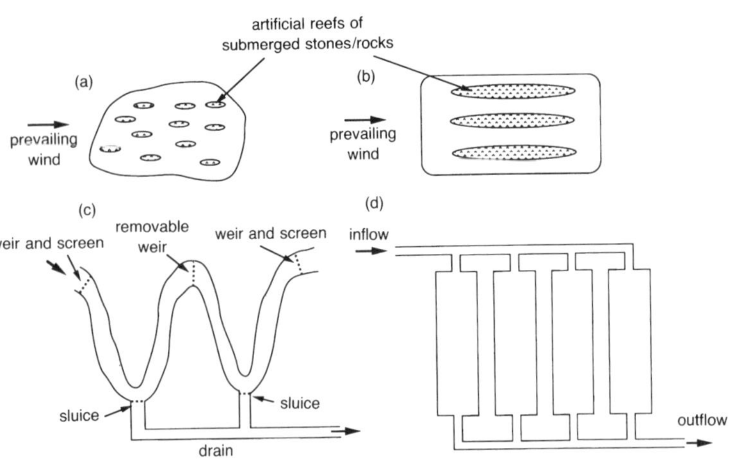

Fig. 7.2 Examples of crayfish farm design: (a) natural lake; (b) earthen pond; (c) simple stream/channel; (d) semi-intensive canal or 'raceway' culture system (see also Plate 7.6).

Plate 7.6 A small freshwater crayfish farm in southern England. The canals were initially stocked with imported signal crayfish.

maintain than canals or raceways (Ackefors 1997) but vegetation and water exchange must be strictly controlled. Stretches of suitable streams may also be used for production. Flow control is accomplished through construction of sluices, screens and by-pass or drainage canals (Fig 7.2c). Yields of signal crayfish in Europe range from 50 to 1000 kg ha^{-1} (Ackefors 2000).

Many commercial ventures still use canals or raceways; designs based on farms built in the 1980s. Some utilise former watercress beds which are 2–3.5 m wide by 10–50 m long but must be deepened from 0.75 m deep to 1.5 m. In Britain semi-intensive farms have been built around specially constructed canals which should give greater yields but only at greater input costs (labour and feed). The canals are typically 1.5–2 m deep with sloping sides to avoid collapse (Fig. 7.2d). The width is 3–10 m and canal lengths range from 50 m to 150 m with banks of 2–3 m width in between. Barriers are required to prevent escapes of crayfish and entry of eels into the farm. Shelter is all-important in most culture systems, both to reduce cannibalism and exposure to light, and to permit increased stocking density. Habitats are provided in the form of rough stone lining, hardcore piles at 10 m intervals or short lengths of weighted plastic pipes. Water flow on most British farms seems likely to give 50% exchange in 54–150 h (Alderman & Wickins 1996), while in Sweden, flows of 150–300 L min^{-1} ha^{-1} give good results.

Stocking may be with juveniles (spring) or one summer old juveniles (late summer/autumn) at three crayfish per metre of canal bank each year for at least 3 years and up to 5 years. The diet of adults in the wild consists of up to 70% vegetable matter, so most farmers encourage good peripheral plant growth. Young crayfish feed on the macro-invertebrates (worms, crustaceans and molluscs in particular) that abound in calcium-rich waters. The effect of supplemental and compounded feeds on farmed stock has yet to be determined.

Production targets are around six crayfish per running metre of canal bank but this could not be achieved without supplemental feeds unless extra habitat is available; natural feed might support three crayfish per running metre and it must be remembered that unless canals are wider than 9 m, only one bank is counted. Sustainable yields from British farms are reported to range from 79 to 396 kg ha^{-1} (Rogers & Holdich 1995). Elsewhere in Europe yields range from 60 to 1000 kg ha^{-1} (Ackefors 2000).

7.6.6.3 Intensive

Possibly only one truly intensive culture system may be in operation (Ackefors 1997). Generally speaking, the term intensive is currently applied to nursery systems for the production of large numbers of juveniles for restocking. About $2–4 \times 10^6$ *Astacus astacus*, *Astacus leptodactylus* and *Pacifastacus leniusculus* are produced annually for this purpose (Westman & Westman 1992). Artificial (e.g. concrete) ponds are typical and can be circular (0.8–1 m diameter × 0.8 m deep), fitted with hides and have a circular water distribution pattern. Initial stocking can be as high as 750–1000 m^{-2}, but after the first winter the density is halved to 300–500 m^{-2}. After the following winter the density is reduced to 50 m^{-2} but this prolonged nursery phase is only practised when large crayfish are required for stocking.

7.6.7 Feeding

In extensive cultures, vegetation and naturally occurring worms, crustaceans and molluscs are consumed, but very little information is available on successful compounded feeds for European crayfish (Hessen 1989). Arrignon (1981) refers to diets made in 1972 that were bound with alginate and based on fish or shrimp meal and cereals. Protein levels were around 18–44%, lipids 1–5%, and minerals 7–10%. Low feeding levels were reported: 1–4% body weight per day for juveniles, 0.3–1% for adults which reduced to less than 1% after mating. More recent research on *Astacus astacus* indicated that in formulated feeds a satisfactory protein content of 30–35% for juveniles exceeded that required by crayfish over 1 year old (20%) (Ackerfors *et al.* 1992) suggesting that diets used for further ongrowing could be cheaper. Optimum protein : energy ratio was in the range 27–37 mg MJ^{-1} and lipid levels above 10% were detrimental (section 2.4.2). Diets based on fenugreek seeds supplemented with sulphur amino acids or saponines supported good growth and survival in 1-year-old *Astacus leptodactylus* (Sevilla *et al.* 1993).

7.6.8 Harvesting

Crayfish may be caught in fyke nets or traps. Of the latter, baited cylindrical (Ackefors 1998) or funnel traps are commonly used and set at 5 m intervals in canals or at 25–50 ha^{-1} in ponds. Traps are set before dusk and fished at dawn, starting 2–3 years after the initial stocking.

7.6.9 Transportation

Fished crayfish are frequently transported live in expanded polystyrene boxes, in layers separated by damp foam, cloths or moist algae and mosses to maintain high humidity. In France, however, live transport of non-European crayfish is not permitted. To comply with this regulation these crayfish are either frozen or the telson and hind gut is simply torn. Because of this bizarre practice, a more humane killing method involving electrocution is being developed (sections 8.7 and 11.2.5). Juveniles (1–2 cm TL) are shipped by air a few weeks after hatching, in 20 cm long perforated plastic tubes with a spiral insert to which they cling. The tubes are placed in a cool, moist container. In another method, 500 juveniles are transported in 12 L containers holding one-third water, two-thirds oxygen and can survive for up to 40 h provided the temperature remains below 10°C. One-summer-old crayfish are carried as described above but at a lower density (300 per container). Adult broodstock are carried in 7 kg boxes with 60–80 adults in two layers in a box (40 × 40 × 15 cm) and will remain in good condition for up to 20 h if kept cool. Ice or cooling packs can be added to the box if required. During the journey they must never be turned on their backs or exposed to chlorinated tap water. At the water's edge crayfish must be allowed to walk backwards into the pond after they have been exposed to air for any length of time.

7.6.10 Processing

Prior to shipment or cooking, crayfish are best held in clean running water for 24–48 h to purge the gut. They are then sold live, frozen or less commonly, cooked and frozen. Edible cooked meat yield is around 9–13% for *Astacus leptodactylus* and 11–15% for *Pacifastacus leniusculus* (Table 4.6g) (Harlioğlu & Holdich 2001).

7.6.11 References

Ackefors H. (1997) The development of crayfish culture in Sweden during the last decade. In: *Freshwater Crayfish 11* (ed. W.T. Momot), pp. 627–654. Louisiana State University, LA, USA.

Ackefors H. (1998) The culture and capture crayfish fisheries in Europe. *World Aquaculture,* **29** (2) 18–24 & 64–67.

Ackefors H.E.G. (2000) Freshwater crayfish farming technology in the 1990s: a European and global perspective. *Fish and Fisheries*, **1**, 337–359.

Ackefors H. & Lindqvist O. (1994) Cultivation of freshwater crayfishes in Europe. In: *Freshwater Crayfish Aquaculture in North America, Europe, and Australia* (ed. J.V. Huner), pp. 157–216. Food Products Press, an imprint of the Haworth Press, Inc., New York.

Ackerfors H., Castell J.D., Boston L.D., Räty P. & Svensson, M. (1992) Standard experimental diets for crustacean nutrition research. II. Growth and survival of juvenile crayfish *Astacus astacus* (Linné) fed diets containing various amounts of protein, carbohydrate and lipid. *Aquaculture,* **104** (3–4) 341–356.

Alderman D.J. & Wickins J.F. (1996) *Crayfish culture*, 22 pp.

Lab. Leafl. (76). MAFF Directorate Fisheries Research, Lowestoft, UK.
Arrignon J. (1981) *L'écrivisse et son élevage*, 178 pp. Gauthier-Villars, Paris.
Arrignon J.C.V. (1993) The development of a *Pacifastacus leniusculus* population in a gravel pit in France. In: *Freshwater Crayfish 9* (eds D.M. Holdich & G.F. Warner), pp. 87–96. Louisiana State University, LA, USA.
Carral J.M., Celada J.D., Gonzàlez J., Gaudioso V.R., Fernández R. & López-Baissón (1992) Artificial incubation of crayfish eggs (*Pasifastacus leniusculus* Dana) from early stages of embryonic development. *Aquaculture,* **104** (3–4) 261–269.
Celada J. D., González J., Carral J. M., Fernández R., Pérez J. R. & Sáez-Royuela M. (2000) Storage and transport of embryonated eggs of the signal crayfish *Pacifastacus leniusculus. North American Journal of Aquaculture*, **62** (4) 308–310.
Foster J. (1993). The relationship between refuge size and body size in the crayfish *Austropotamobius pallipes* (Lereboullet). In: *Freshwater Crayfish 9* (eds D.M. Holdich & G.F. Warner), pp. 345–349. Louisiana State University, LA, USA.
Golubev A. & Bakulin A. (1998) The experience of obtaining and growing up of larvae of crayfish *Astacus leptodactylus* in recycling device system in Belarus. In: *Aquaculture and Water: fish culture, shellfish culture and water usage* (compiled by H. Grizel & P. Kestemont), pp. 99–100. Abstracts presented at Aquaculture Europe '98, Bordeaux, France, 7–10 October 1998, European Aquaculture Society, Special Publication No. 26.
Gonzàlez J., Carral J.M., Celada J.D., *et al.* (1993) Management of crayfish eggs (*Pasifastacus leniusculus*) for intensification of juvenile production. In: *Freshwater Crayfish 9* (eds D.M. Holdich & G.F. Warner), pp. 144–146. Louisiana State University, LA, USA.
Harlioğlu M.M. & Holdich D.M. (2001) Meat yields in the introduced freshwater crayfish, *Pacifastacus leniusculus* and *Astacus leptodactylus*, from British waters. *Aquaculture Research*, **32**, 411–417.
Hessen D.O. (1989) Crayfish food and nutrition. In: *Crayfish Culture in Europe* (eds J. Skurdal, K. Westman & P.I. Bergen), pp. 164–174. Norwegian Directorate for Nature Management, Trondheim, Norway.
Holdich D.M. (1993) A review of astaciculture: freshwater crayfish farming. *Aquatic Living Resources*, **6**, 307–317.
Huner J.V. (1995) Farming freshwater crayfish in Finland. *Fish Farming International*, **22** (3) 34–35.
Keller M. (1988) Finding a profitable population density in rearing summerlings of European crayfish *Astacus astacus* L. In: *Freshwater Crayfish 7* (ed. P. Goeldlin de Tiefenau), pp. 259–266. Musée Zoologique Cantonal, Lausanne, Switzerland.
Keller M.M. & Keller M. (1995) Yield experiments with freshwater crayfish *Astacus astacus* (L.) in aquaculture. In: *Freshwater Crayfish 10* (eds M.C. Geddes, D.R. Fielder & A.M.M. Richardson), pp. 506–511. Louisiana State University, Baton Rouge, LA, USA.
Koksal G. (1988) *Astacus leptodactylus* in Europe. In: *Freshwater Crayfish 7* (ed. P. Goeldlin de Tiefenau), pp. 365–400. Musée Zoologique Cantonal, Lausanne, Switzerland.
Lewis S.D. (2001) *Pacifastacus*. In: *Biology of Freshwater Crayfish* (ed. D.M. Holdich), pp. 511–40. Blackwell Science, Oxford, UK.
Mackevičienė G., Mickėnienė L., Burba A. & Koreiva E. (1997) Aquaculture of the noble crayfish, *Astacus astacus* L., in Lithuania. In: *Freshwater Crayfish 11* (ed. W.T. Momot), pp. 599–607. Louisiana State University, LA, USA.
Mackevičienė G., Mickėnienė L., Burba A. & Mažeika V. (1999) Reproduction of noble crayfish *Astacus astacus* (L.) in semi-intensive culture. In: *Freshwater Crayfish 12* (eds M. Keller, M.M. Keller, B. Oidtmann, R. Hoffmann & G. Vogt), pp. 462–470. Weltbild Verlag, Germany.
Momot W.T. (1992) Stocking and exploitation as management methods for European crayfish. *Finnish Fisheries Research*, **14**, 145–148.
Nylund V. & Westman K. (1992) Crayfish diseases and their control in Finland. *Finnish Fisheries Research*, **14**, 107–118.
Pérez J.R., Carral J.M., Celada J.D., Muñoz C., Sáez-Royuela M. & Antolín J.I. (1999) The possibilities for artificial incubation of white-clawed crayfish (*Austropotamobius pallipes* Lereboullet) eggs: comparison between maternal and artificial incubation. *Aquaculture,* **170** (1) 29–35.
Piwernetz D. & Balg J. (1999) Growth experiments with *Astacus astacus* in a 6,000 m^2 pond previously used for extensive farming of carp. In: *Freshwater Crayfish 12* (eds M. Keller, M.M. Keller, B. Oidtmann, R. Hoffmann & G. Vogt), pp. 535–539. Weltbild Verlag, Germany.
Rogers W.D. & Holdich D.M. (1995) Crayfish production in Britain. In: *Freshwater Crayfish 10* (eds M.C. Geddes, D.R. Fielder & A.M.M. Richardson), pp. 583–595. Baton Rouge, Louisiana State University, LA, USA.
Rogers D., Holdich D.M. & Carter E. (1999) Conservation of the native crayfish *Austropotamobius pallipes* population at Dowdeswell Reservoir during engineering works. In: *Freshwater Crayfish 12* (eds M. Keller, M.M. Keller, B. Oidtmann, R. Hoffmann & G. Vogt), p. 942. Weltbild Verlag, Germany.
Sevilla C., Linden E., Baccou J.C. & Sauvaire Y. (1993) Influence d'une alimentation a base de fenugrec sur la survie, la mue et la croissance d'*Astacus leptodactylus*. In: *La production et l'exploitation des écrevisses en France* (ed. C. Roqueplo), pp. 71–79. Publ. Assoc. Dev. Aquacult. Cestas No. 35.
Skurdal J. & Taugbøl T. (2001) *Astacus*. In: *Biology of Freshwater Crayfish* (ed. D.M. Holdich), pp. 467–510. Blackwell Science, Oxford, UK.
Westman K. (1973) Cultivation of the American crayfish *Pacifastacus leniusculus*. In: *Freshwater Crayfish 1* (ed. S. Abrahamsson), pp. 211–220. Studentlitteratur, Lund, Sweden.
Westman K. & Mannine K. (1996) Institutes, research workers and programmes related to research on crayfish in Europe, 82 pp. Finnish Game and Fisheries Research Institute, *Kala-Ja Riistaraportteja* (65).
Westman K. & Westman P. (1992) Present status of crayfish management in Europe. *Finnish Fisheries Research*, **14**, 1–22.

7.7 Crayfish: Australia

7.7.1 Species of interest

Redclaw crayfish (*Cherax quadricarinatus* and *C. albertsii*, the latter from Papua New Guinea); marron (*C. tenuimanus*); yabby (*C. destructor*, *C. albidus* and *C. rotundus*) and the New Zealand koura (*Paranephrops planifrons* and *P. zealandicus*). Koura were thought to be uneconomic to farm due to their slow growth but at least three farms have been set up in New Zealand (section 7.7.11). Considerable interest is currently being shown in the commercial culture of redclaw both in Australia and elsewhere (Table 4.1b). The semi-intensive ongrowing methods employed in earthen ponds are similar to those used for shrimp and prawns. Marron farming expanded into South and Western Australia during the late 1990s although it has been less successful abroad. Several intensive culture trials have also been tried with battery systems based on those used for clawed lobsters. This section includes information from published and unpublished sources. Further details may be found in Jones (1990); Merrick and Lambert (1991); Morrissy (1992a); Mills *et al.* (1994); O'Sullivan (1995); Semple *et al.* (1995); Hyde (1998); Jones *et al.* (1998); Lawrence *et al.* (1998); Lawrence and Morrissy (2000); Lawrence and Jones (2001).

7.7.2 Broodstock

Broodstock may be obtained from the wild (with permission), from farms and from dealers. However, close seasons and other legislation in Australia prevent the taking of wild spawning or egg-carrying females in some areas and seasons. Populations in permanent waters become self-sustaining and provide a seasonal supply of stock for extensive operations. Larger projects depend upon wild broodstock as well as hatchery-reared juveniles. However, year-round breeding of yabby is possible through temperature and photoperiod control, and while redclaw breeds readily throughout the year, marron generally spawn annually (Merrick & Lambert 1991). Holker (1989) described a marron broodstock facility of circular tanks fitted with individual housing and stocked at up to 12 crayfish m^{-2} and recommended that broodstock marron over 4 years old (over 80 mm CL) be discarded to reduce the risk of poor egg quality or hatch rates. Lawrence and Jones (2001) recommend holding putative broodstock marron in 500 m^2 ponds at 2 m^{-2} with a sex ratio of one male to two females. Mating is accomplished in smaller ponds of 150 m^2. Redclaw broodstock may be held at 20–60 m^{-2} (Rouse 1995; Barki & Karplus 2000). A guide to setting up a selective breeding programme for redclaw has been published (Jones *et al.* 1998; section 8.10.1.2).

7.7.3 Mating and spawning

Males and newly moulted females held separately in hatchery tanks (Mills *et al.* 1994) may be put together for breeding unless fighting occurs, in which case the male should be replaced. The males should be of similar size, within 25 g of each other (Rouse 1995). Mating normally takes place within a few hours or days, after which either the males are removed and the females left undisturbed to spawn and incubate their eggs, or the females are transferred to a similar tank with a mesh false bottom and stocked at 10–12 females m^{-2}. In trials with redclaw high spawning rates were maintained for up to 3 months at densities up to 20 m^{-2} with sex ratios of 1 : 1 to 1 : 5 males to females proving satisfactory (Yeh & Rouse 1995). Females whose young have become independent should be separated from females still carrying eggs or young to avoid cannibalism (Levi *et al.* 1999). Rematuration in yabbies can be triggered by a drop in temperature and/or the presence of males provided internal food reserves are adequate. A practical scheme for sampling and stimulating females for rematuration is presented by McRae and Mitchell (1997).

7.7.4 Incubation and hatching

Egg-bearing females may be held in tanks or submerged cages in ponds, preferably with some form of shelter until the eggs hatch. Eggs can also be stripped from females (from 4 days after spawning in redclaw) and incubated artificially, provided microbial infestations can be controlled. Survival of stripped eggs to the independent juvenile stages may be less in redclaw (48–63%) than in marron (89%) (Henryon & Purvis 2000). Subsequent attempts to store redclaw eggs at low temperatures (5–10°C) to compensate for fluctuations in juvenile supplies and conserve hatchery space, were not successful (King 1993) and the techniques are unlikely to be adopted commercially. In view of the widespread dissemination of redclaw crayfish and the lessons learnt from shrimp culture, artificial incubation may have a key role in producing SPF stocks (Edgerton & Owens 1997; section 8.9.4.4). After hatching, and when the young have left the mother, they are collected or escape through mesh to imitation weed hides. In redclaw and yabby, incubation lasts 20–40 days (30–90 days in marron) and the young stay with the mother for a further 2–4 weeks (Table 4.6g). These small juveniles concentrate in bunches of twigs, rope fibres or straws placed in the ponds, or are harvested directly from the tanks for ongrowing. It is best to keep crayfish of a common size together as cannibalism increases with disparity of sizes, especially in crowded populations. Morrissy (1976) calculated that 1000 marron females of 50 mm CL would release sufficient juveniles to stock a 1 ha pond at around 5–10 m^{-2} after allowing for up to 75% mortality.

7.7.5 Nursery

Juvenile crayfish, in particular redclaw, are sensitive to handling and transportation (Jones & Grady 1997) and some growers recommend farmers in the Americas to eliminate the hatchery phase and stock egg-bearing females directly into nursery ponds (Mattei 1995). It seems likely that many growers and perhaps hatcheries use nursery ponds or tanks to acclimatise, or increase the size of juveniles prior to stocking, to capitalise on the best seasonal temperatures for growth (Mosig 1999). During this phase they are fed commercial crayfish pellets (Meade & Watts 1995), compost red worms, sorghum and lucerne (alfafa). Densities may reach 200 m^{-2} (Holker 1989) and growth, but not survival, of redclaw crayfish can be improved by the addition of substrates such as mesh bags, which provide extra shelter (Karplus *et al.* 1995). In contrast, neither survival nor growth of juvenile yabbies held in indoor tanks was improved by the presence of various kinds of shelter (Verhoef & Austin 1999). The value of providing shelters is probably species dependent and requires further study. Overwintering nurseries for redclaw culture in warm temperate regions could capitalise on short growing seasons but are expensive to operate. However, overwintering at ambient temperatures outside Australia may be possible but could present some ecological problems (Karplus *et al.* 1998; section 11.3.3). Survival of redclaw juveniles stocked in laboratory tanks at 980–1842 m^{-2} ranged from 4% to 84% (mean 46.3%) over 25–50 days (Jones 1990). Growth, however, is better at densities of 100 m^{-2} and below, than at 250 m^{-2}. In Central and South America and the Caribbean, mesh bag collectors are used to trap redclaw juveniles of 3–6 g from spawning-nursery ponds. Collections of up to $0.5–1 \times 10^6$ juveniles ha^{-1} every 4–5 months are reported (Rouse 1995).

7.7.6 Ongrowing

Techniques for ongrowing crayfish in Australia are conveniently described under four categories: extensive (including farm dam) farming; backyard 'hobby' culture; commercial semi-intensive farming and experimental intensive 'battery' or controlled-environment culture. Elsewhere, most modern farms follow the semi-intensive approach, which can be accomplished year-round in the tropics or seasonally, with or without rotation with another crop, in subtropical regions. Yields, however, range from low (<400 kg ha^{-1} similar to those obtained in extensive systems), to those more typical of semi-intensive systems (2000–4000 kg ha^{-1}). Total Australian production is around 100–300 mt yr^{-1}.

7.7.6.1 Extensive

In Australia, this method is more akin to the 'stock and forget' method used in parts of Europe. Either natural or man-made water reservoirs constructed for watering agricultural livestock are utilised. These may be excavated ponds (farm dams) filled with run-off water in winter (stocked in Western Australia with *Cherax albidus* which burrows less destructively than *C. destructor*), or gully dams built in hilly regions. Gully dams are considered best for *C. tenuimanus* (Morrissy 1992a) as they usually have a stream flowing through them, although they can still be polluted by agricultural run-off. On the other hand farm dams may be susceptible to overloading with detritus and organic pollution. It is recommended that ponds have hard clay bottoms and are designed to be completely drainable for cleaning (Morrissy 1992a; Lawrence *et al.* 1998). Stocking levels for yabby are typically around 1–2 m^{-2}, giving yields under 400 kg ha^{-1} yr^{-1}.

7.7.6.2 Backyard culture

This method appears widespread in Australia but is not likely to attract substantial investment. It has merit in that it allows entrepreneurs to experience some of the demands of an aquaculture operation without serious loss of capital. Ponds may be plastic swimming pools or specially constructed tanks (2–4 m diameter, 35–90 cm deep) provided with shelters and shade cloth to control temperature and protect against bird predation. Aeration may be continuous or applied for about 10 minutes, two to three times per day. A shallow area is sometimes provided in case oxygen levels fall to critical levels.

Stocking density is 10 m^{-2}. Feed is commonly chicken or trout pellets at up to 30 g m^{-2} per week but compost red worms are best. The maximum yield is likely to be 300 g m^{-2} with survivals of 2–3 crayfish m^{-2}.

7.7.6.3 Semi-intensive

The transition from extensive to semi-intensive cultures is not a clear-cut process but basically involves the following changes:

Extensive	Semi-intensive
deep ponds	shallow ponds
mixed age, small biomass	single age, large biomass
slow, continuous production	complete harvest every 12–24 months
no aeration	aeration

Ponds suitable for commercial operation are 50–250 × 20 × 1–1.5 m deep, with excavated topsoil lining the banks down to the waterline and planted with Kikuyu grass for stability (Holker 1989). Pond sides are pitched at angles of not less than 45° (1 : 1 – horizontal : vertical) on long sides and over 60° (1 : 1.7) on the short sides. Between 100 and 200 artificial weed bunches may be added per 1000 m^2 for shelter. A U-shaped pond bottom allows crayfish to move into the shallows if oxygen levels fall. An anti-escape barrier of planking or similar material may be necessary, as may shade netting. Many farms employ bird netting or other anti-predator devices (O'Sullivan *et al.* 1994). Dogs in particular can be very cost-effective against birds (Wilson 2000). Poaching can be a significant worry on Western Australian marron farms. The drains must be able to empty the pond in 5 h for harvesting, but the supply should be capable of filling a pond in 24 h. Ponds are cleaned annually and treated with agricultural limestone (dolomite) at 100 kg ha^{-1} to control algae. After 5 years the ponds are dried, allowed to crack and a further 200–300 kg ha^{-1} of dolomite tilled in to neutralise bottom deposits (Curtis & Jones 1995).

Fertilisation may be employed to increase natural productivity (O'Sullivan 1995). Paddlewheel (best) or submerged pipe aeration is used and feeding is done at sunset along pond margins. Cheaper feeds are more often used in ongrowing ponds than in ponds holding early juveniles or potential broodstocks. Formulated feeds (Morrissy 1992b; O'Sullivan & Kiley 1997a), proprietary cattle or chicken pellets, lucerne and sorghum hay are given as food three to four times each week based on crayfish biomass. A water exchange every 2–3 days is recommended but may be increased to maintain quality as required. Adults are stocked at 2 m^{-2} if required; otherwise 5–15 juveniles of uniform size (5–10 g) are initially stocked per square metre. Stocking with redclaw juveniles below 2 g is not advised. Experimental studies in net pens (16 m^2) indicated that both yield and economic return increased when 17 g juveniles were stocked at 15 m^{-2}. Yields equivalent to 5 mt ha^{-1} were obtained in 140 days (Jones & Ruscoe 2000). Some farms in the Americas reduce density by restocking part of the population into other ponds 3–4 months prior to harvesting. Young crayfish can survive up to 80 h out of water

Plate 7.7 Production pond for redclaw crayfish showing typical shelters, including pipe stacks and bundles of mesh. (Photo courtesy Clive Jones, Department of Primary Industries, Queensland, Australia.)

under favourable conditions, but after acclimatisation to pond temperatures they should be released onto the bank so that they can walk in. Yields of about 2000–4000 kg ha^{-1} in two crops are a reasonable expectation after 12 months (40–70 g animals) to 2 years (100–200 g animals) depending on species and temperature (Rouse 1995). Monosex culture of redclaw, yabby and yabby hybrids (sections 2.6.3 and 12.8.3) is under investigation.

Published information (Mills *et al.* 1994; Curtis & Jones 1995; O'Sullivan 1995; Lawrence & Whisson 2000) about commercial practices in Australia for the redclaw, marron and yabby indicates that farms typically operate 4–40 ponds of 0.4–7.9 ha each and 1–2 m deep. In recent years there has been a trend towards using smaller ponds (1200–2500 m^2) for redclaw production. Most also use pumped bore and well water and eight out of ten use aerators, but exchange rates vary widely from 10–15% per day to once per year. Feeding once or twice a day in the late afternoon and early morning is considered best. Six out of ten farms provide shelters. Shelters and aeration are considered essential for semi-intensive redclaw culture and Western Australian marron farmers use three to four shelters per 100 marron stocked. Car tyres were commonly used but they are cumbersome at harvest time and concern over leaching of toxins, e.g. cadmium (section 11.3.1.3), in a confined volume of water has stimulated use of alternative materials (O'Sullivan & Kiley 1997b), for example, bundles of synthetic mesh have proved an effective alternative. Some shelters are suspended off the bottom or form underwater towers. Water hyacinth has been grown in pens in redclaw ponds to provide shelter and improve water quality, but maintenance costs and other problems make this approach unattractive (Tysoe 1997). Ponds should be dried and prepared with lime and fertilised between crops (section 8.3.3). Staff number between one and seven full-time, and up to three part-time. Harvesting is mostly by draining or by pump rather than by trapping. Many farms have specific nursery ponds, since crayfish breed readily in ongrowing ponds which leads to unwanted competition for resources, and a subsequent increase in the disparity of sizes at harvest.

In Ecuador, semi-intensive ponds (0.25 ha) are stocked with 3–4 juveniles m^{-2} and yield 1500–2000 kg ha^{-1} in 5–7 months. Smaller ponds stocked with 7 juveniles m^{-2} can yield 4000 kg ha^{-1} in the same period. Predator control is practised using netting, fencing and dogs, especially where avian pests are abundant.

7.7.6.4 Intensive

There seems little doubt that battery farming of individually confined *Cherax* species is technically possible. Stocking densities of 100–200 juveniles m^{-2} yielding >5000 kg ha^{-1} yr^{-1} are possible. Economic feasibility however is unlikely unless the twin problems of developing cost-effective diets (King 1992; Jussila & Evans 1998; Keefe & Rouse 1999) and low-labour rearing systems can be solved. Several technical investigations have been made (Cogan 1987; Anon. 1988a; O'Sullivan 1990; Du Boulay *et al.* 1993; Whisson 1995) and one commercial operation with redclaw in the USA is reported (Mattei 1995).

7.7.7 Polyculture

Polyculture of redclaw with tilapia has been attempted in Israel and the USA, generally with unsatisfactory results (section 5.5; Rouse & Kahn 1998). The provision of aquatic plants (macrophytes) to provide shelter for marron in the presence of silver perch was beneficial but again growth and survival of the crayfish were poor (Whisson 1997). However, preliminary experiments indicate silver perch cage culture in marron ongrowing ponds may be worth investigating (Whisson 1999).

7.7.8 Harvesting

Most harvesting is accomplished by draining the pond or through the use of baited traps. Egg-bearing females tend not to be caught in traps but a novel multi-level refuge system tested with yabbies effectively harvested populations unselectively (Mitchell *et al.* 1994). Redclaw crayfish, however, respond strongly to a current of water coming from outside the pond and can be caught in aerated flow-traps (Fig. 7.3), which may be portable or built into the pond floor (Jones 1990). A fish pump can be adapted for use harvesting yabbies from ponds built with a sump or deep area at the drain (Walladge 1995). Immediately after harvesting, crayfish are placed in clean water to flush away organic debris from the gill chambers; the gut contents may then be purged.

7.7.9 Transportation

Many attempts have been made to culture marron and redclaw outside Australia using semi-intensive methods similar to those described above. Supplies of juveniles (<10 g) are shipped in double plastic bags contain-

Plate 7.8 A flow-trap containing boxes of harvested redclaw crayfish. The air supply to the trap has been turned off briefly for the photograph. The ramp down which the water flow is directed and up which the crayfish walk from the pond is seen on the right. (Photo courtesy Clive Jones, Department of Primary Industries, Queensland, Australia.)

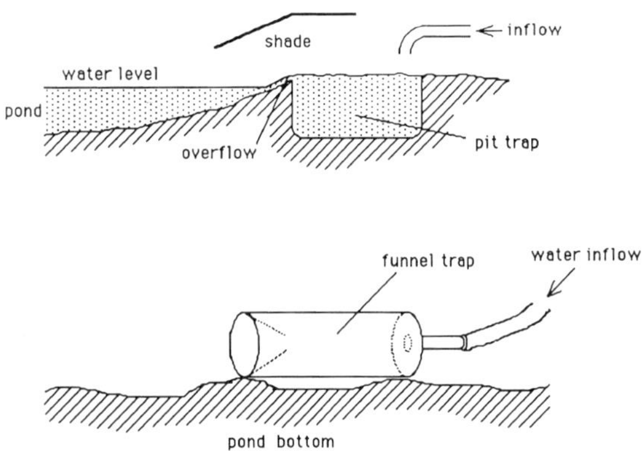

Fig. 7.3 Pit and funnel traps for catching Australian crayfish (*Cherax* spp.) that move towards a flow of water.

ing water (10 L per 100–150 juveniles), some prewashed coconut fibre and oxygen. The bags are placed in cardboard boxes with cooling packs. In one instance, juveniles and adults survived shipment to Dominica, when packed conventionally in polythene bags containing 1 L of water and oxygen. On arrival a quarantine dip in nitrofurazone solution (an antibiotic) and then salt water was given to remove external infestations. Shipping mortality was 50–100% and it was felt that storage at 15–20°C was necessary at transfer points during the flight (Anon. 1988b; Alon *et al.* 1989). Further details of acclimation and disinfection procedures may be found in Rubino *et al.* (1990). Adults are transported live, out of water, in clean, moist coconut fibre or absorbent packing foam material in polystyrene boxes again with cooling packs to keep temperatures below 15°C (O'Sullivan 1995; Jussila *et al.* 1999). Cooling yabbies prior to packing is also advantageous (Roe 1996).

7.7.10 Processing

Australian crayfish are purged in clean, running, fresh water or in a spray system and sold live or cooked and frozen. Some are exported as frozen headless product (Merrick & Lambert 1991; Curtis & Jones 1995). Hand grading of yabbies for example might be done to produce five classes: small (30–50 g), medium (50–70 g), large (70–90 g), extra large (90–110 g) and seconds. The latter are damaged yabbies sold for further processing (Saunders & O'Sullivan 1998).

7.7.11 Koura (*Paranephrops*) culture in New Zealand

Two species of koura (*Paranephrops planifrons* and *P. zealandicus*) are endemic to New Zealand but since commercial exploitation of wild stocks is prohibited by the Freshwater Fisheries Regulations 1983 (McDowall 1995), aquaculture is the only option for producing koura for sale. Most current information on farming methods and production comes from the farmers themselves (e.g. extensive culture of *P. zealandicus* on South Island (Diver 1998); semi-intensive for *P. planifrons* on North Island (Smythe 1998)) (Tables 4.6g and 5.5). In the early 1980s the Fisheries Research Division of the Ministry of Agriculture and Fisheries (MAF) investigated koura aquaculture in tank systems but concluded that farming koura was not likely to be a commercial propo-

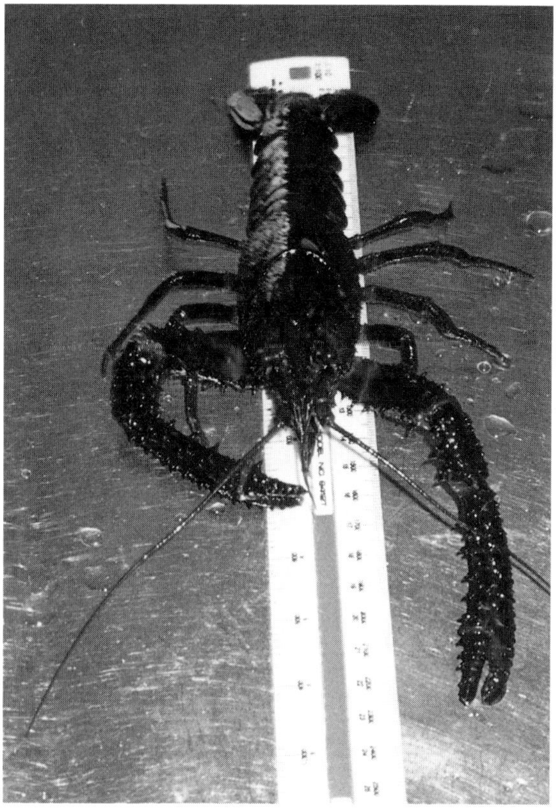

Plate 7.9 Large male Koura (*Paranephrops planifrons*), 135 mm TL, from a pilot farm in New Zealand. (Photo courtesy David Smythe and Peter Wilhelmus, New Zealand Clearwater Crayfish (Koura) Ltd.)

sition (section 10.6.2.3). In the light of this unfavourable outlook, Australian marron were introduced to New Zealand in 1987 with the hopes of farming them. However all stocks were later ordered to be destroyed by the government as there was no information on their possible environmental impact.

Renewed interest in koura aquaculture has resulted in about twelve registered koura farm proposals to date (2000). Most ventures are probably in the very early stages but at least two businesses, one using extensive and the other semi-intensive methods, are currently selling crayfish. One of these farms is in Central Otago (South Island) and since about 1997, has developed a low-input, moderate scale, extensive aquaculture system based on natural conditions. The farm is believed to sell about 11 kg per week of *P. zealandicus* at around \$26 kg^{-1} (S. Parkyn, 2000 pers. comm.). A farm on North Island reported creditable stock densities of between 20 and 30 *P. planifrons* m^{-2} when drain harvesting broodstock (70–130 g individuals) from a pond of 100 m^2 (D. Smythe & P. Wilhelmus, 2001 pers. comm.).

7.7.12 References

Alon N.C., Rubino M.C., Wilson C.A. & Armstrong J.M. (1989) Pond culture of the Australian marron lobster, *Cherax tenuimanus*, in the Eastern Caribbean: survival and growth. *Aquaculture '89 Abstracts,* p. 56, from World Aquaculture Society Meeting 1989, Los Angeles, USA.

Anon. (1988a) Battery culture research in W.A. *Austasia Aquaculture Magazine*, **2** (9) 16.

Anon. (1988b). Lessons from shipping marron overseas. *Austasia Aquaculture Magazine,* **2** (9) 16–17.

Barki A. & Karplus I. (2000) Crowding female red claw crayfish, *Cherax quadricarinatus*, under small-tanks hatchery conditions: what is the limit? *Aquaculture*, **181** (3–4) 235–240.

Cogan P. (1987) Marron battery study proves encouraging. *FINS,* **20** (3) 5–6. Fishing Industry News Service, Perth, W. Australia.

Curtis M.C. & Jones C.M. (1995) Overview of redclaw crayfish, *Cherax quadricarinatus*, farming practices in northern Australia. In: *Freshwater Crayfish 10* (eds M.C. Geddes, D.R. Fielder & A.M.M. Richardson), pp. 447–455. Louisiana State University, Baton Rouge, LA, USA.

Diver P. (1998) *Sweet Koura Enterprises*, 10 pp. (mimeo). Paper presented to the Maori and the Business Aquaculture Conference, 1998, Auckland, New Zealand.

Du Boulay A.J.H., Sayer M.D.J. & Holdich D.M. (1993) Investigations into intensive culture of the Australian red claw crayfish *Cherax quadricarinatus*. In: *Freshwater Crayfish 9* (eds D.M. Holdich & G.F. Warner), pp. 70–78. Louisiana State University, LA, USA.

Edgerton B. & Owens L. (1997) Age at first infection of *Cherax quadricarinatus* by *Cherax quadricarinatus* bacilliform virus and *Cherax Giardiavirus*-like virus, and production of putative virus-free crayfish. *Aquaculture,* **152** (1–4) 1–12.

Henryon M. & Purvis I.W. (2000) Eggs and hatchlings of the freshwater crayfish, marron (*Cherax tenuimanus*), can be successfully incubated artificially. *Aquaculture*, **184** (3–4) 247–254.

Holker D.S. (1989) Marron – *Cherax tenuimanus*, 16 pp. Unpublished report.

Hyde K.W. (1998) *The New Rural Industries: a handbook for farmers and investors*, 152 pp. Rural Industries Research and Development Corporation, Canberra, Australia.

Jones, C.M. (1990) *The Biology and Aquaculture Potential of the Tropical Freshwater Crayfish* Cherax quadricarinatus, 109 pp. Information Series QI 90028, Queensland Department of Primary Industries, Queensland, Australia.

Jones C.M. & Grady J. (1997) Growout of redclaw at two demonstration sites in North Queensland. *Austasia Aquaculture Magazine*, **11** (2) 52–56.

Jones C.M. & Ruscoe I.M. (2000) Assessment of stocking size and density in the production of redclaw crayfish, *Cherax quadricarinatus* (von Martens) Decapoda: Parastacidae), cultured under earthen pond conditions. *Aquaculture,* **189**

(1–2) 63–71.

Jones C.M., McPhee C.P. & Ruscoe I.M. (1998) *Breeding Redclaw: management and selection of broodstock*, 31 pp. Information Series QI 98106. Queensland Department of Primary Industries, Queensland, Australia.

Jussila J. & Evans L.H. (1998) Growth and condition of marron *Cherax tenuimanus* fed pelleted diets of different stability. *Aquaculture Nutrition,* **4**, 143–149.

Jussila J., Paganini M., Mansfield S. & Evans L.H. (1999) On physiological responses, plasma glucose, total hemocyte counts and dehydration, of marron *Cherax tenuimanus* (Smith) to handling and transportation under simulated conditions. In: *Freshwater Crayfish 12* (eds M. Keller, M.M. Keller, B. Oidtmann, R. Hoffmann & G. Vogt), pp. 154–167. Weltbild Verlag, Germany.

Karplus I., Barki A., Levi T., Hulata G. & Harpaz S. (1995) Effects of kinship and shelters on growth and survival of juvenile Australian redclaw crayfish (*Cherax quadricarinatus*). In: *Freshwater Crayfish 10* (eds M.C. Geddes, D.R. Fielder & A.M.M. Richardson), pp. 494–505. Louisiana State University, Baton Rouge, LA, USA.

Karplus I., Zoran M., Milstein A., *et al.* (1998) Culture of the Australian red-claw crayfish (*Cherax quadricarinatus*) in Israel. III. Survival in earthen ponds under ambient winter temperatures. *Aquaculture,* **166** (3–4) 259–267.

Keefe A.M. & Rouse D.B. (1999) Protein requirements for juvenile Australian red claw crayfish, *Cherax quadricarinatus*. In: *Freshwater Crayfish 12* (eds M. Keller, M.M. Keller, B. Oidtmann, R. Hoffmann & G. Vogt), pp. 471–477. Weltbild Verlag, Germany.

King C.R. (1992) Growth of juvenile red-claw crayfish, *Cherax quadricarinatus*. In: *Proceedings of Aquaculture Nutrition Workshop*, Salamander Bay, 15–17 April 1991 (eds G.L. Allen & W. Dall), pp. 100–101. NSW Fisheries, Brackish Water Fish Culture Research Station, Salamander Bay, Australia.

King C.R. (1993) Egg development time and storage for redclaw crayfish *Cherax quadricarinatus* von Martens. *Aquaculture*, **109** (3–4) 275–280.

Lawrence C. & Jones C. (2001) *Cherax*. In: *Biology of Freshwater Crayfish* (ed. D.M. Holdich), pp. 635–69. Blackwell Science, Oxford, UK.

Lawrence C. & Morrissy N. (2000) *Yabby farming – frequently asked questions,* 56 pp. Fisheries Western Australia.

Lawrence C. & Whisson G. (Eds.) (2000) *Proceedings of Australian Crayfish Aquaculture Workshop,* 44 pp. International Association of Astacology, Curtin University of Technology, Perth, Australia.

Lawrence C., Morrissy N., Bellanger J. & Cheng Y.W. (1998) Final report, FRDC Project 94/075: Enhancement of yabby production from Western Australian farm dams. *Fisheries Research Report. Fisheries, Western Australia,* **112**, 1–134.

Levi T., Barki A., Hulata G. & Karplus I. (1999) Mother-offspring relationships in the red-claw crayfish *Cherax quadricarinatus*. *Journal of Crustacean Biology*, **19** (3) 477–484.

Mattei E. (1995) The red claw learning curve. *Aquaculture Magazine,* **21** (6) 20–22.

McDowall R.M. (1995) Koura. In: *Aquaculture Potential of Non-salmonid Fishes and Other Species in New Zealand Fresh Waters: an analysis of the prospects and technical issues*, pp. 30–33. NIWA Science and Technology Series No. 20, Hamilton, New Zealand.

McRae T.G. & Mitchell B.D. (1997) Control of ovarian rematuration in the yabby, *Cherax destructor* (Clark). In: *Freshwater Crayfish 11* (ed. W.T. Momot), pp. 299–310. Louisiana State University, LA, USA.

Meade M.E. & Watts S.A. (1995) Weight gain and survival of juvenile Australian crayfish *Cherax quadricarinatus* fed formulated feeds. *Journal of the World Aquaculture Society,* **26** (4) 469–474.

Merrick J.R. & Lambert C.N. (1991) *The Yabby, Marron and Red Claw: production and marketing*, 185 pp. J.R. Merrick Publications, Artarmon, Australia.

Mills B.J., Morrissy N.M. & Huner J.V. (1994) Cultivation of freshwater crayfishes in Australia. In: *Freshwater Crayfish Aquaculture in North America, Europe, and Australia* (ed. J.V. Huner), pp. 217–289. Food Products Press, an imprint of The Haworth Press, Inc. New York.

Mitchell B.D., Collins R.O. & Austin C.M. (1994) Multi-level refuge utilization by the freshwater crayfish *Cherax destructor* Clark (Decapoda:Parastacidae): a potential harvest and sampling technique. *Aquaculture and Fisheries Management,* **25**, 557–562.

Morrissy N.M. (1976) Aquaculture of marron, *Cherax tenuimanus* (Smith). 1. Site selection and the potential of marron for aquaculture. *Fish. Res. Bull. West. Aust.*, **17** (1) 1–27.

Morrissy N.M. (1992a) *An introduction to marron and other freshwater crayfish farming*, 36 pp. Fisheries Department of Western Australia.

Morrissy N.M. (1992b) Feed development for marron, *Cherax tenuimanus*, in Western Australia. In: *Proceedings of Aquaculture Nutrition Workshop.* Salamander Bay, 15–17 April 1991 (eds G.L. Allen & W. Dall), pp. 57–63. NSW Fisheries, Brackish Water Fish Culture Research Station, Salamander Bay, Australia.

Mosig J. (1999) Hothouse yabbies keep growing all winter. *Austasia Aquaculture Magazine,* **13** (5) 16–18.

O'Sullivan D. (1990) Intensive freshwater crayfish system tested. *Austasia Aquaculture Magazine*, **5** (4) 3–5.

O'Sullivan D. (1995) Techniques for semi-intensive culture of freshwater crayfish in Australia. In: *Freshwater Crayfish 10* (eds M.C. Geddes, D.R. Fielder & A.M.M. Richardson), pp. 569–582. Louisiana State University, Baton Rouge, LA, USA.

O'Sullivan D. & Kiley T. (1997a) Athmaize pellet working for crayfish farmers. *Austasia Aquaculture Magazine*, **11** (2) 27–28.

O'Sullivan D. & Kiley T. (1997b) Redclaw farm has all ingredients for success. *Austasia Aquaculture Magazine,* **11** (2) 20–23.

O'Sullivan D., Camkin J. & Lai-Koon A.C. (1994) *Reducing predation on freshwater crayfish farms*, 60 pp. Aquaculture Sourcebook No. 9, Turtle Press, Hobart, Tasmania, Australia.

Roe J. (1996) Handling and packaging of yabbies for export. *Austasia Aquaculture Magazine,* **10** (2) 58.

Rouse D.B. (1995) Australian crayfish culture in the Americas. *Journal of Shellfish Research,* **14** (2) 569–572.

Rouse D.B. & Kahn B. M. (1998) Production of Australian red

claw *Cherax quadricarinatus* in polyculture with Nile tilapia *Oreochromis niloticus*. *Journal of the World Aquaculture Society,* **29** (3) 340–344.
Rubino M., Alon N., Rouse W.D. & Armstrong J. (1990) Marron aquaculture research in the United States and the Caribbean. *Aquaculture Magazine,* **16** (3) 27–44.
Saunders K. & O'Sullivan D. (1998) Collaboration smooths yabby sales and marketing. *Austasia Aquaculture Magazine,* **12** (2) 28–30.
Semple G.P., Rouse D.B. & McLain K.R. (1995) *Cherax destructor, C. tenuimanus* and *C. quadricarinatus* (Decapoda: Parastacidae): a comparative review of biological traits relating to aquaculture potential. In: *Freshwater Crayfish 8* (ed. R.P. Romaire), pp. 495–503. Louisiana State University, LA, USA.
Smythe D.R. (1998) *New Zealand Clearwater Crayfish Ltd.* 7 pp. (mimeo). Paper presented at the New Zealand Marine Farming Association Inc. 25th Annual Conference 1998. New species forum: freshwater crayfish (koura). 27–30 August, Nelson, New Zealand.
Tysoe A. (1997) Redclaw aquaculture with water hyacinth. *Aquaculture Magazine,* **23** (4) 28–35.
Verhoef G.D. & Austin C.M. (1999) Combined effects of shelter and density on the growth and survival of juveniles of the Australian freshwater crayfish, *Cherax destructor* Clark, Part 2. *Aquaculture,* **170** (1) 49–57.
Walladge P. (1995) Yabby pump works like a dream. *Freshwater Farmer,* **2** (2–3) 90.
Whisson G.J. (1995) Growth and survival as a function of density for marron (*Cherax tenuimanus* (Smith)) stocked in a recirculation system. In: *Freshwater Crayfish 10* (eds M.C. Geddes, D.R. Fielder & A.M.M. Richardson), pp. 630–637. Louisiana State University, Baton Rouge, LA, USA.
Whisson G.J. (1997) Investigating polyculture of marron, *Cherax tenuimanus*, with silver perch, *Bidyanus bidyanus*, and an aquatic macrophyte, *Vallisneria* sp. In: *Freshwater Crayfish 11* (ed. W.T. Momot), pp. 577–586. Louisiana State University, LA, USA.
Whisson G.J. (1999) Interaction between juvenile marron (*Cherax tenuimanus*) and fingerling silver perch (*Bidyanus bidyanus*) in a recirculating tank system. In: *Freshwater Crayfish 12* (eds M. Keller, M.M. Keller, B. Oidtmann, R. Hoffmann & G. Vogt), pp. 593–603. Weltbild Verlag, Germany.
Wilson G. (2000) Good ideas from a pioneer redclaw grower. *Austasia Aquaculture Magazine,* **14** (3) 7–9.
Yeh H.S. & Rouse D.B. (1995) Effects of water temperature, density, and sex ratio on the spawning rate of red claw crayfish *Cherax quadricarinatus* (von Martins). *Journal of the World Aquaculture Society,* **26** (2) 160–164.

7.8 Clawed lobsters

7.8.1 Species of interest

American lobster (*Homarus americanus*); European lobster (*H. gammarus*). Although clawed lobsters are technically straightforward to culture, the need to individually confine them during the ongrowing phase is a serious constraint to economic viability. Current interest centres on the prospects for enhancing or extending wild stocks and ranching hatchery-reared juveniles on natural or artificial reefs.

7.8.2 Broodstock

Substantial trade in live lobsters from the east coast of North America and Western Europe often provides opportunities for the purchase of egg-bearing (berried) females for use as broodstock. In Canada and the USA, however, the possession or sale of berried females is prohibited and special dispensation would be required to utilise this source of material for a commercial hatchery.

The ease with which ovigerous females can be selected from commercial sources, and the extent of the risk of introducing disease to a broodstock facility, are very much dependent on the live handling practices of the fishermen and lobster merchants. Common practices include keeping lobsters at sea in moored keep boxes, in intertidal ponds (lobster pounds) both of which makes access and sorting difficult, or in land-based shallow tanks and tray systems which may or may not utilise recirculated natural or artificial seawater (Beard & McGregor 1991). Claws are either banded (Europe) or held closed by a small wooden peg forced into the articulation joint (parts of North America) to prevent fighting and claw loss during storage. Pegging, however, increases the risk of gaffkaemia, the most important disease of lobsters in captivity (section 2.5.5). Overcrowding and large fluctuations in water quality, e.g. low salinity, high ammonia and turbidity, are factors which can adversely affect the quality of larvae that eventually hatch, without noticeably affecting the marketability of the adult lobsters. Occasionally egg masses are heavily infested with worms and epizootic organisms which can easily spread to larvae culture systems (Aiken & Waddy 1995). Disinfection is possible with hypochlorite – *H. americanus* (Saduski & Bullis 1994) – and iodine – *H. gammarus* – although hatching may be inhibited (Uglem *et al.* 1996).

Transportation of lobsters from the fishery to the storage merchants may be done either in water (vivier transport systems) or 'dry' when lobsters are packed between layers of damp seaweed, foam or sacking (section 8.4.6). Lobsters are best cooled to below 8–10°C before packing. The effects of out-of-water transportation on egg viability are likely to be detrimental. After the eggs hatch the females may be resold for consumption. If lobsters are to be reared from wild-caught females for release in enhancement or ranching programmes (section 7.8.11) the broodstock should be taken from populations as close as practicable to the areas selected for release. This is necessary not only for sound genetic and ecological reasons but also possibly because of differing behavioural patterns (Waddy & Aiken 1998).

7.8.3 Maturation and mating

The feasibility of rearing broodstock from wild, immature and hatchery-bred lobsters to meet the demands of a putative lobster farm has been investigated in North America (Hedgecock 1983; Waddy & Aiken 1984; Aiken & Waddy 1985a; Aiken & Waddy 1995). Lobsters grown rapidly to maturity at 20°C often did not perform well as broodstock. Egg production was poor (only about 5% of *H. americanus* spawned) and attachment was weak. Males produced fewer sperm and spermatophores. Cyclic temperature regimes are essential for normal maturation, spawning and egg attachment (Waddy & Aiken 1991) but most workers conclude that at present better results are obtained from wild-caught egg-bearing females. However, the ability to condition wild pre-ovigerous females with a high degree of reliability has been developed (see below) and minimises the social and legal opposition to taking berried females from the fishery (Aiken & Waddy 1985b).

Control over mating of captive broodstock utilises either natural copulation between selected animals (which sometimes involves mating with intermoult females) or artificial insemination (Talbot & Helluy 1995). The latter offers little control, however, since egg extrusion and fertilisation follow mating, but often not for several months. Additionally, eggs spawned after artificial insemination frequently do not attach well to the female.

7.8.4 Spawning

The environment in which captive females spawn has a marked influence on the success with which the eggs are attached to the pleopods for incubation. For example, if the females do not properly position themselves on their backs, or if they do not remain so long enough for the natural adhesive to set, the newly extruded eggs fail to attach securely (Talbot 1991). Waddy (1988) reported a lengthy conditioning technique for improving egg production and attachment in *H. americanus*. Animals were held for 5 months (December to April) at 0–5°C. Spawning and attachment increased with the time spent at these winter temperatures. Unfortunately 3–4 years of normal summer–winter temperature cycles were required before most females spawned predictably, and additional time was required (3–4 spawning cycles) before proper attachment occurred reliably. No such in-depth study has been made on *H. gammarus,* but in captivity this species frequently displays an annual spawning cycle rather than the biennial rhythm characteristic of *H. americanus*.

7.8.5 Incubation

Incubation may take 4–18 months according to tempera-

Plate 7.10 Inspection of egg-bearing European lobsters housed in a communal broodstock shelter system. The lobsters claws are held closed with rubber bands to prevent damage. The roof of the shelter has been temporarily removed.

ture, and losses can occur at any time, perhaps as a result of abnormal attachment stalk formation (Waddy & Aiken 1991), adverse water quality, infestation or because of abnormal egg aeration and grooming behaviour by the female (Kuris 1991). Work with *H. gammarus* (Beard & Wickins 1992) indicated that shelters designed to be in sympathy with the behavioural requirements of wild females during incubation improved the proportion of eggs carried to full term and hence the numbers of larvae produced per female. The shelter configuration permitted natural grooming and egg care behaviour to occur with minimal disturbance from neighbouring lobsters. Exposure of eggs that are close to hatching to salinity below about 29‰ is also likely to be detrimental (Wickins *et al.* 1995a).

The development of the eggs is monitored by changes in colour and later by the size of the eye of the developing larva (Perkins 1972). By this means, the time of hatching may be predicted to within a few weeks from a small sample of eggs.

As with many other crustacean species, artificial incubation of lobster eggs removed from the female prior to hatching is possible but disease risks and consequent losses are often unacceptably high.

7.8.6 Hatching

The hatching period typically lasts 3–5 days and most, but by no means all, eggs hatch overnight. Differences in viability of larvae hatched in different seasons have been reported and it is likely that larvae hatching first from a brood are more robust than those hatching later (Eagles *et al.* 1986). In practice this indicates that only those larvae hatching during the first 2–3 days should be used in cultures (Wickins *et al.* 1995a).

The number of larvae that hatch from a female varies with female size. For example female *H. gammarus* weighing 450–1500 g may release from 800 to 13 000 larvae each (Beard & Wickins 1992). The free-swimming larvae are easily collected by allowing them to pass in a current of water from the female's incubation chambers to a separate container fitted with a 1.5 mm bar mesh screen. There the larvae may be washed with clean seawater and counted before being transferred to the larvae rearing vessels.

The hatching of eggs from wild-caught broodstock is convenient on a small scale but a major commercial venture would require at least monthly supplies of larvae for year-round production. Studies with *H. gammarus* demonstrated early progress towards this by hatching batches of larvae at regular 3-month intervals over a 4-year period (Beard *et al.* 1985). This was achieved through careful selection of broodstock females (from the north and south of Britain) with eggs at an appropriate stage of development. The females were fed at 2–3% of their body weight per day, with a 1 : 1 mixture of fresh mussel (*Mytilus*) gonad and fresh-frozen shrimp (*Crangon*). Finer manipulation of hatching time was achieved through temperature control during incubation, but some batches 'forced' at higher temperatures or held captive for several months (Wickins *et al.* 1995a) seemed to yield less viable larvae. Similar results were obtained in a larger study with wild-caught *Homarus americanus* (Waddy & Aiken 1984).

An alternative approach was tried by Hedgecock (1983) who, by photoperiod manipulation, attempted to control time of spawning rather than time of hatching. One advantage of this method was that incubation could be done at the culture temperature (provided this was constant), eliminating the need for additional, controlled low-temperature facilities. Disadvantages were a lack of flexibility, poor spawning rate (60%) and excessive egg loss. He calculated that 79 broodstock animals would be needed for the production of 80 000 lb (36 320 kg) of lobsters per month and this figure was used by Coffelt and Wikman-Coffelt (1985) in preparing a model for a battery farm unit capable of producing 1×10^6 1-lb lobsters per year. In a similar exercise based on manipulation of both spawning and incubation, Aiken and Waddy (1985a) suggested 200–800 pre-ovigerous females would be required to produce 1×10^6 marketable lobsters per year.

7.8.7 Larvae culture

The most commonly used rearing containers are based on the Hughes 40 L capacity 'kreisel' (Hughes *et al.* 1974). This is designed to maintain a homogeneous distribution of larvae and their food by means of a spiral, upwelling flow pattern. Sizes range up to 80 L capacity (Beard & Wickins 1992), rarely larger, and several may be linked together in recirculation systems (Fig. 7.4). Other simpler systems have been used with success and one, based on bivalve culture techniques, was adapted for lobsters at Cutler, Maine, USA (Beal *et al.* 1998). The method involves stocking 7500–10 000 newly hatched larvae into a vigorously aerated 400 L conical-bottomed, fibreglass tank containing newly hatched *Artemia* and a suspension of algae. The lobsters are transferred by net to another tank every other day when the seawater is renewed. Large quantities of algae are required which,

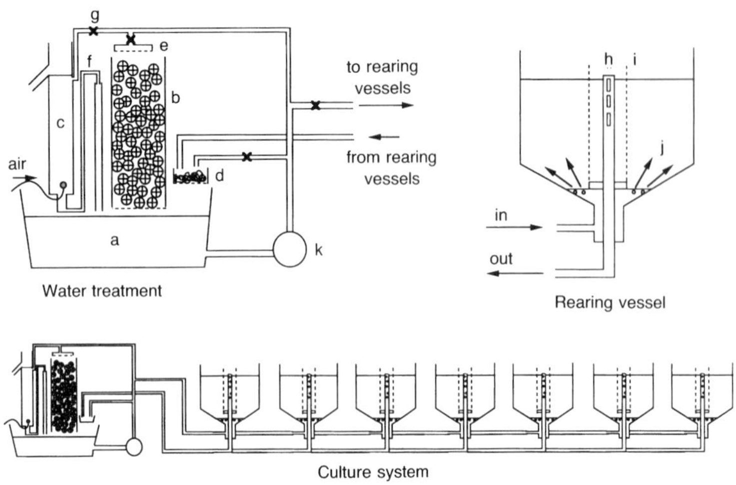

Fig. 7.4 Lobster larvae culture system: (a) reservoir; (b) biological filter; (c) foam fractionation column; (d) washable strainer; (e) sprinkler; (f) constant head pipe; (g) valve; (h) overflow; (i) screen; (j) upwelling water current; (k) pump.

together with expensive *Artemia*, adds considerably to the cost. The approach was feasible in Maine due to free community labour but may not therefore be viable elsewhere except in a research environment. Indeed the system has been successfully adapted for *Homarus gammarus* research in Ireland (Browne & Mercer 1998). Variations on these two themes exist in the UK, one of which involves the use of freeze-dried krill as food in large heavily aerated vats (K. Adamson, 1999 pers. comm.).

The larval phase in *H. gammarus* lasts 14–18 days at about 18°C and survival rates are typically 20–40%; with *H. americanus,* however, survival can be as high as 60–70% at densities of about 40 larvae L^{-1}. Metamorphosis may be spread over a 10-day period in cultured populations of either species. The diet may be live or frozen adult *Artemia* (around 48 mL per 1000 larvae daily (Eagles *et al.* 1986)), frozen mysid shrimp, krill or other zooplankton, chopped molluscs or prepared feeds. However, choices are constrained in practice since fouling, feeding regime and survival rates are intimately linked. Accumulation of suspended and dissolved organic material encourages the growth of epibiotic infestations on larval exoskeletons and can seriously interfere with moulting. This is the main problem associated with the use of prepared and non-living foods. If poor quality food is fed at the start of the culture period, recovery seldom occurs even if a good diet is subsequently fed. In this regard it is usually necessary to enhance the fatty acid (EPA and DHA) content of *Artemia* from some sources prior to use (sections 2.4.2, 7.11.1 and 7.11.2.1). Live adult *Artemia* (maintained in cultures at a concentration of about four *Artemia* per larva) is probably the best food available (Chang & Conklin 1993), although its culture is expensive, particularly when it is fed with live algae or complex compounded feeds.

A convenient alternative has proved to be frozen mysid shrimp (*Neomysis* spp.) supplemented with live, newly-hatched *Artemia* nauplii enriched with EPA and DHA for 6 h per day, on 3 days each week (Beard & Wickins 1992). During the time the nauplii are in the kreisel, flow is reduced and a fine mesh screen is temporarily placed over the outlet screen to retain the nauplii.

One of the most laborious tasks is to separate the newly developed instar 4 lobsters from the remaining instar 3 larvae. If this is not done several times each day during the period of metamorphosis, extensive cannibalism occurs. Lighting also influences survival, dim natural light being better than darkness. Heavy mortality (up to 20%) commonly occurs among larvae and post-larvae between instars 4 and 6, and studies with *Homarus gammarus* have shown that better survival rates among early juveniles can be obtained if only the earliest larvae to reach instar 4 are kept for further culture (Wickins *et al.* 1995a).

7.8.8 Nursery

In captivity, lobsters are cannibalistic throughout their larval and post-larval life. Losses are unavoidable during the larval phase, but after settlement they soon become economically unacceptable. Some authors have calculated that the losses that occur when juveniles are reared communally may be tolerable for a short period (perhaps 1–6 months post-metamorphosis), in relation to the savings made in labour and feed costs. However, communal rearing is unlikely to be commercially acceptable because the intimidation and fighting which occurs, even when shelters are provided, increases size heterogeneity and the proportion of damaged or crippled individuals in the population. Stunting may follow communal rearing, where differences in size of up to three times can occur between the biggest and smallest individuals after just 3 months (Van Olst *et al.* 1980; Aiken & Waddy 1988; Jørstad *et al.* in press). This indicates there would be a need to grade the lobsters at least once during a period of communal rearing so that small individuals could resume rapid growth. Grading, however, has proved difficult and laborious because of the need to first separate all the lobsters from their shelters. Incidentally, if hides are provided, experience indicates they must have two exits so that lobsters cannot get trapped by intruders. Shelters constructed or modified by juvenile lobsters in captivity seldom have entrance and exit aligned linearly (Wickins *et al.* 1996) although older lobsters will shelter under roof tiles.

The drive to solve the problems caused by the lobster's intolerance of crowding has led to experiments considered bizarre in some countries. Indeed, some of these indicated that periodic removal, mutilation or immobilisation of claws may be worthwhile. However, the labour costs and risks of disease following such operations seem unlikely to appeal to investors. The types of communal rearing tanks employed in laboratory and pilot studies included large 59 m^2 outdoor tanks containing a 10 cm deep layer of oyster shells (Henocque 1983), 1 m^2 laboratory tanks and 0.3 m^2 floating cages containing 5 cm shell sand and a variety of mollusc shells and cobbles as shelters (Jørstad *et al.* in press) and small aquaria containing various type of hides configured in two or three dimensions (Van Olst *et al.* 1980; Conklin & Chang 1993). The best survival obtained in Henocque's large tank system was 67% from an initial stocking density of 28 lobsters m^{-2}. After 1 year at ambient seawater temperatures the lobsters had, however, only grown to a mean weight of 5 g. The diet used was a shrimp pellet but there was also much natural production of macroalgae and benthos in the tank. In the Norwegian trials (Jørstad *et al.* in press) size variability and claw losses (16–51% losing one or both claws) increased at elevated temperatures and with disturbance.

The need to reduce production costs of juveniles reared for release in restocking or ranching programmes (sections 7.8.11, 10.6.3.1 and 10.6.3.2) makes communal rearing superficially attractive. However, healthy juveniles with a complete set of appendages will be better equipped to locate and modify defensible spaces (Wickins *et al.* 1995b) and to feed on a wider variety of prey organisms, than those damaged by fighting (Ali & Wickins 1994; Wickins *et al.* 1996).

Diets are at present very expensive and frozen or live adult *Artemia*, if used, would need to be fed in communal rearing tanks at a rate of approximately 10 g per day per 50 g of lobsters. Many natural diets do not produce growth rates as good as those achieved with live adult *Artemia* (Conklin & Chang 1993) although a supplement of lobster larvae (first instar) or fresh mussel gonad tissue significantly improves growth and survival (section 7.8.11.1).

There is great variety in the types of individual holding systems that have been built and tested with juveniles. Several of these are also suitable, when rescaled, for the further on-growing of larger lobsters (section 7.8.9). The most important features of individual compartments are that they must each receive a supply of well-oxygenated water, be self-cleaning and amenable to automatic feeding. Approaches have included blocks of mesh cages suspended in deep tanks of moving seawater containing a suspension of live *Artemia* which can pass through the mesh, and shallow trays containing mesh floored compartments suspended over or under rotating sprinkler bars (Van Olst *et al.* 1980) or subjected to a tidal rise and fall of water (Beard *et al.* 1985).

Survival in individual confinement systems is good (>80%) after about instar 6. Transfer to larger containers becomes necessary when the size of the compartment begins to restrict growth (Richards & Wickins 1979; Table 7.8).

7.8.9 Ongrowing

Most studies of ongrowing made in research and pilot units involved holding each lobster separately. In fact the whole future of battery culture (both for lobsters and for freshwater crayfish; section 7.7.6.4) depends primarily on the advent of cost-effective diets and system designs. Four interrelated factors are critical:

(1) Calculation of the minimum sizes of container which do not inhibit growth in selected sizes of lobster (see Table 7.8), and the minimum number of different container sizes (and hence transfers) that need to be employed for the 2–3 year growth period.
(2) Configuration of, and materials used for, the containers must be cost-effective.

Table 7.8 Suggested container floor areas for individually confined clawed lobsters.

Lobster size (mm CL)	Age (mo)	Area of container (cm^2)
5–10	1	25
11–25	4	115
26–40	12	310
41–60	24	620
61–85	30	1058

(3) Satisfactory removal of waste (and dead lobsters) and adequate water exchange in each container.
(4) Accurate and rapid distribution of food to each container.

The basic requirements for the sizes and numbers of transfers are known, as are the general tolerance limits of lobsters to water quality parameters (section 8.5). In addition, a variety of ingenious container/system configurations and materials have been developed and tested in a number of countries over the past 25 years. Most fulfil the requirements of space per lobster and water exchange but not all cope satisfactorily with the waste removal, feeding and cost criteria. Current thinking is that the containers will be designed in such a way as to save space, and in a limited range of sizes to reduce manufacturing costs. The design must also ensure that lobsters can be moved, or the containers expanded easily (though as infrequently as possible) during the culture period. Some of the experimental and pilot systems were described by Van Olst *et al.* (1980); Beard *et al.* (1985); McCoy (1986); Waddy (1988) but little development has occurred since the mid-1980s (Conklin & Chang 1993; Aiken & Waddy 1995). Examples include:

(1) Simple troughs divided into compartments by wooden or plastic slats and mesh screens through which water flows continuously.
(2) Rectangular tanks or troughs containing blocks of containers fitted with perforated floors through which water is distributed by specialised flushing or tidal systems.
(3) Various experimental systems tested in North America and Norway, for example:

   (a) Deep tanks containing horizontal stacks of tubes or corrugated sheets of plastic, or modular cages fitted to a supporting framework.
   (b) Deep tanks containing vertical or horizontal stacks of perforated trays. These are serviced by sequential removal of the stacks or by continuous slow rotation of trays. According to Waddy (1988), the latter represented the best design available at the time for holding large numbers of large juveniles over 25 mm CL. However, a \$5m plant incorporating this technology was partly constructed on Prince Edward Island, Canada but went into receivership before commercial viability could be established (Campbell 1989).
   (c) Shallow, round tanks containing a revolving group of floating mesh-bottomed containers. Water is jetted upwards or downwards into the containers as they revolve. (A notable example of this existed at Tiedemann's Norwegian lobster plant which had a capacity of 120 000 1-year-old lobsters; Grimsen *et al.* 1987.)
   (d) A novel, cylindrical stacked system built and tested by Sanders Associates, Inc. at Kittery, Maine; Moss Landing, California, and at Hawaii at a cost of \$3m. The company indicated that the initial capital investment would be \$10m (1986 prices) (McCoy 1986).

Of all these systems, those using horizontally stacked shallow trays of containers are perhaps the most easily constructed and serviced. They may also be the least expensive for laboratory and pilot scale studies. A compact version of the shallow tray system incorporating computer controlled tidal flushing, food injection mechanisms and water quality monitoring is being developed in Britain. It is designed to be suitable for both lobsters and freshwater crayfish (McLeod 1998).

The supply of food to individually confined lobsters involves separate techniques for small (20 mm CL, 4 months post-metamorphosis) and larger lobsters. If it can be made economically attractive to grow and feed live adult *Artemia* to juveniles up to 4 months of age, then it should be possible to house the small lobsters in mesh or perforated containers stacked horizontally or vertically in a deep-water tank in such a way that live *Artemia* would swim freely through all the chambers. From an engineer's viewpoint the design of such a system would be straightforward, but care would be needed to ensure even food distribution and to prevent fighting and subsequent limb loss through the meshes. The main consideration would be the cost of culturing the food and the labour of transferring the lobsters to different containers after 4 months. One or two further transfers would also be required.

Lobster culture in North America and Europe remains constrained by lack of commercial compounded feeds for ongrowing lobsters to market size. The development of a purified diet that supported good growth and high survival in *Homarus americanus* represented significant progress in the early 1980s and still serves as a foundation for the formulation of commercial diets (Conklin *et al.* 1983). Nevertheless, some early formulations were thought to be responsible for producing severe moulting difficulties (moult death syndrome; section 8.9); a problem studied on both sides of the Atlantic. Not all poor diets elicit an 'all or nothing' response, however, and in some cases normal moulting can be restored by the addition of frozen or fresh natural foods once or twice each week (Ali & Wickins 1994). Prototype mechanical feeders have facilitated the precise and rapid provision of both wet and dry diets to trays of individually held lobsters (Wickins *et al.* 1987), while computerised food distribution systems have been developed in Norway (Grimsen *et al.* 1987) and Great Britain (McLeod 1998).

It has been calculated (Conklin *et al.* 1983) that a unit producing one million marketable lobsters annually would require 24 mt of food per day. Storage and preparation of this food would be a major operation. As far as is known, no one has reared lobsters from metamorphosis to commercial size in 2–3 years solely, or even largely, on a compounded diet. Claims that growth rates achieved with artificial feeds over a limited period of time can be extrapolated to indicate that market sized lobsters could be obtained in this time, must therefore be regarded as speculative. Eyestalk ablation (section 2.3) and hormonal manipulation possibly through genetic engineering may hold promise for the future but progress will depend heavily on further research in these areas (Conklin & Chang 1993; Aiken & Waddy 1995).

Experiments conducted in the UK suggest that not all farmed lobsters would reach a marketable size at the same time; an important factor when estimating revenues. The observed percentage of European lobsters reaching 80 mm CL in 0.75–3.5 years is shown in Table 4.5.

Most culturists agree that the homarid lobsters are remarkably resistant to disease. Problems chiefly arise when wild-caught lobsters are crowded together (even mutilated to prevent fighting) in storage pounds. Apart from this, few serious losses have been reported among individually held juveniles or indeed larvae, which could not be explained in terms of poor water, diet or husbandry. As with many other species, the true significance of disease will probably not be known until commercial culture becomes a reality. It may be noted here that artificial seawater can be successfully used to store lobsters (Beard & McGregor 1991) but it is unlikely that maximum growth rates would occur in lobsters cultured in the simple salt solutions used in most live storage systems.

A limited form of aquaculture called pounding exists in North America in which recently moulted individuals caught in summer and autumn are held in tidal enclosures (called pounds) until they can be sold at a higher price during winter and spring. Typically they are fed salted herring or fish processing scraps although potential scarcity of these items means alternative diets are being sought (Floreto *et al.* 2000).

7.8.10 Harvesting and processing

The harvesting of cultured lobsters from indoor battery units is likely to be highly mechanised or automated. Grading according to size and appearance would be followed by packing live for shipment to market either in vivier trucks or out of water in boxes for air freight (section 8.4.6). Other processing (shelling, cooking, freezing in brine) would only be necessary for sub-size or sub-standard lobsters (section 3.3.4.1).

7.8.11 Hatchery supported fisheries, ranching

In the absence of demonstrably profitable culture operations, interest in the prospects for ranching and developing hatchery supported lobster fisheries has revived particularly in Europe (Bannister & Addison 1998; Agnalt *et al.* 1999) and North America (Gendron 1998) but also elsewhere (Addison & Bannister 1994) (section 5.7; Table 5.8). An summary of the procedures involved is presented in Fig. 7.5.

7.8.11.1 Juvenile production

In essence the techniques entail the culture of juveniles from wild-caught broodstock to either the first benthic stage (North America) (Beal *et al.* 1998; Waddy & Aiken 1998) or, in European experiments, to between 3 and 12 months from hatching (Grimsen *et al.* 1987; Beard & Wickins 1992; Browne & Mercer 1998; Latrouite 1998). The juveniles are then taken to sea and released onto carefully selected lobster grounds (section 7.8.11.3). It is expensive to grow juveniles for as long as 3 months but trials in which younger, instar 6–8 juveniles (6–10 mm CL; 1–2 months in the hatchery) were tagged and released either showed poor survival (Cook 1995) or took 7 years to reach minimum legal landing size – 85 mm CL

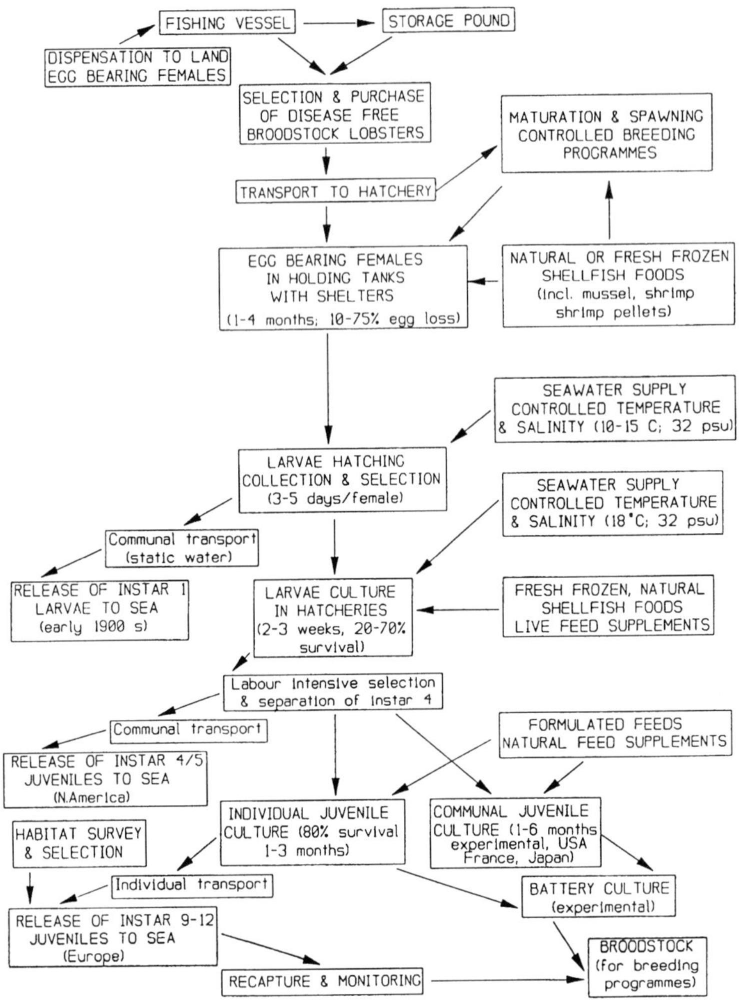

Fig. 7.5 Steps in the production and release of clawed lobsters (from Wickins 1997). Crown copyright, 1997 reproduced with the permission of CEFAS.

– (Burton 1998) compared to the 4–5 years taken by lobsters released at 3 months of age (instar 10–12; 15 mm CL) (Bannister *et al.* 1994). There is clearly a trade-off between the cost of producing juveniles and the time during which the lobsters are exposed to the rigors of the natural environment (section 10.6.3.3).

Culture methods are almost identical to those described above (Wickins *et al.* 1995b), although research in the UK showed juveniles could be given special treatment to ensure the best chance of survival on the seabed. Specifically, a daily diet of frozen mysid shrimp or prawn pellets was supplemented on only 2 days each week with either fresh-frozen lobster larvae or fresh mussel gonad tissue. The supplements improved survival (83–100%) and growth (5–6 mm CL to 10–12 mm CL in 77–88 days), enhanced their natural pigmentation, increased moulting success and reduced their sensitivity to handling (Ali & Wickins 1994). The addition of live bivalve spat (2–4 mm) stimulated rapid crusher claw development, thus increasing the range of food items accessible to each lobster upon release (Wickins 1986). In Norway, shell sand was used to encourage differentiation of the claws (Van der Meeren & Uglem 1998).

7.8.11.2 Tagging

The ability to distinguish hatchery-reared lobsters from their wild counterparts is essential if the success of any

Plate 7.11 The mass culture of clawed lobster larvae in modified Hughes 'kreisels' at the former CEFAS Conwy Laboratory, UK.

pilot release is to be evaluated. Three methods have been used to distinguish released animals:

(1) Release outside the normal geographic range of the species. For example, *Homarus americanus* released in Japanese waters (Henocque 1983). Any lobster subsequently caught is easily recognised and must have come from the release programme.
(2) Culture of hybrids (*H. americanus* × *H. gammarus*) or unusually coloured lobsters (Beal *et al.* 1998). Recaptured hybrids would be moderately easy to identify, colourmorphs more so.
(3) Tagging with an internal, coded microwire tag (Wickins *et al.* 1986; Uglem & Grimsen 1995). With this method, however, it is impossible to recognise a tagged animal without specialised detection equipment, and a rigorous recapture sampling approach is essential (Bannister *et al.* 1994). Microwire tagging requires expensive equipment and until recently extracting the tag to read the code meant killing the animal. A visible, alphanumeric encoded microtag (section 8.10.1.1) designed for insertion in the clear tissues of fish was successfully used in juvenile (10–13 mm CL) spiny lobster *Jasus edwardsii*. Tags were implanted directly beneath the intersternal membranes of the ventral abdomen and could be read under water (Edmunds 1992). However, attempts to use internal visible elastomer tags with clawed lobsters have yet to prove satisfactory (Linnane & Mercer 1998).

Releases under the first two techniques may be made using newly metamorphosed instar 4 or 5 juveniles, but under the third, consistent placement of micro-tags in such small individuals may be hard to achieve. The use of half-sized tags for lobsters of 4–6 mm CL is possible but requires further development (Burton *et al.* 1995).

7.8.11.3 Transport and release

Juveniles are transported individually in perforated tubes placed in tanks or plastic bags of seawater, or in compartmented trays contained in tanks of recirculating seawater (Beard & Wickins 1992; Wickins *et al.* 1995b). Juveniles released in Norwegian waters after transportation in damp seaweed or newspaper did not behave normally for some time after reaching the seabed and suffered undue predation (Van der Meeren 1991). In general, out-of-water transportation prior to release is not advisable (Linnane *et al.* 1997).

For maximum survival, it is vital that lobsters are released directly onto suitable seabed substrate (Howard 1988). Surveys are conducted using high-resolution sonar and divers to locate seabed areas sufficiently heterogeneous to provide the newly released juvenile lobsters with immediate shelter from currents as well as with defensible crevices. This is because a lobster must construct a burrow system that will sustain it for the next 2 or so years so that, provided the burrow meets all its shelter and food requirements, it can remain underground thereby reducing the risk of undue mortality arising from predation and competition (Wickins *et al.* 1996). A cobble-boulder field on cohesive clay or mud is ideal (Wahle & Steneck 1991) (section 7.8.12). It is also advantageous to release juveniles in, or close to, an area of good but sparsely inhabited, adult habitat, e.g. close to a natural boulder/scree outcrop or a new artificial reef so that after about 2–3 years when they increasingly emerge to forage, a wide range of suitably sized crevices will be available to them (Wickins 1999) (sections 7.8.12, 8.11.2 and 12.7).

Releases are most reliably accomplished by divers or, where extensive, contiguous areas of suitable substrate exist, through a pipe reaching from a surface vessel to the sea bed (Fig. 7.6; Burton 1992; Cook 1995; Wickins *et al.* 1995b). In some Norwegian habitats, surface releases in shallow water at times when predator density is low have been successful (Van der Meeren *et al.* 1998) but in the French trials, lobsters were sealed in individual tubes fitted with a soluble paper cap which dissolved soon after reaching the seabed (Latrouite 1998).

The significance of a high level of fitness at release is now widely recognised (Svåsand *et al.* 1998). Small juveniles reared in mesh-bottomed compartments instinctively seek shelter and commence burrowing within minutes of release to a suitable substrate. Laboratory studies indicated that the provision of pre-release burrowing or shelter seeking experience was probably unnecessary for animals of up to 3 months of age (Wickins & Barry 1996). Of greater importance to survival upon release was likely to be bearing a full set of appendages, good pigmentation and high levels of internal food reserves (Ali & Wickins 1994), stress-free transportation (Van der Meeren 1991) and prior acclimation to seabed temperatures, which greatly increased the speed with which they constructed defensible burrows (Van der Meeren 1993; Wickins unpublished data).

7.8.11.4 Monitoring

The use of suction sampling techniques to monitor survival of juveniles within 1–2 years of release has proved successful in North America but not in European waters where neither wild nor released juveniles could be found (Wahle 1995). Thus when tagged lobsters are released during the pilot study phase, a sound strategy for monitoring recaptures five or more years later is essential for project viability to be accurately assessed and to help with deciding the appropriate scale of operation (section 10.6.3.3; Bannister *et al.* 1994; Wickins 1997; Agnalt *et al.* 1999). Once success has been established it would not normally be necessary to tag further juveniles released in that area.

7.8.12 Habitat modification

Habitat modification can be of particular relevance to

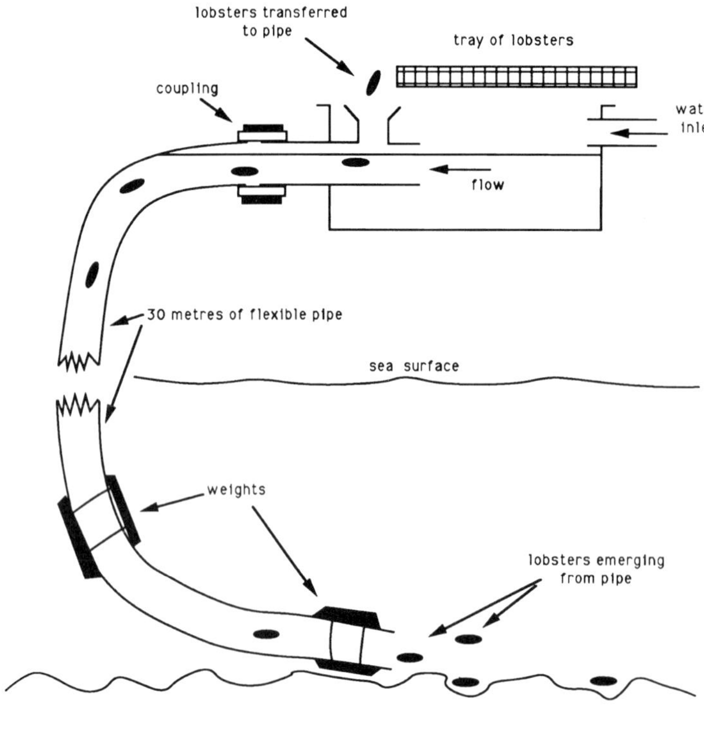

Fig. 7.6 The arrangement used to release juvenile lobsters directly on to suitable seabed habitat through a 30 m flexible pipe. (Redrawn from Beard & Wickins 1992 and adapted from the original idea of W. Cook and P. Oxford, North Western and North Wales Sea Fisheries Committee.)

lobster stock enhancement programmes for two reasons. Firstly, when it is appropriate to release juveniles in, or close to, an area of good but sparsely inhabited, adult habitat, e.g. close to a natural boulder/scree outcrop or a new artificial reef but where there is no nearby natural nursery ground in which juvenile clawed lobsters can safely begin burrow construction. Secondly, and arguably more importantly from an economic standpoint, the deployment of attractive, starting points for burrowing (e.g. artificial shelters) should give much smaller (and hence cheaper to produce) clawed lobster juveniles a better chance of survival than when they are released onto natural habitat already fully occupied by burrowing predators and competitors with similar habitat requirements. If the shelters were made of natural or biodegradable material, many potential objections to their deposition on the seabed might be overcome. Future trials may indicate that it is uneconomic or impractical to provide artificial habitat on a scale sufficient for enhancement. In this case, a less controllable alternative would be to increase environmental complexity over wholesome and penetrable substrates through the addition of natural materials such as cobbles and boulders obtained, for example, from graded channel or harbour dredgings.

Research on the prospects for making natural habitat more suitable for newly released lobsters showed that juvenile lobsters burrowing in cohesive, muddy substrates utilised hard surfaces, such as embedded stones, to act as a starting point for excavation. In one series of experiments, such surfaces were provided both in form of cobbles and as artificial shelters. These shelters were simply short lengths of box section conduit pushed into the mud substrate. One end was cut at an angle to provide a defensible 'letter-box' shaped opening close to the mud surface. The shelters gave the lobsters immediate access to a penetrable but cohesive substrate in which they readily constructed burrows. The shelters proved extremely attractive, with 92% of the 25 burrows built in mixed cobble/shelter habitat being started within the shelters (Wickins 1999). The lobsters also constructed significantly more burrows within shelters situated where the current was likely to be greatest (up to 18 $cm\,s^{-1}$ at the mud surface) thus enhancing the burrow's hydrodynamic features and minimising the lobster's energy expenditure on creating a flow within the burrow.

Studies have identified survival, from the time of settlement (or release) to the time of emergence from the shelter to begin foraging (at 2–3 years of age), as a crucial factor in determining economic stocking and fishery management strategies. Therefore, successful stock enhancement of clawed lobsters requires the identification, and whenever possible optimisation, of the factors affecting the behaviour of juveniles critical to this cryptic period of the life cycle. In any lobster stock enhancement programme conducted on an artificial reef or in an otherwise suitable habitat area where natural settlement does not occur, the fished stock would be continuously replenished by 2–3-year-old juveniles emerging from their burrows to occupy reef crevices and grow to a harvestable size. In turn, these nursery burrows would need to be continuously restocked from a hatchery. For hatchery-reared juveniles to survive and remain in the nursery area, their ecological requirements must be met in full. Any deficit in the number of juveniles recruiting to the fished stock would, in the long term, compromise the harvest.

7.8.13 References

Addison J.T. & Bannister R.C.A. (1994) Re-stocking and enhancement of clawed lobster stocks: a review. *Crustaceana*, **67** (2) 131–155.

Agnalt A.-L., Van der Meeren G.I., Jørstad K.E., *et al.* (1999) Stock enhancement of European lobster (*Homarus gammarus*): a large scale experiment off south-western Norway (Kvitsøy). In: *Stock Enhancement and Sea Ranching* (eds B.R. Howell, E. Moksness & T. Svåsand), pp. 401–419. Fishing News Books, Oxford, UK.

Aiken D.E. & Waddy S.L. (1985a) The uncertain influence of spring photoperiod on spawning in the American lobster (*Homarus americanus*). *Canadian Journal of Fisheries and Aquatic Science*, **42** (1) 194–197.

Aiken D.E. & Waddy S.L. (1985b) Production of seed stock for lobster culture. *Aquaculture*, **44** (2) 103–114.

Aiken D.E. & Waddy S.L. (1988) Strategies for maximizing growth of communally reared juvenile American lobsters. *World Aquaculture*, **19** (3) 61–63.

Aiken D.E. & Waddy S.L. (1995) Aquaculture. In: *Biology of the Lobster* Homarus americanus (ed. J.R. Factor), pp. 153–175. Academic Press, New York.

Ali Y. & Wickins J.F. (1994) The use of fresh food supplements to ameliorate moulting difficulties in lobsters – *Homarus gammarus* (L.) – destined for release to the sea. *Aquaculture and Fisheries Management*, **25**, 483–496.

Bannister R.C.A. & Addison J.T. (1998) Enhancing lobster stocks: a review of recent European methods, results, and future prospects. *Bulletin of Marine Science*, **62** (2) 369–387.

Bannister R.C.A., Addison J.T. & Lovewell S.R.J. (1994) Growth, movement recapture rate and survival of hatchery-reared lobsters (*Homarus gammarus* (Linnaeus, 1758)) released in the wild on the English east coast. *Crustaceana*, **67** (2) 156–172.

Beal B.F., Chapman S.R., Irvine C. & Bayer R.C. (1998) Lobster (*Homarus americanus*) culture in Maine: a community-based, fishermen-sponsored public stock enhancement program. *Canadian Industry Report of Fisheries and Aquatic Sciences,* **244,** 47–54.

Beard T.W. & McGregor D. (1991). Storage and care of live lobsters, 33 pp. *Lab. Leafl.* (66). MAFF Directorate Fisheries Research, Lowestoft, UK.

Beard T.W. & Wickins J.F. (1992) Techniques for the production of juvenile lobsters (*Homarus gammarus* (L.)), 22 pp. *Fisheries Research Technical Report* (92). MAFF Directorate Fisheries Research, Lowestoft, UK.

Beard T.W., Richards P.R. & Wickins J.F. (1985) The techniques and practicability of year-round production of lobsters, *Homarus gammarus* (L.) in laboratory recirculation systems, 22 pp. *Fisheries Research Technical Report* (79). MAFF Directorate Fisheries Research, Lowestoft, UK.

Browne R. & Mercer J.P. (1998) The European clawed lobster (*Homarus gammarus*): stock enhancement in the Republic of Ireland. *Canadian Industry Report of Fisheries and Aquatic Sciences,* **244,** 33–41.

Burton C.A. (1992) *Techniques of lobster stock enhancement*, 51 pp. Seafish Report No. 396. Sea Fish Industry Authority, Edinburgh, UK.

Burton C. (1998) Lobster stocking, update 1998, 2 pp. *Seafish Aquaculture Shellfish Bulletin*. Sea Fish Industry Authority, Edinburgh, UK.

Burton C.A., Beard T.W., Wickins J.F. Cook W. & Bannister R.C.A. (1995) United Kingdom lobster stock enhancement programme. *NMT Network News,* **2** (3) 1–2.

Campbell M. (1989) Prince Edward Island: lobster culture technology on hold. *Atlantic Fish Farming,* 20 August 1989 (page unknown).

Chang E.S. & Conklin D.E. (1993) Larval culture of the American lobster (*Homarus americanus*). In: *Handbook of Mariculture,* 2nd edn, *Vol. 1 Crustacean aquaculture* (ed. J.P. McVey), pp. 489–495. CRC Press, Boca Raton, FL, USA.

Coffelt R.J. & Wikman-Coffelt J. (1985) Lobsters: one million one pounders per year. *Aquacultural Engineering,* **4** (1) 51–58.

Conklin D.E. & Chang E.S. (1993) Culture of juvenile lobsters (*Homarus americanus*). In: *Handbook of Mariculture,* 2nd edition, *Vol. 1 Crustacean aquaculture* (ed. J.P. McVey), pp. 497–510. CRC Press, Boca Raton, FL, USA.

Conklin D.E., D'Abramo L.R. & Norman-Boudreau K. (1983) Lobster nutrition. In: *Handbook of Mariculture, Vol. 1 Crustacean aquaculture* (ed. J.P. McVey), pp. 413–423. CRC Press, Boca Raton, FL, USA.

Cook W. (1995) *A lobster stock enhancement experiment in Cardigan Bay, final report,* 33 pp. North Western and North Wales Sea Fisheries Committee, Lancaster, UK.

Eagles M.D., Aiken D.E. & Waddy S.L. (1986) Influence of light and food on larval American lobsters, *Homarus americanus*. *Canadian Journal of Fisheries and Aquatic Science,* **43**, 2303–2310.

Edmunds M. (1992) Visible implant microtags. *The Lobster Newsletter,* **5** (2) 13.

Floreto E.A.T., Bayer R.C. & Brown P.B. (2000) The effects of soybean-based diets, with and without amino acid supplementation, on growth and biochemical composition of juvenile American lobster, *Homarus americanus*. *Aquaculture,* **189** (3–4) 211–235.

Gendron L. (ed.) (1998) Proceedings of a workshop on lobster stock enhancement held in the Magdalen Islands (Québec) from 29–31 October 1997. *Canadian Industry Report of Fisheries and Aquatic Sciences,* **244** xi & 135 pp.

Grimsen S., Jaques R.N., Erenst V. & Balchen J.G. (1987) Aspects of automation in a lobster farming plant. *Modeling, Identification and Control,* **8** (1) 61–68.

Hedgecock D. (1983) Maturation and spawning of the American lobster, *Homarus americanus*. In: *Handbook of Mariculture, Vol. 1 Crustacean aquaculture* (ed. J.P. McVey), pp. 261–286. CRC Press, Boca Raton, FL, USA.

Henocque Y. (1983) *Techniques d'élevage du homard et experience d'implantation au Japon,* 84 pp. Unpublished report to La Maison Franco-Japonaise.

Howard A.E. (1988) Lobster behaviour, population structure and enhancement. *Symposia. Zoological Society of London,* **59,** 355–364.

Hughes J.T., Shleser R.A. & Tchobanoglous G. (1974) A rearing tank for lobster larvae and other aquatic species. *Progressive Fish-Culturist,* **36** (3) 129–132.

Jørstad K.E., Agnalt A-L., Kristiansen T. & Nøstvold E. (in press) High survival and growth of European lobster juveniles (*Homarus gammarus*), reared communally on a natural bottom substrate. *Marine and Freshwater Research.*

Kuris A.M. (1991) A review of patterns and causes of crustacean brood mortality. In: *Crustacean Egg Production* (eds A. Wenner & A. Kuris), pp. 117–141. A.A. Balkema, Rotterdam, Netherlands.

Latrouite D. (1998) The French experience with enhancement of European lobster *Homarus gammarus*. *Canadian Industry Report of Fisheries and Aquatic Sciences,* **244,** 55–58.

Linnane A. & Mercer J.P. (1998) A comparison of methods for tagging juvenile lobsters (*Homarus gammarus* L.) reared for stock enhancement. *Aquaculture,* **163** (3–4) 195–202.

Linnane A., Uglem I., Grimsen S. & Mercer J.P. (1997) Survival and cheliped loss of juvenile lobsters *Homarus gammarus* during simulated out-of-water transport. *Progressive Fish-Culturist,* **59** (1) 47–53.

McCoy H.D. II (1986) Intensive culture systems past, present and future. Part 1, *Aquaculture Magazine,* **12** (6) 32–35.

McLeod D. A. (1998) Future prospects for crustacean farming in Europe. *FAO Eastfish Magazine,* (5–6) 43–44.

Perkins H.C. (1972) Developmental rates at various temperatures of embryos of the Northern lobster (*Homarus americanus* Milne-Edwards). *Fishery Bulletin of the National Oceanic and Atmos. Adm. (U.S.),* **70,** 95–99.

Richards P.R. & Wickins J.F. (1979) Ministry of Agriculture, Fisheries and Food. *Lobster culture research*, 33 pp. *Lab. Leafl.* (47). MAFF Directorate Fisheries Research, Lowestoft, UK.

Saduski T.J. & Bullis R.A. (1994) Experimental disinfection of lobster eggs infected with *Leuchothrix mucor*. *Biological Bulletin,* **187**, 254–255.

Svåsand T., Skilbrei O.T., Van der Meeren G.I. & Holm M. (1998) Review of morphological differences between reared and wild individuals: implications for sea-ranching of Atlantic salmon, *Salmo salar* L., Atlantic cod, *Gadus morhua* L., and European lobster, *Homarus gammarus*. *Fisheries Management and Ecology,* **5**, 473–490.

Talbot P. (1991) Ovulation, attachment and retention of lobster eggs. In: *Crustacean Egg Production* (eds A. Wenner & A.

Kuris), pp. 9–18. A.A. Balkema, Rotterdam, Netherlands.
Talbot P. & Helluy S. (1995) Reproduction and embryonic development. In: *Biology of the Lobster* Homarus americanus (ed. J.R. Factor), pp. 177–216. Academic Press, New York.
Uglem I. & Grimsen S. (1995) Tag retention and survival for lobster juveniles (*Homarus gammarus*(L.)) marked with coded wire tags. *Aquaculture Research*, **26**, 837–841.
Uglem I., Uksnøy L.E. & Bergh Ø. (1996) Chemical treatment of lobster eggs against epibiotic bacteria. *Aquaculture International,* **4**, 1–8.
Van der Meeren G.I. (1991) Out-of-water transportation effects on behaviour in newly released juvenile Atlantic lobsters *Homarus gammarus*. *Aquacultural Engineering,* **10**, 55–64.
Van der Meeren G.I. (1993) Initial response to physical and biological conditions in naïve juvenile Atlantic lobsters *Homarus gammarus* L. *Marine Behaviour and Physiology,* **24**, 70–92.
Van der Meeren G.I. & Uglem I. (1998) Lobster stock enhancement in Norway, with emphasis on a large-scale release project at Kvitsy. *Fisken og Havet*, (13) 83–89.
Van der Meeren G.I., Agnalt A-L. & Jørstad K.E. (1998) Lobster (*Homarus gammarus*) stock enhancement in Norway, experiences from large-scale release project, from 1990 to 1997. *Canadian Industry Report of Fisheries and Aquatic Sciences,* **244,** 63–68.
Van Olst J.C., Carlberg J.M. & Hughes J.T. (1980) Aquaculture. In: *The Biology and Management of Lobsters, Vol. 2, Ecology and management* (eds J.S. Cobb & B.F. Phillips), pp. 333–384. Academic Press, London.
Waddy S.L. (1988) Farming the Homarid lobsters: state of the art. *World Aquaculture,* **19** (4) 63–71.
Waddy S.L. & Aiken D.E. (1984) Broodstock management for year-round production of larvae for culture of the American lobster, 14 pp. *Canadian Technical Report of Fisheries and Aquatic Sciences* (1272).
Waddy S.L. & Aiken D.E. (1991) Egg production in the American lobster, *Homarus americanus*. In: *Crustacean Egg Production* (eds A. Wenner & A. Kuris), pp. 267–290. A.A. Balkema, Rotterdam, Netherlands.
Waddy S.L. & Aiken D.E. (1998) Lobster (*Homarus americanus*) culture and resource enhancement: the Canadian experience. *Canadian Industry Report of Fisheries and Aquatic Sciences,* **244,** 9–18.
Wahle R.A. (1995) A trans-atlantic perspective on *Homarus* recruitment. *The Lobster Newsletter,* **8** (1) 1–3.
Wahle R.A. & Steneck R.S. (1991) Recruitment habitats and nursery grounds of the American lobster *Homarus*: a demographic bottleneck? *Marine Ecology Progress Series,* **69**, 231–243.
Wickins J.F. (1986) Stimulation of crusher claw development in cultured lobsters, *Homarus gammarus* (L.). *Aquaculture and Fisheries Management*, **17**, 267–273.
Wickins J.F. (1997) Strategies for lobster cultivation. *CEFAS Shellfish News,* (4) 6–10.
Wickins J.F. (1999) *Lobster behaviour and stock enhancement*, 4 pp. CEFAS, Lowestoft, UK.
Wickins J.F. & Barry J. (1996) The effect of previous experience on the motivation to burrow in early benthic phase lobsters (*Homarus gammarus* (L.)). *Marine and Freshwater Behaviour and Physiology,* **28,** 211–228.
Wickins J.F., Beard T.W. & Jones E. (1986) Microtagging cultured lobsters for stock enhancement trials. *Aquaculture and Fisheries Management*, **17**, 259–265.
Wickins J.F., Jones E., Beard T.W. & Edwards D.B. (1987) Food distribution equipment for individually held juvenile lobsters, *Homarus* sp. *Aquacultural Engineering*, **6** (3) 277–288.
Wickins J.F., Beard T.W. & Child A.R. (1995a) Maximizing lobster, *Homarus gammarus* (L.), egg and larval viability. *Aquaculture Research*, **26**, 379–392.
Wickins J.F., Bannister R.C.A., Beard T.W. & Howard A.E. (1995b) A video record of techniques used to rear, tag release and monitor recaptures of hatchery-reared lobsters, 42 min. *Fisheries Research Video Rept.* (1) MAFF Directorate Fisheries Research, Lowestoft, UK.
Wickins J.F., Roberts J.C. & Heasman M.S. (1996) Withinburrow behaviour of juvenile European lobsters *Homarus gammarus* (L.). *Marine and Freshwater Behaviour and Physiology*, **28**, 229–253.

7.9 Spiny lobsters

7.9.1 Species of interest

Caribbean spiny lobster (*Panulirus argus*); western rock lobster (*P. cygnus*); scalloped spiny lobster (*P. homarus*); California spiny lobster (*P. interruptus*); Japanese spiny lobster (*P. japonicus*); ornate rock lobster (*P. ornatus*); mud spiny lobster (*P. polyphagus*); painted spiny lobster (*P. versicolor)*; southern or red rock lobster (*Jasus edwardsii*); packhorse or green rock lobster (*J. verreauxi*); common (European) crawfish (*Palinurus elephas*); Mediterranean slipper lobster (*Scyllarides latus*) and mud lobster or Moreton Bay bug (*Thenus orientalis*). At present the prospects for the commercial culture of most of these lobsters are constrained by two factors: the technical difficulties of rearing the phyllosome larvae (Fig. 2.3c) and the unpredictable or restricted availability of wild pre-juveniles (pueruli) or juveniles for fattening or ongrowing (sections 7.9.4 and 7.9.5). However it has recently been reported (Australian Fresh Corporation 2000) that *Thenus* spp. have been reared from hatchery-produced larvae to market size in a collaborative research project in Queensland, Australia (section 12.8.5). Prospects for attracting spiny lobsters and their juveniles to selected seabed areas by providing artificial habitats are being investigated in several countries interested in developing ranching programmes (Briones-Fourzán *et al.* 2000; section 5.7.2).

7.9.2 Broodstock, incubation and hatching

Little difficulty is expected in obtaining and transporting sufficient wild-caught broodstock to supply future commercial hatcheries. Methods are similar to those used for homarid lobsters (section 7.8.2). Mature spiny lobsters can be successfully maintained on diets of bivalve mollusc (mussels) and crustacean flesh; indeed the general husbandry techniques and conditions that are applied to *Homarus* broodstock are also likely to be suitable for spiny lobsters. Maturation and spawning frequencies may be enhanced in some species by increasing photoperiod and temperature (MacDiarmid & Kittaka 2000). Mating, spawning, incubation and hatching readily occur among captive broodstock of a number of species. Size does matter since large male *Jasus edwardsii* and *Panulirus argus* can fertilise more females than small males, and females mated by large males bear larger clutches of eggs. Male *Jasus edwardsii* can fertilise up to 12 females in a single spawning season (Gibson & Frusher 1997). Multiple copulations are common and in *Jasus*, egg extrusion occurs immediately after successful mating. Delay in mate availability, particularly in *Jasus*, by as little as 5–10 days after a female is ready to mate depresses clutch size and emphasises the need for large, active males to be present in broodstock tanks (MacDiarmid & Butler 1999; MacDiarmid *et al.* 1999a). Hatching occurs over 3–5 days and at this time females need to stand on the tips of the pereopods, often on an elevated structure, to aid larvae dispersal. We are not aware of any detailed studies on which expectations of hatchery performance could be based.

7.9.3 Larvae culture

The main problems with rearing the larvae arise from the prolonged larval life, the unusual larval morphology and a fundamental lack of information on larval diet and feeding habits (Mikami *et al.* 1994). Current knowledge of the phyllosoma digestive system was reviewed by Mikami and Takashima (2000). Japanese researchers were the first to rear a palinurid lobster from egg to puerulus (for historical summary, see Kittaka 2000), but the effort involved during the lengthy larval phase (often over 300 days) suggests that it would be very difficult to maintain hygienic culture conditions for so long in a commercial hatchery. Of particular importance is an ample exchange of good quality water throughout the culture period (Shioda *et al.* 1997; Igarashi & Kittaka 2000).

The species of spiny lobster cultured so far and the duration of their larval lives are: *Panulirus japonicus* (304–391 days), a warm-temperate water species (25°C); *Jasus lalandii* (306 days); *J. novaehollandiae* (319 days); *J. verreauxi* (189–359 days); *J. edwardsii* (212–319 days) and *Palinurus elephas* (64–132 days), all cool temperate species (20°C), and a *Jasus novaehollandiae* x *J. edwardsii* hybrid, both of which are now thought to be the same species (Kittaka 2000). Few cultured phyllosoma larvae have reached or lived beyond the puerulus stage. The larvae are cultured in 40–100 L capacity conical bottomed vessels in upwelling seawater, not unlike the vessels used in the culture of *Homarus* larvae (Kittaka 2000). Some through-flow systems use water purified by filtration and ultraviolet irradiation, others employ recirculation and additions of algae. For example, the alga *Nannochloropsis* is used in Japan at initial densities of $1–2 \times 10^6$ cells mL^{-1} to control bacteria and maintain water quality. Best survival of phyllosomas was obtained when the water was renewed every 13–14 days (Shioda *et al.* 1997). At high culture densities the long appendages of the phyllosomes readily become entangled with debris suspended in the water (cast shells, filamentous bacteria and algae) and particularly high standards of husbandry are essential. Many diets have been tested (Ritar 1999; Kittaka 2000) but the best are newly hatched and partially grown *Artemia*, *Sagitta* spp., various fish larvae and mussel flesh. Phyllosomas often cling to the sides of the vessel but we do not know if they graze when in this position, or if their survival could be improved through the provision of additional flat surfaces. *Jasus edwardsii* larvae are attracted to and prey on jellyfish rich in HUFAs (section 2.4.2) and which have a similar amino acid composition (Kittaka 1997).

The phyllosome larvae of *Thenus* spp. are considerably easier to rear since there are only four instars and a nisto stage which are completed in 40 days (section 12.8.5).

7.9.4 Nursery

The newly metamorphosed pre-juvenile or puerulus stage of spiny lobsters, and the equivalent nisto stage of slipper lobsters, do not usually feed but rely on nutrient reserves built up during the late phyllosoma stages (Mikami & Takashima 1993; Lemmens 1994). The puerulus lasts from about 13–14 days in *Palinurus elephas* and *Panulirus japonicus* to 25 days in *Jasus verreauxi* before metamorphosing to the first juvenile stage. New Zealand

Plate 7.12 *Jasus edwardsii* pueruli collected from the wild for ongrowing. The dark colour is due to the pigmentation of the new exoskeleton developing under the transparent exoskeleton of the puerulus. (Photo courtesy A. Blacklock, National Institute of Water and Atmospheric Research, New Zealand.)

trials with wild-caught *Jasus edwardsii* pueruli grown on in cages indicate growth and survival rates generally similar to those achieved in land-based systems (Jeffs & James 2000). Feeding does not seem to be a problem, for juveniles will accept live *Artemia* between their first and second moults and thereafter will take artificial diets. Growth is from 2 g to 10 g in 4 months (Cray Corporation Ltd 1994; Thomas *et al.* 1998) and captured juveniles can be grown at 90 m^{-2} with up to 98% survival (Table 4.4).

The pueruli of some species, for example *Panulirus argus*, congregate in large numbers in suitable habitat (mangrove roots, seaweeds) and can be caught in artificial floating 'bushes' or 'seaweed habitat traps' that resemble the clumps of algae that the pueruli seek for refuge (Marx & Herrnkind 1985). Various other types of collector have also been developed (Phillips & Booth 1994). In Tasmania, for example, they include mesh collectors consisting of synthetic netting compacted into a cubical frame and crevice collectors which can be a spaced stack of plywood sheets (Thomas *et al.* 1998). In India, Mangalore tiles suspended near the seabed are used. Pueruli also commonly occur in bivalve farms on mussel ropes and in oyster trays. Collectors are widely used for fishery catch predictions and research as well as for aquaculture purposes (Phillips & Booth 1994). Captured pueruli and early juveniles may then be transferred to protected seabed areas or to farms for ongrowing (section 5.7.2). One Australian study (see Van Olst *et al.* 1980) determined that the cost of culture plus the expense of capture of the pueruli made the price of the farmed animal twice that of the fished product. In contrast, a later study (Meagher 1994 – *apud* Linton 1998) claimed that capture and collection of *P. cygnus* could be profitable and provide an internal rate of return of 24.5% (see also section 10.6.4). Experience to date has shown, however, that concentration and reliability of occurrence of the young stages in many regions are inadequate for commercial exploitation and in many cases collection for aquaculture is prohibited. Yet in New Zealand, where much is known of larval recruitment patterns (Booth & Phillips 1994), a quota trade-off agreement has been reached which means that for every tonne of catch quota retired from the fishery, 40 000 juveniles may be taken locally for ongrowing or fattening (Booth *et al.* 1999). A similar scheme considered for Tasmania (Anon. 1997) led to an agreement that 25% of all pueruli taken will be returned to the fishery after one year of growth (van Barneveld 2001). Other such schemes will no doubt be proposed elsewhere in the future.

7.9.5 Ongrowing

In Taiwan wild-caught juveniles ranging in size from 0.5 g to 250 g are grown to 300–800 g in small 200 m^2 ponds. On average, animals stocked at 25 g mean weight reach 330 g in 16 months. Growth rate is heterogeneous and animals are graded every 2–3 months. About five to six ponds are needed to accommodate this practice. Water exchange is about 10% per day and survival about 80%. The largest Taiwanese farm is reported to have 13 ha of ponds and the capacity to produce 150 000 marketable lobsters per year (Chen 1990). Juvenile spiny lobsters readily accept and grow well on natural foods such as abalone, mussels, crab and squid. Hard, compounded pellets are also utilised and a water stable extruded diet based on soybean meal has been developed both for cultured *P. argus* and for use as trap bait (Brown *et al.* 1995).

Several other methods for ongrowing have been proposed elsewhere and in India it is reported that *Panulirus homarus* and *P. ornatus* could be grown profitably in 5–10 mt concrete tanks from 50–100 g to 250 g in 5 months at initial stocking densities of 25 m^{-2} (Rahman & Srikashnadhas 1994). In *P. ornatus*, promising growth rate increases without reduction in survival have been obtained through unilateral eyestalk ablation (sections 2.3 and 11.2.5). Fattening of *P. argus* is being investigated in Brazil to encourage local fishermen to diversify their activities in the face of a declining fishery (Table 5.2c). Sub-legal sized lobsters are held in shallow, 500 m^2 sea pens for 3–4 months and fed locally available mussels and fish. Initial mortality in early trials was high (30%) possibly due to stresses of capture, transport, excessive sunlight and an inadequate diet (Assad 1998); all factors which are characteristic of many preliminary aquaculture trials.

Since 1983 wild-caught juvenile *Panulirus polyphagus* have been increasingly cultured (fattened) in floating fish farm cages in Singapore (Table 5.2c). The cages are typically wooden framed, 2–5 m square and about 2–3 m deep. Regular cleaning of the synthetic mesh netting is necessary to remove fouling organisms. Output is around 24 mt per year and the lobsters are fed daily with chopped trash fish and mussels. The use of live rather than opened mussels reduces the labour required for preparation and feeding as well as reducing fouling (James & Tong 1998). Fattening in pens and cages is also practised in India, Japan and Australia. Behavioural changes may influence performance under culture conditions. For example, while juvenile *Jasus edwardsii* are primarily solitary and cryptic, older animals become increasingly gregarious. It is thus difficult to obtain good, comparative data on the effects of density on growth rate and survival in commercially farmed animals because the sizes at stocking and culture conditions (tanks, sea pens, cages) vary so widely.

7.9.6 Transportation

Adult spiny lobsters may be harvested by seining and are transported live using the same techniques and precautions that are used for clawed lobsters (section 8.4.6). Rough handling and exposure to poor quality water can stress the lobster's immune system (Jussila *et al.* 2000) (sections 2.5.2 and 8.5). Animals being transported live to consumer outlets and processing plants are purged for 3 days in clean, flowing seawater chilled to 12°C before being packed between layers of crushed ice or in polystyrene boxes of damp sawdust with cooling packs. Vivier trucks or tankers equipped with temperature controlled seawater containers and compressed air or oxygen are also used (Sugita & Deguchi 2000).

With regard to restocking projects, little work has been published on methods of transportation and release to seabed habitat of pueruli or juvenile spiny lobsters, nor of monitoring subsequent survival (Butler & Herrnkind 2000).

7.9.7 Processing

Most processed spiny lobsters are beheaded, washed, graded and frozen shell-on. Small spiny lobsters are shelled and canned.

7.9.8 Habitat modification

Fishermen in several Caribbean countries construct fields ('campos') of artificial shelters (called 'casitas', 'casas Cubanas' or 'pesqueros'), within which wandering palinurids take shelter and can be conveniently captured (Miller 1983; Fee 1986; Briones-Fourzán *et al.* 2000). The shelters measure about 1.5–4 m^2, and consist of a wooden frame and a flat or corrugated roof under which the spiny lobsters hide (Cruz & Phillips 2000). They are placed about 20–30 m apart and in some areas density may reach 10 000 casitas in 160 km^2. Up to 200 lobsters take shelter under a traditional (4 m^2) Cuban pesquero although the trend is towards smaller, more easily handled shelters with fibre-cement roofs. Other shelters are made from car tyres (Cruz & Phillips 2000; Spanier 2000). The yield from such an area may be 40–65 mt of spiny lobsters annually (Eggleston *et al.* 1990a).

The main effect of the shelters seems to be to reduce mortality of adults and juveniles due to predation by fish (section 5.7.2). The lobsters prefer shelters with small entrances so that larger predators are excluded. Survival of a range of sizes of lobsters may be enhanced therefore by the provision of a range of different sized shelters in otherwise unprotected areas where lobsters forage for food (Eggleston *et al.* 1990b; Spanier 1994). If the shelter units are small and dispersed when deployed, they are less likely to attract predators than large structures and their use can create useful nursery habitat where none existed previously (Herrnkind *et al.* 1997). While it is uncertain whether such activities would augment the fishery or merely concentrate the stock, it does seem likely that the concept could be utilised to ranch those species of spiny lobster with low migratory instincts (Sosa-Cordero *et al.* 1998). Behavioural studies with *Jasus edwardsii* indicate that in this case juveniles should be deployed in such a way that they can quickly gain individual shelters much like clawed lobsters (section 7.8.11.3) while larger spiny lobsters would be better released *en masse* because their defence lies in their gregariousness (MacDiarmid *et al.* 1999b).

7.9.9 References

Anon. (1997) Tasmania looks to rock lobster culture. *Austasia Aquaculture Magazine,* **11** (2) 40–45.

Assad L.T. (1998) Growout of juvenile *Panulirus argus* in cages. *The Lobster Newsletter,* **11** (1) 11–13.

Australian Fresh Corporation (2000) Aquaculture of Moreton Bay bugs. http://www2.dpi.qld.gov.au/fishweb/research/Welcome.html

van Barneveld R. (2001) Rock lobster enhancement and aquaculture in Australia. In: *Australian Aquaculture Yearbook* (ed. R. Navarro), pp. 50–52. National Aquaculture Council, Executive Media Pty Ltd, Melbourne, Australia.

Booth J.D. & Phillips B.F. (1994) Early life history of spiny lobster. *Crustaceana,* **66** (3) 271–294.

Booth J., Davies P. & Zame C. (1999) Commercial-scale collections of young rock lobster for aquaculture. *The Lobster Newsletter,* **12** (1) 13–14.

Briones-Fourzán P., Lozano-Álvarez E. & Eggleston D.B. (2000) The use of artificial shelters (casitas) in research and harvesting of Caribbean spiny lobsters in Mexico. In: *Spiny Lobsters: fisheries and culture,* 2nd edn (eds B.F. Phillips & J. Kittaka), pp. 420–446. Fishing News Books, Oxford, UK.

Brown P.B., Leader R., Jones S. & Key W. (1995) Preliminary evaluations of a new water-stable feed for culture and trapping of spiny lobster (*Panulirus argus*) and fish in the Bahamas. *Journal of Aquaculture in the Tropics*, **10**, 177–183.

Butler M.J. IV & Herrnkind W.F. (2000) Puerulus and juvenile ecology. In: *Spiny Lobsters: fisheries and culture,* 2nd edn (eds B.F. Phillips & J. Kittaka), pp. 276–301. Fishing News Books, Oxford, UK.

Chen L-C. (1990) *Aquaculture in Taiwan*, 273 pp. Fishing News Books, Blackwell Scientific Publications, Oxford, UK.

Cray Corporation Ltd. (1994) Molting periodicity in a captive population of juvenile spiny lobster (*Jasus edwardsii*). *The Lobster Newsletter,* **7** (1) 6.

Cruz R. & Phillips B.F. (2000) The artificial shelters (pesqueros) used for the spiny lobster (*Panulirus argus*) fisheries in Cuba. In: *Spiny Lobsters: fisheries and culture,* 2nd edn (eds B.F. Phillips & J. Kittaka), pp. 400–419. Fishing News Books, Oxford, UK.

Eggleston D.B., Lipscius R.N. & Miller D.L. (1990a) Stock enhancement of Caribbean spiny lobster. *The Lobster Newsletter,* **3** (1) 10–11.

Eggleston D.B., Lipscius R.N., Miller D.L. & Coba-Letina L. (1990b) Shelter scaling regulates survival of juvenile Caribbean spiny lobster *Panulirus argus*. *Marine Ecology Progress Series*, **62**, 79–88.

Fee R. (1986) Artificial habitats could hike crab and lobster catches. *National Fisherman,* **67** (8) 10–12 & 64.

Gibson I.D. & Frusher S. (1997) How large a harem can one rock lobster handle? *The Lobster Newsletter,* **10** (1) 6–7.

Herrnkind W.F., Butler M.J. IV. & Hunt J.H. (1997) Can artificial habitats that mimic natural structures enhance recruitment of Caribbean spiny lobster? *Fisheries,* **22** (4) 24–27.

Igarashi M.A. & Kittaka J. (2000) Water quality and microflora in the culture water of phyllosomas. In: *Spiny Lobsters: fisheries and culture,* 2nd edn (eds B.F. Phillips & J. Kittaka), pp. 533–555. Fishing News Books, Oxford, UK.

James P.J. & Tong L.J. (1998) Development of a regime for feeding captive NZ red rock lobsters on mussels. In: *Aquaculture and Water: fish culture, shellfish culture and water usage* (compiled by H. Grizel & P. Kestemont), pp. 122–123. Abstracts presented at Aquaculture Europe '98, Bordeaux, France, 7–10 October 1998, European Aquaculture Society, Special Publication No. 26.

Jeffs A. & James P. (2000) Cage culture of the spiny lobster *Jasus edwardsii* in New Zealand. In: *Abstracts, Aqua 2000, Responsible aquaculture in the new millennium* (compiled by R. Flos & L. Creswell), p. 309. European Aquaculture Society, Special Publication No. 28.

Jussila J., Jago J., Tsvetnenko E. & Evans L.H. (2000) Live transport as origin of immune system stress in western rock lobster (*Panulirus cygnus*). In: *Abstracts, Aqua 2000, Responsible aquaculture in the new millennium* (compiled by R. Flos & L. Creswell), p. 323. European Aquaculture Society, Special Publication No. 28.

Kittaka J. (1997) Application of ecosystem culture method for complete development of phyllosomas of spiny lobster. *Aquaculture,* **155** (1–4) 319–331.

Kittaka J. (2000) Culture of larval spiny lobsters. In: *Spiny Lobsters: fisheries and culture,* 2nd edn (eds B.F. Phillips & J. Kittaka), pp. 508–532. Fishing News Books, Oxford, UK.

Lemmens J.W.T.J. (1994) The western rock lobster *Panulirus cygnus* (George, 1962) (Decapoda: Palinuridae): the effect of temperature and developmental stage on energy requirements of pueruli. *Journal of Experimental Marine Biology and Ecology,* **180**, 221–234.

Linton L. (1998) *The potential for tropical rock lobster aquaculture in Queensland,* 22 pp. Information Series QI 98020. Queensland Department of Primary Industries, Queensland, Australia.

MacDiarmid A. & Butler M.J. (1999) Sperm limitation in exploited spiny lobsters. *The Lobster Newsletter,* **12** (1) 2–3.

MacDiarmid A.B. & Kittaka J. (2000) Breeding. In: *Spiny Lobsters: fisheries and culture,* 2nd edn (eds B.F. Phillips & J. Kittaka), pp. 485–507. Fishing News Books, Oxford, UK.

MacDiarmid A., Stewart R. & Oliver M. (1999a) *Jasus* females are vulnerable to mate availability. *The Lobster Newsletter,* **12** (1) 1–2.

MacDiarmid A., Butler M. & Booth J. (1999b) Why do juvenile

red rock lobsters aggregate? *Austasia Aquaculture Magazine,* **12** (5) 50–51.
Marx J. & Herrnkind, W. (1985) Factors regulating microhabitat use by young juvenile spiny lobsters, *Panulirus argus*: food and shelter. *Journal of Crustacean Biology,* **5** (4) 650–657.
Meagher T. (1994) *The practicality of cultivating puerulus of the western rocklobster,* Panulirus cygnus *in Western Australia.* Western Australian Fishing Industry Council, (62). (Not seen, *apud* Linton 1998.)
Mikami S. & Takashima F. (1993) Development of the proventriculus in larvae of the slipper lobster, *Ibacus ciliatus* (Decapoda: Scyllaridae). *Aquaculture,* **116** (2–3) 199–217.
Mikami S. & Takashima F. (2000) Functional morphology of the digestive system. In: *Spiny Lobsters: fisheries and culture,* 2nd edn (eds B.F. Phillips & J. Kittaka), pp. 601–610. Fishing News Books, Oxford, UK.
Mikami S., Greenwood J.G. & Takashima F. (1994) Functional morphology and cytology of the phyllosomal digestive system of *Ibacus ciliatus* and *Panulirus japonicus* (Decapoda, Scyllaridae and Palinuridae). *Crustaceana,* **67** (2) 212–225.
Miller D.L. (1983) Shallow water mariculture of spiny lobster (*Panulirus argus*) in the Western Atlantic. In: *Proceedings of First International Conference on Warm Water Aquaculture – Crustacea.* 9–11 February 1983 (eds G.L. Rogers, R. Day & A. Lim), pp. 238–245. Brigham Young University, HI, USA.
Phillips B.F. & Booth J.D. (1994) Design, use, and effectiveness of collectors for catching the puerulus stage of spiny lobsters. *Reviews in Fisheries Science,* **2** (3) 255–289.
Rahman M.K. & Srikashnadhas B. (1994) The potential for spiny lobster culture in India. *Infofish International,* (1) 51–53.
Ritar A. (1999) Recent research on the propagation of southern rock lobster in Tasmania: Part 1. *Austasia Aquaculture Magazine,* **13** (5) 52–54.
Shioda K., Igarashi M.A. & Kittaka J. (1997) Control of water quality in the culture of early-stage phyllosomas of *Panulirus japonicus*. *Bulletin of Marine Science,* **61** (1) 177–189.
Sosa-Cordero E., Arce A.M., Aguilar-Dávila W. & Ramírez-González A. (1998) Artificial shelters for spiny lobster *Panulirus argus* (Latreille): an evaluation of occupancy in different benthic habitats. *Journal of Experimental Marine Biology and Ecology,* **229**, 1–18.
Spanier E. (1994) What are the characteristics of a good artificial reef for lobsters? *Crustaceana,* **67** (2) 173–186.
Spanier E. (2000) Artificial reefs of the Mediterranean coast of Israel. In: *Artificial reefs in European Seas* (eds A.C. Jensen, K.J. Collins & A.P.M. Lockwood), pp. 1–19. Kluwer Academic, Netherlands.
Sugita H. & Deguchi Y. (2000) Shipping. In: *Spiny Lobsters: fisheries and culture,* 2nd edn (eds B.F. Phillips & J. Kittaka), pp. 633–640. Fishing News Books, Oxford, UK.
Thomas C., Crear B., Ritar A., Mills D. & Hart P. (1998) Research into the aquaculture of the southern rock lobster in Tasmania. *Austasia Aquaculture Magazine,* **12** (4) 24–27.
Van Olst J.C., Carlberg J.M. & Hughes J.T. (1980) Aquaculture. In: *The Biology and Management of Lobsters, Vol. 2, Ecology and management* (eds J.S. Cobb & B.F. Phillips), pp. 333–384. Academic Press, London.

7.10 Crabs

7.10.1 Species of interest

Mud or mangrove crabs (*Scylla* spp.): green mud crabs (*Scylla oceanica, S. olivacea, S. tranquebarica*); red mud crabs (*Scylla paramamosain, S. serrata*); swimming crab (*Portunus trituberculatus*); Chinese mitten, river or hairy crabs (*Eriocheir sinensis, E. japonica*); blue crab (*Callinectes sapidus*); Caribbean king crab (*Mithrax spinosissimus*); Australian giant crab (*Pseudocarcinus gigas*); edible estuarine crab (*Thalamita crenata*).

Traditionally, the culture of crabs for the table (particularly mud crabs in South-east Asia), relies on wild-caught juveniles ranging from 3 g to 150 g in size at stocking. They are grown in enclosures or ponds for 3–8 months to produce marketable adults weighing 250 g or more. In the context of mud crab aquaculture, 'fattening' refers to the process of holding adult crabs for just 2–4 weeks in order for them to acquire advantageous market characteristics such as a hard shell or ripe ovaries. The latter are known as 'egg or red crabs' and command particularly high prices. Some fish or prawn hatcheries also rear crab larvae for restocking (e.g. in Japan). Extensive polyculture as well as semi-intensive monoculture of *Scylla* spp. for the table is practised in Taiwan where the industry has become differentiated into hatchery, nursery, ongrowing and fattening operations. Hatchery technology for *Scylla* spp. now exists in Australia (Williams & Field 1999) where crabs reared from the egg, giving 70–90% survival to megalopa, have been grown to 600 g in commercial ponds when stocked at a size of 15–20 mm carapace width (CW). Densities are around 3 m^{-2} (P. Sorgeloos, 2001, pers. comm.). Similar techniques are being developed elsewhere, e.g. in the Philippines (Quinito *et al.* 2000). Over the last decade, production of cultured *Eriocheir* spp. is reported to have increased from 3305 mt (1989) to 123 249 mt in 1998 (FAO 2000). Methods used include extensive and semi-intensive pond culture, often in disused shrimp ponds, polyculture with rice and stock enhancement (Tables 4.6h, 5.6 and 5.8). We are unsure what proportion of the total production is due to stock enhancement.

7.10.2 Broodstock and larvae culture

In 1981 production of *Scylla* spp. in Taiwan was based on wild-caught seed, but hatchery rearing is now practised. In Australian trials mature females were held in 7000 L

Plate 7.13 A Chinese combined hatchery/nursery facility rearing button-sized (5–25 g) mitten crabs (*Eriocheir sinensis*) for sale to farmers. (Photo courtesy Zhou Xin and Li Kangmin, Asian Pacific Regional Research and Training Centre for Integrated Fish Farming.)

shaded tanks fitted with false, sand bottoms. Water of 28–31°C was exchanged at 200% per day. After spawning females were placed individually into 1000 L tanks with no substrate or feed for the 10 day incubation period. Alternatively they may be held in floating mesh trays to release their larvae. Eyestalk ablation of wild-caught broodstock females has increased the control over larval supplies, and larvae survival rates of up to 60% are claimed in Taiwan at culture densities of six larvae L^{-1}. Higher densities (20–50 L^{-1}) in 7000–10 000 L tanks result in very variable survivals from 3–20%, with an average of 15% reported from pilot, commercial scale studies (Williams & Field 1999). The long spines typical of crab zoeae (Fig. 2.3d) make these larvae particularly vulnerable to entanglement with filamentous material in the water, especially at high culture densities, emphasising the importance of tank hygiene (Williams *et al.* 1999). Daily addition of cultured probiotic bacteria (*Thalassobacter utilis*, strain PM4) improved survival of *Portunus trituberculatus* from 16% to 28% (Nogami *et al.* 1997). First instar zoeae are fed rotifers at 10–15 mL^{-1} and at instar 3, newly hatched *Artemia* nauplii are given initially at a density of 1 mL^{-1}, then increasing to 5 mL^{-1}. Later stages are fed live *Artemia* nauplii at up to 5–10 mL^{-1}. Culture vessels vary in design but modified lobster 'kreisels' (section 7.8.7) can be used in conjunction with recirculation systems (Heasman & Fielder 1983). Other crab species such as *Cancer irroratus* can be reared using very similar methods. For mud crab, salinity is reduced from 32 to 25‰ from fifth instar zoea to megalopa (Quinitio *et al.* 2000). *Eriocheir* larvae are reared in natural or artificial seawater at 12–15‰ for 30–40 days. During the later instars they are acclimated to freshwater over 5–7 days prior to transfer to nursery ponds (Li Kangmin, 2001 pers. comm.).

7.10.3 Nursery

After around 15–25 days, crab zoea metamorphose to the megalopa stage and are then fed minced fish and bivalve flesh. Survival of *Portunus triturbiculatus* megalopa to first crab in 12 days is around 50–60% and similar to that achieved with mud crab, with mortalities occurring largely through cannibalism. The use of suspended plastic or other filamentous artificial substrate to provide shelter and resting places is beneficial (O'Sullivan 1997). In Taiwan mud crab megalopae are held at 2000–3000 m^{-2} in nursery ponds for 2 weeks with survivals of 40–80% expected. Taiwanese farms often have four to five earth-bottomed ponds, each of 15–20 m^2, containing a 10 cm layer of beach sand. Temperature is controlled by shade netting, salinity held between around 20‰ and, if a flow is available, the water exchange rate is up to one pond volume per day. Blended trash fish (1 kg d^{-1} per 30 000 crabs) is given for the 2 weeks of culture after which the crabs have reached 1–2 cm CW. Hatchery-reared mud crab megalopae can also be grown in net cages in ponds until they reach 3 g, a size suitable for stocking (Quinito & Parado-Estapa 2000). They may be further nursed in cages at lower density (120 m^{-2}) until they reach 30–40 g (Castaños 1997).

In China, *Eriocheir sinensis* megalopae (approximately 6–7 mg live weight) are stocked in nursery ponds or rice fields at about 50–120 m^{-2}. The latter are modified by digging peripheral trenches and sumps in which the crabs grow to around 0.2–0.5 g (called 'bean size') in 50 days and to 'button' size, which is anything between 5 g and 25 g in a further 2–3 months. Yields range from about 220–300 kg ha^{-1} after 4–5 months. When food is given it consists of about 40% animal and 60% vegetable material, of which 35% is grass or duckweed. Feeding is initially done eight to ten times a day at 100–150% of body weight which decreases to 7–10% fed four to six times a day. Lime and fertilisers are used in pond preparation and during culture (respectively) and the water is partly renewed every 3–10 days according to conditions (Li 1998; Liu Fengqi, 2001 pers. comm.). The juvenile crabs overwinter in the ponds and are stocked in rice fields or crab ponds for ongrowing in spring.

7.10.4 Ongrowing

Extensive polyculture of *Scylla* spp. is practised in the Philippines where it traditionally forms a secondary, low-density crop in intertidal ponds with shrimp or milkfish (Table 5.6). Juveniles may enter with the tide or are purchased at sizes of 2–3 cm CW from push-net fishermen, from gatherers working on mudflats and mangrove stands or from nursery operators. If crabs are stocked intentionally rather than incidentally, farmers may install an overhanging fence of bamboo stakes around the inside edge of the pond to prevent escapes. Berried females in particular try to escape to the sea to spawn. Ponds are 0.1–1 ha in area; they may have a central raised platform and contain hides. Stocking is usually done once a year (1000–10 000 crabs ha^{-1}) at the start of the growing season (May–August) although continuous stocking is sometimes attempted. Some farmers feed chopped shrimp and trash fish at 6–10% of the body weight per day, often on the rising tide to reduce pollution. Crabs grow to 200+ g or 8 cm CW in 4–6 months and provide yields of around 340 kg ha^{-1} annually. Higher yields up to around 700 kg ha^{-1} are possible with these methods. When stocked with large crabs of 150 g at 2 m^{-2}, market size of 250 g is reached in 3–4 weeks, a practice seen in Sri Lanka, India and Thailand. In Indonesia, crabs of 70–110 g are also grown on in cages (3 m^3). In Indonesia, Malaysia, the Philippines and Vietnam, crabs are being cultured in pens (9 × 18 m), often built in mangrove areas (Overton & Macintosh 1997). This is proving attractive because it involves very low-cost techniques and leads to the protection of existing mangroves, and the establishment of new mangrove stands complete with their inherent resources. In Taiwan *Scylla* spp. are frequently grown with *Gracilaria* (seaweed), shrimp or fish but pond management habitually gives priority to the conditions required by these other species.

In China, 5–25 g juvenile crabs are stocked in modified rice fields at densities equivalent to around 4000–8000 ha^{-1} during February to March and ongrown to 125 g in 6–9 months. The aquaculture area is about

Plate 7.14 A typical, small-scale, tidal, mud crab pond, Mekong Delta, Vietnam. The fences are to prevent crabs escaping or migrating to sea. (Photo courtesy D.J. Macintosh, University of Aarhus, Denmark.)

Plate 7.15 Soft-shell crab production farm in Thailand. Mud crabs are held in individual boxes until they moult, at which time they are harvested. (Photo courtesy D.J. Macintosh, University of Aarhus, Denmark.)

15–20% of the total and fencing is constructed to prevent the crabs migrating to sea. Sometimes aquatic plants are planted to increase shelter availability. Feed is given at 5% body weight three to four times per day. Repeated draining and flooding of the trenches concentrates the crabs for harvesting in October to November and yields are around 300–500 kg ha^{-1} (Li 1998). When crabs are grown in extensive monocultures, either 3–5 g crabs are stocked at 3–4.5 m^{-2} or 10–16 g crabs are stocked at 1.5–2 m^{-2}. Survival from February/March to October is around 40%, at which time crabs will have reached 70–150 g. Yields are usually around 450 kg ha^{-1} but range up to 1500 kg ha^{-1}. In November the largest crabs (200 g) are selected for the next year's broodstock. In southern China some farmers accelerate development by heating the hatchery water and improving feeding regimes during overwintering in order to harvest 100 g crabs after only 1 year. During the late 1990s many crabs (30–70%) began to exhibit precocious sexual maturation having reached only 20–30 g by October. These crabs do not grow but gradually die during the next year. At the time of writing, the cause of this problem is unknown (Liu Fengqi, 2001 pers. comm.).

Crab monoculture is also practised in Taiwan (Chen 1990). Typically the ponds are 0.2–0.5 ha earth ponds with several sluice gates if they are tidal rather than pumped. Crab stocking density is 0.5–3 m^{-2} and feed includes live snails and trash fish (10–15 g m^{-2} daily). Crabs grow to 8–9 cm CW in 3–4 months (summer) or in 5–6 months in winter with 30–70% survival. Yields are estimated to be about 8000–9000 crabs ha^{-1}, or approximately 1800 kg ha^{-1} per crop.

Recent trials in the Philippines showed that monosex culture of male mud crabs was most economical at densities of 0.5 m^{-2} when survivals of 88% were obtained in 150 m^2 ponds – giving yields equivalent to over 3000 kg ha^{-1} (Triño *et al.* 1999). Similarly, monosex culture of *Portunus triturbiculatus* in China gave increases in survival of 1.5 times that of mixed sex culture and increased yields by 54% (Guo *et al.* 1997).

Another form of mud crab monoculture in Taiwan is essentially a holding and fattening operation to supply 'red crabs' (females packed with an internal mass of orange eggs) for gourmet restaurants. Such specialist farms have 5–15 ponds of 50–600 m^2 each stocked at 2–4 crabs m^{-2} with 8–12 cm CW females. They are fed once daily with live snails, shrimp or trash fish (up to 200 g per crab). Survival over the 1–3 month holding period is around 70–90%. Similar entrepreneurial fattening operations are undertaken in Singapore and Hong Kong (section 5.2.2). Elsewhere the operation is conducted in cheaper pens, tidal ponds and floating cages (Chong 1995). Fattening of mitten crabs from 100–150 g to 250 g in China yields 450–750 kg ha^{-1} in about 4–5 months from an initial stocking density of 1200–2000 ha^{-1} (Li Kangmin, 2001 pers. comm.).

7.10.5 Harvesting

Harvesting is done using baited lift nets, bamboo traps, cages, gill nets or when the pond is drained. During high tides mud crabs swim against the current and can be caught by dip nets. Mitten crabs can be caught by draining, trapping or hand collecting at night when they emerge onto the banks.

7.10.6 Transportation

Megalopae are best transported at 50 L^{-1} in double polythene bags inflated with oxygen and surrounded with ice or ice packs in polystyrene containers to maintain 20°C (Quinitio & Parado-Estepa 2000). In Japan seaweed may be provided as a substrate to reduce cannibalism. Juvenile crabs are transported in wicker baskets, straw bags or ventilated containers with moist mangrove leaves. The claws may be cut or bound. Adult crabs are commonly transported live, with the claws bound or glued. They may survive for up to 17 days if held in chilled, humid conditions (for example, between layers of damp synthetic 'cotton wool') in ventilated, insulated polystyrene boxes (Dagoon 1997). Survival can be reduced in damp newspaper, wood shavings, seaweed or other less favourable conditions to around 7 days or less (Rendon & Ronquillo 1999) (section 3.3.5). Additional methods and precautions taken during the live storage and transportation of crabs and other crustaceans are described in section 8.4.6.

7.10.7 Processing

Most cultured and fattened crabs are sold live, unprocessed. Increasing amounts are sold frozen, whole, halved, in blocks or IQF. Crab meats are sold in frozen blocks or canned. Chitin and chitosan are extracted from waste shells and used in non-allergenic contact lenses, artificial skin and waste water filters (section 3.3.7).

7.10.8 Hatchery supported fisheries, ranching

Intensive culture of juvenile *Portunus trituberculatus* for restocking occurs in Japan, where $10–50 \times 10^6$ are released to the sea each year. Smaller quantities of *P. pelagicus* (983 000) and *Scylla* spp. (56 300) were released in 1996 (Imamura 1999) and the recapture rate is around 3–12%. There is very little culture of *Portunus* spp. to market size. The hatcheries involved only rear crabs for 3 weeks to 3 months in a year and rear other species at other times. A 77% survival rate is reported from the stocking of *Eriocheir sinensis* into reservoirs in China (Duan *et al.* 1996) but uncontrolled transplantations between estuarine systems have resulted in reduced growth as two different stocks have become mixed (section 11.3.2). In Indonesia and Vietnam, European marine consultants are examining the prospects for enhancing mud crab stocks.

The following account applies largely but not exclusively to *Portunus trituberculatus*.

7.10.8.1 Broodstock

Mated females are caught from March to August and transported in 5–10 L of water (short 30 min journeys) or in 1000 L vivier vehicles (1–5 h journeys) at one crab per 20 L of water.

7.10.8.2 Spawning and incubation

Hatchery tanks used for incubation and spawning typically contain 25 m^2 of sand, 10 cm deep on a false floor. Water depth is 50 cm and the flow is around 200% per day or 25 000 L per day. Shade covers are used to control algal growth and temperature, and the substrate is kept clean to reduce the risk of egg infections. The temperature is raised from 10–20°C over 8–12 days in March–April and then to 23°C. Stocking density is one to three females m^{-2} but, to reduce fighting injuries, some farmers remove the dactyl of the claws and increase the stocking density to 10 m^{-2}. Careful daily feeding with bivalves or fish (3–5 kg d^{-1} per 20–25 crabs) can result in 80% survival at the time of hatching. Females with a firm round orange egg mass are selected and moved carefully to 500–1000 L hatching tanks in which water is gently aerated and continuously renewed. During the 20–25 day incubation period the eggs change colour from orange to black. The appearance of two red-purple spots in the eggs indicates hatching will occur in 3 days. Fecundity ranges from 1×10^6 to 3×10^6 zoeae for 400–1000 g females. Rotifers (30 mL^{-1}) are added prior to hatching, which usually occurs between 8 PM and midnight. Small or non-phototactic batches of larvae are discarded and few females are retained to spawn a second time. Overall only a small proportion of the purchased broodstock are used.

7.10.8.3 Larvae culture

Newly hatched larvae of *Portunus trituberculatus* and *Scylla* spp. can be treated with 25 ppm formalin to inhibit fungal infections (Hamaski & Hatai 1993). The product of one spawning of *Portunus trituberculatus* is stocked to give a density of 10–50 larvae L^{-1} in 75–300 m^3 capacity culture tanks. Many operators prefer circular 100 m^3 tanks. The tanks are supplied with sufficient water to give a daily exchange of 10% but this may be adjusted to control algal growth and water quality. Salinity is 30–33‰, pH 8.0–8.5, but light levels below 3000 lux (36 μE m^{-2} s^{-1}) are inadequate. Some tanks are fitted with a slow (1 rpm) stirring bar to reduce bottom fouling. The larvae pass through four zoeal instars, each lasting 3–4 days, followed by one megalopa instar lasting 5–7 days at 20–25°C (Cowan 1983). Survival is around 14%. Separately cultured *Nannochloropsis* or *Chaetoceros* or a culture of bacteria and yeast together with rotifers (3–10 mL^{-1}) are fed to the first zoea but live *Artemia* are added for zoea 2 onwards. Sieved clam and shrimp flesh (140 μm fragments) are given later five to six times a day. Hanging nets or plastic mesh are added for the megalopae to cling to, but a 30% mortality is common between zoea 4 and the megalopa instar. The tanks are emptied through a siphon or drain and the postlarvae retained on 520 μm mesh netting.

7.10.8.4 Transportation and release

Transportation to the release site is done in 1000 L capacity vivier trucks at 15–19°C. The crabs are stocked at 150 L^{-1} and provided with frayed rope 'shelters'. The crabs are pumped to smaller containers and released at the sea surface from a boat or from the beach. Stage one crabs (C1) cling to seagrass and only burrow when they reach the C2 stage. The heavy losses following surface releases has led to the construction of specialised nursery facilities. These may be:

- Onshore tanks in which the crabs are stocked at 1–3 L^{-1} and grow to C4 with 20–40% survival. After 1–3 weeks they are released at the sea surface.
- Netted inshore areas fitted with frayed rope shelters and stocked at 100–500 m^{-2}. The crabs are fed, survival is 20–40% over 1–3 weeks and they are released by opening the enclosure to the sea.
- Open inshore areas again fitted with frayed rope shelters but surrounded only by a low (40 cm) underwater fence to prevent early dispersal.

7.10.9 Soft-shelled crabs

Production of newly moulted or soft-shell crabs is really a short-term holding operation in which the wild-caught adult crabs are generally not fed. These crabs are in demand by sport fishermen for bait but also form the basis of a small but expanding and lucrative gourmet food market, e.g. for the blue crab (*Callinectes sapidus*) in the USA. Production in 1993 was more than 772 mt worth $3.5m. The same technology would be applicable with very little modification, to other species, for example *Cancer irroratus*, *C. magister*, *C. borealis* (Oesterling & Provenzano 1985) and *Scylla* spp., although if the soft-shell mud crab market expanded, it could aggravate existing problems of overfishing wild stocks (section 11.3.1.2). The ability to control and make moulting more predictable is being investigated (section 12.8.6).

In North America two types of 'shedding' system are in use:

(1) Floating boxes (3.6 × 1.2 × 0.45 m deep) holding 200–300 crabs, and fitted with an external 20 cm wide lip to give stability. Problems arise from strong currents, water quality fluctuations and predators. They can be time consuming and many producers have switched to on-shore systems.
(2) Onshore shallow trays or tables, which may be provided with a through-flow of water or linked to a recirculation system. The trays may be stacked according to operational convenience and made of wood, fibreglass or other non-toxic materials. Typical dimensions are 2.4 × 1.2 × 0.25 cm deep with 10 cm water depth. Flow is designed to promote a self-cleaning action and may be set to three to four tray volumes per hour. Recirculation systems are now commonplace (Webster 1998), and recent modifications include the incorporation of side-stream protein skimmers which can be operated intermittently when required (D. Webster 2000, pers. comm.).

The stock are caught in traps or pots and inspected for the presence of 'peelers' and 'busters'. Peelers have a red line along the edge of the last two flattened sections of the 'paddle fins' (the last pair of pereopods), indicating that moulting will occur in 1–3 days. When a split develops along the posterior edge of the carapace, the crab becomes a 'buster' and will moult in 2–3 hours. The trays are examined every 4–6 hours and busters removed to other trays to moult. They are then observed every 15 min to a few hours until they have expanded to full size, when they are removed for sale before the shell hardens and their value lost. In some areas where peeler crabs are not abundant, intermoult crabs are held and fed until they moult. The main constraint to the industry appears to be the supply of good quality peeler crabs. In Australia, efforts are being made to develop techniques for rearing soft-shelled blue crabs throughout the year (Navarro 2001).

7.10.9.1 Processing

Soft shelled crabs are chilled or packed and frozen for shipment to restaurants.

7.10.10 References

Castaños M. (1997) Grow mudcrab in ponds. *Asian Aquaculture,* **19** (3) 14–16.

Chen L-C. (1990) *Aquaculture in Taiwan*, 273 pp. Fishing News Books, Blackwell Scientific Publications, Oxford, UK.

Chong L.P. (1995) The culture and fattening of mud crabs. *Infofish International,* (3) 46–49.

Cowan L. (1983) *Crab farming in Japan, Taiwan and the Philippines*, 85 pp. Information series QI 84009, Queensland Dept. of Primary Industries, Queensland, Australia.

Dagoon N.J. (1997) Post-harvest, processing. *Asian Aquaculture,* **19** (3) 23–25.

Duan M., Zhao Y., Sun X., Wang J. & Yang R. (1996) On enhancement of the mitten crab in the small-typed reservoir. *Shandong Fish. Qilu Yuye,* **13** (6) 17–18 [in Chinese – English abstract seen].

FAO (2000) http://www.fao.org/waicent/faoinfo/fishery/statist/fisoft/fishplus.htm *(apud* FAO (2000) *FAO yearbook, Fishery statistics, Capture production 1998*. Vol. 86/1 and FAO (2000) *FAO yearbook, Fishery statistics, Aquaculture production 1998*. Vol. 86/2.)

Guo X., Zhang X. & Li Y. (1997) Technique of isolating female from male culture of *Portunus trituberculatus*. *Trans. Oceanol. Limnol. Haiyang Huzhao Tongbao,* (3) 71–75 [in Chinese – English abstract seen].

Hamaski K. & Hatai K. (1993) Prevention of fungal infection in the eggs and larvae of the swimming crab *Portunus trituberculatus* and the mud crab *Scylla serrata* by bath treatment with formalin. *Bulletin of the Japanese Society of Scientific Fisheries,* **59** (6) 1067–1072.

Heasman M.P. & Fielder D.R. (1983) Laboratory spawning and mass rearing of the mangrove crab, *Scylla serrata* (Forskal), from first zoea to first crab stage. *Aquaculture,* **34** (3–4) 303–316.

Imamura K. (1999) The organisation and development of sea farming in Japan. In: *Stock Enhancement and Sea Ranching* (eds B.R. Howell, E. Moksness & T. Svåsand), pp. 91–102. Fishing News Books, Oxford, UK.

Li K.M. (1998) Rice aquaculture systems in China: a case of rice-fish farming from protein crops to cash crops, 15 pp. In: *Integrated Bio-Systems in Zero Emissions Applications*. Proceedings of the Internet Conference on Integrated Bio-Systems 1998 (eds E-L. Foo & T. Della Senta). http://www.ias.unu.edu/proceedings/icibs

Navarro R. (Ed.) (2001) Major R&D reform heralds new era for Australian prawn sector. *Australian Aquaculture Yearbook,* p. 46. National Aquaculture Council, Executive Media Pty Ltd, Melbourne, Australia.

Nogami K., Hamasaki K., Maeda M. & Hirayama K. (1997) Biocontrol method for rearing the swimming crab larvae *Portunus trituberculatus*. *Hydrobiologia,* **358** (1–3) 291–295.

Oesterling M.J. & Provenzano A.J. (1985) Other crustacean species. In: *Crustacean and Mollusk Aquaculture in the United States* (eds J.V. Huner & E. Evan Brown), pp. 203–234. AVI Inc., Westport, CT, USA.

O'Sullivan D. (1997) Scientists meet on mud crab culture. *Austasia Aquaculture Magazine,* **11** (4) 47–51.

Overton J.L. & Macintosh D.J. (1997) Mud crab culture: prospects for the small-scale Asian farmer. *Infofish International,* (5) 26–32.

Quinitio E.T. & Parado-Estepa F. (2000) Transport of *Scylla serrata* megalopae at various densities and durations. *Aquaculture,* **185** (1–2) 63–71.

Quinitio E.T., Parado-Estepa F., Rodriguez E. & Millamena O.M. (2000) Seed production of mud crab *Scylla* spp. In: *Abstracts, Aqua 2000, Responsible aquaculture in the new millennium* (compiled by R. Flos & L. Creswell), p. 580. European Aquaculture Society, Special Publication No. 28.

Rendon C.G. & Ronquillo J.D. (1999) Tolerance of the mud crab, *Scylla serrata* (Forskål, 1755), to handling, storage, and transport. In: *Book of Abstracts, World Aquaculture '99*, 26 April–2 May 1999, Sydney, Australia, p. 628. World Aquaculture Society, Baton Rouge, LA, USA.

Triño A.T., Millamena O.M. & Keenan C. (1999) Commercial evaluation of monosex pond culture of the mud crab *Scylla* species at three stocking densities in the Philippines. *Aquaculture,* **174** (1–2) 109–118.

Webster D. (1998) Soft crabs and recirculating systems. *Aquaculture Magazine,* **24** (2) 23.

Williams G. & Field D. (1999) Towards commercial rearing of the mud crab. *Austasia Aquaculture Magazine,* **12** (5) 39–41.

Williams G., Wood J. & Shelley C. (1999) Removal of biofilm from culture vessel surfaces improves survival rates of mud crab (*Scylla serrata*, Förskal) larvae from zoea 1 to megalopa. In: *Book of Abstracts, World Aquaculture '99*, 26 April–2 May 1999, Sydney, Australia, p. 689. World Aquaculture Society, Baton Rouge, LA, USA.

7.11 Non-decapod crustaceans

7.11.1 Species of interest

Branchiopods: *Artemia* spp., *Moina* spp. and *Daphnia* spp.; copepods: *Amphiascoides* spp., *Tisbe* spp. *Acartia* spp. and *Centropages* spp., and mysids: *Mysidopsis* and *Leptomysis* spp. The use of the brine shrimp *Artemia* as a live food for the commercial rearing of fish, cephalopod and crustacean larvae, and in some cases juveniles, is almost universal. *Artemia* cysts are, however, expensive and may not always be available to a hatchery at an economically viable price. Also, the newly hatched nauplii often do not provide all the nutrients necessary for complete larval development (Léger & Sorgeloos 1992).

Nevertheless they can be enriched with, for example, essential fatty and amino acids and vitamins (Merchie *et al.* 1995; Han *et al.* 2000) prior to feeding and even used as carriers for medicines (Touraki *et al.* 1999). As the culture of novel species, particularly of fin- and ornamental fish, has increased during the past decade or two, so the demand for specialist live feeds has grown. This need has been partially met through the rearing of other small organisms such as rotifers and nematodes and, increasingly, by culturing non-decapod crustaceans, particularly copepods which form the natural diet of many marine fish larvae and juveniles. Together with formulated microparticulate diets, live and enriched live foods can provide hatchery operators with an attractive array of options. Very often a range of live and prepared foods of different sizes and with specific nutritional content will be supplied, changing as the larvae pass through different developmental stages (sections 2.4.8, 7.2.4 and 7.3.3).

7.11.2 Branchiopods

The two main groups of branchiopod crustaceans reared are the saline anostracan branchiopod *Artemia* (predominantly *Artemia franciscana*) and the fresh- or brackish-water cladocerans *Daphnia* and *Moina*. Marine species of cladoceran do exist, for example *Diaphanosoma aspinosum* and *Moina salina,* but reliable techniques for their mass culture are not yet fully developed. Occasionally, other species of branchiopod can become so numerous in fish nursery ponds that they become pests, competing with the juveniles for food. Eradication is both difficult and costly, involving frequent sediment removal and drain/fill cycles (Thursten 1995).

7.11.2.1 Anostracan branchiopods: Artemia spp.

Artemia are particularly convenient to use in hatcheries. Their cysts are harvested from the wild, cleaned, dried and vacuum-packed in cans in which they can be stored for several months or even years. Nauplii hatch after 18–36 h incubation in seawater ready to be fed to larvae. Originally supplies came from coastal saltworks in San Francisco Bay and the Great Salt Lake in Utah, but sources expanded in the late 1970s to include Argentina, Australia, Canada, China, Colombia, Cuba and France. However, considerable variation in cyst supply, hatch rate and nutritional quality within and between sources stimulated substantial research effort to develop and refine harvesting, processing, hatching and enrichment technologies. By 1997 the global, annual demand had risen to 1500 mt of which over 80% went to shrimp hatcheries. Cysts are also produced commercially in managed ponds and saltworks, e.g. in Brazil (Camara 1996) and Vietnam (Baert *et al.* 1997) but quantities are small by comparison (1–20 mt each) and many units, after initial successes, show much reduced yields. Overall, changing shrimp and fish production strategies and improvements to enrichment technologies (and hence in *Artemia* utilisation), coupled with climatic and anthropologically induced fluctuations in their natural environment, make it difficult to forecast the future demand for, and supply of, *Artemia* cysts (Lavens & Sorgeloos 2000).

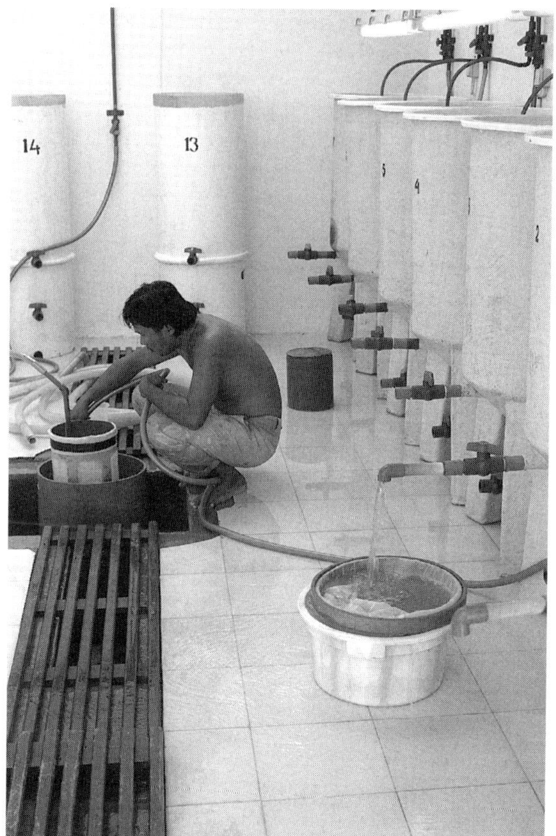

Plate 7.16 Freshly hatched *Artemia* nauplii being drained from a hatching tank into a sieve (foreground). The technician is rinsing an earlier batch of nauplii to reduce bacterial contamination prior to feeding to larval shrimp.

Pond culture systems range from those used in solar salt works to purpose built or adapted earth ponds (Baert *et al.* 1996). In Vietnam, for example, they are shallow (30–50 cm depth), 0.5–2 ha in area and are operated as batch cultures, producing cysts, adults or juveniles according to season and requirements. They are initially stocked with around 60–100 *Artemia* L^{-1} and supplied with algae-rich water from fertilised ponds or intake canals.

Intensive indoor systems are used in some hatcheries and nurseries where juveniles and adults are reared in 1000–5000 L oval or 'D'-ended raceways. These are also operated independently to produce sequential batches. Air lift pumps circulate the water, a proportion of which may be treated and recycled. Newly hatched nauplii are stocked at 2000–8000 L^{-1} according to the final size required (juveniles or adults). The food is often a prepared suspension of rice bran in seawater maintained at concentrations up to 200 g m^{-3} and is monitored two or more times each day by Secchi disk (see Glossary). Harvesting is most simply done by hand net and yields vary from 5 kg to 25 kg m^{-3} after 10–14 days of culture according to individual size required at harvest, and initial stocking density (Dhert *et al.* 1992; Dhont *et al.* 1993; O'Sullivan 1997).

Although optimum conditions for hatching vary between the sources of the cysts, basic requirements are for cysts to be placed at 2 g L^{-1} in conical bottomed vessels of brightly illuminated (2000 lux (= 25 $\mu E\, m^{-2} s^{-1}$) at the water surface) and vigorously aerated, seawater at 25–28°C, 15–35‰ salinity and pH 8.0. After hatching, the nauplii are separated from unhatched cysts and empty shells by stopping the aeration and illuminating the bottom of the vessel. The nauplii concentrate at the bottom and are siphoned or drained off within 5–10 min into a submerged net. They are then washed thoroughly in clean water (while submerged to prevent damage to the nauplii), counted by sub-sampling and fed immediately to the larvae. With every hour's delay the nauplii utilise some of their internal yolk reserves and become less nutritious. If they are required at intervals throughout the day, they may be stored under gentle aeration at up to 8×10^6 L^{-1} at temperatures below 10°C for up to 24 h. More detailed accounts of the hatching, preparation and feeding of *Artemia* include Dhont *et al.* (1993), Lavens and Sorgeloos (1996) and Hoff and Snell (1997).

Disinfection of nauplii with formaldehyde or sodium hypochlorite is achieved by the following steps: hatch cysts in clean seawater as described above, harvest the nauplii on a sterile mesh, discard the hatching water; wash thoroughly with sterile sea- or freshwater, disinfect nauplii with sodium hypochlorite (5 mg L^{-1} for 15 min) or with formaldehyde (50–70 mg L^{-1} for 5 min), wash nauplii in sterile seawater until no disinfectant is detected (GomezGil-RS *et al.* 1994; Sahul Hameed & Balasubramanian 2000). The nauplii can then be fed to the larvae.

Cyst decapsulation is the process of removing the indigestible and potentially dirty outer shell (chorion) of the cyst. The cysts are first hydrated in aerated seawater at 25°C for 60–90 min in a conical bottomed vessel, then collected and washed on 120 μm mesh to remove debris. At this stage they may be stored at 4°C for a few hours if required. The seawater decapsulation solution is made using liquid bleach (sodium hypochlorite solution) or bleaching powder (calcium hypochlorite). Some authorities recommend the addition of alkali to raise the pH. The requirement is for 0.5 g of active chlorine and 14 mL of decapsulation solution per gram of cysts. Decapsulation produces heat so once mixed, the solution and cysts must be aerated and kept cool (20–25°C) in a water bath or with ice. The process takes 2–7 min, after which the decapsulated cysts are filtered off and thoroughly washed. They are then rinsed in a 0.1% sodium thiosulphate solution to remove residual chlorine and washed again. At this stage they may be hatched or stored for up to 7 days at 4°C (Dhont *et al.* 1993; Hoff & Snell 1997). Cysts from some sources (e.g. Turkmenistan) do not decapsulate well under standard conditions. Owing to the increased labour involved and recent advances in disinfection techniques (see below) decapsulation is no longer as widely used as it was in the 1980s.

Enrichment of *Artemia* is often necessary to increase the levels of specific nutritional or medicinal components. After the newly hatched *Artemia* have used up their internal reserves (usually in about 12 h) they are fed on live or spray-dried micro-algae, yeasts, emulsions of HUFA-rich marine oils, liposomes (Wickins 1972; Touraki *et al.* 1995; Evjemo & Olsen 1999) or microcapsules (Polk *et al.* 1994), each selected or prepared to supply the required nutrients. Fatty acids, in particular the highly unsaturated fatty acids (HUFAs) DHA and EPA (section 2.4.2), are important for fish and crustacean larvae, and their stability, and the form in which these components are ultimately presented to the larvae (e.g. as TAG lipid or polar lipid) may be critical (McEvoy & Sargent 1998; Harel *et al.* 1999). For enrichment, the *Artemia* nauplii (12–24 h post-hatch) are placed at densities of 100–250 mL^{-1} in 200–250 L conical-bottomed vessels of vigorously aerated seawater containing a suspension of the enrichment material for 12–24 h at 20–25°C. They are then collected, washed and fed to larvae as soon as possible or stored at 4°C for not more than 12–18 h, since the levels of some nutrients can decline markedly during starvation (Evjemo *et al.* 1997; Estévez *et al.* 1998).

A procedure combining both hatching and enrichment steps with a disinfection process has been developed commercially; this advance claims to reduce tank space and labour requirements (De Wolf *et al.* 1998; Lavens *et al.* 2000).

7.11.2.2 Cladoceran branchiopods: Daphnia, Moina

Daphnia are reared in freshwater containing algae, rice bran or fertilised with animal manure. Yields between 100 g and 1000 g m^{-3} per week have been recorded (Murugan 1989), the highest and most stable yields being obtained when frequent, low-level feeding or fertilisation regimes are used (Jana & Chakrabarti 1997). Even so, *Daphnia* are less suited to intensive culture than *Moina*. Adult *Daphnia* (1400–3000 μm) are roughly twice the length of *Moina* adults and are therefore used to feed larger fish larvae or juveniles. In turn, adult *Moina* (700–1500 μm) are larger than newly hatched *Artemia* (400–500 μm) and are used widely in the ornamental fish trade, as well as for feeding fish and sometimes prawn larvae (Alam *et al.* 1993). They can be cultured in shallow (30–40 cm) aquaria, outdoor concrete tanks or ponds containing fertilised, brackish water (4–20‰). Considerable quantities are produced in Singapore for the aquarium trade, in ponds fertilised with pig wastes. Good results have been achieved in Hawaii (yields of 6–30 *Moina* mL^{-1}) in culture trials using imported *Chlorella* as food at a density of $2–5 \times 10^6$ cells mL^{-1}. Problems can arise however, at high culture densities due to low oxygen and pH levels. Harvesting by net or siphon at regular daily intervals helps to maintain steady production levels. Stopping the aeration causes *Daphnia* to rise to the surface where they can be readily seen and removed. *Daphnia* can be stored under refrigeration and should be enriched with algae, yeast or fatty acid emulsions if stored for more than a day or two.

Typically, populations of *Moina* and *Daphnia* grow well for the first 6–9 days before declining. Population growth is related to food availability and also to individuals switching from a parthenogenetic (see Glossary) to a sexual reproductive mode under adverse environmental conditions. This results in the production of eggs with delayed development (resting eggs). The unpredictability of the switch makes batch culture more feasible than continuous, semi-natural culture (Hoff & Snell 1997; Bonou & Saint-Jean 1998). Further details of production methods and expected yields are given by Mims *et al.* (1991), Delbare and Dhert (1996) and Hoff and Snell (1997).

7.11.3 Copepods

Copepods have excellent nutritional properties for fish larvae and Støttrup (2000) has recently reviewed their culture for use in marine fish hatcheries. For example, wild-caught *Eurytemora velox* have been shown to be a superior diet to enriched *Artemia* for halibut larvae on account of higher levels of DHA and a higher DHA : EPA ratio (Shields *et al.* 1999). The benthic harpacticoid copepods (*Amphioscoides*, *Tigriopus*, *Tisbe*) are generally easier to rear intensively than the pelagic calanoid copepods (*Acartia*, *Centropages*, *Eurytemora*). Because of this, many hatcheries rearing temperate marine fish only use copepods during critical stages of the larval life and revert to other diets, including *Artemia* thereafter. Some Danish turbot hatcheries rely on outdoor cultures of wild-caught copepods and other natural zooplankton (Støttrup *et al.* 1998). These may be reared in large, semi-buried tanks up to 28 m diameter × 4 m deep. Under adverse environmental conditions around 12 species of copepod can produce resting (diapause) eggs which remain in the sediment until conditions are right for hatching. These eggs may be disinfected and used as inocula for new cultures (Næss & Bergh 1994). Additional details of copepod culture requirements are given by Mims *et al.* (1991), Delbare *et al.* (1996) and Hoff and Snell (1997).

7.11.3.1 Harpacticoid copepods: Amphioscoides, Euterpina, Tigriopus, Tisbe

Harpacticoids graze on biological growths attached to surfaces, and when artificial substrates such as biological filtration media are added to culture tanks to increase the surface area available and therefore their food supply, culture densities can be high (e.g. 100 000 nauplii L^{-1}). Artificial foods may also be utilised. *Amphioscoides atopus* for example, was cultured at 25–35‰ in a 4 m^2 tank containing a bottom layer of limestone cobbles which became coated with biofilm following the addition of fish flakes and algae. Yields averaging nearly 3×10^6 individuals per day have been maintained over a period of 5 months, although a method for cleaning between the cobbles will need to be developed if cultures are to be maintained for even longer periods (Sun

& Fleeger 1995). High-density *Tigriopus japonicus* cultures in 200 mt tanks have yielded from 1 kg to 5 kg day^{-1} for 3 months (Fukusho 1980). On a smaller scale, *Tisbe holothuriae* can be either batch (7–10 day cycle) or, with further system development, continuously cultured on daily feeds of microalgae. In the batch system, shallow trays (40 × 60 cm) containing 3 L of 0.22 μm filtered seawater were initially stocked with 40 000 ovigerous females, and subsequently harvested daily. The copepods were fed each day with microalgae (400 mL *Rhodomonas* suspension at 1×10^6 cells mL^{-1}). Average production was 300 000 nauplii daily. In the prototype, through-flow, continuous culture system, the copepods were grown in a 150 L conical bottomed vessel containing biofilter media (4.5 cm polypropylene balls) to increase the surface area available for grazing. Computer controlled valves governed the input of algae, water flow and harvesting. The system yielded 230 000 nauplii daily but production was limited by food availability. Individuals from either system harvested on a 60–80 μm mesh could be stored for up to 3–5 days at <5°C prior to feeding to larvae (Støttrup & Norsker 1997). Many ornamental fish require very small live prey and another harpacticoid, *Euterpina acutifrons*, whose adults are about half the weight of *Artemia* nauplii and whose nauplii are only 40–50 μm in size, is cultured in Hawaii for this purpose. They are grown in 450 L tanks at 26°C, stirred and aerated using large air bubbles and fed on live microalgae. Small bubbles must be avoided since they stick to larvae and carry them to the surface. Production cycle time is about 1 week (Szyper *et al.* 2000).

7.11.3.2 Calanoid copepods: Acartia, Centropages, Eurytemora, Gladioferens

Calanoids are generally filter-feeding planktonic crustaceans which feed readily on suspended microalgae or other microparticulate material. In contrast to harpacticoids, they only tolerate low culture densities (300 adults L^{-1}) and comparatively few successful large-scale cultures have been reported. One large-scale culture of *Acartia tsuensis* in 24 m^3 fertilised tanks produced sustainable exploitation rates of 27% per day (Ohno *et al.* 1990) and a promising new method for the culture of tropical species of *Acartia* has recently been described (Schipp *et al.* 1999). Stock cultures maintained in 60 L of gently aerated seawater (28–32°C and 30–34‰ salinity) were used to inoculate 1000 L conical bottomed culture tanks with 50–100 adults and 150–200 copepodids L^{-1}. Three species of algae (*Rhodomonas, Tetraselmis* and *Isochrysis*, in the ratio 2 : 1 : 1) were fed daily to a final density of 20 000 cells mL^{-1}. This provided a broad range of nutritional components as well as different sizes of food for the different life cycle stages. On average 750 copepodids and 320 adults L^{-1} were harvested every 8 days from the vessels, which were productive for over 6 months. Resting or diapause eggs of *Centropages hamatus* can be produced and stored for several months until required for hatching and feeding to marine fish larvae. Carboy vessels of 19 L capacity yielded 100 000 eggs each, in 2–4 weeks using a diet of mixed dinoflagellate algae (Marcus & Murray 2000).

One temperate, estuarine species (*Gladioferens imparipes*) has been identified that seems to tolerate higher adult densities during culture and does not cannibalise its nauplii. This may be because late stage copepodids and adults adhere to surfaces (rather like harpacticoids) while the nauplii remain pelagic. Further research on its suitability for fish larvae and large-scale culture seems likely (Payne & Rippingale 2000).

7.11.4 Mysids

Mysid shrimp (10–20 mm in length) are much larger than cultured copepods and are a favourite prey of cuttlefish and squid (Hanlon *et al.* 1991; Lee *et al.* 1994). They are also widely used in toxicity testing and bioassay procedures. This has generated interest in developing techniques for their culture.

7.11.4.1 Mysidopsis

The estuarine opossum shrimp *Mysidopsis almyra* is cultured in shallow trays (125 × 50 × 8 cm water depth) of gently aerated seawater at 20‰ salinity and with pH levels maintained above pH 7.6. Temperature is reduced weekly from 28°C through 23°C to 20°C as the mysids grow to adults. The water is renewed every 12–24 h from an integral biological filtration system. The mysids are fed newly hatched *Artemia* nauplii and grow to 4 mm in length in 7 days. Adult densities can be high (50 L^{-1}) and yields can average 133 newly hatched mysids per day for 17 weeks. In trials, *Artemia* were the major cost item (Domingues *et al.* 1998, 1999), but costs could be reduced during *Leptomysis* culture by feeding rotifers for the first 20 days followed by a mixed diet containing 33% *Artemia* nauplii and 66% rotifers (Domingues *et al.* 2000).

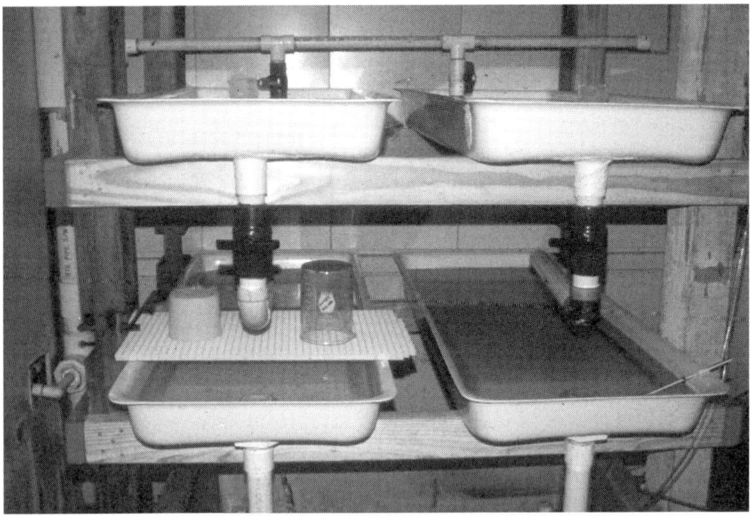

Plate 7.17 Production of mysid shrimp as live food for cuttlefish and squid. The upper trays contain adult broodstock and the lower the hatchlings. The hatchlings are transferred by draining from the top trays where they are spawned to the bottom trays to avoid cannibalism. The large, lower tank (barely visible) contains the filter through which the water is recycled once per day. Mysid production averaged 250 hatchlings per day for over 120 days. (Photo courtesy P.G. Lee, University of Texas Medical Branch, Galveston, USA.)

7.11.5 References

Alam M.J., Ang K.J. & Cheah S.H. (1993) Weaning of *Macrobrachium rosenbergii* (de Man) larvae from *Artemia* to *Moina micrura* (Kurz). *Aquaculture*, **112** (2–3) 187–194.

Baert P., Bosteels T. & Sorgeloos P. (1996) Pond production. In: *Manual on the Production and Use of Live Food for Aquaculture* (eds P. Lavens & P. Sorgeloos), pp. 196–251. FAO Fisheries Technical Paper No. 361. Food and Agriculture Organization, Rome, Italy.

Baert P., Anh N.T.N., Quynh V.D., Hoa N.V. & Sorgeloos P. (1997) Increasing cyst yields in *Artemia* culture ponds in Vietnam: the multi-cycle system. *Aquaculture Research,* **28**, 809–814.

Bonou C.A. & Saint-Jean L. (1998) The regulation mechanisms and yield of brackish water populations of *Moina micura* reared in tanks. *Aquaculture,* **160** (1–4) 69–79.

Camara M.R. (1996) *Artemia* production in coastal saltworks in Brazil: past, current practices, and perspectives. In: *Improvement of the commercial production of marine aquaculture species. Proceedings of a workshop on fish and mollusc larviculture* (eds G. Gajardo & P. Coutteau), pp. 173–178. Impresora Creces, Santiago, Chile.

Delbare D. & Dhert P. (1996) Cladocerans, nematodes and trochophora larvae. In: *Manual on the Production and Use of Live Food for Aquaculture* (eds P. Lavens & P. Sorgeloos), pp. 283–295. FAO Fisheries Technical Paper No. 361. Food and Agriculture Organization, Rome, Italy.

Delbare D., Dhert P. & Lavens P. (1996) Zooplankton. In: *Manual on the Production and Use of Live Food for Aquaculture* (eds P. Lavens & P. Sorgeloos), pp. 252–282. FAO Fisheries Technical Paper No. 361. Food and Agriculture Organization, Rome, Italy.

De Wolf T., Dehasque M. & Coutteau P. (1998) Intensive hygenic *Artemia* production. *Bulletin of the Aquaculture Association of Canada,* **98** (4) 25–26.

Dhert P., Bombeo R.B., Lavens P. & Sorgeloos P. (1992) A simple semi flow-through culture technique for the controlled super-intensive production of *Artemia* juveniles and adults. *Aquacultural Engineering,* **11** (2) 107–119.

Dhont J., Lavens P. & Sorgeloos P. (1993) Preparation and use of *Artemia* as food for shrimp and prawn larvae. In: *Handbook of Mariculture,* 2nd edn, *Vol. 1 Crustacean aquaculture* (ed. J.P. McVey), pp. 61–93. CRC Press, Boca Raton, FL, USA.

Domingues P.M., Turk P.E., Andrade J.P. & Lee P.G. (1998) Pilot-scale production of mysid shrimp in a static water system. *Aquaculture International,* **6**, 387–402.

Domingues P.M., Turk P.E., Andrade J.P. & Lee P.G. (1999) Culture of the mysid, *Mysidopsis almyra* (Bowman), (Crustacea: Mysidacea) in a static water system: effects of density and temperature on production, survival and growth. *Aquaculture Research,* **30**, 135–143.

Domingues P.M., Fores R., Turk P.E., Lee P.G. & Andrade J.P. (2000) Mysid culture: lowering costs with alternative diets. *Aquaculture Research,* **31** (8–9) 719–728.

Estévez A., McEvoy L.A., Bell J.G. & Sargent J.R. (1998) Effects of temperature and starvation time on the pattern and rate of loss of essential fatty acids in *Artemia* nauplii previously enriched using arachidonic acid and eicosapentaenoic acid-rich emulsions. *Aquaculture,* **165** (3–4) 295–311.

Evjemo J.O. & Olsen Y. (1999) Effect of food concentration on the growth and production rate of *Artemia franciscana* feeding on algae (*T.Iso*). *Journal of Experimental Marine Biology and Ecology,* **242** (2) 273–296.

Evjemo J.O., Coutteau P., Olsen Y. & Sorgeloos P. (1997) The stability of docosahexaenoic acid in two *Artemia* species following enrichment and subsequent starvation. *Aquaculture,* **155** (1–4) 135–148.

Fukusho K. (1980) Mass production of a copepod *Trigriopus japonicus* in combination culture with a rotifer, *Brachionus olicatilis*, fed ω-yeast as a food source. *Bulletin of the Japanese Society of Scientific Fisheries*, **46** (5) 625–629.

GomezGil-RS B., Abreu-Grobois F.A. Romero-Jarero & de los Herrera-Vega M. (1994) Chemical disinfection of *Artemia* nauplii. *Journal of the World Aquaculture Society,* **25** (4) 579–583.

Han K., Geurden I. & Sorgeloos P. (2000) Enrichment strategies for *Artemia* using emulsions providing different levels of *n-3* highly unsaturated fatty acids. *Aquaculture,* **183** (3–4) 335–347.

Hanlon R.T., Turk P.E. & Lee P.G. (1991) Squid and cuttlefish mariculture: an updated perspective. *Journal of Cephalopod Biology*, **2,** 31–40.

Harel M., Ozkizilcik S., Lund E., Behrens P. & Place A.R. (1999) Enhanced absorption of docosahexaenoic acid (DHA, 22:6n-3) in *Artemia* nauplii using a dietary combination of DHA-rich phospholipids and DHA-sodium salts. *Comp. Biochem. Physiol., B-Biochem. Mol. Biol.,* **124** (2) 169–176.

Hoff F.H. & Snell T.W. (1997) *Plankton culture manual,* 142 pp. Fourth Edition, Florida Aquafarms Inc., FL, USA.

Jana B.B. & Chakrabarti L. (1997) Effect of manuring on in situ production of zooplankton *Daphnia carinata. Aquaculture,* **156** (1–2) 85–99.

Lavens P. & Sorgeloos P. (1996) Manual on the production and use of live food for aquaculture, 295 pp. *FAO Fisheries Technical Paper* (361). FAO, Rome, Italy.

Lavens P. & Sorgeloos P. (2000) The history, present status and prospects of the availability of *Artemia* cysts for aquaculture. *Aquaculture,* **181** (3–4) 397–403.

Lavens P., Thongrod S. & Sorgeloos P. (2000) Larval prawn feeds and the dietary importance of *Artemia.* In: *Freshwater Prawn Culture: the farming of* Macrobrachium rosenbergii (eds M.B. New & W.C. Valenti), pp. 91–111. Blackwell Science, Oxford, UK.

Lee P.G., Turk P.E., Yang W.T. & Hanlon R.T. (1994) Biological characteristics and biomedical applications of the squid *Sepioteuthis lessoniana* cultured through multiple generations. *Biological Bulletin,* **186**, 328–341.

Léger P. & Sorgeloos P. (1992) Optimized feeding regimes in shrimp hatcheries. In: *Marine Shrimp Culture: principles and practices* (eds A.W. Fast & L.J. Lester), pp. 225–244. Elsvier Science, New York, USA.

Marcus N.H. & Murray M.M. (2000) A reliable source of copepod nauplii for rearing marine fish larvae: the potential of diapause eggs. In: *Abstracts, Aqua 2000, Responsible Aquaculture in the New Millennium* (compiled by R. Flos & L. Creswell), p. 437. European Aquaculture Society, Special Publication No. 28.

McEvoy L.A. & Sargent J.R. (1998) Problems and techniques in live prey enrichment. *Bulletin of the Aquaculture Association of Canada,* (4) 12–16.

Merchie G., Lavens P., Dhert Ph., Dehasque M., Nelis H., De Leenheer A. & Sorgeloos P. (1995) Variation of ascorbic acid content in different live food organisms. *Aquaculture,* **134** (3–4) 325–337.

Mims S.D., Webster C.D., Tidwell J.H. & Yancey D.H. (1991) Fatty acid composition of *Daphnia pulex* cultured by two different methods. *Journal of the World Aquaculture Society,* **22** (2) 153–156.

Murugan N. (1989) The mass culture of a cladoceran, *Daphnia carinata* (King), for use as food in aquaculture. In: *Aquacultural Research in Asia: management techniques and nutrition* (eds E.A. Huisman, N. Zonneveld & A.H.M. Bouwmans), pp. 190–202. Pudoc Wageningen, Netherlands.

Næss T. & Bergh Ø. (1994) Calanoid copepod resting eggs can be surface-disinfected. *Aquacultural Engineering,* **13** (1) 1–9.

Ohno A., Takahashi T. & Taki Y. (1990) Dynamics of exploited populations of the calanoid copepod, *Acartia tsuensis. Aquaculture,* **84** (1) 27–39.

O'Sullivan D. (1997) Brine shrimp success draws overseas interest. *Austasia Aquaculture Magazine,* **11** (1) 40–43.

Payne M.F. & Rippingale R.J. (2000) Evaluation of diets for the culture of the calanoid *Gladioferens imparipes. Aquaculture,* **187** (1–2) 85–96.

Polk A.E., Amsden B., Scarratt D.J., Gonzal A., Okhamafe A.O. & Goosen M.F.A. (1994) Oral delivery in aquaculture: controlled release of proteins from chitosan-alginate microcapsules. *Aquacultural Engineering,* **13** (4) 311–323.

Sahul Hameed A.S. & Balasubramanian G. (2000) Antibiotic resistance in bacteria isolated from *Artemia* nauplii and efficacy of formaldehyde to control bacterial load. *Aquaculture,* **183** (3–4) 195–205.

Schipp G.R., Bosmans J.M.P. & Marshall A.J. (1999) A method for hatchery culture of tropical calanoid copepods, *Acartia* spp. *Aquaculture,* **174** (1–2) 81–88.

Shields R.J., Bell J.G., Luizi F.S., Gara B., Bromage N.R. & Sargent J.R. (1999) Natural copepods are superior to enriched *Artemia* nauplii as feed for halibut larvae (*Hippoglossus hippoglossus*) in terms of survival, pigmentation and retinal morphology: relation to dietary essential fatty acids. *Journal of Nutrition,* **129** (6) 1186–1194.

Støttrup J.G. (2000) The elusive copepods: their production and suitability in marine aquaculture. *Aquaculture Research,* **31** (8–9) 703–711.

Støttrup J.G. & Norsker N.H. (1997) Production and use of copepods in marine fish larviculture. *Aquaculture,* **155** (1/4) 231–247.

Støttrup J.G., Shields R., Gillespie M., *et al.* (1998) The production and use of copepods in larval rearing of halibut, turbot and cod. *Bulletin of the Aquaculture Association of Canada,* (4) 41–45.

Sun B. & Fleeger J.W. (1995) Sustained mass culture of *Amphiascoides atopus* a marine harpacticoid copepod in a recirculating system. *Aquaculture,* **136** (3–4) 313–321.

Szyper J.P., Brittain K., Tamaru C. & Ako H. (2000) Copepod nauplii as food for marine ornamental fish larvae. Excerpt from *Makai Sea Grant College Program,* **22** (1).

Thursten S. (1995) Clam shrimp and fairy shrimp in fish fry ponds can halve production. *Austasia Aquaculture Magazine,* **9** (6) 13–18.

Touraki M., Rigas P. & Kastritis C. (1995) Liposome mediated delivery of water soluble antibiotics to the larvae of aquatic animals. *Aquaculture,* **136** (1–2) 1–10.

Touraki M., Niopas I. & Kastritis C. (1999) Bioaccumulation of trimethoprim, sulfamethoxazole and *N*-acetyl-sulfamethoxazole in *Artemia* nauplii and residual kinetics in seabass larvae after repeated oral dosing of medicated nauplii. *Aquaculture,* **175** (1–2) 15–30.

Wickins J.F. (1972) The food value of brine shrimp *Artemia salina* L., to larvae of the prawn, *Palaemon serratus* Pennant. *Journal of Experimental Marine Biology and Ecology,* **10**, 151–170.

Chapter 8
Techniques: General

8.1 Materials

Suitable materials for the construction of culture vessels and for use in pumps and plumbing are judged by two main criteria, their toxicity to the cultured species and their resistance to corrosion. Other attributes such as strength, weight, ease of working and cost are more easily recognised. As a general rule, toxicity is likely to be more of a problem in recirculation systems where dissolved substances can accumulate to harmful levels in the water, while corrosion is a major problem in marine and brackish-water systems. It is also worth remembering that many species accumulate dissolved substances to toxic levels in their bodies. One example is copper, which is a vital component of crustacean respiratory pigment but which is also toxic when in excess. Wheaton (1977), Hawkins and Lloyd (1981), Dexter (1986), Muir (1988) and Huguenin and Colt (1989) provide useful reviews of materials suitable for aquaculture systems.

The use of materials other than quarried rock (e.g. various concrete mixes, tyres, iron and steel) for the construction of artificial reefs (section 8.11.2) is a contentious issue in some countries because of the possibility of toxic materials leaching into the environment (Collins & Jensen 1997) and the risk of structural disintegration. A number of research programmes to determine the acceptability of potentially useful materials are in progress (Jensen *et al.* 2000a).

8.1.1 Concrete

Concrete is commonly used in aquaculture systems and storage reservoirs. Seawater-resistant concretes are available (e.g. sulphate-resistant Portland cement to British Standard BS 4027: 1972) but two problems frequently arise in use. The first is the elevated pH and alkalinity levels that occur in water in contact with new concrete; the second is the break-up of the concrete caused by corrosion of the reinforcing steel inside. The mix and curing conditions are critical factors in stability in the marine environment (Yuebo & Kwok-Hung 1997). The risk of problems arising due to high alkalinity is reduced if the concrete is thoroughly washed for several weeks with periodic renewals of water. In some cases it may be necessary to coat the concrete with epoxy resin paint before the animals are introduced. Competent mixing and pouring is essential if corrosion is to be prevented.

In Europe, by-products from coal-fired electricity generating stations (pulverised fuel ash (PFA) and flue gas desulphurisation gypsum) are stabilised with cement to form the concrete blocks often used in marine artificial reefs (Jensen *et al.* 2000b). Other cement-stabilised materials include oil and domestic incinerator ash, some furnace slags, phosphogypsum (a by-product from the fertiliser industry) and harbour dredgings. Tyres too may be compressed and consolidated in concrete blocks (Figley 1994). The heavy metal content of these materials varies with their origin, which means that all the potential end products require independent toxicity tests before they can be accepted for deployment.

8.1.2 Metals

As a general rule, and especially in recirculation systems, metals should not be allowed to come into contact with the culture water. In particular, metals to be avoided include copper, zinc, and alloys containing these metals such as brass, gunmetal and bronze that are found in some pumps and valves. Iron and most steels corrode readily in seawater, exceptions being titanium steel and, to a lesser extent, type 316 stainless steel. Newer, corrosion-resistant alloys with higher strength than 316 stainless steel have been developed, but may be too costly for general use. Protective coatings can be applied to most metals but in our experience are rarely satisfactory in the long term. The slightest fault or damage to the coat can lead to corrosion spreading beneath the coating and the unsuspected release of toxic materials into the water. Condensation dripping from galvanised bolts or cadmium-tipped masonry nails is another example of a potential source of contamination. Interestingly, Méndez *et al.* (1997) reported that levels of copper, cadmium and zinc in shrimp hepatopancreas tissue increased with stocking density during extensive/semi-intensive ongrowing in ponds although they were cautious about implications regarding possible toxic effects.

8.1.3 Plastics and other materials

A wide range of plastic, fibreglass and epoxy resin-based materials is used in culture systems. The safest from a toxicity point of view are 'food grade' materials but it is nevertheless advisable to soak all materials in several changes of water for 10–14 days before use, to reduce levels of potentially toxic leachates (Carmignani & Bennett 1976). Additives such as colourants, plasticisers, antioxidants and stabilisers are often present in plastics and leach very readily from recycled plastics in particular. Mould release agents may be present on fibreglass tanks, which again should be well soaked before use. Wood preservatives are usually toxic (Mercaldo-Allen & Kuropat 1994) and can be carried into the water, for example from walled enclosures, cage and pen supports and by drips from tanks held on wooden frames above reservoirs. Marine grade plywood sealed with epoxy resin paint is widely used in indoor installations.

8.1.4 Pond sealing materials

Several methods are used to line ponds built in areas where porosity, inward seepage or other soil inadequacies prevail (Wheaton 1977). All require proper prior pond preparation, removal of vegetation, sticks, stones, etc. The methods include compaction of suitable soil (about 20% clay) to a depth of 20 cm with about six passes of a sheepsfoot roller or, more expensively, liners (Singh 1993) or reinforced membrane systems (Stroethoff & Hovers 1996). Liners may be made of polyethylene, synthetic rubber, polypropylene, or polyvinylchloride (PVC) and ideally they should be UV-stabilised to resist sunlight if and when they are exposed. It is advisable to protect liners with about 15–25 cm of sand or coarse soil (sections 8.2.2.3 and 8.3.9). The same recommendations about soaking for several weeks with frequent water changes apply here as to plastics.

8.2 Pond design and construction

We are concerned with the earthen or predominantly earthen ponds widely used in outdoor crustacean farming and fattening operations. We do not address the design or construction of structures such as tanks or raceways. These aspects can be found in Wheaton (1977).

8.2.1 Layout and configuration

The layout and configuration of ponds within a crustacean farm will be largely determined by the type of culture to be performed (species and intensity) and the characteristics of the site, particularly the topography, soil type and the positioning of the water supply. Before design of the layout, a comprehensive survey is vital to determine the soil type (to a depth at least 1 m below the intended base of the ponds) and detailed levels throughout the whole area involved (sections 6.3.3 and 9.5.1) In addition to these basic considerations, the design of farms is now starting to reflect the need for effective disease management and for reduced environmental impacts. It is no longer usually just a matter of linking production ponds to a water supply and a drainage canal. For example, disease control in many Thai shrimp farms requires closed or semi-closed water management (section 8.3.7) as well as the use of ponds for reservoirs and for effluent treatment prior to water reuse. Many farms now incorporate settlement ponds to receive effluents and these bodies of water act as a buffer between the production ponds and the outside environment. One farm in Saudi Arabia uses round production ponds of 1 ha grouped into blocks of 18, and each block receives water from an 8 ha 'pre-greening' reservoir in which algae blooms are initiated. The effluent flows into a 10 ha settlement pond before returning to the sea. Thus for each 18 ha of ongrowing ponds there are another 18 ha of ponds dedicated to pretreatment and post-treatment of the water (section 7.2.6.5).

If an extensive operation is planned, then the general objective will be to perform a minimum of earthworks to create a maximum surface area of ponds. In the case of semi-intensive culture, attention should also be given to creating straight embankments and roughly rectangular ponds and, if intensive production is intended, then ponds should be restricted in size to a maximum of about 0.5–1 ha. Whereas most sites can be suitable for building small ponds, large extensive culture ponds (5–100 ha) require land with a shallow gradient such as that found in coastal and estuarine margins or on alluvial plains.

If the water reaches the farm site by gravity (freshwater, river or stream) or tidal flow (brackish water or seawater), the elevation of the ponds will be restricted in accordance with the level of the water supply. A pumped water supply, on the other hand, enables land above the level of the water source to be exploited and thus allows more flexibility in the arrangement of the farm and its drainage. The water intake point must be located away from the discharge of the same and other farms, but it may not be feasible to draw water more than about 1 km or so from the site (Muir & Lombardi 2000).

A reservoir can be advantageous for tidally flushed shrimp farms since it allows for water exchange to be performed for longer periods than would be possible on high tides alone. Even if a pumped water supply is provided it may only be efficient to operate the pumps on high tides, so again, a reservoir can be useful to extend the duration of water exchanges. With this in mind, the distribution canal in many Ecuadorian shrimp farms is often widened to 30–60 m so that it acts as a reservoir. If a water quality problem is encountered, for example low dissolved oxygen concentrations at night, the reserve of water can be used to rapidly flush a pond. A reservoir can however act as a large sediment trap, so ideally it should be drainable to enable accumulated silt to be removed. For some crayfish farms with a seasonal or intermittent water supply, a dam or reservoir may be essential.

Prevailing wind direction may need to be taken into account when a farm layout is designed, either to maximise wind induced circulation or, by orientating embankments to interrupt the fetch of the wind, minimising the risks of wave damage.

Some examples of crayfish farm designs are given in Chapter 7 (Fig. 7.2). Further arrangements for crustacean ponds are presented in Fig. 8.1. Ponds with curved

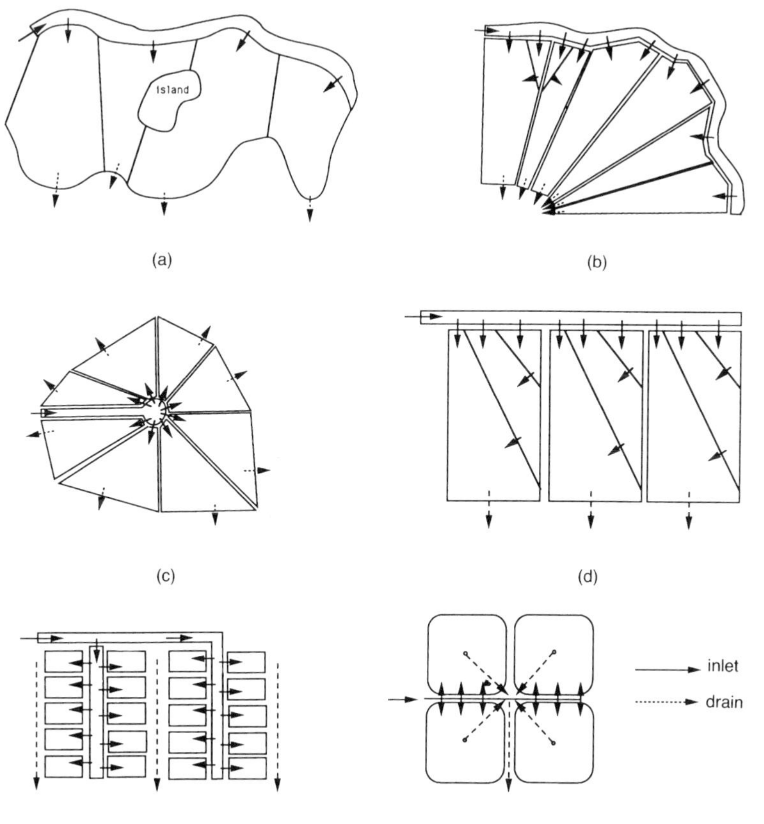

Fig. 8.1 Examples of pond layouts: (a) supply canal and ponds following natural contours; (b) supply canal following natural contours, two small nursery ponds are included; (c) daisywheel; (d) three-phase pond system; (e) rectangular ponds; (f) centrally drained, concrete-walled ponds.

margins are feasible for extensive culture and can be arranged to take advantage of site contours. In the three-phase system (Figure 8.1d) the small triangular ponds act as nurseries, from which the juveniles are transferred by gravity to the larger adjacent ponds for two successive phases of ongrowing. In theory this arrangement helps to maximise the productive use of the available surface area of the farm. In practice, though, it is vulnerable to irregularities in the supply of juveniles and becomes inefficient when the three steps get out of phase. Rectangular or square ponds are most convenient when they are of a uniform size since this allows standard size nets to be used (important for frequent *Macrobrachium* seine harvests; section 7.3.6) and straightforward calculation of drain and fill times. With inlet and outlet points positioned in opposing short embankments, rectangular ponds permit efficient water circulation and exchange. Modern square ponds often have central drain structures. Various other shapes, such as triangles and rhomboids, are often incorporated in farm designs to maximise use of the available land, and these too can be suitable provided no unproductive 'dead' spots are created by restricted water movement. If nursery ponds are included in the farm design then these are best located near to the main centre of activity since this will facilitate management. A farm layout also needs to be designed with security and access needs in mind. On a sloping site it usually makes sense to put the main access and infrastructure on the higher ground and the ongrowing ponds in lower areas (Muir & Lombardi 2000).

The need for efficient water movement has encouraged the use of round ponds and tanks for intensive and super-intensive culture. Aerators or water jets create circular currents that provide efficient mixing and help to sweep accumulated waste to central drains. However there are limits to the size of round ponds, particularly if super-intensive culture is planned. Centre drains work well in small ponds without soil bottoms but they do not work well in larger ponds with soil bottoms. In ponds of 0.25 ha or greater it is not economically practical to produce strong enough water currents to move solid wastes to the centre, and even if it were so, soil eroded by the strong water currents would settle in the centre of the basin and bury the drain.

8.2.2 Construction

The first step in construction may need to be the formation of a perimeter drainage ditch to prevent waterlogging of the construction area. On the other hand, if the earth is too dry, at some stage it may require moistening for the construction of solid embankments. To provide a reference point for surveying work it is usually necessary to establish a primary datum point in the form of a permanent marker or a prominent durable natural or man-made feature. Once the site has been cleared of vegetation and the topsoil put aside (for covering embankments and roadsides later on) the layout of canals, roads and ponds can be marked with stakes. The process of pond construction can then begin, usually by a process of cutting and filling in which soil is skimmed from the pond beds and transferred to form embankments, normally with the aid of earthworking machinery. High-volume earth-moving is most economically done with a scraper (LICA 1988). Alternatively, construction can be performed by excavators, although ponds created in this way are typically small and can be difficult to drain without the aid of pumps. In some developing countries it makes economic sense to use manual labour rather than machinery, particularly for small ponds, where one person might be expected to move on average 1 m^3 of earth per day.

For planning purposes (Estilo 1988), the process of building a series of ponds, complete with drainage and supply canals, can be divided into twelve steps:

(1) Site clearing
(2) Topsoil stripping
(3) Staking of centre lines and templates
(4) Preparation of embankment foundations
(5) Excavation of drainage canals
(6) Construction of embankments
(7) Forming and compaction of embankments
(8) Excavation of pits for water control structures
(9) Levelling of pond bottom
(10) Construction of water control structures and refilling of pits
(11) Construction of elevated canals
(12) Construction of embankment protection

However, it may not always make sense to follow this sequence precisely; if a large number of ponds are to be constructed in phases, it can be advantageous to start with the elevated water supply canal and then build the ponds around it, because in this way some ponds can be brought into production before the whole farm is completed.

Muir and Lombardi (2000) describe the process of earthen pond construction by cutting and filling, with a particular focus on freshwater prawn ponds in areas with gradients of between 2% and 5%. To minimise earth movements and to ensure that embankments are correctly built, the process is usually divided into two operations. Firstly, soil is cut away from the highest areas and transferred to the lowest areas to create a level and well-compacted platform. Secondly, new cuts are made into the platform to provide material for building the embankments (section 8.2.2.1). Freshwater prawn ponds share many similarities with ponds used for the semi-intensive culture of Australian crayfish (section 7.7.6.3).

To ensure efficient drainage, pond beds should be sloped with a minimum gradient of 0.1% or preferably 0.2–0.5%. Good drainage can be essential for harvesting and pond preparation (section 8.3.3). Tree stumps, peat and other organic material should be removed from the pond beds and all holes or depressions must be filled because they can retain animals during harvesting and may harbour pests between crops. Channels cut into the bed can improve the drainage and provide deepened areas that serve as refuges from extremes of ambient temperatures. The channels may be arranged around the periphery of a pond, diagonally across the middle, or in the shape of a fishbone, but to ensure efficient harvesting they must all be sloped to the outlet point. Deep peripheral channels (known as 'prestamos' in parts of Latin America), such as those found in many Ecuadorian and Thai farms, are created to provide material for adjacent embankments. However, these channels readily become fouled with black organic sludge and often prove laborious and difficult to clean.

Sites for pond construction should be chosen with impermeable soils that are suited to embankment construction (section 6.3.3). In some situations permeable soils can be sealed using clay blankets or the addition of a sealant such as bentonite, a fine-grained clay, although this can add greatly to construction costs. When using sealants, laboratory analysis of soil will be necessary to ascertain what type of sealant is appropriate and what quantities will need to be applied. Embankments can be made impervious by using sealants or by incorporating a clay barrier. The latter may take the form of a layer of clay on the embankment's inside surface, or a central clay core (also known as a 'key' or a foundation cut-off) around 0.5 m thick that extends down to an impervious layer of the substrata.

Pond size and shape have an important bearing on the amount of earth that needs to be moved to construct a farm. The relationship between pond size and volume of earthworks for a 40 ha farm is presented in Fig. 8.2a. Larger ponds require less embankment per hectare and consequently they are less expensive per hectare to build. On the other hand, they are unsuited to intensive culture methods because of the difficulty of effectively regulating water quality over such a large area. A size of 8 ha is considered ideal for extensive crayfish ponds in the USA (Avault & Huner 1985). *Macrobrachium* ponds typically measure 0.2–0.5 ha (Muir & Lombardi 2000).

The relationship between pond shape and volume of earthworks is illustrated in Fig. 8.2b. Clearly, square-shaped ponds (length : width ratio 1 : 1) require the smallest volume of earth to be moved per hectare (only rectangular forms considered). All the same, despite their added cost, elongated ponds (length : width ratio 2.5–4 : 1) are often preferred since this shape facilitates feeding, harvesting and pond maintenance. Ponds of maximum width 30–50 m are favoured for freshwater prawn farming, especially where the frequency of seining is high, while narrow, canal-type ponds, 2–20 m wide and 50–150 m long, are popular for semi-intensive on-growing of crayfish because of the bank burrowing habits of these crustaceans. Canal-type ponds in the UK are usually 1.5–2 m deep with embankments 2–3.5 m wide or wider if the soil is unstable (section 7.6.6.2). Different

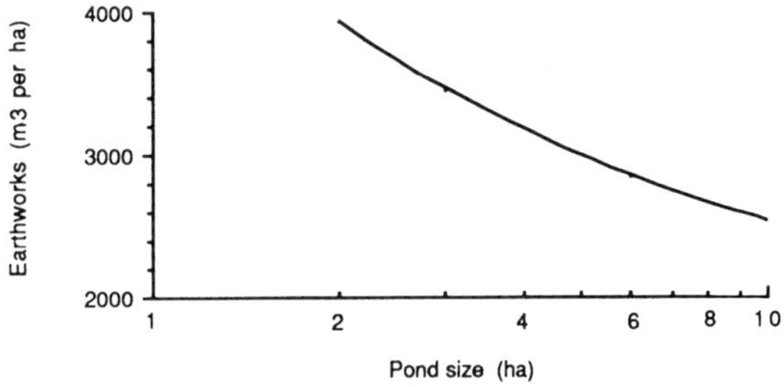

Fig. 8.2a Relationship between the amount of earth to be moved per ha and pond size for a 40 ha farm (based on Yates 1988).

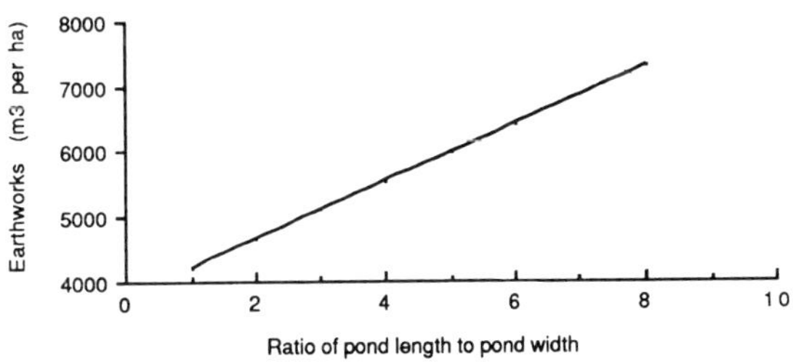

Fig. 8.2b Relationship between pond shape and amount of earth required to be moved for a 40 ha farm with ponds of 2 ha each (modified from Yates 1988).

aspects of pond design affect the profitability of shrimp farms in that pond shape and pond size, for example, have a much greater influence on financial viability than the slope and width of embankments (section 10.6.1.5).

8.2.2.1 Embankments

In general embankments should be overbuilt because they will always be subject to erosion, both from wave action at the edge and rainfall on the surface. Embankments made with sandy soils are particularly at risk and may need to be twice as wide as clay ones. The optimum height of embankments depends partly on the species in culture and the climate involved. Deep ponds can protect a crop against extremes of temperature but they also take longer to warm up than shallow ponds and this may not be desirable in some temperate climates. For extensive crayfish ponds a water depth of 0.3–0.5 m is adequate and enables low-cost levees to be built (Avault & Huner 1985). Minimum water depth in other cases, however, should be 0.8 m, to reduce the growth of vegetation on the pond bed. In brackish-water ponds this minimum depth helps guard against salinity fluctuations caused by heavy rains. Average depths of 0.8–1.5 m are usually ideal for semi-intensive and intensive farming, although even deeper ponds may be desirable where plastic liners are used (sections 7.2.6.5 and 12.8.1). Some satisfactory results have been reported in freshwater prawn ponds with just 0.3–0.4 m of water but only in an area with little diurnal temperature fluctuation (Valenti & New 2000).

Embankment construction should allow for an additional freeboard of 0.3–0.7 m. The width of an embankment at its top depends primarily on access requirements, but should at least equal the height and should never be less than around 1 m. A width of 2–3 m is more usual and 3.5–5.0 m may be required if the embankment is to carry vehicles safely. On exterior dry faces a slope of 1–2 : 1 (horizontal : vertical) is suitable, whereas inside faces in contact with water need shallower slopes with a minimum gradient of 2 : 1 and preferably 3–4 : 1. The shallow slope is especially important if embankments are made of light earth (non-cohesive) or will be subject to strong wave action.

For the construction of ponds for integrated rice and prawn culture in seasonally flooded land, it may be necessary to build perimeter embankments 1 m higher than the surrounding land to avoid inundation. These types of ponds can usually have steep-sided, almost vertical, embankments, because of the heavy clay soils typical in rice-growing areas, but they require frequent maintenance and rebuilding.

It is important to construct embankments correctly at the outset because mistakes are very difficult to remedy and embankment failure can jeopardise the stock of a whole farm. The first step is to excavate down to a watertight, impervious foundation. Embankments should then be constructed in a series of layers of about 20 cm thick. Each layer must be thoroughly compacted, for which the soil may require moistening. Allowances should be made for embankment settlement that can be 10% of height, and for soil shrinkage that can be 10–20% of volume. Spaces can be left in the embankment for the installation of inlet and outlet gates, or alternatively the embankment can be constructed intact and later cut away in the relevant spots using a backhoe. Baffle levees (small embankments) can be positioned in some extensive ponds to direct water flow over a greater area of a pond (Avault & Huner 1985).

Vegetation is valuable for embankment stabilisation and it can be encouraged with a layer of topsoil. Excessive growth however is generally a hindrance to harvesting and should be controlled. Trees should not be planted because their roots will weaken the embankments. To reduce embankment erosion, gravel or porous plastic sheeting can be placed on inside surfaces, and to limit further wave damage on downwind sections an array of vertical wooden stakes can be effective. If one corner of the embankment is widened with a ramp it will allow for vehicle access during pond cleaning or maintenance.

New ponds should be filled slowly over a period of several days to allow the embankments to become fully saturated before they are subjected to the full weight of water.

8.2.2.2 Farm dams

Farm dams are usually built to provide reservoirs for irrigation or for livestock but they can also produce crops of crayfish and, by making use of existing land contours, they are usually straightforward to construct. After site clearance and topsoil stripping, a foundation cut-off is built to prevent seepage under the dam. This involves digging a trench along the centre line where the dam is to be built, down to a depth that extends to an impervious layer. The trench is then filled with impervious material in 10–15 cm layers each compacted with a heavy roller (sheepsfoot or grid roller). The water supply lines for irrigation or livestock and a spillway pipe (minimum diameter 15 cm) are then installed with anti-seepage collars prior to building the dam. To make the dam, soil is built up in layers and compacted, but if the soil is too dry to form a ball when moulded by hand it may be necessary to moisten it by spraying on water. An emergency spillway, generally wider than 3 m, is usually created to safely discharge excess flow round the dam, and the surfaces of the dam and spillway can be planted with grass to limit erosion. The dam width at the top is usually a minimum of 2.5 m, or 4.3 m if there is a roadway planned. The usual slopes for the dam walls are 2 : 1 on the dry side and 3 : 1 on the wet side and a freeboard of at least 30 cm is normal (LICA 1988).

8.2.2.3 Lined ponds

The use of pond liners makes it feasible to build ponds in sandy or acid sulphate soils that would normally prohibit pond aquaculture. By the early 1990s some 700 ha of plastic-lined ponds were reported to be in production in Oman and Indonesia (P. Fuke, 1990 pers. comm.). Today pond liners are widely used in shrimp ponds and are also reported in New Zealand prawn and Australian yabby farms. Lined ponds also have the advantage of isolating pond water from groundwater and preventing cross-contamination by seepage. Prior to installation the pond bed must be levelled and sloped to drain and it is vital to remove all sharp objects such as roots that could puncture the liner. Ponds must be prepared as precise rectangles to minimise waste of liner material. Sometimes a layer of sand or geotextile is laid beneath the liner. Sheets of liner material are either cut and joined in the field, or prefabricated in the factory. The edges of the liner rise up the embankments and are usually buried in a trench for anchorage. Subsoil vents or drains are important to allow the escape of gases and prevent the liner from 'ballooning'. Drainpipe fittings require special flanges to permit watertight joints with the liner material (Singh 1993).

Lined ponds have been used in areas with sandy soils for yabby farming. A 10 cm layer of sand covers the liners and extends up the embankments where it is seeded with grass. A 3 m high, solar powered electrified fence was erected on one farm to keep out poachers, as well as cattle and kangaroos that could damage the liner. In Indonesia shrimp ponds of 0.3 ha with a central drain have also been made using pond liners (Stroethoff & Hovers 1996). Again a layer of sand was placed on top of the liners but in this case the banks of the ponds were covered in an inexpensive and novel form of reinforced concrete employing bamboo frames and open woven bamboo sheets rather than steel. The same construction method was used for the central drain structure.

8.2.2.4 Inlet and outlet structures

In the simplest shallow ponds water may be controlled with plastic, metal or bamboo tubes fitted with valves or turn-down drains. The tubes are buried in the embankment and should ideally be fitted with collars to reduce water seepage. To prevent the passage of shrimp or other animals a staked net may be positioned upstream of the inlet, or a strainer of netting or split bamboo located on the end of the tube.

Monks (see Glossary) and sluices are more specialised water control devices (Fig. 8.3). A sluice may be constructed of wood, brick or concrete and forms an opening in an embankment. Vertical grooves located in the sluice walls accept mesh screens and wooden boards, the latter to control the water flow rate or pond level. Prefabricated ferro-cement sluice gates can be obtained in some countries. A monk is located within the pond or water supply canal and is connected to a tube or channel that passes through the embankment. The advantage of this arrangement is that the embankment is left largely intact and vehicles can easily pass over it. Monks also incorporate grooves for mesh screens and for water control boards (unless water flow is regulated by means of a valve located in the tube). To increase the surface area of screens and thus reduce the frequency of blocking, the upstream end of a monk may be flared or formed in the shape of a Y. Centrally drained intensive ponds are sometimes equipped with a central monk structure or a vertical standpipe surrounded by a mesh screen.

If two vertical stacks of boards are used for water control they can be set 15–25 cm apart and the space between them packed with soil to make a watertight barrier. The monk depicted in Fig. 8.3 is equipped with two sets of grooves for this purpose, another two sets for screens (coarse and fine) and a fifth set in which boards can be located to encourage the exchange of bottom water rather than surface water. To prevent the theft of a crop by draining, the water level control boards can be locked in position with a padlock or a secure lid. If drain harvesting is intended a harvesting basin can be located immediately behind the outlet gate. This will serve to keep the catch submerged as it is collected.

Pumping stations require solid foundations of concrete or of wooden piles and the point where the pumps discharge their water must be protected from erosion with stones, rocks or a concrete spillway.

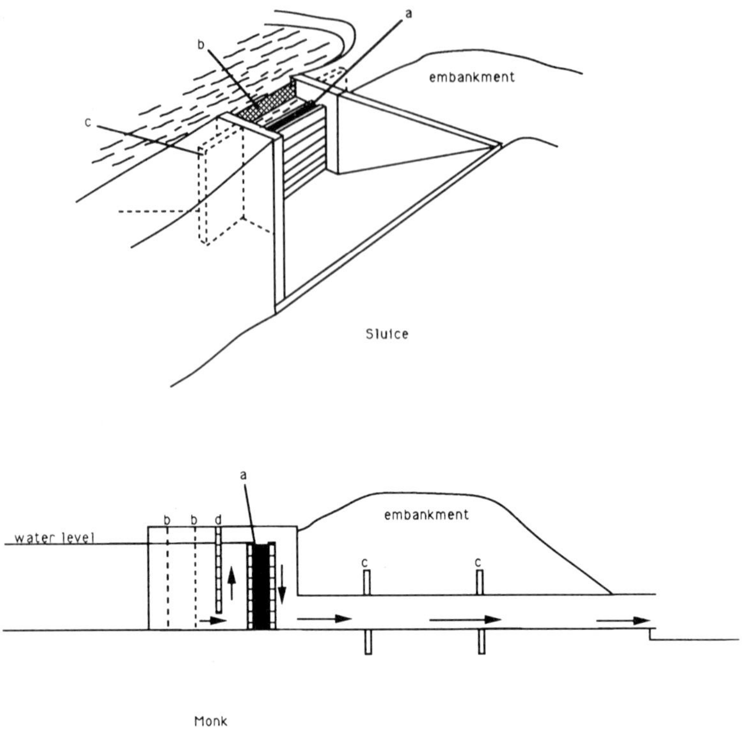

Fig. 8.3 Water control structures: (a) earth packed between two vertical stacks of wooden boards; (b) mesh screen; (c) brickwork or concrete keyed into embankment to stop seepage; (d) stack of boards with gap at base.

8.2.2.5 Construction in areas with acid sulphate soils

The special problems posed by acid sulphate soils are discussed in sections 6.3.3.5 and 8.3.8. Although affected sites should generally be avoided, there are three basic ways of tackling the problem:

(1) Drain the soils and wait until natural oxidation and leaching removes the acidity;
(2) Apply lime to neutralise the acidity;
(3) Prevent the oxidation of the iron pyrites so the acidity is not expressed.

Unfortunately it can take many years for acidity to leach naturally so it is not really economically feasible to build ponds and wait. Similarly, adding the full complement of lime may require 25–150 mt ha^{-1} and such large applications are not usually feasible (Boyd 1995a). Most programs for controlling acidity rely on a combination of techniques and aim to ameliorate rather than cure the problem. Repeated applications of moderate amounts of lime can be beneficial.

Some extensive hand-built fishponds have been constructed in acid sulphate areas by minimising the disturbance of existing soil layers (for example, sometimes roots are left in place). Since embankments are responsible for large amounts of acidity, building larger ponds (less embankment per hectare) can also have some benefits. The construction of small ponds may only be feasible if plastic liners are used, if walls are made of concrete, or if embankments are covered with a deep layer of non-acidic topsoil.

The approach described by Brinkman and Singh (1982) is designed to reclaim acid sulphate sites in a single dry season lasting only 4 months. It basically involves harrowing, drying and filling the ponds, and then allowing the pH of the water to stabilise before draining and repeating the procedure. Up to three or more treatment cycles may need to be performed to raise the water above pH 5. At the same time embankments are leached by repeatedly filling and draining shallow basins constructed along their tops. Before the pond is used agricultural limestone is applied at 500 kg ha^{-1} on the pond bottoms and at 0.5–1.0 kg m^{-2} on the embankments.

8.3 Pond management

8.3.1 Introduction

Every pond system has its own limitations to productivity based on its physical and chemical characteristics and available inputs of water, feeds and juveniles. The pond manager is faced with the challenge of making the most efficient and profitable use of these resources, knowing that conditions will rarely be optimal and that for the most part the crop will remain hidden from view. Successful pond management depends largely on collecting reliable and regular data on the condition of the crop and the status of the culture environment and deciding how this information may best be used in the application of fertilisers and feeds, and in the control of water exchange and aeration. Controlling the entry of predators, competitors and disease carriers is also of fundamental importance and a primary aim of pond management is to maintain adequate pond water and sediment quality to limit the stressful conditions that can precipitate disease outbreaks.

It is worth noting at this point that very significant variations in productivity can arise between different farms and ponds, due to differences in soil characteristics. This emphasises the important role that site selection can have on project viability and the need to investigate soil quality before a site is chosen (section 6.3.3).

As far as possible the following account presents the key aspects of pond management in a generalised manner of relevance to all pond-reared crustaceans, but, to illustrate particular points, many examples specific to particular crustacean species or groups are included. Further specific details of pond culture techniques are included in Chapter 7 along with other ongrowing methods.

8.3.2 Biological processes

An understanding of the basic biological processes at work in a pond, and the chemical changes they bring about, is very helpful to farmers if they are to manage pond conditions effectively and keep water quality within acceptable ranges (section 8.5).

Phytoplankton forms the basis of the natural food chain in a pond ecosystem and, in the process of photosynthesis, uses the energy of sunlight to synthesise organic molecules. The process can have a profound influence on water quality, largely because in the daytime it results in the liberation of oxygen and the consumption of carbon dioxide. Oxygen is essential for crustaceans and nearly all other organisms within a pond, since it is needed for respiration. However it can rise to harmful levels (see below). Carbon dioxide concentration is important primarily because of its influence on pH. It acts as an acid in water, so as it is removed during photosynthesis, acidity declines and pH rises. On sunny days the pH in ponds rich in phytoplankton can rise to pH 9 or 10, at which levels the growth rates of the crustaceans can be impaired.

During darkness the metabolism of the phytoplankton changes from photosynthesis to respiration (the process by which organic molecules are oxidised to obtain energy); oxygen is consumed and carbon dioxide is released to form carbonic acid. The latter adds to the other respiration products (e.g. ammonia) of the crustaceans being farmed (and those of nearly all other non-plant organisms in the pond) with the result that oxygen concentrations and pH fall by night to reach a minimum around dawn. Figure 8.6a illustrates a typical pattern of variation in pH and dissolved oxygen levels within a diurnal cycle in an outdoor pond. Unfortunately an excess of oxygen at one time of day does not compensate for a deficit at another; both high and low concentrations can be harmful to crustaceans. In fact, abrupt changes in any water quality factor are likely to be stressful and will adversely affect growth and susceptibility to disease (section 8.9).

When circulation within a pond is poor, often as a result of calm weather, water can become strongly stratified with regard to temperature, pH and oxygen. In the absence of mixing, conditions deteriorate at the bottom of a pond as oxygen is consumed, pH declines and ammonia concentrations rise. Stratification can be especially bad in brackish-water ponds following rainfall, when a layer of less dense freshwater may form on the surface. This severely impedes the process of gaseous exchange between the water and the atmosphere that normally helps to limit the fluctuations in carbon dioxide and oxygen levels that result from photosynthetic activity and respiration. Stratification can be counteracted by circulation and aeration (sections 8.3.6.5 and 8.3.6.6).

Within the natural food web the phytoplankton are the primary producers that provide food for organisms at higher levels in the food chain. Zooplankton, for example, will graze on phytoplankton, and plankton productivity as a whole will support a community of micro-organisms (bacteria, fungi, protozoa) and invertebrates (worms, molluscs, crustaceans) on the pond bed, principally by providing a rain of nutrient organic material (faecal pellets, dead organisms, exuviae). The monitoring and control of algae populations are critical aspects of pond management (section 8.3.5.2). If very dense phytoplankton populations develop they can rapidly exhaust supplies of inorganic nutrients and undergo a catastrophic mortality known as a 'crash'. The resulting mass of decomposing organic matter can consume much of the available oxygen and endanger the crop once more. Benthic algal mats and uncontrolled growth of macroalgae are also likely to jeopardise crustacean production.

Some farmed crustaceans feed directly on primary producers. Crayfish and prawns frequently feed on aquatic plants and most penaeid shrimp and crayfish will consume benthic micro-organisms. Hence the natural productivity of a pond represents a valuable source of nutrition for crustaceans and, in general, for this reason it should be encouraged. Since phytoplankton growth is often limited by the availability of inorganic nutrients, it can be encouraged by the addition of fertilisers (section 8.3.6.2).

Although natural productivity alone, or enhanced with fertiliser, can be adequate for extensive cultures, semi-intensive and intensive farming operations require the addition of supplementary feeds to maintain rapid growth rates. However, in addition to boosting growth, the application of feeds has important implications for water quality and must be controlled if the dangers of anaerobic conditions and ammonia toxicity are to be avoided (section 8.3.6.3). Feed is not only consumed by the crustacean crop, since uneaten and partially digested fragments are also consumed (decomposed) by bacteria, other micro-organisms and invertebrates on the pond bed. Although this community often provides food for the crop, a major part of its impact relates to its heavy demand for oxygen. In one study on shrimp ponds receiving a daily average of 37 kg of feed ha^{-1}, oxygen consumption in the sediments accounted for 51% of the total oxygen demand of the whole pond system! The shrimp, by comparison, accounted for only 4% with the remainder consumed by organisms within the water column (Madenjian 1990). Comparable results were obtained in freshwater prawn ponds, again emphasising the critical influence of sediment respiration on water quality in ponds receiving supplemental feeds (Moriarty & Pullin 1990).

8.3.3 Pond preparation and rejuvenation

Since crustaceans dwell and forage on the pond bed, the condition of the substrate has a critical influence on their well-being. Pond preparation, both initially and between cycles, has a major impact on the substrate and on water quality, particularly at the early stages of the pond production cycle. Preparation basically involves draining, drying, and turning the soil and chemical treatment.

Drying enables air to penetrate the sediments and thereby assists in the breakdown and mineralisation of organic matter and the release of hydrogen sulphide. The mineralisation of organic matter produces inorganic nutrients (nitrate, phosphate, carbonates) that will improve the fertility of the pond, reduce the oxygen demand of the sediment, and consequently reduce the impact of any previously formed anaerobic decomposition products. Opinions as to the ideal length of the drying period vary. Up to 7 days or until the top centimetre of the soil has dried, has been recommended by ASEAN (1978). Boyd (1995b) considers 2–3 weeks to be suitable in most circumstances and notes that excessive drying should be avoided because it inhibits microbial activity. Turning the top 10–15 cm of the pond bed with a plough exposes more of the sediments to the air, encouraging aerobic decomposition, but it may not be feasible between every crop. Tilling the soil with a disk harrow to 5–10 cm has a similar effect but requires less energy than ploughing and may be preferable. Either process should be performed after the pond bed has dried sufficiently to support a tractor but before all soil moisture is lost. Afterwards it may be necessary to recompact the soil with a roller to give a firm surface.

The deep rich organic sediments that accumulate in some intensive ponds can be pumped out or flushed out with hoses to shorten the drying-out and reoxidation period required to recondition the ponds. However this approach to sludge management can have a severe environmental impact unless the sediments are trapped in a containment pond. If sludge is dumped on raised ground, measures are necessary to stop the sludge washing back into ponds or to the environment following heavy rains. Sludge management is discussed further in section 8.3.6.7.

Treating pond beds with lime (see Glossary) has several beneficial effects but is particularly cost-effective when the bottom soil is acidic (pH <7). It increases the pH of the mud, improves benthic productivity, buffers against large daily fluctuations in the pH of the water, boosts primary productivity by increasing the availability of carbon dioxide for photosynthesis, and improves the availability of nutrients, particularly phosphates. Its impact is most beneficial when accompanied by a programme of fertilisation. The lime should be added 3–4 days after a pond is drained, before the pond bed becomes completely dry. Application rates of agricultural limestone (calcium carbonate) vary according to pH. Those given by Boyd (1999a) are:

pH	Lime (kg ha^{-1})
>7	0
6.5–7	500
6–6.5	1000
5.5–6	2000
<5.5	3000

Limed ponds should be filled with water and left for at least a week, and the pH of the water checked before animals are introduced. Ponds are sometimes treated with other chemicals specifically aimed at eliminating predators and competitors (section 8.3.6.1) or disease. To treat the soil in a pond that has succumbed to disease Boyd (1999a) recommends burnt lime (calcium oxide, 1000 kg ha^{-1}) or hydrated lime (calcium hydroxide, 1500 kg ha^{-1}) to raise the pH above 10. The lime needs to be applied uniformly when pond bottoms are still wet. The alternative of using calcium hypochlorite is sometimes taken but organic matter in pond soils quickly reduces chlorine residuals to non-toxic chloride and as much as 500 mg L^{-1} may be needed for disinfection purposes. This implies the use of around 1000 kg ha^{-1} calcium hypochlorite, which is a more expensive option than burnt or hydrated lime.

After a pond has been drained for harvest, it can be useful to measure the organic carbon levels in the soil. Levels of less than 0.5% are too low because a certain amount of organic matter is beneficial to benthic productivity but levels of 3–4% or more are excessive and can be reduced by fertilising the soil with nitrogen (200–400 kg ha^{-1}) to enhance microbial activity. Nitrates such as sodium nitrate are ideal because they dissolve in soil water, penetrate to anaerobic zones and serve as an oxygen source for bacteria (Boyd 1999a). Basic processes governing soil chemistry are dealt with at length by Boyd (1995a), particularly with regard to freshwater ponds.

When a pond is refilled prior to restocking, a bloom of phytoplankton can be encouraged in a small volume of water (10–30 cm deep) by the addition of fertilisers (section 8.3.6.2). As the phytoplankton density increases, the water level can be raised in steps.

Pond preparation may include routine repairs to drainage channels, embankments and water control structures, and the filling of holes. In the case of extensive crayfish ponds it may be necessary to plant forage crops (section 7.5.4).

8.3.4 Stocking

Although it is advantageous to standardise stocking densities, particularly beyond the level of extensive farming, in practice variations may be necessary depending on season and the availability of juveniles. Periodic review of yield levels, feed conversion ratios and the size range of harvested animals will indicate the most profitable density at which to operate (section 10.5). Since stocking density influences the size of animals at harvest, it can to a certain extent be adjusted so that the product meets the sizes and product flow required by the markets (sections 3.2.4 and 7.3.7). Figure 8.4 illustrates the relationship between stocking density and harvest size for extensive/semi-intensive shrimp culture in Ecuador.

During the stocking process and the period directly afterwards, there is an enhanced risk of mortality due to predation and stress. When animals arrive for stocking they are usually weakened as a result of handling and transport and should be acclimated gradually, minimising

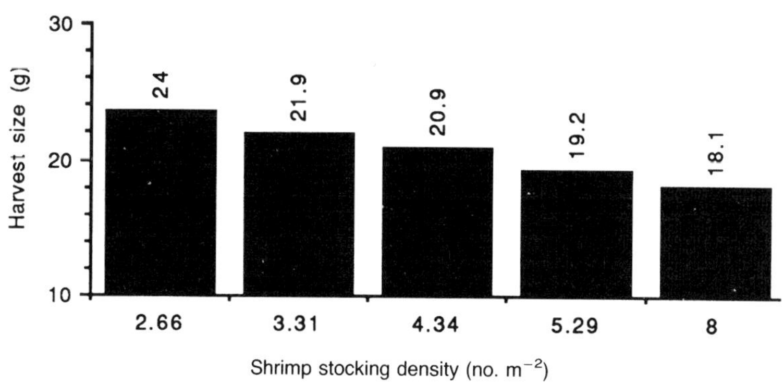

Fig. 8.4 Relationship between stocking density and harvest size for extensive/semi-intensive *Litopenaeus vannamei* farming in Ecuador (based on 152 observations) (Hirono 1986).

exposure to rapid changes in environmental conditions (section 7.2.4). Young post-larvae of hatchery origin are generally more delicate than larger nursery-reared or wild-caught juveniles (section 7.2.5). Requirements for acclimation vary from one situation to the next, but in the tropics stocking during the heat of the day should be avoided.

During stocking it is useful to count a number of juveniles (usually 100 or more) into a test cage placed in the pond where they can be fed if necessary and counted again 2–4 days later to estimate stocking mortality. The results are helpful to pond management since low survival can give an early warning of problems and may help in identifying the cause. Rather than waste time and effort on a pond where high stocking mortality is suspected, it may be better to restock. However if high survival is obtained in a test cage this does not guarantee that other problems, notably predation, will not influence the outcome in the pond at large.

8.3.5 Monitoring

If ponds are to be managed successfully, regular and reliable information must be gathered on pond conditions and the status of the crop. Although certain management procedures can be standardised, many critical decisions must be based on the daily measurements made in the ponds rather than standard tables for growth, feeding and water exchange (section 8.6).

Regularly reviewed, accurate records provide a basis for the understanding of performance trends and the effects of different management strategies. Sometimes as ponds mature over their first 2–4 years of operation they gradually become more productive. Eventually however a steady decline in yield may be detected indicating an overall deterioration in pond conditions. Assuming this is not related to general environmental degradation outside the farm (sections 6.3.1.4 and 11.5.3.1), it can signify the need for improved pond bottom treatment between crops and possibly the need to overhaul the ponds and remove accumulated silt and organic sediment (section 8.3.6.7). All the same, it is clear that careful pond preparation cannot in itself guarantee that a farm's productivity will not decline over time. Lee *et al.* (2000) found that productivity in semi-intensive shrimp ponds dropped by an average of 6% per year after ponds were 3 years old, despite the maintenance of good quality soil through drying, tilling and liming between every crop. Poor harvests were usually linked to white spot syndrome virus, for which the prevalence and virulence appeared to increase over time. In a model of pond productivity Lee *et al.* (2000) found that unpredictable factors such as disease accounted for around half of the variability around the mean yield of 3.8 mt ha^{-1} per crop. The other half of the variability could be accounted for by reference to the age of the pond, the stocking density, the crop duration and the size of the pond.

Accurate records are also essential to understand day-to-day problems and to account for irregularities that only become apparent at harvest time (e.g. unexpected mortality). Any technicians involved in the collection of data should be aware of their importance and be well instructed in the use and calibration of instruments (section 8.6).

8.3.5.1 Crop biomass and growth

Obtaining good estimates of the crop biomass (or standing crop) present in a pond is often essential for the efficient management of feeding rates. The only situation where such estimates are unimportant is when feed is supplied on trays (section 8.3.6.3). For the farmer the process of biomass estimation is complicated by the fact that the crop remains largely invisible and survival rates cannot be determined with any precision until harvesting is completed.

Crop biomass may be estimated by measuring the average size of the crustaceans and estimating the number of individuals present in the pond (a population estimate). The best methods of estimating population density are only accurate to perhaps ± 20%. They rely on laborious, repeated sampling and assume a relatively random distribution of animals within a pond. Accuracy can be improved if samples are taken at various points in a pond and results for several crops are compared with actual harvest numbers (Falguiere *et al.* 1989). In time a rough picture will emerge for each pond of how the animals distribute themselves, but this can vary with pond bottom configuration (section 8.2.2), water temperature, season, size of animals, stage of the tide (in coastal or estuarine ponds) and time of day.

Cast-nets are commonly used for sampling but the process is more of an art than a science (Dugger 2000). The list of factors that can bias the results is extended by water depth, water clarity, crustacean density, design of cast-net, cast-net mesh size and type, the person throwing the net, pond bottom texture, water flow patterns and dissolved oxygen levels. To obtain useful data it becomes important to standardise sampling procedures using particular cast-nets with a known catchment area (or by seining specified areas). Cast-net sampling should be performed at a set time of day (e.g. 2–4 PM), usually from a boat, by the same employee, at numbered and marked stations throughout the pond. Seven to twelve casts per hectare, depending on pond size, are usually adequate. Dugger (2000) has tried to make population estimates using side-scan sonar but found the approach to be inaccurate because the target animals could not be distinguished from rocks and shells. Sometimes it is possible to estimate numbers by snorkel-diving and counting the animals seen while swimming over a given distance. An added advantage of diving is that the amount of uneaten food remaining in a pond can be determined (section 8.3.6.3).

Bearing in mind the general unreliability of sampling methods, the use of a standard mortality curve based on expected losses (low-level cannibalism, predation, escapes) is often the simplest practical way of making a population estimate. This method is likely to be of real value only if it is continually updated in the light of farm yields (numbers and sizes) and takes account of differences in performance between groups of ponds. The unpredictability of initial post-stocking mortality can be overcome somewhat by using a mortality test cage (section 8.3.4). Large well-managed farms usually incorporate all of these methods in order to minimise food wastage and pond fouling.

Regular samples, at 7–14 day intervals, can be taken to estimate growth increments, and, provided no large-scale mortality takes place within the crop, the growth rate can be taken as a fundamental indicator of the success of pond management. Growth that is close to predicted values indicates that pond conditions are adequate and that the crop is healthy and feeding well. Samples of individual weights of crustaceans provide an impression of size variability, which in the case of *Macrobrachium* may assist in the planning of partial harvesting (section 7.3.6). Accurate biomass and average weight estimates allow for more efficient scheduling of harvests and better co-ordination between farmer, processor and buyer.

8.3.5.2 Water quality

Pond water quality measurements are best taken either at the water exit point or where good access can be obtained to deeper water. It may be useful to build a small jetty for this purpose as long as its structure does not interfere with harvesting. The subject of water quality monitoring is of general importance to all phases and types of crustacean culture and is also discussed in section 8.6. Desirable levels and ranges for some of the important water quality factors are given in Table 8.3.

Dissolved oxygen (DO) readings are usually taken at least twice per day for water close to the pond bed. Afternoon measurements can be used to monitor peak DO levels induced by photosynthetic activity. Since the most critical period for low oxygen levels is around dawn, readings can be taken in advance to warn of likely problems. One method relies on plotting measurements taken at 8 PM and 11 PM, and extrapolating in a straight line to obtain a predicted DO level at 6 AM. the next day (Boyd 1990). Remedial action (e.g. activating aerators) can be taken if the predicted level falls below a set minimum. Some shrimp farms aim to keep levels above 4–5 mg L^{-1} which improves both survival and yields (McGrow *et al.* 2001) and Boyd (1990) notes that feed conversion ratios in shrimp ponds increase drastically if DO levels fall below 2–3 mg L^{-1} at night. The oxygen requirements of moulting Crustacea are considerably higher than those of intermoult animals and, since moulting usually occurs at night, it may be doubly important to increase aeration or exchange rates during this period. Fortunately, as long as favourable oxygen levels can be quickly restored, it appears that crustaceans that survive short-term oxygen stress can make a complete recovery. Allan and Maguire (1991) ran experiments to simulate pond conditions in which emergency aeration successfully raised DO levels after an oxygen crisis. Neither the duration (up to 12 h) nor the level of DO stress (down to 0.5 mg L^{-1}) significantly reduced the growth or food conversion ratio of shrimp that were returned to favourable water conditions for 21 days.

Temperature measurements are simple to make and can be used to quantify the diurnal and seasonal variations that influence feeding and growth rates, and to observe any differences between surface and bottom waters that would indicate signs of stratification.

Turbidity can be conveniently measured using a Secchi disc at least once or twice a day, although in intensively managed ponds Wyban *et al.* (1989) recommend three daily observations to check that phytoplankton blooms are stable (sections 8.3.6.2 and 8.3.6.4). If water samples are viewed under a microscope and phytoplankton counted using a haemocytometer, the concentration of microalgae can be compared to Secchi disc readings to establish what part of turbidity is due to suspended organic detritus and sediment and what part is due to the presence of the algae. However in general it is not feasible to routinely count algae in all of a farm's ponds and microscopic observation is more usefully directed towards obtaining information on gross community structure. Observations of pond water colour can suggest which type of algae is predominant. For example, in brackish-water ponds a green coloration is generally indicative of flagellates, while brown generally denotes diatoms. Unusual colours or shades can be useful indicators of the presence of toxic algae blooms similar to 'red tides' (see Glossary).

Salinity can fluctuate widely on estuarine sites and can rise due to evaporation. After heavy rains differences in salinity between surface and bottom water can be used to check for stratification. Salinity can be measured with a refractometer to a precision of ± 1 or 2‰.

Commercial test kits can also be used to measure the concentrations of ammonia, nitrites, nitrates, phosphates and silicates although when testing brackish- or saltwater, it should be remembered that certain kits are designed for use only in freshwater (section 8.6). Usually it is not necessary to measure these concentrations on a regular basis, except for ammonia, which is liable to build up in intensive systems.

8.3.5.3 Other observations

Observations of the amount of uneaten food remaining every day should be made in more intensive pond systems because daily adjustments in feeding rates may be necessary, and the first sign of stress is often a cessation of feeding. Mesh trays with food on them can be lowered to the pond bed, left for a set period, then raised and inspected. Such observations can help establish whether a problem with sluggish growth is related to underfeeding or not. If farmers notice a decline in growth rate and respond by increasing feeding rates without checking that the food is actually being consumed, they risk severely polluting the pond (section 8.3.6.3). Some observation of food remains and the condition of a pond bottom can also be made by snorkel-diving with a torch.

The softness of a pond bottom can be used to locate areas of accumulated organic sediments. Softness can be gauged using a pole from a boat and the information gained can be used to select the best sites for, or to reorientate, aeration devices to prevent or disperse sediments. Care should, however, be taken not to provoke an oxygen crisis by resuspending sediments (but see section 8.3.7), and it may be better to leave accumulated sludge undisturbed while a pond is in production or, in small intensive ponds, pump it out or void it through a central drain (section 8.3.6.7).

It is possible to make simple deductions about the organic load in sediments and the presence of the highly toxic gas hydrogen sulphide by wading into a pond, taking a mud sample and observing odour, texture and colour. Black sediments with a foul smell like bad eggs are indicative of hydrogen sulphide, which forms as the result of sulphide excretion by anaerobic bacteria. The toxic effects of hydrogen sulphide are felt at very low concentrations and are greatest when the pH is low (acidic conditions). The production of this dangerous gas can be minimised by maintaining aerobic conditions throughout the pond and the topmost sediments, by avoiding overfeeding and by allowing the drying and oxidisation of organic sediments between crops (section 8.3.3). Problems with the build-up of hydrogen sulphide and ammonia in more intensive systems have apparently been ameliorated by the addition of zeolite at 250 kg ha^{-1} (Chen 1990). However zeolite may have little effect in brackish water or saltwater and, while it can technically absorb ammonium, very large amounts would be needed to reduce ammonia concentrations (Boyd 1995b).

Observing the behaviour and condition of the crustaceans can provide a timely warning of existing and potential problems. Shrimp, for example, seen circling around the edge of a pond may be suffering from stress due to lack of oxygen, and crayfish under similarly low oxygen conditions may even start to migrate from a pond. The presence of dead animals in population samples or around the margin of a pond is an obvious cause for concern. Carcasses, however, are usually rapidly consumed and a steady mortality over a long period may go unnoticed. Dying animals or those showing signs of abnormality can be preserved for microscopic and histological examination if disease is suspected. Observed softness in shell texture indicates recent moulting or a problem with shell mineralisation (section 8.9). Soft-shelled or pre-moult crustaceans stop feeding, and if moulting is largely synchronous in a pond population (sometimes triggered by the influx of new or freshwater) it may be necessary to reduce feeding rates or delay a planned harvest. In transparent shrimps and prawns, the fullness of the alimentary tract can be observed, to establish whether the animals are feeding and (in support of observations of feeding trays and water quality) to assist in decisions on feeding rates.

Algal growth can be a particular problem in freshwater ponds and routine monitoring and control of weeds in prawn ponds is often worthwhile (Valenti & New 2000). Crayfish, on the other hand, tend to consume plant material and weed growth is seldom a problem on farms.

8.3.6 Control

8.3.6.1 Predators and competitors

Most predators and competitors in crustacean ponds can be eliminated by draining and drying between crops, provided pond beds are well constructed and no pools remain (section 8.2.2). Chemical control may be required if complete draining cannot be achieved. While a pond is in production, mesh screens on inlet and outlet gates prevent the entry of most adult water-borne predators and competitors.

Teaseed cake, the residue after extracting oil from the seeds of *Camellia*, can be applied as a selective fish poison. It usually contains 10–15% saponin that is 50 times more toxic to fish than to shrimp, and biodegrades after a few days. The cake must be dried, ground and soaked in water for 24 h. It is applied to shallow water and puddles at rates of 12–20 g m^{-3} to give 1.2–3.0 mg L^{-1} saponin (ASEAN 1978). The use of higher dosages, 2.5–10 mg L^{-1} saponin, is reported by Chen (1990). Hovers (1999) uses teaseed cake at 10–15 mg L^{-1} to kill fish and also uses the same dose to induce moulting. Rotenone is another selective fish poison that is most effective in fresh or low-salinity water and is often applied in the form of derris root that contains around 5% active ingredient. The recommended application rate is 4 mg L^{-1} of dry root to give 0.2 mg L^{-1} rotenone. After poisoning, dead fish must be removed and the treated pools left for several days for the chemicals to deactivate.

Prior to using teaseed cake or derris root it can be helpful to perform a bioassay on fish in an aquarium because activity can vary between batches of raw material. Sometimes as much as 50 mg L^{-1} of derris root are needed to kill the fish (Wang 1999).

A list of chemical compounds and products used as piscicides and for pond sterilisation has been compiled by Jory (1995a), but the usefulness of certain chemical treatments has been questioned. For example Boyd (1999a) notes that there is no evidence that formalin, chlorine, benzalkonium chloride, povidone iodine or zeolite have significant beneficial effects on soil or water quality, and goes on to recommend that in general no chemicals should be added to pond water at all except for fertiliser and agricultural limestone (500 kg ha^{-1} if alkalinity drops below 60 mg L^{-1}). Boyd (1999a) does however concede that bacterial inoculants (section 8.9.4.2) and grapefruit seed extracts may improve the survival rates of some cultured species. Diesel oil (30 L ha^{-1}) has been used 3 days prior to stocking *Macrobrachium* ponds to eliminate air-breathing predaceous insects (Daniels *et al.* 1995). The use of insecticides, however, is not generally recommended for prawn culture because of the potential risk of bioaccumulation and toxicity to the crop (Boyd & Zimmerman 2000).

Steps may be necessary to control birds, some of which may be protected species. Non-destructive methods

Plate 8.1 Bamboo stakes arranged around the margin of a Thai shrimp pond to deter cast-netting by poachers. Paddlewheel aerators and a hatchery facility are also visible.

should be used since they will be less likely to arouse public ire (section 11.2.5). Scarers can be used, though some types tend to lose their efficacy with time. Hand-held laser projectors (rifles) are advertised as environmentally friendly bird scarers and dogs can be very effective on small to medium-sized farms. The presence of dogs and other livestock around the ponds may allow certain parasites to complete their life cycles but we know of no reports of health problems in this respect. Strings stretched across ponds can be an effective deterrent to diving birds but are only feasible for small units. Wading birds can be discouraged by regular attention to embankments to maintain their slopes and by eliminating shallow areas in ponds. Baited jars sunk in pond banks can be effective traps for land crabs. Fences may be necessary around ponds to keep out various frogs, snakes and toads or to prevent non-native species of crayfish from escaping to natural waters. Night watchmen, guard dogs (or geese on small operations), lighting and perimeter fencing may be essential to guard against theft, and stakes can be positioned to interfere with cast nets. Total security on large farms, however, is very difficult to achieve. Electric fences have been installed to protect crayfish ponds in Australia (section 8.2.2.3).

The control of macrophytes is often necessary in freshwater prawn ponds and herbicides are sometimes applied. The most widely used chemicals include copper sulphate and copper chelates and if they are used in accordance with manufacturers' recommendations they are rarely directly toxic to aquatic animals. However, the decay of the plants killed by the herbicides can reduce dissolved oxygen concentrations. Once the aquatic vegetation has been eliminated phytoplankton should bloom and shade the pond bed and prevent regrowth (Boyd & Zimmermann 2000).

8.3.6.2 Fertilisation

Fertiliser is the principal agent employed for promoting natural productivity in ponds where the concentration of inorganic nutrients is low. The most important components for phytoplankton productivity are nitrogen and phosphorus and there are two types of fertilisers: organic and inorganic. The former represent much less concentrated sources of nutrients and are thus much more bulky to transport. For example, 37 kg of dry chicken manure supplies the same amount of nitrogen as 1 kg of urea. Although organic fertilisers provide additional material to boost benthic productivity, if used excessively, their decomposition can create anaerobic conditions on the pond bed. Despite their disadvantages they are often readily available and cheap and can be ideal for small-scale and extensive aquaculture operations. The manure of geese and ducks is often preferred to others for its relatively high phosphate content. Chicken and cow manures, however, are often available in larger quantities. It should also be remembered that there is a danger of animal wastes being contaminated with pesticides, antibiotics and heavy metals and because of this and potential dissolved oxygen problems some authorities suggest that they should always be avoided; cleaner alternative organic fertilisers are some plant meals (Boyd 1999a).

Fertilisers, particularly manures, are often applied to the pond bed in advance of filling and stocking. Once an initial phytoplankton population has been established, usually after 5–15 days, its maintenance usually requires further fertiliser applications unless a nutrient-rich water supply is employed. For bloom maintenance Boyd (1999a) suggests applying 1–2 kg nitrogen and 0.5–1 kg phosphate per hectare per week. Fertilisers may be slowly leached into the water from floating perforated plastic drums, sunken wicker silos or from porous sacks held in the inflowing water current or tied to stakes within the pond. Alternatively they may be placed on a wooden platform 30 cm below the water surface. It is advisable not to broadcast solid fertilisers over the pond since the nutrients will be deposited on the pond bed rather than used to fuel primary productivity in the water column. This may cause a carpet of benthic algae to develop in ponds where light can reach the bottom, an effect that is undesirable and sometimes necessitates the application of algicides to prevent a reduction in crop yields.

The control of phytoplankton productivity can be difficult, especially in large ponds. It usually requires manipulation of the rates of water exchange and feeding as well as fertilisation. The objective is usually to keep turbidity levels (as measured with a Secchi disc) within set limits. Visibility to a depth of 25–40 cm is generally recommended for shrimp or prawn ponds although 15–35 cm may also be suitable for freshwater prawn units. Supplementary feeds partly act as organic fertilisers, so fertilisation rates usually need reducing as feeding rates build up in the later stages of a culture cycle.

To a certain extent fertilisation regimes can be designed to favour particular types of phytoplankton. Diatoms, which are usually prevalent in moderate or high salinity water, are usually preferred in shrimp ponds and can be encouraged with fertiliser high in nitrogen. In contrast, generally undesirable blue-green algae that often bloom in lower-salinity water are able to fix dissolved atmospheric nitrogen and are thus likely to be favoured by fertilisers high in phosphates. Boyd (1990) recommends roughly a 20 : 1 ratio of nitrogen : phosphorus to maintain diatom-dominated blooms in brackish-water ponds. This equates to a 9 : 1 ratio of urea : triple superphosphate. Although diatoms require silicates, most tropical brackish waters have fairly high silicate concentrations so the value of using silicate fertiliser is likely to be site-specific. There is some evidence that applications of silicate and/or chelated iron can stimulate diatoms in some situations (Boyd 1999a). Interestingly, despite a general preference for diatoms in shrimp ponds, flagellates are preferred in some Taiwanese farms because their concentrations are more stable and easier to control (Chen 1990).

Sodium nitrate is a better fertiliser than urea or ammonium-based fertilisers but is also more costly. Unlike the others, it is a source of oxygen for bacteria, it is non-toxic, it does not form acidity, and it does not have an oxygen demand (Boyd 1999a). In contrast, urea quickly hydrolyses to ammonia, which is toxic.

Although chemical analysis may reveal which nutrients, if any, are limiting to productivity, so assisting in the formulation of a suitable fertilisation programme, the pattern of nutrient availability can be expected to change greatly with rainfall and season. Routines for fertilisation should be established as experience is gained with each pond and at each site. Greater fluctuations in algal populations may occur in plastic-lined ponds compared to earthen ponds since the former may tend to develop less diverse and less stable populations of plankton. As a result, greater control over water exchange and the application of fertiliser become necessary.

Hard and fast rules for the frequency of application of inorganic fertilisers do not exist, and recommendations vary from two or more times per week to once every 2–4 weeks. Decisions should be based on changes in turbidity shown by Secchi disc readings. In response to excessive concentrations of algae, applications should be reduced in quantity rather than eliminated, in order to reduce the risk of a population crash.

Examples of fertilisation regimes (bi-weekly quantities ha^{-1}) include:

(1) For semi-intensive fish/*Macrobrachium* culture in Hawaii (Malecha 1983):
   60 kg single superphosphate
   60 kg ammonium sulphate or liquid ammonia;

(2) For shrimp ponds in Hawaii (Chamberlain 1987):
   6.6 kg urea
   2.7 kg triple superphosphate
   10 kg calcium silicate
   0.7 kg mineral mix (made from zinc oxide).

For further reading, details of 21 different fertilisation schedules applicable to aquaculture ponds have been compiled by Tacon (1990), and Cook and Clifford (1998a) discuss the fertilisation of shrimp ponds and nursery tanks (see also section 8.9.4).

8.3.6.3 Feeding

The productivity of ponds relying on natural productivity alone or natural productivity boosted by fertilisation rarely exceeds 500 kg ha^{-1} per crop. Thus feeding becomes essential if greater yields are required. In shrimp culture the need for feed can be linked to the stocking density employed, placing it at densities greater than 2–5 shrimp m^{-2} for *Penaeus monodon* and above 5–10 shrimp m^{-2} for *Fenneropenaeus indicus*.

Feeds are usually broadcast by hand either from the pond banks or from small boats. Mechanisation is possible using feed blowers towed behind tractors or four-wheel motorbikes, but these have a maximum range of 25 m and are only suited to small ponds. The operators of one very large Ecuadorian shrimp farm (1600 ha) found it convenient to distribute pellets using an adapted crop-spraying aircraft. In contrast to fish farmers, few crustacean farmers rely on automatic or demand feeders. The distribution of food to a large number of individually held crustaceans, for example battery-reared lobsters or crayfish, requires special apparatus (section 7.8.9).

Increasing awareness of the costs and problems caused by overfeeding and the impacts of effluents on the neighbouring environment has led to feeding strategies that promote more efficient feed utilisation and greater use of the nutritious organisms produced in the ponds. Placing feed on numerous submerged trays rather than broadcasting has enabled feeding rates in many semi-intensive shrimp farms to be much more carefully tailored to consumption rates (see below). In reduced and zero water exchange systems feeding strategies have also undergone a fundamental reappraisal, placing emphasis on promoting a pond ecosystem that minimises water quality fluctuations and maximises the efficiency of nutrient assimilation by the crop (section 8.3.7).

Plate 8.2 The routine inspection of feeding trays allows feeding rates to be adjusted in line with consumption rates and is also a convenient opportunity to inspect the crop for abnormalities.

The frequency of feeding and the times of day the food is given are important aspects of husbandry that must be in sympathy with the animal's physiological needs if good growth is to be achieved. The utilisation of feeds can usually be improved by increasing the number of applications per day. Four to six feeds per day are provided in some intensive systems and two to four feeds per day are usual in semi-intensive shrimp and prawn culture. The optimal times for feeding are not well understood, but for shrimp they appear to differ between species. Feeding schedules for *Penaeus monodon* in Asia usually give a larger proportion, 55–65%, in the evening and at night rather than in the morning and afternoon (Jory 1996). In contrast, in trials with *Litopenaeus vannamei*, day feedings produced slightly better growth than feeding at night, suggesting that for this species day feeding is at least as good as night feeding (Robertson *et al.* 1993). For *Farfantepenaeus subtilis*, Nunes and Parsons (1999) found that feed consumption was lower at 6 AM than at 9.30 AM and 2.30 PM and linked this to low oxygen levels at daybreak and subsequent suppression of feeding activity. A list of the chemical compounds known to increase searching behaviour, attract the animal directly to the food or stimulate the actual consumption of the food (section 2.4.6) is given by Lee and Meyers (1997) and a useful internet site is http://www.aquafeed.com.

Standard commercial crustacean diets are usually incomplete in their nutritional profile (section 8.8.2) and animals have to rely on natural productivity in a pond to make up for shortfalls in essential nutrients. As the standing crop in a pond increases, particularly in intensive systems, the supply of essential nutrients becomes limiting to growth unless a higher-quality diet can be used. Improved diets usually contain greater proportions of protein and this is reflected in their higher price. It may be worth increasing the quality of a diet in the later stages of ongrowing, provided that improved growth compensates for the extra expense. In less intensive systems, however, there may be little benefit in raising protein levels (section 2.4.1). For *Litopenaeus vannamei* stocked at 5–11 m^{-2}, Teichert-Coddington and Rodriguez (1995) found that diets containing 20% protein were adequate and that there was no advantage from a diet containing 40% protein.

The food conversion ratio (FCR) relates the weight of feed applied and the weight of crustaceans harvested. For dry diets, low ratios (<2 : 1) are desirable and they indicate that feeds are being converted efficiently into crustacean flesh and/or that natural productivity is making a significant contribution to growth. Higher food conversion ratios (>3 : 1) are suggestive of overfeeding, poor diet quality or slow growth. Efficient conversion of feeds is critical to the economics of crustacean culture and is the primary task of feed management.

The most common approach to feed management relies on the use of standardised feeding tables and regular estimates of crustacean biomass. Daily rations are calculated according to a set schedule that varies in accordance with the average size of the crustaceans and the pond biomass estimate (section 8.3.5.1). Feeding schedules are often supplied initially by feed companies and are then customised to suit individual preferences and farm conditions. The normal feeding rate is usually expressed in terms of percent body weight per day and in most systems this percentage will decline as the average size of the animals increases. Jory (1996) compiled examples of eight different schedules for the culture of *Litopenaeus vannamei* and *L. stylirostris*. The data indicate a feeding rate of 11–25% body wt per day for shrimp of 1 g, declining to 2.8–3.9% body wt per day for shrimp of 10 g, and 1.7–2.5% body wt per day for shrimp of 22 g.

In practice it is dangerous to follow feeding tables blindly and it is important to reduce feeding rates if there are problems with water quality, particularly low dissolved oxygen levels (section 8.3.5.2). One strategy for feed management in freshwater prawn ponds involves cutting feeding rates by 50% if the oxygen concentration falls below 3.5 mg L^{-1} in the early morning or if the temperature drops below 24°C during the day. Feeding is interrupted altogether if oxygen concentrations fall below 2 mg L^{-1} and temperatures fall below 18°C (Valenti & New 2000). If excess feed lies uneaten on the pond bed this can exacerbate water quality problems. On the other hand if too much caution is exercised, underfeeding can sacrifice potential growth. These drawbacks, together with the difficulty of making accurate population estimates (section 8.3.5.1), have led to the increasing popularity of feeding trays both to provide information on actual feed consumption rates and to deliver food to the crop in a more efficient manner.

The use of feeding trays has been developed within the shrimp industry because shrimp, unlike other farmed crustaceans such as *Macrobrachium*, are neither aggressive nor territorial and are tolerant to crowding. If a small proportion of the feed scheduled to be fed to a pond is distributed on such trays and later the trays are inspected, valuable information on feed consumption rates can be obtained. This approach is known as the indicator method and it allows daily rations to be adjusted in line with shrimp appetite. If large numbers of feeding trays are installed in a pond (15–30 ha^{-1}) it becomes feasible to place all the feed on the trays and rations can be even more precisely tailored to demand. This latter approach was first developed to a commercial scale in Peru and is sometimes known as the Peruvian method (Cook & Clifford 1997a). When feeding trays are lifted they also provide a valuable opportunity to inspect the crop and detect any abnormalities.

The indicator method was developed in Taiwan using a limited number of trays (1–6 ha^{-1}) arranged in three rows. Trays either receive a fixed amount (e.g. 150 g each) or a percentage (e.g. 3%) of the total pond ration and are filled after the pond has been fed so that when they are inspected after a set period (1 or 2 hours) they provide information on residual appetite. They can quickly detect when shrimp are particularly hungry and when there is a loss of appetite, for example during synchronous moulting. Cruz (1991) prepared a comprehensive manual describing the indicator method and the use of a tray consumption index (TCI) to summarise the data of feed remaining. The TCI is used to adjust the total pond ration following sets of tables that are either aggressive (rapid and large response to changes in TCI) or conservative (more gradual response to changes in TCI). The use of indicator trays in semi-intensive *Penaeus monodon* farming has revealed how consumption patterns vary with season, temperature and moult cycle and has enabled improvements in FCRs from 2.0 : 1 to 1.8 : 1 and yield increases of 11% (Bador *et al.* 1998). Feed consumption rates dropped by as much as 50% when shrimp were moulting. There is some evidence that the indicator system is less efficient in large ponds with shrimp densities below 9 m^{-2} (Cook & Clifford 1998b). There are also problems when competitors such as crabs, prawns and fish monopolise the trays, consume a lot of feed and thereby interfere with the indicator function.

The Peruvian method, in which all feed is supplied on trays, can overcome some of the problems of the indicator system and lead to yet greater feeding efficiency. Trays are usually placed in a pond in a grid pattern at a density of around 16 ha^{-1} prior to filling. Further trays may be added to provide a density of 25 trays ha^{-1} by the end of the production cycle. Viacava (1995) records the use of one tray per 500 m^2 (i.e. 20 ha^{-1}) in Peru at stocking densities of 15–20 post-larvae m^{-2}. A two-man team in a canoe visits numbered trays in turn to record and remove feed remains and to add fresh feed. The amount of feed given is effectively controlled at two levels – firstly by the feed worker in the canoe who adjusts individual tray rations immediately in response to the level of uneaten remains (using a specially graduated container), and secondly by a feed supervisor who compiles records of feed remains in a pond, calculates a consumption index, and controls the total daily consumption. Decisions to suspend or reduce overall feeding rates are also taken in response to DO levels, for example if early morning levels fall below 3 mg L^{-1}.

The drawback with the feeding tray method is that it requires additional labour and material costs and a period of training before it can succeed. Viacava (1995) records an increase in labour costs by a factor of 1.8. Simple faults, for example lowering trays too quickly such that feed drifts away, or inaccurate recording or estimation of feed remains, can nullify the beneficial impact of the system. One way to counter this is to provide a financial bonus to the pond workers on a pond by pond basis that varies depending not simply on productivity but also on the feed conversion ratio obtained. Thus there is an incentive to perform laborious, repetitive tasks with diligence. According to Viacava (1995), one worker can service 200 trays in a 10 ha pond in 6 h. Other semi-intensive farms use two people per 20 ha and one food supervisor per 60 ha.

Feeding trays have various designs but most consist of mosquito mesh stitched to a square or circular frame, 70–80 cm wide, and have a lip to retain feed. Some have bamboo frames and concrete sinkers, others frames of steel reinforcing bars or plastic pipe filled with sand or concrete. A bridle with three or four lines is attached to the frame and is tied to an anchoring pole by a short length of rope. Most trays lay flat on the bottom but some are raised on short legs of 10 cm.

The benefits of feeding trays usually outweigh the drawbacks. They minimise feed waste, benefiting pond soil, water quality and the environment and allow for the rapid detection of disease problems that are often manifested by a drop in appetite. Feeding trays have led to an 18% increase in annual productivity in semi-intensive *Penaeus monodon* ponds in Indonesia (Ayranto 1998) and to 30% savings in feed in semi-intensive Peruvian farms producing *Litopenaeus vannamei*, with FCRs falling from 1.7 : 1 to 1.2 : 1 (Viacava 1995).

8.3.6.4 Water exchange

Daily water exchange has traditionally been employed to maintain healthy phytoplankton blooms, to flush away toxic metabolites, to make good losses due to seepage, and in brackish-water ponds to limit fluctuations in salinity caused by evaporation. In freshwater prawn ponds it is the most widely used tool of water quality management (Boyd & Zimmermann 2000). However the need for water exchange to control phytoplankton and metabolites has come under close scrutiny, and the development of reduced or zero exchange culture systems has demonstrated that there are viable alternatives to water exchange for controlling water quality (Hopkins *et al.* 1995). Moreover these alternatives (section 8.3.7) can have advantages in terms of disease management and environmental impacts.

Water exchange rates in ponds are often expressed in terms of the inflowing volume as a percentage of the pond volume. This, however, does not represent the true exchange rate obtained (Table 8.1) (section 8.4.3).

Estimates of water requirements differ widely. In extensive crayfish culture in farm dams in Australia, water supply need only be sufficient to keep the reservoir full. A flow of 118 L min^{-1} ha^{-1} from a well is considered adequate for a 32 ha crayfish farm in the USA (Avault & Huner 1985). Crayfish canal ponds in England have been designed with flows giving 50% exchange in 54–150 h (section 7.6.6.2). Ideally, in semi-intensive and intensive systems, the water supply should provide for a minimum exchange rate in each pond with additional capacity for emergency flushing. While average water use in freshwater prawn ponds in Hawaii has been recorded at 94–271 L min^{-1} ha^{-1}, peak demand may be as high as 3700 L min^{-1} ha^{-1} (Malecha 1983). More recently, flows of 120–560 L min^{-1} ha^{-1} for *Macrobrachium* ponds are reported (Muir & Lombardi 2000). To keep water moving freely through the ponds, all screens on water control structures must be regularly cleaned to prevent blocking with debris. If incoming water has a high sediment load it is better to use a settling reservoir first, which will require routine dredging to remain effective.

Calculations for overall water requirements on shrimp farms have usually been based on a maximum daily exchange of 10–20%. Clifford (1994) for example used daily rates of between 2% and 11.6% for the semi-intensive culture of *Litopenaeus vannamei*. However there is mounting evidence of little additional benefit from exchange rates much above 5% per day and that the need for reserve capacity to allow for pond flushing in an emergency can be greatly reduced with careful feed management and the use of aerators to supplement DO levels. Various trials to measure the effects of water exchange on shrimp growth and survival have revealed no difference at exchange rates between 4% and 14% per day and between 2.5% and 25% (Hopkins *et al.* 1993). Allan and Maguire (1993) ran rearing trials for 8–9 weeks and found no difference in shrimp growth and survival in the range 0–40% exchange per day. They concluded that water exchange can reduce nutrient concentrations and phytoplankton densities but that most of the reduction occurs at water exchange rates under 5% per day. Daily water exchange rates are often managed as a simple function of the crop biomass in a pond. However it makes more sense to have a very flexible approach driven by water quality measurements, and linked to the overall feeding rate rather than the crop biomass.

An alternative way of reducing phytoplankton densities, often practised with freshwater prawns, is to introduce filter-feeding fish to the ponds and perform polyculture (section 7.3.5.2).

8.3.6.5 Circulation

The absence of mixing in a pond in calm conditions, and the stratification which often results, are barriers to the transfer of oxygen from the surface layers to deeper areas. A large part of the positive impact of mechanical aeration is due to physical mixing and destratification of the water layers. In addition to improving dissolved oxygen concentrations, mixing also evens out the distribution of phytoplankton and reduces temperature stratification. In US crayfish ponds boat and trapping lanes between forage crops are used to improve water circulation.

Low-energy water circulation devices based on large electrically driven propellers have shown potential to homogenise water quality within prawn ponds (Rogers & Fast 1988). The improved uniformity of prawn distribution that resulted reduced cannibalism and aggression and significantly decreased the heterogeneity of prawn sizes. In these tests, a 0.5 hp circulation unit was credited with moving 8300 L of water per minute.

8.3.6.6 Aeration

Aeration devices have chiefly been used in small (<2 ha) intensive ponds to avoid water quality problems and permit high stocking densities. Their use in larger semi-intensive ponds for shrimp, for example in the USA, has accompanied trends towards intensification. Oxygenation is the most important function of aerators but the water currents they create are very beneficial and move oxygenated water away from the aerator to other parts of the pond and reduce thermal and chemical stratification. In ponds where feeding rates exceed 25–35 kg ha^{-1} day^{-1} dissolved oxygen levels can fall to 2 mg L^{-1} at night (Boyd & Zimmerman 2000), so if large water exchange rates are not feasible, aerators are the ideal option to improve water quality. Chamberlain (1987) recorded the value of aeration in reducing requirements for water renewal in shrimp ponds. In trials in Hawaii, 0.4 ha shrimp ponds aerated with two 1 hp paddlewheels required 62% less water exchange than non-aerated ponds and were also more productive. It is estimated that 1.34 hp (1 kW) of aeration can increase production by 500 kg ha^{-1} $crop^{-1}$, but to register a benefit it needs to be provided at least at a rate of around 2–3 hp ha^{-1}. Aeration is probably not necessary for yields below 2000 kg ha^{-1} $crop^{-1}$ and is only required at night for yields between 3000 and 4000 kg ha^{-1} $crop^{-1}$ (Boyd 1999a). Boyd (1998) reviewed pond aeration systems and concluded that an optimal approach for sustainable shrimp production may be to stock 20–30 animals m^{-2} and to aim for 4–6 mt ha^{-1} $crop^{-1}$ for which an aeration rate of 10–11 hp ha^{-1} would be required.

The usual numbers of paddlewheel aerators (typically 1 hp each) used in intensive Taiwanese shrimp ponds has been put at four or more per hectare for ponds stocked at 10–30 animals m^{-2} and eight or more per hectare for ponds stocked with more than 30 animals m^{-2} (Chiang & Liao 1985). Results in the USA quoted by Chamberlain (1987) are roughly in line with these levels of aeration: 10 hp ha^{-1} of paddlewheel aeration proved adequate for a shrimp density of 40 m^{-2}, and 1.5–3 hp ha^{-1} avoided all but occasional oxygen problems at densities of 20–30 shrimp m^{-2}. Wyban *et al.* (1989) recommend aeration levels of 20 hp ha^{-1} in super-intensive round ponds and AQUACOP (1989) have used pairs of 2 hp propeller-aspirator pumps in super-intensive ponds of 0.1 ha (equivalent to 40 hp ha^{-1}). Australian shrimp farmers use on average 16 hp ha^{-1} at densities averaging 35 shrimp m^{-2} (Peterson 1998).

The increasing demand for oxygen during the on-growing period can be met by bringing more aerators into operation. In Taiwan an initial 2.5 hp ha^{-1} may be increased to 10 hp ha^{-1} by the end of the culture period. Aerators can also be activated during particular critical periods of the day, when oxygen levels are at their lowest during the night or become excessive in the afternoon.

Aerators that combine a strong horizontal flow with some degree of vertical mixing are more effective than aerators that create predominantly horizontal or vertical flow alone. Floating paddlewheels and propeller-aspirator pumps are the most popular aerators and have proved to be the most cost-effective. Both types provide aeration and impart strong mixing currents. A 2 hp propeller-aspirator pump tested by Boyd and Martinson (1984) mixed salt throughout a 0.4 ha pond in 1.5 h whereas the same mixing by wind-driven surface currents alone took 60 h. Aeration performance tests in tanks indicated that paddlewheels are more efficient in transferring oxygen and circulating water than other types of commonly used aerators. But it was noted that, for small ponds not requiring units of 1.3 hp or more, other aerators are often cheaper to buy (Boyd 1998).

Paddlewheel aerators may lower water temperatures by a total of 2°C or 3°C, which may be more than for equivalent propeller-aspirators, and this may be a good or bad thing depending on a farm's location and the season. It may be wise to avoid the use of paddlewheel aerators when problems with high salinities are anticipated because they can accelerate evaporative losses. There are wide variations in the maintenance requirements for different brands of paddlewheels. Those with more paddles on the wheels put less strain on motors and gears and incur lower maintenance costs (Estrada 2000).

Ideally aerators should generate water currents over the whole pond bed with sufficient velocity to suspend organic particles, but not so strong that fresh feed pellets or mineral soil particles are swept away. Usually, however, aerators generate localised areas with strong currents and sediments are suspended only to resettle in mounds in calmer areas. These sediments include organic solids, have a high oxygen demand and become anaerobic (section 8.3.6.7). It is often suspected that propeller aspirators suspend more sediment than paddlewheel aerators but both have a similar impact in this regard (Boyd 1998). Aerators should be positioned to minimise erosion, and not placed too close to the edge of the pond where they can erode embankments (Peterson & Pearson 2000).

8.3.6.7 Sludge

The question of what to do about the organically rich sediments that accumulate in patches in production ponds is a persistent problem in most semi-intensive and intensive systems. Organic matter in the sediment decomposes and causes anaerobic conditions at the soil surface with the release of toxic metabolites such as hydrogen sulphide into the water. It can be better to leave sediments undisturbed during ongrowing so that denitrification processes can release to the atmosphere some of the total nitrogen added as feed. Certainly the resuspension (re-aeration) of anaerobic sludge deposits needs to be avoided because it can provoke an oxygen crisis and may inhibit denitrification (Hopkins *et al.* 1994). Pond water quality can be improved if the sludge is removed, for example by concentration and voiding at a central drain, but the issue of what to do with the sludge then still needs to be resolved. Disposal on higher ground has been proposed, along with spreading on impoverished agricultural or forest land but brackish-water and marine ponds sediment have a high salt content so land disposal can constitute an environmental hazard.

In contrast, in intensive, zero water exchange systems the approach is to provide strong aeration and water currents in order to maintain as much potential organic sediment as possible in suspension where aerobic processes can proceed (section 8.3.7).

Faced with the task of reconditioning pond beds after a drain harvest, farmers often mistakenly assume that accumulated mounds of black sediment comprise mostly organic matter and they attempt to remove them. In fact they usually contain 95–98% mineral soil and only a little organic matter (2–5%) so removal of this waste is bad practice and inefficient (Boyd 1995b). Washing with high-pressure jets is very polluting and is to be avoided. It is better to dry, till and treat sediments (section 8.3.3) and spread them back over eroded areas of the pond bed and try to improve aeration techniques to minimise erosion.

8.3.6.8 Effluent treatment

Crustacean farming generates wastes in the form of particulate and dissolved organic and inorganic material. These arise from uneaten feeds (e.g. 15% of the feed provided), faeces (28%), and excretion and moult casts which together with food used for body maintenance can amount to a further 48% (Primavera 1994). In addition, farm effluents also contain numerous small organisms that feed or live on these materials and minerals from pond soils. It is estimated that in brackish-water shrimp ponds, 10–15% of the organic carbon and 20–70% of the nitrogen and phosphorus are converted to shrimp flesh, while most of the rest contributes to effluent load (Boyd 1995c). Put another way, production of 1 mt of shrimp can release 56.5 kg of nitrogen and 15 kg of phosphorus to the water (Boyd 1999b). Fertilisation also increases levels of nitrogen and phosphorus in effluents but the overall nutrient budgets will, of course, depend greatly on the feed and pond management strategies applied (Teichert-Coddington *et al.* 2000). Careful monitoring of turbidity, shrimp biomass and subsequent adjustment of feeding levels can do much to reduce wastage (Jory 1995b) but when pond water quality deteriorates, farmers often attempt to improve conditions by flushing with new water, thereby increasing their effluent discharges. Under extreme conditions, this can result in the loss of valuable phytoplankton (Hopkins *et al.* 1995). Large discharges may also become necessary at other times such as when ponds are drain-harvested or to maintain salinity after heavy rain.

Effluents can contain very high levels of nitrogen (1900–2600 mg L^{-1}) and phosphorus (40–110 mg L^{-1}) as well as therapeutant chemicals and antibiotics (Sze 1998). Dispersal of dissolved nutrients outside the farm depends on current and tidal regimes and may cause harmful microbial blooms (eutrophication – see Glossary) in the vicinity, but most particulates will sediment out as anaerobic deposits close to the farms. Both are likely to adversely affect the quality of any water pumped into farms from the locality as well as aid the spread of disease organisms. Legislation governing discharges is well established in many countries (sections 11.4.3 and 11.5.3), and for new farms consideration to managing pond drainage and discharges should be given at the design stage (Cook & Clifford 1997b; section 8.2.1).

Several methods for minimising wastes and their impact are practised (Chen 1998; Funge-Smith & Briggs 1998). These often involve increased operating or facility costs which can sometimes be offset, for example, against reduced pumping and fertilisation costs, against benefits from the production of forage organisms, or revenue from bivalves, seaweeds or herbivorous fish cultured in treatment ponds (Hopkins *et al.* 1995). The methods include:

(1) Removing sludge deposits from ponds to dry during or after each production cycle (Hopkins 1994; Hopkins *et al.* 1994) (section 8.3.6.7). Where disposal of sludge on land as fertiliser or landfill is not feasible (e.g. due to salt content or odour), demineralisation and stabilisation in created wetlands of planted grass hedges may be possible (Summerfelt *et al.* 1999) (section 12.6).
(2) Passage of effluents through constructed wetlands (Kadlec & Knight 1996) including mangrove forest, seaweed or salt-tolerant plant (halophyte) filter beds and rock filters. Estimates suggest that as much as 2–22 ha of mangroves (Robertson & Phillips 1995) or 2.5–13 ha of halophyte filter beds (Brown & Glenn 1999) would be required to treat the effluent from a 1 ha shrimp pond depending on farm management practices (section 12.6). If pipework or channels were needed to evenly distribute the effluent, this approach would be prohibitively expensive. Based on the capacity for nitrogen removal alone, however, Rivera-Monroy *et al.* (1999) estimated that 0.04–0.12 ha of mangrove would be sufficient to treat effluent from a 1 ha shrimp pond. On the other hand, partial recycling of effluent from a 340 ha Colombian shrimp farm through a 119 ha mangrove area demonstrated a satisfactory reduction in suspended solids and nutrient levels (Gautier *et al.* 2000). Filter beds of graded rocks placed in channels (Fujii *et al.* 1997), close to sea cages or farm outfalls (Xu *et al.* 1996; Laihonen *et al.* 1997) also have potential to reduce suspended solids and nutrient load, largely by increasing sedimentation rates and by biofiltration. However, reefs becoming fouled with algae (perhaps because they are in too shallow water) may not remove sufficient nutrients to be worthwhile (Antsulevich *et al.* 2000).
(3) Retaining effluent waters in large settlement or oxidation lagoons prior to discharge (section 8.2.1). Although such lagoons may occupy 4–20% of farm area, they can be used to buffer the volumes of effluent released each day. Where annual discharge licence fees are based on maximum daily volume released, such control can save money (Ford & Robertson 1995). The most efficient use of a lagoon may be for treating the last 10–20% of pond water (e.g. at harvest) which usually contains most of the settleable solids. A retention time of 6 h can be adequate for this (Teichert-Coddington *et al.* 1999).
(4) Reusing water from lagoons or a series of reservoirs in production ponds after treatment (Sandifer & Hopkins 1996). Steps involved can include sedimentation followed by the culture of nutrient-absorbing microbes, aquatic plants and seaweeds, filter-feeding bivalves or *Artemia* and/or bottom-feeding organisms such as fish (Kanit 1996; Chin & Ong 1997). Ideally the species chosen should have some commercial value (Chien & Liao 1995; Lin 1995; section 8.3.7). Early recycle systems disadvantageously occupied up to 30–50% of total farm area (Flegel *et al.* 1995).
(5) Reducing or eliminating water exchange and/or culture intensity to take advantage of in-pond digestion processes (Hopkins *et al.* 1993; Horowitz & Horowitz 2000; section 8.3.7). Recent estimates suggest these natural processes (Boyd 1999b) could remove 50% of nitrogen and 65% of phosphorus. The cost of the increased aeration needed in ponds with reduced or zero water exchange (section 8.3.6.6) is usually more than offset by the reduced pumping costs. In low water exchange, super-intensive shrimp ponds, well-formulated, low-protein diets can be used to give additional savings since some shrimp can exploit the natural productivity of the system (McIntosh 2000; sections 7.2.6.6, 8.3.2, 8.3.6.3 and 8.3.7). Species such as *Litopenaeus stylirostris, L. vannamei* and *Penaeus esculentus* seem better at utilising the nutritious detrital/microbial flocs generated and maintained in such systems than do *Penaeus monodon* and *Marsupenaeus japonicus* (McNeil 2000).

Considerable improvement of effluent quality is still possible on the majority of farms, however, through better utilisation of pond natural productivity as feed for the cultured stock, the use of low protein and phosphorus feed formulations (sections 2.4.1 and 2.4.5), better feeding strategies (Cho *et al.* 1994; Jory 1995b), stock monitoring and water management (Nunes & Parsons 1998).

Hatchery effluents are much smaller in volume than those from ponds and treatment methods may additionally include mechanical filtration through natural (slow gravity flow) or artificial sand filters (pumped systems); chlorination reservoirs; activated carbon filters and ozonation towers, or combinations of these methods. The main aims are usually to prevent disease transmission and the release of medicines to the environment.

8.3.7 Reduced and zero water exchange systems

The operation of reduced or zero water exchange systems, also known as semi-closed or closed systems, presents a crustacean farmer with both constraints and opportunities. Essentially the farmer must learn to manage a fixed body of water, without the ability to dilute phytoplankton or ammonia concentrations with an influx of new water and without the option of being able to react to low dissolved oxygen levels with an emergency flushing. At the same time, the farmer must develop an integrated approach to water, feed and soil management, limit the ingress of water-borne disease, minimise environmental impacts and enhance the efficiency of nutrient assimilation by the crop. With typical food conversion ratios in open systems of 1.65–2.4 : 1, only 18–27% of nitrogen is assimilated, leaving considerable room for improvement in the latter aspect (Funge-Smith & Briggs 1998).

Reduced and zero exchange systems predominate in shrimp farms in Thailand and they have been developed as a necessary response to environmental degradation and widespread disease, particularly yellow head virus and white spot syndrome virus. Incoming water may be treated with calcium hypochlorite (60% active) at 300 kg ha^{-1} and then aerated to disperse the chlorine and limed to promote a bloom of phytoplankton. However chlorine does not really function as a disinfectant when targeted at disease-carrying organisms in ponds because it may not reach concentrations high enough to react with all the organic matter present. Insecticides, for example trichlorfon (0.5–0.8 mg L^{-1}), actually represent a safer, cheaper and more carefully targeted option and, if used in a responsible manner, should not present any hazards. Shrimp farmers, much more so than the farmers of terrestrial crops, have a very strong incentive to use insecticides responsibly because any abuse, for example excessive concentrations or the use of products that do not break down quickly, will swiftly kill all the shrimp in a pond.

Once ponds in Thailand have been filled with treated water, no water at all is exchanged for 2 months except maybe for the addition of some freshwater in the dry season to replace losses due to evaporation. From 2 months onwards some topping up may be performed with pretreated water. As the production cycle progresses the level of nutrients in the system increases and the challenge of successful water management intensifies. Ammonia concentrations can reach 3 mg L^{-1} and phytoplankton can become very dense, threatening a mass die-off and possible oxygen crisis. Dinoflagellates and blue-green algae can bloom and these are considered stressful for shrimp. Farmers sometimes attempt to eliminate excess phytoplankton by killing it with benzalkonium chloride, formalin or chlorine and then removing the resulting foam that forms on the pond surface and collects at the pond edge. However this algaecide-based approach to bloom management may be futile and dangerous. Bacterial remediation is also attempted but it is unclear if it is effective or not (section 8.9.4.2). The use of carbon sources such as sugar and molasses has yielded interesting results in terms of modification of microbial communities, but if sugar applications are intermittent then bacteria are liable to bloom and crash and the original ammonia problem can return. Ideally ammonia would be denitrified to nitrate, subsequently converted to nitrogen and volatilised, but this function is not performed by the bacteria present in bacterial remediation products (Funge-Smith & Briggs 1998). In one zero exchange system in Australia, Body (2000) suggests the application of 10 mg L^{-1} of molasses if ammonia levels exceed 0.5 mg L^{-1}. Closed systems often support dense populations of fouling organisms like *Zoothamnium* and this can lead to problems with gill infestations. While closed systems can help to keep diseases out, if disease agents gain entry to the system they can spread very quickly.

Another type of reduced water exchange system is operated in Thailand in low-salinity water. Ponds of 0.1–0.5 ha in areas traditionally used for rice paddy are filled with freshwater and one corner isolated with a makeshift barrier of poles and plastic sacking. Hypersaline water (70‰) from a road tanker is then run into the impounded zone to provide a salinity of around 5‰ in preparation for stocking with *Penaeus monodon* postlarvae at densities of 30–40 m^{-2}. After 2 weeks the sacking material is removed and the shrimp disperse.

Plate 8.3 Paddlewheel aerators awaiting deployment at a semi-intensive shrimp farm in Australia. Environmental imperatives are favouring aerators over water exchange as the primary tool of water quality management.

Freshwater is added each week and the salinity falls to almost zero by the time ponds are harvested (75–120 days). It may be necessary to add lime to improve the pH buffering capacity of the water. Yields of 3–4 mt ha^{-1} crop^{-1} are achieved (Shivappa & Hambrey 1997).

The comparative productivity of closed and open systems is still a matter of debate and investigation. Very good yields have been achieved in super-intensive closed system ponds in Belize (section 7.2.6.6), but it is apparent that there are limits to the intensity and productivity of closed systems without resorting to water exchange or some type of filtration. Hopkins *et al.* (1993) place these limits somewhere between stocking densities of 22 and 44 shrimp m^{-2}, which corresponds to peak feeding rates of between 70 kg and 140 kg ha^{-1} day^{-1}. This is in accordance with estimates of the assimilative capacity of static freshwater catfish ponds (112 kg ha^{-1} day^{-1} of 32% protein feed). Funge-Smith and Briggs (1998) note that growth rates may be slower in closed intensive systems than in open ones. Boyd (1999a) recognises that some water exchange may be useful in intensive systems to reduce ammonia concentrations, but he seriously doubts the need for any water exchange in less intensive systems, except perhaps to reduce salinity in dry seasons. Water exchange can be counterproductive precisely because it flushes away nutrients and plankton. Catfish farmers have successfully adapted to zero exchange systems and harvest their ponds with seine nets to conserve the water in the pond. The average catfish pond is drained only once every 6 years.

Aeration plays an essential role in most closed systems. For intensive ponds Hopkins *et al.* (1995) estimate that the elimination of water exchange may increase supplemental aeration requirements by about 10%. In semi-intensive systems either aeration or water exchange are needed to counteract an oxygen imbalance. More information on aeration is provided in section 8.3.6.6.

In closed systems it is also important to carefully manage feeding rates so as not to overwhelm the assimilative capacity of a pond ecosystem. Reliance on normal feeding tables is usually inappropriate and it may even be better to use constant feeding rates. Hopkins *et al.* (1996) experimented with constant feeding rates with a view to promoting stability and obtained good results. *Litopenaeus vannamei* were reared at 38 m^{-2} in plastic-lined, rectangular ponds of 0.1 ha and 1.3–1.5 m deep. A 26 cm layer of sandy soil covered the liner and ponds were aerated at a rate equivalent to 10 hp ha^{-1} by day and 20 hp ha^{-1} for the 6 hours before dawn. Aerators imparted a circular current that concentrated waste in the centre of the pond. Ponds received a 40% protein diet at a fixed rate equivalent to 57 kg ha^{-1} day^{-1}. Yields after 153 days were 5–6 mt ha^{-1} crop^{-1} of shrimp weighing on average 15–16 g. Water exchange rates of 15% and 0% were tested and the former gave a slightly but not statistically significantly better result. Total ammonia nitrogen levels were higher in the zero exchange ponds (3.16 compared with 0.92 mg L^{-1}).

A logical extension of the zero water exchange approach is to recondition and reuse water for existing and future crops (section 8.3.6.8). This can be done most easily with the aid of treatment ponds, feedback canals and reservoirs and, ideally, new farms should be designed with a closed water management strategy in mind. The effectiveness of treatment is enhanced if a secondary crop of bivalves, herbivorous fish, seaweeds or aquatic plants is produced to assimilate excess nutrients and plankton. But for biological treatment ponds to be effective the secondary crop must be routinely harvested to ensure that biomass is removed from the system, otherwise the treatment pond merely serves to dilute or postpone the eutrophication potential of pond effluents (Hopkins *et al.* 1995).

Some zero exchange farms in Thailand operate recycling systems in which water is taken from shrimp ponds and goes through three phases of treatment before reuse – sedimentation, biological treatment (with fish, mussels, oysters or seaweeds), and finally aeration, in a manner similar to systems used for domestic wastewater treatment (Lin 1995). The successful integration of secondary species for biological treatment is, however, rarely achieved in practice, partly because of the problem of the heavy solids load in effluent water even after passage through a sedimentation pond. Tilapia cause problems because their nest-building habits stir up more sediment; bivalve molluscs generate large amounts of pseudo-faeces that can quickly cause self-fouling; and seaweeds do not thrive because their fronds are quickly smothered with sediment. Ideally the secondary crop should be ready after 4–5 months so that it can be harvested at the same time as the shrimp. Compared to shrimp, secondary crops are low-value species with more limited markets, so there is little financial incentive to persevere. In view of the practical problems of treating and recycling pond effluent, the trend in Thailand is towards the use of a limited water exchange system for the production ponds, replacing water loss from treated reservoirs – rather than from recycled water (Funge-Smith & Briggs 1998).

8.3.8 Ponds with acid sulphate soils

The problems posed to pond culture by acid sulphate soils were discussed in section 6.3.3.5. The successful management of affected ponds requires usual practices to be modified, particularly with regard to pond bottom preparation. Normal rejuvenation procedures involving drying and oxidation in air (section 8.3.3) would serve to exacerbate acidity problems. Indeed, complete draining of ponds may need to be avoided to keep pond bottoms and embankments waterlogged and largely anaerobic. Between crops, repeatedly filling and leaching an affected pond can reduce acidity and will certainly be more beneficial than drying. Routine monitoring of pH is particularly important.

To reduce the leaching of acid from embankments into pond water, pond levels can be kept higher than surrounding waters. This is more easily achieved with a pumped than a tidal water supply. Another advantage of a pumped supply is that water can be exchanged each day to limit the build-up of acid in a pond. Tidal flushing, by comparison, may only be possible during spring tides. If native acid-resistant grasses are planted on embankments they can help to stabilise the soil.

8.3.9 Lined ponds

Lined ponds can extend the range of sites where crustaceans can be grown to include sandy or acid sulphate soil locations but their use usually means abandoning traditional concepts of pond management. Specific differences will include deeper water (1.5–2 m) and increased responsiveness to water management needs in relation to phytoplankton control. Lined ponds are also easier to clean and sterilise between crops than earthen ponds. They are particularly suited to intensive and super-intensive cultures because they prevent the soil erosion that would otherwise arise due to the strong currents generated by the aerators.

Pruder *et al.* (1992) found increased phosphorus in effluents from experimental lined ponds (1.8 m^2) stocked with *Litopenaeus vannamei*, but no other major concerns. However it has been observed that cannibalism and poor feed conversion can arise in shrimp culture if currents sweep feeds too quickly to the centre of the pond (Funge-Smith & Briggs 1998). Another possible problem is gas production beneath liners not laid over subsoil vents or drains, which often necessitates the use of weights. The disposal of old pond liners can present problems and in Thailand has been of particular concern with some cheap liners that only give a service life of two crops.

Opinions vary as to whether it is better to provide a sand substrate in lined ponds or to leave them bare. Trials with shrimp, *Litopenaeus vannamei*, in bare or plastic-lined tanks have produced equal or better growth rates than shrimp grown on sand, clay or mud surfaces, suggesting that burrowing is not a physiological necessity (Pruder *et al.* 1992; section 4.6.4). However, improved

Plate 8.4 Deep plastic-lined pond used for intensive shrimp culture in Oman. The crop is being harvested with a seine net and then packed in ice to be loaded into the truck waiting on the embankment. (Photo courtesy of P. Fuke, Chelmsford, Essex.)

feed conversion and growth rates have been obtained with *Penaeus semisulcatus* cultured on a sand substrate as opposed to a bare fibreglass bottom (Al Ameeri & Cruz 1998). The use of lined ponds in shrimp culture is further discussed in sections 7.2.6.5 and 7.2.6.6.

8.4 Water treatment methods

This section summarises water treatment methods applicable to both fresh and seawater, with any differences indicated. The broad aims of water treatment are to:

- provide a near optimal environment for maximum growth of the cultured crustacean;
- exclude disease-causing organisms;
- economise on the quantity of water to be pumped or heated.

Unfortunately, the moment water is drawn from the natural environment, changes occur in its physicochemical and biological characteristics. Treatment attempts to slow or minimise these changes and to maintain conditions within limits tolerated by the animals (section 8.5). A number of reviews cover the subject in more detail (Wheaton 1977; Wickins & Helm 1981; Brune & Tomasso 1991; Timmons & Losordo 1997) and most authors agree that no amount of treatment will satisfactorily rectify problems arising from a poorly chosen site (section 6.3.1).

8.4.1 Abstraction

Water from natural sources may be used untreated (for example, in ponds); alternatively, if it is to be used in a hatchery, nursery or disease-free (i.e. closed cycle or biosecure) broodstock production unit it may be treated prior to entry (pretreatment) or within the culture facility (treatment).

Large-scale, land-based ongrowing units operated at well-chosen sites generally have minimal pretreatment needs. The water is drawn directly from the sea, river, lake or estuary by tide, gravity or pump with minimal mechanical filtration (Muir & Lombardi 2000). Supplies drawn from streams and rivers are liable to contain plant material that could block intake structures (e.g. leaves during autumn) and specific cleaning devices may be necessary such as the use of compressed air jets under a downstream sloping screen (Ewing & Sheahan 1996). Once on site, the water may be fed directly to the ongrowing facility or to reservoirs, which allow greater control over fluctuations in supply and water quality (Chien & Liao 1995); together with effluent treatment reservoirs the latter may occupy as much as 50% of the farm area (New 1999). If destined for a hatchery or nursery, the water is fed into a storage or pretreatment reservoir where some settlement of solids occurs and phytoplankton growth may be encouraged by the application of fertilisers and inoculation with algae starter cultures.

Plate 8.5 A tower of sticks over which pumped borehole water cascades to remove minerals and unwanted gases and gain oxygen prior to use on a Taiwanese shrimp farm.

Shade netting is sometimes employed to control the amount of light reaching the algae in small reservoirs, while algal levels in ponds can be diluted by increased water exchange (section 8.3.6.4).

Borehole water may be cascaded, or aerated in a reservoir to reduce carbon dioxide, hydrogen sulphide and ammonia or to precipitate minerals that frequently occur in groundwaters (section 6.3.1.4). Smaller quantities of water suitable for hatcheries may be drawn through various types of filter before entry (Colt & Huguenin 1992). Where a suitable substrate exists or can be improvised, sub-sand extraction is a technique for drawing water down through layers of sand and gravel into a buried, screened intake pipe. Once the substrate around the intake point has stabilised, it acts both as a mechanical and a biological filter (see below) provided it is operated frequently (Cansdale 1981).

8.4.2 Primary treatment

Large land-based filters used in pretreatment take many forms (from towers to sunken pits), contain any of a range of materials (sand, gravel, rocks, coral, shells, plastic rings or beads) and are operated in a variety of ways (upflow, downflow, submerged, trickling or pressurised). In circumstances where settlement or sedimentation is necessary, circular or long V-shaped, purpose-built tanks, sedimentation ponds and canals or compact plate separators can be used. Tilted plate or tube separators (also called particle interceptors) are static arrays of inclined parallel plates or tubes up which water passes in a laminar flow. Close packing of the elements permits the minimum of distance for particles to settle out onto the surfaces. Unfortunately these devices do not work efficiently with finely divided solids or 'sticky' organic materials that are reluctant to move off the inclined surfaces into the collection sump (Timmons 2000). Water and pond bottom disinfection with chlorine and quicklime (respectively) is commonly used in closed cycle or biosecure production units for shrimp (especially broodstocks) and time must be allowed for microbial populations to stabilise after treatment (Bratvold *et al.* 1999; sections 8.3.3 and 8.9.4).

8.4.3 Secondary treatment

Adjustment of salinity, temperature, pH and oxygen levels in the water supply are most important, normally straightforward, operations. In calculating the flow required to dilute metabolites, maintain temperature and in some cases add oxygen, it is necessary to compute the displacement time or exchange rate of water in the tank. In practice, providing a given percentage of the pond volume each day does not actually exchange that amount of water. A better idea of exchange (assuming complete mixing) can be derived from the equation:

$$T = -\ln(1 - F) \,.\, V/R$$

which can be rearranged to:

$$R = -\ln(1 - F) \,.\, V/T$$

and:

$$F = 1 - (e^{-TR/V})$$

where T = days needed to get x% exchange; V = pond

volume (m^3); F = fractional water replacement in time T; R = inflow ($m^3\ d^{-1}$).

Table 8.1 shows the approximate number of days to achieve partial replacement of 50%, 75% and 90% of water in culture tanks or ponds. From the above equation ($R = -\ln(1 - F)\ .\ V/T$) the daily water requirement needed to achieve, for example, 10% exchange per day in a 40 ha pond of average water depth 1 m can be calculated:

Table 8.1 The time (days) taken to achieve 50%, 75%, and 90% water exchange in a pond receiving 8%, 10%, 12% and 14% of its volume in new water per day (assuming complete mixing).

% exchanged	Flow rate (% pond volume day^{-1}) 8	10	12	14
50	8.7	6.9	5.8	4.9
75	17.3	13.9	11.5	9.9
90	28.8	23.0	19.2	16.4

Plate 8.6 Side-stream protein skimmers are employed to treat water in some blue crab shedding systems. (Photo courtesy D.W. Webster, University of Maryland, USA.)

4 ha × 10 000 m^2 × 1 m depth = 40 000 m^3 of water in the pond

Rate of inflow to achieve 10% exchange per day = $-\ln(0.9) \times 40\ 000/1 = 4214\ m^3\ d^{-1}$ or around 176 $m^3\ h^{-1}$.

Other than in extensive pond systems, oxygen is added by means of specific aeration or oxygenation methods rather than by the water flow alone, but it is worth noting that crustaceans can be adversely affected by total gas supersaturation, as can fish, and to avoid gas bubble disease, allowances have to be made for the presence of metabolic gases already in the water when calculating the input from an aeration system (EIFAC 1986; Colt 2000). Paddlewheel aerators and propeller-aspirator pumps are widely used, especially in intensive shrimp farms where several units consuming up to 20 kW ha^{-1} may be employed (Boyd 1998; section 8.3.6.6). The increased circulation they produce is beneficial but if too severe commonly causes erosion of banks and sediment accumulation at the pond centre.

It may sometimes be necessary to sterilise water destined for hatcheries or recirculation systems. A number of shrimp hatcheries, for example, add commercial sodium hypochlorite solution to a freshly filled indoor reservoir to give an initial concentration of around 5–20 mg L^{-1} free chlorine. The water is then recycled through a rapid (pressure) sand filter for 24 h, after which any remaining free chlorine is neutralised with sodium thiosulphate. The uses of chlorination and other methods including ozonation and ultraviolet irradiation were reviewed by Rosenthal (1981) while Summerfelt and Hochheimer (1997) provide a more recent review of ozone use in aquaculture.

In specialised research hatcheries, additional treatment facilities for dark storage, ultra-fine filtration and activated charcoal treatment may be included (Wickins & Helm 1981). At certain times of the year phytoplankton blooms or elevated dissolved organic loads in the water may necessitate the use of air or air/ozone foam fractionation treatment (protein skimming) for their breakdown. The process typically involves the upward passage of fine air or air/ozone bubbles through a downward flowing column of water (Timmons *et al.* 1995). It is often used in densely stocked recirculation systems for the breakdown of refractory organic molecules in solution that are not readily oxidised by biological filtration (Rosenthal 1981). Foaming in seawater invariably leads to an increase in suspended particulates and a filtration step generally follows. However a new design (based on a cyclonic counter-current system) is reported to have the capacity to remove 860g of suspended solids and 15 g dissolved organic carbon per day from an intensive marine farm (Hussenot & Lejeune 2000).

Ion exchange resins for the removal of dissolved substances (ammonia, metals and some organic compounds) may be appropriate in fresh- but not in seawater where chelation agents such as disodium EDTA, sodium metasilicate and Fuller's earth are widely used to deactivate a range of growth inhibiting substances (Wickins & Helm 1981). A polymeric heavy metal absorbent (PHMA) has also been shown to be effective in decreasing high concentrations of copper, zinc, lead and cadmium (Yuan *et al.* 1993).

The beneficial bacteriostatic properties of extracts from macro- and microalgae have been recognised since about 1990. They can enhance shrimp larvae survival, especially when microparticulate diets are being fed. However, attempts to sterilise hatchery water supplies can upset the balance of natural microbial populations and allow surviving bacteria to dominate or become more virulent. The use of 5 µm filtered, but otherwise untreated, seawater in shrimp hatcheries has proved advantageous, particularly when artificial diets are used (Alabi *et al.* 1997).

8.4.4 Recirculation systems

Serious losses in the shrimp farming sector caused by environmental degradation and disease outbreaks together with increasing restrictions on effluent discharges have stimulated many farm managers to recycle some or most of their pond water. These aspects of pond (i.e. outdoor) recirculation and water reuse are discussed in sections 7.2.6.5, 7.2.6.6, 8.3.6.8 and 8.3.7.

Indoor recirculation systems are increasingly used in broodstock production units (e.g. Menasveta *et al.* 2000), hatcheries (Mallasen & Valenti 1998), nurseries (Davis & Arnold 1998), overwintering facilities, pilot battery operations (Mattei 1995), for disease-free stock production (Lee *et al.* 1998) and in stock enhancement rearing programmes (Beard & Wickins 1992). They reduce the risk of disease, conserve heat and preserve water that has had expensive treatment or has cost a lot to transport (Valenti & New 2000) or prepare (e.g. artificial seawater; Bidwell & Spotte 1985). The proportion of water renewed varies from a continuous 'bleed-in' to almost closed systems where water is renewed only rarely. During recirculation the water is continuously treated to:

(1) Maintain the required temperature and salinity;
(2) Make good losses due to evaporation and leakage;
(3) Stabilise chemical changes by:
   (a) replacement of depleted components (oxygen, buffering capacity, calcium);
   (b) detoxification and dilution of substances that accumulate (ammonia, nitrite, nitrate, carbon dioxide, dissolved organic materials, suspended solids; Wheaton 1977; Wickins 1985a,b; van Rijn 1996; Hochheimer & Wheaton 1998).

The cost of maintaining temperature in a recirculation system is largely dependent upon the amount and temperature of new water that has to be added to the system in order to maintain water quality. Yet several modern, closed cycle, European eel farms operate at 23–25°C with minimal additional heat, relying mainly on that generated by system pumps and compressors. Higher heating costs arise in controlled environment ongrowing systems than in hatcheries but, in both, good control of chemical changes in the water combined with efficient insulation can go some way towards minimising these costs without the need for extensive use of heat exchangers or heat pumps. However, the high capital costs of providing adequate insulation, reliable water treatment plant and heat transfer equipment militates against a successful demonstration of commercially viable crustacean farming in recirculation systems in cool temperate regions (Van Gorder 1990). Nevertheless, continued refinements to water treatment technology (Timmons & Lorsodo 1994) combined with better understanding of dietary needs and waste management are maintaining commercial optimism as well as research interest (section 7.2.6.6).

Control of salinity is normally by addition of artificial sea salts or freshwater and generally presents few technical problems. Mixing of seawater with some natural groundwaters could cause problems of precipitation unless the latter are well aerated first.

In any aquaculture system oxygen is consumed by the cultured animal, its live food (if any), other heterotrophic organisms in suspension and attached to surfaces, and by nitrifying bacteria. In cloudy, organically rich water considerably more oxygen may be consumed, and ammonia and carbon dioxide produced, by organisms other than those being cultured. This is particularly evident in intensive ongrowing recirculation systems and at night in ponds after photosynthesis has stopped (section

8.3.2). The ways in which the effects can be minimised in recirculation systems are, firstly, rigorous attention to feeding regimes and feeding husbandry (section 8.3.6.3), and secondly, efficient mechanical filtration to remove much of the suspended matter. Aeration at this stage supplies oxygen and removes excess dissolved carbon dioxide, thereby tending to stabilise pH (Wickins 1984a).

During intensive culture, and particularly in recirculation systems, the mineral content of the water may change. For example, in marine recirculation systems, both calcium (essential for shell formation after moulting) and magnesium may be lost by precipitation with phosphate and through uptake by the cultured species. Similar changes thought to occur in shrimp ponds could be responsible for some outbreaks of chronic soft-shell disease (section 8.9.1). Under such circumstances the addition of new water or chemical restoration of the lost minerals may be required (Wickins & Helm 1981; Laurent *et al.* 1997).

8.4.5 Biological filtration

Biological filters have two primary functions: the oxidation of ammonia by autotrophic micro-organisms and the oxidation of dissolved and some fine suspended organic materials by populations of heterotrophic micro-organisms (simply, autotrophic = feeds on inorganic compounds; heterotrophic = feeds on organic compounds). By this definition crustaceans are heterotrophs and, like the microbes, also produce ammonia and carbon dioxide wastes. The autotrophs, on the other hand, feed on the ammonia and produce hydrogen ions and nitrate as waste products. The simplified reactions are:

Nitrosomonas

$$55\ NH_4^+ + 5\ CO_2 + 76\ O_2 \rightarrow C_5H_7O_2N + 54\ NO_2^- + 52\ H_2O + 109\ H^+$$

Nitrobacter

$$400\ NO_2^- + 5\ CO_2 + NH_4^+ + 195\ O_2 + 2\ H_2O \rightarrow C_5H_7O_2N + 400\ NO_3^- + H^+$$

The hydrogen ions (acid) produced by *Nitrosomonas* are normally neutralised or buffered by the alkaline reserve of the water, but in densely stocked recirculation systems can result in a catastrophic loss of buffering capacity as the acid pushes the carbonate/bicarbonate equilibrium to the right:

$$4\ H^+ + 2\ CO_3^{2-} \leftrightarrow 2\ H^+ + 2\ HCO_3^- \leftrightarrow 2\ H_2CO_3 \leftrightarrow 2\ H_2O + 2\ CO_2$$

The loss of bicarbonate and associated rapid decline in pH is likely to prevent proper mineralisation of the crustacean exoskeleton (Wickins 1984b). Additions of sodium bicarbonate to freshwater systems (Loyless & Malone 1997) or sodium hydroxide to marine systems (Wickins 1985a) are made in compensation. Ponds and lakes acidified by industrial wastes (e.g. acid rain) can also result in shell mineralisation and moulting problems in freshwater crayfish but conditions can be improved by the addition of limestone (Iivonen *et al.* 1995). Similarly marine and brackish-water ponds may be rejuvenated by the addition of lime (section 8.3.3).

A biological filter provides a large surface area for colonisation by useful micro-organisms, through the material (the filter medium) with which it is packed. Literally hundreds of different types of biological filter have been described. Commonly used filter media include stone chips which may (e.g. limestone), or may not (e.g. granite) contribute to the calcium content, alkalinity or buffering capacity of the water; plastic rings, spheres, beads or sheets are also used, packed either at random or coherently. The water to be treated may pass downwards or upwards through the filter, and downflow filters may be submerged or percolating. In display aquaria and lightly loaded systems the filter functions may be combined with that of mechanical filtration (e.g. slow sandbed filters), but in heavily loaded systems more consistent and predictable performance is achieved by separating mechanical and biological filter functions. This is because mechanical filters require regular back-flushing and surface raking, processes that can disrupt the performance of the microbial populations. Many modern biological filters contain plastic media designed to provide a large surface area per unit volume while at the same time containing a high percentage of voids so that the filter can never become blocked. Other types of biological filter include compact, rotating biological contactors (RBCs), which are disc or drum systems turning slowly in a sump tank, and fluidised bed filters in which finely divided filter particles (sand or buoyant plastic beads) are held in suspension by an upwelling flow of wastewater (Muir 1982; Hochheimer & Wheaton 1998; Aneshansley 2000). The rotating disc types are very effective at removing ammonia (Rogers & Klemetson 1985; Knösche 1994) and require virtually no pumped head of water but their use in seawater systems requires special attention to the materials and engineering design to avoid corrosion and consequent mechanical breakdown of the moving parts. Modern floating bead filters act not only as biological filters but also as particle traps and can thus also be used as water clarifiers (Malone 2000). They are compact units (Greiner & Timmons 1998) and the beads are designed in such a way as to retain nitrifying bacteria during backwashing to remove trapped solids as well as the faster growing heterotrophic organisms that compete with the nitrifiers for space (Malone *et al.* 1998). Further information on modern biological filters can be found in Timmons and Losordo (1997).

Calculation of the size and number of the biological filters required is one of the most uncertain elements of system design because the relative proportion of the available surface area occupied by heterotrophs and autotrophs in the filter alters as the cultured crustaceans grow, as the amount and composition of feed added to the system changes, and as changes occur in the quantity and composition of incoming make-up water (Hochheimer & Wheaton 1998; Malone *et al.* 1998). Calculations are simplified in systems where much of the particulate organic material in suspension is first filtered out mechanically, since heterotrophs generally grow faster and will colonise a biological filter more quickly than the autotrophs, leaving less capacity for ammonia oxidation (Wickins 1985a,b). Periodic removal of accumulated heterotrophic biofilm (slimes) can be a problem, particularly in submerged filters. Floating bead filters are easily backwashed and cleaning frequency can be readily adjusted to maximise nitrification performance. Examples of filter performance under a range of culture conditions are reported in Table 8.2 and by Wheaton (1977), Wickins and Helm (1981), Muir (1982), Wickins (1983, 1985a,b) and Rogers and Klemetson (1985). Additionally, example calculations to show the size of biological filters required for specific biological loads are presented by Hochheimer and Wheaton (1998) for percolating, plastic ring and RBC types, and by Malone *et al.* (1998) and Malone and Beecher (2000) for floating bead filters.

The initial colonisation of a filter begins as soon as food or animals are put into the system because the micro-organisms are ubiquitous in nearly all water supplies. The crustaceans to be cultured, however, should not be placed in the system until the filter populations have become established (about 3–9 weeks at 20–28°C). This is to avoid exposure to ammonia and also to nitrite that invariably accumulates until the rate of consumption by *Nitrobacter* equals the rate of production by *Nitrosomonas* (Fig. 8.5). Nitrate is much less toxic and levels are reduced either by dilution (effluent regulations permitting) or by denitrification units (see below) in which certain heterotrophic bacteria reduce nitrate to nitrogen in the absence of oxygen (van Rijn 1996). The filter maturation process may be hastened by adding 10% or more media from an established filter, chemical nutrients (e.g. ammonium citrate, ammonium chloride, sodium nitrite), some commercial filter seeding mixtures, or even a few freshly opened filter-feeding bivalve molluscs to the system (Beard & Wickins 1992; Hochheimer & Wheaton 1998). It is reported that stable nitrification may be achieved both in marine and freshwater systems in just 1–2 days, if filter media precoated with selected populations of nitrifying bacteria are used (Horowitz & Horowitz 1998).

Denitrification units are more likely to be found in intensive freshwater fish production units than in crustacean systems because of the stringent regulations governing nitrate discharges in countries where most fish recirculation systems have been built. In addition, the denitrification process occurs under anaerobic conditions that require particularly careful control of nutrient inputs, thereby adding to costs. Denitrification reactors commonly contain populations of bacteria attached to sand or plastic media and are made anaerobic either by purging with nitrogen gas or by maintaining such slow flows that all the oxygen is used up by other microbes (van Rijn 1996). Alcohol or sugars are supplied as a carbon source for the denitrifying bacteria, although the possibility of using degraded fish wastes as a carbon source has also been demonstrated (Aboutboul *et al.* 1995). During denitrification, hydroxyl ions are released, which help to reduce the alkalinity losses caused by hydrogen ion production during nitrification and thus to stabilise system pH. A potentially useful advance has been the development of (experimental) polymeric beads in which denitrifying bacteria are entrapped together with a suitable carbon source. Denitrification starts as soon as the beads are introduced into nitrate-rich, and presumably anaerobic waters (Tal *et al.* 1997). Discharges of phosphorus in effluents are also subject to strict regulation and can be initially reduced by improving the assimilation of dietary phosphorus (section 2.4.5). Further reduction can occur due to uptake by aerobic nitrifying bacteria (Wickins & Helm 1981) and anaerobic denitrifers (van Rijn & Barak 1998), which would need to be periodically harvested from the system. Greater use of denitrifying units could be expected if super-intensive, indoor crustacean farms become more

Table 8.2 Examples of biological filter performance in marine and freshwater recirculation systems.

Input (mg N L^{-1})	Filter type	Hydraulic load (day^{-1})	Temp. (°C)	pH	Daily ammonia removal	Time to establish nitrification (days)	Reference
0.5–2	Marine, plastic, percolating	—	20	8.2	—	37	Wickins & Helm 1981
1 + live lobsters	Marine, plastic percolating	—	24	8.2	—	35	Wickins & Helm 1981
Live shrimp	Marine, plastic percolating	—	26	8.1	—	37	Wickins & Helm 1981
1–4 + 50 g mussels	Marine, 12–25 mm gravel, submerged	20.5(a)	26	—	501(e)	24–35	Forster 1974
—	Marine, 12–25 mm gravel, submerged	82.0(a)	—	—	1112(c)	—	Muir 1982
—	Marine, 12–25 mm gravel, submerged	246.0(a)	—	—	2178(c)	—	Muir 1982
0.1	Marine, plastic, percolating	95(a)	20	7.8–8.2	0.22(d)	—	Richards & Wickins 1979
0.2	Marine, plastic, percolating	182(a)	28	—	0.03–0.38(d)	—	Wickins 1982
0.28	Marine, plastic, percolating	153(a)	28	—	0.08–0.39(d)	—	Wickins 1982
0.39	Marine, gravel, 200 m^2 m^{-3} specific surface area	26(a)	28	—	0.03–0.10(d)	—	Wickins 1982
—	Marine, gravel, 210 m^2 m^{-3} specific surface area	360(a)	20	—	0.84(d)	—	Goldizen 1970
0.5	Freshwater, gravel submerged	—	6	—	0.25(d)	—	Wheaton 1977
0.5	Freshwater, gravel submerged	—	12.5	—	0.64(d)	—	Wheaton 1977
0.5	Freshwater, gravel submerged	—	20	—	1.03(d)	—	Wheaton 1977
0.08–9.3	Freshwater, plastic rotating biodrum	0.006–0.03(b)	25–30.8	7.1–8.4	82–96(e)		Rogers & Klemetson 1985
0.08–9.3	Rotating biological contactor	0.002–0.07(b)	25–30.8	7.1–8.4	69–99(e) 2.83(d)	—	Rogers & Klemetson 1985
0.08–9.3	Slag, percolating	0.003–0.03(b)	25–30.8	7.1–8.4	38–61(e)	—	Rogers & Klemetson 1985
<0.6	Freshwater, downflow submerged gravel	0.4	7	—	0.07(d)	—	Vandenbyllaardt & Foster 1992
<0.6	Freshwater, downflow submerged gravel	1.03	7	—	0.22(d)	—	Vandenbyllaardt & Foster 1992
<0.6	Plastic rings	—	7	—	0.38–0.45(d)	42	Vandenbyllaardt & Foster 1992
1.5	Percolating, plastic rings	50–300(a)	24	6.5–9.0	1.0(d)	—	Hochheimer & Wheaton 1998
1–1.5	Rotating biological contactor	403(b)	24–28	6.5–9.0	0.28–0.94(d)	—	Hochheimer & Wheaton 1998
—	Percolating, plastic media	—	—	—	0.28–0.55(d)	—	van Rijn 1996
<1.0	Floating beads	—	—	—	0.25–0.5(d)	—	van Rijn 1996
16 kg feed m^{-3} beads day^{-1}	Floating beads	576(a)	20–30	6.5–8.0	350–450(c) (i.e. 0.3–0.4(d))	—	Malone *et al.* 1998

(a) m^3 m^{-3}; (b) m^3 m^{-2}; (c) g N m^{-3}; (d) g N m^{-2}; (e) %.

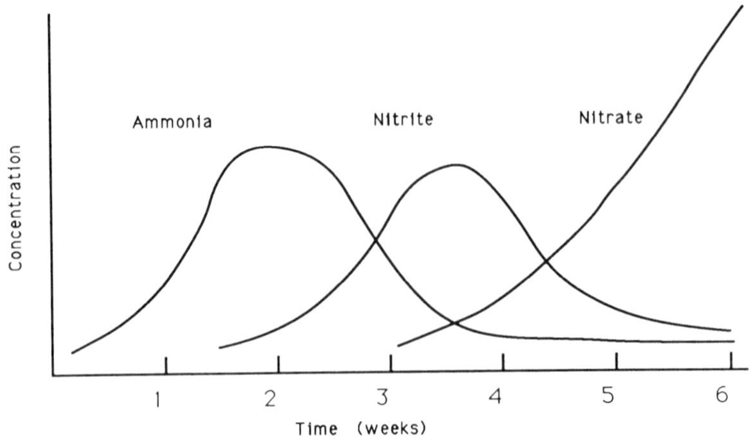

Fig. 8.5 Changes in levels of dissolved nitrogenous waste during maturation (start-up) of a recirculation system. (Note: Time scale will change with temperature.)

widespread, especially in localities where charges are levied according to effluent content and quantity.

Some medication treatments applied to stocks held in recirculation systems could affect the performance of microbes in biological filters and, unless information to the contrary is available, it is advisable to incorporate a filter by-pass system for use during such treatments (Bower & Turner 1982). Formalin treatments to remove marine fish parasites, on the other hand, may have little effect on biofilter performance.

8.4.6 Display, live storage and transportation

Biological filters and water recirculation are also widely used in display aquaria (Anon. 1988) and sometimes in live storage systems for clawed and spiny lobsters and crabs (Beard & McGregor 1991). Vivier transport systems rely on refrigeration and aeration, with or without recirculation, to maintain live crustaceans during transportation by road, sea or air. Descriptions exist of systems for crustaceans in general (Richards-Rajadurai 1989), *Macrobrachium* (Phillips & Lacroix 2000) and spiny lobsters (Sugita & Deguchi 2000). Cascade systems have been developed for road transportation of clawed and spiny lobsters and crabs (Whiteley & Taylor 1989) in which the crustaceans are held in vertically stacked trays continuously sprayed with recycled, chilled seawater. The gills of the animals are adequately covered although the animals are not totally submerged in water. In this system the weight of water carried is considerably less and the space the trays can occupy is considerably greater than in a conventional vivier truck fitted with deeper, aerated tanks. Attempts to develop systems in which crustaceans can be transported out of water but in a cooled, sprayed mist have not been successful.

At the other end of the scale, some crabs, crayfish and kuruma shrimp (*Marsupenaeus japonicus*) are typically transported to market out of water in baskets, damp sacks, or chilled sawdust respectively. A comprehensive review of live holding and transportation methods used for fish, molluscs and crustaceans in South-east Asia was prepared by Macintosh (1987).

Broodstock shrimp and prawns, larvae and small post-larvae are commonly transported by road or air in double-skinned polythene bags containing one-third water and two-thirds oxygen (e.g. Correia *et al.* 2000); Fig. 7.1 and Tables 7.2 and 7.3). The most important factors influencing survival under these conditions are temperature, oxygen and handling. Attempts to improve survival by removing ammonia and buffering pH with chemical additives have not always been successful (section 7.2.2.2). Prior to shipment, animals are acclimated to a suitably low temperature (Samet *et al.* 1996) and starved to reduce their metabolism. The rostrum of adult prawns is sheathed or removed to prevent it puncturing the bags. Survival and reproductive performance of broodstock shrimp (*Fenneropenaeus indicus*) following live transport can be improved, however, by packing them individually in perforated polythene tubes containing coconut mesocarp dust. The dust is held inside the tubes by a sheath of plastic mesh. The tubes are suspended in horizontal layers in a tank of chilled, aerated seawater (Table 7.2). Further details of transportation techniques are given under each species group in Chapter 7.

8.5 Water quality tolerance

In all aquaculture operations, but especially those where the water is treated, it is desirable to know the maximum and minimum acceptable levels of the changes that occur in water chemistry so that effort is not spent trying to achieve unnecessary goals. Unfortunately there is a shortage of such data based on long-term growth studies with crustaceans and much reliance is necessarily placed on values extrapolated from short-term acute tests and from studies with fish. Unlike the crustaceans used in traditional laboratory tolerance tests, farmed crustaceans are exposed to mixtures of metabolites, minerals or toxins in the water that may act synergistically or antagonistically. For example, toxicity of ammonia is exacerbated by high pH, nitrite may be ameliorated by the presence of chloride ions (Meade & Watts 1995) and detrimental effects of elevated total hardness (calcium and magnesium) levels in freshwater are enhanced as alkalinity increases (Latif *et al.* 1994; Vera 2000). In addition, the concentrations of many of the substances will undoubtedly vary throughout each day (see below). The examples of reported 'acceptable' ranges for ongrowing several major species or groups of crustacean shown in Table 8.3 are moderately variable and can thus only provide a guide rather than definitive values for system design. Levels acceptable in a hatchery are generally more restrictive since larvae and small post-larvae are often more sensitive than juveniles and adults.

In the majority of cultivated crustaceans most sub-lethal stressors (excretory products, industrial and agricultural chemical pollutants) affect ionic regulation; mainly of sodium and chloride ions. The effects of a number of such compounds on crustacean osmoregulatory capacity and on the tissue morphology of the major excretory organs, especially the gills, have been reviewed by Lignot *et al.* (2000) and indicate that measurement of a crustacean's ability to maintain its internal ionic balance against the external environment is likely to provide a good early warning system when monitoring animal health during culture. Measurement of acute stress response (usually time to death) to a controlled level of adverse environmental quality (sudden exposure to low salinity, temperature or formalin; usually a combination of two) is often used to test the vigour of a sample of post-larvae or juveniles prior to sale (section 7.2.4) although the results will not necessarily predict performance during ongrowing. A standardised test has been developed for *Macrobrachium* post-larvae based on exposure to ammonia, and is claimed to be more sensitive than salinity stress tests (Lavens *et al.* 2000).

8.6 Monitoring water quality

The metabolic activity of all organisms in an aquaculture system varies throughout each 24 h period, being generally less at night than in the daytime and reaching a maximum during and just after feeding. It has been recommended (EIFAC 1986) that the daily cycle of metabolite levels in culture waters should be determined at least once in order to locate the periods of maximum and minimum levels of vital components and measure selected factors to obtain information relevant to water management. This would include periodic measurements to monitor the condition of pond bottoms on outdoor farms (Boyd 1995a). Monitoring changes in the levels of every factor likely to affect crustacean growth and survival is clearly impracticable on a commercial farm and it is worth considering which are the key factors (Boyd & Fast 1992). In the majority of situations these will include oxygen, temperature, pH, total ammonia nitrogen and turbidity, but priority will vary according to species, life cycle stage and the culture system used. Other factors often of less importance in established systems include salinity, alkalinity, nitrite nitrogen, carbon dioxide, mineral ions and dissolved organic materials. Figures 8.6a and 8.6b show typical changes in key factors over a 24 h period in (a) a shrimp pond and (b) a hatchery recirculation system, and illustrate how samples taken at different times of the day from the same system, or at the same time of day from different systems, will give entirely different estimates of water quality.

Having considered when and what water quality factors to measure, it is appropriate to draw attention to the problems of making the measurements and analyses themselves. For example, ion-sensitive electrodes used for pH and oxygen measurements should be calibrated daily, replaced as soon as performance declines (this can be as often as every 18 months for some pH electrodes) and the standard solutions and buffers used for calibration stored and renewed according to the maker's instructions. These and other examples associated with the calculation and expression of results are presented in detail by EIFAC (1986), Boyd and Fast (1992) and, with special reference to monitoring for disease prevention purposes, by Brock and Main (1994).

It is worth mentioning two points concerning samples sent to analytical laboratories. Firstly, proper collection and preservation of the sample is vital if changes are not

Table 8.3 Desirable ranges and levels of water quality factors.

Species/group	Temp. (°C)	Salinity (‰)	Oxygen (mg L^{-1})	pH	Un-ionised ammonia (NH_3-N mg L^{-1})	Nitrite (NO_2-N mg L^{-1})	Hardness and alkalinity ($CaCO_3$ mg L^{-1})	Other
Penaeids	26–30(aa)	5–35(p,aa)	>5(p) 85–103%(p)	7.8–8.3(w)	0.09–0.11(t) (<0.02–0.07 in presence of 1.5 mg L^{-1} nitrite(y))	<0.1–0.25(p,t)	150–200 mg L^{-1} alkalinity(p)	160–400 mg L^{-1} Ca^{2+}(i); 100–200 mg L^{-1} NO_3-N; <0.002 mg L^{-1} H_2S(b); <0.45 mg L^{-1} Cu^{2+}(ab); <20 mg L^{-1} CO_2(p); <10 mg L^{-1} ferrous Fe(i); 2–14 mg L^{-1} suspended solids(i)
Macro-brachium	26–30; 25–32(a)	0; 12 for larvae	>4.5; >75% 3–7(a)	7–8.5(a)	<0.1; 0.1–0.3(a)	<1.4; <0.1(ad)	30–50 mg L^{-1}(a); 20–60 mg L^{-1} alkalinity(a); <53 mg L^{-1}@ <50 mg L^{-1} alkalinity(r,s); 25–100 mg L^{-1} @25–100 mg L^{-1} alkalinity(e)	3–8 mg L^{-1} SO_4(c)
Crayfish, temperate	14–23	0, <5	>6(min. 3(v))	6.7–8.5 (v,ac) (min. 6.0)	<0.1	<0.5(x)	50–200 (min. 40)	>5–44 mg L^{-1} Ca^{2+}; <0.1 mg L^{-1} H_2S(g); <0.1 mg L^{-1} ferrous Fe(g); <5 mg L^{-1} free CO_2(v,ac); <3 mg L^{-1} iron(v); 50–100 mg L^{-1} Ca^{2+}(q)
Crayfish, tropical	23–28(f) (min 14)	<1.5	>7.8(h); >80 %	7.0–8.5(ac)	<0.1	< 0.2(g)	60–100(f); >50 mg L^{-1} total alkalinity(l)	<0.1 mg L^{-1} ferrous Fe(g)
Lobsters, clawed	18–22	28–35(d)	6.4(d)	7.8–8.2	<0.014(d)	—	—	<10 µg L^{-1} copper(z)
Lobsters, spiny	23–30	32–36	>70% <105% (min. 3)	8.0–8.6	<0.1(n)	<1.0(n)	—	max. 1.2 mg L^{-1} COD(m); NO_3-N <100(n)
Crabs	23–30(j)	18–34(u) range 0–40(k)	>70% <130%	8.0–8.5(j)	—	—	—	<0.062 mW cm^{-2} UVB radiation(o)

(a) Boyd & Zimmermann 2000; (b) Kuo 1988; (c) New & Singholka 1982; (d) Van Olst *et al.* 1980; (e) Vera 2000; (f) O'Sullivan 1988; (g) Culley & Duobinis-Gray 1989; (h) Sammy 1988; (i) Wickins 1981; (j) Cowan 1983; (k) Oesterling & Provenzano 1985; (l) Jones & Burke 1990; (m) Kittaka 1997; (n) Booth & Kittaka 2000; (o) Hovel & Morgan 1999; (p) Brock & Main 1994; (q) Koksal 1988; (r) Brown *et al.* 1991; (s) Latif *et al.* 1994; (t) Olivares & Yule 2000; (u) Castaños 1997; (v) Eversole & Brune 1995; (w) Bray & Lawrence 1992; (x) Liu & Avault 1996; (y) Cavalli *et al.* 1998; (z) Mercaldo-Allen & Kuropat 1994; (aa) Lester & Pante 1992; (ab) Chen & Lin 2001; (ac) Lawrence & Jones 2001; (ad) Correia *et al.* 2000.

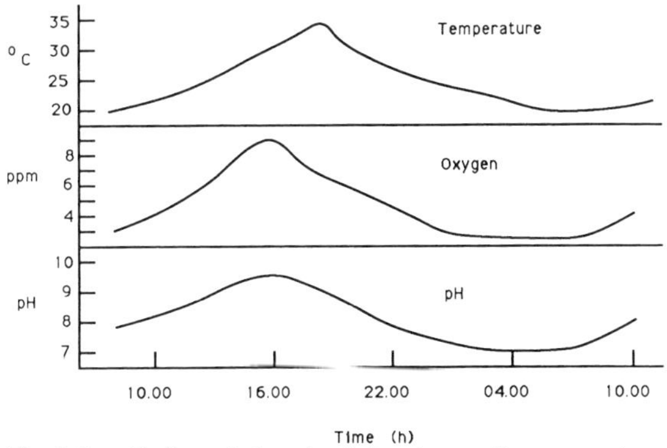

Fig. 8.6a Daily variations in vital water quality parameters in an outdoor earth pond.

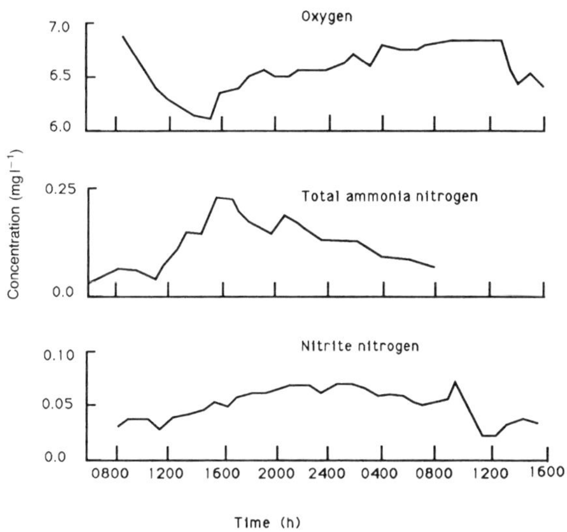

Fig. 8.6b Daily variations in selected water quality parameters in a marine recirculation system containing lobsters. Food was given between 0800 and 1000 h daily. Note the coincidence of adverse conditions just before 1600 h and the slow recovery of oxygen levels despite vigorous aeration in the system.

to occur in transit, and secondly, the measurements required must be specified in case the techniques normally used by the laboratory are unsuitable. For example, a laboratory routinely engaged in freshwater analysis may not be equipped to deal with ionic interferences from substances found in brackish- or saltwater. Similar precautions are taken with soil samples (Boyd 1995a). Commercial portable test kits have been found suitable for aquaculture use in fresh- and seawater, e.g. Boyd and Daniels (1988).

Additional factors that arise when monitoring recycled water are discussed by Rosenthal *et al.* (1980); Wickins and Helm (1981) and Wickins (1985a,b). In brief, these concern short-term changes in the organic and inorganic load on the biological filters due to normal husbandry operations and the cyclic water quality variations induced by this and by the natural, periodic sloughing of biological growths from the filter media. In modern floating bead filters such growths are controlled by the standard operational washing cycles (Malone *et al.* 1998). Long term changes include the accumulation of refractory organic materials that cannot easily be broken down by biological filtration alone, a gradual decline in pH, and in some cases a loss of the buffering contribution normally expected from limestone or oystershell filter media (Wickins 1985a). The latter may prevent normal mineralisation of the crustacean exoskeleton after moulting (Wickins 1984b) and, in systems employing plastic filter media but no denitrification unit, monitored doses of hydroxyl or carbonate ions may be required to maintain buffer capacity (Hochheimer & Wheaton 1998).

8.7 Humane slaughter

Public debate on animal welfare matters, particularly in developed countries, is increasingly being extended to include fish and shellfish (Breen 1995). Researchers too are being affected as animal ethics committees start to include these groups in new codes of laboratory practice. Methods used in the trade for killing crustaceans attract the most attention from the public and typically involve plunging the animals into boiling water (either as individuals or in bulk), beheading, or, in the case of some crabs, by piercing the nerve centres with a spike. Concern naturally arises, with the first method, if too large an individual or too many animals are added to the water at once, causing it to go off the boil. This can result in those at the top living for an unnecessarily long time (UFAW 1978). Several methods intended to render crustaceans insensible prior to killing by conventional means have been advocated, one of the simplest being to quickly chill them in a saltwater-ice slurry for 20 min prior to boiling (NSW Agriculture, undated pamphlet). Any adulteration of the flesh by anaesthetics or, for some lobsters and crayfish, by deep freezing, would, of course, be unacceptable to the consumer. Other methods intended to numb their senses that are perceived by some to be more humane, include placing marine crustaceans in freshwater or the converse, placing them in deoxygenated water or bringing them slowly to the boil in a covered pan; covered presumably so that their struggles cannot be seen. Adoption of any of these additional procedures would incur additional time and financial costs to the industry. Electrocution or electro-stunning devices have been developed for individual freshwater crayfish (Ducruet *et al.* 1993; sections 7.6.9, 11.2.5 and 12.1) and marine lobsters (Anon. 1999a). It is believed that commercial models capable of batch operation could be developed given sufficient consumer demand (C. Buckhaven, 2000 pers. comm.).

8.8 Food preparation and storage

8.8.1 Larvae feeds

The larvae of all farmed crustaceans grow and survive best on living foods or at least with a supplement of live food (sections 2.2 and 2.4.8). Examples include single-celled algae which are followed by rotifers or newly-hatched *Artemia* (section 7.11.2.1) for penaeid shrimp and crabs with small larvae, and newly-hatched or partially grown *Artemia* for caridean prawns, other crabs and the clawed and spiny lobsters. Live foods are, however, expensive and their culture incurs additional facilities and management costs.

Algae are cultured in a variety of ways. Traditional shrimp hatcheries encourage a 'bloom' of algae in outdoor, or illuminated indoor tanks prior to spawning, by the addition of fertilisers. Fish or chicken manures, or preferably, clean agricultural fertilisers containing silicate as well as the usual phosphate and nitrogen compounds are widely used (e.g. 50 µg at N L^{-1}, 50 µg at Si L^{-1} and 10 µg at P L^{-1}; µg at = microgram atoms); the silicate encourages the growth of desirable diatom rather than flagellate species of algae. Sometimes it is necessary to inoculate the water from a stock culture of the preferred alga. Generally, however, a mixture of several endemic species develops. Under favourable conditions it is possible to obtain peak densities of $3–4 \times 10^6$ diatom cells mL^{-1} after 3–4 days; however, lower densities of 50 000 to 150 000 cells mL^{-1} seem preferable in the larval tanks. Advanced shrimp hatcheries culture algae at very high densities in nutrient enriched, sterilised seawater in illuminated culture vessels and feed controlled amounts at regular intervals during the protozoea and early mysis stages of culture (section 7.2.4).

Rotifers are sometimes cultured in penaeid hatcheries to provide an intermediate food during the transition from the filter feeding protozoea stage to the raptorial feeding mysis stage, however the brine shrimp *Artemia* is the most widely used food for crustacean larvae. Details of its culture, hatching, decapsulation and enrichment are given in section 7.11.2.1 and the culture of other crustacean species suitable for live food is discussed in sections 7.11.2.2, 7.11.3 and 7.11.4. Useful reviews and descriptions of the techniques and facilities used in the various approaches to live food culture are described for marine algae (Laing & Ayala 1990; Smith *et al.* 1993; Wikfors & Smith 1998), algae, rotifers and *Artemia* (Liao *et al.* 1993) and these plus other micro-crustaceans in section 7.11.

Attempts to develop microparticulate diets capable of replacing live foods commenced some 25 years ago and today a wide range of products exists. For the most part these diets work well in commercial hatcheries as a supplement to live foods, but some have been shown capable of completely replacing live foods, at least for some species under laboratory conditions. At one end of the scale, microparticulate diets include centrifuged, preserved microalgae, microalgae pastes and spray-dried microalgae (Laing *et al.* 1990); at the other are included microencapsulated, microbound, complexed and liquid diets. Microencapsulated diets contain the nutrients within a shell or capsule of cross-linked protein or lipid, while in microbound feeds a binding agent (e.g. gelatin, alginate) within the nutrient mix holds all the constituents together. Recently lipid spray-beads have been developed using a process simpler than that required for producing lipid-walled microcapsules (Langdon 2000). Complexed particles are a blend of two or more other particle types (Villamar & Langdon 1993), for example, a water-soluble vitamin supplement in a microcapsule that is itself embedded in a microbound particle containing the main dietary components. Liquid diets are among the more recent developments and are essentially a slurry of particles in a suitable suspension medium. They are marketed by at least three companies at the time of writing and, although generally more expensive, they are claimed to cause less fouling and can be continuously dosed into larvae cultures using peristaltic pumps. Additional ingredients such as probiotic bacteria and lipid emulsions are also readily added.

The simplest microcapsules to prepare are those with a gelatin–acacia wall (for method see Southgate and Lou 1995) but longer-lasting cross-linked protein and nylon-protein walled microcapsules are made commercially. Liposomes are phospholipid-walled vesicles that can be used to enrich *Artemia*, for example, with water-soluble vitamins, antibiotics or nutrients so that they may be delivered more effectively to shrimp and fish larvae (Touraki *et al.* 1995). Methods for the production of microbound particles are outlined by Barrows (2000). Most dry diets are packed under vacuum or in an atmosphere of nitrogen to prolong storage life. Liposomes can be freeze-dried and remain stable for several years in the absence of oxygen, while wet diets such as microalgae pastes have added food-grade preservatives to give a frozen shelf life of up to 2 years. Further information on non-living foods and their use for the larvae of the relevant species groups is given in section 2.4.8 and Chapter 7 respectively.

8.8.2 Juvenile and adult feeds

In extensive and lower-density semi-intensive ongrowing systems, the natural production of food sources is encouraged by the controlled addition of fertilisers (chicken, duck, cattle manures or agricultural chemicals) or, in the case of crayfish, by the addition of hay (lucerne, sorghum) or the planting of forage grasses (low-yield rice varieties). Specific examples are given in Chapter 7 for each species group and in the section on pond management (section 8.3.6.2). As the intensity of ongrowing increases, so does the reliance on compounded feeds whose composition must be tailored to the nutritional demands of the species being cultured if rapid and economical growth is to be achieved (Houser & Akiyama 1997). For example, it would not be economic to feed a high-protein (60%) diet designed for *Marsupenaeus japonicus* to *Fenneropenaeus merguiensis* or *Macrobrachium rosenbergii* which are able to thrive on low-protein (25%) diets.

Three main types of feed are produced for juvenile and adult stages; microparticulate diets for post-larvae, pelleted or extruded diets in various sizes for ongrowing and predominantly fresh, natural invertebrate tissues for broodstocks. Pelleted diets are often provided as supplements (5–25%) to the fresh diets fed to broodstock (Wouters *et al.* 2000). Many farms in South-east Asia manufacture their own feeds, particularly those growing *Macrobrachium* (D'Abramo & New 2000), and this practice was reviewed by New *et al.* (1993). The best formulated diets currently available are the pelleted feeds manufactured for intensive and super-intensive penaeid shrimp farms. Since the production of freshwater prawns is generally under semi-intensive or extensive conditions, there has been less incentive to manufacture diets of similar quality for *Macrobrachium* species, especially since natural productivity often plays a major role in their nutrition (D'Abramo & New 2000). It is not yet clear if the best of the penaeid diets would support good growth and survival during the prolonged ongrowing period for clawed and spiny lobsters without fresh food supplements (Booth & Kittaka 2000). At present, research diets specifically formulated for lobsters allow only 80% of the growth achieved when natural diets are fed.

In the past few years considerable attention has been given to the development of diets that minimise the impact of farm effluents on the receiving waters and surrounding ecosystems (Cho *et al.* 1994; sections 8.3.6.8, 11.4.3 and 12.6). In essence the diets are highly nutrient-dense, contain easily digestible components (sometimes with added enzymes), and are formulated with particular regard to the digestibility, assimilability and energetic interactions between the various components (section 2.4). The quality of the raw materials used in diets is also important. This is especially true of fishmeal, which can vary considerably from batch to batch. For example, the growth of young shrimp improved with the freshness of the fishmeal used in their diets, i.e. meal processed just 12, 25 or 36 h after capture. A similar effect was also observed with the more carnivorous species of older shrimp that require higher dietary protein levels (Ricque-Marie *et al.* 1998).

8.8.2.1 Diet preparation

It seems unlikely that good performance could be obtained if crustaceans were fed solely on proprietary chicken or trout pellets. Such pellets, together with trash fish, are used as supplements on some extensive or semi-intensive shrimp and crayfish farms where they also provide food for small aquatic organisms upon which the farmed species feeds. In other words, the pellets fertilise the pond water in the same way as additions of cheaper poultry or cattle manure. Any thoughts of using unprocessed chicken offal, vegetable or slaughterhouse wastes, particularly in semi-intensive and intensive operations, should be dismissed, as they are likely to cause considerable fouling, increased oxygen demand and contain unsuitable quantities or imbalances of micronutrients. In addition, some animal wastes can be contaminated with medicants, hormones or growth promoters that might be harmful or illegal if found in crustacean flesh (see also section 12.5).

Crustacean diets must be physically stable in water to prevent premature disintegration caused during repeated manipulation by the animal during feeding. Binding agents such as some glutens, starches or gums are commonly used, while manufacturing techniques that involve moisture and heat tend to gelatinise natural starches and increase their binding capacity. Industrial scale manufacture of compounded feeds for aquaculture generally involves using conventional plant but additional steps (finer grinding, activation of binding agents) are required to control pellet stability (Fig. 8.7).

Pellets for early shrimp ongrowing are about 1.8–3.0 mm diameter and sink when added to water. As with all formulated ongrowing diets, the ingredients must be finely ground and bound together well to be physically stable in water. Particle size reduction by grinders (pulverisers) or hammer mills followed by screening is the most time-consuming and expensive step in feed production. Of the ingredients, 95% should be about 250 μm or less, the remainder no larger than 400 μm (Tan & Dominy 1997). Laboratory studies showed that optimum ingredient particle size for shrimp pellets containing fishmeal, wheat, shrimp and soya meals was 124 μm in terms of water stability, durability and in producing good growth in 1.7 g *Litopenaeus vannamei*. However the energy costs of grinding and pulverising increased exponentially with decrease in the size of particle required from 0.3 kW h mt^{-1} at 586 μm, through 2.3 kW h mt^{-1} at 272 μm, to 23.8 kW h mt^{-1} at 69 μm (Obaldo *et al.* 1998).

Proper mixing of the finely ground materials is essential and confers two advantages. It facilitates the even distribution of each dietary component throughout all the pellets produced and in doing so allows a more uniform reaction with the natural or added binding agent (Cuzon *et al.* 1994). The sequence in which ingredients are added to the mixer is also important, for example, the binder should not be rendered ineffective by a coating of fat or oil. Types of mixer and test protocols are described by Behnke *et al.* (1992). The degree of binding is increased by additional heat and moisture in a conditioning process (usually involving steam injection), which aids starch gelatinisation and bonding with protein ingredients. A post-pelleting, conditioning step may also be employed prior to the drying and cooling phase of production. The final steps involve screening to remove fines (dust) and oversized pellets, passage through crumble rollers to produce any different sizes of crumb required, and bagging in 20–25 kg lots (pelleted feeds) or 5–10 kg for crumbles. For further reading, Tan and Dominy (1997) describe the equipment used in a shrimp feed plant producing 1.5 mt h^{-1}, give guidelines for all critical steps in the production process and indicate solutions to common problems.

Steam compaction pelleting and extrusion cooking are the most widespread methods for manufacturing

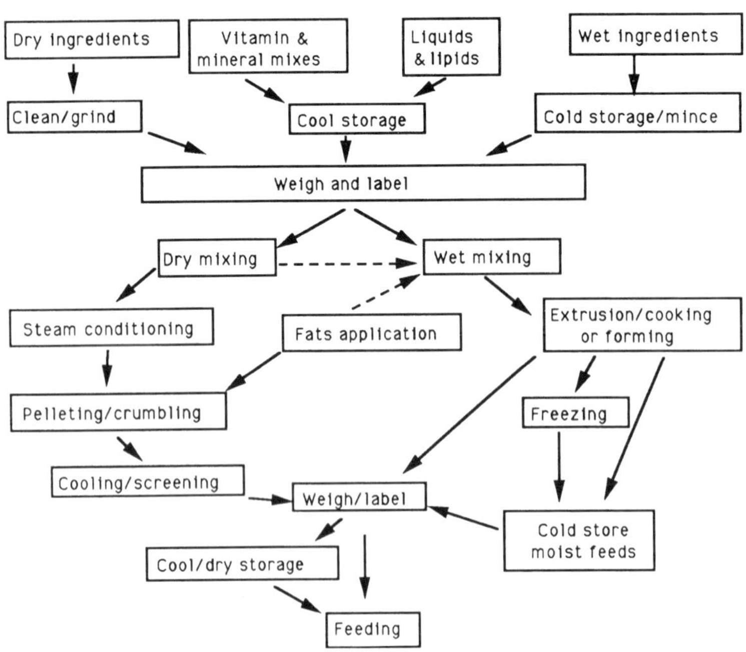

Fig. 8.7 Steps in feed preparation.

crustacean diets, the latter being more versatile and energy-efficient but otherwise more expensive. In conventional pelleting, the conditions during processing (heat pretreatment of the ingredients and the pellet drying temperature) are likely to have greater effect on water stability than the binding agent used (Flores & Martinez 1993). Conditions for optimising the reaction between the binding agent and the rest of the dietary ingredients are critical. In the case of vegetable starches too little water or steam prevents adequate gelatinisation while too much makes the mix too viscous to pass through pelleting dies. Thick dies (e.g. 2.2 mm diameter × 55 mm length) operate at a compression ratio of 25 : 1 or higher and give more shear and heat to the pellet, the latter aiding the subsequent drying process (Devresse 1998a). The interaction between the binding agent and temperature during drying of the pellets also seems particularly critical (Flores & Martinez 1993).

Extrusion is a process in which the feed ingredients are plasticised and cooked by a combination of pressure, heat, mechanical shear and friction forces in the extruder barrel (Kearns 1998). Generally, a specific binding agent is not required during extrusion cooking (cereals in the diet can provide sufficient starch) and the process can be adjusted to produce wet or dry pellets by controlling the moisture flow to the extruder. Studies indicate that although extrusion increases the apparent digestible energy in cereal grains poorly utilised by shrimp, no single extrusion condition (wet or dry) gives optimal gelatinisation and feed digestibility for the sources of starch tested (wheat, rice, corn, milo; Davis & Arnold 1995). Shear, however, has the greatest influence on starch gelatinisation and water stability although shrimp growth has been shown to be better on extruded diets having less than the maximum gelatinisation and stability (Obaldo *et al.* 1999). It therefore seems best to optimise steps in the extrusion process to suit both the ingredients and the end use.

Extruded feeds, while physically more stable, have a sponge-like structure which can take up water and increase losses of some water-soluble vitamins and micronutrients due to leaching (Gadient & Schai 1994). Fat-coating of water-soluble vitamins can reduce leaching losses by 50% in both extruded and pelleted feeds and additionally provide protection during processing and storage. The additional cost of coating with fat can be offset to some extent against the reduced requirement for vitamins added to compensate for leaching losses (Marchetti *et al.* 1999). The incorporation of chemical attractants (e.g. free amino acids such as taurine) and feeding stimulants can also be advantageous in reducing the time available for nutrient leaching. However, detection of the chemical does not necessarily imply the diet will be acceptable or consumed and assimilated efficiently; indeed the attractiveness of the diet may become attenuated with time (Lee & Meyers 1997). The best method of incorporating such substances in the diet (e.g. before pelleting, coating after pelleting or just prior to feeding) has not yet been defined and it is likely that water quality in the pond will affect both the crustacean's ability to detect, and its response to, the chemicals.

8.8.2.2 Storage

Significant losses or deterioration of feed can occur during storage because of theft or damage by insect, rodent and bird pests, fungal, mould and mite infestations, and chemical changes in the feed due to enzymic action and oxidative rancidity. Lipids are particularly vulnerable to degradation in poorly prepared diets, especially if they are stored at tropical temperatures (30–40°C) without added antioxidants. Vitamins and heat-sensitive additives are also at risk. Poor diet preparation and storage conditions rapidly lead to the development of rancidity, which renders the diets useless, and predisposes towards the growth of moulds and fungi that produce toxins, especially in diets containing high levels of starch and lipids (Sarac & Swindlehurst 1992). Most common are perhaps the fungi *Aspergillus* spp. which produce potent aflatoxins (e.g. type AF B_1) that cause histopathological damage to the hepatopancreas and antennal gland of shrimp at levels as low as 50 parts per billion (ppb), and have reduced growth and digestibility coefficients after 8 weeks exposure to 400 ppb (Ostrowski-Meissner *et al.* 1995). Mould inhibitors (e.g. propionic acid) are sometimes added to dry diets but in general effective storage conditions and stock turnover rates (3 weeks to less than 12 weeks) will minimise contamination risks.

Dry pelleted feeds should be stacked no more than ten high on wooden pallets in dry, cool and well-ventilated storage areas out of direct sunlight. Stock turnover should be carefully controlled to avoid storage of more than 3 months from date of manufacture. Different feeds will require different storage conditions. Critical features in the construction of feed stores (ventilation, insulation and pest exclusion) are described by Goddard (1996). Bulk storage systems (usually hoppers) are also used and enable farmers to economise by buying in bulk. Disadvantages of hoppers include reduced control over temperature and the break-up of pellets at the bottom of the hopper due to the weight of those above. Moist diets (e.g. for broodstock) and labile ingredients for late incorporation into prepared diets will require refrigeration, preferably at –30°C if storage is to be as long as 3–6 months.

8.9 Disease diagnosis, transmission, prevention and control

Diseases, or abnormal states of health, can arise as a result of non-infectious and infectious agencies. They can be of genetic origin, due to dietary inadequacies or adverse environmental conditions as well as to pathogenic organisms (section 2.5). Any factor predisposing an animal to stress can increase its vulnerability to diseases, especially invasion by pathogens. One estimate has put the cost of disease to the Asian shrimp industry between 1994 and 1998 at over $1bn (D. Lightner, *apud* Rosenberry 1998).

8.9.1 Non-infectious diseases

With regard to non-infectious or abiotic diseases, concerns up to the mid-1990s were for the increasing incidence of the reversible conditions known as soft-shell or crinkle-shell and blue-shell that arose during ongrowing in several South-east Asian countries and Australia (Sarac & Rose 1995). The indications are that soft-shelled shrimp result from an inability to store and mobilise calcium and phosphorus properly, while carotenoid metabolism seems suspect in blue shrimp (Menasveta *et al.* 1993) (section 3.3.1.1). In some cases, notably in Taiwan, the Philippines and India, occurrence of both conditions was linked to the increased use of variable quality feed ingredients, the incorporation of substitute feed materials, and to feed formulation changes hastily made in an attempt to survive in an increasingly competitive industry (Sheeks 1989). In other situations, in India, Malaysia and Taiwan, for example, pesticides or the chemical composition of both pond bottom and water were also suspected and seemed particularly important when the ponds were used very intensively. Saponin, a plant-derived toxicant used to eliminate fish from shrimp ponds prior to stocking (section 8.3.6.1), also produced soft- and crinkle-shelled animals at concentrations above 20 mg L^{-1} (Nagesh *et al.* 1999). Poor shell mineralisation also occurs in freshwater crayfish in acidified waters (section 8.4.5).

Moult death or exuvia entrapment syndrome is reported in larvae and juveniles of a number of species including *Homarus* (Bowser & Rosemark 1981), *Panulirus* (Booth & Kittaka 2000), *Macrobrachium* (Johnson & Bueno 2000) and *Palaemon* (Wickins 1972). It is associated mainly with an inadequate diet (e.g. lacking some essential fatty acids) but can arise from other stressful conditions (Conklin 1990). In shrimp, vitamin C deficiency can result in a reversible blackening of the abdominal cuticle edges. Black staining of redclaw crayfish cuticle, which constrains product marketability, also occurs on some farms in Australia, but for a different reason. This condition seems linked to iron and manganese oxides in ponds (iron and manganese frequently appear in groundwaters) but, interestingly, a 15 min dip in a molasses solution removes the blemishes. The gills of most crustaceans are also readily discoloured by iron (Nash *et al.* 1988). In addition to shell discoloration, the account of crayfish abiotic diseases given by Evans and Edgerton (2001) included:

- Conditions of muscle wasting and necrosis sometimes associated with gill fouling or to exposure to acute and chronic stressors such as low pH, handling and holding out of water.
- Histopathological and reproductive changes following exposure to heavy metals (mercury, lead, aluminium, iron).
- Acute sensitivities to insecticides, herbicides, fungicides and other chemicals from agriculture, forestry, mining and industrial urban development that generally decreased with advancing crayfish maturity and increasing body size. Overall, insecticides were found to be more toxic to crayfish than herbicides, which in turn were more toxic than fungicides.

The responses of clawed lobsters to heavy metal and organic contaminants and of crayfish to organic toxicants have been reviewed by Harding (1992) and Mercaldo-Allen and Kuropat (1994) (lobsters) and Eversole and Seller (1996) (crayfish). Another environmental disorder, gas bubble disease, most frequently affects larvae, causing them to float uncontrollably, but it can also affect older animals in systems receiving a warmed, pumped water supply, e.g. power station effluent. It commonly arises after water becomes supersaturated with gases due to pump cavitation, pump intake-side air leaks, after backwashing a pressure sand filter or from elevated, heated, water header tanks. The excess gases can be driven off by vigorous aeration with large air bubbles.

It is generally accepted that animals kept in artificial (culture) conditions experience stress at some time or another and become more susceptible to infectious diseases. It is safe to assume that potentially pathogenic organisms are present in all culture systems and will create disease or infestation problems in weak and stressed animals. In the hatchery the first lines of defence are quarantine of imported stocks and hygiene. In pilot and newly established farms stressful periods may occur while operators gain experience with husbandry and water management. Almost inevitably some diseases and infestations will be encountered at this stage, but will not necessarily be a major threat later when more experience has been gained and sound working practices established. However, shrimp diseases become a greater constraint to progress as semi-intensive and intensive farms strive to become more competitive. Outbreaks are increased by the stresses induced when culture densities are increased and when the quality of farm inputs is reduced to save production costs (Kautsky *et al.* 2000).

8.9.2 Diagnosis

For all practical purposes, routine monitoring of the general condition and state of health of the crustaceans being cultured is considered essential. Additional examinations are worthwhile following sudden changes in environmental conditions. For example, outbreaks of viral diseases have been associated with an abrupt increase in water hardness, calcium levels or a decrease in temperature and salinity (Flegel *et al.* 1997). An excellent account of diagnostic methodology incorporating lists of equipment required, signs and symptoms to look for, examination and diagnostic techniques has been published for shrimp but provides good principles for monitoring other species (Lightner 1996). Quality of larvae can be assessed using criteria based on those recommended for *Macrobrachium* (Tayaman & Brown 1999) and for post-larval shrimp on those reported in Table 7.4.

The first signs of stress or disease are often reduced appetite, abnormal swimming or postural behaviour and a continuous low level of mortality. Obvious external signs include infestations of epibiotic growths on the cuticle (a sign of reduced cleaning activity or moulting frequency) or on the eggs of brooding females (Fisher 1986), increased cannibalism and moulting difficulties, increased individual size variation and a high prevalence of deformities (Wyban *et al.* 1993), discoloration, lesions and, finally, mass mortalities. Chronic diseases such as 'black spot' (El-Gamal *et al.* 1986) while not necessarily fatal, reduce market acceptability and can result in considerable financial loss. In marron, as in many shrimp and prawn species (Johnson & Bueno 2000), exposure to stressful conditions, e.g. low pH, rough handling or prolonged holding out of water, results in a characteristic whitening of the abdominal muscle (idiopathic muscle myopathy) and elevated blood haemocyte levels (Evans *et al.* 1999). A better non-specific test of present condition is probably osmoregulatory capacity (see Glossary; Lignot *et al.* 2000; sections 7.2.4, 8.5 and 12.2).

Where knowledge of the culture conditions and history indicates an infectious disease, the diagnostic methods available, in addition to the gross and clinical signs described above, include: microscopy (smears, wet-mounts, histological sectioning and histochemistry); microbiology (isolation and culture of pathogens); electron microscopy; serological tests with immune sera and molecular techniques. Classical microscopic methods allow recognition of the acute phase of infection but lack the sensitivity to detect latent or carrier states of infection. In conventional microbiological tests it often takes 2–3 days to ensure accurate identification. Most molecular techniques however are much more rapid and some can be cheaper and simpler to conduct (Mialhe *et al.* 1992; Lightner & Redman 1998). They include serological tests with monoclonal antibodies (e.g. fluorescent antibody and ELISA; see Glossary), and gene probes that may be labelled with radioisotopes, enzymes, antigens or chemoluminescent molecules (e.g. dot blot hybridisations and in situ hybridisation assays on histological sections; Lightner 1996). Sensitivity of these techniques can be increased through application of polymerase chain reaction (PCR) methods to replicate small segments of nucleic acids, specific to the identification of particular pathogens. Commercial diagnostic kits based on gene probe methods are now available for all the important shrimp viruses (Lightner 1999).

8.9.3 Transmission

Some pathogens are found naturally in many different host populations over a wide geographical range; others are more limited in their distributions. The widespread practice of shipping shrimp broodstock, nauplii and post-larvae and crayfish juveniles and adults between farms and countries has led to the introduction and reintroduction of serious diseases in a number of areas (sections 2.5 and 11.3.3). Several countries also conduct trade in live wild or farmed crustaceans for the table or aquaria, some of which may escape, be released or even discarded untreated, into natural waters or landfill sites. Primary pathogens often present more serious problems for crustacean farmers and are less easily prevented from entering outdoor culture operations than indoor controlled environment systems where incoming water and (some) feeds may be sterilised prior to use.

Translocation of live crustaceans is not the only means by which diseases are spread. The Asian shrimp viruses (WSSV and YHV; for abbreviations, see section 2.5.4) have been found in consignments of frozen shrimp imported into the USA and demonstrated to be infectious (Nunan *et al.* 1998). Mechanisms by which they can be transmitted to both cultured and wild stocks include the discharge of untreated effluents from shrimp importing, processing and repacking plants into coastal waters, improper disposal of solid wastes in landfill sites accessible to scavenging birds and vermin, and the use of imported shrimp as bait for anglers or as food for other captive crustaceans in zoos, home aquaria and research laboratories. It seems probable they can be carried between ponds and farms by predatory birds, other crustaceans and insects such as the water boatman (corixid beetles; Lightner *et al.* 1997). The possibility that insidious introduction of alien virus diseases into local fished stocks has already occurred cannot be discounted.

Cross-species transmission of viral and other pathogens is also possible in some circumstances and in this regard the use of mammalian and avian slaughterhouse wastes as ingredients in crustacean aquaculture diets, for example to replace fishmeal, should be subject to meticulous examination (section 12.5).

8.9.4 Prevention and control

Management and containment of disease need careful site selection (clean water supplies, away from discharges from other farms, good quality soil) in the first instance followed by the best husbandry, water and pond management practices available to ensure minimisation of environmental and dietary stresses during culture. In hatcheries and nurseries attention centres on hygiene, feed and water quality (section 8.4.3); in ponds on proper bottom treatments, and on fertilisation, feed content and feeding regimes in relation to water management (section 8.3.6). Of critical importance, however, is the strict enforcement of recommended quarantine procedures whenever animals have to be imported (ICES 1995; FAO/NACA/AAHRI 1996). Escapes of animals must be prevented and disinfection of water and equipment with which imported stocks have been in contact must be rigorously enforced. Similar control over quality and cleanliness of other farm or hatchery inputs such as water, feedstuffs and personnel is equally vital (sections 9.7.2, 11.3.3 and 11.3.4). Today, the use of sensitive, rapid diagnostic tools coupled with efficient enforcement of import and quarantine regulations constitutes the primary focus in the USA and parts of Europe for controlling the introduction of diseases from both live and processed crustaceans.

In the absence of effective controls, several countries rely on the use of commercially available, high-health, specific pathogen free (SPF) or specific pathogen resistant (SPR) strains of shrimp (sections 8.9.4.4 and 8.10.1.3). Indeed, it is thought that some wild stocks of *Litopenaeus vannamei* and *L. stylirostris* are already developing natural resistance to TSV and IHHN (respectively), in areas where the viruses have become enzootic for several years (Lightner 1999). SPF and SPR strains of other farmed crustaceans have yet to be developed but advances are being made towards this with Australian crayfish.

If an outbreak of disease cannot be controlled many operators prefer to kill any remaining stock, disinfect the facility and restart after a sanitary 'dry-out' period (sections 7.2.4 and 11.3.4). This is, of course, more readily practised in indoor hatcheries than in ongrowing systems, but even so there is no guarantee that the outbreak will not occur again. Although destruction of all susceptible stock once infected animals are found is common practice among farmers of mammalian and avian domestic stock, no financial compensation is available following the destruction of crustacean stock. One big Ecuadorian hatchery decided it was better to live with the viruses endemic in wild broodstock, and minimise the risks by improved husbandry, than to keep killing expensive broodstock. Disease control in recirculation systems presents special problems since many treatments designed to kill infectious or infesting organisms will also kill beneficial microbial populations resident in biological filters. Few published data are available on the detailed costs and frequency of occurrence of disease outbreaks to farmers of particular species or groups of crustaceans, and it would thus seem prudent to make allowances for production losses at several levels during project appraisal (sections 9.3.4 and 10.4.2).

A wide range of chemicals (collectively termed 'drugs' when used in aquaculture) is being used for disinfection and treatment of animals, water and pond bottoms in all types of aquaculture operation (Massaut *et al.* 2000). Possible mechanisms by which some of these may disrupt (to a greater or lesser extent) crustacean metabolic processes are described by Bainy (2000). Very few drugs, however, are approved for use in aquaculture by national authorities in the USA, Europe and Japan (section 11.5.3.2). Some may only work satisfactorily in freshwater, others function equally well in marine, brackish- and freshwater environments. Since biological films form on most solid surfaces in contact with culture waters (pipes, biofilters, rearing vessels), their removal physically or with cleaning agents (detergents – sodium hydroxide and carbonate, silicates, organic acids, surfactants) is often necessary for effective use of disinfectants (Flick 1998). Inorganic fertilisers are important in pond management to maintain adequate levels of primary nutrients (nitrogen, phosphorus, potassium and silicate) in the water in order to sustain desired densities of phytoplankton populations (section 8.3.6.2). For simplicity, some examples of the chemical compounds commonly used in fish and shellfish hatcheries and ponds (GESAMP 1996) are grouped below according to their intended function:

- Disinfectants (formaldehyde, glutaraldehyde, hypochlorite, chloramine T, chlorine dioxide, iodine preparations – iodophors, quaternary ammonium compounds, ozone). Some disinfectants are used on surfaces, including floors, tanks and inside pipes as well as for sterilising water. The action and use of cleaning agents and disinfectants in aquaculture is described by Flick (1998).
- Water treatment (hypochlorite, ozone for breakdown of refractory organic compounds – section 8.4.3; chemicals for pH and buffering control – section 8.4.4; metasilicates and other metal chelating agents – EDTA – section 8.4.3; agricultural fertilisers – ammonium phosphate, urea, calcium nitrate, alum and gypsum to remove pond turbidity – section 8.3.6.2). Introductions to calculating correct dosages for ponds, tanks and raceways are given in imperial and metric units by Mitchell (1996) and Avault (1997).
- Therapeutants (formalin, various antibiotics – sulphonamides, tetracyclines, 4-quinolones, nitrofurans, erythromycin and 'phenicols' – sections 11.3.4 and 11.5.3.2).
- Algicides and pesticides (ammonia, saponin, rotenone, organotin compounds and nicotine – tobacco dust – to control fish and snails in ponds, organophosphates for infestations in hatcheries, copper compounds and formalin to control algae blooms and external infestations of protozoans and filamentous bacteria, trifluralin used as a prophylactic fungicide in hatcheries, also, though not advisable because of flavour tainting risks, diesel is sometimes used to control aquatic insects – section 8.3.6.1).
- Pond bottom treatments (hydrated lime, calcium/sodium bicarbonates, gypsum – unless total alkalinity is below 40 mg L^{-1} – section 8.3.3).
- Anaesthetics (clove oil in ethanol, ice; note that many compounds used for fish are ineffective on Crustacea; section 11.2.5).

There is a significant risk that many if not all of the chemicals used in crustacean farming constitute a health hazard to employees and other site workers. Inhalation of dust by unprotected staff during the preparation of medicated feeds, or particularly when using rotenone powder, can produce severe respiratory distress. Exposure to antimicrobial agents can cause skin irritation and other hypersensitivity reactions. Fertilisers can be corrosive and some are explosive; liming chemicals and cleaning agents can also be corrosive. Hypochlorite reacts with organic matter to form carcinogenic trihalomethanes, although we know of no problems in this regard reported from crustacean farms. Often the labelling on packaged or repackaged chemical products does not give sufficient instruction as to dosage and use in different environmental conditions (tropics or temperate zones, fresh- or saltwater), nor information on the proportion of active ingredient contained (chlorine, iodine), on storage conditions and expiry dates, or on correct disposal of unused chemicals and containers. Farmers may combine several chemical treatments, perhaps in a desperate attempt to control mortalities without regard to the potential production of harmful gases, by-products or of inactivation of the one or more ingredients (Mishra & Singh 1999). Boyd and Massaut (1999) assessed qualitatively levels of risk to food, environment and those handling the main chemicals used in pond preparation and management, including pest and bactericidal compounds. They concluded that most present little food safety risk although some can cause environmental pollution. Other substances present handing risks which, they emphasised, could be significantly increased by the unauthorised mixing of two or more compounds. The example given was the mixing of the fertiliser, ammonium nitrate, with diesel fuel to make a substitute industrial explosive. More rigorous assessment of the risks involved is made difficult by the lack of quantitative information on the amounts of chemicals used and how they are being applied. As a result, comprehensive impact assessments for new projects are constrained. Even so, the risks to personnel can be substantially reduced by the implementation of standard health and safety precautions, such as following correct handling procedures, using protective clothing and masks, post-application washing and accurate record keeping for all chemicals used, in line with HACCP protocols (sections 3.2.2 and 9.6).

Other issues of concern regarding the use of chemicals and drugs in crustacean farming are discussed elsewhere in this book: residues in marketed product (section 3.2.1), effects on biological filter organisms (section 8.4.5), development of resistance to antibiotics (section 11.3.4) and eutrophication (section 11.4.3).

8.9.4.1 Vaccines

Since the 1980s attempts have been made to develop vaccines for shrimp (Itami *et al.* 1989) and at least two companies offered vaccines for use in shrimp larvae cultures. The effectiveness of such vaccines under commercial conditions remains equivocal. Bath treatment of shrimp with formalin-killed *Vibrio* spp. and β-glucan provides some protection (30–50 days) against vibriosis (section 2.5.5) and it is now thought that crustaceans may possess some form of adaptive immunity mechanism, albeit one quite unlike that of vertebrates (section 12.2). A vaccine against Gaffkaemia for clawed lobsters has been effective in field trials (Keith *et al.* 1992) but there are reports that it may suppress moulting (Aiken & Waddy 1995).

8.9.4.2 Probiotics

Concern over antibiotic residues in crustacean flesh and over the development of increasing resistance of pathogens (Inglis *et al.* 1997) has stimulated widespread interest in the use of probiotics. As yet, however, that interest is mainly confined to penaeid shrimp production. A probiotic (as originally defined) is a live microbial feed supplement, which benefits the host animal by improving its intestinal microbial balance. The observed increased resistance to pathogens may be due to competitive exclusion of potentially harmful bacteria from the gut, to enzymes that improve digestibility and nutrient availability or to stimulation of the immune system. The term probiotic is now also being used in the aquaculture industry when live microbial inoculations (usually *Bacillus* spp., also *Lactobacillus*, *Pseudomonas*, nitrifying bacteria and some *Vibrio* spp.) are made to culture waters (hatchery, nursery or ongrowing ponds) as biocontrol and bioremediation agents (Gatesoupe 1999). These additions are made to improve survival of the animals by promoting nitrification, organic oxidation, the reduction of blue-green algae or by excluding potentially pathogenic bacteria either by competition or antagonism. The definition in this context is much broader and embraces the use of live microbes to improve the internal and/or external microbial balance and environment of the cultured stocks.

In crustacean aquaculture probiotics are most widely used in shrimp hatcheries (section 7.2.4) and in Ecuador, for example, additions of cultured *Vibrio alginolyticus* among other species are reported to have reduced the use of antibiotics by 95%, increased production by 35% and reduced shrimp hatchery 'down-time' from 21 days to 7 days per year (Devresse 1998b).

The selection of a probiotic species is largely empirical. Adding a sucrose substrate to larvae cultures to promote the growth of 'good' bacteria is one simple method; another is to culture selected bacteria separately and add them to the culture medium as if they were microalgae (section 7.2.4). Similarly, ongrowing diets with high C : N ratios may be used to encourage the development of beneficial bacterial flocs in low water exchange ongrowing ponds (section 8.3.7). Manufactured probiotic mixes are applied to ponds in the form of a series of doses or inoculations (e.g. 5 days per week) repeated throughout the culture period. There is, however, little evidence of consistent, beneficial results from such treatments in ponds (Sonnenholzner & Boyd 2000), the main difficulty being to create a pond environment in which the desired microbial population will outcompete others and become dominant as well as stable.

8.9.4.3 Immunostimulants

A number of commercially available compounds are claimed to activate crustacean defence systems (section 2.5.3). They may contain a single, active component or a mixture of two or more. In use they may also be combined with other therapeutants including antibiotics. The stimulants may be administered orally, by immersion bath or by injection. Oral administration gives a good non-specific immune response and is generally the most cost-effective. Injection can induce a stronger response but is only practical for large individuals (lobsters, crayfish) and particularly valuable shrimp broodstock. Immersion produces a weaker response, usually requires extra handling and crowding of the stock, but is more cost-effective than injection. Post-larval shrimp, for example, can be treated by immersion for at least 2 h either at the hatchery or on arriving at the pond side during standard acclimation treatment. When given with the feed, the immunostimulant may be added as a top dressing in combination with a fish oil, although effectiveness varies according to how well the compound adheres to the feed. Heat-stable immunostimulants such as β-1,3-D glucan can be directly and cost-effectively incorporated into a diet (e.g. via a commercial feed premix), and will remain effective after pellet formation. For larvae, some immunostimulants can be fed in microparticulate form, either directly or by prior feeding to a live food such as *Artemia*.

Published accounts reveal variable degrees of non-specific immune responses to immunostimulant compounds. Some variation may be due to inadequate purification, to contamination of the compound with cellular debris during preparation or to inadequate dosages. Another source can be blocking of the crustacean's haemocyte receptor sites with dietary ingredients such as carrageenan (Dugger & Jory 1999). There may also be a short refractory period (possibly 1–48 h) after initial stimulation due to over-utilisation of cellular resources, leaving the treated crustacean temporarily vulnerable until the full complement of circulating haemocytes and defence mechanisms is restored (Lorenzon *et al.* 1999). Trials made in poor environmental conditions such as overcrowded or dirty ponds would also tend to minimise the effectiveness of any treatment. Further research is required to develop consistently effective treatment protocols (section 12.2).

8.9.4.4 SPF stock production

To produce shrimp that are free of specified pathogens, primary, secondary and sometimes tertiary quarantine facilities and operating procedures must be established (Lotz 1997). Several such systems have been designed in the USA and operate in strict biological isolation. They utilise recirculation technologies with little or no water exchange to minimise the risks of introducing pathogens, and are often referred to as biosecure units. Prototypes and preliminary economic analyses of such systems have been prepared by Browdy and Bratvold (1998); Moss *et al.* (1998); Ogle and Lotz (1998) and Samocha and Lawrence (1998). The primary systems are stocked with clean larvae or post-larvae and regularly checked (every 30–45 days) for pathogens. The pathogen-free shrimp are then raised to maturity and spawned in the secondary system. Only after the consequent F_1 generation is certified free of the specified pathogens are they introduced into the main breeding programme and passed to the broodstock production (multiplication) systems. Further rigorous testing for the pathogens to be specifically excluded is done at every step. SPF sentinel animals may also be held in bioassay systems and fed moults or limbs from the broodstock that are, by now, too valuable to be killed for routine diagnostic purposes. Subsequent offspring are raised in the mass production systems that provide 'high-health' post-larvae for sale to farms (Lotz *et al.* 1995). High-health shrimp are those originating from SPF stocks that have been transferred to commercial facilities for mass production and where testing to confirm their SPF status is no longer practicable (Wyban *et al.* 1993). Crustaceans certified as 'disease free' for shipment abroad are not uncommon, but caution should be exercised in accepting the validity of some of the claims.

Several misconceptions exist concerning SPF shrimp stocks and it must be emphasised that these shrimp can still succumb to diseases, especially if transplanted and exposed to pathogens not previously encountered (Lotz 1997; Bédier *et al.* 1998). They perform best in protected environments, for example, in super-intensive and recirculation systems (sections 7.2.5 and 7.2.6.6) and in other specialist biosecure systems used in disease and genetic research (Browdy & Bratvold 1998; section 12.3). SPF stocks are, however, susceptible to disease in conventional ponds where specific pathogen-resistant (SPR) stocks tend to do better.

8.10 Genetics

8.10.1 Selective breeding programmes

Selective breeding programmes use both individual and family selection. A large number of crosses must be produced for each generation and the number of families used per generation must be maximised in order to minimise loss of genetic variation (Gjedrem & Fimland 1995; Browdy 1998). Many families from defined matings are now being reared from *Litopenaeus vannamei* (Wyban *et al.* 1993), *L. stylirostris* (Bédier *et al.* 1996) and *Penaeus monodon* (Benzie *et al.* 1997).

8.10.1.1 Tagging and stock monitoring

When individuals with desirable phenotypes are selected from populations to be potential founders of new genetic lines, they must be identifiable if credible pedigree records are to be maintained. Identification of specific individuals is facilitated if a crustacean can be labelled with a readily visible tag that is not lost at moult (section 7.2.2.2). The eyestalk ring tag developed for large penaeid shrimp by Rodriguez (1976) is worth noting but is not easily applied to lobsters because of their much shorter eye peduncle. Internally placed tags that are visible or readable from the outside are being developed. They include small (1.0 mm × 2.5 mm), visible implant fluorescent (VIF) tags made of medical-grade elastomer pigmented with non-toxic fluorescent material, biocompatible, alphanumeric visible implant tags (Jerry *et al.* 2001), and microchip, passive integrated transponder (PIT) tags (Caceci *et al.* 1999). The latter are, as yet, too big for use with all but the largest shrimp or prawns but might be better tolerated by lobsters and the larger crayfish species.

8.10.1.2 Improving growth rates

Repeated inbreeding from one stock may result in a reduction of genetic variation or loss of culture performance. Indeed, husbandry practices that seem likely to select unintentionally for adverse traits have been reported. Most prevalent are the selection of broodstock from among the first pond-raised prawns or shrimp to mature or spawn regardless of size (Doyle *et al.* 1983; Sbordoni *et al.* 1986), and the practice of leaving slow-growing, red swamp crayfish in the ponds to become next season's broodstock (Lutz & Wolters 1989). Similarly, farmers of Australian crayfish may, by selling all the largest and fastest growing individuals first, be inadvertently selecting for slower growth by using the smaller, less valuable animals as broodstock. With this in mind, a guide to setting up a simple selective breeding programme to improve growth rates on commercial redclaw crayfish farms (where tagging of individuals may not be practicable) has been prepared (Jones *et al.* 1998) and recommends the following procedures:

(1) Stock and rear under good conditions (at 5–10 m^{-2}), healthy, uniformly sized, juveniles (5–10 g) obtained from good quality egg-bearing females, preferably those taken from different ponds to ensure adequate genetic variability.
(2) After about 6 months select the largest 5–10% of each sex (preferably several hundred animals >85 g in weight) and set aside, for subsequent mating, sufficient numbers of these to generate the required numbers of juveniles for the farm's normal stocking programme. Using a large number of small tanks or ponds in the mating programme reduces the chance of inbreeding and allows better control and monitoring of broodstocks than a few large ponds. Good record keeping is essential (particularly if animals are not tagged) and can later be used to confirm the status of animals sold.
(3) Transfer egg-bearing females to juvenile production ponds and allow juveniles to grow under optimum conditions to about 5–10 g.
(4) Harvest and sort juveniles to stock into ongrowing ponds.
(5) Repeat the selection process, occasionally introducing new genetic material. This may be in the form of selectively improved stock from another

Plate 8.7 Tanks used for mating redclaw crayfish (*Cherax quadricarinatus*) in selective breeding programmes. Once berried, females are removed and stocked into cage enclosures in ponds where juveniles are released. (Photo courtesy Clive Jones, Department of Primary Industries, Queensland, Australia.)

farm provided the performance of the new stock is comparable to, or better than, the original. An upward trend in growth rates should be apparent in 3–5 years, provided no changes in environment or management practices mask the results.

8.10.1.3 SPR breeding programmes

One difficulty of using conventional breeding techniques to select for disease resistance is that the criterion of success is survival, which is under the control of many genes as well as environmental factors (section 2.6). The degree of genotype–environmental interaction will affect the usefulness of stock selected from breeding programmes intended to supply a wide range of culture environments since a specific difference in the environment does not have the same effect on all genotypes (Coman *et al.* 2000). The ability to rear stock from an appropriate founder population under environmental conditions similar to those of the target farm's locality is therefore likely to be highly advantageous (Tave 1994). Crosses can also cause the transfer of genes that may be of little benefit or even harmful. Because of the environmentally induced variability in some characteristics, studies of nuclear DNA and mitochondrial DNA (mtDNA) are being made for their usefulness in establishing pedigrees, linkage mapping and identifying quantitative trait loci (QTL) that influence commercially important traits (Benzie 1998; Lai *et al.* 2000). Various techniques are employed including restriction fragment length polymorphisms – RFLPs, randomly amplified polymorphic DNA – RAPDs, amplified fragment length polymorphisms – AFLPs, and microsatellite techniques. Gene mapping has already commenced in *Marsupenaeus japonicus* (Moore *et al.* 1999), *Litopenaeus vannamei* (Alcivar-Warren *et al.* 1997), crayfish (Fetzner & Crandall 2001) and *Homarus* (Tam & Kornfield 1996). These techniques are also useful in identifying genetically distinct populations (e.g. of *Cherax tenuimanus* – Imgrund *et al.* 1997, and *Jasus* spp. – Ovenden & Brasher 2000) and for informing breeders when genetic diversity has been reduced in a population so that genetically new parent stock can be introduced. Using these techniques to construct a genome map of a species will greatly facilitate international progress towards domestication (Alcivar-Warren *et al.* 1997).

Gaining from a selection programme depends upon the amount of genetic variability within the initial population and the extent to which this can be exploited. Current research on crustaceans indicates that selection for disease resistance is possible (Tang *et al.* 2000), at least in shrimp (*Litopenaeus stylirostris* and *L. vannamei*), either by the selection of animals naturally more resistant to a specific pathogen or from an acquired immunoresistance (Bédier *et al.* 1998; section 12.2). In these particular cases, the selective breeding of SPR shrimp populations resulted in loss of genetic variability and, although growth was not adversely affected, when the shrimp experienced environmental stress they became vulnerable to other pathogens. In this they were like SPF shrimp (section 8.9.4.4). Commercial use and production of SPR shrimp is well established, for example in Mexico (Anon. 1999b; Clifford 2000).

8.10.1.4 Artificial insemination

Artificial insemination can be useful in pairing specific individuals in breeding and hybridisation programmes, for example for lobsters (Talbot *et al.* 1983), crabs (Lee & Yamazaki 1989) and shrimp (Ting *et al.* 1991; Benzie *et al.* 1997). The technique is used frequently in some commercial shrimp hatcheries but otherwise is largely limited to research hatcheries. Extraction of the spermatophores is achieved by dissection, or non-destructively by electrical stimulation (electro-ejaculation) (Samuel *et al.* 1998) or by manually squeezing the terminal ampullae at the bases of the fifth pereopods (Redón *et al.* 1997). Electro-ejaculation can, however, cause damage to the protective spermatophore wall resulting in uptake of water and swelling. The method of implantation varies with species according to the type of thelycum present (sections 7.2.2.5 and 7.8.3) and good results are reported with shrimp by implanting the sperm mass from a spermatophore into the thelycum together with an artificial fluid containing 62.5 mg mL^{-1} trypsin (Lin & Hanyu 1990).

In vitro fertilisation is reported for a number of shrimp species (Bray & Lawrence 1992) but one of the main constraints has been that the penaeid egg membrane hardens within 12–15 min after contact with seawater. Upwelling a concentrated sperm suspension under a spawning female (Misamore & Browdy 1997) can alleviate the problem.

8.10.2 Genetic manipulation

Genetic manipulation is done either at the chromosome level, to increase or decrease chromosome number in cells in order to produce sterile or monosex populations (section 2.6.3), or at the level of the gene to introduce a beneficial characteristic from another species (section 2.6.4). The techniques used for increasing the number of chromosomes in crustaceans include, for example in *Fenneropenaeus indicus*, the application of a heat shock (30–44°C) of 1–7 min duration to shrimp eggs between 6 and 46 min after they were spawned or a combination of heat or cold shock and exposure to chemicals, commonly cytochalasin B (Benzie 1998). Similarly, 39–75% triploids were induced in *Fenneropenaeus chinensis* by heat shocks (28–32°C) to fertilised eggs for 8–16 min starting 8–20 min after spawning. Eight populations of 29–118 shrimp were reared to about 10 cm total length and many exhibited ovarian abnormalities (Li *et al.* 1999). Eggs of *F. chinensis* have also been induced to form viable (presumably haploid) embryos by activation with irradiated sperm. When subsequently subjected to cold or cytochalasin B shocks, the embryos yielded 15–37% diploid gynogenetic nauplii (Cai & Feng 1993).

Transgenic crustaceans contain a gene within their chromosomal DNA, usually transplanted from another species, that is intended to improve production traits such as growth rate, disease resistance or cold tolerance (Lutz 1999). Techniques that are suitable for mass insertion of a gene sequence into embryos include electroporation (fertilised eggs are placed in a solution containing the construct and the permeability of their membranes temporarily increased by a pulse of high voltage), biolistic methods in which tissues or embryos are bombarded with gold or tungsten microparticles coated with material containing the DNA construct, and the use of viral vectors (sections 2.6.4 and 12.3). However, microinjection of DNA into individual embryos, despite being more labour-intensive and time-consuming, can be more efficient (Preston *et al.* 2000).

8.11 Hatchery supported fisheries, ranching and habitat modification

The priority for any stock enhancement or habitat modification scheme must firstly be to define the aims of the programme, conduct cost–benefit analysis and develop suitable monitoring methods to judge the effectiveness of the project. The scale of the planned releases can then be made appropriate for the defined objectives. For example, in the UK stock enhancement experiments, the minimum numbers of juvenile lobsters needed to yield a scientifically useful number of returns from a specific fishery was estimated to be 10 000 lobsters per year for 5 years, which in the event proved to be satisfactory (Bannister & Addison 1998). A typical UK lobster fishery along 10–20 miles of coastline involves 10–20 fishermen landing collectively around 45 000 lobsters annually. To create a new fishery of this size would be a very substantial, long-term undertaking and a more tenable objective might be to aim for a 10% increase in annual landings (Anon. 1995). Alternatively, the objective might be to restore an ailing breeding stock or maintain activity in a rural fishing community regardless of cost (sections 7.2.9 and 10.6.1.9). Techniques used to achieve such aims may include the placement of structures that modify sea or lake bed habitat specifically for the enhancement of crustacean fisheries (sections 7.6.6.1, 7.8.12, 7.9.8 and 7.10.8.4). They may also include the redesign of conventional structures planned for non-fishery purposes (coastal defence reefs, breakwaters and harbours) in order to maximise the number of micro- and macrohabitats available and hence increase biodiversity and shellfish ranching and fishing opportunities (Jensen *et al.* 1998; section 8.11.2).

8.11.1 Restocking and ranching

Comprehensive ecological and hydrodynamic surveys are prerequisites of any release programme and must indicate suitable habitat and season, both for release and subsequent growth, as well as provide data on predators and natural recruitment. These surveys may include the use of side-scan sonar and scuba divers. For example, on a cohesive clay/mud substrate, many burrowing crustaceans including juvenile clawed lobsters (Wickins 1999), create hydrodynamically advantageous burrow systems that facilitate the exchange of oxygenated water (Ziebis *et al.* 1996). Optimum habitats for *Homarus* will therefore include mature, heterogeneous cobble-boulder layers overlaying a fertile, penetrable substrate. Juvenile spiny and slipper lobsters, on the other hand, require an environment containing clumps or bushes of macroalgae (seaweeds and sea grasses) and a similarly heterogeneous substrate of coral rubble or rocky outcrops containing appropriately sized crevices in which to conceal themselves as they grow larger (Nonaka *et al.* 2000) (section 7.9.4).

Production of the juveniles for release may be from licenced collectors (spiny lobsters) or contracted from small- or large-scale, private or publicly funded hatcheries (clawed lobsters, crayfish, crabs; Wickins 1997) but in most cases the released animals will be indistinguishable from their wild counterparts when captured. This means that fishermen are unlikely to contribute willingly towards the cost of their production, and also that no private restocking schemes will be implemented without the security of ownership rights (section 11.5.3.1). The management of large national programmes may involve a blend of levies on the fishermen and subsidies for the producers or a system of transferable share quotas.

8.11.2 Habitat modification

Artificial reefs are widely perceived to be effective in aggregating fish and, in some countries, have become an important fishery management tool. Radically different approaches to reef construction and deployment, however, exist between countries. In Japan, artificial reefs are primarily intended to benefit commercial fishermen. They are designed and built by engineers from non-waste materials and are placed on carefully selected sites. By contrast, reefs in the USA are often large, constructed from low-cost 'materials of opportunity' and are deposited in deep waters, typically to improve recreational fisheries. Some European reefs are placed to control, for example, inshore trawling, and may, additionally, increase fishing opportunities for rural communities. In Britain, however, the term 'artificial reef' is widely understood to include any man-made structure built below high water. Such reefs are usually built for non-fishery purposes such as coastal protection, harbour training walls, pipeline protection, tidal power generation or recreation (Collins *et al.* 1994a).

Artificial reefs are known to affect the abundance or exploitation of commercially valuable crustaceans. For example, permanently submerged reefs of rock rubble (Todd *et al.* 1992) or stabilised, pulverised fuel ash (Collins *et al.* 1994b) can provide new habitat for immigrant, wild lobsters and crabs. On a smaller scale, reefs of stones or cobbles are often deployed in European freshwater crayfish fisheries and farms to provide increased shelter, thereby improving opportunities for survival at high stock densities (section 7.6.6.1). In a slightly different vein, internal levees constructed in large Louisiana crayfish ponds increase the area available for broodstock burrows and hence the next season's crop (section 7.5.2). While most underwater constructions will attract fish and shellfish, specific designs have been developed for particular purposes (Grove *et al.* 1991), e.g. to provide increased shelter for spiny lobsters (Briones-Fourzán *et al.* 2000) and slipper lobsters (Spanier & Almog-Shtayer 1992); attract settlement of pueruli (Nonaka *et al.* 2000); increase habitat diversity (Haroun & Herrera 1995) and food availability (Bailey-Brock 1989), or for improving water quality (section 8.3.6.8). In general, species diversity and perhaps 'productivity' increase with reef complexity (Seaman 1997). The orientation and area of surfaces and the internal spaces available for colonisation by attaching organisms are critical factors (Wickins & Barker 1997; Jensen *et al.* 1998), and influence the species, diversity and rate of growth of reef biomass (Hatcher 1995). Crevices too, play an important role in providing shelter for prey and predator alike as well as for commercially important shellfish species such as slipper (Spanier 1994), spiny (Norman *et al.* 1994; Hotta *et al.* 1995) and clawed lobsters (Jensen & Collins 1997). Studies are now needed to extend knowledge of the detailed spatial (tolerable nearest neighbour distances, foraging behaviour) and habitat needs (crevice size and shape preferences, food availability) and thus carrying capacity of a reef structure for lobsters of different sizes to survive and grow within a defined area (section 12.7).

8.12 References

Aboutboul Y., Arviv R. & van Rijn J. (1995) Anaerobic treatment of intensive fish culture effluents: volatile fatty acid mediated denitrification. *Aquaculture*, **133** (1) 21–32.

Aiken D.E. & Waddy S.L. (1995) Aquaculture. In: *Biology of the Lobster* Homarus americanus (ed. J.R. Factor), pp. 153–175. Academic Press, New York.

Alabi A.O., Yudiati E., Jones D.A. & Latchford J.W. (1997) Bacterial levels in penaeid larval cultures. In: *Diseases in Asian Aquaculture III* (eds T.W. Flegel & I.H. MacRae), pp. 381–388. Asian Fisheries Society, Manila, Philippines.

Al Ameeri A.A. & Cruz E.M. (1998) Effect of sand substrate on growth and survival of *Penaeus semisulcatus* de Haan juveniles. *Journal of Aquaculture in the Tropics*, **13** (4) 239–244.

Alcivar-Warren A., Garcia D.K., Dhar A.K., Wolfus G.M & Astrofsky K.M. (1997) Efforts towards mapping the shrimp genome: a new approach to animal health. In: *Diseases in Asian Aquaculture III* (eds T.W. Flegal & I.H. MacRae), pp. 255–263. Asian Fisheries Society, Manila, Philippines.

Allan G.L. & Maguire G.B. (1991) Lethal levels of low dissolved oxygen and effect of short-term oxygen stress on subsequent growth of juvenile *Penaeus monodon*. *Aquaculture*, **94**, 27–37.

Allan G.L. & Maguire G.B. (1993) The effect of water exchange on production of *Metapenaeus macleayi* and water quality in experimental ponds. *Journal of the World Aquaculture Society*, **24** (3) 321–328.

Aneshansley E.D. (2000) Advances in fluidized sand bed biofilters. *Global Aquaculture Advocate*, **3** (3) 35–36.

Anon. (1988) Selling live seafood. *Seafood International*, June 1988, pp. 37–43.

Anon. (1995) *Lobster stocking: progress and potential. Significant results from UK lobster restocking studies 1982–1995*,

12 pp. MAFF Directorate Fisheries Research, Lowestoft.

Anon. (1999a) 'Humane' lobster stunner prototype. *Fish Farming International*, **26** (12) 42.

Anon. (1999b) Seed firm hits Pacific heights. *Fish Farming International*, **26** (10) 13.

Antsulevich A., Laihonen P. & Vuorinen I. (2000) Employment of artificial reefs for environmental maintenance in the Gulf of Finland. In: *Artificial Reefs in European Seas* (eds A.C. Jensen, K.J. Collins & A.P.M. Lockwood), pp. 319–329. Kluwer Academic, Netherlands.

AQUACOP (1989) Production results and operating costs of the first super-intensive shrimp farm in Tahiti. *Aquaculture '89 Abstracts*, World Aquaculture Society Meeting 1989, p. 56. Los Angeles, USA.

ASEAN (1978) *Manual on Pond Culture of Penaeid Shrimp*, 131 pp. ASEAN National Coordinating Agency of the Philippines. Ministry of Foreign Affairs, Manila, Philippines.

Avault J.W. Jr (1997) Control of diseases and calculations for chemical treatments, some fundamentals reviewed. *Aquaculture Magazine*, **23** (3) 80–83.

Avault J.W. Jr & Huner J.V. (1985) Crawfish culture in the United States. In: *Crustacean and Mollusk Aquaculture in the United States* (eds J.V. Huner & E. Evan Brown), pp. 1–61. AVI Inc., Westport, Connecticut, USA.

Ayranto M.J. (1998) *A summary report on the use of feeding trays*, 15 pp. Banggai Sentral Shrimp, Surabaya, Indonesia.

Bador R., Scura E.D. & Naivosoa R. (1998) The use of feeding trays in the semi-intensive grow-out of *Penaeus monodon*: a tool to better understand shrimp feeding behaviour in ponds. *World Aquaculture*, **29** (1) 27.

Bailey-Brock J.H. (1989). Fouling community development on an artificial reef in Hawaiian waters. *Bulletin of Marine Science*, **44** (2) 580–591.

Bainy A.C.D. (2000) Biochemical responses in penaeids caused by contaminants. *Aquaculture*, **191** (1–3) 163–168.

Bannister R.C.A. & Addison J.T. (1998) Enhancing lobster stocks: a review of recent European methods, results, and future prospects. *Bulletin of Marine Science*, **62** (2) 369–387.

Barrows F.T. (2000) Larval feeds: two methods for production of on-size, microbound particles. *Global Aquaculture Advocate*, **3** (1) 61–63.

Beard T.W. & McGregor D. (1991) Care and storage of live lobsters, 33 pp. *Lab. Leafl., (66).* MAFF Directorate Fisheries Research, Lowestoft, UK.

Beard T.W. & Wickins J.F. (1992) Techniques for the production of juvenile lobsters 22 pp. *Fish. Res. Tech. Rept.*, (92). MAFF Directorate Fisheries Research, Lowestoft, UK.

Bédier E., Ottogalli L., Patrois J., Weppe M. & AQUACOP. (1996) Development of the *Penaeus stylirostris* AQUACOP SPR43 strain: in situ experiments and genetic enhancement. In: *World Aquaculture '96. Book of abstracts* (ed. R.L. Creswell), pp. 32–33. World Aquaculture Society, Baton Rouge, LA, USA.

Bédier E., Cochard J.C., Le Moullac G., Patrois J. & AQUACOP (1998) Selective breeding and pathology in penaeid shrimp culture: the genetic approach to pathogen resistance. *World Aquaculture*, **29** (2) 46–51.

Behnke K.C., Fahrenholz C., Bortone E.J., Eisenberg S. & Dominy W.G. (1992) Feed mixing and mixer testing of shrimp feeds. In: *Proceedings of the Special Session on Shrimp Farming* (ed. J. Wyban), pp. 206–211. World Aquaculture Society, Baton Rouge, LA, USA.

Benzie J.A.H. (1998) Penaeid genetics and biotechnology. *Aquaculture*, **164** (1–4) 23–47.

Benzie J.A.H., Kenway M. & Trott L. (1997) Estimates for the heritability of size in juvenile *Penaeus monodon* prawns from half-sib matings. *Aquaculture*, **152** (1–4) 49–53.

Bidwell J.P. & Spotte S. (1985) *Artificial Seawaters: formulas and methods*, 349 pp. Jones & Bartlett Publishers, Boston, Woods Hole, USA.

Body A. (2000) *Sugar/molasses, C:N ratios*. Correspondence, 23/09/00. shrimp@egroups.com

Booth J.D. & Kittaka J. (2000) Spiny lobster growout. In: *Spiny Lobsters: fisheries and culture,* 2nd edn (eds B.F. Phillips & J. Kittaka), pp. 556–585. Fishing News Books, Oxford, UK.

Bower C.E. & Turner D.T. (1982) Effects of seven chemotherapeutic agents on nitrification in closed seawater culture systems. *Aquaculture*, **29** (3–4) 331–345.

Bowser P.R. & Rosemark R. (1981) Mortalities of cultured lobsters, *Homarus*, associated with a molt death syndrome. *Aquaculture*, **23** (1–4) 11–18.

Boyd C.E. (1990) *Water quality management and aeration in shrimp farming*, 70 pp. American Soybean Association, Singapore.

Boyd C.E. (1995a) *Bottom Soils, Sediment, and Pond Aquaculture*, 348 pp. Chapman & Hall, New York.

Boyd C.E. (1995b) Soil and water quality management in aquaculture ponds. *Infofish International*, (5) 29–36.

Boyd C.E. (1995c) Chemistry and efficacy of amendments used to treat water and soil quality imbalances in shrimp ponds. In: *Swimming through troubled water. Proceedings of the special session on shrimp farming* (eds C.L. Browdy & J.S. Hopkins), pp. 183–199. Aquaculture '95. World Aquaculture Society, Baton Rouge, LA, USA.

Boyd C.E. (1998) Pond water aeration systems. *Aquacultural Engineering*, **18** (1) 9–40.

Boyd C.E. (1999a) *Pond water and soil management procedures to minimize the effects of disease epidemics in shrimp farming*, 16 pp. (mimeo). Presented at Conferencia Regional de Camaronicultura, 7–8 July 1999, Hotel Plaza Paitilla Inn, Panama, Grupo FCE, Panama.

Boyd C.E. (1999b) Management of shrimp ponds to reduce the eutrophication potential of effluents. *Global Aquaculture Advocate*, **2** (6) 12–13.

Boyd C.E. & Daniels H.V. (1988) Evaluation of Hach fish farmer's water quality test kits for saline water. *Journal of the World Aquaculture Society*, **19** (2) 21–26.

Boyd C.E. & Fast A.W. (1992) Pond monitoring and management. In: *Marine Shrimp Culture: principles and practices* (eds A.W. Fast & L.J. Lester), pp. 497–513. Elsevier Science, New York.

Boyd C.E. & Martinson D.J. (1984) Evaluation of propeller-aspirator-pump aerators. *Aquaculture,* **36** (3) 283–292.

Boyd C.E. & Massaut L. (1999) Risks associated with the use of chemicals in pond aquaculture. *Aquacultural Engineering*, **20**, 113–132.

Boyd C.E. & Zimmermann S. (2000) Grow-out systems – water quality and soil management. In: *Freshwater Prawn Culture: the farming of* Macrobrachium rosenbergii (eds

M.B. New & W.C. Valenti), pp. 221–238. Blackwell Science, Oxford, UK.

Bratvold D., Lu J. & Browdy C.L. (1999) Disinfection, microbial community establishment and shrimp production in a prototype biosecure pond. *Journal of the World Aquaculture Society*, **30** (4) 422–432.

Bray W.A. & Lawrence A.L. (1992) Reproduction of *Penaeus* species in captivity. In: *Marine Shrimp Culture: principles and practices* (eds A.W. Fast & L.J. Lester), pp. 93–170. Elsevier Science, New York.

Breen P.A. (1995) Killing lobsters humanely: call for discussion. *The Lobster Newsletter*, **8** (2) 5.

Brinkman R. & Sing V.P. (1982) Rapid reclamation of brackish water fishponds in acid sulphate soils. In: *Proceedings of Bangkok Symposium on acid sulphate soils* (eds H. Dost & N. van Breeman), pp. 318–330. Int. Inst. Land Reclamation and Improvement, Publ. 18. Wageningen, Netherlands.

Briones-Fourzán P., Lozano-Álvarez E. & Eggleston D.B. (2000) The use of artificial shelters (casitas) in research and harvesting of Caribbean spiny lobsters in Mexico. In: *Spiny Lobsters: fisheries and culture,* 2nd edn (eds B.F. Phillips & J. Kittaka), pp. 420–446. Fishing News Books, Oxford, UK.

Brock J.A. & Main K.L. (1994) *A Guide to the Common Problems and Diseases of Cultured* Penaeus vannamei, 242 pp. World Aquaculture Society, Baton Rouge, LA, USA.

Browdy C.L. (1998) Recent developments in penaeid broodstock and seed production technologies: improving the outlook for superior captive stocks. *Aquaculture* **164** (1–4) 3–21.

Browdy C.L. & Bratvold D. (1998) Preliminary development of a biosecure shrimp production system. In: *Proceedings of the US marine shrimp farming program biosecure workshop*, 14 February 1998 (ed. S.M. Moss), pp. 19–38. The Oceanic Institute, HI, USA.

Brown J.J. & Glenn E.P. (1999) Management of saline aquaculture effluent through the production of halophyte crops. *World Aquaculture*, **30** (4) 44–49.

Brown J.H., Wickins J.F. & Maclean M.H. (1991) The effect of water hardness on growth and carapace mineralization in juvenile freshwater prawns, *Macrobrachium rosenbergii* (de Man). *Aquaculture*, **95**, 329–345.

Brune D.E. & Tomasso J.R. (Eds.) (1991) Aquaculture and water quality, 606 pp. *Advances in World Aquaculture*, **3**. World Aquaculture Society, Baton Rouge, LA, USA.

Caceci T., Smith S.A., Toth T.E., Duncan R.B. & Walker S.C. (1999) Identification of individual prawns with implanted microchip transponders. *Aquaculture*, **180** (1–2) 41–51.

Cai N. & Feng L. (1993) Artificial induction of gynogenesis in Chinese shrimp, *Penaeus orientalis* Kishinouye. *Aquaculture*, **111** (1–4) 314.

Cansdale G. (1981) Sea water abstraction. In: *Aquarium Systems* (ed. A.D. Hawkins), pp. 47–62. Academic Press, London.

Carmignani G.M. & Bennett J.P. (1976) Leaching of plastics used in closed aquaculture systems. *Aquaculture*, **7** (1) 89–91.

Castaños M. (1997) Grow mudcrab in ponds. *Asian Aquaculture*, **19** (3) 14–16.

Cavalli R.O., Peixoto S.M. & Wasielesky W. (1998). Performance of *Penaeus paulensis* (Perez-Farfante) broodstock under long-term exposure to ammonia. *Aquaculture Research*, **29**, 815–822.

Chamberlain G. (1987) Status report: intensive pond management. *Coastal Aquaculture*, **4** (1) 1–7.

Chen J-C. & Lin C-H. (2001) Toxicity of copper sulfate for survival, growth, moulting and feeding of juveniles of the tiger shrimp, *Penaeus monodon*. *Aquaculture*, **192** (1) 55–65.

Chen L-C. (1990) *Aquaculture in Taiwan*, 273 pp. Fishing News Books, Blackwell Scientific Publications, Oxford, UK.

Chen S. (1998) Aquacultural waste management. *Aquaculture Magazine*, **24** (4) 63–69.

Chiang P. & Liao I. C. (1985) The practice of grass prawn (*Penaeus monodon*) culture in Taiwan from 1968 to 1984. *Journal of the World Mariculture Society*, **16**, 297–315.

Chien Y.-H. & Liao I. C. (1995) Integrated approach to prawn growout system design. In: *Swimming through troubled water. Proceedings of the special session on shrimp farming* (eds C.L. Browdy & J.S. Hopkins), pp. 167–182. Aquaculture '95. World Aquaculture Society, Baton Rouge, LA, USA.

Chin K.K. & Ong S.L. (1997) Water conservation and pollution control for intensive prawn farms. In: Water quality and pollution control in Asia (ed. T. Matsuo). *Water Science and Technology*, **35** (8) 77–81.

Cho C.Y., Hynes J.D., Wood K.R. & Yoshida H.K. (1994) Development of high-nutrient-dense, low-pollution diets and prediction of aquaculture wastes using biological approaches. *Aquaculture*, **124** (1–4) 293–305.

Clifford H.C. III (1994) Semi-intensive sensation. *World Aquaculture*, **25** (3) 6–12 & 98–104.

Clifford H.C. III. (2000) Shrimp farming in Mexico: recent developments. *Global Aquaculture Advocate*, **3** (2) 79–81.

Collins K. & Jensen A. (1997) Acceptable use of waste materials. In: *Proceedings of First Conference of European artificial reef research network*, 26–30 March 1996, Ancona, Italy (ed. A.C. Jensen), pp. 377–390. Southampton Oceanography Centre, Southampton, UK.

Collins K.J., Jensen A.C., Lockwood A.P. M. & Lockwood S.J. (1994a) Coastal structures, waste materials and fishery enhancement. *Bulletin of Marine Science*, **55** (2–3) 1240–1250.

Collins K.J., Jensen A.C., Lockwood A.P.M. & Turnpenny W.H. (1994b) Evaluation of stabilised coal fired power station waste for artificial reef construction. *Bulletin of Marine Science*, **55** (2–3) 1251–1262.

Colt J. (2000) Design of gas transfer systems part 1: site and water supply considerations. *Aquaculture Magazine*, **26** (4) 55–61.

Colt J. & Huguenin J. (1992) Shrimp hatchery design: engineering considerations. In: *Marine Shrimp Culture: principles and practices* (eds A.W. Fast & L.J. Lester), pp. 245–285. Elsevier Science, New York.

Coman G.J., Crocos P.J. & Preston N.P. (2000) Effect of the interaction of genotype and environment on the survival and growth of the kuruma shrimp *Penaeus japonicus*. In: *Abstracts, Aqua 2000, Responsible aquaculture in the new millennium* (compiled by R. Flos & L. Creswell), p. 136. European Aquaculture Society, Special Publication No. 28.

Conklin D.E. (1990) Nutritional factors in "molt death" of juvenile lobsters (*Homarus americanus*). *Crustacean Nutrition Newsletter*, **6** (1) 71–72. World Aquaculture Society, Baton Rouge, LA, USA.

Cook H. & Clifford H. (1997a) Feed management for semi-intensive shrimp culture: part 2. *Aquaculture Magazine*, **23** (6) 37–42.

Cook H.L. & Clifford H.C. (1997b) Managing water discharges from semi-intensive shrimp ponds. *Aquaculture Magazine*, **23** (2) 31–38.

Cook H.L. & Clifford H.C. (1998a) Fertilization of shrimp ponds and nursery tanks. *Aquaculture Magazine*, **24** (3) 52–54 & 58–62.

Cook H. & Clifford H. (1998b) Feed management for semi-intensive shrimp culture. *Aquaculture Magazine*, **24** (1) 30, 32 & 34–37.

Correia E.S., Suwannatous S. & New M.B. (2000) Flow-through hatchery systems and management. In: *Freshwater Prawn Culture: the farming of* Macrobrachium rosenbergii (eds M.B. New & W.C. Valenti), pp. 52–68. Blackwell Science, Oxford, UK.

Cowan L. (1983) *Crab farming in Japan, Taiwan and the Philippines*, 85 pp. Information Series QI 84009. Queensland Department of Primary Industries, Queensland, Australia.

Cruz P.S. (1991) *Shrimp feed management*, 58 pp. Kabukiran Enterprises Inc., Davao City, Philippines.

Culley D.D. & Duobinis-Gray L. (1989) Soft-shell crawfish production technology. *Journal of Shellfish Research*, **8** (1) 287–291.

Cuzon G., Guillaume J. & Cahu C. (1994) Composition, preparation and utilization of feeds for Crustacea. *Aquaculture*, **124** (1–4) 253–267.

D'Abramo L.R. & New M.B. (2000) Nutrition, feeds and feeding. In: *Freshwater Prawn Culture: the farming of* Macrobrachium rosenbergii (eds M.B. New & W.C. Valenti), pp. 203–220. Blackwell Science, Oxford, UK.

Daniels W.H., D'Abramo L.R., Fondren M.W. & Durant M.D. (1995) Effects of stocking density and feed on pond production characteristics and revenue of harvested freshwater prawns *Macrobrachium rosenbergii* stocked as size-graded juveniles. *Journal of the World Aquaculture Society*, **26** (1) 38–47.

Davis D.A. & Arnold C.R. (1995) Effects of two extrusion processing conditions on the digestibility of four cereal grains for *Penaeus vannamei*. *Aquaculture*, **133** (3–4) 287–294.

Davis D.A & Arnold C.R. (1998) The design, management and production of a recirculating raceway system for the production of marine shrimp. *Aquacultural Engineering*, **17**, 193–211.

Devresse B. (1998a) Production of water stable shrimp feeds, 11 pp. (mimeo). *Proceedings of IV international symposium on aquatic nutrition*, 15–18 November 1998, La Paz, Mexico.

Devresse B. (1998b) Nutrition and health: the nutriceutical approach. *International Aquafeed Directory, 1997/8*, pp. 51–59.

Dexter S.C. (1986) Materials science in aquacultural engineering. *Aquacultural Engineering*, **5** (2–4) 333–345.

Doyle R.W., Singholka S. & New M.B. (1983) 'Indirect selection' for genetic change: a quantitative analysis illustrated with *Macrobrachium rosenbergii*. *Aquaculture*, **30** (1–4) 237–247.

Ducruet S., Rojat G., Nicolas L. & Laurent P.J. (1993) A new device for electrocution of crayfish. In: *Freshwater Crayfish 9* (eds D.M. Holdich & G.F. Warner), pp. 154–157. Louisiana State University, LA, USA.

Dugger D.M. (2000) *Cast net sampling*. Correspondence, 07/06/2000. Shrimp@egroups.com

Dugger D.M. & Jory D.E. (1999) Bio-modulation of the non-specific immune response in marine shrimp with beta-glucan. *Aquaculture Magazine*, **25** (1) 81–89.

EIFAC (1986) Flow-through and recirculation systems. Report of the working group on terminology, format and units of measurement. *EIFAC Technical Paper* (49) 1–100.

El-Gamal A.A., Alderman D.J., Rodgers C.J., Polglase J.L. & Macintosh D.J. (1986) A scanning electron microscope study of oxolinic acid treatment of burn spot lesions of *Macrobrachium rosenbergii*. *Aquaculture*, **52** (3) 157–171.

Estilo V.E.J. (1988) Pond construction, design and facilities. In: *Technical Considerations for the Management and Operation of Intensive Prawn Farms* (eds Y.N. Chiu, L.M. Santos & R.O. Juliano), pp. 30–79. U.P. Aquaculture Society, Iloilo City, Philippines.

Estrada J. (2000) Air aspirator vs paddlewheel. Correspondence. 4/12/2000. Shrimp@egroups.com

Evans L.H. & Edgerton B.F. (2001) Pathogens, parasites and commensals. In: *Biology of Freshwater Crayfish* (ed. D.M. Holdich), pp. 377–438. Blackwell Science, Oxford, UK.

Evans L.H., Fotedar S, Fan A. & Jones B. (1999) Investigation of idiopathic muscle necrosis and circulating hemocytes in the freshwater crayfish *Cherax tenuimanus* exposed to acute and chronic stressors. In: *Freshwater Crayfish 12* (eds M. Keller, M.M. Keller, B. Oidtmann, R. Hoffmann & G. Vogt), pp. 356–370. Weltbild Verlag, Germany.

Eversole A.G. & Brune D.E. (1995) Water quality and management in crawfish culture ponds. *Journal of Shellfish Research*, **14** (1) 264–265.

Eversole A.G. & Seller B.C. (1996) Comparison of relative crayfish toxicity values. In: *Freshwater Crayfish 11* (ed. W.T. Momot), pp. 274–275. Louisiana State University, LA, USA.

Ewing R.D. & Sheahan J.E. (1996) Airlift debris removal system for water intake structures. *Progressive Fish-Culturist*, **58** (4) 284–285.

Falguiere J.C., Mer G., Gondouin P.H. & Defossez J. (1989) Evaluation of three population sampling methods for freshwater prawn *Macrobrachium rosenbergii* culture in earthen ponds. *Aquaculture '89 Abstracts*, from World Aquaculture Society Meeting 1989, p. 81. Los Angeles, USA.

FAO/NACA/AAHRI (1996) *Revised guidelines for the responsible movement (introduction and transfer) of living aquatic organisms,* 48 pp. OBJ1.Doc Revision 3. FAO/NACA/AAHRI, Workshop on health and quarantine, 28 January 1996. Bangkok, Thailand [seen in draft].

Fetzner J.W. Jr. & Crandall K.A. (2001) Genetic variation. In: *Biology of Freshwater Crayfish* (ed. D.M. Holdich), pp. 291–326. Blackwell Science, Oxford, UK.

Figley W.K. (1994) Developing public and private tire reef unit construction facilities. *Bulletin of Marine Science*, **55** (2–3)

1334 [summary only].

Fisher W.S. (1986) Defences of brooding decapod embryos against aquatic bacteria and fungi. In: *Pathology in Marine Aquaculture* (eds C.P. Vivares, J.R. Bonami & E. Jaspers), pp. 357–363. European Aquaculture Society, Special Publication No. 9.

Flegel T.W., Sriurairatana S., Wongteerasupaya C., Boonsaeng V., Panyim S. & Withyachumnarnkul B. (1995). Progress in characterization and control of yellow-head virus of *Penaeus monodon*. In: *Swimming through troubled water. Proceedings of the special session on shrimp farming* (eds C.L. Browdy & J.S. Hopkins), pp. 76–83. Aquaculture '95. World Aquaculture Society, Baton Rouge, LA, USA.

Flegel T.W., Boonyaratpalin S. & Withyachumnarnkul B. (1997). Progress in research on yellow-head virus and white spot virus in Thailand. In: *Diseases in Asian Aquaculture III* (eds T.W. Flegel & I.H. MacRae), pp. 285–295. Asian Fisheries Society, Manila, Philippines.

Flick G.J. Jr. (1998) Common chemicals for cleaning and disinfecting aquaculture facilities. In: *Proceedings of Second International Conference on Recirculating Aquaculture*, 16–19 July 1998, pp. 7–19. Virginia Tech, VA, USA.

Flores S.E. & Martinez S.E.V. (1993) Critical operations on the manufacture of pelleted feeds for crustaceans. *Aquaculture*, **114** (1–2) 83–92.

Ford K. & Robertson C. (1995) Prawn farm effluent: yes it can be successfully managed. *Austasia Aquaculture Magazine*, **9** (2) 33–34.

Forster J.R.M. (1974) Studies on nitrification in marine biological filters. *Aquaculture*, **4** (4) 387–397.

Fujii S., Niwa C., Mouri M. & Jindal R. (1997) Pilot-plant experiments for improvement of polluted canal/klong water by rock-bed filtration. In: Water quality and pollution control in Asia (ed. T. Matsuo), *Water Science and Technology*, **35** (8) 83–90.

Funge-Smith S.J. & Briggs M.R.P. (1998) Nutrient budgets in intensive shrimp ponds: implications for sustainability. *Aquaculture*, **164** (1–4) 117–133.

Gadient M. & Schai E. (1994) Leaching of vitamins from shrimp feed. *Aquaculture*, **124** (1–4) 201–205.

Gatesoupe F.J. (1999) The use of probiotics in aquaculture. *Aquaculture*, **180** (1–2) 147–165.

Gautier D., Mogollón J.-V., Ruiz M. & Amador J. (2000) The integration of mangrove and shrimp farming: a case study. In: *Abstracts, Aqua 2000, Responsible aquaculture in the new millennium* (compiled by R. Flos & L. Creswell), p. 232. European Aquaculture Society, Special Publication No. 28.

GESAMP (1996) *Chemicals in coastal aquaculture*. Report of GESAMP working group 31. 24–28 May 1996. SEAFDEC, Tigbauan, Iloilo, Philippines [seen in draft].

Gjedrem T. & Fimland E. (1995) Potential benefits from high health and genetically improved shrimp stocks. In: *Swimming through troubled water. Proceedings of the special session on shrimp farming* (eds C.L. Browdy & J.S. Hopkins), pp. 60–65. Aquaculture '95. World Aquaculture Society, Baton Rouge, LA, USA.

Goddard S. (1996) Feed handling and storage. In: *Feed Management in Intensive Aquaculture,* Chapter 6, pp. 99–116. Chapman & Hall, New York.

Goldizen V.C. (1970) Management of closed system marine aquariums. *Helgoländer Wiss. Meeresunters.*, **20**, 637–641.

Greiner A.D. & Timmons M.B. (1998) Evaluation of the nitrification rates of microbead and trickling filters in an intensive recirculating tilapia production facility. *Aquacultural Engineering*, **18** (3) 189–200.

Grove R.S., Sonu C.J. & Nakamura M. (1991) Design and engineering of manufactured habitats for fisheries enhancement. In: *Artificial Habitats for Marine and Freshwater Fisheries* (eds W. Seaman & L.M. Sprague), pp. 109–152. Academic Press, San Diego, USA.

Harding G.C. (1992) American lobster (*Homarus americanus* Milne Edwards): a discussion paper on their environmental requirements and the known anthropogenic effects on their populations. *Canadian Technical Report of Fisheries and Aquatic Sciences*, (1887) vi & 1–16.

Haroun R. & Herrera R. (1995) Artificial reefs in Canary Islands: an overview of their present situation. In: *ECOSET '95, Proceedings of International Conference on ecological system enhancement technology for aquatic environments*, 30 October–2 November 1995. **2**, 727–731. Tokyo, Japan.

Hatcher A.M. (1995) Trends in the sessile epibiotic biomass of an artificial reef. In: *ECOSET '95, Proceedings of International Conference on ecological system enhancement technology for aquatic environments*, 30 October–2 November 1995. **2**, 125–30. Tokyo, Japan.

Hawkins A.D. & Lloyd R. (1981) Materials for the aquarium. In: *Aquarium Systems* (ed. A.D. Hawkins), pp. 171–196. Academic Press, London.

Hirono Y. (1986) Shrimp pond management. In: *Acuacultura del Ecuador*, 49 pp. [in Spanish and English]. Cámara de Productores de Camarón, Guayaquil, Ecuador.

Hochheimer J.N. & Wheaton F.W. (1998) Biological filters: trickling and RBC design. In: *Proceedings of Second International Conference on Recirculating Aquaculture*, 16–19 July 1998, pp. 291–318. Virginia Tech, VA, USA.

Hopkins J.S. (1994) An apparatus for continuous removal of sludge and foam fractions in intensive shrimp culture ponds. *Progressive Fish-Culturist*, **56** (2) 135–139.

Hopkins J.S., Hamilton R.D. III., Sandifer P.A., Browdy C.L. & Stokes A.D. (1993) Effect of water exchange rate on production, water quality, effluent characteristics and nitrogen budgets of intensive shrimp ponds. *Journal of the World Aquaculture Society*, **24** (3) 304–320.

Hopkins J.S., Sandifer P.A. & Browdy C.L. (1994) Sludge management in intensive pond culture of shrimp: effect of management regime on water quality, sludge characteristics, nitrogen extinction, and shrimp production. *Aquacultural Engineering*, **13** (1) 11–30.

Hopkins J.S., Sandifer P.A. & Browdy C.L. (1995) A review of water management regimes which abate the environmental impacts of shrimp farming. In: *Swimming through troubled water. Proceedings of the special session on shrimp farming* (eds C.L. Browdy & J.S. Hopkins), pp. 157–166. Aquaculture '95. World Aquaculture Society, Baton Rouge, LA, USA.

Hopkins J.S., Sandifer P.A., Browdy C.L. & Holloway J.D. (1996) Comparison of exchange and no-exchange water management strategies for the intensive pond culture of marine shrimp. *Journal of Shellfish Research*, **15** (2) 441–445.

Horowitz A. & Horowitz S. (1998) Immediate and stable nitrification in biofilters by microbial manipulation. In: *Proceedings of Second International Conference on Recirculating Aquaculture*, 16–19 July 1998, pp. 359–64. Virginia Tech, VA, USA.

Horowitz A. & Horowitz S. (2000) Micro-organisms and feed management in aquaculture. *Global Aquaculture Advocate*, **3** (2) 33–34.

Hotta, K., Suzuki, T. and Ohno, M. (1995). Technology for creating an artificial seaweed community using ferrous sulphate. In: *ECOSET '95, Proceedings of International Conference on ecological system enhancement technology for aquatic environments*, 30 October–2 November 1995. **2**, 487–494. Tokyo, Japan.

Houser R.H. & Akiyama D.M. (1997) Feed formulation principles. In: *Crustacean Nutrition* (eds L.R. D'Abramo, D.E. Conklin & D.M. Akiyama), pp. 492–518. Advances in World Aquaculture 6, World Aquaculture Society, LA, USA.

Hovel K.A. & Morgan S.G. (1999) Susceptibility of estuarine crab larvae to ultraviolet radiation. *Journal of Experimental Marine Biology and Ecology*, **237**, 107–125.

Hovers A.P. (1999) Teaseed cake/powder. Correspondence 2/8/2000. shrimp@egroups.com

Huguenin J.E. & Colt J. (1989) Design and operating guide for aquaculture seawater systems. *Developments in Aquaculture and Fisheries Science*, **20**, 1–264.

Hussenot J. & Lejeune A. (2000) Use of foam fractionation equipment to improve water quality in ponds: a new experimental equipment. In: *Abstracts, Aqua 2000, Responsible aquaculture in the new millennium* (compiled by R. Flos & L. Creswell), p. 298. European Aquaculture Society, Special Publication No. 28.

ICES (1995) *Code of Practice on the Introductions and Transfers of Marine Organisms 1994*, 12 pp. A slightly amended bilingual English/French version of ICES (1994) *Report of the ICES Advisory Committee on the Marine Environment, 1994, Annex 3*, 122 pp. ICES Cooperative Research Report No. 204.

Iivonen P., Jaervenpaeae T., Lappalainen A., Mannio J. & Rask M. (1995) Chemical, biological and socio-economic approaches to the liming of Lake Alinenjaervi in southern Finland. In: Fifth International Conference on Acidic Deposition: Science and policy, 26–30 June 1995, Goteborg, Sweden (eds P. Grennfelt, H. Rodhe, E. Thoerneloef & J. Wisniewski). *Water, Air, Soil Pollution,* **85** (2) 937–942.

Imgrund J., Groth D. & Wetherall J. (1997) Genetic analysis of the freshwater crayfish *Cherax tenuimanus*. *Electrophoresis*, **18**, 1660–1665.

Inglis V., Abdullah S.Z., Angka S.L., *et al.* (1997) Survey of resistance to antibacterial agents used in aquaculture in five South-east Asian countries. In: *Diseases in Asian Aquaculture III* (eds T.W. Flegel & I.H. MacRae), pp. 331–337. Asian Fisheries Society, Manila, Philippines.

Itami T., Takahashi Y. & Nakamura Y. (1989) Efficacy of vaccination against vibriosis in cultured kuruma prawns *Penaeus japonicus*. *Journal of Aquatic Animal Health*, **1** (3) 238–242.

Jensen A.C. & Collins K.J. (1997) The use of artificial reefs in crustacean fisheries enhancement. In: *Proceedings of First Conference of European artificial reef research network*, 26–30 March 1996, Ancona, Italy (ed. A.J. Jensen), pp. 115–121. Southampton Oceanography Centre, UK.

Jensen A.C., Hamer B.A. & Wickins J.F. (1998) Ecological implications of developing coastal protection structures. In: *Coastlines, Structures and Breakwaters* (ed. W. Allsop), pp. 70–81. Thomas Telford, London.

Jensen A., Collins K. & Lockwood P. (2000a) Current issues relating to artificial reefs in European seas. In: *Artificial Reefs in European Seas* (eds A.C. Jensen, K.J. Collins & A.P.M. Lockwood), pp. 489–499. Kluwer Academic, Netherlands.

Jensen A.C., Collins K.J. & Lockwood A.P.M. (eds) (2000b) *Artificial Reefs in European Seas*. Kluwer Academic, Netherlands, 508 pp.

Jerry D.R., Stewart T., Purvis I.W. & Piper L.R. (2001) Evaluation of visual implant elastomer and alphanumeric internal tags as a method to identify juveniles of the freshwater crayfish, *Cherax destructor*. *Aquaculture*, **193** (1–2) 149–154.

Johnson S.K. & Bueno S.L.S. (2000) Health management. In: *Freshwater Prawn Culture: the farming of* Macrobrachium rosenbergii (eds M.B. New & W.C. Valenti), pp. 239–258. Blackwell Science, Oxford, UK.

Jones C. & Burke J. (1990) Water quality crucial to production. *Australian Fisheries*, **49** (11) 21–25.

Jones C.M., McPhee C.P. & Ruscoe I.M. (1998) *Breeding redclaw. Management and selection of broodstock*, 31 pp. Information Series QI 98106. Queensland Department of Primary Industries, Queensland, Australia.

Jørstad K.E. & Farestveit E. (1999) Population genetic structure of lobster (*Homarus gammarus*) in Norway, and implications for enhancement and sea-ranching operation. *Aquaculture*, **173** (1–4) 447–57.

Jory D.E. (1995a) Management of natural productivity in marine shrimp: semi-intensive ponds. *Aquaculture Magazine*, **21** (6) 90–100.

Jory D.E (1995b) Feed management practices for a healthy pond environment. In: *Swimming through troubled water. Proceedings of the special session on shrimp farming* (eds C.L. Browdy & J.S. Hopkins), pp. 118–143. Aquaculture '95. World Aquaculture Society, Baton Rouge, LA, USA.

Jory D.E. (1996) Management of commercial and farm-made feeds in marine shrimp ponds. *Aquaculture Magazine*, **22** (1) 86–97.

Kadlec R.H. & Knight R.L. (1996) *Treatment Wetlands*. CRC Press, Lewis Publishers, Boca Raton, Florida, USA.

Kanit C. (1996) Phytosanitation – utilisation of *Gracilaria* in reclaimation of shrimp pond effluents. Report on a regional study and workshop on the taxonomy, ecology and processing of economically important red seaweeds. Bangkok, Thailand FAO *NACA Environ. Aquacult. Dev. Ser.,* (3) 331.

Kautsky N., Rönnbäck P., Tedenaren M. & Troell M. (2000) Ecosystem perspectives on management of disease in shrimp pond farming. *Aquaculture*, **191** (1–3) 145–161.

Kearns J.P. (1998) Extrusion reviewed. *Aquafeed International*, (3) 33–37.

Keith I.R., Paterson W.D., Airorie D. & Boston L.D. (1992) Defence machanisms of the American lobster (*Homarus americanus*): vaccination provided protection against gaffkemia infections in laboratory and field trials. *Fish and Shellfish Immunology,* **2**, 109–119.

Kittaka J. (1997) Application of ecosystem culture method

for complete development of phyllosomas of spiny lobster. *Aquaculture*, **155** (1–4) 319–331.

Knösche R. (1994) An effective biofilter type for eel culture in recirculating systems. *Aquacultural Engineering*, **13** (1) 71–82.

Koksal G. (1988) *Astacus leptodactylus* in Europe. In: *Freshwater Crayfish: biology and exploitation* (eds D.M. Holdich and R.S. Lowery), pp. 365–400. Croom Helm, London.

Kuo J. C-M. (1988) Shrimp farming management aspects. In: *Shrimp '88, Conference proceedings*, 26–28 January 1988, Bangkok, Thailand, pp. 161–174. Infofish, Kuala Lumpur, Malaysia.

Lai L., Boeing P. & Jones K. (2000) Developing and using DNA markers to improve aquaculture species. In: *Abstracts, Aqua 2000, Responsible aquaculture in the new millennium* (compiled by R. Flos & L. Creswell), p. 362. European Aquaculture Society, Special Publication No. 28.

Laihonen P., Hänninen J., Chojnacki J.C. & Vuorinen I. (1997). Some prospects of nutrient removal with artificial reefs. In: *Proceedings of First Conference of European artificial reef research network*, 26–30 March 1996, Ancona, Italy (ed. A.J. Jensen), pp. 85–96. Southampton Oceanography Centre, UK.

Laing I. & Ayala F. (1990) Commercial mass culture techniques for producing microalgae. In: *Introduction to Applied Phycology* (ed. I. Akatsuka), pp. 447–477. SPB Academic Publishing, The Hague, Netherlands.

Laing I., Child A.R. & Janke A. (1990) Nutritional value of dried algae diets for larvae of Manila clam (*Tapes philippinarum*). *Journal of the Marine Biology Association UK*, **70**, 1–12.

Langdon C. (2000) Artificial microparticles for delivery of nutrients to marine suspension feeders. *Global Aquaculture Advocate*, **3** (1) 40–41.

Latif M.A., Brown J.H. & Wickins J.F. (1994) Effects of environmental alkalinity on calcium-stimulated dephosphorylating enzyme activity in the gills of postmoult and intermoult giant freshwater prawns *Macrobrachium rosenbergii* (de Man). *Comparative Biochemistry and Physiology*, **107A** (4) 597–601.

Laurent P.J., Nicolas J. & Paris L. (1997) Noble crayfish restockings, synthesis of the experiments in Lorraine and Morvan, lessons to learn. *Association des Astaciculteurs de France*, (51) 34–58.

Lavens P., Thongrod S. & Sorgeloos P. (2000) Larval prawn feeds and the dietary importance of *Artemia*. In: *Freshwater Prawn Culture: the farming of* Macrobrachium rosenbergii (eds M.B. New & W.C. Valenti), pp. 91–111. Blackwell Science, Oxford, UK.

Lawrence C. & Jones C. (2001) *Cherax*. In: *Biology of Freshwater Crayfish* (ed. D.M. Holdich), pp. 635–69. Blackwell Science, Oxford, UK.

Lee D.O'C., Hillion B. & Letessier L. (2000) An analysis of factors influencing pond yields in a semi-intensive shrimp farm. In: *Abstracts Aqua 2000, Responsible aquaculture in the new millennium* (compiled by R. Flos & L. Creswell), p. 379. European Aquaculture Society, Special Publication No. 28.

Lee P.G. & Meyers S.P. (1997) Chemoattraction and feeding stimulation. In: *Crustacean Nutrition* (eds L.R. D'Abramo, D.E. Conklin & D.M. Akiyama), pp. 292–352. Advances in World Aquaculture 6, World Aquaculture Society, Baton Rouge, LA, USA.

Lee P.G., Turk P.E. & Lawrence A.L. (1998) Design and construction of a commercial biosecure, closed, recirculating shrimp production system. In: *Proceedings of Second International Conference on Recirculating Aquaculture*, 16–19 July 1998, p. 387. Virginia Tech, VA, USA.

Lee T.H. & Yamazaki F. (1989) Cytological observations on fertilization in the Chinese fresh-water crab, *Eriocheir sinensis*, by artificial insemination (in vitro) and incubation. *Aquaculture*, **76** (3–4) 347–360.

Lester L.J. & Pante Ma. J. R. (1992) Penaeid temperature and salinity responses. In: *Marine Shrimp Culture: principles and practices* (eds A.W. Fast & L.J. Lester), pp. 515–34. Elsevier, Amsterdam, Netherlands.

Li F., Zhou L., Xiang J., Liu X. & Zhu J. (1999) Triploidy induction with heat shocks to *Penaeus chinensis* and their effects on gonad development. *Chinese Journal of Oceanology Limnology*, **17** (1) 57–61.

Liao I-C., Su H-M. & Lin J-H. (1993) Larval foods for penaeid prawns. In: *Handbook of Mariculture*, 2nd edn, *Vol. 1 Crustacean aquaculture* (ed. J.P. McVey), pp. 29–59. CRC Press, Boca Raton, FL, USA.

LICA (1988) *Application handbook for land improvement contractors*, 25 pp. Land Improvement Contractors of America.

Lightner D.V. (1996) *A handbook of shrimp pathology and diagnostic procedures for disease of cultured penaeid shrimp*. (*Pag. var.)* World Aquaculture Society, Baton Rouge, LA, USA.

Lightner D.V. (1999) The penaeid shrimp viruses TSV, IHHNV, WSSV, and YHV: current status in the Americas, available diagnostic methods, and management strategies. *Journal of Applied Aquaculture*, **9** (2) 27–52.

Lightner D.V. & Redman R.M (1998) Shrimp diseases and current diagnostic methods. *Aquaculture*, **164** (1–4) 201–220.

Lightner D.V., Redman R.M., Poulos B.T., Nunan L.M., Mari J.L. & Hasson K.W. (1997) Risk of spread of penaeid shrimp viruses in the Americas by the international movement of live and frozen shrimp. *Rev. sci. tech. Off. Int. Epiz.*, **16** (1) 146–160.

Lignot J.-H., Spanings-Pierrot C. & Charmantier G. (2000) Osmoregulatory capacity as a tool in monitoring the physiological condition and the effect of stress in crustaceans. *Aquaculture*, **191** (1–3) 209–245.

Lin C.K. (1995) Progression of intensive marine shrimp culture in Thailand. In: *Swimming through troubled water. Proceedings of the special session on shrimp farming* (eds C.L. Browdy & J.S. Hopkins), pp. 13–23. Aquaculture '95. World Aquaculture Society, Baton Rouge, LA, USA.

Lin M.N. & Hanyu I. (1990) Improvements on artificial insemination in gravid females of close thelycum *Penaeus penicillatus*. In: *Proceedings of Second Asian Fisheries Forum*, 17–22 April 1989 (eds R. Hirano & I. Hanyu), pp. 627–630. Tokyo, Japan.

Liu H. & Avault J.W. (1996) Effect of nitrite on growth of juvenile red swamp crawfish, *Procambarus clarkii*. *Journal of Shellfish Research*, **15** (3) 759–761.

Lorenzon S., De Guarrini S., Smith V.J. & Ferrero E.A. (1999)

Effects of LPS on circulating haemocytes in crustaceans *in vivo*. *Fish and Shellfish Immunology*, **9**, 31–50.

Lotz J.M. (1997) Disease control and pathogen status assurance in an SPF-based shrimp aquaculture industry, with particular reference to the United States. In: *Diseases in Asian Aquaculture III* (eds T.W. Flegal & I.H. MacRae), pp. 243–254. Asian Fisheries Society, Manila, Philippines.

Lotz J.M., Browdy C.L., Carr W.H., Frelier P.F. & Lightner D.V. (1995) USMSFP suggested procedures and guidelines for assuring the specific pathogen status of shrimp broodstock and seed. In: *Swimming through troubled water. Proceedings of the special session on shrimp farming* (eds C.L. Browdy & J.S. Hopkins), pp. 66–75. Aquaculture '95. World Aquaculture Society, Baton Rouge, LA, USA.

Loyless J.C. & Malone R.F. (1997) A sodium bicarbonate dosing methodology for pH management in freshwater-recirculating aquaculture systems. *Progressive Fish-Culturist*, **59**, 198–205.

Lutz C.G. (1999) Transgenic organisms in aquaculture: a little bit of this, little bit of that. *Aquaculture Magazine*, **25** (5) 83–85.

Lutz C.G. & Wolters W.R. (1989) Estimation of heritabilities for growth, body size, and processing traits in red swamp crawfish *Procambarus clarkii* (Girard). *Aquaculture*, **78** (1) 21–33.

Macintosh D.J. (1987) *An overview on the handling of aquaculture products in ASEAN (Thailand, Malaysia, Singapore, Indonesia and the Philippines)*, 365 pp. ASEAN Food Handling Bureau, Kuala Lumpur, Malaysia.

Madenjian C.P. (1990) Patterns of oxygen production and consumption in intensively managed marine shrimp ponds. *Aquaculture and Fisheries Management*, **21** (4) 407–417.

Malecha S.R. (1983) Commercial pond production of the freshwater prawn, *Macrobrachium rosenbergii*, in Hawaii. In: *Handbook of Mariculture, Vol. 1 Crustacean aquaculture* (ed. J.P. McVey), pp. 231–260. CRC Press, Boca Raton, FL, USA.

Mallasen M. & Valenti W.C. (1998) Comparison of artificial and natural, new and reused, brackish water for the larviculture of the freshwater prawn *Macrobrachium rosenbergii* in a recirculating system. *Journal of the World Aquaculture Society*, **29** (3) 245–350.

Malone R.F. (2000) Treatment of recirculating waters with floating bead bioclarifiers. *Global Aquaculture Advocate*, **3** (3) 65–66.

Malone R. F. & Beecher L. E. (2000) Use of floating bead filters to recondition recirculating waters in warmwater aquaculture production systems. *Aquacultural Engineering*, **22**, 57–73.

Malone R.F., Beecher L.E. & DeLosReyes A.A. Jr. (1998) Sizing and management of floating bead filters. In: *Proceedings of Second International Conference on Recirculating Aquaculture*, 16–19 July 1998, pp. 319–41. Virginia Tech, VA, USA.

Marchetti M., Tossani N., Marchetti S. & Bauce G. (1999) Leaching of crystalline and coated vitamins in pelleted and extruded feeds. *Aquaculture*, **171** (1–2) 83–91.

Massaut L., Sonnenholzner S. & Boyd C.E. (2000) Risks associated with chemicals and other agents used in attempts to control white spot syndrome. *Global Aquaculture Advocate*, **3** (3) 26–29.

Mattei E. (1995) The redclaw learning curve. *Aquaculture Magazine* **21** (6) 20–22.

McGraw W., Teichert-Coddington D.R., Rouse D.B. & Boyd C.E. (2001) Higher minimum dissolved oxygen concentrations increase penaeid shrimp yields in earthen ponds. *Aquaculture*, **199** (3–4) 311–21.

McIntosh R.P. (2000) Changing paradigms in shrimp farming: IV. Low protein feeds and feeding strategies. *Global Aquaculture Advocate*, **3** (2) 44–50.

McNeil R. (2000) Zero exchange, aerobic, heterotrophic systems: key considerations. *Global Aquaculture Advocate*, **3** (3) 72–76.

Meade M.E. & Watts S.A. (1995) Toxicity of ammonia, nitrite, and nitrate to juvenile Australian crayfish, *Cherax quadricarinatus*. *Journal of Shellfish Research*, **14** (2) 341–346.

Menasveta P., Worawattanamateekul W., Latscha T. & Clark J.S. (1993) Correction of black tiger prawn (*Penaeus monodon* Fabricius) coloration by astaxanthin. *Aquacultural Engineering*, **12**, 203–213.

Menasveta P., Panritdam T., Sihanonth P., Powtongsook S., Chuntapa B. & Lee P. (2000) Design and function of closed, recirculating seawater system with denitrification for culturing black tiger shrimp broodstock. In: *Abstracts, Aqua 2000, Responsible aquaculture in the new millennium* (compiled by R. Flos & L. Creswell), p. 467. European Aquaculture Society, Special Publication No. 28.

Méndez L., Acosta B., Palacios E. & Magallón F. (1997) Effect of stocking densities on trace metal concentration in three tissues of the brown shrimp *Penaeus californiensis*. *Aquaculture*, **156** (1–2) 21–34.

Mercaldo-Allen R. & Kuropat C.A. (1994) Review of American lobster (*Homarus americanus*) habitat requirements and responses to contaminant exposures, 52 pp. *NOAA Technical Memorandum*, NMFS-NE-105.

Mialhe E., Boulo V., Bachère E., *et al.* (1992) Development of new methodologies for diagnosis of infectious diseases in mollusc and shrimp aquaculture. *Aquaculture*, **107** (2–3) 155–164.

Misamore M. & Browdy C.L. (1997) Evaluating hybridization potential between *Penaeus setiferus* and *Penaeus vannamei* through natural mating, artificial insemination and in vitro fertilization. *Aquaculture*, **150** (1–2) 1–10.

Mishra A.K. & Singh M. (1999) Drug, vaccine and pesticide use in aquaculture. *Infofish International*, (5) 25–30.

Mitchell A.J. (1996) Treatment calculations for ponds, tanks and raceways, Part 1. *Aquaculture Magazine*, **22** (4) 90–94.

Moore S.S., Whan V., Davis G.P., Byrne K., Hetzel D.J.S. & Preston N. (1999) The development and application of genetic markers for the Kuruma prawn *Penaeus japonicus*. *Aquaculture*, **173** (1–4) 19–32.

Moriarty D.J.W. & Pullin R.S.V. (eds) (1990) Detritus and microbial ecology in aquaculture, 420 pp. *Proceedings of the conference on detrital food systems for aquaculture in 1985*. Bellagio, Como, Italy.

Moss S.M., Reynolds W.J. & Mahler L.E. (1998) Design and economic analysis of a prototype biosecure shrimp growout facility. In: *Proceedings of the US marine shrimp farming program biosecure workshop*, 14 February 1998 (ed. S.M. Moss), pp. 5–17. The Oceanic Institute, HI, USA.

Muir J.F. (1982) Recirculated water systems in aquaculture. In: *Recent Advances in Aquaculture* (eds J.F. Muir & R.J. Roberts), pp. 357–446. Croom Helm, London.

Muir J.F. (1988) Shrimp farming: engineering and equipment. In: *Shrimp '88, Conference proceedings*, 26–28 January 1988, Bangkok, Thailand, pp. 136–160. Infofish, Kuala Lumpur, Malaysia.

Muir J.F. & Lombardi J.V. (2000) Grow-out systems – site selection and pond construction. In: *Freshwater Prawn Culture: the farming of* Macrobrachium rosenbergii (eds M.B. New & W.C. Valenti), pp. 126–156. Blackwell Science, Oxford, UK.

Nagesh T.S., Jayabalan N., Mohan C.V., Annappaswamy T.S. & Anil T.M. (1999) Survival and histological alterations in juvenile tiger shrimp exposed to saponin. *Aquaculture International*, **7**, 159–167.

Nash G., Anderson I.G. & Shariff M. (1988) Pathological changes in the tiger prawn, *Penaeus monodon* Fabricius, associated with culture in brackishwater ponds developed from potentially acid sulphate soils. *Journal of Fish Diseases*, **11** (2) 113–123.

New M.B. (1999) Sustainable shrimp farming in the dry tropics: third generation shrimp farming technology. *World Aquaculture*, **30** (4) 34–43 & 62.

New M.B. & Singholka S. (1982) *Freshwater prawn farming. A manual for the culture of* Macrobrachium rosenbergii, 116 pp. FAO Fisheries Technical Paper 225. FAO, Rome, Italy.

New M.B., Tacon A.G.J. & Csavas I. (eds) (1993) Farm-made aquafeeds, 434 pp. *Proceedings of FAO/AADCP Regional expert consultation on farm-made aquafeeds*, 14–18 December 1992, FAO-RAPA/AADCP. Bangkok, Thailand.

Nonaka M., Fushimi H. & Yamakawa T. (2000) The spiny lobster fishery in Japan and restocking. In: *Spiny Lobsters: fisheries and culture*, 2nd edn (eds B.F. Phillips & J. Kittaka), pp. 221–242. Fishing News Books, Oxford, UK.

Norman C.P., Yamakawa H. & Yoshimura T. (1994) Habitat selection, growth rate and density of juvenile *Panulirus japonicus* (Von Siebold, 1824) (Decapoda, Palinuridae) at Banda, Chiba Prefecture, Japan. *Crustaceana*, **66** (3) 365–383.

NSW Agriculture (Undated pamphlet) *Guidelines for avoiding cruelty in shellfish preparation*, 6 pp. Animal Welfare Unit, New South Wales Agriculture, New South Wales, Australia.

Nunan L.M., Poulos B.T. & Lightner D.V. (1998) The detection of white spot syndrome virus (WSSV) and yellow head virus (YHV) in imported commodity shrimp. *Aquaculture*, **160** (1–2) 19–30.

Nunes A.J.P. & Parsons G.J. (1998) Dynamics of tropical coastal aquaculture systems and the consequences to waste production. *World Aquaculture*, **29** (2) 27–36.

Nunes A.J.P. & Parsons G.J. (1999) Feeding levels of the southern brown shrimp *Penaeus subtilis* in response to food dispersal. *Journal of the World Aquaculture Society*, **30** (3) 331–348.

Obaldo L.G., Dominy W.G., Terpstra J.H., Cody J.J. & Behnke K.C. (1998) The impact of ingredient particle size on shrimp feed. *Journal of Applied Aquaculture*, **8** (4) 55–66.

Obaldo L.G., Dominy W.G. & Ryu G.H. (1999) Growth, survival, and diet utilization of Pacific white shrimp, *Penaeus vannamei*, fed dry extruded diets with different levels of shear and moisture. *Journal of Applied Aquaculture*, **9** (1) 47–57.

Oesterling M.J. & Provenzano A.J. (1985) Other crustacean species. In: *Crustacean and Mollusk Aquaculture in the United States* (eds J.V. Huner & E. Evan Brown), pp. 203–234. AVI Inc., Westport, CT, USA.

Ogle J.T. & Lotz J.M. (1998) Preliminary design of a closed, biosecure shrimp growout system. In: *Proceedings of the US marine shrimp farming program biosecure workshop*, 14 February 1998 (ed. S.M. Moss), pp. 39–47. The Oceanic Institute, HI, USA.

Olivares E. & Yule A. (2000) Effect of nitrite and ammonia on growth of *Penaeus indicus*. In: *Abstracts, Aqua 2000, Responsible aquaculture in the new millennium* (compiled by R. Flos & L. Creswell), p. 521. European Aquaculture Society, Special Publication No. 28.

Ostrowski-Meissner H.T., LeaMaster B.R., Duerr E.O. & Walsh W.A. (1995) Sensitivity of the Pacific white shrimp, *Penaeus vannamei*, to aflatoxin B_1. *Aquaculture*, **131** (3–4) 155–164.

O'Sullivan D. (1988) Queensland cray farmers opt for local species. *Aquaculture Magazine*, **14** (5) 46–49.

Ovenden J.R. & Brasher D.J. (2000) Stock identity of the red (*Jasus edwardsii*) and green (*Jasus verreauxi*) rock lobsters inferred from mitochondrial DNA analysis. In: *Spiny Lobsters: fisheries and culture*, 2nd edn (eds B.F. Phillips & J. Kittaka), pp. 302–320. Fishing News Books, Oxford, UK.

Peterson E. (1998) The effect of aerators on the benthic shear stress in a pond. *Austasia Aquaculture Magazine*, **12** (1) 64 66.

Peterson E. & Pearson D. (2000) Round peg in a square hole, aeration in a square shrimp pond. *Global Aquaculture Advocate*, **3** (5) 44–46.

Phillips H. & Lacroix D. (2000) Marketing and preparation for consumption. In: *Freshwater Prawn Culture: the farming of* Macrobrachium rosenbergii (eds M.B. New & W.C. Valenti), pp. 345–368. Blackwell Science, Oxford, UK.

Preston N.P., Baule V.J., Leopold R., Henderling J., Atkinson P.W. & Whyard S. (2000) Delivery of DNA to early embryos of the kuruma prawn, *Penaeus japonicus*. *Aquaculture*, **181** (3–4) 225–234.

Primavera J.H. (1994) Environmental and socioeconomic effects of shrimp farming: the Philippine experience. *Infofish International*, (1) 44–49.

Pruder G.D., Duerr E.O., Walsh W.A., Lawrence A.L. & Bray W.A. (1992) The technical feasibility of pond liners for rearing Pacific white shrimp (*Penaeus vannamei*) in terms of survival, water exchange rate and effluent water quality. *Aquacultural Engineering*, **11**, 183–201.

Redón M.J., Ros R.M., Rielo J.A. & San Feliu J.M. (1997) First attempt of interspecific hybridization between the shrimps *Penaeus kerathurus* Forskäl, 1775 and *Penaeus japonicus* Bate, 1888. *Aquaculture Research*, **28**, 271–277.

Richards P.R. & Wickins J.F. (1979) Ministry of Agriculture, Fisheries and Food. Lobster culture research, 33 pp. *Lab. Leafl., (47)* MAFF Directorate Fisheries Research, Lowestoft, UK.

Richards-Rajadurai P.N. (1989) Live storage and distribution of fish and shellfish. *Infofish International*, (3) 23–28.

Ricque-Marie D., Abdo-de La Parra Ma. I., Cruz-Suarez L.E., Cuzon G., Cousin M., AQUACOP & Pike I.H. (1998) Raw material freshness, a quality criterion for fish meal fed to shrimp. *Aquaculture*, **165** (1–2) 95–109.

van Rijn J. (1996) The potential for integrated biological treatment systems in recirculating fish culture – a review. *Aquaculture*, **139** (3–4) 181–201.

van Rijn J. & Barak Y. (1998) Denitrification in recirculating aquaculture systems: from biochemistry to biofilters. In: *Proceedings of Second International Conference on Recirculating Aquaculture*, 16–19 July 1998, pp. 179–187. Virginia Tech, VA, USA.

Rivera-Monroy V.H., Torres L.A., Bahamon N., Newmark F. & Twilley R.R. (1999) The potential use of mangrove forest as nitrogen sinks of shrimp aquaculture pond effluents: the role of denitrification. *Journal of the World Aquaculture Society*, **30** (1) 12–25.

Robertson A.I. & Phillips M.J. (1995) Mangroves as filters of shrimp pond effluent: predictions and biogeochemical research needs. *Hydrobiologia*, **295** (1–3) 311–321.

Robertson L., Lawrence A.L. & Castille F.L. (1993) Effect of feeding frequency and feeding time on growth of *Penaeus vannamei* (Boone). *Aquaculture and Fisheries Management*, **24**, 1–6.

Rodriguez L.M. (1976) A simple method of tagging prawns. *Natural and Applied Science Bulletin, University of the Philippines*, **28**, 303–308.

Rogers G.L. & Fast A.W. (1988) Potential benefits of low energy water circulation in Hawaiian prawn ponds. *Aquacultural Engineering*, **7** (3) 155–165.

Rogers G.L. & Klemetson S.L. (1985) Ammonia removal in selected aquaculture water reuse biofilters. *Aquacultural Engineering*, **4** (2) 135–154.

Rosenberry R. (1998) *World Shrimp Farming 1998.* Rosenberry, Shrimp News International, San Diego, USA, **11**, 1–328.

Rosenthal H. (1981) Ozonation and sterilization. In: *Aquaculture in Heated Effluents and Recirculation Systems.* (ed. K. Tiews), **1**, 219–274. Heenemann Verlagsgesellschaft, Berlin.

Rosenthal H., Anjus R. & Kruner G. (1980) Daily variations of water quality parameters under intensive culture conditions in a recycling system. In: *Aquaculture in Heated Effluents and Recirculation Systems* (ed. K. Tiews), **1**, 113–120. Heenemann Verlagsgesellschaft, Berlin.

Samet M., Makamura K. & Nagayama T. (1996) Tolerance and respiration of the prawn (*Penaeus japonicus*) under cold air conditions. *Aquaculture*, **143** (2) 205–214.

Sammy N. (1988) Breeding biology of *Cherax quadricarinatus* in the Northern Territory. In: *Proceedings of First Australian Shellfish Aquaculture Conference* (eds L.H. Evans & D. O'Sullivan), pp. 79–88. Perth, 1988, Curtin University of Technology, Perth, Australia.

Samocha T.M. & Lawrence A.L. (1998) Preliminary design and operating specifications for a biosecure shrimp growout facility in Texas. In: *Proceedings of the US marine shrimp farming program biosecure workshop*, 14 February 1998 (ed. S.M. Moss), pp. 49–58. The Oceanic Institute, HI, USA.

Samuel M.J., Kannupandi T. & Soundarapandian P. (1998) Assessing the suitable position for electrode placement and electroejaculation on different size males of cultivable prawn *Macrobrachium malcolmsonii* (H. Milne Edwards). *Indian Journal of Experimental Biology*, **36** (1) 115–117.

Sandifer P.A. & Hopkins J.S. (1996) Conceptual design of a sustainable pond-based shrimp culture system. *Aquacultural Engineering*, **15** (1) 41–52.

Sarac H.Z. & Rose C. (1995) The chronic soft-shell syndrome in prawns. *Austasia Aquaculture Magazine*, **9** (2) 37–40.

Sarac H.Z. & Swindlehurst R. (1992) Prawn feed – The rules of storage. *Austasia Aquaculture Magazine*, **6** (3) 35–40.

Sbordoni V., De Matthaeis E., Sbordoni M.C., La Rosa G. & Mattoccia M. (1986) Bottleneck effects and the depression of genetic variability in hatchery stocks of *Penaeus japonicus* (Crustacea, Decapoda). *Aquaculture*, **57** (1–4) 239–251.

Seaman, W. (1997) Does the level of design influence success of an artificial reef? In: *Proceedings of First Conference of European artificial reef research network*, 26–30 March 1996, Ancona, Italy (ed. A.J. Jensen), pp. 359–376. Southampton Oceanography Centre, UK.

Sheeks R.B. (1989) Taiwan's aquaculture- at the crossroads. *Infofish International*, (6) 38–43.

Shivappa R.B. & Hambrey J.B. (1997) Tiger shrimp culture in freshwater? *Infofish International*, (4) 32–36.

Singh T. (1993) The use of pond liners in aquaculture. *Infofish International*, (5) 50–53.

Smith L.L., Fox J.M. & Treece G.D. (1993) Intensive algae culture techniques. In: *Handbook of Mariculture,* 2nd edn *Vol. 1 Crustacean aquaculture* (ed. J.P. McVey), pp. 3–13. CRC Press, Boca Raton, FL, USA.

Sonnenholzner S. & Boyd C.E. (2000) Managing the accumulation of organic matter deposited on the bottom of shrimp ponds… Do chemical and biological probiotics really work? *World Aquaculture*, **31** (3) 24–28.

Southgate P.C. & Lou D.C. (1995) Improving the n-3 HUFA composition of *Artemia* using microcapsules containing marine oils. *Aquaculture*, **134** (1–2) 91–9.

Spanier E. (1994) What are the characteristics of a good artificial reef for lobsters? *Crustaceana*, **67** (2–3) 173–186.

Spanier E. & Almog-Shtayer G. (1992) Shelter preferences in the Mediterranean slipper lobster: effects of physical properties. *Journal of Experimental Marine Biology and Ecology*, **164** (1) 103–116.

Stroethoff K.H. & Hovers A.P.H.M. (1996) Shrimp farming in sandy areas. *Infofish International*, (5) 24–29.

Sugita H. & Deguchi Y. (2000) Shipping. In: *Spiny Lobsters: fisheries and culture,* 2nd edn (eds B.F. Phillips & J. Kittaka), pp. 633–640. Fishing News Books, Oxford, UK.

Summerfelt S.T. & Hochheimer J.N. (1997) Review of ozone processes and applications as an oxidizing agent in aquaculture. *Progressive Fish-Culturist*, **59** (2) 94–105.

Summerfelt S.T., Adler P.R., Glenn D.M. & Kretschmann R.N. (1999) Aquaculture sludge removal and stabilization within created wetlands. *Aquacultural Engineering*, **19** (2) 81–92.

Sze C.P. (1998) Shrimp pond effluents and the environment: the Malaysian scenario. *Infofish International*, (6) 28–34.

Tacon A.G.J. (1990) *Standard methods for the nutrition and feeding of farmed fish and shrimp.* **1**, 1–117; **2**, 1–129; **3**, 1–208. Argent Laboratories Press, Redmond, USA.

Tal Y., van Rijn J. & Nussinovitch A. (1997) Improvement of structural and mechanical properties of denitrifying al-

questions. These will relate to where and in what form the products will be marketed; whether the product can be sold to a processor or directly to consumers or buyers at the farm gate; and whether any special arrangements or facilities will be necessary for handling or distribution. Key aspects of crustacean markets are discussed in Chapter 3.

If the culture technology appears to be appropriate and reliable (Chapters 7 and 8), the emphasis of the prefeasibility study can shift to finance and economics. Proposed sources and types of funding, and the availability of grants and aid need to be investigated, bearing in mind that at this early stage only rough costings for a project will be available. Some idea of likely investment and operating costs for a range of different crustacean farming projects is provided in Chapter 10. Financial feasibility analysis should not be introduced as an afterthought for justifying the financial feasibility of the proposed enterprise (Rhodes 2000). If specific rates of return on capital are expected it should be determined whether the project has a reasonable chance of achieving them. It may be necessary to eliminate some high-risk or low-profit schemes at the prefeasibility stage unless these characteristics are compatible with the stated objectives.

Perhaps the most important pre-investment decision is the selection of a suitable site. At the prefeasibility stage it should be established whether the area being considered for the project is appropriate with regard to climate, availability of broodstock or seedstock, and infrastructure. Although the final site selection, including surveying and soil and water analysis, will usually be performed later as an initial part of the feasibility study, sites can be assessed at this stage to see if they conform to basic requirements of topography, soil characteristics and water supply. Some of the most important topics to be considered during site selection are described in Chapter 6. Additional sources of information that will assist prefeasibility studies are considered later in section 9.3.5.

If the prefeasibility study concludes that the project shows good potential to achieve the desired yields and objectives, the decision can be taken to proceed with the more thorough and costly full feasibility study. In such cases the prefeasibility study should be used to highlight the most important factors influencing the project's viability, and indicate the areas (e.g. marketing, water supplies) requiring more detailed investigation in the full feasibility analysis. The identification of potential problems and weaknesses in underlying assumptions concerning productivity is a primary objective and, if achieved satisfactorily, will greatly enhance the effectiveness and value of the full feasibility study. If on the other hand the outcome of the prefeasibility study is negative and the proposed project shows little promise, the findings may facilitate the redefinition of more realistic objectives, the selection of different culture options, or indicate that interest in crustacean farming should be abandoned altogether (or at least held back until more favourable market or economic conditions prevail).

9.3.2 Feasibility study

In a full feasibility study a detailed analysis and assessment of a project is carried out to enable the levels of risks and rewards to be more fully quantified. The study will contain full technical details of processes involved in the project, and all assumptions underlying anticipated markets, yield predictions and cost estimates. To achieve this, all available information on the site, water, soil and infrastructure must be collected and assessed. Any gaps in the data must be exposed to establish if they are likely to undermine confidence in the technical assumptions. In a comprehensive study, it may be possible to include comparisons with the performance of similar, viable projects elsewhere. In addition, it is usually at this stage that possible social and environmental implications are considered in depth (Chapter 11). The question of how to go about implementing the construction phase of the project is then addressed and choices made regarding the possible purchase of turnkey packages; if and how consultants, technical assistance and government assistance will be used; how risk will be managed; and what safeguards will be sought (sections 9.3.4 and 9.3.6).

Part of the feasibility study will consider how to finance the total operation and will include the proposed arrangements for loans and equity, taking into account the overall investment costs, the period for start-up, and the expected running costs. The study should convince not only those who prepare it but also the project's backers (bankers, landowners and government officials) of the overall viability of the operation. To this end an investment appraisal is essential, preferably supported by an analysis of the risks involved (sections 10.3.3, 10.4 and 10.4.1). If a feasibility study is prepared well it should be easy to identify *bona fide* objectives. However, it should be remembered that, even with an accurate and thorough feasibility analysis, success can never be guaranteed. It is difficult to be certain about all the elements required to make a project function as planned.

For example, weather and longer-term climatic change can be hard to predict and water quality can deteriorate, especially if aquaculture activities or other developments are expanding in the chosen area.

9.3.3 Managerial control

If the process of implementation is to proceed smoothly, it is essential to establish clear overall managerial responsibility for the project by named individuals. The key figure among these will be the project manager, perhaps advised by project financial controllers, but other members of the project team will also need to be identified and given clear, delegated responsibilities to carry out specified tasks on time. The project manager takes the central co-ordinating role and ensures that the different elements of the project are implemented without any disruptions. In effect, the primary function of the project manager is to manage risk.

9.3.4 Project risk management

Risk management is the process of identifying and minimising the many problems, or risks, that can arise and undermine a project and prevent it from fulfilling expectations. Essentially it is the same activity as project management because both processes involve taking steps to avoid the project being late, over budget or substandard in terms of any or all quality aspects. Risk management should begin in the validation phase because it has implications for the project design. Later on in the course of project implementation, an emphasis on detailed planning is a central requirement of effective risk management.

To start the process, risks need to be identified and evaluated. This requires skill and knowledge of similar projects in the past, and is facilitated by an understanding of project management techniques (Verzuh 1999). Input from key people who will physically implement the project is also important. Risks can initially be identified on the broad basis that 'what can go wrong, will go wrong'. Subsequent evaluation will involve asking – what is the probability of each risk happening and what is the likely impact?

Once the risks have been identified and evaluated, action plans can be formulated to contain them. In this regard it is instructive to consider the kinds of problems that have caused projects to fail in the past and to look at some general strategies that would have improved their chances of success. These include:

- Forming the project team from people who already have experience of similar projects.
- Ensuring that the design is suited to actual site conditions and is not purely a product of an isolated design office.
- Improving reliability by keeping systems as simple as possible and by avoiding unnecessary equipment and procedures – if a system contains many critical components the chance of system failure increases exponentially with the number of components.
- Being aware of the impact of legislation on design – e.g. requirements for effluent treatment or for avoiding the escape of exotic and genetically altered species.
- Avoiding late changes in project specifications.
- Seeking early approval from regulating authorities since they are traditionally notorious for making late decisions.
- Anticipating delays in the arrival of materials and equipment, particularly if imported goods, customs clearance, and remote sites are involved.
- Preparing contingency plans for unusually bad weather.
- Considering the potential impact of macro-economic factors such as exchange rates and inflation.

As project implementation proceeds, risks will need to be monitored and contingency plans prepared. For example, if there is a significant possibility that a particular contractor will fall behind schedule, it may be necessary to closely compare progress with the work schedule to provide an early warning of any delays. Drawing up contracts that keep contractors on time and within budget is an important means of containing or transferring risks and is discussed in section 9.4.2. A very common contingency plan is to retain some spare money outside the main budget, but this should be viewed as a last resort. To reduce risk it may be necessary to eliminate a particularly uncertain part of the project entirely. But if it turns out that the overall risk is unacceptably high, i.e. there is some likelihood of a major impact on project schedule and costs, and there are no opportunities for mitigation, then the project's viability may come into question.

9.3.5 Sources of information

A range of information sources can be effectively exploited to expand the basic material provided in this book. Market information at the beginning of the validation phase will usually come from second-hand sources rather than from performing actual market studies. Information can be sought in trade journals, from the Internet and also from existing producers, processors, retailers and restaurants. From these sources a basic idea of sales potential and the best product forms can be established. A market study, involving a market survey or the test marketing of sample products, may be required as part of the full feasibility analysis, in which case, specialised help may be needed from a company or consultant with marketing expertise.

Information about culture techniques can be obtained from scientific and technical literature including Chapters 5, 7 and 8 of this book and the references therein. Consulting industry sources, university departments, government departments and extension services in the chosen area can be a valuable aid to assessing the 'state of the art' and to establishing whether any special technical adaptations may be necessary for local conditions.

Preliminary information on possible sites can be obtained from maps, preferably ones that indicate relief and vegetation type as well as access points (roads, tracks). Detailed maps can be most useful once suitable general areas have been identified; however, in some countries they may either not exist or may only be available to military personnel. Marine charts sometimes provide the most accurate information on the form of coastal margins and indicate the extent of areas such as salt flats, mangroves or swamps. Oceanographic institutes, national government resource and fishery departments or universities can be good sources of information on prevailing sea currents and temperatures. Climatic information, particularly rainfall and temperature data, may be more easily found through government or university departments concerned with agriculture rather than aquaculture, and these sources can also be useful for information on dominant soil types. Mineral resource surveys can provide valuable information on soil types and may be available from commercial resource surveying reports, government agencies or departments concerned with mining, and as a result of gas and oil exploration and geographical surveys. Land registry details are usually available through local government and district offices.

Light aircraft can be chartered to obtain an overview of large areas of land or coastline and aerial photographs can be taken and analysed. Some satellite images are ideal for identifying zones with different types of vegetation and land usage and can provide up-to-date information on the extent and location of existing aquaculture operations such as coastal shrimp farms (section 6.3).

Economic and financial information can be obtained through foreign trade missions and government departments concerned with industrial, agricultural, fisheries and aquaculture development. Trade associations may also be of some assistance along with aid agencies and various types of commercial, co-operative and development banks.

In the past decade the Internet has become a particularly valuable source of information. Notable sites dealing with aquaculture and fisheries in general include the Aquaculture Network Information Centre (http://ag.ansc.purdue.edu/aquanic/) and World Wide Web Sources for Aquaculture, Fisheries, Aquaria and Fish Diseases (http://www.stir.ac.uk/Departments/NaturalSciences/Aquaculture/fishing/fish/f_web.htm) and they both have dozens of links to more specialised sites that deal with particular crustacean species or the activities of particular aquaculture companies or government funded organisations. Just to give a taste of the wide variety of sources: information on *Macrobrachium* farming in the USA is available at http://ext.msstate.edu/pubs/pub2003.htm; Australian crayfish farming at http://www.wa.gov.au/westfish/aqua/broc/aqwa/marron/index.html; and an e-mail discussion forum for people active in or interested in the shrimp industry is operated by egroups.com (http://www.groups.yahoo.com/group/shrimp).

9.3.6 Consulting services

9.3.6.1 Types of organisations

Consulting services may be procured from individuals or from groups of consultants working in many different types of organisations. Individual consultants may offer particular specialist help, for example with marketing, economics, biology or engineering, or alternatively they may provide more general expertise covering a range of such disciplines. Aquaculture consulting groups often consist of in-house or subcontracted specialists and may be organised as partnerships or private companies. Some groups, however, are government enterprises or receive government backing and are likely to promote equipment and technologies only from organisations within their own countries. Yet other groups are attached to universities. The manufacturers and distributors of aquaculture feeds, equipment or other supplies often also provide consulting services in support of their main business. Some shrimp processing companies provide free technical advice to farmers, generally on the condition that all harvested product is sent to their operation.

Assistance with the technical aspects of crustacean farming may be provided as an integral part of a turnkey package (section 9.3.8).

9.3.6.2 How and where to locate consultants

Many consultants and consulting companies advertise in aquaculture trade publications; others can be contacted through universities, government research stations, trade associations, professional institutions or through people already engaged in crustacean farming. Recommendations can also be obtained from aquaculture supply companies but if they offer their own consulting services their advice may tend to be centred on the use of their own products. The advantage of such companies is that they are often in routine contact with many operators and are in a good position to compare results at different operations and recommend consultants accordingly.

9.3.6.3 How to use consultants

Consultants can be of value for their specialised skills and general advice right from prefeasibility analysis though to construction and operation. Their importance generally increases with the scale of the project. In the validation phase they can be employed to put together complete prefeasibility and feasibility analyses or brought in to tackle particular tasks such as site surveys, market surveys or environmental impact assessments. For the last of these, they should be aware of current guidelines for responsible aquaculture development (e.g. FAO 1997) and asked to pay special attention to likely changes in regulations governing effluent volume, effluent treatment requirements and the escape of exotic and genetically selected or altered crustaceans.

When it comes to the construction phase and the start-up period of operations, it is generally beneficial for supervision of the work to be undertaken by the same people who performed the feasibility study. In this way the benefits of continuity are retained. However, some tasks may require independent and specialised help, such as pond construction and site management, staff recruitment and training, accounting and financial planning.

Technical advice can be essential during the start-up period when inexperienced staff require instruction and training. Advisors may need to be retained until smooth running is established with all routines defined and documented, including technical operations, sanitation, quality control, data collection, monitoring and reporting formats. Assistance may be needed on a long-term basis, either continuously or at regular intervals although, by emphasising the importance of the training role and by encouraging the delegation of responsibilities to other technical employees, reliance on individual outside experts can gradually be reduced.

During the operational phase consultants may be brought in as 'trouble-shooters' to identify and resolve production problems, particularly with regard to persistent disease problems. However, taking the advice of consultants who offer 'quick-fix' solutions may not be advisable because these solutions tend to be very short term in their effectiveness (Muir 1988). Valuable advice about production problems can be obtained from aquacultural suppliers, especially if they have encountered similar problems in other operations. Nevertheless, it should be ensured that they do not simply take advantage of the opportunity to promote their products, such as therapeutants, in ways that treat the symptoms of problems rather than the causes. Some of the larger processing companies may offer practical assistance and advice to client farms because of mutual interest in maintaining productivity levels.

At any stage of a project's development, the use of a second technical opinion to provide a detached viewpoint can be very valuable, especially in critical areas of production. During the operational phase consultants can assist technical staff in identifying problems and in communicating with the project owner. In some situations it may even be worthwhile to consider a confrontational approach: Mock (1987) recommended that 'you (a project owner) bring someone in once in a while unannounced that can review and challenge your so-called experts'.

9.3.6.4 Choosing a consultant

Larger aquaculture projects often require substantial and complex engineering design work and it may be advantageous to employ an engineering design or consulting engineer group to prepare proper design specifications, engineering drawings and tender documents. Since the interpretation from biology to engineering must be done with great care, the engineering group must have aquaculture experience; otherwise a bioengineer should be hired to shape the design to the needs of the species to be cultured.

Small aquaculture consulting companies may be competent with regard to biology but may not possess the skills needed for all aspects of planning and implementation. Some recruit additional experts as and when they need them for a specific project, in effect acting like recruiting agencies and providing back-up to the experts and a link with the clients. In all cases it is advisable to verify the suitability of the experts that are being hired.

Consulting roles can be broadly divided into specialist and general. The value of a general aquaculture consultant may be greatest in the early phases of planning when project options must be chosen to suit desired objectives, and a prefeasibility study is required to highlight the areas where more specialised inputs may be needed for full feasibility analysis and implementation work. The general consultant may also play a key role in compiling the feasibility study and co-ordinating and supervising the work of the more specialised consultants. Specialist consultants are most valuable in the later stages of planning after the needs for specific expertise have been clearly defined and when the implementation phase has advanced towards construction. When specialists are given precise roles it is usually a straightforward matter to maintain control, monitor progress and obtain value for money.

During the operational phase, both specialists and general consultants can be useful. If a production problem can be identified as relating to a particular subject, such as nutrition, disease or water quality, specialist help may be appropriate. However, production problems are often caused by a series of interrelated factors and an aquaculture consultant with broad experience may be able to identify the most likely causes, suggest general improvements in areas such as husbandry, and recommend specialist help only if it is needed. Alternatively, if a production system is in general disarray and a rapid resolution of problems is required, it may be advisable to hire a team of experts each with different specialisations. This may be costly initially but can result in considerable time savings.

9.3.6.5 How to get the best from consulting services

Since the project backers pay for the services of consultants in order to obtain guidance in areas outside their direct experience, it is of fundamental importance that they be convinced of the consultants' expertise and be fully prepared to take the advice in the end. There is no sense in paying for advice simply to ignore it. If at any stage confidence in consultants is lost, then they should be replaced rather than allowed to carry on and reach unreliable conclusions. The danger of expensive mistakes is at its greatest when over-optimistic or overambitious project developers combine with incompetent or inexperienced advisors. A good consultant will always be prepared to recommend that a project be stopped or redirected if it looks unfavourable.

To be sure of the competence of consultants, it is necessary to investigate and discuss with them their past activities. People with whom they have worked previously should also be questioned to establish whether the consultant or consulting group do in fact perform the services that they claim, or whether their skills in the relevant field are only limited. Experience should cover the chosen species and culture system and also, ideally, its application in the relevant location or at least the same type of location. Entrepreneurs and investors should be aware that, as in any industry, there are some individuals and companies that may make false claims and others that lack sufficient experience to be credible.

Muir (1988) recommends studying the attitude of the consultants and considering whether they are too slick and fast in their approach; i.e. very efficient at promoting and selling their services but possibly less proficient at performing them. Muir (1988) also considers that consultants should be prepared to admit that they have made mistakes in the past because this is an essential aspect of learning. Their methods, style of work and personal approach must be acceptable to the employer if efficient communications and working relationships are to be established and maintained, particularly over long-term contracts.

An initial statement of the work that is required from the consultant, the terms of reference (TOR), should be drawn up, including a timescale and if possible a budget framework for the project. This can be especially critical for validation studies where the scope and form of the assignment are open to different interpretations. The TOR should specify the information that the consultant is required to produce, along with relevant outline formats for reports, plans, design drawings, budgets and recommendations. Details of training assignments may need to be included, along with requests for background documentation in support of reports.

Ideally more than one consultant or consulting group should be approached so that price comparisons can then be made, making sure to take into account any hidden extras above the specified daily or monthly rates. However, experience and the quality of the services offered should be the overriding factors used in the selection of consultants, rather than the cost.

Although most internationally active consultants are from developed countries, there are many situations where suitably qualified people in developing countries can provide the best advice and good value for money. Such experts can take advantage of cultural affinities and may be particularly familiar with local climatic or physical conditions, design practices, and legal and bureaucratic complexities.

Contracts can be drawn up on several bases. Time-based contracts, relating to a specified number of man-months, are commonly used and are favoured, for example, by the World Bank for general planning and feasibility studies, for design, detailed engineering and construction, and for technical assistance assignments (World Bank 1981). The time-based rate includes salary, social costs, firm's overheads, fee or profit and usually an overseas allowance. Other types of contracts include lump sum (fixed fee), percentage (based on a percentage of construction costs), and cost-plus-fixed-fee. It may be necessary to include a contingency allowance in the consultant's contract to allow for the cost of any unforeseen work.

Production targets and incentives can be used to encourage commitment from technical advisors as well as production workers. Incentives should not, however, be allowed to promote imbalance in the production schedule by encouraging 'storming' – i.e. reaching a production target by the end of a set time without regard to the quality of the work, its impact on the infrastructure or environment, or the quality of the final product. If an incentive scheme extends to the whole workforce, bonuses should be paid in addition to set salaries, and salary levels should not be restricted because of the earning potential of the bonus scheme. Performance-related clauses can also be included in the contracts of technical advisors to allow for termination in the event of continually bad results. However, the root causes of problems are not always the fault of the technical assistants and initially, during start-up operations, some legitimate delays in reaching production targets may need to be accommodated.

9.3.6.6 Indemnity insurance

It is worth establishing whether, and up to what level, a consultant or consulting company is insured against problems that arise from their negligence and incorrect advice or information. If they have indemnity insurance from a reputable source it usually means that they have an established track record and are recognised as an authority in their field. Indemnity insurance is more commonly possessed by consulting companies than by individuals.

9.3.7 Contract growing (nucleus/plasma) schemes

In contract growing (nucleus/plasma) schemes, a nucleus operation is responsible for providing individual contract farmers (the plasma) with technical assistance. Farmers also usually receive pond construction services, feed, seed and preliminary training, and in return are required to provide or buy the land for their ponds, manage the ongrowing phase and then sell their product back to the central processing plant. In one such scheme for shrimp, Aquastar in Thailand, the nucleus operated a hatchery, nursery, feed mill and processing plant. However, this project and another major one in Indonesia failed for reasons discussed elsewhere in this book (sections 10.4 and 11.2.3) and the potential of nucleus/plasma schemes has yet to be realised.

In theory a nucleus/plasma scheme can provide a more even distribution of wealth than the establishment of a single farm on the same scale, because it involves local people in the role of small-scale farmers rather than simply as farm labourers. The technical back-up from the nucleus is usually of a far higher quality than that normally available to small-scale operators, and the improved dissemination of good advice can be beneficial to the industry in general.

Success with small-scale farms relies greatly on the diligence, motivation and innovation of the farmers and a nucleus/plasma scheme must encourage the development and expression of these same positive attributes. Good results are likely to rely on very skilful management of the nucleus operation. The system must be flexible enough to meet the requirements of a large number of individual farmers and maintain an atmosphere of mutual co-operation with equal priority given to each grower, even when complications arise as a result of such problems as a shortage of seed or water. Contract growers, for their part, must learn to co-ordinate stocking and harvesting operations with the nucleus operation. One potential difficulty arises because of the need for protection against incompetent contract growers. Some form of dismissal may be a reasonable sanction, but it would not be proper to expel a farmer from his land if he had previously owned it, or to deprive his pond of water from the company supply system.

9.3.8 Turnkey projects

When a project developer acquires a turnkey package he usually receives comprehensive design and management services and in return provides the funding and sometimes the land and water. Many consulting companies, particularly those whose *forte* is engineering, offer turnkey options. The package may be arranged to cover the initial prefeasibility analysis right through the planning stages to the provision of technical assistance during the operational phase. Sometimes, however, the decision to use a turnkey package is taken after independent prefeasibility or feasibility studies have concluded that this is the best approach.

The turnkey approach is only advantageous if very high-quality services are being offered. Great trust will be placed in the consulting company and the owner will relinquish much of the control over the project. The results of buying a bad or inappropriate package may not become apparent until the project is installed and then fails to perform to specifications, by which time it may be extremely expensive or impossible to take corrective measures. To establish the quality of the services that are offered and to judge the likelihood of success, the track record of the consulting company should be investigated. Ideally, more than one company should be approached and the final choice based on results obtained previously with similar projects.

The advantage of employing a single company in the turnkey approach is that continuity is provided to assist in the precise translation of an idea into a viable operation. Responsibility for good site selection, functional design and operational success rests with a single group that has unique control. The approach can be especially useful for particular specialised projects or newly emerging technologies where expertise is not widely available. However, as crustacean farming technology becomes more widespread, the turnkey approach may become increasingly less attractive because it is often far less expensive to hire the services of individuals with wide experience in the industry.

The use of a performance-related contract can help obtain full commitment but disputes can arise over the payment of royalties on production. In one Ecuadorian hatchery project the owners refused to pay agreed royalties to the company which provided the technology, maintaining that it was only the development of their own management and technical inputs which had secured a viable production level. Because of such problems, it is often in the interest of the technology provider to receive the bulk of remuneration on completion of the project rather than as royalties on production. Once an operation is finished and running successfully the owners have effectively acquired the product they paid for and the technology provider is in a weak bargaining position since technical developments and improvements are a feature of present-day crustacean farming. Contracts may need to be carefully formulated to prevent disputes and ensure mutual interest in the success of a project. If contractual disputes do arise they can be very expensive, time-consuming and, in some countries, impractical to resolve.

Some large aquaculture engineering companies (with access to financial resources) acquire equity in the projects they are setting up, either by injecting cash or exchanging technical inputs for a percentage share of the project ownership. This joint venture approach encourages commitment from both partners.

9.3.9 Government assistance

Governments often offer technical and financial assistance for crustacean farming in relation to national interests. Support goes to training, research and extension services associated with universities, agricultural colleges and research institutes. Demonstration hatcheries and production units may be established commercially with national government grants in order to extend new technologies to traditional farmers and to train new technicians and farm managers.

Governments may also back promotional campaigns for crustacean products on home and export markets, and encourage auxiliary industries such as those producing seed and feed. In Hawaii during the mid-1980s some prawn farmers relied on a supply of juveniles from a government-sponsored *Macrobrachium* hatchery and would otherwise probably not have remained in business. When government involvement is sought in the development of a new project, the process of obtaining land concessions, leases and other permits can be greatly simplified.

Extension services, research stations and universities can be very useful for providing scientific and technical information on crustacean farming. Experienced extension workers, available to make routine and emergency visits, can provide a most valuable service, particularly for small-scale farmers with limited access to other types of assistance (section 11.5.2.1).

The Malaysian government promoted its shrimp culture industry by providing advice and technical assistance through the Department of Fisheries (USDC 1989). It also offered tax rebates and tax relief to this and other developing industries. In Malaysia and more recently in Mexico and Cuba, state-backed, economic development corporations have become active in shrimp farming in conjunction with foreign (joint venture) partners. As part of the Shrimp Farming Industry Development Plan in Taiwan, low interest loans and infrastructure grants were made available and water resources re-allocated (Lee 1988). However, these measures, together with a paddy conversion plan and subsidies on electricity, helped to boost shrimp farming to levels that resulted in severe environmental damage (section 11.4.2). Government-owned or sponsored crayfish hatcheries exist in many northern European countries to provide juveniles for national restocking programmes. Also Japanese aquaculture has long enjoyed government support for both land-based and stock enhancement operations deemed to be in the national interest.

9.4 Detailed planning phase

Once the potential viability of a project has been validated, planning can proceed in much greater detail. For this to be done efficiently, the technical specifications of the project must be fixed as soon as possible because the cost of changing the scope or design of a project escalates rapidly through the final phases of implementation. Late design changes, particularly once construction is under way, may mean that construction contracts need to be renegotiated and this usually leads to serious delays and cost overruns.

9.4.1 Time planning and control

The project manager and his team are responsible for planning and controlling the time at which key components of the project are carried out. To do this requires identifying what the key components are, what they involve, how long they are likely to take and what are their earliest start dates, taking into account their interdependence with the completion of other components of the project. A work breakdown structure (WBS) divides the project into its component tasks (or work packages) and then a project planning schedule (Gantt chart) is created to relate the components of the WBS to a time schedule. An example of a Gantt chart for the simplified construction phase of a crustacean hatchery is given in Fig. 9.3 with the component tasks of the WBS listed down the left-hand side. The use of the techniques of critical path analysis, network analysis and the programme evaluation and review technique can then be used to optimise the time scheduling of different tasks within the project (Verzuh 1999). Once the project planning schedule has

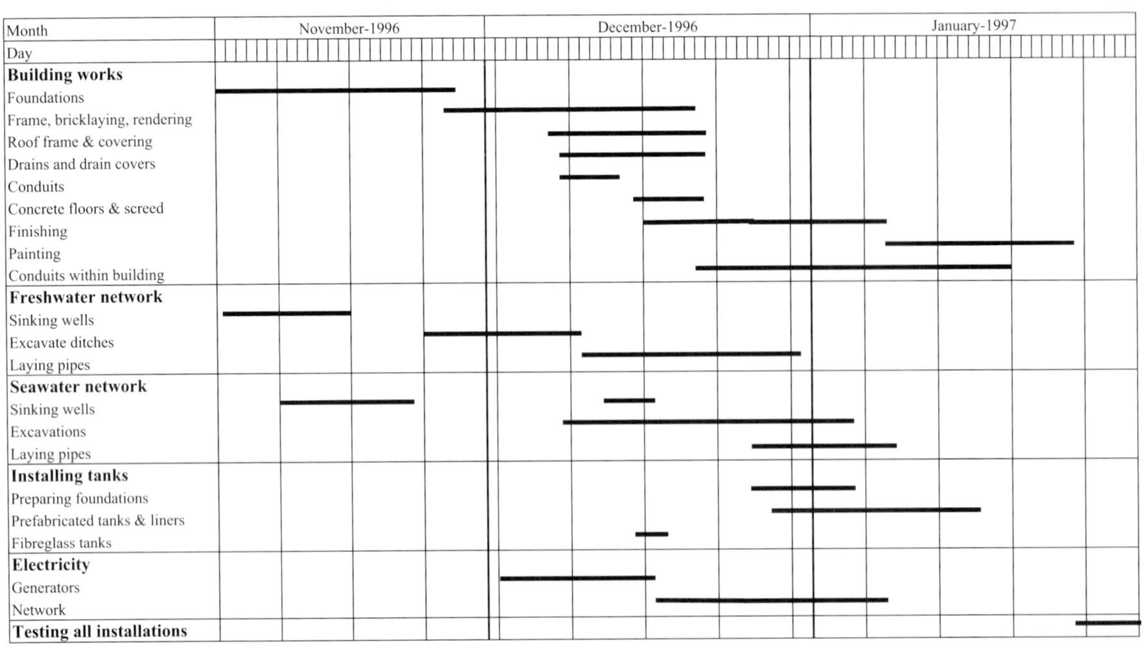

Fig. 9.3 Example of a work planning schedule (Gantt chart) for the construction of a crustacean hatchery project.

been optimised, inputs of labour, specialist services, materials and equipment can then be planned accordingly.

9.4.2 Cost control

Effective cost control is a dynamic process in which the project manager must adopt a forward thinking approach and anticipate any events that may cause delays or involve extra expense. Careful monitoring of costs is important in this process, but eventual success will also rely very heavily on good communication between the project manager, his team and the contractors who actually incur the costs, so that timely feedback is received on actual progress with individual tasks. Only by this means can advance warning of possible problems be gained so that the project manager can react in time to mitigate negative impacts.

Cost control begins well before full-scale investment. An initial detailed costing of a project relies on establishing the cost of the tasks in the work breakdown structure described above (section 9.4.1 and Fig. 9.3). Cost planning then involves identifying key stages in the detailed WBS and adding the costs of all relevant tasks to obtain a key stage cost. A plot of cumulative key stage costs over time will then represent a series of reference points for cost control and also provide a schedule for the input of investment funds. Figure 9.4, explained below, incorporates this plot of key stage costs over time, which is also referred to as the budgeted cost of work scheduled (BCWS).

Cost monitoring techniques can evaluate the extent to which an actual project is advancing at the planned rate or is costing the amount predicted in the budget. To do this involves calculating the schedule variance, the slippage, and the cost variance. The schedule variance involves comparing the budgeted and actual costs over time, and is calculated as

Schedule variance = BCWP − BCWS

where BCWP is the budgeted cost of work performed.

Figure 9.4 shows a plot of the BCWS and BCWP over time for a project in which progress has fallen behind schedule, a process called slippage. In this illustration, after 10 weeks the project has a slippage of 3.5 weeks and the schedule variance is –$28 000. If the schedule variance were positive it would mean that the project was ahead of schedule. Although schedule variance is expressed in money units it is, like slippage, really a measure of progress rather than cost. In the example given, after 10 weeks the budgeted cost of the work performed is $86 000 whereas it was anticipated that $114 000 of work would have been completed at this stage. It is a simple matter then to compare these figures and calculate that project progress after 10 weeks is only 75% of what it should be.

Schedule variance is calculated on the basis of budgeted costs alone but the calculation of cost variance requires actual expenditure data. It is calculated as

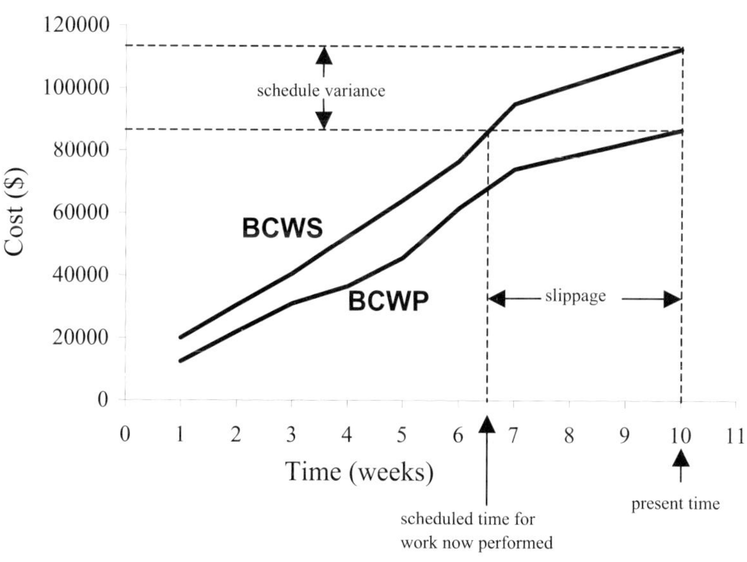

Fig. 9.4 Example plot of budgeted costs for a project that is behind schedule, to illustrate the concepts of slippage and schedule variance. (Refer to text for explanation.)

Cost variance = BCWP – ACWP

where ACWP = actual cost of the work performed.

Negative cost variance implies the project is costing more than anticipated. The project illustrated in Fig. 9.5 shows actual costs exceeding budgeted costs and hence the cost variance in negative. In this case the work performed has cost $120 000 when it was scheduled to cost only $86 000. The project is thus 40% over budget after 10 weeks. By calculating schedule and cost variance at regular intervals during the implementation phase, the project manager can see if his efforts are succeeding in keeping the project on target.

One important means of cost control is the use of contracts to try to transfer the risk of cost overruns and delays to contractors. In theory the simplest approach is to draw up fixed cost contracts but in reality this requires a highly detailed document in which there is no room for unexpected work. If something new crops up during implementation that is not specified in the contract then the project may be exposed to additional costs and the risks are not contained. If a fixed cost contract is carefully drawn up so that the contractor is liable for unforeseen costs, the risk is effectively transferred but the contractor usually raises his price to reflect this. As a result the contractor can generate excessive profits at the expense of the project. There are alternatives to fixed price contracts. Cost-plus contracts involve the client reimbursing the actual costs of the work and then adding an additional percentage to provide a return for the contractor. However, under this scheme the contractor has a big incentive to escalate or exaggerate inputs and costs and it can result in unnecessary 'gold-plating' of a project. In a cost-plus fixed fee contract there is no incentive to maximise costs but there is also no incentive to minimise them either. This problem is partly addressed in target-cost contracts in which there are penalties for time and cost overruns to encourage a contractor to hurry up. In most cases it is useful to keep a retention of perhaps 10% until contractors finish to the satisfaction of the client. Half of this amount can be paid on completion of the work and then the remainder after 6 months of satisfactory operation. One general problem, particularly with large construction projects, is that contractors often give low bids to secure the work and then are at risk of bankruptcy. Some prior idea of likely project costs (section 10.6) is useful to help identify realistic bids.

9.5 Implementation phase

The final implementation phase of a project usually begins with the production of the final design drawings and construction documents and proceeds through construction and start-up to full operation.

9.5.1 Acquisition of site and construction

Before a site is acquired, and as a final part of the site selection process, detailed surveying is usually necessary in order to prepare a topographic map on a scale of 1 : 1000 or 1 : 2000. In addition to marking eleva-

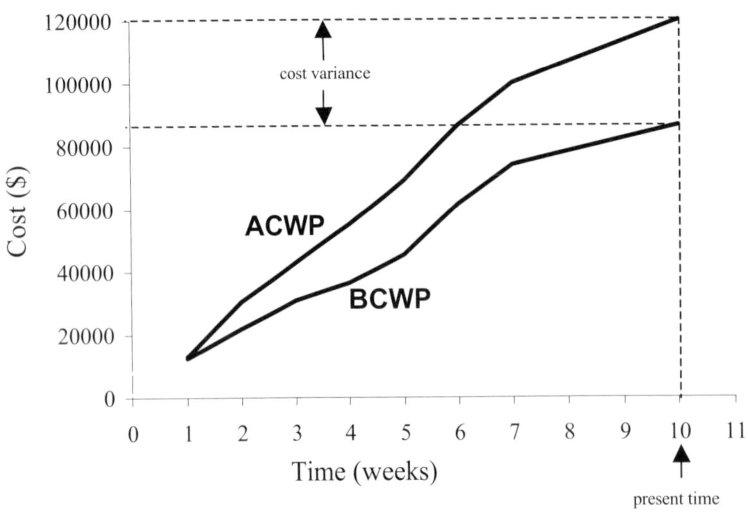

Fig. 9.5 Example plot of actual and budgeted costs for a project that is more expensive than anticipated, to illustrate the concept of cost variance. (Refer to text for explanation.)

Plate 9.2 Manual labour employed during the construction of shrimp ponds in India.

tions, buildings, access points, paths, channels, bodies of water, vegetation types and the positions of soil sample cores, the map will need to define the precise location of boundaries. Once these have been established, arrangements can be made to take possession of the land, either by purchase or lease agreement.

For remote sites road access may need to be constructed and accommodation provided for a construction workforce. For large projects a temporary site office and store within a lockable compound may be needed so that tools, equipment and materials can be properly guarded. Shipping containers are often convenient (Muir & Lombardi 2000). A temporary power supply can be essential and if freshwater is not available reservoir tanks may be needed to provide drinking water and for mixing concrete and mortar. Water may also be needed for compacting pond embankments (section 8.2.2).

Construction work can be managed by a site engineer but the project manager should supervise and monitor progress. It is important to proceed with construction as swiftly as possible because any delays can severely influence the economic viability of a project (section 10.4.1.1). Careful planning and good logistical support are essential and may be aided by employing a quantity surveyor. Cheap, locally available materials should be used for buildings, providing the overall quality of construction suits the planned life of the project (section 8.1) and there are no special planning regulations (for example, in earthquake zones). A large hatchery may be expected to last a minimum of 10 years. After construction, time must be allocated for the installation of equipment such as pumps and generators; suitably qualified personnel may be required for supervision and to ensure all electrical and water systems function correctly.

9.5.2 Start-up

Start-up operations usually begin after construction has been completed and all equipment has been installed and tested. However, to generate early income in some farms it may be possible to start production before all the ponds have been built. During the start-up of hatcheries, defects in plumbing and electrical installations often come to light and need to be resolved before production can commence. It usually takes some months or even years before a reliable production routine can be established, and investors and project backers must allow for the learning curve as new staff are trained and gradually gain practical experience.

If an integrated operation is planned, to incorporate a hatchery, nursery, farm, processing unit and possibly feed mill, then it may be beneficial initially to construct only the farm and rely on purchasing juveniles and feed from outside sources. This will enable production assumptions to be tested, site suitability to be confirmed, and income to be generated before investment in the additional operations. Ideally, each operation in a vertically integrated set-up should be economically viable in its own right and be able to compete in outside markets with its products or services. However, if one operation such as a hatchery is newly built and unique in a particular area, its value as a vital link in an integrated process may justify subsidisation.

9.5.3 Consolidation

From an operational base there may be many opportunities to sell alternative products and skills and provide advice to other people involved or interested in the crustacean farming business. The demand may be especially great in a pioneer industry. Indeed, some early marron and redclaw crayfish growers in Australia made more money from selling juveniles than from retaining them for ongrowing.

Despite the existence of opportunities to sell technology and the need to restrict the number of visitors to a busy installation, there are dangers in maintaining 'secrecy' in order to retain some perceived commercial advantage. If technical staff and other members of the workforce are convinced that they are engaged on a very special project they will expect special salaries in recognition or otherwise will be encouraged to go elsewhere and sell their 'secrets'. Mayo (1988) considered people who were new to aquaculture and who kept building fences around their projects and research. He warned that fences work both ways, so that while trying to retain some advantage through secrecy, the influx of ideas and solutions from outside could easily be stifled. In the long run free exchange of information and ideas between operators and researchers alike will have a much greater impact on success. Good results in aquaculture are much more closely related to hard work, diligence and trial and error than to the application of 'secret' or 'magical' formulae.

9.5.4 Operational phase

In order to monitor and control operational costs and maintain profitability in an increasingly competitive industry, strict accounting procedures may need to be introduced. Costs for materials, energy and manpower can be collated for each centre of operations in order to more easily identify areas where economies can be made. For example, when assessing a hatchery operation it may be worth splitting costs between different operational units, e.g. broodstock and maturation, larvae culture and live feed production. Problems with very detailed cost accounting systems can arise, however, due to unpredictability in the performance of aquaculture systems.

On a farm it can be advantageous to keep records of the amounts and costs of all inputs, such as labour, feed, seed, energy and fertiliser, destined for each pond. After a harvest the revenue can then be compared to total production cost, and results interpreted in terms of the management practice applied to that pond. This approach may be particularly useful in nucleus/plasma schemes where individual operators or individual ponds may vary in performance.

At some stage the decision to scale up operations may need to be taken and a choice faced between expanding an existing operation or building new facilities elsewhere. For an ongrowing operation, the limitations of the site and its water supply will strongly influence the choice. For a hatchery, overall size has implications for the flexibility of operations as well as the cost of production (section 10.6.1.1) and, rather than continually increase the capacity of an original facility, it is sometimes preferable to establish an independent operation at a new site. Most hatcheries suspend production for cleaning and sterilisation, so independent operations may be able to provide a more consistent supply of seed.

9.6 Food safety and HACCP

To gain access to the world's most valuable food markets, crustacean farmers and processors need to ensure that their products are safe to eat. The most reliable management tool for this purpose is known as Hazard Analysis, Critical Control Points (HACCP) and it is being incorporated into food safety regulations worldwide. The principles are introduced in sections 3.2.1 and 3.2.2, and important practicalities are discussed here.

Although HACCP largely aims to confront food safety hazards at the processing stage, it has significant implications for aquaculture operations because certain important hazards can only be controlled effectively at the farm. Some of the most serious potential hazards are best addressed during a farm's early planning stages through careful selection of the site and the water supply. For example, contamination with heavy metals, pesticides and with certain pathogenic bacteria and viruses can be avoided by choosing sites that are distant from industrial and intensive agriculture activities and sewage pollution (section 6.3.1). When HACCP is applied to a farm in operation, the main areas of concern are the use of drugs (medicants, growth promoters), chemicals, probiotics and the feed supply. Antibiotic residues pose a particular danger to consumer health and for this and other reasons (section 11.3.4) the value of applying antibiotics to crustacean ponds and feeds is highly questionable.

If they are used, minimal withdrawal periods must be respected to ensure that no residues remain at harvest time. To avoid unwittingly contaminating a crop during ongrowing, the use of uncertified feeds is also to be avoided. Sulphiting agents are anti-oxidants that are mixed with water and applied to shrimp during or after harvest to control melanosis, an enzymatic blackening process. However, sulphite residues in food can trigger allergic-type reactions in certain consumers, so residues need to be kept below 100 mg L^{-1} of edible flesh and all treated product requires labelling. Since for most farmers and processors antibiotic and pesticide testing is not feasible on a regular basis, the best approach to these hazards is preventive. For example, in another branch of aquaculture (catfish production), farmers draft plans for the entire production procedure and specify how and when approved chemicals are used throughout all stages of pond production. Any deviations are justified and recorded. This approach, which covers standards for many other aspects of farm management as well, is known as Total Quality Assurance (TQA) and it can be incorporated within or adopted as a supplement to HACCP (Otwell & Flick 1995).

HACCP-based regulations in the USA, Canada and the EU stipulate that the importer is responsible for ensuring that imported fishery products are processed under conditions equivalent to those prevailing in the importing country. This firmly puts the onus on importers to check that effective HACCP plans operate in exporting plants and it has the effect of forcing close co-operation between exporters and importers. Exporters have an incentive to comply with HACCP because it can give them a competitive edge over non-compliant exporters. Despite the advantages of HACCP, Japan still relies heavily on the traditional approach of random sampling and finished product testing (Sophonphong & Lima dos Santos 1998).

In a processing plant, the implications of HACCP differ greatly depending on the complexity and type of processes involved. For instance, the supply of live crustaceans can safely be performed with relatively few controls whereas the production of value-added foods, such as ready to cook, ready to serve, or vacuum-packed items, necessitates strict controls. Many types of operators and middlemen are not required to operate under HACCP plans at all. In the USA, people who manage holding systems for soft-shell crab production are required by the Food and Drug Administration (FDA) to conduct a hazard analysis but the preparation and implementation of an HACCP plan is only needed if the analysis reveals a hazard other than one directly related to sanitation (FDA 1995; Rippen *et al.* 1999).

Processors usually maintain basic sanitary standards by preparing and following a written set of procedures known as sanitation standard operating procedures (SSOPs). SSOPs are a vital component of the good manufacturing practices (GMPs) that define measures of general hygiene for the plant and the personnel. GMPs and SSOPs in combination are considered by regulators to be essential prerequisites to an effective HACCP plan because they create the environment in which safe food can be produced. They allow the HACCP plan to focus on the food and the processing steps rather than on basic aspects of sanitation.

The control of sanitation is a basic requirement in all cases and it involves monitoring and record keeping in the following eight areas:

(1) Safety of water that comes into contact with food or that is used for ice;
(2) Condition and cleanliness of food contact surfaces including utensils, gloves and outer garments;
(3) Prevention of cross-contamination from unsanitary objects to food, food packaging material and other food contact surfaces;
(4) Maintenance of hand washing, hand sanitising and toilet facilities;
(5) Protection of food, food packaging material and food contact surfaces from adulteration with lubricants, fuel, pesticides, cleaning compounds, sterilising agents, condensate and other chemical, physical and biological contaminants;
(6) Proper labelling, storage and use of toxic compounds and other chemicals;
(7) Control of employee health conditions that could result in the microbiological contamination of food, food packaging materials and food contact surfaces;
(8) Exclusion of pests from the food plant.

When new HACCP-based regulations are adopted, there is often a period of confusion as information is disseminated and interpreted, and precise concepts are translated between languages. Much of the confusion is due to the fact that, while the regulations deal with generalities, each individual processing operation has many unique features that do not fit readily into a generalised scheme. This situation is not helped by the fact that there are at least four major sets of regulations and recommendations (from the USA, Canada, the European Union and the Codex Alimentarius Commission (CAC) of the

FAO) and each is subject to revision. Thus a producer who wishes to provide quality assured product for several markets is faced with multiple and moving targets. There is also potential for confusion on the basis of whether the objective is quality or safety since while the USA, EU and CAC guidelines only address food safety, the Canadian Quality Management Programme has wider quality objectives that also include avoiding economic fraud and preventing the production of substandard batches. To reduce the confusion, the CAC guidelines are being promoted as the international norm. Understandably, some operators, who become exasperated by the requirements of different regulators, come to view HACCP merely as a bureaucratic obstacle and are tempted to produce different HACCP plans to suit different markets (Brooks 1997). Luckily however most regulators are aware that for any particular plant and process there are a range of different HACCP plans that can be acceptable as long as they are carefully implemented and can deliver the desired food safety objectives. The CAC accepts the capability of different inspection and certification systems to achieve the same objectives regardless of precise details of different methods and the FDA in the USA will accept dissimilar measures that can achieve the same level of health protection.

9.6.1 Implementation of an HACCP plan

An essential starting point in the preparation and implementation of an HACCP plan is to ensure commitment from the top of an organisation. The adoption of HACCP is a reflection of a company's commitment to producing safe food, so its implementation will fail without the backing of top management. Three inaugural steps to the development of an HACCP plan are necessary:

(1) *Assemble a team of people to devise and implement the HACCP plan*. The team needs to include key personnel who will be responsible for running and maintaining the processing operation and for assuring product quality. Ideally some of the team will have attended HACCP training courses. Outside HACCP expertise may also be needed.

(2) *Describe the products and their intended use by the purchaser*. This is a key reference point for the HACCP plan. For example, crustacean products which are intended to be fully cooked by the consumer before eating will pose fewer hazards than those which are destined to be eaten raw or only partially cooked.

(3) *Prepare a process flow diagram*. This diagram usually starts with the inputs of raw material, including all ingredients and packaging materials, and ends with the storage and/or shipment of finished product to the buyers. A process flow diagram is helpful because it enables hazards and processing steps to be considered in a logical sequence. An example is given in Fig. 9.6, based on individually quick frozen product.

The next phase is setting up the plan itself following the seven principles outlined in section 3.3.2.

(1) *Conduct hazard analysis and determine control methods*. The analysis aims to determine whether there are food safety hazards for the particular products and processes involved. Food safety hazards fall into three categories: biological, chemical and physical. Biological hazards are harmful bacteria, viruses or parasites that may either be naturally present in the culture environment or that are introduced through contamination by domestic or industrial waste. In a processing plant, bacteria and viruses of human origin can also contaminate product during handling. Measures to eliminate or to prevent the spread of biological hazards comprise a combination of the following:

- Attention to basic plant and personal hygiene.
- Attention to temperature during storage and duration of storage. Appropriate storage temperatures are 0–5°C for chilled product and below –18°C for frozen product.
- Preventing contamination of water supplies by human wastes.
- Preventing cross-contamination between cooked and raw product.
- Thorough cooking. If not carried out during processing then clear instructions to cook before eating are provided to the consumer.

Chemical hazards are elements or compounds that can cause illness or injury and include sulphiting agents, unapproved colourants, pesticides, antibiotic residues and heavy metals. Physical hazards are foreign objects in food that can cause harm when eaten and include fragments of glass or metal that may contaminate the product before or during processing. Metal detectors are sometimes deployed at the ends of processing lines to eliminate metal fragments. Food safety hazards of rel-

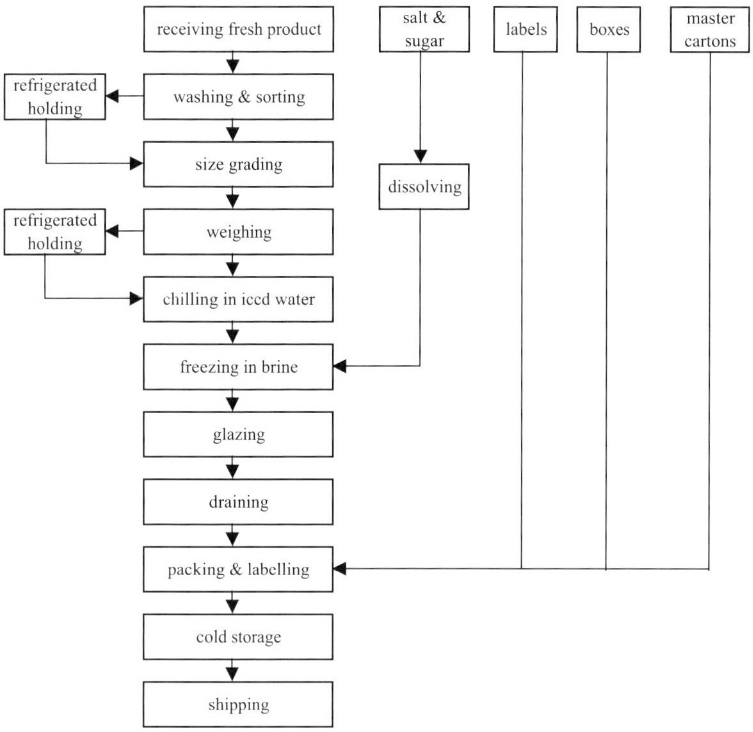

Fig. 9.6 Process flow diagram for a plant producing whole crustaceans individually quick frozen in brine.

evance to aquaculture and seafood are considered in more detail by Reilly and Käferstein (1997) and Seafood HACCP Alliance (1997).

There are other quality defects that are not directly related to food safety and these are eliminated from the hazard analysis. They include spoilage, insects, hair and economic fraud and are usually countered by SSOPs or GMPs. Food spoilage is unpleasant but it does not automatically represent a food safety hazard unless it is caused by pathogens or by toxic microbial by-products that can make a person sick.

(2) *Identify critical control points (CCPs).* A CCP is defined as a point, step or procedure at which control can be applied and a food safety hazard can be prevented, eliminated or reduced to acceptable levels. This is fine in principle but in practice it is often difficult to decide if and when a CCP is needed. At first there is a tendency to identify too many CCPs, but if HACCP is to be sufficiently focused and effective CCPs need to be kept to a minimum. HACCP guidelines (e.g. Seafood HACCP Alliance 1997) indicate that a hazard should only merit a CCP if (a) it is reasonably likely to occur and (b) if not properly controlled, it is likely to result in an unacceptable health risk to consumers. In this manner, CCPs can be restricted to the most significant hazards. In addition, CCPs need to be restricted to the point in the process at which they will be most effective. To clarify the situation, decision trees are available in HACCP guidelines but, even with these, precise interpretation is difficult and input from experienced people may often be needed. Experts can more quickly distinguish between those hazards that can effectively be dealt with by SSOPs and those that require CCPs. In the final plan, hazards will typically be controlled by a combination of SSOPs and CCPs so the HACCP plan and the SSOPs will be intimately linked. An example of a simple decision tree to identify CCPs (for example at each of the processing steps in Fig. 9.6) is shown in Fig. 9.7. The responses to the questions in such a tree can be usefully compiled in a document known as a Hazard Analysis Worksheet in which hazards at each step can be systematically dealt with. Justifications for the decisions are usually given, and for hazards

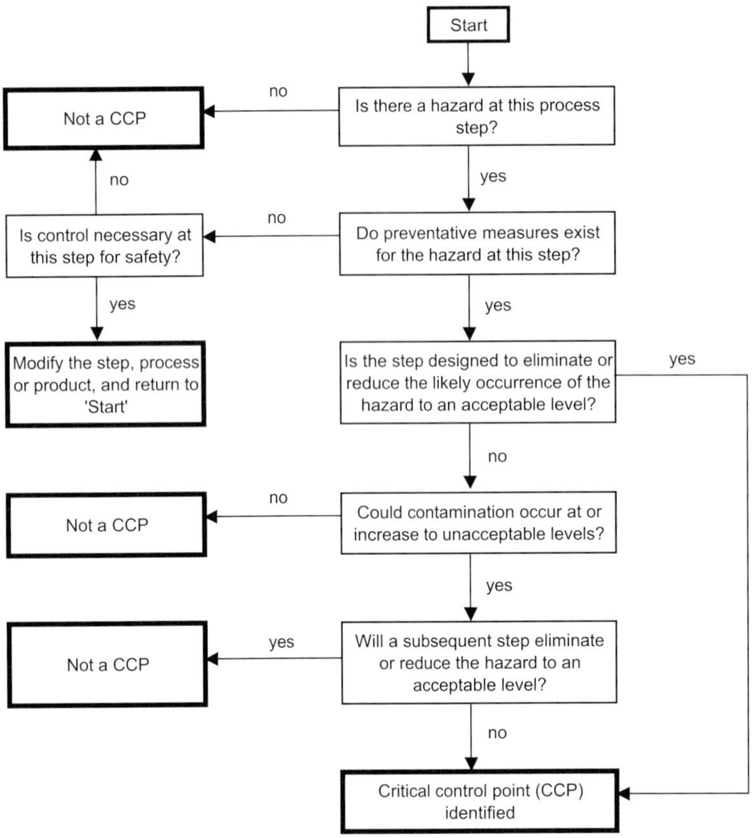

Fig. 9.7 Example of a simple decision tree for the identification of critical control points.

that will be addressed by SSOPs, GMPs and TQA rather than CCPs, relevant links can be shown.

(3) *Establish critical limits.* In many cases, the appropriate critical limits may not be readily apparent or available. Tests may need to be conducted or information gathered from sources such as scientific publications, regulatory guidelines, experts or experimental studies. Selection of the best critical limits is often driven by practicality and experience and if information is not available, conservative values should be chosen (Seafood HACCP Alliance 1997). Although some of the main food hazards are microbial, setting critical limits in terms of the results of microbiological tests is not suited to a CCP because results may not be available for a matter of days. If CCPs are to be effective, feedback is needed more quickly. For example, appropriate critical limits for the control of pathogenic bacteria in cooked crustaceans usually relate to the duration and temperature of a cooking process because these variables can be measured directly. A minimum cooking time and temperature will combine to eliminate the bacteria in question.

(4) *Establish monitoring and checking procedures.* These procedures are necessary to establish whether a critical control point is actually under control. Monitoring devices, the intended frequency of observations and the staff responsible for making them are determined and recorded.

(5) *Establish corrective actions.* When critical limits are exceeded, processors are required to modify the process and bring it back under control. Records of corrective actions are kept and used in subsequent verification procedures. If product is made during a period of non-compliance with critical limits and the potential food safety hazards are deemed significant, the product will either need to be reworked or destroyed.

(6) *Establish record keeping.* Full documentation is needed during the preparation of the HACCP plan,

including details of hazard analysis and the justification for critical limits. Following implementation of the HACCP plan, information on the monitoring of CCPs and on the effectiveness of corrective actions will also be needed.

(7) *Establish verification procedures*. This initially involves a validation of the HACCP plan prior to implementation to confirm that it has the potential to deliver the desired food safety benefits. Later on, verification involves confirmation that CCPs and critical limits are satisfactory. An important detail is to ensure that monitoring devices such as thermometers, scales, pH meters, chemical analytical equipment and test-kits are routinely calibrated (De Beer & McLachlan 1998). It is often necessary to conduct routine microbiological testing of end-product, of potable water supplies and of swab samples from work surfaces to check that biological hazards are under control. Finally, audits may be needed in which an unbiased person conducts systematic evaluations, including on-site observations and record reviews (Lupin 2000).

9.7 Management

The earlier sections of this chapter have addressed the management and planning issues arising during the establishment of successful projects. This final section looks at some of the main areas where sound management is needed to keep such projects running smoothly.

9.7.1 Husbandry and management practice

An aquaculture operation must allocate sufficient resources (manpower and equipment) to maintain good husbandry and staff management practices. Some general examples applicable to all species at all stages of the life cycle are given below.

Good husbandry practices

- The frequent, regular inspection of animals, water flow and aeration rates, and keeping a watch for unusual behaviour, animals not feeding, and signs of infestations and diseases.
- The adoption of a husbandry regime (stocking densities and handling methods) sympathetic to the life cycle stage, especially for maturing broodstock and incubating females.
- Frequent, regular feeding, monitoring of feeding rates and removal of uneaten food, detritus and sludge.
- Regular monitoring of food quality in both live and stored feeds.
- The frequent, regular inspection of water flow and aeration rates, immediate attention to leaks.
- Rationalised, regular water quality monitoring (see section 8.6).
- Attention to the control of temperature, lighting, weeds and predators, and minimisation of disturbance and vibrations.
- Awareness of the precautions to be taken regarding use of chemicals, especially antibiotics.

Plate 9.3 Predator-proof fencing around redclaw crayfish ponds in Australia. (Photo courtesy Clive Jones, Department of Primary Industries, Queensland, Australia.)

- Planned reporting and documentation of all of the above.

Good management practices

- The provision of clear instructions for and adequate supervision of staff.
- Good training with proper attention to the transfer of technological know-how from technical advisors/ consultants to farm staff at all levels as appropriate.
- The installation of comfortable working practices (jobs involving a struggle do not get done properly).
- The definition of clear lines of communication and areas of responsibility.
- Consideration of staff welfare (health, cultural and social isolation, particularly on remote sites, and incentives; e.g. Anon. 1999).
- Protection of the site from poachers and vandals.
- Provision of clearly defined and regularly updated emergency procedures covering both staff welfare and culture operations. Shepherd and Morris (1987) review practical emergency procedures.

9.7.2 Health management

The need for greater emphasis on health management in crustacean farming has been brought into sharp focus by the formidable disease problems facing much of the global shrimp farming industry. Many of the principles of health management can be applied at the level of individual farms but it is increasingly clear that governments have a critical role to play (sections 11.3.3, 11.3.4 and 11.5.3.2). 'Off-farm' management involves management of the ecosystems that support aquaculture and this requires a co-ordinated effort at national and regional levels to integrate aquaculture into watershed and coastal management strategies (Hotta & Dutton 1995). Specific areas of concern are effluent control, the introduction of pathogens and exotic species, and the need to reduce user conflicts. Governments can regulate resource use, particularly of land and water and provide legal and institutional arrangements to minimise resource use conflicts. Disease prevention in aquaculture is a formidable challenge in itself without the added burden of environmental degradation caused by unregulated coastal development (section 11.4).

An example from the shrimp industry in India illustrates how action at the farm level alone is inadequate for health management. Individual farms have implemented crop 'holidays' and pond disinfection procedures but they have not, in themselves, succeeded in containing disease (Mohan 1996). As a result, more action has been urged at a national level to:

- restrict stocking densities to 30 shrimp m^{-2};
- carry out routine health monitoring of farms;
- enforce mandatory disease notification and restrictions on the movement of broodstock and seedstock;
- encourage the use of settlement tanks for the control of waste discharges;
- develop closed and semi-closed production systems.

The growing emphasis on health management is essentially in recognition of the basic truism that prevention is better than cure. Effective disease control entails the adoption of a systems approach because aquatic animal disease is the end result of a series of linked events, and because treatment of disease goes beyond a mere consideration of the pathogen (Phillips 1996). If the focus remains only on the pathogens, this tends to lead to ineffective 'cures' and the inappropriate use of chemicals for a quick 'fix'. There is only a limited understanding of the relationships between disease, stress and the environment but some diseases are clearly associated with poor environmental conditions and the stress this places on the crustaceans being cultured (section 8.9). Shrimp, for example, readily succumb to vibriosis in a stressful environment. On the other hand some primary pathogens seem to cause disease even when there is no apparent stressor, e.g. white spot syndrome and yellow head viruses in shrimp (section 12.2).

One important aspect of health management is the need to set up effective quarantine procedures (Arthur 1996). However in many countries this remains a distant and elusive goal because of a lack of political will at the national level. Quarantine programmes form part of a first line of defence against the possible adverse effects of the introduction of exotic species. Codes of practice are a starting point for the design of national legislation and for the drafting of international agreements. They need to incorporate regionally agreed-upon lists of certifiable pathogens, standard diagnostic procedures and provision for health certificates of unambiguous meaning. To cut the spread of diseases in general it is necessary to reduce dependence on imported broodstock and seedstock (sections 8.9.4.4 and 8.10.1.3).

Very often the practicalities of health management on a particular crustacean farm involve the adoption of good management practices with a particular emphasis on maintaining a suitable environment and reducing the risks of introducing infectious agents. Disease prevention, particularly of opportunistic pathogens, requires careful attention to the management of water quality, pond bottom conditions, feeding rates and algae blooms. These are all important aspects of pond management and good management practice and are covered in detail in other parts of this book (sections 8.3 and 9.7.1). The critical role of pond management practice in disease control is emphasised in the comprehensive manual on shrimp health management by Chanratchakool *et al.* (1998). Strategies to limit disease problems in shrimp farming, which may also be applicable for the culture for certain other crustaceans, include:

- The use of inland freshwater ponds, as an alternative to brackish-water ponds, into which hypersaline water is added from tankers (although this has particular environmental risks; sections 11.4.1, 11.4.3 and 11.5.3.3).
- Avoiding the introduction of fish and invertebrates with the water supply by using rotating drum filters or through the treatment of incoming water, either in the reservoir, the intake canal, or the pond, with calcium hypochlorite, ozone or a non-persistent insecticide;
- Careful selection and acclimation of post-larvae and the use of SPF or SPR stocks.
- Reducing stocking densities and shortening the duration of the ongrowing cycle.
- Reduction or elimination of water exchange to reduce the risks of introducing pathogens and their vectors (i.e. improving biosecurity) and the development of culture systems suited to these limitations (section 8.3.7).
- Incorporating waste management systems into project design at an early stage to limit environmental impacts and the potential spread of pathogens.

Further details of these and other pond management strategies are given in section 8.3 and species specific information is given in Chapter 7. Aspects of health management in freshwater prawn culture are dealt with by Johnson and Bueno (2000). They note that important diseases have been associated with freshwater prawn culture but that there have been no reports of virus diseases that would pose a large-scale threat to the industry. One approach to managing prawn health would be to obtain SPR stocks through a carefully controlled programme in a similar way as for penaeid shrimp.

The intensification of culture systems has important implications for health management – some negative and some positive. Not only does it create the high-density conditions that can induce water quality problems and favour the transmission of diseases, but at the same time it provides environments over which greater control can be exercised, both to minimise these negative effects and to reduce the risk that pathogens will enter the production system in the first place. The net impact of intensification on disease control is thus difficult to discern and is often masked by the more dominant factor of the overall quality of management prevailing at a particular farm.

Another means that is believed to promote a healthy production environment is to perform polyculture rather than monoculture (sections 7.2.6.2, 7.3.5.2, 7.7.7 and 7.10.4). Clear evidence in support of this viewpoint is not yet available but it is an attractive hypothesis, based on the idea that polyculture systems increase diversity and stability and thus decrease the risks of soil and water quality problems. The polyculture of tilapia with shrimp has attracted particular attention in this regard for its potential to limit the effects of white spot syndrome virus (section 7.2.6.2).

9.7.3 Management of crop risk

In a discussion of how best to evaluate and contain crop risk it is worthwhile to consider the views of people in the insurance industry because it is their business to deal in such risks. Wiley (1993) described the main areas of interest in the risk evaluation of a shrimp farm from the point of view of a firm specialising in risk assessment. The account focused on the farming of *Penaeus monodon* but the approach was based on aquaculture in general so it has relevance beyond shrimp farming alone. Basically Wiley (1993) divided risk into primary and secondary categories, as shown in Table 9.1, on the assumption that enough basic knowledge of shrimp farm design and culture techniques exists to identify the most important characteristics of well-operated farms and their sites. Categories were weighted to indicate relative importance and the resulting framework was then used as the basis of a points scoring system to be applied at individual farms to see how well they were equipped to contain risks. The overall score of a particular farm would be used to grade it on a six-level scale ranging from 'excellent' to 'reject'. Thus crop insurance could be offered at a discount to 'excellent' farms and at higher rates for inferior operations. Looking in more detail at this framework gives an insight into how farms might best reduce the risks of production problems and at the same time minimise the cost of insurance.

Table 9.1 Categories of risks influencing the success of a crustacean farm and their relative importance (based on Wiley 1993).

Primary category	Relative weighting*	Secondary category	Relative weighting*
Competence	I	Personnel and management	I
		Security	II
		Production and financial history	III
Environment	II	Site characteristics	I
		Farm design	I
		Land use	II
		Water quality	II
Technical	III	Water supply and management	I
		General operations	I
		Stocking density	II
		Quality of post-larvae/juveniles	II

*Relative weighting is based on a descending scale of importance, with I corresponding to the highest contribution to the overall risk evaluation.

The three primary risk categories laid down by Wiley (1993) are competence, biophysical environment and technical aspects.

(1) *Competence* is the most important of these primary categories, and it largely refers to the quality of management. Specific aspects cover the number of years of operation, the level of management experience, the crop production history and financial performance. Interruptions in farm performance and frequent management changes reflect badly on the overall ability of the organisation to succeed technically and financially: companies that can retain staff are better equipped to manage risk. The farm must also be secure to minimise crop loss due to theft, sabotage, vandalism or other malicious acts and this requires a physical security programme of fences, checkpoints and patrols, backed up by good relations with the local community.

(2) The *biophysical environment*. This refers to the air, water and land conditions on and around the farm. Essentially it highlights the importance of good site selection and farm design. Shortcomings in the design features such as dikes, pumps, canals, water handling, access and building structures will increase susceptibility to pollution, natural catastrophes, and disease. The surrounding land use is a good indicator of environmental quality. Conflicts over land and resources, e.g. with urban, industrial and tourist developments, need to be avoided as well as pollution and nutrient enrichment caused by other shrimp farms. Rural sites need to be developed with 'reasonable' space and adequate dikes between neighbouring farms. To protect the quality of incoming water, the intake should be distant from effluent discharge of other farms.

(3) *Technical aspects*. These relate basically to good animal husbandry and sound pond management practice, with particular attention to the selection and acclimation of post-larvae, pond preparation, feeding regimes, water supply and intake and waste water quality management. If there are significant deviations from accepted 'best' practices this is considered to place the crop at a higher risk of loss. General operations management should be established with good routines because these are indicative of sound technical capabilities for preventing stress and disease. Key aspects are population sampling, feed rationing, personnel management, aeration, acclimation and stocking of post-larvae, and pond preparation. Good technical protocols are a necessary, but not a totally sufficient entity in themselves for effective risk management. They must be combined with qualified personnel and good environmental conditions.

In summary, minimising the risk of crop losses depends on providing effective management, a healthy environment, and appropriate culture techniques, but the most crucial of these three is effective management.

9.8 References

Anon. (1999) Mexico, teamwork the key at Aquastrat – shrimp farm puts success down to staff. *Fish Farming International*, **26** (10) 18–19.

Arthur J.R. (1996) Fish and shellfish quarantine: the reality for Asia-Pacific. In: *Health Management in Asian Aquaculture* (eds R. Subasinghe, J.R. Arthur & M. Shariff), pp. 11–28. FAO fisheries technical paper No. 360, FAO, Rome.

Brooks E. (1997) Application of HACCP in developing countries – problems and prospects. *Infofish International*, (6) 51–54.

Chanratchakool P., Turnbull J.F., Funge-Smith S.J., MacRae I. H. & Limsuwan C. (1998) *Health Management in Shrimp Ponds*, 152 pp. Aquatic Animal Health Research Institute, Bangkok, Thailand.

De Beer J. & McLachlan R.E. (1998) HACCP implementation in United States and developing countries. *Infofish International*, (1) 46–52.

FDA (1995) *Procedures for the safe and sanitary processing and importing of fish and fishery products; final rule*. FDA-CFSAN, Federal Register: 18 December 1995, **60** (242) 65096–65202.

FAO (1988) *Planning an aquaculture facility – Guidelines for bioprogramming and design*. ADCP/REP/87/24, UNDP, FAO, Rome.

FAO (1997) *Aquaculture development. FAO Technical Guidelines for Responsible Fisheries*. (5) 1–40. Food and Agriculture Organisation, Rome.

Hotta K. & Dutton I.M. (eds) (1995) *Coastal Management in the Asia-Pacific Region: issues and approaches*, 421 pp. Japan International Marine Science and Technology Federation, Tokyo, Japan.

Johnson S.K. & Bueno S.L.S. (2000) Health management. In: *Freshwater Prawn Culture: the farming of* Macrobrachium rosenbergii (eds M.B. New & W.C. Valenti), pp. 239–258. Blackwell Science, Oxford, UK.

Lee J.C. (1988) Government policies and support for shrimp farming. In: *Shrimp '88, Conference proceedings*, Bangkok, Thailand, 26–28 January 1988, pp. 175–181. Infofish, Kuala Lumpur, Malaysia.

Lupin H.M. (2000) Internal auditing of HACCP-based systems in the fishery industry. *Infofish International*, (4) 56–64.

Mayo R.D. (1988) *The Birdsong Junction Handbook*, 100 pp. JM Montgomery Consulting Engineers, Bellevue, WA, USA.

Mock C.R. (1987) A penaeid shrimp farm in Ecuador. *Aquaculture Magazine*, **13** (3) 36–43.

Mohan C.V. (1996) Health management strategy for a rapidly developing shrimp industry: an Indian perspective. In: *Health Management in Asian Aquaculture* (eds R. Subasinghe, J.R. Arthur & M. Shariff), pp. 75–87. FAO fisheries technical paper No. 360, FAO, Rome.

Muir J. F. (1988) Aquaculture consultancy services – the realities (Parts 1 and 2). *Infofish International*, (5) 19–20; (6) 48–50.

Muir J.F. & Lombardi J.V. (2000) Grow-out systems – site selection and pond construction. In: *Freshwater Prawn Culture: the farming of* Macrobrachium rosenbergii (eds M.B. New & W.C. Valenti), pp. 126–156. Blackwell Science, Oxford, UK.

Otwell W.S. & Flick G.J. Jr. (1995) A HACCP program for raw, cultured penaeid shrimp. In: *Proceedings of the Special Session on Shrimp Farming* (eds C.L. Browdy & J.S. Hopkins), pp. 218–226. World Aquaculture Society, Baton Rouge, LA, USA.

Philips M.J. (1996) Better health management in the Asia-Pacific through systems management. In: *Health Management in Asian Aquaculture* (eds R. Subasinghe, J.R. Arthur & M. Shariff), pp. 1–10. FAO fisheries technical paper No. 360, FAO, Rome.

Reilly A. & Käferstein F. (1997) Food safety hazards and the application of the principles of the hazards analysis and critical control point (HACCP) system for their control in aquaculture production. *Aquaculture Research*, **28**, 735–752.

Rhodes R.J. (2000) Economics and business management. In: *Freshwater Prawn Culture: the farming of* Macrobrachium rosenbergii (eds M.B. New & W.C. Valenti), pp. 369–392. Blackwell Science, Oxford, UK.

Rippen T.E., Fisher R.A. & Oesterling M.J. (1999) HACCP for soft crab producers. *Aquaculture Magazine*, **25** (3) 18–20.

Seafood HACCP Alliance (1997) *HACCP: Hazard analysis and critical control point training curriculum*, second edn, 228 pp. Publication UNC-SG-96–02, North Carolina Sea Grant, North Carolina State University, NC, USA.

Shepherd B.G. & Morris J.G. (1987) A review of practical emergency procedures for fish culturists. *Aquacultural Engineering*, **6** (3) 155–169.

Sophonphong K. & Lima dos Santos C.A. (1998) Fish inspection equivalence agreements: overview and current developments. *Infofish International*, (2) 42–49.

USDC (1989) *Malaysian shrimp culture*, 3 pp. National Marine Fisheries Service, F/IA23: KK, PN, IFR-89/50, National Oceanographic and Atmospheric Administration, US Dept. of Commerce.

Verzuh E. (1999) *The Fast Forward MBA in Project Management*, 332 pp. J. Wiley & Sons, New York.

Wiley K. (1993) Environmental risk assessment in shrimp aquaculture. *Infofish International*, (2) 49–55.

World Bank, 1981. *Guidelines for the use of consultants by world bank borrowers and by the World Bank as executing agency*, 38 pp. World Bank, Washington, DC, USA.

Chapter 10
Economics

10.1 Introduction

People often invest in crustacean farming projects without fully appreciating the costs and risks involved. Initially they may be captivated by euphoric press accounts of the industry or be deluded by overenthusiastic promoters. But optimism soon turns to disillusionment or despair when ill-conceived projects sink into financial crisis after suffering from start-up delays, cost overruns and missed productivity targets. Because of the frequency of such disasters, it is worth remembering that most companies go bankrupt for a lack of cash rather than a lack of profits, and that aquaculture-based operations are no exception to this rule. A financial crisis will lead to abrupt failure, even while most problems of a biological or technical nature can, in time at least, be solved. Problems are particularly common during the start-up phase when insufficient cash is available to cover operating expenses. For most projects positive cash flows should not be expected for a minimum of 3–5 years after initial investments have been made. Even after this period financial problems are still common. Many if not most companies in the aquaculture industry remain undercapitalised and are unable to fulfil their potential.

Despite such constraints there are many success stories underpinning the global expansion of crustacean farming. This chapter assesses the main financial and economic factors influencing the viability of projects and, by reference to examples of budgets, identifies likely investment and operating costs. Techniques for investment appraisal are described along with methods of assessing risk. The factors that influence risk levels, such as intensification, are also discussed. Since the calculations required for investment appraisal are based on numerous species-specific assumptions, these also need to be fully understood in advance. If the assumptions for key factors such as productivity levels are based on false or misleading estimates, incorrect conclusions will be drawn. It is thus essential to evaluate the technical aspects of a project (Chapter 7) to gain an understanding of its likely performance prior to undertaking a financial or economic appraisal. Eventual success with a project will also rely critically on effective management (Chapter 9).

While financial appraisal focuses on the viability of a commercial venture from the point of view of its owners, economic appraisal can be applied to tackle the wider question of whether society as a whole will be better off or worse off by undertaking a particular project (section 10.3.3.2). The technique attempts to correct for situations in which the actual prices paid for goods and services do not represent their true value, for example as a result of market distortions or government policies. Economic appraisal can be extended to cost–benefit analysis, which includes externalities – indirect costs and benefits that accrue to third parties and fall outside the immediate financial concerns of a project (section 10.3.3.3). The negative externalities of aquaculture projects include pollution, loss of natural habitat and reduction in biodiversity. Putting monetary values on such effects is notoriously difficult but, if aquaculture is to make progress towards the elusive goal of sustainability, such impacts will need to be carefully considered before projects are set up (Chapter 11). In one important respect the impacts of aquaculture may be longer term than those of agriculture because, if ponds are built, changes in the nature of the land may make it prohibitively expensive for subsequent conversion to other crops. In this regard, an environmental fund or other financial safeguard may be imposed as a condition when the project is approved (sections 10.3.4 and 11.5.3).

Despite the usual focus on the negative externalities of crustacean farming, the activity also generates positive externalities and their importance should not be underestimated. An increase in the supply of crustaceans leads to lower prices and significant consumer benefits. Lower prices also slowly undermine the financial viability of wild fisheries, thereby reducing dependence on over-exploited wild stocks and cutting the wastage associated with open access to these common resources. Examples of these effects have already been identified with crayfish (section 3.3.3) and with shrimp. For the latter it was estimated that prices in the USA would be 70% higher but for imports of farmed shrimp (Keithly *et al.* 1992). Although this may be an unwelcome development for fishermen, it has always been a fundamental ideal of aquaculture to reduce pressure on natural fisheries by farming, so evidence of progress in this regard can be viewed as a positive contribution to world development.

10.2 Finance

Obtaining finance for crustacean farming projects can be a major hurdle. Investors, bankers and the directors of aid agencies are becoming increasingly wary of supporting new projects because of high-profile project failures and because of adverse environmental impacts. Essentially the sources of funding for aquaculture projects can be divided into two categories – private investment and capital assistance.

10.2.1 Private investment

Private investment can be of foreign or domestic origin, may involve the private or public sale of equity, and is usually backed up by finance from commercial banks. Finance sometimes comes from specialised venture capital companies that generate funds through the sale of equity. But these companies usually consider the capital needs of aquaculture firms too small to be of interest and they have very demanding investment criteria, including high rates of return. McCoy (1986), for example, lists the following requirements:

- annual growth rates of at least 25%, preferably 50%;
- pre-tax margins of at least 35%, preferably 50–60%;
- minimum return on investment of 30%;
- management sufficiently 'attractive' to make a public offering within 5–8 years;
- unique patentable technology is preferred;
- up to 30% ownership of the project;
- minimum investment level around $1m.

Funds for aquaculture operations are rarely generated directly through the public sale of equity although some companies, usually those involved in salmon production, have made public offerings on the Vancouver Stock Exchange in Canada. The shortage of this type of funding is reflected in the shrimp industry, which, despite a turnover of some $25bn per year, is almost entirely financed by private rather than public capital. This lack of public finance, i.e. from companies whose shares are listed on a major stock exchange, is given as one reason why the shrimp industry appears to be stuck on a plateau (Woodhouse 2000). Several large publicly traded companies (e.g. British Petroleum, Unilever, SANOFI and the International Proteins Corporation) diversified into shrimp farming in the 1980s only to pull out later.

Loans for crustacean farming operations are particularly difficult to obtain in areas where the activity does not have a successful track record. In Africa, for example, a shortage of funding has been blamed on general ignorance of the aquaculture industry and on a lack of confidence from commercial banks, foreign aid organisations and international finance institutions (Skabo 1988). Such attitudes have, however, only been reinforced by the failure of major shrimp projects such as those in the Gambia in the 1980s and more recently in Guinea. The financial arrangements of the latter have come under scrutiny by a commission set up to investigate public finances (Anon. 1999) and are based on concerns about the viability of industrial-scale projects in many parts of the continent. In other parts of the world, such as Hawaii, problems in obtaining capital for expansion and operating expenses have also been linked to an unfavourable investment climate left in the wake of some large but short-lived projects (Main & Deupree 1986).

10.2.2 Capital assistance

Capital assistance, in the case of developing nations, usually comes from external sources and is provided in the form of loans from institutions such as the World Bank, and grant aid packages from donor countries and organisations. In the case of developed countries, capital assistance is made available in the form of government enterprise grants and regional development subsidies, sometimes linked to providing employment.

The bulk of external capital assistance is provided in the form of 'soft' concessionary loans with low interest, extended repayment terms. The Asian Development Bank (ADB) has played a leading role in this area, promoting aquaculture, particularly of shrimp, in countries such as Indonesia, the Philippines, Pakistan and Sri Lanka. Some World Bank (WB) loans have been provided for schemes in which aquaculture was one of many components. For example, within a total WB package of $200m, $5–30m may have been aimed at aquaculture (Nash 1988). In the late 1990s an aid package was put together in Bangladesh to upgrade shrimp and fish production. It comprised $33m from the World Bank, $15.5m from the UK and $9.5m from the Bangladesh government (Griffin 1999). Other organisations that have provided 'soft' loans to aquaculture include the Inter-American Development Bank (IDB), the International Development Agency (IDA) and United Nations organisations such as the International Fund for Agriculture Development (IFAD). According to data from the Food and Agriculture Organization (FAO) of the United Nations, official aid to aquaculture research and development averaged $124m per year between 1988 and 1995. Development banks were the main source of funding (69%) followed by bilateral sources (17%) and multilateral sources (7%). Asian (65%) and African (16%) regions consistently got the most support, and four countries, India, China, Bangladesh and Mexico, accounted for 64% of all external aid to aquaculture (Shehadeh & Orzeszko 1997). Governments and various organisations including the EU and the United Nations provide assistance in the form of grants that, of course, do not require repayment. While part of this grant aid is aimed directly at the construction of research facilities, hatcheries and farms, part is usually provided in the form of technical assistance to help maximise the beneficial impact of the whole aid package. In Cuba, for example, the FAO has provided both financial backing and technical assistance for the farming of shrimp and prawns.

External capital assistance is channelled to individual projects via government departments, commercial or development banks and specialised credit institutions. Domestic government inputs will usually involve the provision of infrastructure and the funding of support activities such as extension services, hatcheries, training and research facilities or grants. Domestic and international commercial or development banks such as the International Finance Corporation (IFC), a US-based development bank linked to the World Bank, has provided credit for large-scale projects over $1–4m and has thus tended to favour large vertically integrated projects incorporating hatchery, farm, processing plant and sometimes a feed mill.

Specialised credit institutions, often in the form of agricultural, fishery or co-operative banks, are geared to supply loans to numerous scattered smallholders because, unlike commercial banks, they usually have a wide network of branches in rural areas. As such their role can be central to the development of community-based, small-scale aquaculture. If the specialised needs of small-scale operators are to be met, however, loans need to be provided promptly, with a minimum of bureaucracy and preferably with the back-up of technical extension and marketing services. In addition, as part of a co-ordinated effort, briefing also needs to be provided on the purpose of credit and how to make best use of it. Development banks have been urged to follow this approach to the promotion of aquaculture, and the argument that they cannot bear the added cost of supervising numerous small loans is considered spurious, bearing in mind that they are ready to provide subsidised credit for large-scale projects (Smith 1984). Certainly, in the parallel case of fisheries development, success in the past has relied upon the provision of appropriate, flexible and timely credit (Tietze 1989).

In developed nations, governments often provide support for aquaculture in the form of enterprise grants and subsidies, particularly if a proposed project is likely to stimulate the economy of a depressed area and provide jobs. Examples of this include Spanish government and EU grants towards the capital expense of constructing shrimp hatcheries and farms in Andalucia. Other schemes to assist businesses in rural areas are aimed primarily at agriculture but can equally be applied to crustacean culture.

10.2.3 Joint ventures

In many developing tropical nations the profit potential of crustacean culture and its support industries has attracted foreign investment. The attraction is based primarily on the presence of favourable climatic conditions and the availability of relatively cheap land and labour. Activity so far has been centred on shrimp culture in Asia and Latin America, with production aimed at export markets because of the limited purchasing power of domestic consumers. Regulations regarding the establishment of joint ventures vary from country to country.

Sometimes a certain minimum level of local participation is required on the basis of equal or majority domestic shareholding. In some cases the foreign party is required to finance local stock acquisition while in others, if the requirements for foreign capital are high, the foreign partners take the majority interest.

Joint venture agreements often involve the provision of technical expertise by the foreign partner. In such instances it is usually worthwhile establishing pilot-scale operations before expanding to full production unless the foreign partner has an excellent record of technical success and the proposed site for the project is considered to be completely satisfactory. The subject of technical assistance and how get the best from it is discussed in section 9.3.6.

Many opportunities for joint venture agreements exist, and interested parties often advertise in trade publications such as *Infofish International*. Great care needs to be taken in choosing a partner. Ideally, selection should be made from a number of candidates and preference given to parties with reliable histories and appropriate experience. Some of the most successful ventures in shrimp culture involve large established companies with experience in the export of fishery or agricultural products.

10.3 Investment appraisal

10.3.1 Objectives

The results of the appraisal of any project or scheme can only be interpreted by reference to specified objectives so these need to be clearly established as a first step in project planning.

10.3.1.1 Private sector

Private crustacean farming ventures can range in size from vertically integrated, multinational concerns down to smallholdings and backyard hatcheries, but each will ultimately be interested in generating profits. Objectives and investment decisions, however, will be based on different criteria at different scales of operation. For example, the largest companies may view involvement in crustacean culture as part of a long-term strategy to exploit trends in increasing seafood consumption. As such, significant investment in unprofitable new technology and production on a pilot or research scale is sometimes justified in the light of potential long-term profits and a leading position in the industry. Smallholders, on the other hand, may be attracted to crustacean culture as an alternative to other activities such as agriculture or the production of fish or salt. In such cases, the potentially greater profits and risks of crustacean farming must be weighed against the immediate loss of income from an established crop. Proven 'low-risk' culture practices are sometimes employed to keep the impact of the critical start-up and learning period to a bare minimum. 'Back yard' hatchery operators, who produce juveniles only in response to the seasonal availability of broodstock and the demand for juveniles, are primarily concerned with the profitability of each production run. This type of operation allows great flexibility but alternative occupations or sources of income are needed to cover periods of inactivity.

Individual farmers and private investors can usually express their objectives in terms of desirable levels of personal income or rates of financial return on investments. The returns that are required by a private investor reflect the sources and cost of the capital employed. The cost of risk capital (equity) will depend on the return expected by the shareholders, while the cost of funds borrowed from a financial institution will vary depending on interest rates and the duration and size of the loan. Since the cost of loans is usually lower than cost of risk equity, the average cost of capital for a particular project will depend on the proportions (the gearing ratio) of the two different types of capital employed. Achieving the correct balance between equity and loan capital can be critical to the success of a project. Though heavy reliance on loans may be an attractive option when interest rates are low, when they are high a highly geared company can be placed in an impossible situation due to the excessive cost of servicing its debt. In such situations, a venture that relies mainly on equity capital can be better placed to survive.

When deciding what rate of return is suitable for a particular type of project it is also important to take account of the level of risk involved. Investment in crustacean culture is very hazardous when compared, for example, to the purchase of secure treasury bonds, and in order for it to be a worthwhile proposition it should demonstrate a potential for generating higher returns. Just what rates of return might be acceptable is a subjective matter, but one method of evaluation involves adding a risk premium to the return possible from an alternative low-risk investment. For example, if a risk premium of 10% were added to the yield from US corporate bonds, at the time of writing (2000) around 7%, then the resulting minimum rate of return would be 17%. Risk premiums of between 5% and 15% are appropriate for most crustacean culture projects, although the highest premiums should be used when intensive or untried technology are to be employed. Ongrowing, nursery and hatchery operations are more risky than projects centred on processing, marketing or feed production because there are more factors influencing output levels and because it is harder to predict the period between making initial investments and attaining profitable output levels. Risk, a major concern of crustacean farmers, is discussed further in section 10.4.

10.3.1.2 Public sector

In the case of governments or aid agencies, the policy objectives for developing aquaculture schemes will usually be based on broad issues of social and economic development. They may be described in terms of benefits to the national economy (e.g. attracting foreign investment, or improving the balance of trade through exports or import substitution) or with reference to rural development schemes (benefiting the local economy, providing employment, improving local incomes, avoiding population drift to urban centres). The possible socio-economic benefits derived from an aquaculture project are summarised in Fig. 10.1. In general, the objective of economic development can be defined as providing the greatest possible increase in the standard of living of a population. In specific terms, the objective of a particular public sector project may be stated in terms of a desired minimum rate of economic return. The exact level of this minimum rate, sometimes referred to as the test discount rate, can be a contentious issue. It will usually vary over time and will depend on the point of view of the particular agency or government involved.

In poor countries, the need to attract foreign investment is usually a crucial consideration in aquaculture development plans. Several methods are applied by governments to achieve this objective, including:

- relaxing exchange controls;
- offering flexibility in regulations regarding national involvement in joint ventures;
- simplifying legal requirements and bureaucratic 'red tape';
- providing training, information, extension and other support services;
- promoting auxiliary industries, e.g. feed production;
- cutting duties on imports of equipment and raw materials;
- providing 'tax holidays' during start-up;
- permitting repatriation of profits and assets;
- providing infrastructure (e.g. designated areas for aquaculture, 'aquaparks').

In the case of Panama, the government is intent on stimulating foreign investment in aquaculture, and restrictions have been lifted on the movement of capital, on mergers and acquisitions and on limits to foreign participation in joint or wholly owned ventures. The paperwork needed to obtain coastal land has been simplified and concessions granted that enable this land to be used as collateral for obtaining loans from banks. Such loans benefit from subsidised bank interest rates as a result of aquaculture being reclassified as an agriculture/husbandry practice. In addition, support industries producing nauplii and post-larvae have been granted exemption from income taxes and are able to import raw materials and equipment duty-free (Jory & Martinez 1998).

The introduction or rehabilitation of crustacean farming on the scale of smallholders or artisanal farmers has been an objective of some governments in developing nations. If supplies of wild or hatchery juveniles are adequate and markets for crustacean products are accessible, this approach has obvious potential for raising incomes in rural communities. However, as a means of producing food to supplement local diets, crustacean farming does not provide the more immediate benefits of finfish farming, which can generate greater yields with lower input costs using simpler and more easily introduced techniques (section 11.2). In addition, the lower value of fish does not tempt the producer to export his crop and it thus becomes available locally.

The considerable overlap between government and private sector objectives allows for co-operation and co-ordination of effort. Examples of this include the establishment of government hatcheries to supply juveniles, and the promotion of contract growing 'nucleus/plasma' schemes in which a privately owned, vertically integrated 'nucleus' operation supplies numerous smallholders ('the plasma') with materials, technical assistance and a guaranteed market. Further aspects of government assistance to the aquaculture sector are discussed in sections 9.3.9 and 11.5.3.

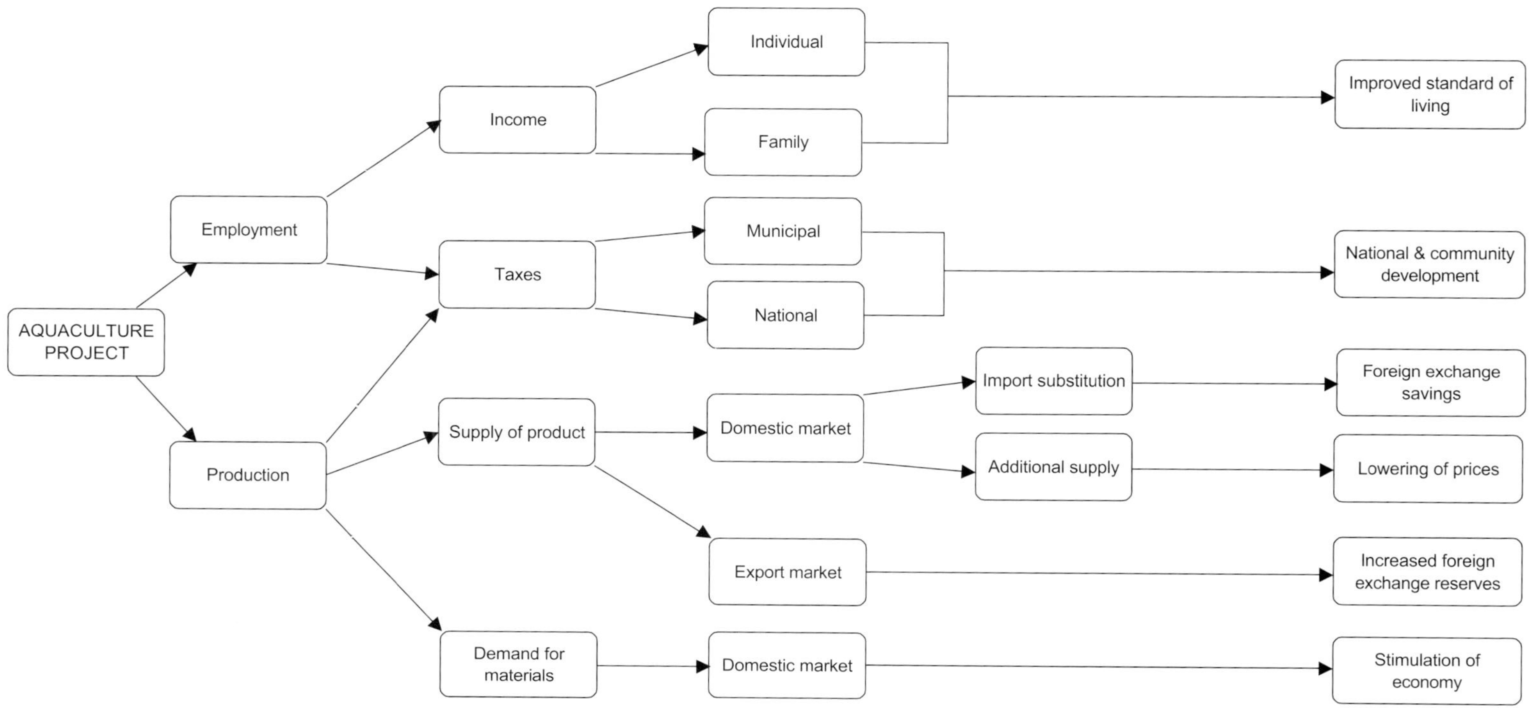

Fig. 10.1 Socio-economic benefits derived from an aquaculture project (modified from Israel 1987a).

10.3.2 Assumptions

A whole series of assumptions are made when a project is formulated and appraised, and the significance of these assumptions must be understood in order to reach correct conclusions. In financial appraisal some assumptions are made to simplify calculations and permit some general conclusions to be drawn. For example, it is often assumed that unlimited capital is available and that tax payments and concessions do not apply. Clearly in any specific situation allowances must be made for these factors. Taxation effects in particular can have a strong influence on profitability.

Other key assumptions relate to the market prices obtained for products, the dimensions and operational characteristics of the project, and the cost and amounts of materials, labour and utilities required for construction and operation. As few of these can be estimated with absolute precision, feasibility studies should always take into account the likely accuracy of underlying assumptions. The techniques of risk analysis can be applied to assess the effects of variation in basic assumptions (section 10.4.1). For most projects the market prices for the final products will have the greatest and most direct effect on profitability. They can only be predicted on the basis of knowledge about existing markets and market trends (Chapter 3).

The projected output level of any operation rests upon a string of assumptions, some of which are under the control of the designer or operator and some of which can only be guessed from experience. An example of some key design and operational assumptions for a shrimp nursery, ongrowing and processing operation is given in Table 10.1. On actual farms, average rates for survival, growth and feed conversion will depend on management skills, environmental conditions and the

Table 10.1 Example of assumptions affecting productivity, associated with the design and operation of a shrimp nursery, ongrowing unit and processing plant (based on IFC 1987).

Phase	Factor	Assumed value	Units
Nursery			
	Number of ponds	11	
	Average pond size	1.4	ha
	Nursery pond area (total)	15.7	ha
	Stocking density	400 000	post-larvae ha^{-1}
	Duration of cycle	35	days
	*Survival rate	70	%
Ongrowing			
	Ongrowing pond area (total)	300	ha
	Average pond size	10	ha
	Number of ponds	30	
	Stocking density	40 000	juveniles ha^{-1}
	Duration of ongrowing period	125	days
	Cycles per year	2.6	
	Number of harvests per year	78	
	*Feed conversion ratio	2.5 : 1	wt. feed : wt. shrimp harvested
	*Survival rate	70	%
	*Average harvest weight	28	g
	*Average harvest	784	kg ha^{-1} crop^{-1}
Processing			
	Capacity	5	mt day^{-1}
	Processing yield	65	%

*Expected outcome rather than design or operational choice.

possibility of disease outbreaks. They can be predicted most reliably when proven species and farming techniques are applied in established farming areas. Special care should be taken to avoid multiplicative errors when the output level of a proposed project is calculated from a series of assumed values. Using figures in Table 10.1, the annual production per hectare from a crustacean farm can be calculated as the product of stocking density (no. ha^{-1}), survival rate, average harvest weight, and number of crops per year. A 20% over-optimistic estimate for one of these factors will exaggerate the annual production per hectare (2038 kg) by 20% (2446 kg), but if all four are overestimated by 20% the productivity will be exaggerated by 107% (4227 kg).

10.3.2.1 Project life

For investment appraisal the life of a project is usually set between 7 and 10 years. Although this is somewhat arbitrary, if profitability is not foreseen within 10 years or less, the project is unlikely to be an attractive investment proposition. Some assessments, for example for development projects, are made on the basis of longer project lives of 15–20 years, or in certain cases even longer. In an evaluation of a project to create an artificial reef and stock it with hatchery-reared lobsters in the UK, project life was set at 100 years to reflect the long economic life of the reef structure (Whitmarsh & Pickering 2000). However, unless discount rates are low (4% or less) the results of investment appraisal for a project with a steady stream of revenue are only weakly influenced by cash flows after 30 years. In general, there are three bases upon which the project life can be determined:

- The market life of the output (not a major concern for established food products).
- The technical life of the major replaceable assets, generally machinery.
- The economic life of the major replaceable assets (this is shorter than the technical life because machinery is replaced before repair and maintenance costs become excessive).

10.3.2.2 Inflation

It is usually assumed that the returns generated by a project will keep pace with inflation, i.e. the inflation of production costs will be compensated by inflation in the prices obtained for the final products. While it is advisable to test possible variations in the market prices for products using sensitivity analysis (section 10.4.1.1), the above assumption is generally valid and it means that the IRR appraisal technique (section 10.3.3.1) will yield a value for the true rate of return (exclusive of inflation). However, because the interest rates of loans and the cost of equity are expressed as nominal rates (inclusive of inflation) rather than true rates, it is sometimes useful to also express the rate of return from a project as a nominal rate. This can be done most simply by adding the expected rate of inflation to the estimated true rate of return. Thus if the inflation rate is expected to be 5%, a project with an estimated IRR value of 20% has a nominal rate of return of 25% and should be able to pay interest at this higher rate.

10.3.3 Appraisal methods

Various techniques are available to evaluate and compare the potential benefits of different investment opportunities. To evaluate purely financial returns, the methods include payback period (PB), accounting rate of return (ARR, also known as return on investment), net present value (NPV) and internal rate of return (IRR). To evaluate returns to society as a whole it is possible to extend the NPV and IRR methods and calculate a project's net social present value (NSPV) or its economic internal rate of return (EIRR).

10.3.3.1 Financial appraisal

The *payback* method is widely used and it simply calculates the time it will take for a project to recover the cost of the original investment. A time horizon is chosen and the decision rule is to accept projects with payback periods shorter than this interval. Strictly speaking, however, it is a poor method because it does not consider cash flows after the end of the payback period and it does not calculate the overall profitability of a project over its expected lifetime. It can thus lead to the selection of short-lived projects over better, longer-term ones. It also fails to take account of the 'time value' of money (simply put, a dollar today is worth more than a dollar in a year's time because the dollar today can be invested immediately to start earning interest). Because of its limitations the payback method is best avoided or used only in conjunction with other appraisal methods. Its advantages are its simplicity and the fact that it can reflect the 'robustness' or risk level of a project, i.e. shorter payback periods usually imply greater 'robustness' and lower risk.

The *accounting rate of return* (ARR, or return on investment) method calculates the average yearly accounting return over the life of the project and expresses this as a percentage of the average investment. However it does not adjust future revenues to take into account yearly incidence (the 'time value' of money) and as such it is inferior to the NPV and IRR methods. Its advantages are that it can be readily calculated from accounting data and that often there are a lot of comparative data available from other projects.

The *net present value* (NPV) technique provides a figure for the overall value of a project by totalling all the expected annual cash flows over its intended life. However, in order to account for the 'time value' of money, each yearly cash flow is adjusted by a discount factor calculated on the basis of the discount rate – the minimum rate of return desired from the project. A project that yields a positive NPV will in theory generate sufficient returns to pay interest at or above the discount rate as well as repay the original investment. This approach leads to the simple decision rule of accepting all projects with positive NPVs and, if projects are mutually exclusive, accepting the one with the highest NPV. All viable projects thus identified should in theory be able to raise finance but, if capital is rationed, a refinement of the NPV approach is to calculate the profitability index, $PI = (NPV + I_0)/I_0$, where I_0 is the initial investment, and to invest in the projects with the greatest indices, up to the capital constraint.

The *internal rate of return* (IRR) method takes the NPV technique one stage further and calculates the discount rate at which discounted revenues, totalled over the life of the project, just balance the original investment, i.e. it calculates the discount rate at which the NPV is zero. The IRR represents the true rate of return from the investment and the decision rule is to accept the project with the highest IRR. IRR is preferable to payback (PB) and ARR because it takes into account the 'time value' of money. It does, however, have a fundamental drawback because it is quite possible for there to be more than one IRR that will generate an NPV of zero. This can occur when a project has an initial cash outflow, a series of positive cash inflows, and then at least one additional outflow. The IRR method can also give misleading results when projects are mutually exclusive. When such projects differ in scale, it is possible for the IRR method to select a small high-return project instead of a larger project with a more modest IRR, even if the larger project has a higher NPV.

NPV is strictly the best method because it selects projects that maximise net present value, or wealth. Its disadvantage is that many business and non-technical people have difficulty understanding the concept. The 'time value' of money and the present value of future sums are not intuitively obvious concepts to most people and this can lead to misunderstandings. The NPV method also entails computing the appropriate discount rate and this may involve some guesswork. The discount rate should reflect either the cost at which capital can be borrowed to finance the project or, if the funds are already available, the return that the capital could expect to earn in the next best alternative investment. The NPV technique also gives no impression of 'robustness'. For these reasons many companies use the alternative approaches because they are easier to understand. IRR is probably the most popular method because in general people are more comfortable with the concept of a rate of return than they are with the idea of a sum of discounted money. In summary: NPV is the best method but it gives no impression of robustness; IRR is respectable but flawed; ARR is good for comparative purposes only; payback is useful for uncertainty; and, all four techniques can be used in parallel to give a more complete picture of financial viability.

10.3.3.2 Economic appraisal

The appraisal of a project by calculating the net social present value (NSPV) or the economic internal rate of return (EIRR) is closely related to the cash flow analysis methods of NPV and IRR. But in order to estimate economic returns to society as a whole (as opposed to solely financial benefits) the prices of goods, services and labour are adjusted so that they more closely represent their true economic value rather than simply their financial cost (this approach can be extended to include externalities; section 10.3.3.3). Distortions in a country's domestic price structures usually result from the imposition of trade barriers and are often most marked in poor nations where there is an acute shortage of foreign exchange. Wage rates can also become distorted as a result of a large pool of unemployed labour and rigidities in labour markets, and so inadequately reflect the cost to the economy of employing additional labour. In economic appraisal, 'shadow' wage rates compensate for such distortions and conversion factors adjust domestic prices to more representative world-equivalent prices. Conversion factors vary between countries and between different categories of goods, and together with shadow wage rates can be obtained from aid agencies.

Although it is difficult to establish the true economic values of certain goods and services such as power, construction, internal transport and the consumption of environmental goods (section 10.3.4), economic appraisal provides a valuable indication to governments and development agencies of a project's overall worth and it serves to expose schemes which are only viable on a financial basis because of protectionist policies. Nowadays, the need for price adjustments in an economic appraisal is declining because many developing countries have introduced economic reforms that result in many more goods reflecting truer, world prices. Trade liberalisation has been particularly important in narrowing the gap between domestic and world prices but, as long as divergences persist, economic appraisal remains valid and it will continue to have an important role in investment decisions.

10.3.3.3 Cost–benefit analysis

To provide a more complete assessment of a project's value to society, economic appraisal can be extended to the more complex approach known as cost–benefit analysis (CBA). In addition to the analysis of resource flows, CBA attempts to evaluate indirect effects (externalities), such as the costs of pollution (section 10.3.4), and it may also take into account distributional elements, i.e. the distribution of costs and benefits between poor and wealthy people and between current and future generations. CBA formalises a commonsense concept of rationality and is often used as the basis for formulating public policy. However, it is often criticised when it is used in this way. Environmentalists who believe that environmental protection is a social good, transcending cost–benefit calculations, are automatically opposed to it (Cropper & Oates 1992). While it is clear that very real problems arise with CBA when costs and benefits cannot be identified, quantified or expressed in money terms, such problems can largely be attributed to the complex nature of the issues involved. As Sylvia (1997) puts it – the management of aquaculture development is a wicked policy problem due to: the fugitive characteristics of aquatic resources; the complexity and uncertainty of environmental interactions; the number and diversity of user groups; the absence of market prices for many aquatic resources; and uncertainties associated with industry impacts. Faced with such difficulties CBA at least has the attribute of exposing its underlying assumptions. Most alternatives to CBA, such as the imposition of zoning or performance standards, really just hide costs and benefits and involve the adoption of arbitrary standards that can be manipulated to fit the needs of any interest group.

10.3.4 Environmental costs

The long-term viability of aquaculture depends on its ability to function within the constraints of the natural environment. Aquaculture becomes unsustainable where there is overexploitation of natural resources and where the capacity of the environment to replenish oxygen and assimilate wastes is exceeded (Chapter 11). Problems with sustainability may be self-inflicted by the aquaculture industry or may arise from competition for resources with non-aquaculture users because population growth, intensification of agriculture and industrialisation all result in consumption of the same environmental goods and services that aquaculture requires (Beveridge *et al.* 1997).

To tackle these problems, incentive-based solutions are increasingly being sought. In one analysis of environmental costs, involving the construction of ponds in tropical coastal areas, it was found that the externalities induced by mangrove destruction could be efficiently internalised by the imposition of a per unit area mangrove development tax (Parks & Bonifaz 1994). Such a tax, if rigorously applied, would have the effect of discouraging aquaculture in mangrove areas that are often exploited in ignorance of the fact that mangrove soils make poor ponds (sections 6.3.3.5 and 8.2.2.5). Estimates of the economic value of the goods and services arising from mangrove ecosystems have been less than \$100 $ha^{-1} yr^{-1}$ when only marketed fishery and forestry products are considered. But when complete systems including maintenance of fauna and air/water purification are included, such valuations rise to between \$1000 and \$11 000 $ha^{-1} yr^{-1}$, strongly reinforcing the view that mangroves should not be destroyed for aquaculture. On the basis of such evidence, Primavera (1997) has proposed increases in annual lease payments for shrimp ponds in the Philippines to fund mangrove planting and rehabilitation.

The 1992 Rio Declaration promotes the view that national authorities should actively encourage the internalisation of environmental costs and the use of economic instruments to achieve this – i.e. the polluter should in principle bear the cost of externalities (Sylvia 1997). This approach is being followed in some western countries that are adopting 'polluter pays' policies in which licences to discharge wastes and monitoring are used to control effluents (Beveridge *et al.* 1997). Non-governmental organisations such as the Worldwide Fund for Nature have also stressed the need for aquaculture ventures to face up to environmental costs directly, for example through the imposition of royalty charges for the use of land and water, and through the establishment of restoration funds that are used to return land to some productive use if and when ponds are abandoned (Clay 1997). The cost of complying with such environmental regulations is starting to be felt in Queensland, Australia, where potential shrimp farmers face licence costs of as much as US\$115 000 before they start construction. Such measures however have generated controversy because the small and nascent aquaculture industry feels it is being unfairly targeted compared to more powerful, long-established industries such as sugar cane and cattle that also create significant environmental impacts (O'Sullivan 2000).

Other strategies for pollution control include:

- The imposition of water quality standards or uniform emission standards.
- Black lists and grey lists (black listed chemicals are not allowed to enter aquatic environment and grey listed ones can be discharged if precautions are taken and/or effluent treatment is performed).
- The adoption of the best available technology or best practicable means available to minimise pollution.
- The precautionary principle – i.e. if it is suspected there is a hazard, even if there is no scientific proof of cause and effect, then take precautionary action (Barg 1992). This is often interpreted as implying no discharge at all is permitted except of uncontaminated 'natural' substances. For the creation of artificial reefs, this implies that quarried rock would be acceptable but not necessarily tyres or concrete blocks incorporating pulverised fuel ash (sections 5.7.2 and 8.11.2).

The rigid adoption of pollution controls, in a one-size-fits-all approach, however does not make economic sense because it does not take account of the extent to which the environment can assimilate waste. The adoption of the precautionary principle can be particularly problematic because strict application on the basis of suspicion alone can block all discharges and thus prevent all new developments, except perhaps certain closed systems.

10.4 Risk

The different risks facing aquaculture ventures can be split into four basic types – economic, market, physical and production. Economic risk covers external macro-economic factors such as inflation, recession and exchange risk. Of these, recession is a particular concern because crustaceans are luxury goods and demand is very sensitive to changes in levels of disposable income (section 3.1). Market risk refers to variability in prices usually arising from changes in supply and demand. Physical risks include acts of nature such as flooding, drought and earthquakes. Production risk or crop risk refers to factors with a direct influence over the success of a crop such as disease, problems with the supply of seed, water quality effects and low-growth or survival rates.

Specific ways of reducing or eliminating risks include taking out insurance against crop failures, storm damage or earthquakes (sections 10.4.2 and 11.5.1.3), diversifying production, and dealing in shrimp futures (section 3.3.1). The management of crop risk is discussed in section 9.7.3 and the techniques available for analysing the potential financial and economic impacts of risk are presented below in section 10.4.1.

An individual's attitude to risk will depend on their personality and their circumstances. Crustacean farmers may be optimistic in general, but small-scale operators in poor countries are often very risk-averse because they have limited access to capital and because a series of crop failures can quickly result in financial ruin. A small shrimp farm of 0.6 ha can be set up in Sri Lanka for just over \$2000, but for each crop the owner may be exposed to an additional \$3000 of debt to cover working capital for feed, fuel and post-larvae (De Silva & Jayasinghe 1993). Thus if a crop fails the farmer may be unable to repay the debt and risks losing the farm. To minimise risks in Vietnam shrimp farmers keep total production costs to a bare minimum so that they can make profits even when survival rates fall to 10%. Unfortunately outbreaks of white spot syndrome virus can result in survival rates as low as 1–3% so the risks are still considerable (Johnston *et al.* 1999).

Larger operations with many ponds are in a better position to survive occasional bad crops but they cannot avoid risks that have an impact on the whole farm or the whole region. The way that different risks (in this case, economic, production and physical) can combine to undermine a large crustacean farm is illustrated by the fate of one project, Dipasena Citra Darmaja, a shrimp farm located in southern Sumatra, Indonesia. A massive enterprise comprising 18 000 ponds and covering an area of 16 000 ha, it is the world's largest aquaculture operation. It was set up with money from BDNI, an Indonesian bank, and with support from the World Bank and the Export-Import Bank of Japan. Ambitious in organisational structure as well as size, the project attempted to capture vast economies of scale while at the same time benefiting from the hard work and diligence of thousands of individual farmers, each with a real stake in a small part of the operation. Individual farmers were lent $65 000 to buy land, a house and two shrimp ponds. The company was the sole provider of feed, post-larvae, power and other basic supplies and was also the only purchaser of the harvested shrimp. In its central co-ordinating role, the mother company had to perform a delicate balancing act between recouping enough income from each farmer to cover inputs and financial costs, and at the same time leaving farmers with enough income and an incentive to carry on. Unfortunately this balance was not achieved. The situation was aggravated by initial meagre harvests and, in 1996, the effects of drought. Struggling to stay afloat, many farmers took out further loans but these loans were denominated in US dollars and became grossly inflated in local currency terms following the Asian financial crisis of 1997. Even when harvests were good, farmers complained that the company made unexpected deductions from their payments. Falling deeper into debt, the farmers became increasingly restless and started to blame the mother company for all their production and financial problems. Tension built up and eventually erupted in violent clashes (section 11.2.3). The company is now valued at $400m, a mere fraction of its debts totalling $2.5bn, and the dreams of its owners and 7750 workers lie in tatters (Murphy 2000). Many of the problems could be blamed on the catastrophic loss of confidence embodied in the Asian financial crisis. However, many outside analysts would say that this loss of confidence was caused by the very existence of such projects that were worth far less than the funds invested in them. In this respect, the failure of banks to properly scrutinise their loans has also been heavily criticised.

Interestingly the Asian financial crisis has also assisted many shrimp farmers. For example, when the Thai currency lost 60% of its value, the price of shrimp, an export commodity, effectively doubled and farms that had been left idle for fear of disease and crop failure were brought back into production because the potential rewards were so much better than before (Polioudakis 2000).

A further example of project failure is illustrative of the production risks inherent in highly intensive systems that push production technology to its limits. Over 2000 people bought shares valued at $13m in Penbur Farms, a shrimp culture system with 0.55 ha of indoor tanks located in Texas. Although some of the critical techniques needed to manage feeding and water quality, at biomass densities of 4 kg m^{-2} or more, were eventually developed on site, the improvements came too late to avoid catastrophic mortalities. Financial crisis led to the project's collapse with debts of $1.5m (Boeing 1998).

To minimise production risks low-density, low-input systems can be chosen. They may not have the allure of highly intensive systems but they have shorter pay-back periods because less working capital is tied up in the crop and less money needs to be invested in equipment (Millamena & Triño 1997). The subject of intensification and its implications for risk are discussed further in section 10.5.

10.4.1 Risk analysis

The investment appraisal methods in section 10.3.3 provide an impression of the viability of an operation in the unlikely event that cash flows are entirely predictable. Risk analysis improves on these methods by taking account of the potential for variable cash flows in an uncertain future. The simplest approach is to use a discount rate modified to reflect the risk level inherent in a particular type of project (section 10.3.1.1). Other ways of dealing with risk involve modelling the performance of a project while allowing key variables to fluctuate as anticipated in the real world. These techniques are sensitivity analysis, Monte Carlo simulation and break-even analysis and they provide a more complete picture of possible outcomes.

10.4.1.1 Sensitivity analysis

In sensitivity analysis a whole range of different internal and external factors can be selected and varied to see how they influence the viability of a project. The selection of appropriate factors can be based on past experience of similar projects, on case studies or on

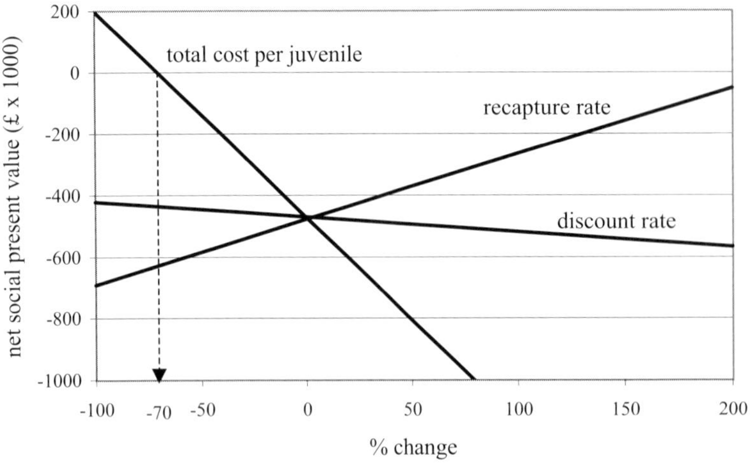

Fig. 10.2 Sensitivity analysis of a hypothetical lobster stock enhancement project (Lee 1994).

experiments. Viability is expressed in terms of financial indicators such as the payback period, NPV or IRR. Conveniently, the sensitivity relationship has a linear property and it can be represented graphically with the percentage change in the factor on the X-axis and a financial indicator on the Y-axis. Steeper lines indicate greater sensitivity. An example is given in Fig. 10.2, based on a hypothetical large-scale lobster stock enhancement project in the UK. Baseline assumptions in this analysis include a total number of 500 000 juveniles released per year; a juvenile cost of £1.38 each (about $2.00); a recapture rate of 10%; and a discount rate of 5%. Under these assumptions the NSPV is negative. The slopes of the lines reveal that viability is most sensitive to the cost of juveniles and that such a project would be viable (NSPV >0) if this cost could be cut by 70% or more (Lee 1994). In fact, progress is already being made in reducing juvenile production costs (section 10.6.3.2), particularly for small juveniles, but in this sensitivity analysis it was assumed that relatively advanced juveniles would be released (instar 10–11) because they have a better chance of surviving in the wild.

Sensitivity analysis is a popular tool because it is flexible and easy to interpret. The method is most useful when there is uncertainty about markets and costs and a quick impression of the robustness of a project is required. Table 10.2 lists various factors that have been included in sensitivity analyses to test their impact on the profitability of crustacean farms. Most analyses should, at the very least, test the assumptions regarding market prices and productivity levels. The selection of other factors would ideally relate to scenarios that are considered probable or possible in any particular case. For example, a prawn ongrowing operation in a temperate climate may investigate the effect of variations in the length of the growing season, and an intensive operation may need to test assumptions regarding energy costs, particularly if based in heated indoor facilities. Sensitivity analysis is particularly valuable for highlighting the areas where research should be focused to have the most beneficial impact on the economic viability of culture systems.

Some general conclusions can be drawn from published cost and sensitivity analyses:

(1) *Economy of scale.* As the size of a system increases, production costs are reduced because many fixed costs can be spread over greater output levels. In one study, Lambregts *et al.* (1993) found that increasing the size of an intensive shrimp farm from 10 ha of ponds to 160 ha of ponds reduced the total investment costs per hectare by 50% and cut production costs per kg by 25%. Other published studies include those for shrimp (Hanson *et al.* 1985), prawns (Sandifer & Smith 1985) and crayfish (Staniford & Kuznecovs 1988). Super-intensive farming systems are usually of a modular design so there are fewer possibilities for shared costs and economies of scale.

(2) *Pond size.* Smaller ponds are more controllable and more productive per hectare (Lee *et al.* 2000) but are more expensive to build and maintain and hence have greater investment and operating costs. In the study by Hanson *et al.* (1985) large shrimp farms and large ponds gave higher rates of return,

Table 10.2 Factors typically used in sensitivity analyses.

(1) Market price
(2) Productivity level
(3) Factors affecting productivity
species or combination of species cultured
availability of juveniles for stocking
stocking density
survival rate
growth rate
harvest weight
length of growing season
crops produced per year
environmental temperature
operating temperature
salinity
dissolved oxygen levels
random events, e.g. hurricanes, crop failures
(4) Design features
size of operation
size of ponds/tanks/containers
water exchange rate
water recirculation rate
(5) Major cost components
Investment
land
construction
duration of start-up period
Operation
stocking/broodstock
feed
labour
energy
interest rates

although increasing the number of ponds in a facility resulted in less variability in IRR values. Not everyone agrees that smaller ponds are more productive – McIntosh (2000), for example, found no differences in yield per unit area over a 2-year period from ponds in Belize ranging in size from 0.065 ha to 1.6 ha.

(3) *Start-up time.* Ideally this should be kept to a minimum. Staniford and Kuznecovs (1988) found it was very important that output in a new yabby farm reached 100% in its second year, and Hanson *et al.* (1985) found that a 100 ha shrimp farm built in 2 years was more profitable than a 200 ha farm built in 3 years.

(4) *Site selection.* Factors affecting yield and annual production potential have the most significant effects on capital and operating costs. In an assessment of brackish-water pond systems, Muir and Kapetsky (1988) found these factors to include water temperature, salinity and silt content, and soil acidity – characteristics that also proved to be especially critical in more extensive pond systems (sections 6.3 and 8.2). Hajek and Boyd (1994) developed a system for rating soil and water quality characteristics according to their likely impact on the costs of establishing and running earthen aquaculture ponds. Their account provides a useful demonstration of the approach by determining limitation ratings for pond construction from a soil survey report.

10.4.1.2 Monte Carlo simulation

Monte Carlo simulation is a more powerful and realistic technique than sensitivity analysis. While sensitivity analysis enables the impact of variations in only one factor to be investigated at a time, in real situations it is possible for a whole range of factors to fluctuate simultaneously. Monte Carlo simulation can deal with such situations and describe the whole spread of possible outcomes for a project. To begin with, the different variables are assigned ranges and probability distributions. The latter are usually assumed to be rectangular, triangular or normal. Particular values for the variables are then generated within the specified distributions using random numbers, and the resulting project NPV or IRR is calculated. The process is repeated to generate another value for the financial indicator, and so on, until a complete probability distribution is derived. Risk is then reflected in the dispersion of this distribution.

The virtue of Monte Carlo simulation is its flexibility and the fact that it can easily be performed on computer spreadsheet packages. Unfortunately such techniques usually require a great deal of data, much of which may be unobtainable, and they may also assume that variables are independent, when many are probably correlated. Correlations can be included in the models but it is difficult enough to predict probability distributions, let alone levels of correlation. Another drawback with simulation is that attempts to be realistic tend to be complex and this usually means that the decision makers will delegate the task to specialists or consultants. Thus, although the makers of the model may understand it, the decision makers may not and thus they may not fully trust it or rely on it. Interpretation of the results may not be a simple matter because someone still has to decide what is an acceptable average outcome and distribution of outcomes.

In one example of Monte Carlo simulation, Staniford (1988) ran 500 simulations of the profitability of a hypothetical 10 ha integrated crayfish (*Cherax destructor*) farming operation in which market prices varied between A$8 and A$12 (US$5.77–8.65) per kg, yield varied between 2000 kg and 3800 kg ha^{-1}, and periodic crop failures occurred due to mismanagement. Results showed that an IRR of around 15% was most probable and that there was only an 18% chance of exceeding an IRR of 20%. Simulations of this type (e.g. for shrimp (Hanson *et al.* 1985), redclaw crayfish (Medley *et al.* 1994) and artificial reefs for lobsters (Whitmarsh *et al.* 1995)) are useful for quantifying levels of risk and can be used to study the impact of random events such as hurricanes. A modified form of Monte Carlo simulation, involving a more efficient stratified sampling technique (Latin hypercube), is often preferred nowadays because of its greater computational efficiency.

10.4.1.3 Break-even analysis

Break-even analysis is also closely related to sensitivity analysis but it addresses viability in a different manner by asking the question – what level of output or sales revenue would be needed to cover costs? The answer is the break-even point: the point where NPV = 0. If the present value of cash inflows (sales revenue) and outflows (total costs) are plotted against output level (Fig. 10.3) the intersection of the two lines indicates the break-even point. The analysis, when properly applied to the present values (i.e. discounted values) of costs and revenues, rather than accounting costs and revenues, gives a production or sales benchmark against which actual figures can be compared. As such it is a very useful management tool. The break-even quantity can be compared with data from existing projects and pilot-scale trials to assess the likelihood of achieving viable productivity levels, and it thereby gives an impression of the 'riskiness' of a project.

10.4.2 Crop insurance

Crustacean farming is a capital-intensive activity in which the crop is often the most valuable asset. Yet this asset is exposed to a significant risk of failure and a major financial loss can result. To guard against this type of shock, many branches of agriculture and some well-established branches of aquaculture, such as salmon farming, turn to insurance companies. Insurers rely on the fact that, in the long run and over a large number of farms, crop losses usually represent a predictable fraction of total output. On this basis of pooled risk, they are willing to take on risk from many individual farms in return for payment of a premium representing a percentage of the value of the crop, or the value of inputs such as feed, seed and labour. Individual farms can benefit from insurance because an insured asset can be used to obtain credit from banks or from feed companies, such that the

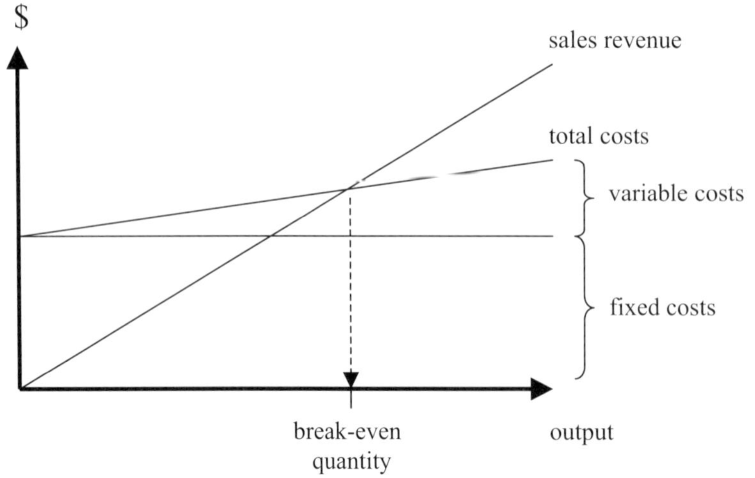

Fig. 10.3 Estimation of break-even quantity: the output level at which sales revenue balances total costs.

expenses occurred in producing a crop do not have to be paid until the crop has been harvested and sold, or, in the event of crop failure, the insurers pay out compensation. With their crops insured, farmers reduce the risk they are exposed to, minimise the need for operating cash and can concentrate on maximising yields.

Despite its advantages, crop insurance is rarely taken out by crustacean farmers (section 11.5.1.3). For most shrimp farmers the main problem is that no protection is offered for the major killers, such as white spot syndrome virus (WSSV) and existing policies are limited to cyclones, power failure, flood damage and some lesser diseases. In Madagascar some farms have been able to obtain fuller disease cover but this is not an area currently infected with WSSV. Lloyds of London have provided insurance for the specific hazards of power loss and hurricanes in the USA at a premium of 2.5% of insured value. There is a possibility that broader cover will be available at some time in the future through US federal crop insurance (Cannon 2000). Farms in Peru have received compensation for flood damage (D. Currie, 2000 pers. comm.).

10.5 Intensification

The intensification of crustacean culture involves increasing the input of capital and/or labour to a given area of land and raising the stocking densities. Its main attraction is its ability to maximise productivity and hence financial returns. However not everyone chooses to culture crustaceans intensively. Some people simply lack the resources or technological know-how and others view it as a risky strategy that increases the chance of crop failure due to disease or sharp fluctuations in water quality. Evidence from the shrimp farming industry in the 1980s tended to support this view because production in extensive and semi-intensive systems in Ecuador continued to expand while the intensive Taiwanese industry collapsed as a result of disease and environmental degradation. At the same time super-intensive systems, such as Marine Culture Enterprises in Hawaii (section 7.2.6.6), also displayed vulnerability to disease outbreaks. More recent patterns in the shrimp industry indicate that it is an oversimplification to categorise intensive operations as risky and extensive systems as safe. Disease appears to strike farming regions in waves, regardless of whether they practise extensive or intensive methods. Ecuador's shrimp farms have themselves suffered heavily from white spot syndrome virus, while output from Thailand, mostly based on smaller but more intensive ponds, has shown remarkable resilience to the same disease. The factors determining the reliability of a farming system are more closely linked to its management, particularly health management, than they are to its intensity level. Also in some important respects, smaller, more carefully managed, intensive production units (especially those based on low or zero water exchange – sections 7.2.6.6 and 8.3.7) are more suited to the control or elimination of disease than large extensive and semi-intensive systems which appear more susceptible to environmental degradation and disease.

Intensification has important implications for the underlying economics of aquaculture. Though it can certainly boost the output of culture systems, changes in the costs of production do not necessarily improve overall profitability. For a given surface area, the additional investments associated with intensification mean that fixed costs are increased and the system must deliver a higher output to cover total costs. This point is illustrated with break-even charts for extensive and intensive systems in Fig. 10.4. In both charts variable costs are proportional to output levels. Yet while an ongrowing

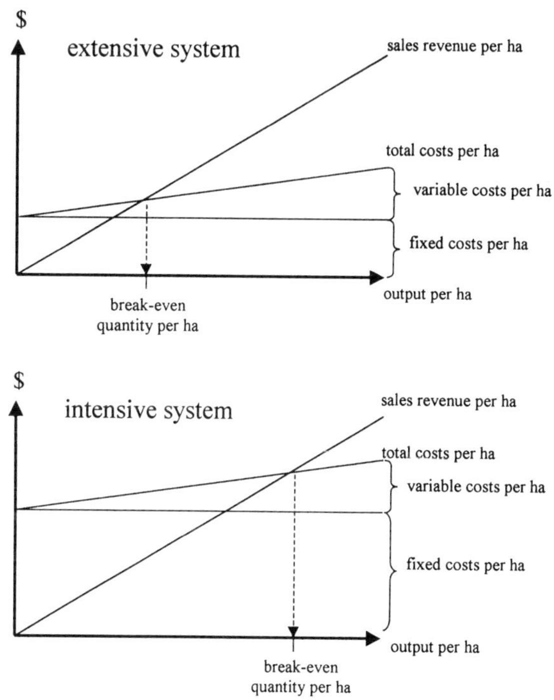

Fig. 10.4 Break-even charts for extensive and intensive culture systems showing the effect of increased fixed costs on the break-even quantity.

system is intensified and the potential output tends to increase as a linear function of the stocking density, the cost of the system tends to increase as an exponential function of the stocking density. These two underlying patterns are apparent in Figs 10.5 and 10.6, based on the data listed in Table 10.7 for shrimp and prawn farms. As a result, as a system is intensified, fixed costs increase faster than output and the total cost of producing a kilogram of crustaceans increases. This effect is apparent with the intensification of shrimp culture systems shown in Fig. 10.7. Although this figure does not extend to densities beyond 50 shrimp m^{-2}, the general trend may hold true for super-intensive culture also, where the data are more variable. The average cost of producing a kilogram of shrimp is $3.70 for extensive and semi-intensive systems while for intensive and super-intensive systems it is $5.75, some 55% higher. Note that these figures were calculated without including the very high production costs of intensive and super-intensive Japanese systems.

Intensive systems may be able to maximise the returns per unit area but the profit margin per kilogram is actually lower than in more extensive systems. This places an upper limit on the viability of intensive systems and it also means they can be vulnerable to drops in the market price of their output. The super-intensive shrimp farm of Marine Culture Enterprises in Hawaii may have been close to this upper limit. The target biomass densities were set as high as 5 kg m^{-2} to try to compensate for the high capital cost of this greenhouse-enclosed raceway system, which were around $1m ha^{-1}. But failure to attain these levels, largely as a result of disease, eventually forced the venture to close (section 7.2.6.6). From the evidence in Fig. 10.6, the upper limit for economically viable intensive shrimp culture may lie in the stocking density range 100–150 shrimp m^{-2} because in this range the investment per unit area appears to rise very sharply. However, pilot-scale trials at a super-intensive system in Michigan, USA, have suggested that the limit may be higher, at 200 shrimp m^{-2} (section 7.2.6.6).

So far, the commercial viability of highly intensive crustacean culture has been limited to specific situations, where:

- a premium price market can be exploited;
- high land prices or a shortage of suitable sites preclude the construction of extensive or semi-intensive ponds;
- the culture of a particular species requires controlled-environment facilities.

In Japan, high land costs and the high price market for live *Marsupenaeus japonicus* combine to support high-intensity farming operations. Australian farmers are having some success penetrating this Japanese market with shrimp grown at lower cost in less intensive systems. Also reliant on a premium price market are the intensive systems for producing soft-shell crabs and crayfish, although strictly speaking these are holding operations rather than farming enterprises (sections 7.5.8 and 7.10.9).

A greater understanding of how ongrowing systems respond to intensification can be gleaned by the estimation of production functions. Production functions are equations that attempt to model the output of a system as a function of the inputs of basic resources such as capital, labour and land. Usha Rani *et al.* (1993) studied farms in India and concluded that shrimp farming is highly capital-intensive rather than labour-intensive. Their production functions revealed that there was plenty of scope to increase investment in the industry, with positive returns (per unit area) from inputs of seed, labour, establishment costs and power costs. However they found that feed was in general excessively used, and that cuts could be made without reducing output levels. Chen (1994) developed production functions for different aquaculture systems in Taiwan. The overall results showed positive returns to inputs of land and labour, as expected, but returns to abalone and grouper farming were better than for *Penaeus monodon* farming. This was a reflection of low survival rates and disease problems in Taiwanese shrimp culture.

Another technique to analyse the productivity of culture systems, commonly applied in agriculture, is bio-economic modelling. Bio-economic models incorporate the biological and physical elements of a culture system and have been used to optimise production management, particularly with respect to the scheduling of stocking and harvesting. Early investigations with crustaceans focused on the potential of lobster culture, but nowadays shrimp systems attract more attention (Leung 1994). The construction of the biological part of the model is the usually the most difficult part of the process, particularly for pond culture because of the complexity of the pond ecosystems.

10.6 Costs

In the sections that follow, published details of the costs of culturing shrimp, prawn, crayfish, lobster and crab are

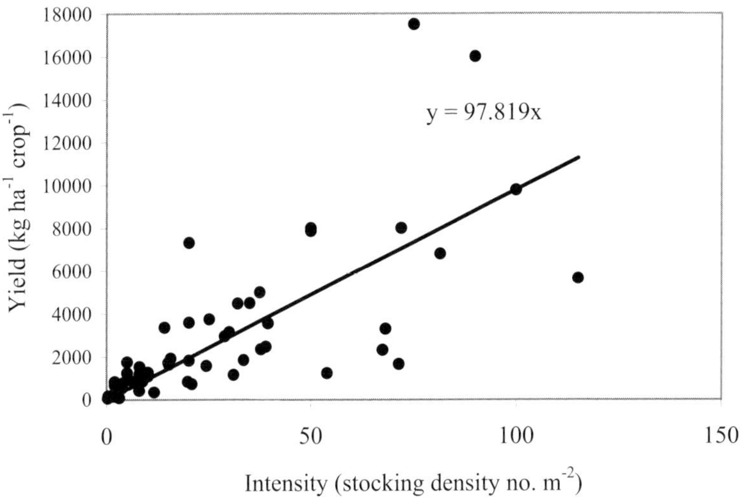

Fig. 10.5 Linear relationship between intensity and yield (data from Table 10.7 with one outlying point removed).

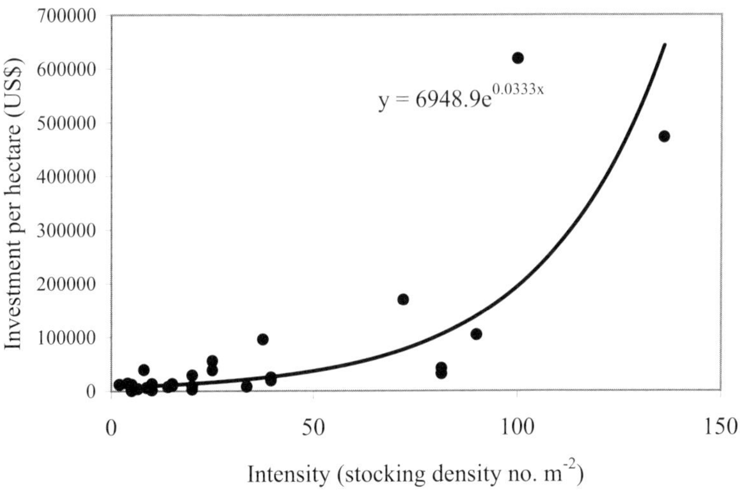

Fig. 10.6 Exponential relationship between intensity and investment per hectare (data from Table 10.7 with one outlying point removed).

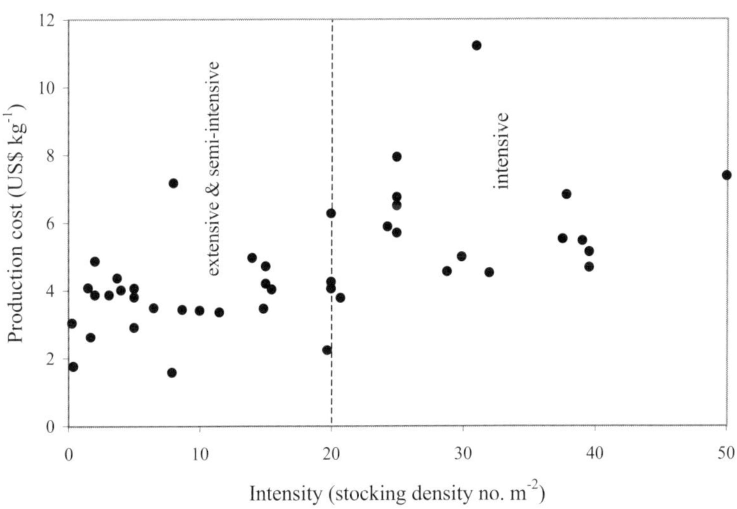

Fig. 10.7 General trend of increasing production cost with increasing intensity (data from Table 10.7 but excluding extensive culture in warm temperate regions).

reviewed and, where possible, compiled to illustrate the major components of investment and operating costs. Although most individual studies necessarily relate to a particular situation and are based on a specific set of underlying assumptions, taken together they serve as useful reference material for the financial appraisal of crustacean farming proposals and may help project sponsors gauge whether they are getting value for money. Cost data obtained in currencies other than US$ have been converted at the exchange rate prevailing when the data were released or published, except in Table 10.10 where the same exchange rate is applied to all listed data. Unless otherwise stated, figures are not adjusted for inflation and are best interpreted with reference to the date of the source.

10.6.1 Shrimp and prawns

10.6.1.1 Hatcheries

The cost of setting up a simple backyard hatchery for *Macrobrachium* or penaeids with the capacity to produce $0.5–5 \times 10^6$ post-larvae per year can be less than \$10 000. Valenti and Daniels (2000) gave a figure of \$4000 for a backyard hatchery in Brazil designed to produce 100 000 *M. rosenbergi* post-larvae per month in two tanks of 1000 L each. For a larger hatchery (6.3×10^6 *M. rosenbergii* post-larvae per year) Fuller *et al.* (1992) quoted a figure of \$83 523. Rhodes (2000a) estimated that investment in prawn hatcheries averaged \$14.41 per 1000 post-larvae produced per year, on the basis of data adjusted for inflation to 1997 values. The lowest costs are possible when cheap or locally available materials, such as bamboo, wood, cement blocks and plastic sheeting, are used in construction. Inexpensive tanks can be made from up-ended sections of concrete drainage pipes or plastic pool liners within wood or bamboo frames. Such structures, however, may only be temporary. In backyard hatcheries equipment is kept to a minimum; where possible, submersible pumps and hoses are used in preference to fixed pumps and rigid pipework, and standpipes are fitted in place of tank valves. All the same, if the hatchery is dependent upon mains electricity, provision must be made for emergency power to maintain aeration in case of power cuts. This may take the form of a small generator or a petrol motor that can directly operate an air compressor.

Since *Macrobrachium* hatcheries can usually rely on supplies of berried female broodstock from local farms, they do not require separate facilities for broodstock production or maturation. Generally speaking, penaeid hatcheries that receive supplies of broodstock or nauplii from elsewhere are also able to concentrate resources on rearing larvae and post-larvae.

Large penaeid hatcheries with the capacity to produce more than 200×10^6 post-larvae per year can easily represent investments of more than \$2m. The expense can be attributed to the construction of facilities with a planned life of 10–20 years, and investment in multiple back-up systems. The latter may include broodstock ponds and a controlled-environment maturation unit (section 10.6.1.2) to provide a supplementary supply of nauplii, as well as spare generators, pumps and compressors to provide security for the electricity, water and air supplies respectively. Plumbing is usually fixed, and some hatcheries incorporate dual seawater distribution lines that enable pipework to be alternated between normal usage and disinfection. A controlled indoor environment is usually provided for the maintenance of algae stock cultures and the early stages of algae culture. The largest projects often include buildings for the accommodation of staff and as much as 20% of the total investment may be spent on acquiring a design and technical assistance package (section 9.3.8).

If a hatchery is to operate above ambient temperatures, investment costs are increased by the need to provide a heating system and sometimes insulation. A *Macrobrachium* hatchery designed by AQUACOP (1982) included a solar heating system that represented 20.5% of the total investment. Many hatcheries in Ecuador have boilers and heat exchangers to raise water temperatures during the cool season.

Table 10.3 lists examples of investment costs for a range of different shrimp and prawn hatcheries. The proportionate components of these investment costs are summarised in Fig. 10.8, and the proportionate components of hatchery production costs, based on the same sources, are shown in Fig. 10.9. From these data it is clear that construction costs (buildings) are the major outlay in establishing a hatchery and that labour is the greatest component of production costs. Some examples of post-larvae production costs are given in Table 10.4. On the basis of data from 11 *Macrobrachium* hatcheries, Rhodes (2000a) calculated that the average cost for a thousand post-larvae was \$10.56 (data adjusted to 1997 values). Table 10.5 lists the works, equipment and services needed to set up a representative medium-sized operation with a production capacity of $4–8 \times 10^6$ penaeid post-larvae ($PL_{7–10}$) per month.

Table 10.3 Examples of investment costs for shrimp and prawn hatcheries.

Species	Location	Output (million per month)	Post-larvae age (a)	Investment (US$) (b)	Facilities included	Reference/source
P. monodon	Philippines	0.05	35	6 000	0.1 ha land, nursery	SEAFDEC 1985
P. monodon	Indonesia	0.25	16	1 400(c)		Yap 1990
M. rosenbergii	Dominica	0.26		215 000(c)		New *et al.* 1978
P. monodon	Vietnam	0.50	10–15	2 900		Quynh 1990
M. rosenbergii	Mississippi	0.53		72 100		Fuller *et al.* 1992
P. monodon	Philippines	0.83	21	113 000		NACA 1986
M. rosenbergii	Tahiti	0.83		336 000	0.3 ha land, broodstock ponds	AQUACOP 1979
M. rosenbergii	unspecified	1.25		412 000(c)	solar heating system	AQUACOP 1982
P. semisulcatus	Kuwait	1.67		394 000	land free	Farmer 1981
not specified	USA Texas	3.75	5	554 000	2 ha land	Johns *et al.* 1981a
P. monodon	SE Asia	4.17	15	540 000(c)	broodstock ponds, maturation, nursery	IFC 1987
L. vannamei	Ecuador	4.58	7	766 000(c)	broodstock ponds, maturation	Ecuador, 1985, unpublished data
L. vannamei	Ecuador	10.00	7?	500 000	asking price (built) + land	A. Al-Hajj, 1989, pers. comm.
P. monodon	Thailand	10.00	20	1 040 000(c)	broodstock ponds, maturation	Thailand, 1988, unpublished data
not specified	USA Texas	15.00	5	1 149 000	2 ha land	Johns *et al.* 1981a
L. vannamei	Ecuador	16.70	7	1 326 000(c)	broodstock ponds, maturation	Ecuador, 1986, unpublished data
L. vannamei	USA Texas	16.90		1 628 000(c)	maturation	Lawrence 1985
L. vannamei	Ecuador	20.00	7?	2 150 000	asking price (built), 21 ha land, maturation	A. Al-Hajj, 1989, pers. comm.
L. vannamei	Ecuador	30.00	15	3 660 000	broodstock ponds, maturation, nursery	Ecuador, 1987, unpublished data
not specified	USA Texas	45.00	5	3 562 000	2 ha land	Johns *et al.* 1981a

(a) Days from metamorphosis. (b) Adjusted for inflation to (1989) US$. (c) Excluding land.

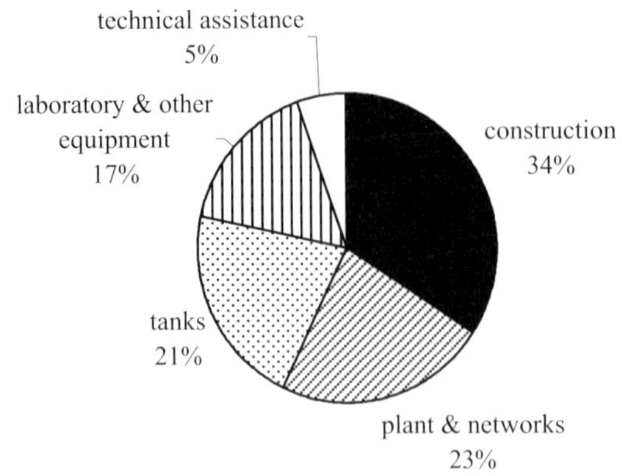

Fig. 10.8 Proportionate investment costs for a shrimp or prawn hatchery (based on sources in Table 10.3).

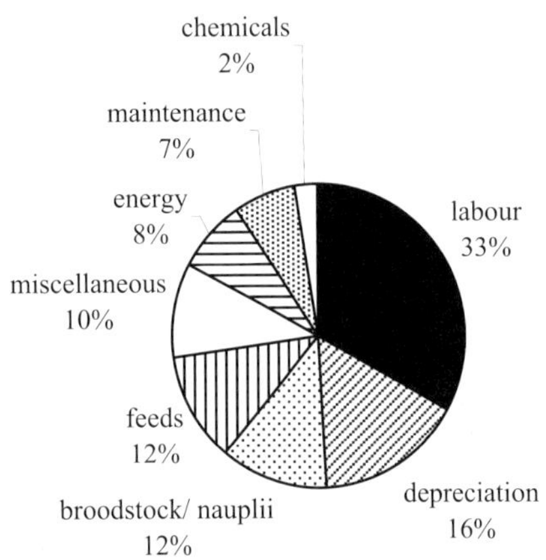

Fig. 10.9 Proportionate production costs for a shrimp or prawn hatchery (based on sources in Table 10.3).

If broodstock ponds are required, their inclusion will constitute a major additional expense. If complete reliance is to be placed on pond-reared stock, IFC (1987) estimate that 1 ha of broodstock ponds are needed for each 10×10^6 post-larvae produced annually. Broodstock ponds in biosecure units in Hawaii are stocked at a density of 2.7 shrimp m^{-2} (Wyban *et al.* 1992).

There are significant economies of scale in hatcheries and it seems that these can be captured in operations producing at least $5–10 \times 10^6$ post-larvae per month (Fig. 10.10). Economy of scale, however, is not the only factor influencing profitability. Indeed Israel (1987b) studied data from hatcheries in the Philippines and established that small-scale and backyard hatcheries gave better financial rates of return than medium and large-sized operations, despite the lower unit production costs prevalent in large hatcheries. This conclusion may have been influenced by accounting methods in which small family concerns tend to exclude labour costs, but it would account for the proliferation of backyard hatcheries in South-east Asia. The advantage of small hatcheries is their flexibility; they are able to operate in response to favourable conditions such as a seasonal abundance of wild spawners. Larger operations, on the other hand, can only take full advantage of economies of scale when able to maintain year-round production.

10.6.1.2 Penaeid maturation units

A significant expense for most penaeid hatcheries is the purchase of broodstock. Table 10.6 gives examples of prices paid for broodstock. In order to avoid or reduce dependence on wild spawners, some hatcheries

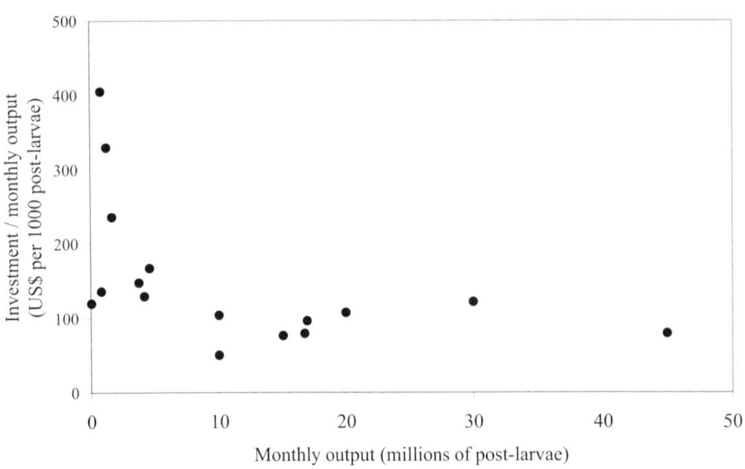

Fig. 10.10 Economies of scale in hatcheries: effect of scale (monthly output) on investment needed to produce 1000 post-larvae monthly (based on data in Table 10.3).

Table 10.4 Examples of post-larvae production costs.

Species	Location	Cost (US$ per 1000)	Age or size	Reference
L. vannamei	Ecuador	3.24	PL_{10-15}	Wilkenfield 1992
L. vannamei	Colombia	3.50	PL_{10-15}	Wilkenfield 1992
L. vannamei	Panama	3.66	PL_{10-15}	Wilkenfield 1992
L. vannamei	Texas	5.20	PL_{4-9}	Wilkenfield 1992
L. vannamei	Costa Rica	6.50	PL_{10-15}	Wilkenfield 1992
L. vannamei	Hawaii	8.00	PL_{4-9}	Wilkenfield 1992
L. vannamei	S. Carolina	9.00	PL_{4-9}	Wilkenfield 1992
P. monodon	Indonesia	2.63	$>PL_{16}$	Wilkenfield 1992
P. monodon	Thailand	5.40	$>PL_{16}$	Wilkenfield 1992
P. monodon	Philippines	5.89	$>PL_{16}$	Wilkenfield 1992
P. monodon	Sri Lanka	7.00	$>PL_{16}$	Wilkenfield 1992
P. monodon	Philippines	3.32	—	Parado-Estepa 1995
M. japonicus	France	25.00	$>PL_{16}$	Wilkenfield 1992
M. rosenbergii	Vietnam	0.68–2.00	26 d (post-hatch)	Nghia 1991
M. rosenbergii	Brazil	4.00	—	Valenti & Daniels 2000 (1998 data)
M. rosenbergii	Mississippi	7.87	—	Fuller *et al.* 1992
M. rosenbergii	various	10.56	—	Rhodes 2000a (1997 data)
M. rosenbergii	Mississippi	24.00	0.25 g	Daniels *et al.* 1995

use maturation facilities to produce their nauplii. In the 1980s the cost of setting up a maturation facility to produce 1.5×10^6 nauplii daily (*Litopenaeus vannamei* – assumed), was estimated at $400 000 (Lawrence *et al.* 1985). The 557 m^2 facility envisaged, formed part of an integrated project including a larvae rearing unit costing $700 000. Further data from the 1980s relate to a maturation unit with 16 circular tanks of 3.66 m in diameter designed to produce on average 3.36×10^6 nauplii (*Litopenaeus stylirostris*) per day. The investment cost was put at $331 772 and the annual operating costs at $243 000 (Johns *et al.* 1981b).

The operating costs of any maturation unit will consist of the major elements illustrated in Fig. 10.11. Labour costs include the salaries of suitably trained or experienced managerial staff, and feed costs are high because a high-quality expensive diet is usually required (sections 2.4.7 and 7.2.2.5). Broodstock are also a major expense (especially if imported or pond-reared) and require routine replacement at a rate of anything between 8% and 60% per month. There have been concerns about the quality of the nauplii produced in some maturation systems. As a result a sharp price differential developed in Ecuador between nauplii from artificially matured females ($0.5–0.6 per 1000) and nauplii from wild-caught gravid females ($1.2–1.4 per 1000) (Rosenberry 1989). However, evidence from elsewhere in the Americas indicates that the quality of nauplii rapidly improves in captive stocks within a few generations of selective breeding. Nauplii produced in Florida have been sold in the Caribbean region for $1.8 per 1000 (D. Lee, 1993 unpublished data).

In addition to a maturation unit, shrimp hatcheries operating in subtropical areas may need to overwinter broodstock from one season to the next in a controlled environment facility. Preston *et al.* (1999) looked at the costs of such a unit in Australia with a capacity for 1440 *Marsupenaeus japonicus* in ten tanks. An investment of $42 000 was needed and the total cost per shrimp was estimated to be $6.30 which was a considerable saving compared to the $43 cost of wild-caught animals. However, despite the possible savings, initial trials revealed quality concerns: the overwintered broodstock were less fecund than wild-caught females and the egg hatch rate was inferior.

10.6.1.3 Nurseries

The stocking of nursery-reared juveniles rather than young post-larvae into ongrowing ponds can improve survival rates, reduce the duration of the ongrowing cycle, and maximise returns, particularly in areas where the growth season is short (section 5.3). The extra value attached to nursed juveniles can be illustrated by data

Table 10.5 Listing of works, equipment and services needed to establish a representative medium-sized penaeid hatchery with a monthly output of 4–8 × 10^6 post-larvae (PL_{7-10}).

Construction/Buildings
- site preparation, inc. roads
- drainage
- hatchery buildings
- ancillary buildings, e.g. canteen, restroom, accommodation
- fencing

Plant, Equipment & Networks
- generator & fuel tank
- electric network & installations
- seawater intake
- seawater pumps & distribution network
- freshwater pump & network
- filters & water treatment apparatus
- air compressors & network
- air conditioning units
- effluent treatment

Tanks
- reservoirs
- broodstock holding/maturation
- spawning/hatching
- larvae rearing
- algae culture
- *Artemia* hatching/holding/enrichment
- rotifer culture/holding/enrichment

Laboratory & Other Equipment
- low power binocular microscope
- stage microscope 40–400 ×
- salinity refractometer
- pH meter
- haemacytometers
- balance, 0.01 g precision
- balance, 1 g precision
- assorted general laboratory items
- hand nets, mesh screens, sieves, air diffusers
- transport vessel, air/oxygen supply
- autoclave for algae culture materials
- refrigerator
- freezer
- food preparation apparatus, e.g. blender, fish cutter
- cleaning equipment, e.g. sponges, brushes
- calculators, computer
- telephone/radio
- road vehicles for personnel/delivery
- fire extinguishers
- furniture
- office equipment
- tools for maintenance and repairs

Services
- surveying
- construction
- design/supervision
- consulting/technical assistance

for *Macrobrachium rosenbergii* in Brazil: whereas post-larvae were sold for $7–25 per thousand, juveniles of 2–3 cm (TL) fetched $40 per thousand (Valenti 1993). These price differences reflect differences in production costs. For example the cost of producing 1000 *Macrobrachium* juveniles of 2 g in Brazil and Costa Rica was estimated to be $6–10 but production costs for larger 10 g juveniles in Thailand were estimated at $50 per thousand (1997 data; Rhodes 2000a). Similarly, data from the 1980s for *Penaeus monodon* show that in the Philippines the price of PL_{40} was roughly double that of PL_{20} (Primavera 1983), and in Taiwan PL_{27-29} fetched around 80% more than PL_{13-15} (Mock 1983). In these and other countries in South-east Asia many nurseries are still operated as independent concerns and their viability is assisted by the short culture period (12–25 days) of each batch of juveniles, which permits many production cycles per year. Nursery ponds incorporated into extensive and semi-intensive farms are typically stocked for longer periods of 20–60 days. They are operated to suit the production and harvesting schedule of ongrowing, and often serve as juvenile holding ponds. In a semi-intensive shrimp farm, nursery ponds usually represent around 5–10% of the total surface area and so a similar proportion of the total farm investment costs can be attributed to them (although allowance needs to be made for the extra cost per hectare of building smaller ponds; section 8.2.2).

In Texas and other locations with warm temperate or Mediterranean climates, the use of nurseries enclosed within greenhouses has been investigated as a means of producing juveniles in advance of the normal ongrowing season. In this way it is possible to boost overall productivity and sometimes obtain two ongrowing cycles per year. Smith *et al.* (1983) used enclosed nurseries for the production of *Macrobrachium* juveniles. Although actual commercial data were not available, they calculated that an ongrowing operation could reasonably afford to pay $30 per 1000 for juveniles of 0.3–0.4 g and up to $45 and $60 per 1000, respectively, for larger animals of 1 g and 2 g. In another system Juan *et al.* (1988) calculated the cost of producing 1 g penaeid juveniles to be $15.1–17.21 per 1000. However, they concluded that using these juveniles to obtain two crops of shrimp per year was not as profitable as stocking young post-larvae (PL_5, which cost only $7.50 per 1000) directly into outdoor ongrowing ponds and obtaining a single crop per year.

The greatest expense of running a nursery is the initial cost of the post-larvae, so the survival rate is critical

Table 10.6 Examples of prices for penaeid broodstock.

Species	Location	Price (US$ each)	Notes	Year	Reference/source
M. japonicus	Australia	43		1999	Preston *et al.* 1999
M. japonicus	Australia	10.5	Pond-reared	1999	Preston *et al.* 1999
L. vannamei	Bahamas	25	Import from Panama	1993	Lee, D. 1993 unpubl.
F. chinensis	China	4.05		1988	Rosenberry 1988
P. monodon	Vietnam	5.5–11.0	Low demand	1989	Quynh 1990
P. monodon	Vietnam	25–30	High demand	1989	Quynh 1990
P. monodon	Indonesia	8.5–22.5		1989	Yap 1990
P. monodon	Thailand	12–50	Gravid & non-gravid	1989	Currie, D. 1989 pers. comm.
P. monodon	Thailand	70	Large	1989	Currie, D. 1989 pers. comm.
P. monodon	Taiwan	60–150	Gravid & non-gravid	1987	Chen 1990
P. monodon	Taiwan	50–500		1980	Chen 1990

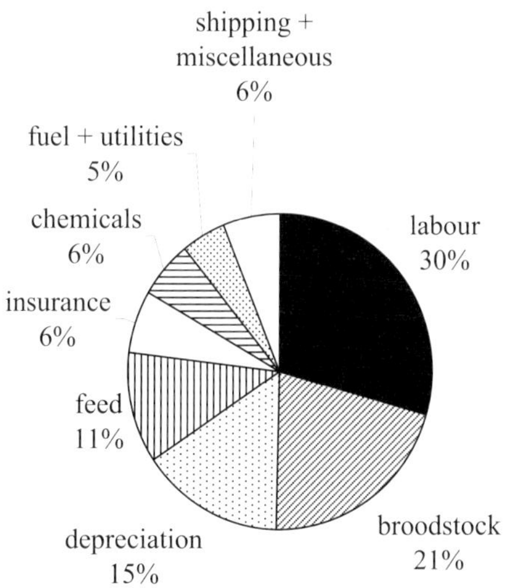

Fig. 10.11 Proportionate operating costs for a shrimp maturation facility (based on Johns *et al.* 1981a, interest and taxes excluded).

to profitability. At least 50–70% survival is desirable. If survival falls to 30% the selling price (or value) of the nursed juveniles would need to be more than triple the purchase cost, simply to cover the initial purchase of post-larvae.

10.6.1.4 Bait shrimp

The market for live bait for fishermen in places such as the southern USA has stimulated interest in the production of small (6.5–8.1 g) shrimp in high-density systems. Samocha *et al.* (1998) calculated that an outdoor pond system with plastic liners and a blower driven aeration grid could be economically feasible (IRR of 14.4%) based on two crops per year of the native species *Litopenaeus setiferus* and a high-quality, 40% protein, diet. A very high stocking density of 700 shrimp m^{-2} was also assumed.

10.6.1.5 Farm investment costs

Table 10.7 shows investment and operating costs for a wide range of shrimp and prawn farms and is based upon actual case studies as well as published estimates. Examples are arranged in order of increasing intensity on the basis of stocking density (Table 7.6). The data range from low-input extensive systems in Asia producing a few tonnes of shrimp per year to super-intensive systems in the USA capable of producing hundreds of tonnes per year. The relationship between the scale of a project and the necessary investment is shown in Fig. 10.12. The slope of the trend line indicates that, for all but the smallest operations, around $5200 is invested per tonne of annual output. Surprisingly this data set does not reveal any economies of scale. This may be partly due to the fact that some of the data relate to super-intensive systems that are modular in design – i.e. large operations simply consist of multiples of small independent production units and there are few opportunities for shared costs.

Construction is normally by far the greatest expense in establishing a shrimp or prawn farm. Data from a number of the semi-intensive farms listed in Table 10.7 indicated that construction accounted for on average 61% of total investment costs (range 44–81%).

Table 10.7 Examples of investment and operating costs for shrimp and prawn farms.

Species	Location	Output (mt yr^{-1}) (a)	Yield (kg ha^{-1} crop^{-1})	Stocking density (no. ha^{-1})	System	Farm size (ha ponds)	Operating costs (US$ yr^{-1})	Production cost (US$ kg^{-1}) (a)	Total investment (US$) (excluding land)	Reference/source
P. monodon	Vietnam	0.8	61	3 000	extensive	10.3	2 466	3.03	—	Ling *et al.* 1999
P. monodon	Thailand	4.8	164	4 000	extensive	12.2	8 364	1.74	—	Ling *et al.* 1999
M. japonicus	Italy	—	189	14 000	extensive			14.88	—	Lumare *et al.* 1989
P. monodon	Bangladesh	3.6	180	15 000	extensive	16.6	14 593	4.07	—	Ling *et al.* 1999
P. monodon	Philippines	2.8	153	17 000	extensive	10.8	7 329	2.61	—	Ling *et al.* 1999
M. rosenbergii	Egypt	0.6	844	20 000	rice polyculture	0.7	2 321	3.86	—	Sadek & Moreau 1998
M. rosenbergii	Egypt	3.1	684	20 000	extensive	2.3	15 030	4.85	27 754	Sadek & Moreau 1996
M. japonicus	Italy	—	525	30 000	extensive	—	—	9.62	—	Lumare *et al.* 1989
P. monodon	Indonesia	0.8	81	31 000	extensive	5.0	3 127	3.86	—	Ling *et al.* 1999
P. monodon	India	0.8	535	37 000	extensive	1.2	3 633	4.35	—	Ling *et al.* 1999
P. monodon	SE Asia	612.0	784	40 000	extensive	300.0	2 445 324	4.00	4 292 296	IFC 1987
L. vannamei	Ecuador	113.0	500	—	extensive	90.0	—	—	1 000 000 (f) (g)	A. Al-Hajj, 1989 pers. comm.
L. vannamei	Ecuador	124.0	500	—	extensive	99.0	—	—	700 000 (f) (g)	A. Al-Hajj, 1989 pers. comm.
L. vannamei	Ecuador	313.0	500	—	extensive	250.0	—	—	1 500 000 (f)	A. Al-Hajj, 1989 pers. comm.
L. vannamei	Ecuador	319.0	500	—	extensive	255.0	—	—	2 500 000 (f)	A. Al-Hajj, 1989 pers. comm.
L. vannamei	Ecuador	434.0	500	—	extensive	347.0	—	—	2 500 000 (f) (g)	A. Al-Hajj, 1989 pers. comm.
L. vannamei	Ecuador	500.0	500	—	extensive	400.0	—	—	4 500 000 (f)	A. Al-Hajj, 1989 pers. comm.
L. vannamei	Ecuador	625.0	500	—	extensive	500.0	—	—	5 000 000 (f)	A. Al-Hajj, 1989 pers. comm.
L. vannamei	Ecuador	775.0	500	—	extensive	620.0	—	—	4 350 000 (f)	A. Al-Hajj, 1989 pers. comm.
P. monodon	Sri Lanka	1.5	1 250	50 000	semi-intensive	0.6	6 070	4.05	2 196 (b)	De Silva & Jayasinghe 1993
P. monodon	Philippines	1.9	935	50 000	semi-intensive	1.0	5 505	2.90	480 (c)	Millamena & Triño 1997
M. rosenbergii	Egypt	7.9	1 760	50 000	semi-intensive	2.3	30 038	3.79	27 754	Sadek & Moreau 1996
L. vannamei	Ecuador	2721.0	853	65 000	semi-intensive	1418.0	9 471 250	3.48	5 672 000	Hirono 1989
F. chinensis	China	16.6	421	79 000	semi-intensive	39.5	25 942	1.56	—	Ling *et al.* 1999
P. monodon	NSW, Australia	7.2	1 200	80 000	semi-intensive	6.0	86 100	11.96	235 400	Hardman *et al.* 1990
P. monodon	Qslnd, Aust.	18.5	1 540	80 000	semi-intensive	6.0	132 400	7.16	235 400	Hardman *et al.* 1990
L. vannamei	Ecuador	295.0	864	86 800	semi-intensive	200.0	1 007 500	3.42	1 200 000	Villalon 1993
P. monodon		1000.0	1 111	100 000	semi-intensive	400.0	—	—	5 295 000	D. Lee, 1987 unpubl. data
P. monodon	Philippines	2.6	1 275	100 000	semi-intensive	1.0	8 841	3.40	1 720	Millamena & Triño 1997
P. monodon	Vietnam	0.9	348	115 000	semi-intensive	1.4	3 096	3.34	—	Ling *et al.* 1999
L. vannamei	Texas, USA	273.0	3 366	140 000	semi-intensive	81.0	1 354 200	4.95	627 280	Parker 1990
P. monodon	Sri Lanka	14.7	1 732	149 000	semi-intensive	5.0	50 784	3.45	—	Ling *et al.* 1999
L. vannamei	Texas	132.0	1 650	150 670	semi-intensive	80.0	618 477	4.69	1 063 426 (d)	Lambregts *et al.* 1993
L. vannamei	Texas	528.0	1 650	150 670	semi-intensive	320.0	2 214 541	4.19	3 092 493 (d)	Lambregts *et al.* 1993
P. monodon	Philippines	20.3	1 929	155 000	semi-intensive	7.5	81 232	4.01	—	Ling *et al.* 1999
F. chinensis	China	21.1	848	197 000	semi-intensive	24.9	46 876	2.22	—	Ling *et al.* 1999
P. monodon	Asia	1000.0	3 000	—	semi-intensive	112.0	—	—	4 672 000	D. Lee, 1987 unpubl. data
P. monodon	Taiwan	24.8	7 333	200 000	intensive	2.3	155 140	6.25	18 821	Chen 1990 (1987 data)

P. monodon	Thailand	583.0	3 600	200 000	intensive	90.0	3 776 000	4.04	2 608 000		Thailand 1988 unpubl. data
P. monodon	Philippines	3.7	1 835	200 000	intensive	1.0	15 698	4.24	2 880		Millamena & Triño 1997
P. monodon	Indonesia	3.0	740	207 000	intensive	2.0	11 122	3.76	—		Ling *et al.* 1999
P. monodon	India	15.2	1 583	243 000	intensive	6.4	89 034	5.86	—		Ling *et al.* 1999
P. monodon	NSW, Australia	22.5	3 750	250 000	intensive	6.0	178 100	7.92	332 700		Hardman *et al.* 1990
P. monodon	Qslnd, Aust.	45.0	3 750	250 000	intensive	6.0	303 200	6.74	332 700		Hardman *et al.* 1990
P. monodon	NSW Australia	75.0	3 750	250 000	intensive	20.0	486 900	6.49	769 600		Hardman *et al.* 1990
P. monodon	Qslnd, Aust.	150.0	3 750	250 000	intensive	20.0	852 100	5.68	769 600		Hardman *et al.* 1990
P. monodon	Sri Lanka	12.6	2 965	288 000	intensive	2.5	57 204	4.54	—		Ling *et al.* 1999
F. indicus	India	100.0	3 155	299 000	intensive	19.8	496 753	4.97	—		Ling *et al.* 1999
P. monodon	Bangladesh	20.7	1 166	310 000	intensive	12.7	232 278	11.20	—		Ling *et al.* 1999
P. monodon	Sri Lanka	15.8	4 486	320 000	intensive	2.2	71 062	4.50	—		Ling *et al.* 1999
L. vannamei	Texas, USA	518.0	1 850	335 000	intensive	280.0	—	—	2 220 400		Jaenike 1989
M. japonicus	Japan	—	4 500	350 000	partial embankments	—	—	34.24	—		Yamaha 1989
P. monodon	Indonesia	1152.0	5 000	375 000	plastic lined ponds	115.0	6 336 000	5.50	10 972 168		P. Fuke, 1990 pers. comm.
P. monodon	Philippines	27.2	2 352	378 000	intensive	8.9	185 282	6.81	—		Ling *et al.* 1999
P. monodon	Malaysia	9.9	2 470	390 000	intensive	2.1	53 613	5.44	—		Ling *et al.* 1999
L. vannamei	Texas	142.4	3 560	395 200	intensive	40.0	729 545	5.12	989 004	(d)	Lambregts *et al.* 1993
L. vannamei	Texas	569.6	3 560	395 200	intensive	160.0	2 656 303	4.66	3 081 175	(d)	Lambregts *et al.* 1993
M. japonicus	Japan	—	8 000	500 000	super-intensive	—	—	37.67	—		Yamaha 1989
L. vannamei	Texas	2.8	7 867	500 000	super-intensive	0.1	20 844	7.36	120 000		Samocha & Lawrence 1998
F. chinensis	China	9.0	1 229	540 000	super-intensive	6.9	41 213	4.86	—		Ling *et al.* 1999
P. monodon	Indonesia	8.8	2 312	675 000	super-intensive	2.0	38 913	4.43	—		Ling *et al.* 1999
P. monodon	Malaysia	18.8	3 293	682 000	super-intensive	3.0	90 462	4.82	—		Ling *et al.* 1999
P. monodon	Taiwan	7.3	1 652	714 000	super-intensive	2.6	53 515	7.33	—		Ling *et al.* 1999
L. vannamei	Brazil	55.4	8 000	720 000	cages	2.8	333	6.01	474 000		Paquotte *et al.* 1998
M. japonicus	Japan		17 500	750 000	Shigueno	—	—	44.52	—		Yamaha 1989
L. vannamei	Texas	217.7	6 803	815 100	super-intensive	32.0	1 168 195	5.37	1 331 077	(d)	Lambregts *et al.* 1993
L. vannamei	Texas	870.8	6 803	815 100	super-intensive	128.0	4 308 326	4.95	4 058 539	(d)	Lambregts *et al.* 1993
P. monodon	Malaysia	497.0	16 000	900 000	super-intensive	10.0	1 392 000	2.80	1 038 000		Malaysia 1988 unpubl. data
L. vannamei	Mississippi	174.3	9 770	1 000 000	super-intensive	8.9	1 870 239	10.73	5 500 000		Ogle & Lotz 1998
P. monodon	Thailand	21.5	5 646	1 151 000	super-intensive	2.0	89 034	4.15	—		Ling *et al.* 1999
L. vannamei	Texas	625.0	25 000	1 360 000	super-intensive	7.1	3 867 500	6.19	3 350 000		Moss *et al.* 1998

(a) Whole shrimp or prawns. (b) Includes land. (c) Ponds rented not constructed. (d) Includes land, (15–37% of total). (e) Does not include depreciation. (f) Asking price already built, including land. (g) Price includes extra land for development.

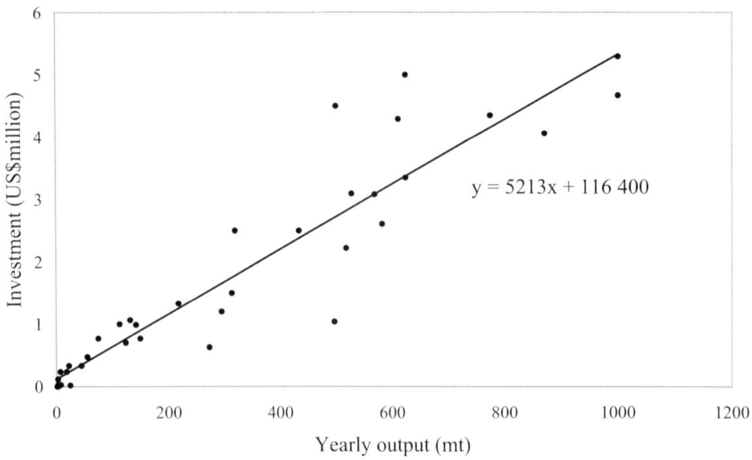

Fig. 10.12 Relationship between scale and investment cost for shrimp and prawn farms (three outlying points removed).

Griffin *et al.* (1993) investigated the costs of building aquaculture ponds in Texas and found that engineering and construction costs constituted 50–60% of total capital costs. The most important component of this cost is earth moving. IFC (1987) calculated that 442 720 m^3 of earth would need to be moved, at an estimated cost of \$2.50 m^{-3}, in order to construct 321 ha of ponds and supply and drain canals. The cost of earth moving can vary greatly with locality and with soil type and the clearance of vegetation, roots and rocks can add greatly to the expense. In some large projects it has proved to be worth buying earth-working machinery and employ crews as an alternative to hiring the services of a contractor for the duration of the construction phase. In a review of investment costs for *Macrobrachium* farms around the world, Rhodes (2000a) found that the average investment cost per hectare in the USA was five times the average for other countries. This was attributed to higher construction labour wages and to more complex designs. The average level of investment for the 13 prawn farms in the dataset was \$18 675 ha^{-1} (1997 prices) but the range was wide, with as little as \$400 ha^{-1} invested in farms in Tamil Nadu India, and as much as \$63 371 ha^{-1} invested in a small, 0.4 ha, farm in Hawaii.

The impact of engineering design on the construction costs and the profitability of a hypothetical 40 ha farm in Texas were analysed by Griffin *et al.* (1993). The design factors with greatest influence were pond shape and pond size and the slope and width of embankments were less important. Square ponds of 2 ha led to construction costs of \$3721 ha^{-1} and an IRR of 17%, but if the ponds were long and thin (20 : 1 length to width) rather than square, the costs increased to \$12 297 ha^{-1} and the IRR dropped to 8%. Increasing the size of square ponds from 2 ha to 10 ha decreased construction costs to \$2345 ha^{-1} and raised the IRR to 21%. Thus building square ponds larger can offer significant savings. All the same, it should be remembered that small ponds are generally more productive per unit area than large ponds. Lee *et al.* (2000) for example found that 2 ha ponds were on average 30% more productive per hectare than 4 ha ponds. The total cost of building a farm in Saudi Arabia with about 100 ha of circular production ponds and another 100 ha of pretreatment reservoirs was \$13m (Falaise & Boël 2000). Of this sum, 30% represented earthworks and concrete works, 30% buildings, 23% equipment, 12% the power network, and the remaining 5% the initial site and feasibility studies.

If ponds are to be made with concrete walls rather than earthen dikes, construction costs per hectare will be far higher – by a factor of six in the case of Taiwan (Chiang & Liao 1985). The cost of constructing small ponds (0.36 ha) with plastic liners is yet higher and has been put at around \$31 500 ha^{-1} (P. Fuke, 1990 pers. comm.). Another estimate for ponds of 0.2 ha put the cost at \$50 000 ha^{-1}, of which 54% represented the cost of the polyethylene liners (Singh 1993). This type of pond has been used in permeable sandy areas in Oman, Indonesia and the Bahamas and is suited to intensive culture (section 7.2.6.5). If too much money is spent building a farm it can be difficult subsequently to recoup the costs. Some intensive shrimp farmers in the Philippines have struggled to operate profitably because of the burden of amortising farm development costs, which can account for 25–35% of total production costs (Cruz 1993). Intensive shrimp farming usually implies increased investment in reinforced pond embankments, water control structures, aerators and electrical installations.

Land purchase can also be a major financial outlay. Prices per hectare vary depending on competing land usage, and may increase in areas where shrimp or prawn culture is successful. In Thailand land for rice may cost \$120–400 ha^{-1} compared to \$12 000–24 000 for shrimp (Primavera 1997). Values ranging between \$2000 ha^{-1} (IFC 1987) and \$40 800 ha^{-1} (Chiang & Liao 1985) have been used in various cost analyses, though in many situations land can be leased at economical rates. In case studies from six different countries (Shang 1983), most land costs appeared as annual lease payments, representing only around 1% or less of the annual operating budget. The highest figure of 4% came from Hawaii. Land costs in Texas, USA, represented 30–38% of total investment for a semi-intensive farm, 16–23% for an intensive farm and 11–16% on a super-intensive farm (Lambregts *et al.* 1993).

A listing of works, equipment and services required to set up a shrimp or prawn farm is provided in Table 10.8. Dredging is sometimes required to enable water from a nearby creek or other body of water to reach the site of a pumping station. Expenditure on the water supply system will include installation and purchase of pumps and piping. Sometimes a well is needed to supply freshwater and, in the case of a prawn farm relying on a borehole, well costs can be considerable. For a *Macrobrachium* farm in South Carolina, Bauer *et al.* (1983) estimated a well would represent 36% of the investment costs. Obviously, because of the cost implications, the proximity of the water supply is one of the major considerations in site selection (section 6.3.1.2).

Super-intensive systems have been developed for use in temperate and Mediterranean climates. They are usually enclosed within greenhouses and incorporate water treatment and recirculation systems, and on a per unit area basis they are some of the most expensive farms to establish. Moss *et al.* (1998) proposed a design for a modular system of paired raceways in which a total water surface area of 7.14 ha represented an investment of \$3.35m, equivalent to \$469 000 ha^{-1}. The key to economic success with super-intensive culture will be to

Table 10.8 Listing of works, equipment and services needed to establish a semi-intensive shrimp or prawn farm.

Construction	
	earthworks
	inlet/outlet gates
	culverts
	harvest basins
	dike stabilisation
	access roads
	dredging
Water supply	
	pumps
	piping
	well
Buildings	
	feed storage
	office
	workshop
	accommodation
	guard huts
Electrical supply/installations	
	generator(s), fuel tank
	network
	lighting
Equipment	
	tools
	ice machine
	communications equipment
	office equipment
	excavator
	harvest pump, nets and other equipment
	aerators
Instruments	
	oxygen meter
	pH meter
	salinity refractometer
	low-power binocular microscope
	balance, 500 × 0.1 g
	weighing scales, 200 × 0.5 kg (for harvesting)
Vehicles	
	boats, outboard motors
	tractor, trailer
	pick-up truck
	truck
	cars
	motorcycles
Miscellaneous	
	fencing
Services	
	design, engineering
	consulting/technical assistance
	surveying
	supervision
	legal

develop systems that are not too expensive but that can deliver consistently high yields with simple and reliable techniques for the management of feeding and water quality. Pilot-scale trials continue and one project has been set up in a barn in Michigan, USA, for around \$0.5m. It consists of reinforced concrete raceways stacked nine high. The raceways are sloped for easy cleaning and water cascades between the different levels (section 7.2.6.6). In Europe self-cleaning raceways are being used that incorporate conveyor belts to remove settled waste (P. Wood, 2000 pers. comm.).

10.6.1.6 Farm operating costs

In most cases the major components of the cost of running a shrimp or prawn farm are feed and juveniles, which between them account for 50–70% of expenditure. Figures 10.13 and 10.14 show proportionate costs for an intensive shrimp farm in Thailand and an extensive prawn farm in Egypt. The particular dominance of seed costs in the latter example is due to the high price of the large post-larvae used in this particular case. Examples of different prices paid for shrimp and prawn seed in various locations are listed in Table 10.9. In *Macrobrachium* farms, Rhodes (2000a) found that feed on average accounted for 30% of operating costs and seed for 20%. The operating costs for the 17 farms in Rhodes' sample on average totalled \$7.08 kg^{-1} (1997 prices).

Shrimp farming is a capital-intensive rather than a labour-intensive activity (section 10.5) and this is reflected in cost studies in which labour costs, though important, rarely exceed 10% of total operating costs. Similarly in prawn farms (in Asia excepting Taiwan) Rhodes (2000a) found that labour costs were less than 12% of total operating costs. Manning levels for semi-intensive farms in Ecuador are around 15–25 people per 100 ha (CPC 1989; Hirono 1989). A small 2–5 ha shrimp farm in Taiwan may be operated by as few as one skilled technician and one permanent worker, with operations such as harvesting and pond bottom cleaning performed by labourers on temporary contracts (Chien & Liao 1987). Labour productivity in the Australian shrimp industry has been put at 9.5 mt per employee per year (O'Sullivan 1999) and in freshwater prawn farms the labour requirements have been estimated to range between 0.5 and 2 workers per hectare, depending on the intensity of management (Valenti & New 2000).

Of the factors that have a clear influence on production costs, location is among the most important (Fig. 10.15).

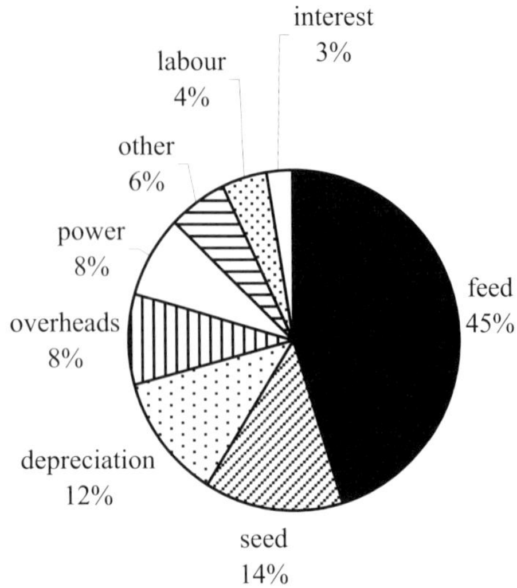

Fig. 10.13 Proportionate operating costs for an intensive shrimp farm in Thailand (based on Ling *et al.* 1999).

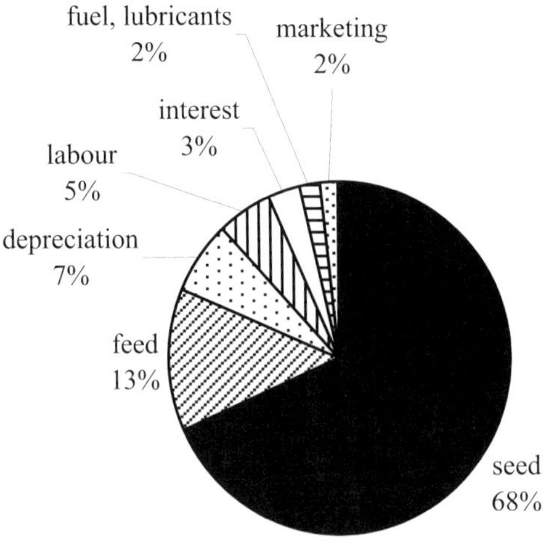

Fig. 10.14 Proportionate operating costs for an extensive *Macrobrachium* farm in Egypt (based on Sadek & Moreau 1996).

This is due to the relatively high operating costs that prevail in industrialised and newly industrialised countries and to restrictions on productivity imposed by the limited ongrowing season in warm temperate and Mediterranean climates. Production costs are particularly high in Japan where there is a shortage of available land and where a decline in shrimp farming has also been linked to disease, ageing ponds and water pollution. In Taiwan the

Table 10.9 Examples of prices paid for shrimp and prawn seed.

Species	Location	Price (US$ per 1000)	Age or size	Reference
L. vannamei	Ecuador	3.63	PL_{10-15}	Wilkenfield 1992
L. vannamei	Ecuador	4.50	—	Villalon 1993
L. vannamei	Brazil	4.60	—	Paquotte *et al.* 1998
L. vannamei	Panama	5.00	PL_{5-10}	D. Lee, 1993 unpubl. data
L. vannamei	Belize	6.00	—	D. Lee, 1993 unpubl. data
L. vannamei	Costa Rica	6.00	PL_{18}	Roma 1997
L. vannamei	Florida	6.50	PL_{4-9}	Wilkenfield 1992
L. vannamei	Colombia	7.00	PL_{10-15}	Wilkenfield 1992
L. vannamei	Panama	7.34	PL_{10-15}	Wilkenfield 1992
L. vannamei	Texas	7.50	—	Griffin *et al.* 1993
L. vannamei	Costa Rica	8.00	PL_{10-15}	Wilkenfield 1992
L. vannamei	USA	8.50	—	Samocha & Lawrence 1998
L. vannamei	Texas	8.60	PL_{4-9}	Wilkenfield 1992
L. vannamei	Texas	9.00	PL_{5}	Lambregts *et al.* 1993
L. vannamei	S. Carolina	9.00	PL_{4-9}	Wilkenfield 1992
L. vannamei	Hawaii	9–11	—	Rosenberry 1994
L. vannamei	Hawaii	12.00	PL_{4-9}	Wilkenfield 1992
L. vannamei	USA	20.00	—	Ogle & Lotz 1998
L. vannamei	USA	25.00	1 g	Moss *et al.* 1998
P. monodon	Taiwan	3.43	—	Kongkeo 1997
P. monodon	Philippines	3.84	—	Parado-Estepa 1995
P. monodon	Indonesia	4.50	—	Kongkeo 1997
P. monodon	Vietnam	5.00	—	Johnston *et al.* 1999
P. monodon	Thailand	5.43	—	Kongkeo 1997
P. monodon	Indonesia	5.77	$>PL_{16}$	Wilkenfield 1992
P. monodon	Thailand	7.00	$>PL_{16}$	Wilkenfield 1992
P. monodon	Philippines	7.20	PL_{18}	Triño & Sarroza 1995
P. monodon	Philippines	7.20	PL_{20} (20 mg)	Millamena & Triño 1997
P. monodon	Philippines	9.06	$>PL_{16}$	Wilkenfield 1992
P. monodon	Philippines	10.21	—	Kongkeo 1997
P. monodon	Sri Lanka	10.60	$>PL_{16}$	Wilkenfield 1992
P. monodon	Sri Lanka	10.90	PL_{20-35}	De Silva & Jayasinghe 1993
M. japonicus	France	33.00	$>PL_{16}$	Wilkenfield 1992
M. japonicus	Spain	35.00	$>PL_{16}$	Wilkenfield 1992
M. rosenbergii	Brazil	7–25	PL	Valenti 1993
M. rosenbergii	Bangladesh	8–12	PL	Angell 1994
M. rosenbergii	USA	8–12	—	Rosenberry 1994
M. rosenbergii	Dominican Rep.	14.52	—	D. Lee, 1993 unpubl. data
M. rosenbergii	Dominican Rep.	16.13	—	D. Lee, 1993 unpubl. data
M. rosenbergii	Dominican Rep.	20.16	45 d (post hatch)	D. Lee, 1993 unpubl. data
M. rosenbergii	Bangladesh	20–30	PL	New (2000) (1999 data)
M. rosenbergii	China	23.00	PL	New (2000) (1999 data)
M. rosenbergii	Guadeloupe	42.00	PL	Lacroix *et al.* 1993
M. rosenbergii	Martinique	51.00	PL	Lacroix *et al.* 1993
M. rosenbergii	Egypt	60.00	PL_{30} (0.3 g)	Sadek & Moreau 1996
M. rosenbergii	Mississippi	60.00	—	D'Abramo *et al.* 1998
M. rosenbergii	Bangladesh	60–100	5–7.5 cm TL	New (2000) (1999 data)
M. rosenbergii	Egypt	88.80	3.0–3.4 g	Sadek & Moreau 1998
M. rosenbergii	Egypt	90.00	2 g	Sadek & Moreau 1996
M. rosenbergii	China	115.00	2–3 cm TL	New (2000) (1999 data)

shrimp industry has been constrained by the high costs of labour, land and water, in combination with a shortage of healthy fry (Ovenden 1994). Shrimp farmers in the USA face high land, labour and materials costs and as a result

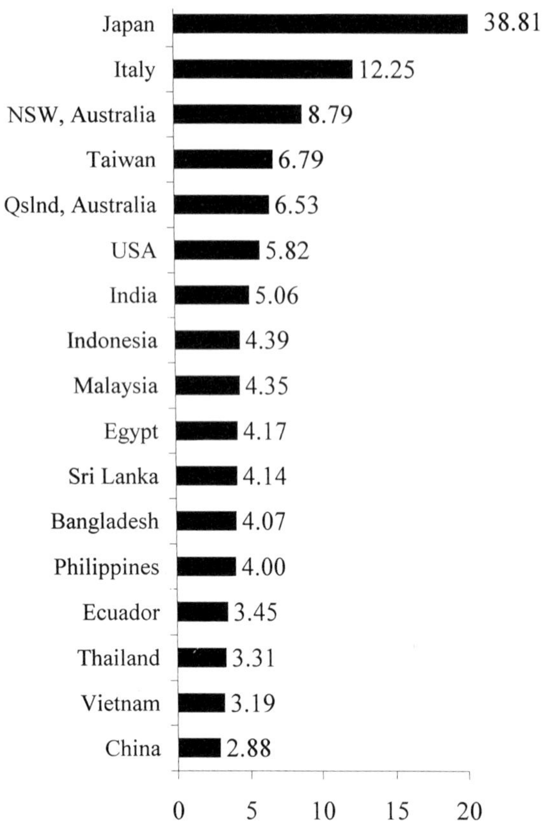

Fig. 10.15 Estimated production cost per kg for shrimp and prawns in various locations (data from Table 10.7).

they have been largely uncompetitive on world markets (Leung *et al.* 1992). Similar factors have greatly limited the commercial success of USA-based *Macrobrachium* farms (Rhodes 2000b).

Energy rarely appears as major component of shrimp and prawn farming production costs, even in systems that require heating. If water is heated it is usually recirculated so that most heat can be retained. An indoor system in Michigan, USA, for example, estimated that heating costs added only \$0.15 kg^{-1} to production costs (section 7.2.6.6). The relatively low fuel requirements of shrimp farming give it a competitive advantage over shrimp trawling which is highly dependent on diesel fuel. Boyd (1998) noted that the cost of aeration is small relative to the value of the crop. For a shrimp pond of 1 ha using 8 kW of aeration and yielding 5000 kg of shrimp, aeration costs including depreciation, maintenance and electricity were 9.5% of crop value. This figure probably overstates typical costs somewhat because it assumed that all aerators were operated for 24 h per day throughout the ongrowing cycle. Many farms reduce energy costs by using fewer aerators at the beginning of the production cycle and some restrict aeration to night-time only.

Despite problems with economic viability in the past, super-intensive recirculating systems are attracting attention because they still hold out potential for shrimp production in highly controlled, disease-free environments close to key markets. A viable modular system has been outlined by Moss *et al.* (1998) on the basis of growth of 1.6 g wk^{-1}, 80% survival, 3.5 crops yr^{-1}, and a resulting production cost of \$6.19 kg^{-1}. However, production costs are highly sensitive to underlying assumptions about productivity. The investigators assumed 2.5 kg m^{-2} $crop^{-1}$ but, in another indoor raceway system, Ogle and Lotz (1998) assumed more conservative values for growth (0.8 g wk^{-1}), survival (50%) and productivity (0.98 kg m^{-2} $crop^{-1}$) with the result that the production cost, at \$10.17 kg^{-1}, was probably too high to form the basis of a profitable operation. The high investment costs of super-intensive systems, when depreciated over the life of the project, have a very strong impact on total production costs. Of the \$10.17 kg^{-1} production cost, for example, 56% represented depreciation of the capital investment (over 5 years). To compensate for such high costs, super-intensive systems must deliver very high yields (section 10.5). Samocha and Lawrence (1998) outlined another super-intensive system comprising sets of four round ponds in greenhouses. They assumed yields of 0.64–0.93 kg m^{-2} $crop^{-1}$, but this productivity level would probably not compensate for the high investment costs of the envisaged system. The intensive outdoor round pond system described by Falaise and Boël (2000) for *Penaeus monodon* farming in Saudi Arabia does not offer the same levels of environmental control but has yielded 0.5–0.75 kg m^{-2} $crop^{-1}$ at a cost of \$6.1 kg^{-1}. Of this total, feed and farm costs account for 54%, general expenses 29%, sales costs 9% and processing 8%. At a market price of \$11 kg^{-1} the IRR of the project is estimated to be a respectable 27%.

10.6.1.7 Cage culture

An alternative system for ongrowing shrimp is to use cages rather than ponds or tanks (section 7.2.6.4). The economics of cage culture trials in Brazil were investigated by Paquotte *et al.* (1998). They found that at the experimental scale employed, ongrowing in cages was not cheaper than in ponds. This was partly attributed to high capital costs – the expense of nets and their short life – and to low labour productivity of 1.3 mt per employee per year. Production costs amounted to $6 kg^{-1} and were higher than the $3.5–4.5 kg^{-1} typical for ponds in Latin America. We are not aware of any published costs for cage farming of shrimp in South-east Asia.

10.6.1.8 Polyculture

Ideally polyculture involves the rearing of compatible species at densities that favour complementary rather than competitive interactions. In reality some competition is often unavoidable, but polyculture can still be financially attractive if the profit from the combined crop exceeds that possible from monoculture alone. Apart from increasing farm output, the advantage of mixed species culture is that it can diversify production and reduce market risk (section 10.4). In addition, when fish are added to crustacean ponds water quality is often improved and the risk of crop loss reduced (Rhodes 2000a).

Some economic data are available for the farming of freshwater prawns with tilapia and carp and with rice. One set of trials in Egypt (Sadek & Moreau 1996) reveals the importance of finding favourable stocking densities. When prawn post-larvae (0.3 g) were reared in monoculture at a density of 2 m^{-2} they yielded an annual financial return of 8%. When prawn post-larvae were reared at the same density in polyculture with tilapia (*Oreochromis niloticus*) (1 m^{-2}) and carp (*Cypinus carpio*) (0.25 m^{-2}) the rate of return increased to 31%, clearly revealing how a mixed species culture can boost profitability. However when the density of tilapia and carp was doubled the fish yield improved greatly but the yield of the more valuable prawns was depressed and the rate of return dropped to 19%. As it turned out, the most profitable strategy in this set of trials was prawn monoculture using larger prawn juveniles (2 g) stocked at a higher density (4 m^{-2}) to obtain a return rate of 58%.

The culture of prawns in rice paddies in Vietnam and Bangladesh can clearly lead to improved returns compared to rice monoculture and the practice would be more common but for shortages of wild or hatchery-reared post-larvae. Such farms have been estimated to generate net revenues of $710 ha^{-1} (rice plus prawns) based on prawn yields of 100–300 kg ha^{-1} crop^{-1} (Hung 1992; section 7.3.5.2).

10.6.1.9 Stock enhancement

Hatchery-reared shrimp and prawn post-larvae have been released into the wild to replenish natural stocks and to boost capture fisheries. The economic success of such activities rests on the fulfilment of two key conditions: firstly, the released animals must make a significant contribution to catches and, secondly, fishing activity must be restrained to prevent any benefits being dissipated by excessive fishing effort. The second condition can be met by assigning and enforcing property rights over the enhanced stocks or by imposing other means of controlling access or fishing effort. Stock enhancement trials with several species have shown that the first of these conditions can be met if nursery-reared juveniles, rather than smaller post-larvae, are released, onto suitable substrate habitat (e.g. crayfish, lobsters, crabs) or into enclosed or partly enclosed bodies of water (shrimps, prawns) (sections 5.7 and 8.11). The second condition, which is really the crux of fishery management the world over, is much harder to satisfy, since if fishing effort could be managed effectively, there would be little call for remedial stock enhancement in the first place.

Shrimp enhancement trials in a 250 ha lagoon fishery in Sri Lanka provide perhaps the clearest evidence that recapture rates can reach a level where the returns to fishermen compensate for the costs of restocking (section 7.2.9). Hatchery-reared *Penaeus monodon* post-larvae were reared to 20 mm (TL) in cages within the lagoon prior to release, and recapture rates of 3% were achieved in two separate experiments. The positive impact on the fishery could be clearly identified because of two factors: (1) the lagoon was cut off from the sea for all but a few weeks each year; and (2) because releases were made out of phase with the periods when the lagoon mouth was open and when wild shrimp were able to recruit naturally to the fishery. The recapture rates justified the investment in restocking but, if they could have been increased marginally to 4%, perhaps by rearing the juveniles to 30 mm before release, then the returns to fishermen would have risen substantially. Despite the success of the trials, the issue of property rights remained unresolved. The lagoon supports a tightly knit fishing community and it was possible to enforce a voluntary closed season with the help of a co-operative association. This agreement enabled the released shrimp to reach more valuable larger sizes before capture, but if exclusive rights to the fishery cannot be assigned, the benefits of any future stock enhancement work are likely to be quickly dissipated (section 11.2.5).

In Japan, inland sea areas and bays have been stocked with hatchery-reared juvenile *Marsupenaeus japonicus*. The costs of the stocking operations were largely borne by the state or the individual prefectures involved, with the long-term objective of stabilising or boosting the coastal shrimp fishing industry, rather than making short-term financial gains. Henocque (1984) reviewed the economics of Japanese stock enhancement programmes. Post-larvae of 10 mm, costing around \$5.95 per 1000 to produce, were reared in government hatcheries and sold to fishermen at an 85% discount. The post-larvae were either released directly to open waters or held in nurseries for around 15 days until they reached 25–30 mm in length. A survival rate of 5.5% was estimated following direct release, and a rate of 15.4–22.7% following nursery rearing. Of these survivors, 35% were estimated to be captured in the fishery, giving overall recapture rates (from early post-larvae) of 1.9% for direct release and 5.4–8% if nursery rearing was used. Obviously each release area has different characteristics, and recapture rates in open fisheries can never be calculated with any precision, but an overall recapture rate of 1.5–2% would justify investment in the long-term objectives. And indeed it does appear that restocking has had beneficial effects on Japanese shrimp fisheries through the stabilisation of recruitment levels. Whether the fishing co-operatives would be keen to absorb 100% of the costs is doubtful unless protectable or enclosed areas were established which offered security for the investment. The future of these stocking programmes has now come into question because the Japanese financial authorities have stated that they cannot continue to invest national funds forever and fishermen, who are the beneficiaries, should start sea farming by themselves (Imamura 1999).

The concepts of stock ownership and security of investment are also of fundamental importance for commercially based ranching operations. In one analysis a profitable balance sheet for shrimp ranching was drawn up on the basis of releasing 25 mm nursed juveniles (approx. 0.2 g). The juveniles were estimated to cost \$22.8–38.0 per 1000 and assumed to be recaptured at 30 g with a fishing cost of \$5.24 kg^{-1} (Oshima 1984). The viability of this hypothetical enterprise rested on the assumed overall recapture rate of 35% (from nursed juveniles, based on 70% survival and 50% capture rate in the fishery); if it were only half this rate, profits would disappear.

Figures for shrimp releases in China indicate that very significant efforts have been made to enhance *Fenneropenaeus chinensis* stocks in and around the Yellow Sea and Gulf of Bohai; for example, in 1986 an estimated total of 4×10^9 hatchery-reared shrimp were liberated (Shang 1989). In the semi-enclosed Jiaozhou Bay, the release of over 350×10^6 juveniles from 1984 to 1986 appears to have had a dramatic effect on the depleted local fishery. Stocks increased by factors of between 4.7 and 7.3 during the 3 year release period, and declined immediately afterwards. Estimated average survival was a very impressive 32% (Liu 1990). In another programme, up to 224×10^6 3–4 cm TL juvenile *F. chinensis* were released per year into Xiang Shan Bay (Xu *et al.* 1997). Prior to restocking, there was almost no fisheries catch, but afterwards catches (of smallish shrimp at 10–12 cm TL) increased and there was a strong correlation between the number released and total fishery landings. The recapture rate was estimated at 6.8–13.3% but the state funding of releases was eventually stopped because there was no practical way to recover the costs of the operation. The programme failed to establish a breeding population and catches swiftly returned to zero. Xu *et al.* (1997) consider that the Xiang Shan Bay scheme was the only Chinese stock enhancement programme to have had a significant effect. It was estimated that an extra tonne of shrimp could be harvested from the fishery for the release of 596 000 juveniles of 3–4 cm TL and on this basis the operation generated a very favourable benefit : cost ratio.

10.6.2 Crayfish

10.6.2.1 Hatchery and nursery

Hatcheries play little role in the US crayfish industry and so economic data come mainly from Australia and Europe. In Europe the cost of establishing a nursery and simple prefabricated hatchery capable of producing 100 000 juvenile crayfish per year was estimated to be \$21 400 (Arrignon 1991). This unit, which was based on the production of the white-clawed crayfish (*Austropotamobius pallipes*) for restocking or ongrowing, would require 5000 broodstock per year. Annual running expenses were estimated at a mere \$1780 and the operation was considered viable at a selling price of \$0.18 per juvenile. Other estimates place the cost of producing juvenile crayfish (*Astacus astacus*) at \$0.04 each for hatchlings, and at \$0.25 each for juveniles of one summer old

(Huner *et al.* 1987), and in Spain it is possible to produce and sell hatchlings of *Pacifastacus leniusculus* at a profit of \$0.10 each. However, in one outdoor recirculating tank system in Germany, the cost of producing 3.5-month-old *Astacus astacus* summerlings was found to be around \$0.63 each, and there was thought to be little potential for profit in this nursery system since the prices obtained for juveniles in Europe usually fell in the range \$0.5–0.75 each (Keller 1988). The price of individuals for stocking generally varies with the quantity ordered and with their size, which reflects the duration of the culture phase. At the time of writing, small (10–20 mm TL) juvenile *A. astacus* were being sold at \$0.14–0.23 each, overwintered juveniles (20–30 mm TL) at \$0.46–0.56, while broodstock were priced at \$1.4–2.3 each (M. Keller, 2001 pers. comm.).

In Australia, the capital cost of a hatchery designed to produce 2×10^6 yabby (*Cherax destructor*) juveniles annually, was estimated to be \$21 680 (Staniford & Kuznecovs 1988). This sum represented 5.7% of the total cost of an integrated project incorporating 10 ha of semi-intensive ongrowing ponds. The proposed nursery for the same operation consisted of 475 above-ground pools of 3.66 m in diameter and was costed at \$50 900 (13.7% of the project's total cost).

10.6.2.2 Restocking and ranching

It has been estimated that an investment of \$7.10 ha^{-1} would be necessary to establish a viable population of *Pacifastacus leniusculus* in an extensive fishery in Sweden. Six to seven years later worthwhile catches of 5–10 kg ha^{-1} could be expected and would generate a revenue of around \$142 ha^{-1} (Huner *et al.* 1987). Waters established for such purposes as flood control and irrigation offer potential sites for stocking crayfish. However, the returns from ranching alone would not justify the costs of creating such waters.

10.6.2.3 Ongrowing

USA

Typically the investment and operating costs of crayfish farms in the USA are kept low. Most farmers make use of marginal land already in their possession and create ponds simply by adapting and raising existing drainage levees (bunds). Pond construction costs, including water control structures, were estimated to be \$499 ha^{-1} in Louisiana (de la Bretonne & Romaire 1989) and \$611 ha^{-1} in Texas (Avault & Huner 1985). There is a shortage of more recent cost data but these figures serve to illustrate how low investment costs are when compared to an estimated \$3013 ha^{-1} invested in the construction of deeper, more substantial semi-intensive shrimp ponds in the same country (Parker 1990).

Most operators are engaged in farming rice, soybeans or sugar cane as well as crayfish, and the costs of vehicles, farm facilities and equipment can be spread among the different crops. Some new investment in wells, pumps, pipes and traps is, however, usually necessary, and a crayfish combine – a specialised open workboat propelled by a spiked wheel – may also be required. Figure 10.16 shows the proportionate investment costs for an extensive 16 ha crayfish pond in southwestern Louisiana, expected to produce between 800 kg and 1700 kg $ha^{-1}yr^{-1}$. The operating costs associated with producing crayfish in extensive ponds using rice as forage are around \$362 ha^{-1}, of which 59% are associated with harvest (Romaire 1995). The major elements, summarised in Fig. 10.17, are bait, labour, fertilisation and forage, water management and maintenance. Forage crops are not always provided, though they are an effective way of enhancing productivity and serve as an alternative to the use of feed or fertiliser (section 7.5.4). Farm owners usually perform most of the labour themselves or utilise an existing farm workforce. However, outside labour may be required to meet the extra demands of harvesting, and this expense can have a major impact on profitability. Huner (1997) confirms that harvesting costs, including bait and labour, account for well over 50% of annual operating costs in crayfish culture systems and natural fisheries. Thus profitability could be increased if the effectiveness of traps, baits and harvesting strategies could be improved.

Crayfish stocking in the USA is usually only performed in the first year of a pond's operation, after which a self-sustaining population becomes established. Stocking costs are thus only incurred once, in stark contrast to most shrimp and prawn farms where expenditure on juveniles can represent 10–50% of an annual operating budget.

In a normal year, half of the crayfish produced in Louisiana are hand-peeled for tail meat but this product is only marginally profitable because labour costs are very high relative to the wholesale price of \$9.00–13.00 kg^{-1}. The activity is seasonal, and cost-effective mechanical crayfish processing systems are not yet available. As a

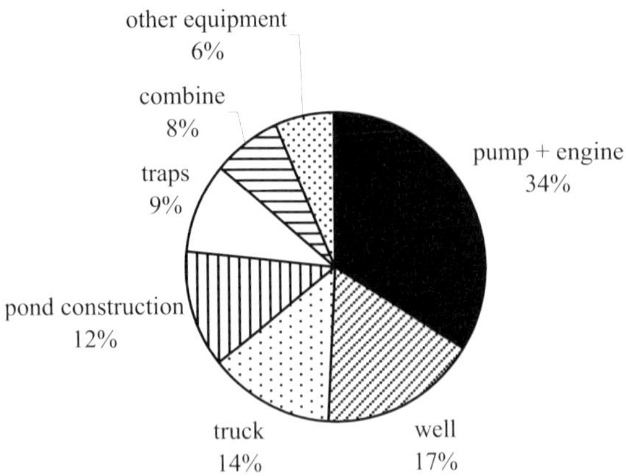

Fig. 10.16 Proportionate investment costs for a 16 ha crayfish farm in southwestern Louisiana (based on de la Bretonne & Romaire 1989).

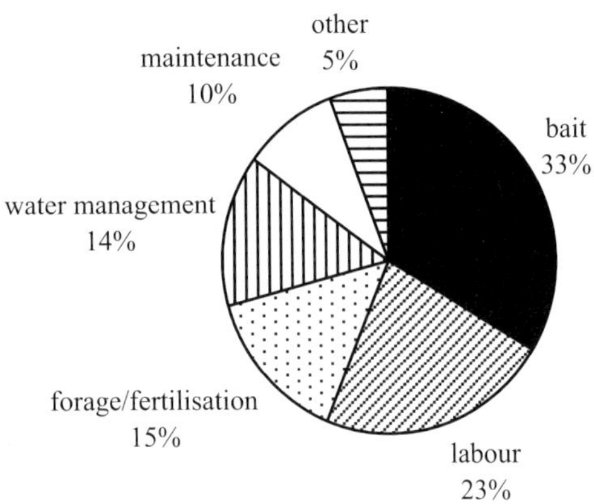

Fig. 10.17 Proportionate operating costs for a 16 ha crayfish farm in Louisiana (based on Romaire 1995).

result, processors cannot match the low cost of imported Chinese product (Huner 1997).

In South Carolina the average yield of crayfish farms is an estimated 840 kg ha^{-1} yr^{-1} and production costs in upland farms of 2–32 ha have been put at \$1.80–2.00 kg^{-1} (Eversole & Pomeroy 1989). The total capital investment needed for a crayfish farm with a single 8 ha pond was calculated at \$24 385, excluding land. Some farmers in South Carolina improve their financial returns by raising waterfowl as well as crayfish.

Interest in the cultivation of Australian redclaw crayfish (*Cherax quadricarinatus*) has led to its introduction to the USA and some theoretical analyses of its profit potential. A relatively optimistic assessment was made by Rubino (1992) for the prospects of a 100 ha farm financed with venture capital. The total investment needed was \$1 637 166 and it generated an IRR of 74% on the basis of a market price of \$8.82 kg^{-1}, a mean harvest weight of 80 g and an annual yield of 2000 kg ha^{-1}. Rubino stressed that this IRR was highly sensitive to changes in yield and market prices and this fact was born out by Medley *et al.* (1994) who found that redclaw farming would not be financially viable on the basis of a more conservative set of assumptions used in a Monte Carlo simulation. The latter authors based market prices on those of native red swamp crayfish, increased by 10–15% to reflect the superior meat yield of redclaw. The prices for the largest specimens were estimated with reference to prices for similar sized *Macrobrachium*, *Nephrops* and small spiny lobsters. A mean productivity of 1029 kg ha^{-1} was assumed on the basis of a stocking density of 3 juveniles m^{-2} and on the results of research trials in the USA and Australia. A mean size at harvest of 48.3 g led to an average market price of \$4.83 kg^{-1}. Other assumptions included a juvenile cost of \$0.05–0.15 each, a diet of corn silage and hay at \$50 mt^{-1} and commercial crayfish feed at \$330 mt^{-1}, investment costs of \$672 525 for a 50 ha farm, and total operating costs of \$390 300 yr^{-1}. The proportionate components of this operating cost are presented in Fig. 10.18. To achieve

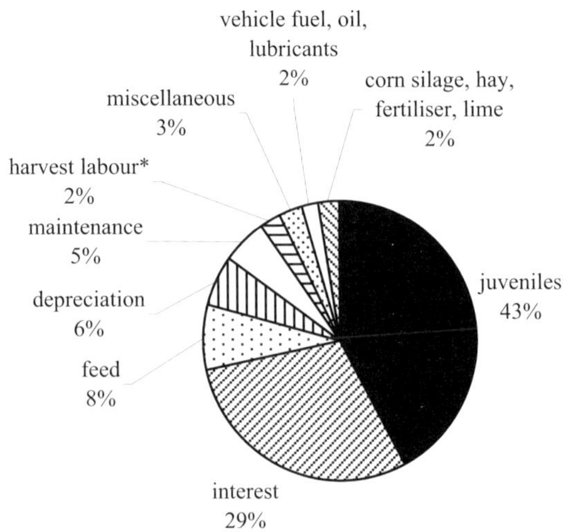

Fig. 10.18 Proportionate operating costs for a redclaw crayfish farm in the USA (based on Medley *et al.*1994). *All other labour and management provided by owner operators.

profitability on the basis of such assumptions it would be necessary to have a longer growing season of 180 days or more and to cut juvenile costs, which account for 42% of the operating budget. Simply increasing the stocking density to 5 m^{-2} would not improve returns because, although an increased yield of 1414 kg ha^{-1} would result, the mean size at harvest would drop to 38 g and the market price would fall to just \$3.85 kg^{-1}.

Europe

In Europe semi-intensive farming is conducted in small ponds along with the stocking of natural or extensive fisheries. Both types of operation rely on a supply of juveniles hatched and reared in captivity. In small semi-intensive canal-type ponds the operating costs comprise labour (30–60%), food (5–30%) and energy (0–15%). They vary depending on the size and intensity of the operation (Clarke 1989). In the UK the cost of excavating a canal-type pond of 0.1 ha and 1.5 m deep was estimated to be \$4700–6250 and the possible gross revenue from ten such ponds put at \$7800–9400 per year (Alderman & Wickins 1996). In France the cost of setting up a farm and hatchery to produce 10 mt yr^{-1} of *Astacus leptodactylus* has been estimated at \$43 000 (Arrignon 1991). Annual running costs were put at \$9000 with gross sales revenues of \$36 000 yr^{-1} due after the second year of operation.

Australasia

Crayfish farming in Australia is performed at various levels of intensity with extensive 'farm dam'-type operations and semi-intensive farms being the most common. Investment costs for extensive yabby farming in existing farm dams are very low. Only traps for harvesting are needed along with containers for gill flushing and transportation. One case study based on a farm with 46 dams identified investment costs of \$3000, annual operating costs (feed, fuel, bait, ice) of \$3400, and a gross annual income of \$11 700 which represented a good return to labour (7 hours per week) of \$23 h^{-1} (Fisheries WA 1999). However, the profitability of extensive yabby farming is far from assured and lower returns to labour of \$6.22 h^{-1} have been identified elsewhere (Mosig 1999). One farmer tried producing yabbies in farm dams and purpose-built ponds but found it impossible to produce substantial harvests and concluded that the activity could only provide a supplementary income. For a sustainable operation, a minimum of 15 ha of dams or ponds are thought to be needed but it is rarely possible to reach this scale because of the cost and limited availability of suitable land and water. Other assessments of the economics of extensive yabby farming draw attention to the high labour requirements for feeding, catching, grading and restocking, and conclude that at a cost of \$9.75 h^{-1}, this can greatly reduce profitability (O'Sullivan 1998). Some extensive yabby farmers simply call in commercial harvesters and arrange to receive a percentage of the crop value.

Semi-intensive operations make use of purpose-built ponds usually stocked with nursery-reared juveniles and require greater financial inputs than extensive systems. For example, Lawrence (1998) put the cost of building a semi-intensive marron farm at \$32 500–39 000 ha^{-1} and estimated annual operating costs at \$9750 ha^{-1}. The estimated gross income from such an enterprise was \$26 000 ha^{-1}. Mills *et al.* (1994) looked at the capital sums needed to set up a 15 ha farm for different crayfish species, including initial stocking costs, and estimated the totals per hectare to be \$35 500, \$50 000 and \$53 000 for yabbies, marron and redclaw, respectively. Annual operating costs per hectare were \$12 800, \$17 600 and \$22 800, respectively, and on the basis of yields of 1.5, 2 and 3 mt ha^{-1} yr^{-1}, IRRs of 9.3%, 10.3% and 19.4%, respectively, were calculated.

The profitability of redclaw farming has also been modelled on the basis of results at two demonstration farms set up to illustrate best practice. A 4 ha farm comprising 40 ponds of 0.1 ha was considered to be the minimum size for commercial viability and was costed at \$195 000. The production costs, at a yield of 3940 kg ha^{-1} yr^{-1}, were estimated to be \$4.61 kg^{-1} and the discounted payback period was 4 years (Lawrence & Jones 2001).

There is some interest in cultivating the two species of crayfish found in New Zealand, *Paranephrops planifrons* and *P. zealandicus* (section 7.7.11), and farms have been set up despite financial appraisals in the 1980s that suggested the activity had little commercial potential (Jones 1985). The study estimated marginal profitability for a farm with 6000 m^2 of ponds stocked at rates of 80 crayfish m^{-2}, and with a market price of NZ\$20 kg^{-1} (about US\$8.8 kg^{-1} at 2001 exchange rates). In practice however the researchers were unable to sustain densities above 20 m^{-2}.

10.6.2.4 Soft-shelled crayfish

Systems for producing soft-shelled crayfish were developed in the USA as a sideline to crayfish farming to exploit the elevated market prices paid for this highly esteemed product. Unfortunately a gold rush mentality spawned over 150 centres of production between 1986 and 1988 and oversupply led to wholesale prices slumping to just \$6 kg^{-1}. As a result the industry collapsed and only the largest producers, who were able to capture economies of scale, have remained viable. The economics of production are not helped by the short, 5 month production season and the fact that fixed costs must be met whether or not there is any production (Huner 1999).

The cost of setting up a unit with the capacity to hold 454 kg of animals and produce 1960 kg yr^{-1} of soft crayfish was estimated at \$15 000, assuming that construction labour was supplied by the owner-operator (Culley & Duobinis-Gray 1989). Production costs, of which 40% comprised the purchase of immature crayfish, were estimated to range between \$9.2 and \$13.0 kg^{-1}, depending on how well an operation was managed. Achieving a moulting rate of around 2% or more per day was critical to profitability, and overall productivity was limited by the length of the growing season which could vary between 4 months and 7 months depending on the weather.

Posadas and Homziak (1993) ran trials to compare open and closed production systems in Mississippi and found that both were marginally unprofitable, particularly when the owner's labour was added to total production costs. For an open (flow-through) system in an existing building an investment of \$7964 was needed (\$18.14 m^{-2} of tray surface), while for a closed (recirculating) system in a new building \$22 731 was needed (\$50.12 m^{-2} of tray surface). The open system could achieve 1.9% moulting rate for 5 months of the year and produce 1780 kg per season at a cost of \$5.93 kg^{-1}. The closed system delivered a lower moulting rate of 1.2% but a longer season of 7 months and yielded 1643 kg at \$9.61 kg^{-1}. The authors concurred that a minimum moulting rate of 2% was needed for financial viability.

To accelerate moulting rates eyestalk ablation can be performed and Gunderson *et al.* (1997) concluded that this approach could be a profitable on the basis of laboratory experiments and spreadsheet modelling. However viability would be sensitive to survival rate following ablation, the length of time till moulting, the number of days the facility was operated, the price paid for the output, and the number of crayfish introduced to the system each day.

10.6.3 Clawed lobsters

Data on the economics of clawed lobster rearing come mainly from Europe and relate to integrated units for producing juveniles for release (see below). Most of the examples considered relate to real operations and all were based on the purchase of wild egg-bearing females from fishermen or live-storage merchants.

The cost of establishing and operating a production unit for juvenile lobsters suitable for restocking has been investigated in the UK by the Sea Fish Industry Authority (C. Burton, 1990 pers. comm.; Burton & Adamson (in prep.)), the North Western and North Wales Sea Fisheries Committee (W. Cook, 1990 pers. comm.) and private organisations (P. Franklin, 1990 pers. comm.; Burton & Adamson, in prep.), and in Norway by the Tiedemanns Group and the Institute of Marine Research (G. van der Meeren, 1990 pers. comm.; T. Kristiansen, 2000 pers. comm.). Data on investment for broodstock/hatchery units and for investment and operating costs for integrated units producing 5000–2500 000 juveniles per year are summarised in Tables 10.10 and 10.11. No account was taken of transport and release costs, or of site surveys. In the absence of large-scale commercial production, these figures represent some of the best cost estimates available.

10.6.3.1 Broodstock and hatchery

The costings for lobster broodstock and hatchery units are usually incorporated within figures relating to integrated units for producing nursed juveniles (comprising broodstock, larvae culture and nursery facilities). However, it is useful to consider them separately since any future ranching or restocking programmes could use instar 5–8 juveniles, the cost of which will be crucial.

Actual data on the costs of setting up a lobster hatchery and nursery by modifying and extending existing buildings have been compiled by Burton and Adamson (in prep.) (Table 10.11). Of the total investment cost (\$55 768), 49% was attributed to nursery rearing, 28% to larvae rearing, 19% to broodstock holding, and the remaining 4% to general equipment needed for all three parts of the operation. Thus the cost of a broodstock and larvae rearing unit alone can be estimated at \$28 189.

Table 10.10 Estimated investment and operating costs for lobster hatchery/nursery units.

Output juveniles yr^{-1}	Stage at release	Investment cost (a) (US$)	Operating cost (b) (US$ yr^{-1})	Operating regime (temperature, no. of batches per year)	Cost per juvenile (US$)	Reference/source
4 980	>8	—	23 000 (c)	18–20°C, 2	4.66	W. Cook, 1990 pers. comm. (d)
10 000–15 000	9–12	150 000–200 000	40 000–50 000	18–21°C, 2	2.67–5.00	Anon. 1995
16 000	11	63 000	28 000	>18°C, 2	1.77	C. Burton, 1990 pers. comm.
25 200	11	67 000	28 000	>18°C, 2	1.12	C. Burton, 1990 pers. comm.
30 000	10–11	189 000	98 000	ambient, 1	3.27	P. Franklin, 1990 pers. comm.
30 000	10–11	124 000	43 000	>18°C, 2	1.45	C. Burton, 1990 pers. comm.
30 149	8	55 768	21 962	17–21°C	0.73	Burton & Adamson (in prep.)
23 000–46 000	10–12	—	—	—	2.9–4.35	T. Kristiansen, 2000 pers. comm
80 000	8	725 000 (e)	194 300	?, 4	2.43	E. Derriman, 2000 pers. comm.; Anon. 2000
100 000	8	65 250	43 645	17–21°C	0.44	Burton & Adamson (in prep.)
100 000	1-yr-old	—	350 100 (f)	20°C, 1	2.67	based on G. van der Meeren, 1990 pers. comm.
299 376	5	65 250	43 645	17–21°C	0.14	Burton & Adamson (in prep.)
2 500 000	1-yr-old	—	— (f)	—	0.73	Dragland 1999; T. Kristiansen, 2000 pers. comm

(a) Exclusive of tax and cost of site. (b) Including depreciation, excluding interest payments. (c) Includes release cost. (d) Refer to text. (e) Combined hatchery and visitor centre. (f) Heat provided by industrial effluent. (UK£1 = US$1.45)

Table 10.11 Investment costs for a 30 000 juveniles per year lobster hatchery/nursery unit based on the modification and extension of existing buildings (Burton & Adamson, in preparation).

	US$*
Building works	21 750
Tanks and stands	12 261
Containers	8 120
Filters	4 314
Pipework	3 915
Electrics and generator	2 175
Instruments	1 160
Aerators	1 088
Various equipment	986
Total	55 768

*UK£1 = US$1.45.

The output of the whole unit has reached around 30 000 instar 8 juveniles per year, but more animals could undoubtedly be produced if they were released at an earlier instar. This is illustrated by further data from Burton and Adamson (*loc. cit.*) that show how a unit capable of producing 100 000 instar 8 juveniles would be capable of producing around 300 000 instar 5 juveniles. The cost of setting up such a unit was put at $65 250 on the basis of data from integrated hatchery/nursery units in Scotland.

10.6.3.2 Integrated juvenile production units

We now revert to a consideration of integrated juvenile

production facilities capable of producing juveniles up to 1 year old from wild-caught broodstock. These larger juveniles were actually produced for restocking trials but would be equally suitable for ongrowing in battery farms.

Estimates for the cost of building a unit to produce 30 000 instar 10–11 juveniles per year have been put at \$124 000 and \$189 000 (Table 10.10). These figures assume that entirely new facilities are constructed. Considerable savings are, however, possible if existing marine hatchery facilities or other suitable buildings can be leased and adapted to lobster production. The costs of setting up a hatchery in Orkney (Scotland) were reduced by relying partly on the conversion or expansion of existing buildings used for seafood trading and partly on a temporary plastic covered greenhouse structure. The investment cost of this unit was estimated at \$55 768 (Burton & Adamson, in prep.) and it has produced around 30 000 instar 8 juveniles per year. An alternative approach, which involves greater investment, is to combine a hatchery and nursery with a visitor centre to generate additional income from paying tourists. This approach is showing great promise in Padstow, Cornwall where a unit to produce 80 000 instar 8 juveniles per year was set up for about \$725 000. Around 70% of this sum was invested in buildings (Anon. 2000).

The proportionate operating costs of a representative juvenile production unit are shown in Fig. 10.19 based on data from four British sources. Labour costs are particularly high and result from the individual holding requirements of juvenile lobsters. Attempts to reduce labour input through communal rearing have resulted in much reduced survival rates, are likely to increase the heterogeneity of growth rates and compromise survival upon release (section 7.8.8).

The estimates of the cost per juvenile, included in Table 10.10, are influenced by the overall approach to production. For example, in addition to considering the building of a completely new and independent facility, the option of setting up an operation in a similar style to a 'backyard' shrimp or prawn hatchery has also been investigated (P. Franklin, 1990 pers. comm.). In this case, an existing hatchery facility equipped with a seawater supply was occupied for a low rent, and all equipment such as pumps, tanks and piping were purchased second-hand or fabricated on site. The owner/operator performed all installation work, received no salary, and operated the finished unit at ambient temperatures. Thus a unit for producing 3600 instar 10–11 juveniles per year was set up for a mere \$2900 to provide animals at a production cost of just \$0.37 each. Although this approach is not sustainable (the operator will eventually require an income) it serves to illustrate the potential of this activity as a part-time occupation requiring a relatively low initial investment.

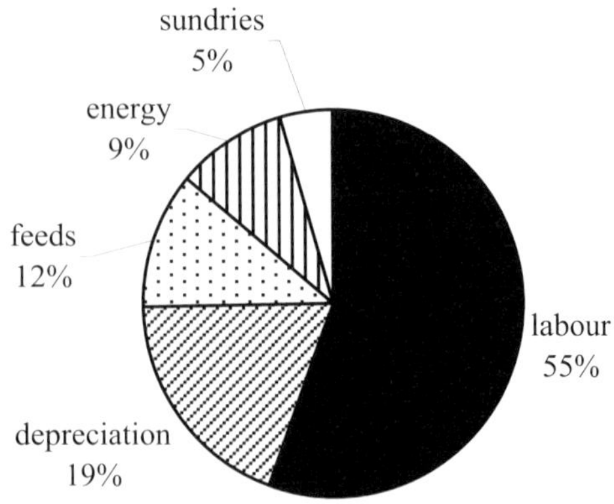

Fig. 10.19 Proportionate operating costs for a lobster hatchery/nursery unit producing 5000–30 000 juveniles per year (based on sources in Table 10.10).

Another approach with potential to cut the cost per juvenile is to increase the scale of production. A small British unit operated by the North Western and North Wales Sea Fisheries Committee, which produced almost 5000 juveniles (of minimum instar 8) in two batches per year, calculated the cost per juvenile to be \$4.66 (including release costs, at around 5–10% of this total; W. Cook, 1990 pers. comm.). However, taking into account the possibility of making considerable savings in a larger, more efficient operation, it was estimated that the cost could be reduced to around \$1.45 each (including release). The savings would result from economies of scale, output of three batches per year, increased energy efficiency and reduced labour requirements through less frequent cleaning and the greater use of an artificial diet. Other calculated UK cost figures ranged between \$1.22 and \$1.95 per juvenile and were based on establishing a juvenile production unit within an existing aquarium facility where space was rented (C. Burton, 1990 pers. comm.). When these figures were modified by including the cost of an independent building to house the same operations, investment and depreciation costs rose but overall production costs fell (to \$1.12–1.77 each) as a result of the elimination of rental charges.

A Norwegian group plans to build a hatchery/nursery unit for 2.5×10^6 1-year-old juveniles per year (Dragland 1999), and on the basis of automation and economies of scale they estimate that the cost per juvenile could be cut to \$0.73 (T. Kristiansen, 2000 pers. comm.). Other cost estimates from Norway are much higher: on the basis of actual data, Grimsen *et al.* (1987) put the cost per juvenile at \$2.6–3.25 (section 7.8.9) and more recently Uglem *et al.* (1998) have estimated \$1.95 each. In Scotland, on the basis of actual data and some cost projections, Burton and Adamson (in prep.) found that increasing a hatchery's output of instar 8 juveniles from 30 000 to 100 000 yr^{-1} could cut the cost per juvenile by 40%, to as little as \$0.44. If production proves to be sustainable at this level of costs, such a reduction would certainly improve the prognosis of the analyses described in section 10.6.3.3.

10.6.3.3 Restocking and ranching

The economic viability of stock enhancement proposals based on the release of hatchery- or nursery-reared juvenile lobsters is impossible to assess without knowledge of survival and recapture rates (section 5.7). Even though billions of instar 4 lobsters have been reared and released in North American waters, the impact of this type of programme remains unquantified. In Britain, instar 10–12 hatchery-reared juveniles were microtagged and released onto good lobster ground over a 5-year period from 1983 to 1988 (section 7.8.11). The monitoring of catches for a further 5 years showed, for the first time, that the 3-month-old juveniles survived in the wild, tended to remain at or close to their release site, contributed to the local fishery and, through egg production some 4–5 years later, probably also enhanced the breeding stock (Bannister & Addison 1998). It is usually not possible to say if genuine enhancement rather than substitution has occurred except where natural recruitment to the fishery has failed, as in Norway, or where releases are made onto hitherto unoccupied ground, as might be the case on an artificial reef.

Estimates can be made of the scale of releases required in a fishery (e.g. *Homarus gammarus* in Bridlington Bay, UK, where annual landings range from 12 mt to 50 mt) in order to justify a programme to achieve a 10% increase in catches (in this case an additional 5000–10 000 0.5 kg lobsters). Assuming a 5% recapture rate (10 000 lobsters to be caught), a 50% survival rate from release to minimum landing size (200 000 to survive at sea), 80% survival from instar 6 to release (400 000 to be released), 20% survival from hatch to instar 6 (500 000 reared to instar 6 from 2.5×10^6 eggs) and at 4000 larvae from each berried female, then the number of females required is 625. Waddy and Aiken (1998) estimated that in a fishery for *H. americanus* where annual landings were 1800 mt, to get a 10% increase in recruitment, the annual releases would need to be 4×10^6 juveniles reared from 16×10^6 instar 1 larvae obtained from 4000 females. This scale of operation would require 4000 80-L kreisels as well as holding trays for 6×10^6 juveniles prior to release, a decidedly non-trivial investment in equipment and labour.

Instar 10–12 juveniles were chosen for release in the UK trials as the smallest size suitable for consistent micro-tagging and because the nursery rearing could be completed in a single phase without the need for transfer to larger individual rearing compartments. Lobsters of this size are, however, far more expensive to produce than instar 5–8 juveniles and it has not been determined whether this extra expense is fully offset by higher survival rates in the sea (Bannister & Addison 1998). The optimum release size is still unknown, though it is critical to the economics of a release programme (section 12.7). For example, Burton and Adamson (in prep.) estimate that instar 5 juveniles could be produced for \$0.14 each, a fraction of the cost of instar 8 juveniles from the same basic hatchery installation. Trials in the UK involving the release of instar 6–8 juveniles, after approximately 8 weeks in the hatchery, have given encouraging results but, as yet, no clear indication of recapture rates. One drawback with these smaller juveniles is they take around 7 years to reach market size compared to 5 years for instar 10–12 juveniles (Seafish 1998).

An ambitious Norwegian stock enhancement project is planned on the basis of a hatchery with an annual output of 2.5×10^6 1-year-old juveniles. The target recapture rate is 50%, which it is hoped can be achieved through careful attention to the selection of suitable release sites and through keeping predators away at the time of release (Dragland 1999). Apart from the cost of producing juvenile lobsters, restocking and ranching programmes would incur significant additional expenditure on-site surveys, habitat modification or protection (ranching projects), transportation, and the release operation itself. On the basis of a bio-economic model and a juvenile cost of \$1.52 each, Borthen *et al.* (1999) identified the need for a recapture rate of 23% to achieve viability. They also noted the requirement for modifications to fishery regulations and the fact that greater benefits would be realised if released lobsters could breed before recapture. Moksness *et al.* (1998) also reviewed the economics of lobster stock enhancement in Norway and found that it showed little promise (negative NPV) on the basis of an assumed 10% discount rate, the same price per juvenile of $1.52, and a 6% recapture rate. However, if the cost per juvenile could be cut by 50% and the recapture rate boosted to 15%, a positive NPV would result. Actual recapture rates in trials in Norway have reached 14% (Agnalt *et al.* 1999). Much would therefore seem to depend on the ability to attain in practice the lower juvenile production costs (<$0.73) estimated by the British and Norwegian researchers.

The potential for releasing lobsters on to an artificially created reef has been investigated in detail in the UK (sections 5.7.2, 7.8.12 and 8.11.2). Economic appraisals arose from research initiated to consider ways of utilising fuel ash waste from power stations to avoid the usual disposal costs (Whitmarsh *et al.* 1995; Whitmarsh & Pickering 2000). The aquaculture component of the scheme envisaged the production of blocks of stabilised, pulverised fly ash and gypsum, the transfer of the blocks to the seabed and subsequent stocking with hatchery-reared juvenile lobsters. The economic analysis was made from the point of view of society as a whole and it was concluded that the net social present value would be negative and that the economic internal rate of return (EIRR) would be 5.1%, just below the minimum rate of return required for public sector projects (the test discount rate), which at the time was 6%. It was also found that if the cost of establishing the reef could be cut by using quarried stone rather than fabricated blocks, the EIRR would rise to 8.0% and exceed the test discount rate. However whether block or quarry stone were employed it was assumed that on average, from the sixth year onwards, 43% of the released lobsters would be recaptured, and it is still a matter of speculation just what recapture rate could be achieved. The analysis showed that viability was highly sensitive to lobster market prices and the cost per juvenile, which in this case was based on $1.45; a value somewhat higher than many of those given in section 10.6.3.2. Prospects for economic viability would be enhanced if additional revenue could be collected from externalities such as sport fishermen paying to fish on the reef.

The regulatory and legislative issues accompanying an artificial reef scheme in the UK are considerable because any viable project, whether privately or publicly funded, would rely on either the establishment of exclusive property rights over the reef and exclusive rights to harvest the lobsters, or strict control of fishing effort and reef access to prevent the benefits of stocking being dissipated. A valuable step in this direction has been taken through a revision of the Sea Fisheries (Shellfish) Act (1967) which provides the means for a licence holder to have exclusive rights to deposit, propagate and harvest lobsters and other crustaceans on the seabed (section 11.5.3.1). A licence to construct a reef for the purpose of ranching, however, is covered by an entirely different set of regulations governing the materials to be used and factors relating to their location in particular (Pickering 2000).

10.6.3.4 Ongrowing

Despite advances in lobster rearing technology, significant problems remain with the very high cost of the necessary ongrowing installations both on- and offshore (sections 5.4 and 7.8.9), and the absence of a cost-effective diet. As a result, lobster farming has yet to become established commercially and real data on economic aspects are unavailable.

10.6.3.5 Holding and fattening

The possibility of holding and fattening under-sized or soft-shell wild-caught lobsters to enhance their value and take advantage of seasonal price fluctuations has stimulated commercial activity and some research. In an economic analysis of a projected operation for fattening 30 000 lobsters of 1–1.5 lb (454–681 g), Bishop and Castell (1978) estimated that positive cash flows could be generated based on:

- post-moult weight gain of 70%, boosted by unilateral eyestalk ablation
- selling price enhancement of 62.5% per pound above purchase price
- mortality rate of 10%
- feeding rate of 1% body wt per day.

They also concluded that two 6-month crops per year were more profitable than a single holding operation lasting 9 months. However, a trial with 3535 lobsters failed because of heavy mortalities following interruptions in the water supply.

10.6.3.6 *Intensive (battery) culture*

The systems envisaged for intensive lobster farming are land-based and rely on individual containment at least during the final year or so of captivity. Despite the lack of real commercial data, the most important factors affecting the economics of such an approach have been identified. Unfortunately, very high capital and labour costs result from the need for individual confinement, and although automation has the potential to reduce labour costs, any increasing sophistication would incur increased development and capital costs.

The amount of space given to each lobster is important. While excess space increases growth rate, it reduces the number of animals that can be held in a production unit. On the other hand, if conditions are cramped depressed growth will extend the rearing period and thereby increase overall costs (sections 7.8.8 and 7.8.9).

The operating temperature is also critical because it largely determines growth rate and hence the duration of the farming cycle. Temperatures between 20° and 22°C are considered optimal, and because of the prohibitive expense of heating large volumes of water, the availability of a water supply at or very close to this temperature is considered to be a virtual necessity for any commercial venture. Allen and Johnston (1976) estimated that production costs would be more than doubled if the incoming water temperature were 12°C rather than 20°C. If a unit does incur significant heating costs, then the use of a recirculation system and/or heat recovery apparatus can result in significant savings. Allen and Johnston (1976) estimated that the expense of installing and operating a recirculation system would be offset by heat savings if the difference between the operating water temperature and the water supply temperature were greater than 5–7°C; and Herdman (1988) estimated that using heat exchangers to transfer effluent heat to incoming water was economically justifiable if the temperature difference was greater than 2°C. However, modern fish recirculation systems function at temperatures up to 25°C, largely on the basis of heat generated by the farm's pumps and compressors (section 8.4.4).

The optimum output level for a production unit has been estimated at around 80 000 animals (of 500 g) per month (Allen & Johnston 1976) and, though operations of this size capture most economies of scale, they represent very large investments. For example, Coffelt and Wikman-Coffelt (1985) estimated that an initial investment of $31.2m would be required for a system with a similar output level. The annual operating costs of this unit, producing 1×10^6 450 g lobsters per year, were projected to be $3.3m, of which labour represented 38% and feed 30%. However, assumptions underlying these calculations have been questioned, notably because of the absence of a suitable formulated diet (section 7.8.9).

It is estimated that it would take on average 27–30 months to produce a lobster of 500 g. Putative investors in lobster culture should expect to wait a minimum of 4–5 years before a project provides any significant returns on investment. However, the economics of culture could potentially be greatly improved through the use of unilateral eyestalk ablation to reduce the growth period from 27–30 months to about 20 months (section 2.3). While 450–500 g lobsters represent the established market size, if 200–250 g animals could be successfully marketed, production costs for these smaller animals would be significantly lower. This is because the time for ongrowing is reduced and because growth rate in captivity declines markedly beyond 250 g (Aiken & Waddy 1995). Also the risk of losses for a particular batch due to disease or other problems is reduced by the shorter ongrowing period. If success were achieved with 250 g lobsters, further reductions in target harvest size might follow. In all cases, however, culture operations will need to make allowances for large variations in individual growth rates (Table 4.5). The cost of producing a 300 g, lobster has recently been estimated at $5.8–7.25, on the basis of a 2 year rearing period (T. Kristiansen, 2000 pers. comm.).

10.6.4 Spiny lobsters

Some preliminary investigations into the economics of spiny lobster farming have been undertaken but as yet there has been little progress in surmounting the fundamental problem of obtaining adequate supplies of juveniles (pueruli). Jeffs and Hooker (2000) do not anticipate any significant improvements in hatchery techniques in New Zealand for at least a decade and have therefore focused attention on catching more pueruli in the wild, using improved collector designs. The current cost of collecting pueruli in New Zealand is estimated at $0.2–0.7 each.

Some of the constraints facing the ongrowing of spiny lobsters overlap with those facing clawed lobster culture, i.e. viability rests heavily on cutting the infrastructure and operating expenses of land-based farming systems including lowering feed and labour costs. Spiny lobsters however have one distinct advantage over clawed lobsters in that they can be held communally with only minimal cannibalism. Production trials in New Zealand with *Jasus edwardsii* have achieved survival rates of 81% (Table 4.6h). Lobsters were fed fresh mussels (costing \$300 mt^{-1}; \$1 kg^{-1} of meat) and with a feed conversion rate of 10 : 1 (on a wet weight basis) they reached 300 g in 4 years at temperatures of 15–23°C. A financial appraisal based on the stocking of 20 000 pueruli per year and production of 5 mt yr^{-1} in 2000 m^2 of tanks was not very favourable because production costs at \$31.1 kg^{-1} exceeded the market price (live sales to Japan). An investment of \$19 3600 was needed of which 64% was for the building and 24% for pumps and filters.

Spiny lobster culture in the tropics can achieve faster growth rates than in temperate regions. By rearing relatively large juveniles (50–100 g) of the native Indian species *Panulirus homarus* and *Panulirus ornatus*, Rahman and Srikrishnadhas (1994) were able to produce 250 g lobsters after just 5–6 months and thereby obtain two crops per year. They calculated the cost of production to be just \$7.22 kg^{-1} which compared favourably to the market price of \$11.7 kg^{-1}, but the studies were severely hampered by a lack of juveniles.

10.6.5 Crabs

Little detailed information on the economics of crab culture has been published, probably because few farms raise crabs in monoculture and even fewer are of significant size. Some Asian shrimp and fish hatcheries may also produce post-larval crabs when the market is favourable, but again cost data for this particular activity are scarce.

In South-east Asia, mud crabs (*Scylla* spp.) are farmed both in extensive polyculture (with shrimp and milkfish) and to a lesser extent in monoculture (section 7.10.4). Extensive systems that rely entirely on the ingress of wild juveniles as tidal ponds are filled often count crabs only as a by-crop. In other operations, in order to boost the yield of crabs, farmers purchase locally netted wild crab juveniles and thus incur significant stocking costs. Feed costs are usually minimal since only trash fish or sometimes rice bran are added to the pond. Labour costs are typically low since owner-operators perform the bulk of the work. Generally, the greater overall productivity possible with polyculture in extensive systems results in greater profits than does crab monoculture (Lapie & Librero 1979). In one analysis of a crab (*Scylla* spp.) and milkfish polyculture project in the Philippines (Anon. 1996) a payback period of 1.2 years was calculated on the basis of yields of 1800 kg crab ha^{-1} yr^{-1} and 1800 kg milkfish ha^{-1} yr^{-1}. The proportionate costs of this operation are presented in Fig. 10.20 and show the dominance of wild crab seed in overall costs. Another analysis from the Philippines (Agbayani *et al.* 1990) has shown that monoculture of mud crab is also potentially lucrative if yields of around 1000 kg ha^{-1} crop^{-1} can be sustained. Samonte & Agbayani (1991) confirm that mud crab culture is viable with short payback periods of 1–2 years.

Agbayani *et al.* (1990) based their economic projections on results from experimental ongrowing trials performed in a shallow (50 cm) earthen pond. Mud crabs were fed with chopped tilapia and it was established that the best results (survival 88%; growth 2.28 g d^{-1}; feed conversion ratio 1.72 : 1; yield 1019 kg ha^{-1} crop^{-1}; size at harvest 232 g) and financial return (return on investment 124%) could be obtained at a stocking density of 5000 juvenile crabs per hectare (mean weight 25.3 g). Higher densities were less profitable because they resulted in significantly lower survival and growth rates and less efficient feed conversion. The cost of setting up a crab farm of 1 ha was put at \$1140, of which 59% represented construction of the pond, and another 28% represented the cost of a perimeter fence. The installations and pond were expected to last for only 2–5 years before replacement and reconstruction would be necessary. Production costs were projected at \$1.10 kg^{-1}, with the major com-

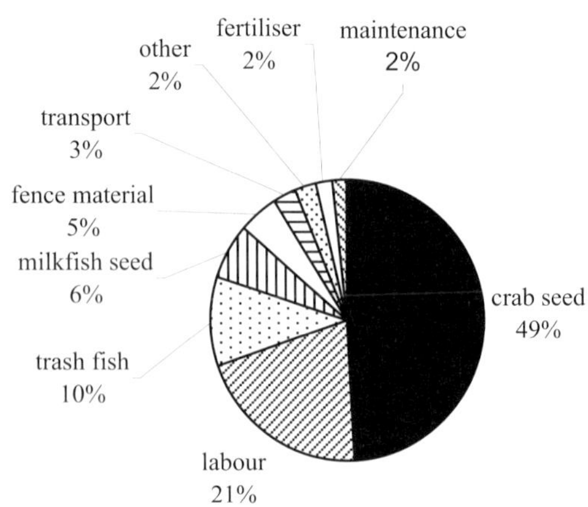

Fig. 10.20 Proportionate operating costs for a 1 ha crab and milkfish farm in the Philippines (based on Anon. 1996).

ponents comprising feed, labour, depreciation and seed. The attractive financial returns, however, relied on the assumption that the yield of 1019 kg ha^{-1} after 90 days, obtained experimentally, could be achieved three times per year in a commercial venture. The authors pointed out that, traditionally, yields per crop from crab monoculture in the Philippines are only around one-third of this figure. Further analysis using mud crabs in the Philippines involved culturing monosex populations at varying densities (Surtida 1997). Male monoculture proved to be more profitable than female monoculture and the optimum stocking density was around 15 000 juveniles ha^{-1}.

The majority of crab farmers are almost totally dependent on supplies of juveniles from the wild, but such supplies are unreliable and this has tended to restrict crab farming. Extensive culture in the Philippines, in which juveniles and small crabs (anything below 150 g) are held for 4–6 months to attain a size of 200 g, can be profitable but is still subject to this constraint (Chong 1993). Crab culture in cages and pens can also be profitable. Tabigoon (1998) calculated that cage culture could achieve payback in 1.8 years on the basis of an investment of \$8800 in 1 ha of cages, operating costs of \$4400 yr^{-1}, two crops per year of 600 kg ha^{-1}, growth from 30–40 g to 250 g in 3–4 months, and 70% survival. Culture in pens was calculated to have an even shorter payback period of 1.6 years. It was assumed that \$7700 ha^{-1} would be invested in net enclosures in mangroves and that operating costs would be \$4200 yr^{-1}. Seed stocked at 9–22 g would reach 275 g in 6 months with a survival rate of 86% and a yield of 485 kg ha^{-1} yr^{-1}. Aldon (1997) has also provided a favourable economic analysis for mud crab farming within netted enclosures in mangrove areas. A 1 ha enclosure would cost \$8260 to set up and operate for a year and with three crops of 200 kg would generate a gross revenue of \$14 400 yr^{-1}.

Crab farmers in Taiwan obtain seed either from fishermen, nursery operators or middlemen, and the prices they pay varying depending on size. Juveniles of 2–3 cm CW may fetch two to five times more than smaller animals of 0.5–1.0 cm (Cowan 1983).

Juvenile crabs (*Portunus trituberculatus*) are produced in Japanese hatcheries at a cost of between \$7.70 and \$21.10 per 1000 (Cowan 1983) and are sold to fishing co-operatives for restocking (section 7.10.8). In one hatchery, energy and food accounted for 83% of variable expenses and 68% of overall production costs. The exact cost of releasing juveniles in the wild, however, varies depending on whether nursery rearing is used and on the type of release method. Nursery rearing in onshore tanks made of concrete or canvas can more than double the cost per juvenile, and release in open water using artificial habitats of branches or frayed rope suspended from rafts can increase overall costs by a factor of 2.5.

In common with all stock enhancement programmes, economic viability depends upon recapture rates and, although these are notoriously difficult to establish, some estimates have been made. For example, in one small bay in Hiroshima, where no previous fishery existed, a recapture rate of 3.2% has been recorded. In other situations, size frequency analysis has been able to identify restocked cohorts, since hatchery seed tend to be larger than their wild counterparts. One such study placed recapture rates at between 3% and 12% (Table 5.8). Economic viability also hinges on crab market prices. At recapture rates of between 1.2 and 2.9% and stocking costs of between \$14.2 and \$34.5 per 1000 juveniles, economically viable returns were considered possible at a selling price of \$1.18 per crab.

Crab holding operations in Taiwan are often run by dealers who trade in both farmed and fished crabs. Fertilised females are retained for fattening and sold at a profit when they reach the highly prized 'red' (full roe) stage (section 7.10.4). In the USA, shedding systems for blue crabs (*Callinectes sapidus*) can take advantage of the big price differential between soft- and hard-shelled animals. Nevertheless, while it is economically viable to take wild crabs on the verge of moulting ('peelers') and hold them for short periods, it is not feasible to take intermoult animals and feed them until they moult. Large production units may have around 100–150 trays and employ about 12 people (section 7.10.9). The turnover period is about 1 week.

10.6.6 Processing plant

Although data on the costs of establishing and running crustacean processing plants are generally scarce, some useful information is available regarding the production of frozen shrimp. The cost of a plant producing 2000 mt per year has been estimated at \$1m for a basic design and \$3m for a high-technology design. The former would require about 400 employees and the latter about 100 (Christensen & Ziegler 1992).

One detailed study was prepared by Waits and Dillard (1987), based on processing shrimp or *Macrobrachium* in Mississippi, USA, in the form of raw frozen shell-on tails, packed in 5 lb (2.27 kg) cartons. The total cost of a plant equipped to produce a daily maximum of 6.5 mt was calculated to be $781 184 (excluding land). The components of this cost are illustrated in Fig. 10.21. The cost of the building (1040 m^2) represented 40% of the total, and the blast and storage freezers together accounted for another 27%. An area of land measuring 0.35 ha would be required for the operation, and the purchase of a site in Mississippi with water frontage (to allow boat-caught shrimp to be landed) was estimated to add another $120 000 to the total investment costs. However, a less expensive inland site ($35 000) could be occupied if incoming product were to be delivered by trucks (usually the case with supplies from aquaculture).

The total operating costs for the processing plant, at an annual output of 1139 mt, were put at $1 165 018 (excluding taxes and interest), giving a total processing cost of $1.02 kg^{-1}. Proportionate costs are shown in Fig.10.22 and illustrate the dominance of labour costs (81%). The cost of producing whole shrimp rather than tails would be significantly lower because no labour would be required for deheading, and because the processing yield for whole product is 100% (compared to 40% for *Macrobrachium* tails and 57–68% for shrimp tails). Transporting product by truck from farms to the processing plant was estimated to add another $0.065 kg^{-1} to the overall costs.

An example of the proportionate costs of operating a shrimp processing plant in Indonesia is given for comparative purposes in Fig. 10.23. The main difference occurs in labour costs that are dominant in the USA (81%) but far lower in Indonesia (22%). In Indonesia the costs of imported products such as packaging (26%)

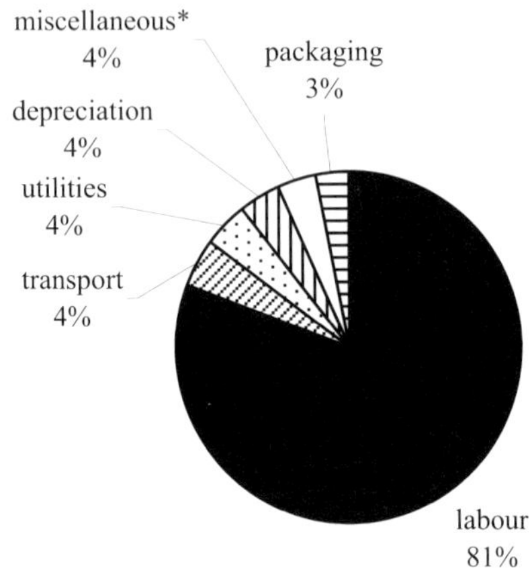

Fig. 10.22 Proportionate operating costs for a shrimp processing plant in the USA (based on Waits & Dillard 1987, excluding interest and taxes). *Includes maintenance, insurance, waste disposal.

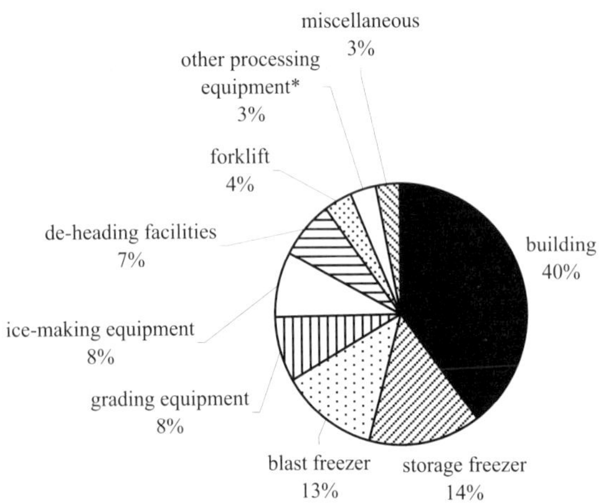

Fig. 10.21 Proportionate investment costs for a shrimp processing plant (max. daily output 6.5 mt frozen headless shrimp or prawns) (based on Waits & Dillard 1987, excluding land, including full cost of ice-making equipment). *For weighing, packing, glazing and strapping.

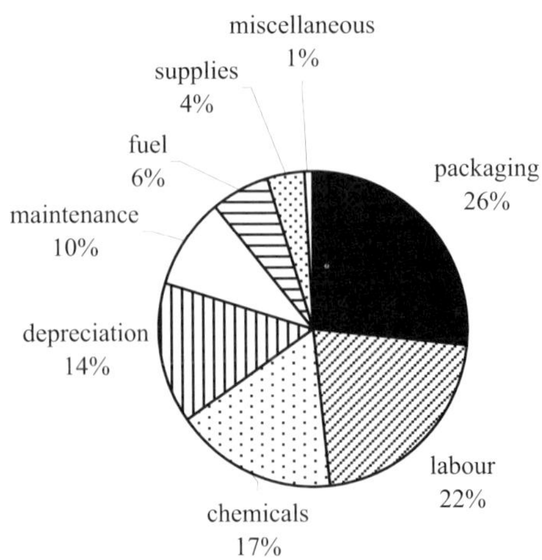

Fig. 10.23 Proportionate operating costs for a shrimp processing plant in Indonesia. Excluding administrative overheads and management costs (D. Lee, 1999 unpubl. data).

and chemicals (17%) are relatively much more important than in the USA.

To help minimise the processing cost, the full capacity of a plant must be used. This becomes difficult, however, when output from fisheries or aquaculture is seasonal. The costs given for the US plant above, relate to a plant operating for a total of 9 months a year with 2 months at maximum capacity, and were based on processing wild-caught shrimp as well as product from aquaculture. No doubt cost savings could be made if consistent quantities of product could be processed all year round. On the other hand, if the same plant were operated for only 2 months a year rather than 9 months, to produce just 314 mt of tails, total processing costs would increase to $\$1.53\,kg^{-1}$.

The bulk of processed crayfish produced in the USA is in the form of fresh and frozen tail meat. Crayfish are peeled after being blanched. The meat yield is a mere 15% and the major processing cost by far is the labour required for peeling by hand. This cost, estimated at around $\$2.75\,kg^{-1}$ of meat (Roberts & Dellenbarger 1989), has been identified as a constraint to crayfish marketing since it results in overall processing costs greater than those of competing shrimp products. Peeling machines have been developed to cut labour requirements but have not been successful. Accounts of US crayfish processing methods are given in sections 3.3.3.1 and 7.5.7.

10.6.7 Feed mill

Since only limited information on the costs of setting up and operating a crustacean feed mill is available, a useful indication of likely expenditure levels can be obtained by considering the situation for an animal (livestock or poultry) feed mill and then taking into account the special requirements of crustacean feeds and their cost implications.

A technical and investment guide for those interested in producing compound feeds on a small scale in developing countries has been compiled by Parr *et al.* (1988). For a plant producing 4200 mt of pellets for poultry or livestock annually, an investment equivalent to around \$371 000 was identified, and annual operating costs were put at \$800 000. Although these totals underestimate the costs for a crustacean feed mill with an equal capacity, Figs 10.24 and 10.25, which illustrate the component costs, provide summaries that can be more generally applied.

The special requirements for aquaculture feeds, as compared to animal feeds, are discussed by Barbi (1987).

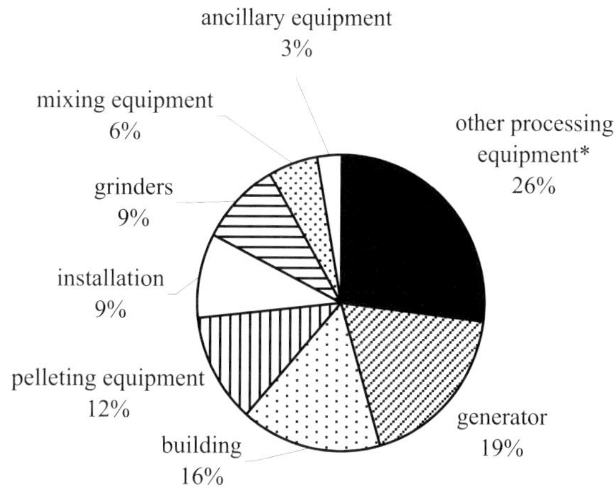

Fig. 10.24 Proportionate investment costs for a feedmill with an annual capacity of 4200 mt (based on Parr *et al.* 1988). *Weighing, elevators, augers, holding bins, grinders, steam production, pellet cooling, bagging, electrical control.

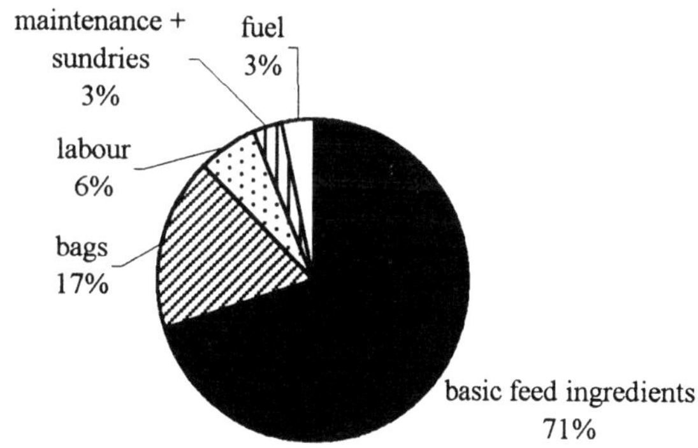

Fig. 10.25 Proportionate operating costs for a feedmill with an annual capacity of 4200 mt (based on Parr *et al.* 1988).

Compounded aquaculture feeds in general, and crustacean feeds in particular, require ingredients that are very finely and uniformly ground and this results in increased investment and operating costs for the necessary milling equipment (section 8.8.2.1). In addition, pelleting machines for crustacean diets are fitted with thicker dies with smaller hole diameters. This improves compaction as the feed mixture is extruded to form pellets, and the extra frictional heat helps to gelatinise starches and improve the binding qualities. However, at the same time the output of the machine is significantly reduced. For example, a pelleter, equipped to produce 2 mt h^{-1} of 3.5 mm poultry pellets, would only be able to produce 1.3 mt h^{-1} of 2.5 mm shrimp pellets (J. Wood, 1990 pers. comm.).

In addition to the conventional pelleting process, high-quality aquaculture feeds, including crustacean diets, are increasingly being produced by a method known as extrusion cooking (section 8.8.2.1). The features and costs of the two alternative processes are compared by Kearns (1989). Extrusion cooking equipment has higher investment and operating costs than conventional pelleting equipment: Barbi (1987) estimated that the investment for an extrusion cooking and drying system would be about 7.5 times greater than for a typical pelleting system, and that the operation and maintenance costs would be about 91% higher. But the more expensive equipment does allow more flexible and lower cost feed formulations to be utilised that can generate significant overall savings.

Devresse (1998) has also made comparisons between the costs of producing feeds using different technologies. He looked at production costs associated with equipment for pelleting or extrusion together with drying and cooling, and included costs for depreciation, maintenance and energy consumption (steam and electricity). He found that the cost of making shrimp pellets was three times the amount of making traditional pellets for land animals due to the energy requirements in post-conditioning and drying. Extrusion was seven times more expensive than traditional pelleting due to the energy involved in extrusion and drying. Devresse (1998) recommended that wherever possible savings should be made by avoiding expensive steps such as drying or post-conditioning and he noted that it is cost-effective to improve pellet stability by modifying the pelleting line to include an injector for gelatine binder.

Despite the extra costs of investing in equipment for producing crustacean diets, and the associated increase in maintenance and utility costs, feed ingredients represent some 71% of operating costs (Fig. 10.25) and it is the cost of these ingredients that has the most significant impact on the overall cost of producing a compounded crustacean feed. Crustacean diets, especially those with high-protein formulations, usually rely heavily on relatively expensive ingredients such as fishmeal, and this is one of the main reasons why they tend to be more costly to produce than other animal feeds.

There is scope for significant cost savings through alterations in the mixture of ingredients of all compounded feeds. Indeed, in the manufacture of poultry and livestock diets, the use of computers to generate the lowest cost formulations has become routine. This enables feed mill operators to take advantage of fluctuations in the price of raw materials to maximise profits, while keeping the overall nutritional profile of their final product largely unchanged. By comparison, the lowest cost formula approach for crustacean feeds is not well developed, principally because of inadequate knowledge of crustacean nutrient requirements and the digestibility of different ingredients (section 2.4).

To produce a stable pellet a conventional pelleting machine requires a formula containing around 25–35% starch ingredients and binders, and this dictates that high-cost aquatic meals (largely fish) be incorporated to make up the desired levels of protein and other nutrients. An extrusion cooked mixture, on the other hand, need contain only 10% starch ingredients or binders to form a stable pellet and this allows more flexibility in the diet formulation, and a range of protein sources, usually cheaper than fishmeal, can be included. Figures provided by Kearns (1989) indicate that an initial investment of \$2m would be needed for a complete plant equipped with an extrusion cooking system to produce 24 000 mt of feed annually. An equivalent plant with more conventional pelleting equipment would cost $\$1.35 \times 10^6$ – significantly less. Although the cost of utilities for the first system was calculated to be greater (\$21.53 mt^{-1} against \$11.93 mt^{-1}), for a diet containing 40.7% protein, cost savings made possible by the cheaper extrusion formula (largely through the reduction of the fishmeal content from 40% to 10%) amounted to over \$100 mt^{-1}.

10.7 References

Agbayani R.F., Baliao D.D., Samonte G.P.B., Tumaliuan R.E. & Caturao R.D. (1990) Economic feasibility analysis of the monoculture of the mud crab (*Scylla serrata*) Forskol. *Aquaculture*, **91** (3–4) 223–232.

Agnalt A.-L., Van der Meeren G.I., Jørstad K.E., *et al.* (1999) Stock enhancement of European lobster (*Homarus gammarus*): a large scale experiment off south-western Norway (Kvitsøy). In: *Stock Enhancement and Sea Ranching* (eds B.R. Howell, E. Moksness & T. Svåsand), pp. 401–419. Fishing News Books, Oxford, UK.

Aiken D.E. & Waddy S.L. (1995) Aquaculture. In: *Biology of the Lobster* Homarus americanus (ed. J.R. Factor), pp. 153–75. Academic Press, New York.

Alderman D.J. & Wickins J.F. (1996) *Crayfish Culture,* 22 pp. Lab. Leafl., (76). MAFF Directorate Fisheries Research,

Lowestoft, UK.
Aldon E.T. (1997) Mudcrab culture in mangrove areas. *Asian Aquaculture,* **19** (3) 21–22.
Allen P.G. & Johnston W.E. (1976) Research direction and economic feasibility: an example of systems analysis for lobster aquaculture. *Aquaculture*, **9**, 155–180.
Angell C.L. (1994) Promotion of small-scale shrimp and prawn hatcheries in India and Bangladesh, 24 pp. *Bay of Bengal Programme Report*, **66**. BOBP, Madras, India.
Anon. (1995) *Lobster stocking: progress and potential. Significant results from UK lobster restocking studies 1982–1995*, 12 pp. MAFF Directorate Fisheries Research, Lowestoft, UK.
Anon. (1996) Mudcrab culture (with milkfish). *Aquafarm News*, **14** (3–4) 23–24.
Anon. (1999) *Scandales financiers. Le second rapport de la Commission Hervé Vincent Bangoura.* http://www.guinee.net/pressrtv/independent/151/HVBRapport2.html
Anon. (2000) Hatchery boost for SW lobster fishery. *Fishing News,* 22 Sept. 2000, p. 7.
AQUACOP (1979) Intensive larval rearing of *Macrobrachium rosenbergii*. A cost study. *Proceedings of the World Mariculture Society*, **10**, 429–434.
AQUACOP (1982) Mass production of juveniles of freshwater prawn *Macrobrachium rosenbergii* in French Polynesia: Predevelopment phase results. In: *Proceedings of Symposium on Coastal Aquaculture 1982*, Part 1, pp. 71–75. Marine Biological Association of India, Cochin, India.
Arrignon J. (1991) *L'écrevisse et son élevage,* 2nd edition. *Technique et Documentation*, 200 pp. Lavoisier, Paris [in French].
Avault J.W. Jr. & Huner J.V. (1985) Crawfish culture in the United States. In: *Crustacean and Mollusk Aquaculture in the United States* (eds J.V. Huner & E. Evan Brown), pp. 1–54. AVI Inc., Westport, CT, USA.
Bannister R.C.A. & Addison J.T. (1998) Enhancing lobster stocks: a review of recent European methods, results, and future prospects. *Bulletin of Marine Science,* **62** (2) 369–387.
Barbi J.W. (1987) *Shrimp and fish feed processing*, 18 pp. (mimeo). Presented at Aquaculture Europe '87 International Conference, 2–5 June 1987. RAI Amsterdam, Netherlands.
Barg U.C. (1992) Guidelines for the promotion of environmental management of coastal aquaculture development, 122 pp. *FAO Fisheries Technical Paper* (328). FAO, Rome, Italy.
Bauer L.L., Sandifer P.A., Smith T.I.J. & Jenkins W.E. (1983) Economic feasibility of *Macrobrachium* production in South Carolina. *Aquacultural Engineering*, **2** (3) 181–201.
Beveridge M.C.M., Philips M.J. & Macintosh D.J. (1997) Aquaculture and the environment: the supply of and demand for environmental goods and services by Asian aquaculture and the implications for sustainability. *Aquaculture Research*, **28**, 797–807.
Bishop, F.J. & Castell, J.D. (1978) 1977 commercial lobster culture feasibility study, Clark's Harbour, Nova Scotia, 18 pp. *Fisheries and Marine Service Industry Report 102*.
Boeing P. (1998) The AquaTech Seafood closed system shrimp farm: a central Texas operation and management perspective. In: *Proceedings of the US Marine Shrimp Farming Program Biosecure Workshop*, 14 February 1998 (ed. S.M. Moss) pp. 59–64. The Oceanic Institute, HI, USA.
Borthen J., Agnalt A.L. & van der Meeren G.I. (1999) A bioeconomic evaluation of a stock enhancement project of European lobster: the simulation model LOBST.ECO with some preliminary results. In: *Stock Enhancement and Sea Ranching* (eds B.R. Howell, E. Moksness & T. Svåsand), pp. 583–596. Fishing News Books, Oxford, UK.
Boyd C.E. (1998) *Mechanical aeration in pond aquaculture*. In: *Proceedings of FEDSM98*: Second International Symposium on Aeration Technology, 21–25 June 1998, Washington, DC. 6 pp. (mimeo).
de la Bretonne L.W. Jr. & Romaire R.P. (1989) Commercial crawfish cultivation practices: A review. *Journal of Shellfish Research*, **8** (1) 267–285.
Burton C. A. & Adamson K. (In prep.) *A preliminary study of the costs of operating a lobster hatchery in Orkney and the development of an economic model for future hatchery programmes*, 17 pp. Seafish Technical Report. Seafish, Ardtoe, Scotland.
Cannon D. (2000) Re: [shrimp] crop insurance. Correspondence, 11/11/2000. shrimp@egroups.com
Chen C.C. (1994) An economic analysis of major species in brackishwater aquaculture. In: *Socioeconomics of Aquaculture* (eds Y.C. Shang, P.S. Leung, C.S. Lee, M.S. Su. & I.C. Liao), pp. 173–191. Tungkang Marine Laboratory Conference Proceedings No. 4.
Chen L.C. (1990) *Aquaculture in Taiwan*, 273 pp. Fishing News Books, Blackwell Scientific Publications, Oxford, UK.
Chiang P. & Liao I.C. (1985) The practice of grass prawn (*Penaeus monodon*) culture in Taiwan from 1968 to 1984. *Journal of the World Mariculture Society,* **16**, 297–315.
Chien Y.H. & Liao I.C. (1987) *Bioeconomic consideration of prawn farming*, 13 pp. Dept. of Aquaculture, National Taiwan College of Marine Science and Technology, Keelung, Taiwan.
Chong L.P. (1993) The culture and fattening of mud crabs. *Infofish International*, (3) 46–49.
Christensen P. & Ziegler H (1992) Ergonomics and economics of automation in the shrimp industry. In: *Proceedings of Third Global Conference on Shrimp Industry*, Hong Kong, 14–16 September 1992 (eds H. de Saram H. & T. Singh), pp. 165–173. Infofish, Kuala Lumpur, Malaysia.
Clarke S. (1989) Freshwater crustacean farming: the world scene. S. Australian Dept Fisheries, *SAFISH*, **13** (4) 10–12.
Clay J.W. (1997) Toward sustainable shrimp aquaculture. *World Aquaculture*, **28** (3) 32–37.
Coffelt R.J. & Wikman-Coffelt J. (1985) Lobsters: one million one pounders per year. *Aquacultural Engineering*, **4** (1) 51–58.
Cowan L. (1983) *Crab farming in Japan, Taiwan and the Philippines*, 85 pp. Information Series QI 84009, Queensland Department of Primary Industries, Queensland, Australia.
CPC (1989) *Libro blanco del camaron*, 79 pp. Camara de Productores de Camaron, Guayaquil, Ecuador [in Spanish].
Cropper M.L. & Oates W.E. (1992) Environmental economics: a survey. *Journal of Economic Literature*, **30,** 675–740.
Cruz P.S. (1993) *Shrimp farming in the Philippines: culture practices and problems,* 30 pp. (mimeo). Kabukiran Enterprises Inc., Davao City, Philippines.

Culley D.D. & Duobinis-Gray L. (1989) Soft-shell crawfish production technology. *Journal of Shellfish Research*, **8** (1) 287–291.

D'Abramo L.R., Brunson M.W. & Daniels W.H. (1998) Freshwater prawns pond production and grow-out. *Mississippi State Extension Service Publication 2003*, 7 pp. http://ext.msstate.edu/pubs/pub2003.htm

Daniels W.H., D'Abramo L.R., Fondren M.W. & Durant M.D. (1995) Effects of stocking density and feed on pond production characteristics and revenue of harvested freshwater prawns *Macrobrachium rosenbergii* stocked as size-graded juveniles. *Journal of the World Aquaculture Society*, **26** (1) 38–47.

De Silva J.A. & Jayasinghe J.M.P.K. (1993) The technology and economics of small-scale commercial shrimp farms in the west coast of Sri Lanka. *Journal of Aquaculture in the Tropics*, **8** (2) 141–149.

Devresse B. (1998) *Production of water stable shrimp feeds*, 3 pp. (mimeo). Fourth International Symposium on Aquatic Nutrition, 15–18 November 1998, La Paz, Mexico.

Dragland Å.(1999) 'Baby-boom' at Tjelderboggen. *Gemini*, (1). http://www.ntnu.no/gemini/1999–01/34b.html

Eversole A.G. & Pomeroy R.S. (1989) Crawfish culture in South Carolina: an emerging aquaculture industry. *Journal of Shellfish Research,* **8** (1) 309–313.

Falaise F. & Boël L. (2000) Economics of sustainable shrimp farming on the Red Sea coast of Saudi Arabia. In: *Abstracts, Aqua 2000, Responsible Aquaculture in the new millennium* (compiled by R. Flos & L. Creswell), p. 210. European Aquaculture Society, Special Publication No. 28.

Farmer A.S.D. (1981) Prospects for penaeid shrimp culture in arid lands. In: *Advances in Food Producing Systems for Arid and Semi-Arid Lands*, pp. 859–897. Academic Press, London.

Fisheries WA (1999) *Farming yabbies*. http:/www.wa.gov.au/westfish/aqua/broc/aqwa/yabby/index.html

Fuller M.J., Kelly R.A. & Smith A.P. (1992) Economic analysis of commercial production of freshwater prawn, *Macrobrachium rosenbergii* postlarvae using a recirculating 'clear-water' culture system. *Journal of Shellfish Research*, **11**, 75–80.

Griffin N. (1999) Shrimp aid for Bangladesh. *Seafood International*, **14,** 15.

Griffin W.L., Lambregts J.A.D., Yates M.W. & Garcia A. (1993) The impact of aquaculture pond engineering design on the returns to shrimp farms. *Journal of the World Aquaculture Society*, **24** (1) 23–30.

Grimsen S., Jaques R.N., Erenst V. & Balchen J.G. (1987) Aspects of automation in a lobster farming plant. *Modelling, Identification and Control*, **8** (1) 61–68.

Gunderson J.L., Richards C. & McDonald M. (1997) Soft crayfish production by eyestalk ablation: can it be profitable? In: *Freshwater Crayfish 11* (ed. W.T. Momot), pp. 567–576. Louisiana State University, LA, USA.

Hajek B.F. & Boyd C.E. (1994) Rating soil and water information for aquaculture. *Aquacultural Engineering*, **13** (2) 115–128.

Hanson J.S., Griffin W.L., Richardson J.W. & Nixon C.J. (1985) Economic feasibility of shrimp farming in Texas: an investment analysis for semi-intensive pond grow-out. *Journal of the World Mariculture Society*, **16**, 129–150.

Hardman P.J.R., Treadwell R. & Maguire, G. (1990) *Economics of prawn farming in Australia*, 20 pp. Presented at International Crustacean Conference, 2–6 July 1990, Brisbane, Australia.

Henocque Y. (1984) Amenagement de la ressource cotiere au Japon: effet des repeuplements marins, 137 pp. *Rapports Techniques 11–1984*, Institute Scientifique et Technique des Peches Maritimes [in French].

Herdman A. (1988) Heating of hatchery water supplies. In: *Aquaculture Engineering Technologies for the Future*. Institution of Chemical Engineers Symposium series No. 111, pp. 343–356. EFCE Publication series No. 66, Hemisphere, London.

Hirono Y. (1989) Extensive shrimp farm management. In: *Proceedings of the Southeast Asia Shrimp Farm Management Workshop*, 26 July–11 August 1989 (ed. D.M. Akiyama), pp. 2–10, Philippines, Indonesia, Thailand. American Soybean Association, Singapore.

Huner J.V. (1997) The capture and culture fisheries of North American crawfish. *World Aquaculture*, **28** (4) 44–50.

Huner J.V. (1999) The fate of the Louisiana soft-shell crawfish. *Aquaculture Magazine,* **25** (3) 46–51.

Huner J., Gydemo R., Haug J., Jarvenpåa T. & Taugbøl T. (1987) Trade, marketing and economics. In: *Crayfish Culture in Europe*, pp. 54–62. Report from the workshop on crayfish culture, 16–19 November, Trondheim, Norway.

Hung L.T. (1992) Integration of crustacean aquaculture with coastal rice farming in Vietnam. *Naga, The ICLARM Quarterly*, **15** (2) 27–29.

IFC (1987) *Marine Shrimp Farming: a guide to feasibility study preparation*, 90 pp. Aquafood Business Associates, International Finance Corp., Charleston, SC 29412, USA.

Imamura K. (1999) The organization and development of sea farming in Japan. In: *Stock Enhancement and Sea Ranching* (eds B.R. Howell, E. Moksness & T. Svåsand), pp. 91–131. Blackwell Science, Oxford.

Israel D.C. (1987a) Economic feasibility of aquaculture projects: a review. *Asian Aquaculture*, **9** (1) 6–9.

Israel D.C. (1987b) Comparative economic analysis of prawn hatcheries. *Asian Aquaculture*, **9** (2) 3–4 & 12.

Jaenike F. (1989) Management of a shrimp farm in Texas. In: *Proceedings of the Southeast Asia Shrimp Farm Management Workshop*, 26 July–11 August 1989 (ed. D.M. Akiyama), pp. 11–21, Philippines, Indonesia, Thailand. American Soybean Association, Singapore.

Jeffs A. & Hooker S. (2000) Economic feasibility of aquaculture of spiny lobsters *Jasus edwardsii* in temperate waters. *Journal of the World Aquaculture Society*, **31** (1) 30–41.

Johns M., Griffin W., Lawrence A. & Fox J. (1981a) Budget analysis of penaeid shrimp hatchery facilities. *Journal of the World Mariculture Society*, **12** (2) 305–321.

Johns M., Griffin W., Lawrence A. & Hutchins D. (1981b) Budget analysis of a shrimp maturation facility. *Journal of the World Mariculture Society*, **12** (1) 104–109.

Johnston D., Clough B., Xuan T.T. & Phillips M. (1999) Mixed shrimp-mangrove forestry farming systems in Cau Mau Province, Vietnam. *Aquaculture Asia*, **4** (2) 6–12.

Jones J.B. (1985) The economics of freshwater crayfish farming in New Zealand. *NZ Agricultural Science*, **19** (2) 64–67.

Jory D.E. & Martinez J.B. (1998) Shrimp culture in Panama: a success story continues. *Aquaculture Magazine*, **24** (4) 91–9.

Juan Y., Griffin W.L. & Lawrence A.L. (1988) Production costs of juvenile penaeid shrimp in an intensive greenhouse raceway nursery system. *Journal of the World Aquaculture Society*, **19** (3) 149–60.

Kearns J.P. (1989) Advantages of extrusion cooking and comparisons with the pelleting process for aquatic feeds. In *Proceedings of the People's Republic of China aquaculture and feed workshop*, 17–30 Sept., 1989 (ed. D.M. Akiyama), pp. 245–69. American Soybean Assoc., Singapore.

Keithly W.R. Jr., Roberts K.J. & Ward J.M. (1992) Effects of shrimp aquaculture on the US market: an econometric analysis. In: *Aquaculture Models and Economics* (eds U. Hatch & H. Kinnucan), pp. 125–156. Westview Press, Colorado, USA.

Keller M. (1988) Finding a profitable population density in rearing summerlings of European crayfish *Astacus astacus* L. In: *Freshwater Crayfish 7* (ed. P. Goedlin de Tiefenau), pp. 259–266. Musée Zoologique Cantonal, Lausanne, Switzerland.

Kongkeo H. (1997) Comparison of intensive shrimp farming systems in Indonesia, Philippines, Taiwan and Thailand. *Aquaculture Research*, **28**, 789–796.

Lacroix D.G., Glude J., Thomas J.E. & Le Menn H. (1993) From research to commercialisation: lessons from four different strategies in the development of freshwater prawn culture (*Macrobrachium rosenbergii*) in America since 1977. In: *World Aquaculture '93. From discovery to commercialization* (eds M. Carrillo, L. Dahle, J. Morales, P. Sorgeloos, N. Svennevig & J. Wyban), p. 101. European Aquaculture Society, Special Publication No. 19.

Lambregts J.A.D., Thacker S.G. & Griffin W.L. (1993) Economic evaluation of different stocking densities for various sized shrimp farms in Texas. *Journal of the World Aquaculture Society*, **24** (1) 12–22.

Lapie L.P.& Librero A.R. (1979) *Crab farming in the Philippines: a socio-economic study*, 56 pp. Research paper series/ Socio-economic survey of the Aquaculture Industry, (21). SEAFDEC-PCARR Research Programme, Los Banos, Laguna, Philippines.

Lawrence A.L. (1985) Marine shrimp culture in the western hemisphere. In: *Second Australian National Prawn Seminar* (eds P.C. Rothlisberg, B.J. Hill & D.J. Staples), pp. 327–36. NPS2, Cleveland, Australia.

Lawrence A.L., McVey J.P. & Huner J.V. (1985) Penaeid shrimp culture. In: *Crustacean and Mollusk Aquaculture in the United States* (eds J.V. Huner & E. Evan Brown), pp. 127–157. AVI Inc., Westport, CT, USA.

Lawrence C. (1998) Marron. In: *The New Rural Industries* (ed. K.W. Hyde), pp. 114–119. Rural Industries Research and Development Corporation, Canberra, Australia.

Lawrence C. & Jones C. (2001) *Cherax*. In: *Biology of Freshwater Crayfish* (ed. D.M. Holdich). Blackwell Science, Oxford, UK.

Lee D. O'C. (1994) *The potential economic impact of lobster stock enhancement,* 27 pp. MSc. Thesis, Department of Economics and Related Studies, University of York, UK.

Lee D. O'C., Hillion B. & Letessier L. (2000) An analysis of factors influencing pond yields in a semi-intensive shrimp farm. In: *Abstracts, Aqua 2000, Responsible Aquaculture in the new millennium* (compiled by R. Flos & L. Creswell), p. 379. European Aquaculture Society, Special Publication No. 28.

Leung P.S. (1994) Bioeconomic modelling in aquaculture after two decades. In: *Socioeconomics of Aquaculture* (eds Y.C. Shang, P.S. Leung, C.S. Lee, M.S. Su & I.C. Liao), pp. 115–137. Tungkang Marine Laboratory Conference Proceedings (4).

Leung P.S., Lee D.J. & Hochman E. (1992) Bioeconomic modelling of shrimp and prawn: a methodological comparison. In: *Aquaculture Models and Economics* (eds U. Hatch & H. Kinnucan), pp. 39–53. Westview Press, Colorado, USA.

Ling B.H., Leung P.S.& Shang Y.C. (1999) Comparing Asian shrimp farming: the domestic resource cost approach. *Aquaculture*, **175** (1–2) 31–48.

Liu J.Y. (1990) Resource enhancement of Chinese shrimp, *Penaeus orientalis*. *Bulletin of Marine Science*, **47** (1) 124–133.

Lumare F., Amerio M., Arata P., *et al.* (1989) Semi-intensive culture of the kuruma shrimp *Penaeus japonicus* by fertilizer and feed applications in Italy. In: *Aquaculture, a biotechnology in progress* (eds N. de Pauw, E. Jaspers, H. Ackefors & N. Wilkins), pp. 401–407. European Aquaculture Society, Bredene, Belgium.

Main K.L. & Deupree R.H. Jr (1986) Commercial aquaculture in Hawaii, 1986, 25 pp. *Information Text Series 031*, College of Tropical Agriculture & Human Resources, HI, USA.

McCoy H.D. II (1986) Intensive culture systems past, present and future Parts I, II and III. *Aquaculture Magazine*, **12** (6) 32–35; **13** (1) 36–40; **13** (2) 24–29.

McIntosh R.P. (2000) Changing paradigms in shrimp farming: III Pond design and operation considerations. *Global Aquaculture Advocate,* **3** (1) 42–45.

Medley P.B., Nelson R.G., Upton Hatch L., Rouse D.B. & Pinto G.F. (1994) Economic feasibility and risk analysis of Australian red claw crayfish *Cherax quadricarinatus* aquaculture in the southeastern United States. *Journal of the World Aquaculture Society*, **25** (1) 135–146.

Millamena O.M. & Triño A.T. (1997) Low-cost feed for *Penaeus monodon* reared in tanks and under semi-intensive and intensive conditions in brackishwater ponds. *Aquaculture*, **154** (1–2) 69–78.

Mills B.J., Morissy N.M. & Huner J.V. (1994) Cultivation of freshwater crayfishes in Australia. In: *Freshwater Crayfish Aquaculture in North America, Europe and Australia* (ed. J.V. Huner), pp. 217–289. Food Products Press, New York.

Mock C.R. (1983) *Penaeid shrimp culture consultation Taiwan,* 51 pp. (mimeo). Trip report, 1–11 December 1983.

Moksness E., Støle R. & Van der Meeren G. (1998) Profitability analysis of sea ranching with atlantic salmon (*Salmo salar*), Arctic charr (*Salvelinus alpinus*), and European lobster (*Homarus gammarus*) in Norway. *Bulletin of Marine Science*, **62** (2) 689–699.

Mosig J. (1999) Grower questions the economics of yabbies. *Austasia Aquaculture Magazine*, **13** (2) 37–39.

Moss S.M., Reynolds W.J. & Mahler L.E. (1998) Design and economic analysis of a prototype biosecure shrimp growout facility. In: *Proceedings of the US Marine Shrimp Farming*

Program Biosecure Workshop, 14 February 1998 (ed. S.M. Moss), pp. 5–17. The Oceanic Institute, HI, USA.

Muir J.F. & Kapetsky J.M. (1988) Site selection decisions and project cost: the case of brackish water pond systems. In: *Aquaculture Engineering Technologies for the Future*. Institution of Chemical Engineers Symposium series No. 111, pp. 45–63. EFCE Publication series No. 66, Hemisphere, London.

Murphy D. (2000) Deeper into the morass. *Far Eastern Economic Review*, 1 June 2000. http://www.feer.com/_0006_01/p58nmoney.html

NACA (1986) A prototype warm water hatchery, 32 pp. *Technology series, selected publication No. 4*. NACA, Bangkok, Thailand.

Nash C.E. (1988) *Investment in aquaculture*, 13 pp. Aquaculture Development and Coordination Programme, UNDP, FAO, Plenary paper, presented at Aquaculture International Congress, 9 September 1988, Vancouver, Canada.

New M.B. (2000) Commercial freshwater prawn farming around the world. In: *Freshwater Prawn Culture: the farming of* Macrobrachium rosenbergii (eds M.B. New & W.C. Valenti), pp. 290–325. Blackwell Science, Oxford, UK.

New M.B., Sanders S., Brown R.L. & Cole R.C. (1978) The feasibility of farming *Macrobrachium* in Dominica, West Indies. *Proceedings of 9th Annual Meeting of the World Mariculture Society*, 3–6 January 1978, pp. 67–90, Atlanta, GA, USA.

Nghia T.T. (1991) Larviculture techniques and economics of small-scale *Macrobrachium rosenbergii* hatcheries in the Mekong Delta, Vietnam. *Larvi '91 – Fish & Crustacean Larviculture Symposium* (eds P. Lavens, P. Sorgeloos, E. Jaspers & F. Ollivier), pp. 368–371. European Aquaculture Society, Special Publication No. 15.

Ogle J.T. & Lotz J.M. (1998) Preliminary design of a closed, biosecure shrimp growout system. In: *Proceedings of the US Marine Shrimp Farming Program Biosecure Workshop*, 14 February 1998 (ed. S.M. Moss), pp. 39–47. The Oceanic Institute, HI, USA.

O'Sullivan D. (1998) Integrating water storage and extensive yabby production. *Austasia Aquaculture Magazine*, **12** (4) 42–44.

O'Sullivan D. (1999) Queensland production returns 1: prawns, barramundi and oysters. *Austasia Aquaculture Magazine*, **13** (5) 27–28.

O'Sullivan D. (2000) New federal regulations on Great Barrier Reef protection to apply to Queensland aquaculture farms. *Austasia Aquaculture Magazine*, **14** (2) 34–39.

Oshima Y. (1984) Status of fish farming and related technological development in the cultivation of aquatic resources in Japan. *TML Conference Proceedings, No. 1*, pp. 1–11. Tungkang Marine Laboratory, Pingtung, Taiwan.

Ovenden C. (1994) *Japan's kuruma prawn market*, 21 pp. Information Series QI 94041, Queensland Department of Primary Industries, Queensland, Australia.

Paquotte P., Chim L., Martin J-L.M., Lemos E., Stern M. & Tosta G. (1998) Intensive culture of shrimp *Penaeus vannamei* in floating cages: zootechnical, economic and environmental aspects. *Aquaculture*, **164** (1–4) 151–166.

Parado-Estepa F.D. (1995) Research on crustaceans. In: *Towards Sustainable Aquaculture in Southeast Asia and Japan* (eds T.U. Bagarinao & E.E.C. Flores), pp. 187–198. SEAFDEC Aquaculture Department, Iloilo, Philippines.

Parker J. (1990) Shrimp farm investment analysis for outdoor ponds in Texas. In: *Texas A&M University shrimp farming short course materials, article 47* (ed. G.D. Treece), pp. 1–793. Texas A & M Sea Grant College Program, College Station, TX, USA.

Parks P. & Bonifaz M. (1994) Nonsustainable use of renewable resources: mangrove deforestation and mariculture in Ecuador. *Marine Resource Economics*, (9) 1–18.

Parr W.H., Capper B.S., Cox D.R.S., *et al.* (1988) The small-scale manufacture of compound animal feed, 87 pp. *Bulletin No. 9*, Overseas Development Natural Resources Institute, Chatham, UK.

Pickering H. (2000) Legal framework governing artificial reefs in the European Union. In: *Artificial Reefs in European Seas* (eds A.C. Jensen, K.J. Collins & A.P.M. Lockwood), pp. 469–487. Kluwer Academic, Netherlands.

Polioudakis M. (2000) *A recent history of shrimp farming in Thailand*. http://members.aol.com/brosenberr/Polioudakis.html

Posadas B.C. & Homziak J. (1993) Costs and returns of open versus closed systems for producing soft-shell crawfish in Mississippi. *Journal of Applied Aquaculture*, **2** (1) 85–103.

Preston N.P., Brennan D.C. & Crocos P.J. (1999) Comparative costs of postlarval production from wild or domesticated Kuruma shrimp, *Penaeus japonicus* (Bate), broodstock. *Aquaculture Research*, **30,** 191–197.

Primavera J.H. (1983) Prawn hatcheries in the Philippines. *Asian Aquaculture*, **5** (3) 5–7 & 8.

Primavera J.H. (1997) Socio-economic impacts of shrimp culture. *Aquaculture Research,* **28**, 815–827.

Quynh V.D. (1990) Shrimp larviculture in Viet Nam. *Larviculture and Artemia Newsletter*, (16) 22–26. State University of Ghent, Belgium.

Rahman M.K. & Srikrishnadhas B. (1994) The potential for spiny lobster culture in India. *Infofish International,* (1) 51–53.

Rhodes R.J. (2000a) Economics and business management. In: *Freshwater Prawn Culture: the farming of* Macrobrachium rosenbergii (eds M.B. New & W.C. Valenti), pp. 369–392. Blackwell Science, Oxford, UK.

Rhodes R.J (2000b) Comparative grow-out costs of freshwater prawn (*Macrobrachium rosenbergii*) farming: a literature review. In: *Abstracts, Aqua 2000, Responsible Aquaculture in the New Millennium* (compiled by R. Flos & L. Creswell), p. 598. European Aquaculture Society, Special Publication No. 28.

Roberts K.J. & Dellenbarger L. (1989) Louisiana crawfish product markets and marketing. *Journal of Shellfish Research*, **8** (1) 303–307.

Roma C. (1997) Costa Rica gaining ground in shrimp culture. *Aquaculture Magazine*, **23** (4) 59–66.

Romaire R.P. (1995) Harvesting methods and strategies used in commercial procambarid crawfish culture. *Journal of Shellfish Research*, **14** (2) 545–551.

Rosenberry R. (1988) Shrimp farming in China. *Aquaculture Digest*, **13** (9) 2–4.

Rosenberry, R. (1989) Ecuador: No seedstock! No money! Exports slipping. *Aquaculture Digest*, **14** (7) 1–2.

Rosenberry R. (1994) Hawaii Aqua-Seed sells prawn (*Macro-*

brachium rosenbergii) and shrimp (*P. vannamei*) postlarvae. *Shrimp News International*, **19** (7) 15.

Rubino M.C. (1992) Economics of red claw (*Cherax quadricarinatus* von Martens, 1868) aquaculture. *Journal of Shellfish Research*, **11** (1) 157–162.

Sadek S. & Moreau J. (1996) Prawn (*Macrobrachium rosenbergii*) culture in earthen ponds in the Nile delta, Egypt: culture parameters and cost benefits. *Israeli Journal of Aquaculture – Bamidgeh*, **48** (4) 201–218.

Sadek S. & Moreau J. (1998) Culture of *Macrobrachium rosenbergii* in monoculture and polyculture with *Oreochromis niloticus* in paddies in Egypt. *Israeli Journal of Aquaculture – Bamidgeh*, **50** (1) 33–42.

Samocha T.M. & Lawrence A.L. (1998) Preliminary design and operating specifications for a biosecure shrimp grow-out facility in Texas. In: *Proceedings of the US Marine Shrimp Farming Program Biosecure Workshop*, 14 February 1998 (ed. S.M. Moss), pp. 49–58. The Oceanic Institute, HI, USA.

Samocha T.M., Burkott, B.J., Lawrence A.L., Yuan Y.S., Jones E.R. & McKee D.A. (1998) Management strategies for production of the Atlantic white shrimp *Penaeus setiferus* as bait shrimp in outdoor ponds. *Journal of the World Aquaculture Society*, **29** (2) 211–220.

Samonte G.P.B. & Agbayani R.F. (1991) Pond culture of mud crab (*Scylla serrata*): an economic analysis. In: *The Mud Crab. Report of the Seminar on mud crab culture and trade* (ed. A.C. Angell), pp. 225–234. Bay of Bengal Programme, Madras, India.

Sandifer P.A. & Smith T.I.J. (1985) Freshwater prawns. In: *Crustacean and Mollusk Aquaculture in the United States* (eds J.V. Huner & E. Evan Brown), pp. 63–125. AVI Inc., Westport, CT, USA.

SEAFDEC (1985) *A guide to prawn hatchery design and operation*, 49 pp. Extension manual No. 9, South East Asia Fisheries Development Centre, Tigbauan, Iloilo, Philippines.

Seafish (1998) Lobster stocking, update 1998. *Lobster Bulletin*. http://www.seafish.co.uk/aquaculture_lobster.htm

Shang Y.C. (1983) The economics of marine shrimp farming: A survey. In: *Proceedings of First International Conference on Warm Water Aquaculture – Crustacea*, 9–11 February 1983 (eds G.L. Rogers, R. Day & A. Lim), pp. 7–18. Brigham Young University, HI, USA.

Shang Y.C. (1989) Marine shrimp farming in China. *Infofish International*, (2) 16–17.

Shehadeh Z.H. & Orzeszko J (1997) External assistance. In: *Review of the state of world aquaculture* (ed. Z. Shehadeh), 163 pp. FAO Fisheries circular (886). Fisheries Department, FAO, Rome.

Singh T. (1993) The use of pond liners in aquaculture. *Infofish International*, (5) 50–53.

Skabo H. (1988) Shrimp farming developments in West Africa. In: *Shrimp '88, Conference proceedings*, 26–28 January 1988, Bangkok, Thailand, pp. 95–102. Infofish, Kuala Lumpur, Malaysia.

Smith I.R. (1984) *Social Feasibility of Coastal Aquaculture: packaged technology from above or participatory rural development*, 42 pp. ICLARM cont. 225, prepared for Consultation on Social Feasibility of Coastal Aquaculture, 26 November–1 December 1984, Madras, India.

Smith T.I.J., Jenkins E.W., & Sandifer P.A. (1983) Enclosed prawn nursery systems and effects of stocking juvenile *Macrobrachium rosenbergii* in ponds. *Journal of the World Mariculture Society*, **14**, 111–125.

Staniford A. (1988) More on the economics of yabbie farming. South Australian Dept. Fisheries, *SAFISH*, **13** (2) 10–12.

Staniford A.J. & Kuznecovs J. (1988) Aquaculture of the yabbie, *Cherax destructor* Clark (Decapoda: Parastacidae): an economic evaluation. *Aquaculture and Fisheries Management*, **19**, 325–340.

Surtida M. (1997) AQD's mudcrab R&D. *Asian Aquaculture*, **19** (3) 17–20.

Sylvia G. (1997) Generating information for aquaculture development: the art and science of economic policy modelling. *Aquaculture Economics and Management*, **1** (1) 87–98.

Tabigoon L. Jr. (1998) Should you use cages and pens instead of ponds? *Asian Aquaculture*, **20** (5) 21–30.

Tietze U. (1989) No fish without credit. *Infofish International*, (5) 46–47.

Triño A.T. & Sarozza J.C. (1995) Effect of a diet lacking in vitamin and mineral supplements on growth and survival of *Penaeus monodon* juveniles in a modified extensive culture system. *Aquaculture*, **136** (3–4) 323–330.

Uglem I., Knutsen H. & Hansen S.H. (1998) Culture of juvenile lobsters (*Homarus gammarus*) in Norway. In: Proceedings of a workshop on lobster stock enhancement held in the Magdalen Islands (Quebec), 29–31 October 1997 (ed. L. Gendron), *Canadian Industry Report of Fisheries and Aquatic Sciences*, **244**, 59–62.

Usha Rani G., Chandra Reddy T. & Ravindranath K. (1993) Economics of brackishwater prawn farming in Nellore district of Andhra Pradesh state, India. *Journal of Aquaculture in the Tropics*, **8**, 221–230.

Valenti W.C. (1993) Freshwater prawn culture in Brazil. *World Aquaculture*, **24** (1) 30–34.

Valenti W.C. & Daniels W.H. (2000) Recirculation hatchery systems and management. In: *Freshwater Prawn Culture: the farming of* Macrobrachium rosenbergii (eds M.B. New & W.C. Valenti), pp. 69–90. Blackwell Science, Oxford, UK.

Valenti W.C. & New M.B. (2000) Grow-out systems – monoculture. In: *Freshwater Prawn Culture: the farming of* Macrobrachium rosenbergii (eds M.B. New & W.C. Valenti) pp. 157–176. Blackwell Science, Oxford, UK.

Villalon J.R. (1993) Commercial semi-intensive penaeid grow-out techniques in Ecuador. In: *Handbook of Mariculture*, 2nd edn, *Vol. 1 Crustacean aquaculture* (ed. J.P. McVey), pp. 237–348. CRC Press, Boca Raton, FL, USA.

Waddy S.L. & Aiken D.E. (1998) Lobster (*Homarus americanus*) culture and resource enhancement: the Canadian experience. In: Proceedings of a workshop on lobster stock enhancement held in the Magdalen Islands (Quebec), 29–31 October 1997 (ed. L. Gendron), *Canadian Industry Report of Fisheries and Aquatic Sciences*, **244**, 9–18.

Waits J.W. III & Dillard J.G. (1987) *The costs of processing and hauling freshwater shrimp in Mississippi*, 22 pp. Bulletin 953, Mississippi Agricultural and Forestry Experiment Station, Mississippi, USA.

Whitmarsh D. & Pickering H. (2000) Investing in artificial reefs. In: *Artificial Reefs in European Seas* (eds A.C. Jensen,

K.J. Collins & A.P.M. Lockwood), pp. 451–467. Kluwer Academic, Netherlands.

Whitmarsh D., Pickering H. & Sarch M.T. (1995) *Economic Appraisal of Artificial Reef Structures for Lobster Production*, 130 pp. Final Report. Centre for Coastal Zone Management, University of Portsmouth, UK.

Wilkenfield J.S. (1992) Commercial hatchery status report: an industry panel viewpoint. In: *Proceedings of the Special Session on Shrimp Farming* (ed. J. Wyban), pp. 71–86. World Aquaculture Society, Baton Rouge, LA, USA.

Woodhouse C. (2000) Huge demand: where's the investment? *Fish Farming International*, **27** (4) 7.

Wyban J.A., Swingle J.S., Sweeney J.N. & Pruder G.D. (1992) Development and commercial performance of high health shrimp using specific pathogen free (SPF) broodstock *Penaeus vannamei*. In: *Proceedings of the Special Session on Shrimp Farming* (ed. J. Wyban), pp. 254–260. World Aquaculture Society, Baton Rouge, LA, USA.

Xu J., Xia M., Ning X. & Mathews C.P. (1997) Stocking, enhancement, and mariculture of *Penaeus orientalis* and other species in Shanghai and Zhejiang Provinces, China. *Marine Fisheries Review*, **59** (4) 8–14.

Yamaha (1989) Prawn culture, 8 pp. *Yamaha Fishery Journal*, (30). Yamaha Motor Co., Shizuoka-ken, Japan.

Yap W.G. (1990) Backyard hatcheries take off in Jepara. *Infofish International*, (2) 42–47.

Chapter 11
Impact of Crustacean Aquaculture

11.1 Introduction

The preceding chapters of this book have dealt with the biological and technical factors that promote or constrain consistent performance and productivity in crustacean aquaculture projects. There are, however, additional considerations equally vital to the long-term success of a venture, that are not always readily appreciated by entrepreneurs, business administrators and their advisers. These are concerned with the social and environmental consequences and the interactions that can arise with institutions as a result of rapid movement into crustacean aquaculture in general, and into tropical shrimp farming in particular. Indeed, success is more likely if attention is paid to the socio-economic factors that have a major influence on the choice of species and even method of production (Tisdell 1994a). Until the early 1980s, the consequences of crustacean farming were widely believed to include a beneficial socio-economic impact with only marginal environmental disruption brought about by implanting a new industry onto otherwise unproductive rural land (NOAA 1988). While this may be true in some instances (e.g. Wilson 1999), a number of popular misconceptions persist, for example, that crustacean aquaculture development automatically produces:

- better standards of living in rural areas;
- more equitable distribution of income, wealth and profits;
- more jobs and job opportunities;
- easily marketable, high-value products;
- food for rural communities;
- improved local nutritional standards;
- significant, sustainable rewards for investors.

Failure to achieve these highly desirable objectives can lead to local disillusionment expressed by increased poaching, vandalism and banditry, perhaps even resulting in project failure (sections 10.4 and 11.2.3). Violent protests ending in fatalities are far from uncommon (Primavera 1997). The blame has sometimes been placed on government agencies, developers, scientists or extension workers for not understanding or giving adequate priority to local needs and conditions, or at other times on farmers and local authorities criticised for resisting change, or lacking initiative (Weeks 1989). In this context it is also relevant to remember that the legislation behind many of the government regulations affecting crustacean farming is established by politicians who are particularly sensitive to public opinion and special interest groups (Boyd 1999).

Small-scale, artisanal crustacean farming generally has little impact on its surroundings. In contrast, modern developments, particularly in the shrimp, prawn and crayfish industries have had significant impact both in ecological and environmental terms. Pressures on the environment and the consequent frequency of failure calls into question the sustainability of many enterprises. The provision of ingredients for feed manufacture, especially fishmeals and oils, also gives cause for concern (New 1999a). In addition, climatic changes are expected to have both positive and negative impacts on crustacean aquaculture worldwide.

The most often reported detrimental impacts caused by crustacean farming are:

Ecological

- Pressure on natural stocks (broodstock, juveniles, also fish used in feed ingredients)
- Transplantations and escapes
- Disease
- Treatment chemicals
- Genetic change

Environmental

- Site clearance (loss of natural storm barriers)
- Loss of nursery habitat
- Water supplies
- Effluents

Social

- Unequal benefit distribution
- Loss of living resources
- Population displacement

In this regard, a number of organisations have taken particular stances on the issues of socially and environmentally responsible crustacean farming. For example, on the one hand, in 1996, 21 non-governmental and community organisations from Latin America, North America, Europe and Asia met in Choluteca, Honduras, and demanded of the international community a global moratorium on any further expansion of shrimp aquaculture in coastal areas that did not meet their criteria for 'sustainability' (Choluteca Declaration 1997). In addition to demanding actions to protect local communities and the environment from exploitation (many aspects of which are discussed later in this chapter), the Declaration also reflected the public unease that exists about genetically modified crustaceans by demanding an outright ban on their use. Indeed, the farming of such animals remains the subject of increasing debate, much of which is ill-informed and over-emotional. On the other hand, and as a direct response to attacks from non-governmental organisations (NGOs) and unfair trade restrictions affecting the shrimp farming industry in particular, a newly formed trade organisation, the Global Aquaculture Alliance (GAA), prepared a draft Code of General Principles (drawn from the aquaculture section of the *FAO Code of Conduct for Responsible Fisheries*; FAO 1995, 2000). The code was designed to assist producers to raise aquaculture products (initially shrimp) in a socially and environmentally responsible manner (Anon. 1997). The Alliance produces a magazine, the *Global Aquaculture Advocate*, covering topics on responsible aquaculture and, in close liaison with industry and researchers, is preparing codes of practice on many specific issues; included among the first were mangrove protection, viruses in shrimp processing wastes and pond effluent controls. Similar, national organisations representing crustacean producers elsewhere (e.g. in Australia and Thailand) are also signing up to codes of conduct for responsible aquaculture and encouraging a more proactive stance among their members.

Conservationists frequently argue that any ecological change brought about by the establishment of a new industry, or the intensification of an existing enterprise, is undesirable. However, it is important to recognise that clear distinctions can be made between the harmful, disruptive consequences of some crustacean farming enterprises and the other, often marginal, changes that arise whenever species interact with their surroundings. Sociologists often fail to give adequate acknowledgement to the benefits gained by whole communities through improved communication networks, health care and employment opportunities. Failure to identify or objectively assess such differences in impact during the planning or feasibility study phases of projects can unduly stifle legitimate and worthwhile developments. Doubtless both the NGOs and the trade organisation mentioned above represent factions in global society that seek the same end result, a socially equitable and environmentally benign aquaculture industry, despite their differences in approach. Continuing dialogue that leads to objectively considered and fairly implemented actions will be essential if these aims are to be met (Hargreaves 1997; Lassen 1997).

For further reading, a summary of the general impacts of inland and coastal zone aquaculture throughout Southeast Asia has been compiled by FAO/NACA (1995). In addition to country overviews, the report provides suggestions regarding the implementation of identified management options, at both government and local levels, designed to minimise adverse effects and promote the industry's sustainability. Accounts have also been published of western society perspectives on conservation issues affecting the shrimp industry (Salmon 1992) and the socio-cultural impacts of shrimp farming and soft-shell crab production in the USA (Fiske & Plé 1992).

11.2 Social impact

'The task of government is to liberate technology from its closed class structure and make it accessible to society at large' (Hayashi 1984).

11.2.1 Institutional involvement

National and international policy-makers often appear to equate development with increased productivity and economic efficiency when planning aquaculture strategies. It is easier and less costly for a nation to develop aquaculture through large-scale, corporate undertakings financed by development banks at subsidised or below market rates, than through participatory rural development by coastal communities (Smith 1984). Capital-intensive technologies are favoured since there is usually a need to increase foreign exchange. Partly as a result of this, governments and international aid agencies have invested heavily in shrimp farming. During the 1980s, the Asian Development Bank and World Bank supported coastal shrimp projects in at least nine Asian countries with Japan, Belgium, the UK, the USA and other developed nations providing bilateral support (Bailey & Skladany 1991). Much less investment was placed in inland freshwater aquaculture operations generally because of restricted resource availability and because the known cultivable freshwater crustacean species usually commanded lower prices. In many of these inland areas fish and rice are produced for local consumption by low-capital, high-labour methods. The major aquaculture policy shifts in South-east Asia (Skladany 1992), that resulted in greater emphasis on attracting investment for producing luxury crustaceans rather than developing fish and molluscan aquaculture to meet the needs of local populations, generated an important but entirely separate debate (Bailey 1997).

Where there is serious concern for the impact that the aquaculture development will have on society, one of the first questions to be asked is whether or not it will reinforce existing socio-cultural and institutional power structures that keep the majority of society members in poverty, or provide opportunities for a wider spread of benefits (Smith 1984). Many of the areas designated by governments for aquaculture development are often economically depressed coastal zones, without local capital and which cover ground inherently unsuitable or difficult to work. Coastal communities are poor because they have no access to alternative employment, and are likely to remain so because existing community and national structures allow local élites to monopolise the benefits of new enterprises (section 10.4). It is also recognised that institutions tend to adapt slowly to changing technologies and may actually inhibit a just distribution of benefits when the technologies are applied. Major social changes thus seem inevitable, particularly when resources are put into the more intensive enterprises that promote rapid economic growth. They will not, however, necessarily lead to the fairer distribution of income. It is left to governments and aid agencies to proactively initiate and support programmes that will ensure community needs are included in investment proposals.

11.2.2 Land ownership and common resources

Conflict between crustacean farmers and others frequently arises because rights to use resources are not well defined or enforced. In Australia, for example, where shrimp and crayfish farming is expanding, resource use is determined by traditional market demand (Holland & Brown 1999). Yet the resources required by aquaculture are not necessarily those easily identifiable in traditional terms and many problems occur either because of poor resource allocation systems or because of the consequences of resource use (seed collection, effluent discharges into waterways). Marked differences may exist between countries in respect of allocation of, for example, water supplies. In many cases access to water comes with the site. In Mexico, water abstraction is prioritised, with aquaculture interests coming seventh after domestic and livestock requirements. In parts of Europe a volume-based licence system operates, while in parts of Australia, water trading encourages efficient utilisation of this essential resource.

In most countries coastal land is state owned but can often be transferred to private ownership if the implementation of a large aquaculture project is judged to be in the national interest. When this happens, local communities may be deprived of traditionally common resources vital to their subsistence (Skladany 1992). Their concept of security changes and they are likely to become increasingly dependent on seasonal jobs in the new industry, which require few skills. Intensive aquaculture operations often employ fewer local people than traditional rice or sugar farms for areas of equivalent size (Bailey 1988). If the project is successful land prices will rise, further enhancing the power of the local landlords, and may encourage alteration of tenancy terms in the landowner's favour. Increased land and seed prices could also arise following overseas investment in a neighbouring country's export industry, for example, if Chinese entrepreneurs interested in North Vietnam's mud crab production became involved in crab farming (Overton & Macintosh 1997). Wages and living standards might also be forced down because workers cannot get employment elsewhere or get enough alternative sustenance from the altered local environment.

Inland rural populations are often far less well nourished than those on the coast that enjoy access to seafish, and those in towns where seafish are marketed. Inland, ownership is clearly defined and freshwater aquacultures tend to be more integrated, family affairs. Because of this, large-scale developments can be sustained although

Plate 11.1 Traditional use of mangrove resources for the construction of barrier traps (kraals) in a lagoon in southern Sri Lanka. The area was being considered for shrimp farm development despite the opposition of the local community.

opportunities are often more limited because neighbouring land or water supplies are already owned. The overzealous conversion of substantial areas of agricultural or aquacultural land, traditionally used for producing staple crops, to the production of high-value, exportable commodities such as coffee, shrimp or prawns, is likely to increase the impact of external market forces on local marketing systems and could further exacerbate local nutrition problems. From the 'cash crop' point of view it is a pity that better endemic crustaceans than *Macrobrachium* (the larvae of the largest species require brackish water) are not more widely available for culture in inland tropical freshwaters. This lack has resulted not only in the widespread introduction of *M. rosenbergii* beyond its natural range but also a growing interest in farming Australian redclaw crayfish, which has no need of saltwater. It is worth noting that one smaller species of *Macrobrachium, M. nipponense,* can be reared throughout its life in freshwater and is now widely grown in China (sections 4.5.2 and 7.3.5.3).

In reality then, the large-scale development of crustacean aquaculture can have considerable social impact. The rural poor frequently depend on a variety of subsistence activities ranging from small-scale trading, manufacturing and farming to fishing, fry collection for fish farms and gathering other aquatic organisms for food, the latter particularly in mangrove areas where building materials, lumber, firewood, charcoal, thatch and a variety of foodstuffs are also sought. When large-scale shrimp culture moves in, vast areas are cleared and local people may be deprived of these traditional resources or access to them. If the soil is later found to be unsuitable, as it often is in cleared mangrove areas, the site is abandoned and the people left with neither employment nor their traditional resources to fall back on. For these reasons alone shrimp farming in the tropics may seem less socially acceptable than farming fish, seaweed, bivalves or crabs.

Nevertheless, success stories do exist. In Venezuela for example, the industry is being developed through close co-operation between public and private sectors with sustainability issues being given the highest priority (Jory 2000). Also, in many developing countries, creation of new freshwater reservoirs is increasing rapidly to meet growing demands for power, irrigation and municipal water supplies but considerate planning can ameliorate the inevitable physical and socio-economic disruption. For example, the introduction of cage aquaculture can provide communities displaced by reservoir construction with a livelihood. Yet at the same time safeguards must also be implemented to prevent wealthy urbanites from entering the new, lucrative industry (Costa-Pierce 1992).

Often no institution exists to protect a village or community from development. Smith (1984) maintained that the allocation of use rights in tropical coastal regions needs to be administered by decentralised institutions, and claimed the lack of decentralisation could be equated directly with lack of effective control over land use and with environmental deterioration. However, to prevent the rise of local tyrants it would be desirable to maintain public accountability at all levels. Furthermore, it is argued that, at the outset, large-scale projects should make provision for payments into environmental funds to ensure some form of restoration can be accomplished if and when ponds are abandoned. Such an approach already exists in industries such as timber and mining and could be enforced through conditions set in operating licences and permits (Clay 1997).

11.2.3 Community relationships

During the initial phases of project implementation some disruption of communities may occur, such as that seen in Ecuador when villagers from the highlands migrated to work on the construction of the coastal shrimp ponds. Although much of the initial construction was mechanised, labourers are required almost continuously since 80% repair and reconstruction is necessary every 4 years (CPC 1989). Similarly, the problems that occurred in the Taiwanese and Thai shrimp farming industries during the late 1980s led to a greater involvement of expatriate technicians in projects elsewhere in South-east Asia and the Americas (section 11.2.6).

A large project may employ a significant number of local people at some time or another but it may also decrease opportunities for alternative employment in the region. One result is wage suppression and increased dependence of tenants and smallholders on landlords and project facilities. Beneficial smallholder programmes have been implemented alongside or within large projects (also called nucleus/plasma schemes) but there is a danger of tying the smallholder restrictively into the main company which may, for example, be the only source of seedstock supply or market outlet (section 9.3.7). Two flagship enterprises which attempted to integrate small-scale farmers with large-scale, private investment in Indonesia (section 10.4) and Thailand (Fegan 1995) were recently (2000) reported to have failed, but in doing so provide useful lessons.

The Indonesian enterprise was huge (covering 16 000 ha and involving some 7750 farmers) while the Thai company was somewhat smaller, employing about 800 people and worked with 700 farmers. Both projects had their own hatcheries, feed mills and processing plants and, in each, the farmers owned or were given loans to buy their own ponds. The parent companies were financially reliant on selling post-larvae, feed and other basic services to the participants and also on buying each farmer's harvest for processing and onward sale. Although there were differences in scale and detail, both aimed to incorporate a considerable degree of social responsibility and were reasonably ecologically sound – both unique features at the time. The Thai project in particular took a lot of trouble to get permission from those affected by the projects, and to minimise impacts on those not involved.

Although there is always more than one point of view regarding failure, factors involved in the demise of the Indonesian project included poor harvests after the El Niño drought of 1997 followed by local currency devaluation, escalating debts, increased feed and seed costs and farmers being paid 70% below the regional average for their crop. Factors involved in the collapse of the Thai project included participants failing to honour their contracts by buying some post-larvae and feed, as well as selling part of their crop outside the contract farming system, and often failing to pay service fees for site maintenance. This increased the farmers' debts to the banks as their loan repayments were linked to the sale of shrimp to the company. Outbreaks of YHV and WSSV diseases caused financial losses to both the company and the farmers, many of whom failed to set aside sufficient contingency funds. The parent company could not recoup its investment (which included emergency loans to farmers to cover disease losses) through the contract farming system, through its export and feed sales activities or by buying shrimp for processing from non-participants. Consequently, it was never able to make the expected return on invested capital (D. Fegan & M. Polioudakis, 2000 pers. comm.).

These stories are regrettable but salutary. If we compare the two projects, one perception might be that, whereas in Indonesia the individual farmers were too tightly bound to the parent company and were eventually driven to violence, the Thai project allowed too much independence (or did not sufficiently enforce the farmers' contractual obligations) such that individual farmers pursued their own short-term, profit-maximising strategies to the financial detriment of the company. Perhaps there is a balance to be struck somewhere between the two, or perhaps the nucleus/plasma model is unworkable or even too idealistic, given the usual free-for-all of the business world and the difficulty and cost of enforcing contracts in some countries.

It is important to remember that establishment of large farms in new areas also brings significant infrastructure development to remote, often neglected regions. Roads, power supplies, telecommunications, even clinics, schools, harbours and jetties may be provided either by the developers or by public authorities. This in itself can bring new employment opportunities for some members of the local community but it is the labour required in processing plants, feed mills and cold stores that possibly has the longest-term impact on local employment patterns, particularly for women (section 11.2.5). Surplus or seasonally spare capacity in cold stores can often be used for the benefit of the local capture fisheries and agriculture operations while some feed mills may be able to support fish and other farmed animal production.

11.2.4 Integration

Prime examples of vertical integration in crustacean aquaculture industries are to be seen in Taiwan and Thailand where, as a first step in the integration process, local fishermen provide 'seed' crustaceans to new farms. The evolution of the 'semillero' industry of Ecuador is perhaps representative. Fishermen in Goa, who traditionally fished post-larval prawns and shrimp from salt works intakes and irrigation sluices for paste manufacture, benefited when the Goan state government developed 500 farms in the area. The fishermen were able to get a better price for their catch by selling them as seed to farmers. Another instance where the advent of crustacean farming brought advantages to local fishermen arose in Louisiana, where unpredictable wild crayfish catches meant that no solid markets could be established. When large numbers of crayfish were cultivated in ponds the increased reliability of production coupled with the earlier harvesting time allowed the marketing to be improved.

The ultimate in vertical integration can be found in some large shrimp farming enterprises that have their own 'closed cycle' broodstock production and maturation facility (section 7.2.2.3), a hatchery, nursery, ongrowing ponds, and often a feed mill and processing plant as well.

Participation in a new industry, however, may also mean that debt is incurred because of the need to borrow to purchase materials, feeds, fertilisers, equipment or medication. Servicing the loan becomes especially difficult if yields do not come up to expectation, or if there are crop failures while the necessary experience is being gained (section 10.2).

11.2.5 Customs, conflicts and sensitivities

Quite apart from the impact of aquaculture on societies, local customs and national or religious sensitivities can have a strong impact on aquaculture. Severe problems might be created for the unwary investor who did not know about religious prohibitions against killing or eating animals (Shang 1990), castes and status rulings affecting labour and activity, demand fluctuations caused by major festivals, and even anti-aquaculture lobbies (Anon. 1990c). Protection of investment may be difficult in areas where pacifist religions predominate or where threatening someone with a weapon is illegal (Chamberlain 1985). Animal welfare issues too are beginning to have an impact upon crustacean slaughter methods (section 8.7), eyestalk ablation practices (section 2.3), and the use of live crustacea in teaching and research. An increasing number of countries are now legislating that invertebrates (previously exempt from most welfare regulations) must be anaesthetised prior to certain manipulative procedures conducted in the laboratory. Chilling in ice may not always provide satisfactory anaesthesia, for example in cold-adapted yabbies. Yet research has shown that a slow (4 hours) but convenient bath treatment using clove oil in ethanol was suitable for specimens of 5–60 g live weight (McRae *et al.* 1999). While the debate about whether or not crustaceans suffer pain is scientifically unresolvable, it would seem that the most humane way of killing them is that which involves the shortest possible time between life and death (section 8.7). Account must therefore be taken of the cost of suitable equipment (e.g. boiling vats, electro-stunners, blast freezers) needed to achieve acceptably brief killing times for a particular species group. Additionally, consumer resistance, particularly to live sales, could temper the financially attractive prospect of increasing growth rate of, for example spiny lobsters, by eyestalk ablation.

It is interesting to note the potential for harm to crustacean farming that can be inflicted by activists engaged in a different sphere of operations (Salmon 1992). A topical example is that of the pressure from animal protection groups that gave rise to the US embargo on nations exporting shrimp fished by vessels not equipped with devices to prevent turtles from becoming caught in nets and drowning (turtle excluder devices – TEDs). Shrimp imports were banned from some 70 countries that did not have a turtle-protection programme comparable to the one in the US. The groups went on to gain a further court order banning all shrimp from the embargoed countries, including farmed shrimp. The consequences to many shrimp growers would have been severe had not the court order been changed several months later, to exclude their crops, before the ban was actually enforced (Gutting 1997). After prolonged litigation, an adverse ruling from the World Trade Organization in 1998 compelled the US State Department to weaken the guidelines of the law to allow the export of shrimp from nations that do not have national laws in place provided individual vessels claim they are using TEDs. The new guidelines exempted shrimp harvested in aquaculture facilities, shrimp harvested with certain gear and trawls equipped with TEDs, and those harvested by small-boat fishermen who pull their nets by hand.

Crustacean aquaculture may also affect, sometimes advantageously, the traditional roles of men and women in the community and in the household. This too needs to be recognised at the planning stage. For example, in Spain the new crayfish industry that developed in the early 1980s changed the lives of many women who were able to add to the family income by making nets and traps, and by selling gear and souvenirs (Lorena 1983). In Western Australia about 90% of yabby farmers are women, for whom living in a farming community affords few opportunities for employment or commercial enterprise (Nenke & Nenke 2000). Women have also come to play a major role in Ecuador's shrimp farming industry over the past 20 years, particularly in catching, sorting and cleaning post-larvae (Flores 1997). This pattern of change and increasing involvement of women is repeated throughout the world although their contribution has been seriously undervalued. International meetings (e.g. Nash *et al.* 1987; Nandeesha 1996) repeatedly identified key problem areas such as the tendency to exclude women from contributing to policy and programme planning as well as from simple decision-making processes at the local or sometimes even at the household level. Some of the discrimination arises from deeprooted traditions but illiteracy and poor education opportunities also play a major part. Training and extension programmes are now being more specifically targeted to include women in a more integrated approach to community and family involvement in aquaculture, but much still remains to be done.

A new crustacean farming enterprise will stand a greater chance of success if it interacts harmoniously with the surrounding community. Fishermen are frequently among the first to feel threatened by aquaculture developments. Several examples of real or potential conflict have been reported and they illustrate the diversity of difficulties that can arise. In Laguna de Bay lake, Philippines, the idea of shrimp and fish culture in bamboo pens was introduced to lessen fishing pressure and give alternative employment to the fishermen. The fishermen who could not afford to build cages were competitively excluded from any benefits. The cages grew so numerous that they hampered navigation, altered the natural water circulation and increased weed growth to such an extent that fish and human health began to suffer. The conflict that developed between cage operators and fishermen was foreseen but decision-makers did not heed the management and control measures proposed by the project initiators (Delmendo 1995). In Kenya, imported red swamp crayfish (*Procambarus clarkii*) proliferated and destroyed the vegetation on which a local fishery depended. The same species also caused conflict in Spain and Japan when it destroyed rice paddy bunds by virtue of its natural burrowing habits. In the USA shrimp fishermen objected to the taking of shrimp post-larvae for Louisiana State University research projects (Avault 1989), and it is highly likely that East Coast lobster fishermen would react similarly if berried, clawed lobsters were taken in any quantity for commercial aquaculture (Aiken & Waddy 1985). Conversely, local Indonesian and Filipino fishermen sided with shrimp farmers when large foreign trawlers entered coastal waters to trawl for large shrimp, and as a result these countries banned inshore trawling. The Spanish crayfish industry began after the eel fishery declined because of pollution from agricultural pesticides. One farmer stocked some ponds with *P. clarkii* and subsequently made a profit. Indiscriminate stocking of the crayfish in public waters followed and many people made profits, but friction developed when the fishery became overcrowded. Louisiana's rice and crayfish growing areas are home to large numbers of waterfowl but in the past 20 years the numbers of colonial wading birds predating on the farmed crayfish have also increased. Attempts to control the avian pests, e.g. by shooting and harassment of roosts (Avault 1995), have brought farmers into conflict with environmentalists (Huner 1995). The situation can only deteriorate as the wetland areas where crayfish are grown contract as a result of climatic, anthropogenic and economic change (J.V. Huner, 2001 pers. comm.).

In an alternative approach, a localised shrimp stock enhancement programme in Sri Lanka (Ekaratne *et al.* 1998) induced and empowered the local village fishermen (as stakeholders) to manage their own natural resources (coral lime, mangrove timber, the fishery itself) and to benefit from them equitably and with minimal social disruption. The project was initiated in a coastal lagoon situated within an area identified for possible future expansion of shrimp farming: an anathema to the villagers, who were well aware of the débâcle caused by such practices in north-west Sri Lanka. The success of the project (sections 5.7.1 and 7.2.9) depended to a large extent on the formation of a lagoon fishermen's association, concurrent implementation of fishery management measures and a high level of family participation. However, continued success may depend on gaining legal empowerment for the community to enforce the exclusion of outsiders.

11.2.6 Expatriate influence

Private entrepreneurs motivated largely by financial considerations undertake much of the aquaculture development in the tropics. To ensure smooth implementation and commissioning of large and costly enterprises, expatriate managers and technicians are usually employed. These may stay on site just long enough to forge good working relationships before moving on to other projects, leaving their local counterparts to contend with the different personalities and approaches to work of their replacements. Foreign staff living in alien or isolated communities nearly always experience some form of sociological stress that can adversely affect their work or personal relationships after a period of time. Groups of staff tend to form cliques and these may be resented locally and result in reactions that could jeopardise the security of the project. On the other hand, the expatriate community might unwittingly introduce alien moral and materialistic values to the detriment of the indigenous society.

Plate 11.2 Discussing the value of a fortuitous catch of aquatic insect larvae as food for juvenile shrimp in northern China.

11.2.7 Summary

In general then, large-scale shrimp production makes little or no contribution to local food availability, and the opportunities for many coastal communities to survive change in a world that may be entering a period of climatic and ecological uncertainty are being severely eroded.

Superficially it might appear that improvements could be made with a substantial commitment by the authorities to improving extension services and the availability of credit to local people, but this often results in dealings made primarily or solely with educated and well-financed businesses or landowners in the area. To achieve the wider objectives during project implementation – including improvements to local nutrition and income (Lambert 1986) – it is first necessary to identify the social aspirations and expectations of the communities involved. Experience shows that in many tropical regions this should lead to the introduction of approaches that emphasise labour-intensive, low-cost technology that is far more in sympathy with local social structures and environmental resources than large imported turnkey farm businesses enclosing wide areas of land. That is not to say that large intensive or semi-intensive farms do not have their place, but it highlights the need for public accountability of those responsible for allowing their implementation. Even so, experience shows that it may take as long as 10 years to transfer appropriate technology for responsible aquaculture to small-scale farmers.

Now that crustacean farming is established, increased accountability could go some way towards ensuring that more rural communities profit equitably from it at the family level and are able to manage their own natural coastal resources. Both the central and local institutions involved would thereby play a greater role in maintaining national social integrity and promoting stability for the future.

11.3 Ecological impact

Most crustacean farms depend upon a number of natural ecosystems to provide clean water, broodstocks, juveniles, ingredients for feed (e.g. fishmeals and oils), to detoxify and convert metabolic wastes into additional nutritious biota within the ponds and to treat effluents outside the ponds. The spatial dimensions of the supporting ecosystem will of course vary with farming intensity, use of wild seed stocks, feeding, water exchange strategies and effluent treatment practices (section 8.3.6.8). For example, estimates of the area required to support a semi-intensive shrimp farm in Colombia with an annual production of 4 mt ha^{-1} range from 35 to 190 times the surface area of the farm. The calculation included the mangrove nursery area needed to provide 10–50% of post-larvae stocked (10–160 times the farm's pond area), and the marine area needed to catch the fish component in the diet (14.5 times the pond area). Intensification to 14 mt ha^{-1} yr^{-1} increased the fishery area by a factor of five and the area for effluent treatment eight times (Kautsky *et al.* 2000). This concept of the ecosystem support area or 'ecological footprint' may have merit in communicating the importance of not exceeding the carrying capacity of local ecosystems and in highlighting the impacts on distant ecosystems (e.g. fisheries) to both farmers and policy-makers alike (section 12.6).

11.3.1 Pressure on natural stocks

Several kinds of activity in coastal regions (agriculture, sewage disposal, industrial discharges, shipping, mining; Brodie 1995) threaten wild stocks of crustaceans, and in many cases this is exacerbated by the increasing demand for wild broodstock or juveniles to support aquaculture. Natural crayfish stocks in Louisiana, for example, are affected by construction of water control structures, roads and anti-flood levees as well as by crop pesticides. However, crustaceans are not the only aquatic group to suffer. Indeed, Phillips (1995) reports estimates of 10 kg of fish and shrimp larvae being killed in by-catches during the collection of 1 kg of tiger shrimp post-larvae for farming and up to 5000 individuals dying as a result of collecting 100 saleable *Macrobrachium* post-larvae. Hatchery production in some areas could remove the need for wild seed collections (Ninawe 1999) but presumably could also jeopardise a subsistence activity involving thousands of women and children living in coastal communities.

11.3.1.1 Broodstock

Penaeid shrimp of larger, and hence more valuable, size than those produced by farms are increasingly sought by fishermen. Also many hatchery operators are convinced that larvae from wild-caught broodstock are more viable that those spawned by pond-raised animals (sections 7.2.2.3 and 12.4). This places significant pressure on many shrimp stocks and it is a sad fact that in several countries the resource is used very extravagantly. In the Philippines and Indonesia, large, breeding shrimp occur in inshore waters and are captured for both fishery and aquaculture. Inshore trawling for shrimp once reached such intensity that foreign vessels were banned, to protect the local industries. In Japan, the established fishery for large, live shrimp provides a convenient source of broodstock for hatcheries, but only 10–50% of the selected females spawn. In the 1980s the demand for broodstock *Penaeus monodon* in Taiwan was substantially increased by the wasteful use of nauplii. This occurred, partly because low survival rates resulted from the use of *Skeletonema*, a poor larvae food, and partly because several broods might be discarded whenever there were not enough nauplii to fill a typical, large Taiwanese rearing tank (section 7.2.4).

11.3.1.2 Wild-caught juveniles

Dependence on wild-caught juvenile shrimp remains commonplace, for example in Ecuador and Bangladesh, and the continued harvesting of very large numbers is reported not only to have depleted some local populations but also to have changed the dominant species normally caught by fishermen (Landesman 1994). Spiny lobster culture is even more dependent on the use of wild-caught pueruli and juveniles but in many countries their collection is prohibited or tightly controlled to protect fished stocks (section 7.9.4).

Owing to its applicability to a wide range of economic circumstances, artisanal mud crab culture and fattening operations (section 7.10.4) became widespread during the mid-to-late 1990s and reliance on wild-caught individuals for stocking has resulted in serious depletion of supplies of juveniles and subadults. The problem is likely to be exacerbated by the potential development of a larger market for soft-shell crabs. In addition, following outbreaks of disease and environmental problems in the shrimp farming sector, there is growing interest in converting to crab culture. An expansion of the crab industry beyond the artisanal level to large-scale extensive and semi-intensive farms could have serious implications for artisanal farmers and would not necessarily rectify any existing environmental degradation arising from shrimp farming (Overton & Macintosh 1997). As a result, there is now considerable interest in the development of hatchery techniques for crabs (section 7.10.1).

11.3.1.3 Habitat

Destruction of nursery grounds or other habitat through industrial or agricultural pollution, pond construction or other change of use (timber and mineral extraction, harbour schemes) can have considerable impacts on crustacean aquaculture. In the case of shrimp, which are primarily an annual crop, loss of nursery habitat could cause the sudden collapse of stocks (Gulland & Rothschild 1984) with serious implications for post-larvae and broodstock collection. Land reclamation in Japan, for example, made serious inroads into shrimp nursery grounds and has stimulated several restocking programmes. Research showed it was important to release juveniles large enough (>30 mm TL) to be independent of the lost habitats if fishing was to be maintained (Fushimi 1999). The demise of the European crayfish stocks over the past 100 years was due to increased modification and pollution of waterways as well as to plague fungus. It has resulted in the widespread culture of non-endemic as well as native species for national restocking programmes (section 7.6.6.1). Collection of ornamental shrimp from the wild for the lucrative aquarium trade is now extensive and methods used for their capture (chemicals and crowbars or dynamite to break open corals and dislodge rocks) cause considerable damage to their coral reef habitats (Jonasson 1987). Concern arising from both reef destruction and the continued removal of these ecologically important species (Dayton 1995) has prompted research into their culture techniques to reduce pressure on natural stocks and ecosystems (section 7.4.4.1). On a positive note, programmes also now exist for the restoration or creation of new crustacean nursery habitats (sections 5.7.2, 7.8.12 and 7.9.8). Many include attempts to design submarine structures built for non-fishery purposes so that they will enhance local biodiversity and provide food and shelter for breeding populations of lobsters and crabs (section 8.11.2). The risk of toxins leaching from some artificial habitat (e.g. tyres in crayfish ponds) or reef materials (including some quarried rocks) cannot, however, be discounted (sections 7.7.6.3, 8.1 and 8.1.1).

11.3.1.4 Incidental fishing

Substantial losses of post-larvae and juveniles are caused when they are taken incidentally into salt evaporation ponds or, in the case of freshwater prawns, sluice controls for irrigation. Of all the examples given above, the pressures on habitat are likely to cause the greatest risks to shrimp, crayfish and freshwater prawn species, while at present broodstock overfishing is probably a greater threat to spiny lobsters than to many other species.

11.3.2 Transplantations

Much reliance is nowadays placed on the ability to choose the species to be farmed regardless of its natural origin (section 4.2). The advantages such choice brings are that markets may be already established and the product known; the farmer can select the best species for overall commercial gain; the culture requirements will be known and, provided the species breeds readily in captivity, stock can easily be obtained. Several examples are given in Table 4.1b. The introduction of non-native crayfish species to restock natural waters in Europe has produced a number of socio-economic advantages including restoration of traditional fishing activities and supportive rural industries, as well as increased opportunities for diversification in agriculture (Ackefors 1999).

The disadvantages of transplantations, however, are that escapes are inevitable and can result in the establishment of breeding populations in the wild (e.g. *Macrobrachium rosenbergii* in Venezuela; Pereira *et al.* 1996). Crossbreeding could theoretically occur in the wild (Misamore & Browdy 1997) but for most crustaceans seems unlikely. Another problem arises from the alien species competing with, and sometimes displacing, ecological homologues (animals occupying the same habitat in the home environment). Examples include the replacement of crayfish *Astacus astacus* by *Astacus leptodactylus* in some European waters (Holdich *et al.* 1999) and *Artemia franciscana* which has been indiscriminately introduced worldwide over many years and in many cases has ousted the native species or strains (Barata *et al.* 1996). Chinese mitten crabs have been introduced both accidentally and intentionally to many regions of the world including Europe and the USA where they now form breeding populations. Their mass seaward migrations to breed have blocked power station cooling water intake screens, while their burrowing habits have weakened flood defences and eroded river banks. Damage to fishing nets is also common (Veldhuizen & Stanish 1999). When these crabs were transplanted intentionally from one Chinese estuary to another to improve the fishery, subtle population changes occurred after they interbred with the local population and resulted in enhanced catches of crabs bearing morphological characteristics intermediate between the two stocks (Li *et al.* 1993). It is clear that further studies are needed on the impacts of transplantations, even simply increasing the abundance of endemic species in a stock enhancement or ranching project could have important repercussions.

Crayfish transplantation and escapes in particular have caused a number of other disruptions (Holdich 1999a) especially to local environments, such as the destruction of weed beds, levees and bunds that has sometimes resulted in fishery or crop losses. In addition, the importation of the American signal crayfish to Great Britain for commercial exploitation resulted in the industry being subjected to new and highly restrictive legislation as well as to competition from the feral signal crayfish fishery that has developed (Holdich 1999b). On the other hand, introductions of red swamp crayfish in China and Spain have unquestionably resulted in commercial harvests of considerable value. Crayfish have also been introduced to advantageously open up and maintain water bodies previously choked with weed (Ackefors 1999). Such positive results and the inevitable publicity that follows put pressure on resource managers not only to utilise the new resource wisely but also to make further introductions elsewhere. This may result in additional, unforeseen management dilemmas, for example when the crayfish become an integral part of local ecosystems, perhaps being the major source of food for endangered birds and mammals (Huner 2001).

Repeated breeding from the offspring of a small number of parents can adversely affect culture performance (Rothlisberg 1998; sections 2.6.1 and 8.10.1.2). It has also fuelled concern over the accidental escape or intentional release of animals (e.g. in stock enhancement programmes) with a different genetic constitution to that of local stocks (Cross 1999; sections 5.7 and 12.7). Particular disquiet can arise in conservationists if the parental stock came from a different location, had been subjected to selection (whether inadvertently or on purpose) or had been genetically engineered either for stock improvement or tagging purposes (Rothlisberg 1998; Jørstad & Farestveit 1999; section 5.7.1). The establishment of the different stock in the new environment brings risks of diminishing the diversity of a species' gene pool, introducing transferred genes into wild gene pools, altering the structure of ecosystem communities and the introduction of diseases against which there may be little or no natural resistance (Lutz 2000a).

11.3.3 Disease transmission

Several serious diseases of Crustacea have been spread through transplantation of live animals for commercial farming operations, national restocking programmes, and for research as well as in frozen product and processing wastes (section 8.9.3). Economically the most significant has been the rapid spread of viral diseases, which from the mid-1980s has resulted in widespread crop failures and slowed the growth of shrimp farming internationally. Viruses caused major losses in Taiwan (1987–88), China (1993–94), Indonesia (1994–95) and India (1994–96), as well as significant problems almost everywhere else, including the USA (1995), Honduras (1994–97) and Ecuador (1993–96 and 1998–99). Taura syndrome virus (TSV) killed farm-raised shrimp throughout the Gulf of Guayaquil in 1992, then spread to every shrimp-producing country in the western hemisphere with the exception of Venezuela where hatcheries maintained captive broodstock and strictly controlled the introduction of new broodstock. The Office International des Épizooties (OIE) maintains a list of notifiable animal diseases, to which in May 1999 was added three shrimp viruses, WSSV, TSV and YHV (section 2.5.4). In recognition of the seriousness of diseases spread through the translocation of species and products, the OIE has developed guidelines for import risk analysis (IRA) for use by importing countries (Edgerton 1999).

A number of crustacean species seem good candidates for culture outside their normal range for various reasons, but their transfer may render them susceptible to potentially pathogenic organisms to which they have little or no natural resistance. They may also bring pathogens into the new environment. For example, the Chinese white shrimp (*Fenneropenaeus chinensis*) has already been transplanted to New Zealand, Europe and the USA for culture trials because it grows well in subtropical and warm temperate zones and has greater tolerance to lower temperatures than many tropical shrimp. No significant production seems to have resulted but it is worth noting that the species is susceptible to HPV and TSV, virus diseases of penaeids in the USA that are not endemic in South-east Asia (Chamberlain 1988a; Lightner *et al.* 1997). Similarly, the Australian redclaw crayfish (*Cherax quadricarinatus*), now widely cultured in tropical regions outside Australia, is susceptible to crayfish plague fungus and has carried the baculovirus CqBV into the USA and Ecuador (Evans & Edgerton 2001). Fortunately this virus is not known to be pathogenic to the crayfish, but recent harmful cross-infections with other baculoviruses in crustaceans have been reported (Hauck *et al.* 2000). It is now well known that crayfish plague fungus can be carried by at least three species of North American crayfish (*Pacifastacus leniusculus, Orconectes limosus* and *Procambarus clarkii*) and has been widely spread by past introductions throughout Europe to most native species of crayfish (*Astacus* and *Austropotamobius*; Söderhäll & Cerenius 1999).

Of course, import regulations exist to control the spread of alien species and diseases; however, in practice they may be difficult to enforce. One reason is that customs personnel may not be trained in the identification of species, another is that there is no guarantee that live crayfish imported for consumption or ornamental use do not escape. The situation within the European Community was exacerbated in 1997 by the abolition of former border veterinary controls which meant that live crayfish for consumption could enter a country unexamined (Skurdal & Taugbøl 2001). Since world trade rules often prohibit import bans on live animals historically traded live as food, legislation was introduced in Great Britain to control the keeping (but not strictly the importation) of non-native crayfish. This legislation was effected in order to prevent free introduction and movement of these animals around the country. Only where crayfish are to be kept for a licensed purpose will their import from a non-EU country be allowed. The licences issued for imports for human consumption can place restrictions on the quantity imported, the facilities in which they are held, and the time over which they are expected to reach the retail market. Only one species, the tropical redclaw crayfish (*Cherax quadricarinatus*), can be imported for the British ornamental fish trade, a concession designed to prevent the transfer of other, primarily temperate water species imported for consumption, to the ornamental market (Scott 2000). Presumably, therefore, the implications are that aquarists could import this species live from countries other than Australia, e.g. Israel, China or Ecuador, where the species is now cultured but where it may also have had contact with unanticipated diseases. Other UK legislation exists to minimise the risk of Gaffkaemia, a bacterial infection of lobsters, which can infect European lobsters held in live storage facilities, primarily by preventing contact with imported North American lobsters (MAFF 1996). However, a global control policy to govern the spread of crustacean diseases seems a long way off (Johnson 1994; section 8.9.4).

11.3.4 Disease treatment chemicals

Managers of many commercial shrimp hatcheries are often under such pressure to meet production targets that they resort to the routine use of prophylactic doses of antibiotics in larvae culture vessels. One attraction of antibiotics, or indeed any chemotherapeutant, is that they can help overcome bad husbandry. A typical scenario in an Ecuadorian hatchery during the 1980s would be to put a low dose of a broad-spectrum antibiotic into the larvae culture water during the protozoeal stages of early season cultures; increase the concentration as the resistance of the disease organism rises; change to a new antibiotic, and later, combinations of antibiotics, as the season advances, until all the treatments become ineffective. At that stage the whole hatchery is closed down, chlorinated and dried out (Chamberlain 1988a). It is now widely recognised that this practice may give rise to more disease-susceptible post-larvae because many will have developed from weak larvae that have been artificially protected from disease. Antibiotics are also administered in feed during ongrowing and, although some may be undetectable in crustacean flesh a week or so after medication has stopped (Mohney *et al.* 1997), residues may still persist in pond bottom sediments (Ruangpan *et al.* 1997).

Predictably, the extensive continued use and misuse of antibiotics has already encouraged the development of lasting resistance in many groups of pathogens (GESAMP 1996; Chythanya *et al.* 1999). It is also known that resistance can be transmitted by direct transfer of genetic material between cells through plasmids and bacteriophages. Indeed, nearly half of the bacteria isolated from aquaculture facilities during a survey of five South-east Asian countries in 1993 were resistant to oxytetracycline and several isolates also showed resistance to other antibiotics (Inglis *et al.* 1997). Similar levels of bacterial resistance (54%) have been reported in isolates from frozen shrimp (Berry *et al.* 1994). Attempts to reduce the risk of single drug resistance arising are confounded by two factors: the relatively small number of medicines licensed or available for use (Schnick 1992; Smith 1998; section 11.5.3.2) and the reluctance of farmers to vary successful treatments by rotating the medicines that are available (Alderman & Michel 1992). Many farm managers do not have time, facilities or the money to identify the causative organism before choosing a course of treatment. It is ironic that, even though broad-spectrum antibiotics are used, they may not be effective against the disease, especially if it is caused by a virus. Misuse of chemicals by farmers often arises from inadequate information or labelling of the products, especially regarding storage, usage under specific environmental conditions, expiry date and disposal of unused product (GESAMP 1996).

Many of the antibiotics used prophylactically or curatively are also used in the treatment of human diseases. Some may cause allergic reactions in the small number of susceptible or pre-sensitised people. The latter may react to the extremely small amounts that could be present, for example, as residues in crustacean tissues (Yndestad 1992). Some may be hazardous to humans during their application but all prophylactic and therapeutic chemicals should be handled in a responsible and safe way (section 8.9.4). In several countries, hatchery effluents, which may contain antibiotics, resistant bacteria or virulent disease organisms, have been freely discharged onto beaches where they present a potential hazard to neighbouring hatcheries and possibly to the public (Brown 1989). It may be feasible to remove some chemotherapeutants from freshwater prior to discharge by adsorption in activated carbon filters (Aitcheson *et al.* 2000). Farm effluents containing aquaculture chemicals may also contaminate bivalve shellfish or fish growing in the vicinity, or poison local fauna and flora. The lower doses encountered by potentially pathogenic organisms, when effluent containing antibiotics is diluted by the receiving waters, further increases the risks of resistance developing.

The use of therapeutic and maturation compounds is widely practised in both commercial and research units (GESAMP 1996; sections 2.4.6, 8.3.6.1 and 8.9.4), but very few chemical agents are approved, for example by the US Food and Drug Administration, for use in animals destined for human consumption. One growth-promoting agent – human growth hormone – seems effective in lobsters (section 12.8.4), but care should be taken at the project planning stage that the intended and potential markets will be ready to accept the cultured product should the use of such agents become necessary during production.

11.4 Environmental impact

Environmental management systems and associated risk analyses are being proposed as a way to integrate awareness of each step in the production process with its impact on the environment and resources involved (Phillips 1995; Jory & Dixon 1999; sections 9.7.2 and 10.3.4). Initial recommendations are aimed at shrimp but the general principles as set out in the Global Aquaculture Alliance's *Guiding Principles for Responsible Aquaculture* are applicable to all crustacean farms (section 11.1). Yet, because of the differences in type and size of production or processing operation, their resource requirements and their local environmental situations, each management system will probably be unique.

Concentrations of farms, particularly shrimp farms, often occur in areas of limited water exchange or close to potentially polluting urban and industrial developments. Integration of crustacean farming into coastal zone management plans at national and regional level is therefore vital but in too many cases has not occurred in time to prevent user conflicts (Hotta & Dutton 1995; Phillips 1995). It is also argued that security of local food supplies are threatened by large-scale uptake of shrimp farming in some regions due to conversion of traditional fish farms, rice paddy and land for other basic food crops to shrimp farming, to salinisation of farming lands and to declining fish and shellfish catches following mangrove destruction (Primavera 1997). This may be true, but man's other activities (mining, forestry, urbanisation and industrial developments) extract a similar if not greater toll on food-growing lands.

11.4.1 Site clearance

Often the first sites to be considered for crustacean farming are those popularly regarded as wastelands. Coastal alluvial plains are one example. With careful design and management, they can support efficient shrimp farming (New 1999b) but when farms are developed haphazardly, ecological and social disasters can result (Hong 1996). Mangrove zones were once viewed in this light and, being near to water supplies, vast areas were destroyed in the construction of aquaculture ponds. Many farms, however, were constructed in mangrove areas previously cleared for non-aquaculture reasons like salt production, mining and urbanisation. It is now widely accepted that destruction of mangroves may destroy not only fish and crustacean nursery grounds but also increasingly important natural flood and storm protection barriers (Sakthivel 1985). Not all mangrove habitats are consequential in these respects, however, and proper development may sometimes be prevented by dogmatic adherence to conservationist principles (New & Rabanal 1985). Often, habitats that contribute much less to

Plate 11.3 Adverse environmental impact caused by clearing mangrove forest for shrimp ponds in Kenya but then abandoning the scheme as an uneconomic proposition.

mangrove ecosystem productivity can be found 100 m or so behind the coastal mangrove belt. These areas can be more appropriate for farm development provided salinisation of adjacent areas is prevented, e.g. by constructing suitable drainage canals or establishing buffer zones (Sze 1997). Here, pond construction is generally cheaper, drainage more efficient and soils of better quality (Phillips 1995). Most governments are now regulating the use of, or protecting, mangrove and other wetland areas (Boyd *et al.* 1998). In addition, many reforestation projects are under way with some specifically designed to ameliorate adverse impacts from farm effluent discharges (section 8.3.6.8), while others attempt to integrate crab or shrimp farming with mangrove silviculture (sections 5.5 and 7.10.4). Although a promising method for restoring timber supplies in impoverished areas, deviations from the specified 30 : 70 pond to forest ratio by some farmers are jeopardising productivity of both activities.

In some areas the conversion of salt flats or unprofitable salinas, rice or sugar fields to ponds may be preferable (Ninawe 1999) as construction is again easier, mangrove nursery and common community resources are not destroyed, and the soil and water quality is better. Alternatively, ponds may be dug inland and a pumped water supply installed. In Thailand, shrimp have been increasingly cultured in seasonally saline areas, sometimes using brine brought in from coastal salt flats to extend the season for cultivation. This practice escalated to such an extent that in 1998 the government banned inland culture of *Penaeus monodon* in 13 provinces. The ban was in direct response to pressure from environmentalists claiming that rice and fruit-growing lands were being salinised (Phillips 1995). However, while some small farms undoubtedly caused damage, many larger operations incorporated buffer zones, dikes and ditches that prevented salinisation of neighbouring agricultural land (Jory 2000). It should be noted that the salinisation of soils used in marine and brackish-water shrimp ponds may prevent them being reconverted for agriculture for several years if the project becomes uneconomic or fails for other reasons (Kayasseh & Schenck 1989). Nevertheless, it may also be possible to mitigate the damage in some such areas, including abandoned shrimp ponds, by growing salt-tolerant plants (halophytes) as forage for ruminants (Brown & Glenn 1999). In the longer term, environmental and sociological changes must be expected when new roads are built to service an isolated project. These can open up an area to change through development of peripheral service businesses and, later on, permanent communities.

11.4.2 Water supplies

Extraction of groundwater to lower salinity in shrimp ponds was common practice and occurred over a wide area in Taiwan in the 1980s. In several areas it eventually led to land subsidence. Partly in response to this major

Plate 11.4 A sunken and abandoned Taiwanese house displaying the effects of land subsidence caused by excessive abstraction of groundwater for shrimp culture.

impact, an Environmental Protection Agency was established. Excessive groundwater abstraction may also cause saltwater intrusions into domestic freshwater sources. Today few farmers find it necessary to add freshwater to cultures of *Penaeus monodon*, and indeed abstraction of freshwater for marine shrimp farming is now banned in Taiwan and parts of Thailand (Phillips 1995).

11.4.3 Effluents

Effluents can take the form of discharges from ponds and hatcheries, the water passing through culture pens and cages and discharges from processing plants. Generally, pond effluents are rich in organic nutrients (phosphates, nitrates) from feed, fertilisers and crustacean metabolic wastes and exuviae (cast shells); substances used on pond bottoms (lime, gypsum); pesticides (saponin and nicotine) and antibacterial agents. Hatchery effluents may also contain disinfectants (formalin, sodium or calcium hypochlorite) and antibiotics (GESAMP 1996). When discharges occur in confined areas or where mixing and dispersion are reduced, there are increased risks of low oxygen levels, eutrophication, toxic algal blooms and disease transfer between farms and hatcheries (Teichert-Coddington 1995; Xu 2000). In addition, specific disease-resistant organisms or antibiotic-resistant bacteria may accumulate (Brown 1989). Discharges of saline water from inland *Macrobrachium* hatcheries, and recently from inland shrimp farms, for example in Thailand, into public water canals (*klongs*) have also had undesirable effects. In contrast, Jory (2000) reports the beneficial contribution of shrimp farm effluent to the reclamation of 350 ha of mangrove in Venezuela.

Wastes arising from the processing of farmed (and captured) crustaceans are also potential pollutants. Liquid effluent results from simple washing, grading and defrosting operations as well as from compaction and other treatments of solid processing wastes (heads, shells, claws) prior to disposal or conversion into usable products (section 3.3.7). Feed manufacturers (particularly in developed countries) encounter strict controls on gaseous emissions and odours from drying and cooling plants. Several methods exist to reduce odours including chemical scrubbing, biological filtration, oxidative particle treatment (ultraviolet light, ozone) and charcoal adsorption, and all add to the cost of feed production (Simonsen 1999).

Monitoring of effluents to ensure regulatory compliance is made costly because the discharges, and their contents, vary with stages in the production or processing cycles, e.g. during pond flushing or draining (Boyd 2000). It is probably more practical, in developing countries at least, to encourage, proactively, the adoption of good management practices that minimise environmental disruption before regulations are imposed by government agencies (Boyd *et al.* 1998). Many countries already have legislation to govern aquaculture wastes and new enterprises are expected to include waste management plans in all proposals (sections 8.3.6.8, 9.3.4, 9.3.6.3, 10.3.4 and 12.6).

11.4.4 Climate

Unusual climatic variations are apparent in many areas of the world today (Weaver & Green 1998) (sections 1.3.1 and 1.4). They pose threats ranging from increased flooding risks in Bangladesh to failure of rains causing higher salinities in ponds, for example in Burma, Sri Lanka and Kenya (Chamberlain 1988b), or droughts affecting crayfish production in Spain (Lorena 1986) and Louisiana (Lutz 2000b; see also section 11.2.5). Some of the effects may be beneficial, especially when warmer pond temperatures allow longer growing seasons. Unpredictability, however, is likely to be the most significant factor, both in terms of short-term changes affecting project appraisal and the longer-term effects that must be considered by those formulating coastal zone management plans (Holmes 1995). It will be impossible, for example, to predict changes in salinity patterns that may occur in estuaries and mangrove areas and their effect on the distribution and availability of seed and broodstocks. One example of the unpredictable effects of weather patterns was the decline in harvests of raw, wet *Artemia* cysts from the Great Salt Lake, Utah, from around 2800 mt in 1997–98 to 1180 mt in 1999–2000 that was followed by an unexpected, record yield of almost 8800 mt in late 2000.

On farms, increased engineering (and pumping) costs may also be incurred if correct salinities and water exchange rates are to be maintained (section 6.2.1). A wise precaution, already taken in several countries, is to investigate the temperature and salinity responses of a range of cultivable species other than those currently farmed, and to pay particular attention to the existence of different physiological strains of valuable species that might perform well under different salinity and temperature regimes (section 4.6.2). In view of current climatic change, protecting the natural genetic diversity of wild crustacean stocks should therefore be recognised as an ecological principle of paramount importance to all involved with the industry (see also section 12.8.3).

11.5 Institutional interactions

At their broadest, the problems facing policy-makers and managers of coastal areas are primarily those concerning resource allocation and the effects of 'externalities', i.e. the impact of one user's activity on another's (FAO 1996). Planners and assessors of crustacean farming investments or development strategies will quickly become aware of the necessity to interact productively with numerous institutions, national and international, commercial and private, if their objectives are to be achieved (Lee 1997). Aspects of some of these interactions are discussed here, to give the reader a feel for the variety and scope of problems and benefits that may arise. For convenience the aspects are grouped under financial, managerial and legislative considerations.

11.5.1 Financial considerations

11.5.1.1 Land/water costs

It was once thought that 'waste' land unsuitable for agriculture would be suitable for modern aquaculture developments, but this is not necessarily so. Fertile clay or loam soils are needed for semi-intensive and intensive farms and usually have a high value. Water (especially freshwater) supplies are also competitively sought. Several developing nations have positive commitments to crustacean aquaculture and have made land available for leasing to local fishermen and farmers, frequently in conjunction with supportive extension and training schemes. In the tropics this is often marginal land, formerly considered to be of low value. Even though it may only be suitable for the less intensive farming strategies, estimates indicate it could be worth from \$1000 to \$11 000 $ha^{-1} yr^{-1}$ to those who depend upon it collectively for subsistence (Primavera 1997). Attempts to intensify production from such areas may yield unreliable harvests and will not make best use of the land.

In contrast, most coastal land in developed countries is difficult to obtain and expensive because it is competitively used and is usually considered to be a public amenity. Sites near unpolluted water supplies and those with access to geothermal or industrial waste heat can be particularly valuable. The use of heated supplies to extend the growing season or make possible the culture of warm-water species in more temperate climates has been discussed in sections 5.3 and 5.4. The point to be considered during the planning of such projects is the financial undesirability of having capital equipment for heat transfer, heat exchange or back-up lying idle for large parts of the year. The same consideration applies to other installations, such as processing plant, which may be under-utilised during the growing season (section 10.6.6).

In the context of using industrial 'waste' heat for aquaculture, it cannot be too strongly emphasised that, firstly, the temperature, flow and chemical composition of the heated medium will be manipulated to suit the primary industry's purposes, not those of the culture enterprise, and secondly, many industrialists may tolerate the presence of an aquafarm simply to alleviate public fears of pollution or to project a 'green' image. Thirdly 'waste' heat is not free; it has to be piped, monitored and transferred directly or indirectly to the farm. Finally, and perhaps most revealingly, there is at least one report in the trade press of charges (royalties) suddenly being demanded for 'waste' heat after several years of aquaculture operation (Anon. 1988).

11.5.1.2 Credit/loans

Whether it is the commencement of a new business, the conversion to Crustacea from a different crop, or the upgrading of an existing crustacean farm, adequate finance will be vital (section 10.2). In developing countries long-term loans at reduced interest rates, together with free technical and management training, are helping many to participate in small-scale crustacean farming (section 10.2.2). Where there is a lack of local capital, co-operatives of perhaps eight to ten farmers may be formed and become eligible for government subsidies (Lee 1997). However, in many countries low interest loans and grants may be difficult to obtain, especially if land or other form of security is not owned. Too often credit is seen to be the monopoly of big business, and in the west, paradoxically, the minimum size of grant or loan on offer may be too large for the individual or family business. Some support may be available in a number of developed countries for small to medium-sized enterprises too far from the marketplace to survive unaided. Publicity and advice about the schemes vary from good to poor, and gaining acceptance for high-risk projects like crustacean farming will be difficult. For example, investors offering venture capital (section 10.2.1) to support small, innovative businesses generally look for annual returns of 50% and withdraw after 5 or so years, a timescale usually incompatible with new crustacean aquaculture projects (Lockwood 1998).

11.5.1.3 Investment and insurance

The high risks inherent in all crustacean farming operations have not deterred entrepreneurs in the past, and crustacean aquaculture has long enjoyed a high degree of 'investor appeal'. An analysis of 25 failed North American aquaculture enterprises revealed that culture technologies *per se* were not to blame except where a flawed technology or business design was given undue promotion to gullible investors. Half of the total estimated losses of $300m was considered to be due to the replacement of committed entrepreneurial leadership with corporate financial managers and the imposition of overbearing central control (Lockwood 1998). Even leading aquaculture research institutions have suffered a similar demise in the name of financial rationalisation.

Significant or repeated failures, however, will give any industry a bad name and tend to discourage support, both financial and occasionally from publicly funded research programmes. The problems are not always due to disease or mechanical failure but commonly include over- or undercapitalisation, over-optimistic expectations of scale and timing of returns, and no allowance made for an adequate learning period or for early crop failures. The need to attract investors in developing countries has led some governments to grant worthwhile tax holidays during the early years of a project. In Brazil, however, the high subsidies for shrimp farms in the 1980s were reported to have been made at the expense of the funding of research and development so vital to the future success of the industry (Chamberlain 1988b).

Protection of investment by appropriate insurance is common practice in many established industrial activities. In crustacean farming and aquaculture, however, it is rare. Most firms and many companies either do not insure or cannot afford insurance. By nature all crustacean farmers are optimists, but reputable insurance companies exist and will offer reasonable terms provided they are satisfied that adequate working practices and precautions are taken to prevent losses (Macfarlane 1997). Usually insurers will undertake a risk management exercise (Hatch *et al.* 1987; Wiley 1992), in which they visit farms to see where and how severe the weaknesses are (Secretan 1986; 1988). In view of the current climatic instabilities, perhaps all should be encouraged to insure! However, in the late 1980s, insurance against loss of stock in the fish farming industry was already being reported as reaching the point of non-viability following increased storm and disease outbreaks (Anon. 1990a). Shrimp crop insurance is available (Anon. 1990b) but it is probable that only projects with stable, committed and appropriately qualified management, the best husbandry practices and best risk management programmes will be insurable at acceptable rates (Wiley 1992; sections 9.7.2, 9.7.3 and 10.4.2).

11.5.1.4 Markets and production costs

Farmed crustaceans, particularly shrimp and prawns, probably contribute little to the diets of the poor since the majority are more profitably sold to Japan, USA and Europe. Freshwater prawns, crayfish and crabs, together with some shrimp, may, however, be sold for good prices in city markets in the country of origin, particularly where there is a tourist trade. Most shrimp farmers, however, whether smallholders or large corporate enterprises, are at the mercy of world commodity markets and fluctuations in international exchange rates. They are vulnerable regardless of how efficient they are in production. The value of the crop may also be affected by temporary oversupply of a specific size group, as was the case with medium-sized shrimp in the late 1980s, or of a specific market. Examples where overproduction caused dramatic price decreases during the 1980s include freshwater prawns in Thailand and hatchery-reared shrimp post-larvae in Taiwan. Similar effects are still apparent today (sections 1.3.1 and 1.3.3). In the mid-1990s, oversupply of the US market with imports of cheap Chinese crayfish tails led to the International Trade Commission voting to impose a 123% tariff on the product. Other, sometimes unjustifiable, tariffs and trade barriers can adversely affect sales and ultimately, the farmers themselves (sections 1.4 and 11.1).

When a decline in market value of the crop or an increase in interest rates or running costs erodes profit margins, the intensive farm with its narrow profit margins will lose profitability first. The extensive shrimp farm run at a low productivity rate may also suffer, and it may be only the semi-intensive farms that combine moderate productivity (yields of 1–3 mt ha^{-1} yr^{-1}) with adequate profit margins that can survive (Kusumastanto *et al.* 1998; section 10.5).

World production of fishmeal, a major ingredient of crustacean diets, stands at about $6.0–6.6 \times 10^6$ mt with fish oil, another key component and source of essential fatty acids (section 2.4.2), standing at 1.3×10^6 mt. Peru is currently the main producer and harvests from 'industrial fishing' are now recovering from the effects of El Niño in 1998. Many former exporting nations (Chile, Japan and Norway) now also import Peruvian fishmeal. Estimates that aquaculture will use 15–17% of world fishmeal supply by the year 2000 (New & Wijkstrom 1990) have been revised upwards (to 40–45%), with shrimp farming alone expected to use 20–30% of that (i.e. 5–7% of global production) during the next 10 years. Fish oil demand for shrimp feeds is expected to increase from 29 000 to 73 000 mt (i.e. 2–5% of global supply) in the present decade (Barlow 2000). Crustacean farming, then, has a small but non-trivial impact on fishmeal and fish oil availability, and intensive and semi-intensive producers are the most likely to feel the effects of price rises caused by increased competition for supplies (Tacon 1998). Investment in value-added crustacean products is likely to provide some advantage as feed costs rise but their success requires the creation of an individual market image (sections 3.2.5 and 3.3.1).

Extending legislation that allows the use of by-catches of fish, normally thrown back into the sea to die, may result in an additional source of material for fishmeal production and could compensate for the anticipated further transfer of some pelagic species (sardines, pilchards, capelin) to the human food market (Barlow 2000). A point of concern raised by New and Wijkstrom (1990) is that shrimp farming within Asian nations has increased the demand for home-produced fishmeal and local supplies of fish for its manufacture. This has reduced the availability of low-cost fish for human consumption, and also inflated the price of important local foods such as poultry that partly rely on fishmeal for their diets. Substitute ingredients (soya, other oil seeds, cereals, single-cell proteins, meat, bone and poultry by-product meals) are being investigated to replace a proportion of the fishmeal used in diets (Lewin 1997; Tacon 2000; but see section 12.5) although substitutes for fish oil are harder to develop (Tacon 1998; section 3.4.1). However, improvements in fish flesh stripping have made fishmeal production more efficient, while new low-temperature processing and steam drying are providing diets that are more effectively assimilated and cause less nitrogen to be excreted. Such diets may considerably reduce pollution levels in the ponds and farm effluent (section 8.3.6.8).

11.5.2 Managerial considerations

The rapidity with which crustacean culture has evolved throughout the world is unprecedented. Entrepreneurial enthusiasm has combined traditional skills with modern research results to produce viable culture technologies that, when applied in developing countries, can satisfy national goals of generating export revenue and increasing employment. The new technologies have been extended to traditional farmers and fishermen with the aid of demonstration shrimp hatcheries and production units that have been established both commercially and with national or overseas government aid. In a number of areas, however, events appear to have moved too quickly and have revealed what Smith (1984) described as 'a lack of institutional preparedness'.

For example, the uncontrolled encroachment of Taiwanese shrimp farmers onto public land, and their profligate use of groundwater as the development of the industry peaked, are well-documented instances that have only been properly addressed in recent years (Chiau 1998). In Laguna de Bay in the Philippines, no agencies stopped the extensive construction of fish and shrimp pens that eventually led to severe eutrophication and restricted navigation in the lake (section 11.2.5). In Great Britain uncontrolled importation of signal crayfish was, predictably, followed by outbreaks of crayfish plague fungus in stocks of the native white-clawed crayfish (section 11.3.3). Many new farm and hatchery projects still place great reliance on technological packages from overseas, which may not always be sustainable in the long term. In a review of constraints to the development of aquaculture in developing countries, Lee (1997) emphasises the importance of government planning and, in particular, monitoring, for predicting and subsequently developing infrastructure needs, for example marketing, rural financial assistance and extension services. Throughout Asia there remain areas where training in both technical and management skills and extension services are insufficient to ensure that maximum benefits will be obtained from crustacean culture (section 11.2).

11.5.2.1 Extension services

Many governments and some of the larger projects and feed companies recognise the value of providing rural extension services to improve communications between the policy makers, researchers or company sales agents and the farmer. To be successful, systematic transfer of information must occur in both directions so that the farmer is aware of new developments and the donors gain feedback concerning the impact and results of their ideas or products. Many extension services, however, do not pay sufficient attention to acquiring feedback information and the exchange becomes unbalanced and unsatisfactory for all concerned (Blakely & Hrusa 1989). Not all societies or communities react in the same way when confronted with new ideas. Industrialised societies seem able to accept and assimilate new ideas more rapidly than traditional societies, where new ideas may be perceived as threatening stability.

The approach adopted by the extension agent will differ according to whether the task is to increase public awareness of a farming opportunity or to encourage the adoption of improved management or a new technique in an existing practice. The approach must be in sympathy with the state of aquaculture development within the locality and properly targeted so that some sections of the community are not ignored in favour of others (Naegel 1995). Changes that increase a farmer's risks (real or imaginary), labour requirements, competition for resources or competition with other enterprises are unlikely to be adopted.

11.5.2.2 Consultants/researchers

The services of skilled specialists may be engaged at any or all stages of project implementation from the initial surveys to the completed application of the culture technology (section 9.3.6). Few specialists deal solely with crustacean ventures and will typically return to more conventional activities after each appointment. Although the specialists will have considerable expertise in their own field, they may not have much expertise or experience of applying it, in crustacean farming projects.

In spite of the present profusion of aquaculture training programmes, many established researchers and consultants in crustacean aquaculture were originally trained in non-aquaculture subjects such as zoology, fisheries or oceanography and have spent much of their lives engaged more in fundamental (primarily academic) research than in applied studies. Some may have difficulty seeing beyond the bounds of their professional training and sometimes fail to appreciate all the long-term implications of their advice. Consequently, their findings are often more applicable to the larger, more technically orientated projects. This leaves many resource-poor farmers unable to identify solutions to their problems in the results coming out of research institutes and universities (Naegel 1995). Also, since the work of consultants (and increasingly, researchers) is likely to be subject to commercial confidences or financial competition, they are prevented from sharing valuable experiences. This reduces the opportunities to learn from past mistakes.

Progress is further impeded because there is insufficient feedback from pilot or commercial operations to those in research and development (R&D). The vast majority of farms cannot support or pay for R&D programmes and would not, in any case, wish to reveal their results and difficulties to others for fear competitors might take advantage of the knowledge. The result is 'research by crisis', and is hardly a satisfactory way of supporting a farming industry. Few scientists see spending time on farms trying to identify the real problems as the kind of productive research that will improve their career prospects. Yet there is a recognised need for an integrated approach to research that will address the needs of farmers in the context of the site-specific circumstances in which they live and work. This is known as farming systems research. It involves the farmer at all stages from problem diagnosis to practicalities of implementation and provides results that can be synthesised and interpreted according to the target audience's perspective. Successful implementation at the farmer's level is of paramount importance and may rely heavily on the communication between extension, technical and aid workers, especially where these act as intermediaries between the farmers and the researchers. It also relies on there being clear financial advantages to adopt any proposed changes (Naegel 1995).

Unethical practices exist in the crustacean farming business as in many other businesses. There are reports that some international consultancy groups have actively recruited skilled and even semi-skilled hatchery operators and hired them back to the developing industry at exorbitant rates. The potential farmer or investor should of course beware of organisations and individuals making false claims and unrealistic proposals (Rosenberry 1984, 1985). It is always wise to check the credentials and track record of advisers before they are employed, and note to what extent they are covered by professional indemnity insurance (Secretan 1988).

11.5.2.3 Managers

There are numerous reasons why projects fail to meet targets or come up to expectations, but many are simply due to poor or inexperienced management (section 9.7). Common problems include:

- Ineptitude at anticipating requirements during project implementation or during operation.
- Lack of appreciation of start-up problems, especially by financiers.
- Lack of awareness of market forces and the desirability to diversify to new species.
- Replacement of skilled, motivated managers with corporate financial controllers (section 11.5.1.3).
- Ignoring the advice of hired specialists.
- Paying insufficient attention to staff needs and expectations.
- Failure to modify jobs in the light of experience.
- Failure to motivate staff and provide suitable incentives (Haughton 1990).
- Inadequate liaison with surrounding businesses, for example those involving pesticide spray programmes against crop pests (Frese 2000) and for the eradication of bilharzia and malaria.
- Overconfidence in computerised pond management and alarm systems (section 8.6).
- Failure to look after monitoring equipment and properly standardise procedures and chemicals used in making water quality measurements.

11.5.3 Legislative considerations

The task of policy-makers and legislators is to reconcile the need to maintain an acceptable environmental and social situation with the economic and efficient functioning of crustacean farming activities. However, many laws governing aquaculture activities were not designed for aquaculture, and those responsible for project feasibility studies and assessments should investigate fully the relevant national and local legislative implications at an early stage. Often the laws applied relate to agriculture or fisheries and are not suitable for aquaculture, let alone crustacean farming. Effluent laws, for example, may not distinguish between biodegradable fish farm discharges and non- or slowly-degradable chemical discharges. An unnecessary burden of compliance may result (Howarth 1990).

Howarth (1995) and Van Houtte (1995) review the essential objectives of aquaculture law and indicate mechanisms by which they may best be achieved. Regulatory instruments, such as imposed standards, operating procedures and prohibited practices provide authorities with the means of good control. But to be credible and realistic, they often require non-trivial amounts of expenditure to gain the information for their formulation and very costly monitoring and enforcement programmes if they are to be effective. Alternatively, a tax or charge may be placed on resource use such as water abstraction and use of a receiving water into which a farm's effluent flows (Tisdell 1994a), or tax incentives offered for adopting approved operational practices, e.g. for effluent treatment (Holland & Brown 1999).

In some countries no laws may be available to protect investment in a crustacean farm or hatchery. Spanish law, for instance, states that the first two metres of land above the high-water mark are public land, so if the Guadalquivir river floods, the entire delta becomes public land and no private aquaculture ventures can be sustained (Lorena 1986).

11.5.3.1 Ownership

The clear delimitation and subsequent protection of investment is not always as straightforward in developing as in developed countries. Moslem inheritance laws, for example, decree that land is divided among descendants but, since a pond is not readily divided, multiple ownership frequently becomes a constraint to development (Shang 1990). In general terms, the ownership of inland freshwater ponds is readily defined and, although limited, the prospects for large-scale development generally seem reasonable. On the coast, however, there is often no security of tenure and one frequently needs political contacts or bribes to get the necessary permits to build ponds. Family ties and land ownership may seem excessively complex to the foreign investor, who probably will not be able to own land directly in any case. Legislation for the protection of traditional rights and custom lands may or may not exist, but the existence of traditional rights will doubtless be perceived as a reality by local inhabitants who will justifiably expect them to be honoured.

Even when a crustacean farm has defined rights, enforcement of those rights could be impossible or at least uneconomic. For example, if there is an agreed entitlement to estuarine or river water containing no more than a specified level of a pollutant, and if this level became exceeded due to the combined discharges from several different sources, each individual polluter could argue they were not to blame for the excess and a successful claim would be unlikely (Tisdell 1994b).

Protection of an investment in developing techniques, trade secrets or 'know-how' comes under intellectual property law, which also covers patents, copyrights and trade marks. The law allows due reward to be obtained for the transfer or sale of protectable ideas and technologies developed, for example from research, or during the pilot or full-scale operational phase of a farm or hatchery. An overview of the situation regarding intellectual property rights in the USA is given by McCoy (1997).

Crustacean ranching involves the release of cultured stock into natural bodies of water and presents particular problems in terms of legal ownership and policing of the site to protect the investment. Attempts to ranch crustaceans are only likely to be of commercial interest in areas where legislation permits leasing or ownership of the submerged land, and where the majority of the stock display limited migratory habits, can be maintained within a defined boundary or can be indisputably distinguished from wild stocks. In other circumstances, releases of hatchery-reared or transplanted (wild) juveniles may be done 'for the public good' in mitigation for habitat change, to augment or extend natural stocks or in support of rural communities. As with pond cultures, marine or brackish-water projects are less easily upheld from a legal viewpoint than inland waters. The legal status of marine ranching including harvesting methods, in international and selected national laws, has been reviewed by Howarth and Lería (1999) but, although primarily concerned with fish, many of the issues identified (introductions, property rights, incentives) are pertinent to crustaceans.

The production of juveniles for national, noncommercial restocking and ranching programmes may of course be conducted commercially. Also, the maintenance of a hatchery and a programme of releases may be imposed as conditions when granting permission for certain engineering or construction developments, harbour dredging operations and specific habitat modification. The justification is generally to compensate for losses due to spoiled habitat or to create new or extend existing fishable resources for the public good. For the latter purpose, a variety of structures commonly known as artificial reefs (section 5.7.2) have been deployed in Japan and the USA for many years. In Europe, with the possible exception of Spain where explicit legal provision exists for artificial reefs (Revenga *et al.* 2000), the rapidly growing interest in artificial reefs has outpaced the development of appropriate law governing property and use rights, controls and incentives. In addition, a multitude of national and international regulations governs their composition and deployment, many of which seriously constrain research and pilot studies aimed at ranching and fishery enhancement (Pickering 1997).

11.5.3.2 Protection or constraint?

The lack of a national policy in the USA and elsewhere during the early days of aquaculture development resulted in difficulty in getting permits and dedicated land. Many conflicts arose with environmental and recreational interests since, while all parties wanted protection, none wanted constraint. In Texas, USA, for example, effluent discharge limits were often specified on a site-by-site basis and potential shrimp farmers could not therefore predict the future costs of compliance (Mattei 1995). The application, to prawn farms, of new regulations to protect the Great Barrier Reef in Australia were perceived by many in the prawn farming industry as unfair or inappropriate, since it was claimed discharges of sewage and from agriculture, forestry and marina developments were far more threatening (O'Sullivan 2000). The duty of government is to reconcile the expectations of the aquaculturists with those of other commercial, amenity and national interests using natural resources. Sometimes, however, the resulting laws become so complex or numerous that not only is compliance difficult but it may even be difficult to get an enterprise going (section 6.2.5.7).

In developed nations like the UK (Howarth 1990), Australia (O'Sullivan 1999), New Zealand and the USA there is a plethora of departments and organisations that become involved with aquaculture applications and project implementation. In one US state it is reported that around 25 permits may be required to establish a shrimp farm (McCoy 1996a). Consulting with, and reviewing, all the interests concerned with yabby harvesting and fisheries in New South Wales, Australia, is delaying the development of a potentially worthwhile crayfish culture industry (Mosig 1999). Together this surfeit of organisations presents formidable difficulties for the newcomer, who has often to make separate applications to each bureaucracy dealing with land use, water abstraction, disease and quarantine measures, effluent discharges, building plans and health and safety, to name but a few. The applicant will also have to comply with numerous regulations that may even have been set with a non-aquaculture activity in mind, but his or her ability to test or challenge those regulations is likely to be severely constrained by the costs involved (McCoy 1996b). An applicant may also be liable for ongoing monitoring costs to verify regulatory compliance once the project becomes operational. In Britain, attempts have even been made to turn such regulatory confusion to personal advantage by using a crayfish farm proposal as a means of obtaining planning permission for a house in the countryside!

Considerable improvements in co-ordination between all those involved are needed. In addition there will be objections from private interests and conservation and amenity groups, many of whom regard themselves as guardians of the national heritage and who see it as their duty to object to any change in the environment whatsoever. Aquaculturists in the US have been put out of business for killing predatory, but legally protected, migratory and endangered birds. In India, the Supreme Court ruling in 1996 that all shrimp farms (except traditional, i.e. extensive, systems) within 500 m of the high-tide line must cease operations, was made in response to environmentalists' concerns for disaffected coastal communities (Murthy 1997). However, the ban does not seem to have been implemented (Rosenberry 1998) and farms that conform to a 'code of conduct' for sustainable fisheries in India or adopt improved technology may be allowed to continue (Govind 1999; Sakthivel 2000).

Yet another area where laws create conflicting perceptions of protection or constraint between consumer and farmer is that of chemical treatments (Alderman & Michel 1992). The long history of unchallenged use of a chemical does not indicate that the substance is safe or has been properly registered by the US Food and Drug Administration (FDA) (or other national authority) for use on crustaceans intended for human consumption. Indeed, compounds widely used in fish culture may even be toxic or harmful to crustaceans regardless of whether they have been approved. Control of substances used in aquaculture is strict in North America, Europe and Japan but less so in parts of Asia, South America and Africa (Schnick 1992). The US FDA terms substances (other than food) intended for use in the diagnosis, cure, treatment or prevention of disease or to affect body form or function, drugs. Some seemingly innocuous compounds (salt, ice – when used in live transport – and sodium bicarbonate) fall under this definition, are afforded low regulatory priority (LRP) status but, nevertheless, must be correctly labelled, of an appropriate quality and used in a prescribed manner (Anon. 1994). Main-line permitted therapeutants are few and an inability to vary, for example, the antibiotics used in the treatment of a disease, increases the risk of resistance developing (section 11.3.4). Excessive regulation may thus appear counterproductive (de Kinkelin & Michel 1992). The status of FDA registrations affecting crustacean farmers was described by Schnick (1988) at a time when it could take 6 years and \$4m to get FDA approval for a therapeutant or other additive. If the compound had already been approved for fish, approval for crustaceans could still cost \$1m and take 3–4 years to process (Idyll 1986). A decade later, it was still not easy to obtain approval to use a chemical in the culture of an animal destined for the table (McCoy 1996c). However, perhaps because of increasing criticism (Meyer 1992) and the growing number of applications by aquaculturists to gain approval to use new drugs (rather than pharmaceutical companies who perceive the sales potential as below the threshold for research investment approval), federal (US) government began working energetically with the industry to facilitate the process (Bell 1995). Obtaining approval for compounds that can be used in crustacean farming is difficult because the relatively small market and high research costs involved deters pharmaceutical companies from seeking approvals. Although it is expected that approval for at least one use for all the eight drugs given priority by the US fish farming industry will be in place by 2001, the extent of approved drug usage will depend on the co-operation of the industry in providing efficacy data (Griffin 2000).

The unwillingness or inability of shrimp farmers and farming companies in the past to embrace practices that protect the environment has precipitated community and consumer outrage. Since producers have not yet taken a sufficiently robust lead in adopting more responsible environmental management methods, legislation to set standards for production, backed up by penalties for non-compliance, are increasingly being demanded by consumers and imposed by governments. Charges might also be levied for the use of land, waterways, wild-caught seed and broodstock as well as for water abstraction and discharge, the latter two already being common practice in developed countries. Indeed, not charging could be perceived as, and possibly declared, illegal, as an unfair but hidden subsidy by the World Trade Organization whose remit includes eliminating unfair trade advantages (Clay 1997). Incentives, such as price premiums paid for shrimp independently certified as being grown using best management practices, are being actively considered by the GAA as a way of encouraging 'sustainable', i.e. responsible, farming.

In western societies, the 'industry', that is to say large-scale shrimp farms (but also some pharmaceutical, feed and biotechnology companies researching GMOs), are widely perceived as uncaring and profit-motivated to the point of being untrustworthy (Boyd 1999). This has led to the development of some successful NGO strategies against crustacean (primarily shrimp) farming such as the incitement of consumer-led boycotts of the products. This tactic uses the press and other information media to lower public indignation thresholds and represents the use of a very blunt instrument to achieve environmental and social goals. Perversely, if such boycotts are implemented extensively, they could have drastic consequences on hundreds of thousands of livelihoods; environmental degradation and poverty could actually increase as farms collapse, with concurrent reduction in the economic health of some developing countries (Lockwood 1997).

11.5.3.3 Positive attitudes and legislation

Some governments, notably in Taiwan (Lee 1988), Thailand (Akrasanee 1988), Spain (Santaella 1989) and the Philippines (Idyll 1986), have long given aquaculture preferential status in a number of areas such as investment, grants and loans and tax holiday incentives. In the Philippines valuable incentives were granted to agriculture and fishery industries (which included aquaculture) because of the greater risks to investment due to the vagaries of the weather and the high incidence of spoilage and pest damage. Good publicity was given and publications distributed to promote awareness of documentary requirements to start aquaculture operations, and much of the income from the legislation goes back to the

Plate 11.5 Clearance of a coconut plantation in Indonesia for the construction of small-scale shrimp farms within the mangrove fringe. Such sites typically provide better quality soils than mangroves but they require a pumped water supply and if they are poorly planned they can lead to the salinisation of surrounding agricultural land.

industry as funding for extension and research support. Today, positive governmental attitudes remain (section 10.3.1.2), despite the numerous adverse impacts attributed to shrimp farming, although policies are now usually geared to providing incentives for more responsible farming. In 1999 the Thai government announced an allocation of research and development funds to improve shrimp feeds; it also reduced electricity charges and relaxed loan requirements for rural farmers to allow uptake of more environmentally benign practices. The industry responded in early 2000 by signing up to a code of conduct covering environmentally and socially responsible farming practices (Anon. 2000). Considerable encouragement was given to the farming (fattening) of spiny lobsters in New Zealand without jeopardising wild stocks by allowing juveniles to be captured under a fishery quota trade-off agreement (section 7.9.4). In Great Britain the scientific demonstration (for the first time) that hatchery-reared lobsters released at sea could form a significant part of fishermen's catches some 5–10 years later (sections 5.7 and 7.8.11), led to a revision of the Sea Fisheries (Shellfish) Act (1967) that provided the means for a licence holder to have exclusive rights to deposit, propagate and harvest lobsters and other crustaceans. This legislative change gave investors in hatchery-reared lobster juveniles similar protection to that traditionally enjoyed by molluscan shellfish growers. In Spain since 1994 artificial reefs to promote stock enhancement (fish and shellfish) have been recognised as worthwhile and have been publicly funded (using EU and Spanish funds) without the need for private participation. The substantial funds made available since 1996 for artificial reef development amount to about $12m and illustrate a positive commitment to the approach (Revenga *et al.* 2000).

New laws or improvements to existing legislation may be implemented in the light of aquaculture developments or may affect aquaculturists even though they may be implemented primarily for other reasons. For example, in many Indian coastal states laws exist to extend and protect agricultural land by preventing the ingress of seawater. To this end, the construction of bunds not only effectively prevented shrimp culture but also in one case led to the destruction of a fishery for *Macrobrachium*. The impounded land eventually proved to be inadequate for profitable rice production and the law had to be amended to allow brackish-water aquaculture development (Sakthivel 1985). The problems associated with that development, particularly shrimp farming, led, in the mid-1990s, to the establishment of the Aquaculture Authority of India to regulate the industry (Sakthivel 2000). Laws to curb the importation of disease organisms with live shrimp, and the over-exploitation of natural stocks of large broodstock *P. monodon*, led to smuggling between countries like Malaysia, Thailand, Taiwan and the Philippines, and prices of up to $1800 for one live female have been reported (Chiang & Liao 1985).

The Ecuadorian government attempted to reduce the country's dependence on US markets by introducing exchange control and import restriction policies. While the broad objective had some merit, the effect was to cause hardship to many shrimp farmers (NOAA 1988). Potentially advantageous legislative changes have been implemented in Cuba where, traditionally, all aquaculture enterprises were government owned. Since the late 1990s, foreign partners have become eligible for consideration as joint venture partners in shrimp, *Macrobrachium* and, presumably, spiny lobster projects. In other countries, such as Mexico, shrimp culture permits were reserved, at least up to 1981, for fishermen's co-operatives, so inhibiting applications from large industrial corporations. Exclusion of private enterprises from direct involvement led to uneasy partnerships between the co-operatives and the private sector. The more successful shrimp farms were almost invariably those operated by private investors, as were most hatcheries, and as a consequence Mexico lagged behind in shrimp culture. In 1990 the Mexican government initiated a national plan for fisheries development and passed legislation that permitted private sector shrimp farming and 49% foreign ownership of shrimp farms (Rosenberry 1990), although some co-operatives feared they would become disadvantaged if the new enterprises monopolised available resources (Cruz 1992). In the event, both co-operative and private shrimp farm output has grown rapidly from 1989, when 104 farms containing 6513 ha of ponds produced an average yield of 582 kg ha^{-1} (Garmendia & Nuñez 1990), to 1999 when 31 000 mt were produced with annual yields reaching 900–2300 kg ha^{-1} (Clifford 2000).

The FAO's voluntary Code of Conduct for Responsible Fisheries (FAO 1995) exhorts countries throughout the world to establish, maintain and develop appropriate legal and administrative frameworks for responsible aquaculture. Industry, governments, government agencies and NGOs are only now recognising the need to work together to create what New (1999a) describes as an 'enabling environment' that will ensure the promotion of aquaculture in general and crustacean farming in particular.

11.6 References

Ackefors H. (1999) The positive effects of established crayfish populations in Europe. In: *Crayfish in Europe as alien species. How to make the best of a bad situation?* (eds F. Gherardi & D.M. Holdich), pp. 49–61. A.A. Balkema, Rotterdam, Netherlands.

Aiken D.E. & Waddy S.L. (1985) Production of seed stock for lobster culture. *Aquaculture,* **44** (2) 103–114.

Aitcheson S.J., Arnett J., Murray K.R. & Zhang J. (2000) Removal of aquaculture therapeutants by carbon adsorption. 1. Equilibrium adsorption behaviour of single components. *Aquaculture,* **183** (3–4) 269–284.

Akrasanee N. (1988) Investment outlook for shrimp farming. In: *Shrimp '88, Conference proceedings,* 26–28 January 1988, Bangkok, Thailand, pp. 186–190. Infofish, Kuala Lumpur, Malaysia

Alderman D.J. & Michel C. (1992) Chemotherapy in aquaculture today. In: *Chemotherapy in Aquaculture: from theory to reality* (eds C. Michel & D.J. Alderman), pp. 3–24. IOE, Paris, France.

Anon. (1988) UK power plant farm closes: pioneer project hit by royalties demand. *Fish Farming International,* **15** (6) 1 & 5.

Anon. (1990a) Risks and claims on the upward path. *Fish Farmer,* **13** (2) 9.

Anon. (1990b) Shrimp crop insurance. *Fish Farmer,* **13** (2) 31.

Anon. (1990c) Monks want ban. *Fish Farming International,* **17** (4) 44.

Anon. (1994) FDA lists low regulatory drugs. *Aquaculture Magazine,* **20** (6) 10–14.

Anon. (1997) Global Aquaculture Alliance formed to guide industry toward environmental sustainability. *World Aquaculture,* **28** (3) 48.

Anon. (2000) Thailand: shrimp farmers sign code of conduct. *Infofish International,* (4) 50.

Avault J.W. Jr. (1989) Social/political aspects of aquaculture. *Aquaculture Magazine,* **15** (4) 70–73.

Avault J.W. Jr. (1995) Insect and bird predators and pests of fish and crustaceans. *Aquaculture Magazine,* **21** (2) 64–70.

Bailey C. (1988) The social consequences of tropical shrimp mariculture development. *Ocean and Shoreline Management,* **11**, 31–44.

Bailey C. (1997) Aquaculture and basic human needs. *World Aquaculture,* **28** (3) 28–31.

Bailey C. & Skladany M. (1991) Aquaculture development in tropical Asia: a re-evaluation. *Natural Resources Forum,* **15**, 66–73.

Barata C., Hontoria F. & Amat F. (1996) Estimation of the biomass production of *Artemia* with regard to its use in aquaculture: temperature and strain effects. *Aquaculture,* **142** (3–4) 171–189.

Barlow S. (2000) Fishmeal and fish oil: sustainable ingredients for aquafeeds. *Global Aquaculture Advocate,* **3** (2) 85–88.

Bell T.A. (1995) New animal drug approvals and the United States aquaculture industry: a partnership for growth. *Aquaculture Research,* **26**, 679–685.

Berry T.M., Park D.L. & Lightner D.V. (1994) Comparison of the microbial quality of raw shrimp from China, Ecuador, or Mexico at both wholesale and retail levels. *Journal of Food Protection,* **57** (2) 150–153.

Blakely D.R. & Hrusa C.T. (1989) *Inland Aquaculture Development Handbook*, 184 pp. Fishing News Books, Blackwell Scientific Publications, Oxford, UK.

Boyd C.E. (1999) Aquaculture and sustainability issues. *World Aquaculture,* **30** (2) 10–13 & 71–72.

Boyd C.E. (2000) Formulating standards for effluent management. *Global Aquaculture Advocate,* **3** (1) 10–11.

Boyd C.E., Massaut L. & Weddig L.J. (1998) Towards reducing environmental impacts of pond aquaculture. *Infofish International,* (2) 27–33.

Brodie J. (1995) Water quality and pollution. In: *Coastal Management in the Asia-Pacific Region: issues and approaches* (eds K. Hotta & I.M. Dutton), pp. 39–56. Japan International Marine Science and Technology Federation, Tokyo, Japan.

Brown J.H. (1989) Antibiotics: their use and abuse in aquaculture. *World Aquaculture,* **20** (2) 34–35, 38–39 & 42–43.

Brown J.J. & Glenn E.P. (1999) Management of saline aquaculture effluent through the production of halophyte crops. *World Aquaculture,* **30** (4) 44–49.

Chamberlain G.W. (ed.) (1985) Sociological factors limit yields from Asian farms. *Coastal Aquaculture,* **2** (4) 11–12.

Chamberlain G.W. (ed.) (1988a) Disease control. *Coastal Aquaculture,* **5** (1) 6–9.

Chamberlain G.W. (ed.) (1988b) Shrimp culture news from around the world. *Coastal Aquaculture,* **5** (1) 13–17.

Chiang P. & Liao I.C. (1985) The practice of grass prawn (*Penaeus monodon*) culture in Taiwan from 1968 to 1984. *Journal of the World Mariculture Society,* **16**, 297–315.

Chiau W-Y. (1998) Coastal zone management in Taiwan: a review. *Ocean and Coastal Management,* **38**, 119–132.

Choluteca Declaration (1997) Choluteca Declaration, Choluteca, Honduras 16 October 1996. *World Aquaculture,* **28** (3) 38–39.

Chythanya R., Nayak D.K. & Venugopal M.N. (1999) Antibiotic resistance in aquaculture. *Infofish International,* (6) 30–32.

Clay J.W. (1997) Towards sustainable shrimp aquaculture. *World Aquaculture,* **28** (3) 32–37.

Clifford H.C. III (2000) Shrimp farming in Mexico: recent developments. *Global Aquaculture Advocate,* **3** (2) 79–81.

Costa-Pierce B.A. (1992) Aquaculture development and large-scale resettlement in Indonesia. *World Aquaculture,* **23** (1) 33–39.

CPC (1989) Libro blanco del camarón, 79 pp. *Cámara de productores de camarón,* May 1989, Guayaquil, Ecuador.

Cross T.F. (1999) Genetic considerations in enhancement and ranching of marine and anadromous species. In: *Stock Enhancement and Sea Ranching* (eds B.R. Howell, E. Moksness & T. Svåsand), pp. 37–48. Fishing News Books, Oxford, UK.

Cruz M.L. (1992) Shrimp mariculture in Mexico. *World Aquaculture,* **23** (1) 49–51.

Dayton L. (1995) The killing reefs. *New Scientist,* **11**, 11–15.

Delmendo M.N. (1995) Fishpen development on Laguna de Bay: a boon or bane to the social, economic and environmental concerns of the area. In: *Ecoset '95 International conference on ecological system enhancement technology for aquatic environments. The 6th International Conference on Aquatic Habitat Enhancement.* 29 October–2 November

1995, **2**: 854–862. Tokyo, Japan.
Edgerton B.F. (1999) A review of freshwater crayfish viruses. In: *Freshwater Crayfish 12* (eds M. Keller, M.M. Keller, B. Oidtmann, R. Hoffmann & G. Vogt), pp. 261–278. Weltbild Verlag, Germany.
Ekaratne S.U.K., Davenport J., Lee D. & Walgama R.S. (1998) Communities protecting coastal resources: Rekawa Lagoon, Sri Lanka. In: *Communities and Conservation, natural resource management in south and central Asia* (eds A. Kothari, N. Pathak, B. Anuradha & B. Taneja), pp. 481–491. Sage Publications, New Delhi, India.
Evans L.H. & Edgerton B.F. (2001) Pathogens, parasites and commensals. In: *Biology of Freshwater Crayfish* (ed. D.M. Holdich), pp. 377–438. Blackwell Science, Oxford, UK.
FAO (1995) *Code of conduct for responsible fisheries,* 41 pp. Food and Agriculture Organization, Rome, Italy.
FAO (1996) Integration of fisheries into coastal area management, 17 pp. *FAO Technical Guidelines for Responsible Fisheries.* (3). Food and Agriculture Organization, Rome, Italy.
FAO (2000) Code of conduct for responsible fisheries. http://www.fao.org/waicent/faoinfo/fishery/agreem/codecond/codecon.htm
FAO/NACA (1995) *Report on a regional study and workshop on the environmental assessment and management of aquaculture development*, 492 pp. TCP/RAS/2253. Food and Agriculture Organization, Network of Aquaculture Centres in Asia-Pacific, Bangkok, Thailand.
Fegan D.F. (1995) Environmental assessment and management of aquaculture development: an industry perspective. In: *Report on a regional study and workshop on the environmental assessment and management of aquaculture development.* TCP/RAS/2253, 437–445. Food and Agriculture Organization, Network of Aquaculture Centres in Asia-Pacific, Bangkok, Thailand.
Fiske S.J. & Plé J-P. (1992) Sociocultural aspects of domestic marine aquaculture, 290 pp. In: *Marine Aquaculture – opportunities for growth.* National Academy Press, Washington, DC, USA.
Flores P.E. (1997) Women in Ecuador's fishing families. *Intercoast Network,* (28) 7–8.
Frese T. (2000) *Ecuadorian shrimp farmers win $10 million verdict against chemical giant DuPont, announces Krupnick Campbell.* Correspondence. 05/12/00. shrimp@egroups.com
Fushimi H. (1999) How to detect the effect in releasing operation of hatchery raised Kuruma prawn postlarvae? Case study of the operation in the Hamana Lake. *Bulletin of Tohoku National Fisheries Research Institute*, **62**, 1–12.
Garmendia E.A. & Nuñez A. L. G-S. (1990) Shrimp culture status in Mexico 1989. *Abstracts from World Aquaculture, 90,* 10–14 June 1990, p. 115. Halifax, Nova Scotia, Canada.
GESAMP (1996) *Chemicals in coastal aquaculture.* Report of GESAMP working group 31. 24–28 May 1996. SEAFDEC, Tigbauan, Iloilo, Philippines [seen in draft].
Govind H. (1999) Chance of reprieve for Indian coastal aquaculture. *Fish Farmer,* **22** (3) 25.
Griffin B.R. (2000) Update on the Federal-State aquaculture drug approval partnership. *Aquaculture Magazine,* **26** (3) 56–58.
Gulland J.A. & Rothschild B.J. (1984) *Penaeid Shrimps: their biology and management*, 308 pp. Fishing News Books, Blackwell Scientific Publications, Oxford, UK.
Gutting R.E. Jr. (1997) The US shrimp embargo. *Infofish International,* (2) 19–22.
Hargreaves J.A. (1997) The quest for sustainable shrimp aquaculture: a moderator's perspective. *World Aquaculture,* **28** (3) 43–47.
Hatch U., Sindelar S., Rouse D. & Perez H. (1987) Demonstrating the use of risk programming for aquacultural farm management: the case of penaeid shrimp in Panama. *Journal of the World Aquaculture Socieity,* **18** (4) 260–269.
Hauck A.K., Marshall M.R., Li J.K-K. & Lee R.A. (2000) A new finding and range extension of baculovirus in the freshwater crayfish *Cherax quadricarinatus* in Utah, North America. In: *Abstracts, Aqua 2000, Responsible Aquaculture in the New Millennium* (compiled by R. Flos & L. Creswell), p. 271. European Aquaculture Society, Special Publication No. 28.
Haughton P. (1990) Spread a little happiness among the 'wage slaves'. *Fish Farmer,* **13** (2) 63–64.
Hayashi T. (1984) Manpower in the diffusion of technology. *United Nations University Newsletter,* **8** (1) 8.
Holdich D.M. (1999a) The negative effects of established crayfish introductions. In: *Crayfish in Europe as Alien Species: How to make the best of a bad situation?* (eds F. Gherardi & D.M. Holdich), pp. 31–48. A.A. Balkema, Rotterdam, Netherlands.
Holdich D.M. (1999b) The introduction of alien crayfish species into Britain for commercial exploitation – an own goal? In: *Crayfish in Europe as Alien Species: How to make the best of a bad situation?* (eds F. Gherardi & D.M. Holdich), pp. 85–97. A.A. Balkema, Rotterdam, Netherlands.
Holdich D.M., Ackefors H., Gherardi F., Rogers W.D. & Skurdal J. (1999) Native and alien crayfish in Europe: some conclusions. In: *Crayfish in Europe as Alien Species: How to make the best of a bad situation?* (eds F. Gherardi & D.M. Holdich), pp. 281–292. A.A. Balkema, Rotterdam, Netherlands.
Holland P. & Brown D. (1999) Resource conflicts in aquaculture: international experiences. *Aquaculture Magazine,* **25** (6) 39–48.
Holmes N. (1995) Coastal dynamics and global change: implications for coastal management. In: *Coastal Management in the Asia-Pacific Region: issues and approaches* (eds K. Hotta & I.M. Dutton), pp. 81–93. Japan International Marine Science and Technology Federation, Tokyo, Japan.
Hong Sc.P.N. (1996) The impact of shrimp pond construction along the mangrove coastal accretion at south-west Ca Mau Cape, Viet Nam. *Asian Aquaculture,* **18** (4) 3–7.
Hotta K. & Dutton I.M. (1995) *Coastal Management in the Asia-Pacific Region: issues and approaches*, 421 pp. Japan International Marine Science and Technology Federation, Tokyo, Japan.
Howarth W. (1990) *The Law of Aquaculture*, 271 pp. Fishing News Books, Blackwell Scientific Publications, Oxford, UK.
Howarth W. (1995) The essentials of aquaculture regulation. In: *Report on a regional study and workshop on the environmental assessment and management of aquaculture development.* TCP/RAS/2253, 459–465. Food and Agricul-

ture Organization, Network of Aquaculture Centres in Asia-Pacific, April 1995, Bangkok, Thailand.

Howarth W. & Lería C. (1999) Legal issues relating to stock enhancement and marine ranching. In: *Stock Enhancement and Sea Ranching* (eds B.R. Howell, E. Moksness & T. Svåsand), pp. 509–525. Fishing News Books, Oxford, UK.

Huner J.V. (1995) An overview of the status of freshwater crawfish culture. *Journal of Shellfish Research*, **14** (2) 539–543.

Huner J.V. (2001) *Procambarus*. In: *Biology of Freshwater Crayfish* (ed. D.M. Holdich), pp. 541–84. Blackwell Science, Oxford, UK.

Idyll C.P. (1986) Aquaculture legislation and regulation in selected countries. In: *Realism in Aquaculture: achievements, constraints, perspectives* (eds M. Bilio, H. Rosenthal & C.J. Sindermann), pp. 545–563. European Aquaculture Society, Bredene, Belgium.

Inglis V., Abdullah S.Z., Angka S.L., *et al.* (1997) Survey of resistance to antibacterial agents used in aquaculture in five South-east Asian countries. In: *Diseases in Asian Aquaculture III* (eds T.W. Flegel & I.H. MacRae), pp. 331–337. Fish Health section, Asian Fisheries Society, Manila, Philippines.

Johnson S.K. (1994) Dissemination of exotic diseases of shrimp: is a control policy attainable? *World Aquaculture*, **25** (4) 30–38.

Jonasson M. (1987) Fish cleaning behaviour of shrimp. *Journal of Zoology (London)*, **213**, 117–131.

Jørstad K.E. & Farestveit E. (1999) Population genetic structure of lobster (*Homarus gammarus*) in Norway, and implications for enhancement and sea ranching operation. *Aquaculture*, **173** (1–4) 447–457.

Jory D.E. (2000) Status of shrimp aquaculture 2000. *Aquaculture Magazine Buyer's Guide 2000*, 29th Annual Edition, pp. 49–60.

Jory D.E. & Dixon H.M. (1999) Shrimp farming and integrated environmental management systems. *Aquaculture Magazine*, **25** (4) 77–82.

Kautsky N., Rönnbäck P., Tedenaren M. & Troell M. (2000) Ecosystem perspectives on management of disease in shrimp pond farming. *Aquaculture*, **191** (1–3) 145–161.

Kayasseh M. & Schenck C. (1989) Reclaimation of saline soils using calcium sulphate residues from the titanium industry. *Ambio*, **18** (2) 124–127.

de Kinkelin P. & Michel C. (1992) The use of drugs in aquaculture. *Infofish International*, (4) 45–49.

Kusumastanto T., Jolly C.M. & Bailey C. (1998) A multiperiod programming evaluation of brackishwater shrimp aquaculture development in Indonesia 1989/1990–1998/1999. *Aquaculture*, **159** (3–4) 317–331.

Lambert G. (1986) Social indicators of the aquaculture-based social laboratory project. *Asian Aquaculture*, **8** (3) 6–8.

Landesman L. (1994) Negative impacts of coastal aquaculture development. *World Aquaculture*, **25** (2) 12–17.

Lassen T.J. (1997) Environmental extremes versus sustainable policies in aquaculture. *World Aquaculture*, **28** (3) 49–51.

Lee C.S. (1997) Constraints and government intervention for the development of aquaculture in developing countries. *Aquaculture Economics and Management*, **1** (1) 65–71.

Lee J-C. (1988) Government policies and support for shrimp-farming. In: *Shrimp '88, Conference proceedings*, 26–28 January 1988, Bangkok, Thailand, pp. 175–181. Infofish, Kuala Lumpur, Malaysia.

Lewin D. (1997) Raw material ups and downs. *Fish Farmer*, **20** (1) 43–44.

Li G., Shen Q. & Xu Z. (1993) Morphometric and biochemical genetic variation of the mitten crab, *Eriocheir*, in southern China. *Aquaculture*, **111** (1–4) 103–115.

Lightner D.V., Redman R.M., Poulos B.T., Nunan L.M., Mari J.L. & Hasson K.W. (1997) Risk of spread of penaeid shrimp viruses in the Americas by the international movement of live and frozen shrimp. *Revue Scientifique et Technique Office International des Epizooties*, **16** (1) 146–160.

Lockwood G.S. (1997) World shrimp production with environmental and social accountability: a perspective and a proposal. *World Aquaculture*, **28** (3) 52–55.

Lockwood G.S. (1998) Capital markets and aquaculture failures. *Aquaculture Magazine*, **24** (6) 37–45.

Lorena A.S.H. (1983) Socio-economic aspects of the crawfish industry in Spain. In: *Freshwater Crayfish 5* (ed. C.R. Goldman), pp. 552–554. AVI Inc., Westport, CT, USA.

Lorena A.S.H. (1986) The status of the *Procambarus clarkii* population in Spain. In: *Freshwater Crayfish 6* (ed. P. Brinck), pp. 131–133. International Association of Astacology, Lund, Sweden.

Lutz C.G. (2000a) Aquatic GMOs: policy questions and considerations. *Aquaculture Magazine*, **26** (3) 83–87.

Lutz C.G. (2000b) The suffering Louisiana crawfish industry. *Aquaculture Magazine*, **26** (4) 48–54.

Macfarlane I. (1997) Strategies for developing shrimp farms. *Fish Farmer*, **20** (1) 33–34.

MAFF (1996) *A guide to shellfish health controls*, 15 pp. + appendices. Ministry of Agriculture, Fisheries and Food, London.

Mattei E. (1995) Shrimp farming in a regulatory climate: south Texas focus. *Aquaculture Magazine*, **21** (3) 54–63.

McCoy H.D. II (1996a) Aquaculture and the law: a primer. *Aquaculture Magazine*, **22** (1) 38–45.

McCoy H.D. II (1996b) Aquaculture and the law: a primer – article two of a series. *Aquaculture Magazine*, **22** (2) 68–80.

McCoy H.D. II (1996c) Aquaculture and the law: a primer – article three of a series. *Aquaculture Magazine*, **22** (3) 72–79.

McCoy H.D. II (1997) Aquaculture and the law: a primer – article four of a series. *Aquaculture Magazine*, **23** (1) 60–66.

McRae T., Horsley K-L. & McKenzie B. (1999) Evaluation of anaesthetic agents for crayfish. In: *Freshwater Crayfish 12* (eds M. Keller, M.M. Keller, B. Oidtmann, R. Hoffmann & G. Vogt), pp. 221–232. Weltbild Verlag, Germany.

Meyer F.P. (1992) Problems of chemotherapy in aquaculture regulatory aspects: an overview. In: *Chemotherapy in Aquaculture: from theory to reality* (eds C. Michel & D.J. Alderman), pp. 80–87. IOE, Paris, France.

Misamore M. & Browdy C.L. (1997) Evaluating hybridization potential between *Penaeus setiferus* and *Penaeus vannamei* through natural mating, artificial insemination and in vitro fertilization. *Aquaculture*, **150** (1–2) 1–10.

Mohney L.L., Williams R.R., Bell T.A. & Lightner D.V. (1997) Residues of oxytetracycline in cultured juvenile blue shrimp, *Penaeus stylirostris* (Crustacea: Decapoda), fed medicated feed for 14 days. *Aquaculture*, **149** (3–4) 193–202.

Mosig J. (1999) Yabbie permits await NSW review. *Austasia*

Aquaculture Magazine, **13** (1) 35–38.

Murthy H.S. (1997) Impact of the Supreme Court judgement on shrimp culture in India. *Infofish International,* (3) 30–34.

Naegel L.C.A. (1995) Research with a farming systems perspective needed for the development of small-scale aquaculture in non-industrialized countries. *Aquaculture International,* **3**, 277–291.

Nandeesha M.C. (1996) Women in fisheries in Indo-China countries. *Infofish International*, (6) 15–21.

Nash C.E., Engle C.R. & Crosetti D. (1987) Women in aquaculture. *Proceedings of the ADCP/NORAD workshop on women in aquaculture.* FAO/UNDP/NORAD ADCP/REP/87/28. FAO, Rome, Italy.

Nenke M. & Nenke M. (2000) The Western Australian yabby industry. In: *Proceedings of Australian Crayfish Aquaculture Workshop* (eds C. Lawerence & G. Whisson), pp. 14–19. International Association of Astacology, Curtin University of Technology, Perth, Australia.

New M.B. (1999a) Global aquaculture: current trends and challenges for the 21st century. *World Aquaculture,* **30** (1) 8–13 & 63–79.

New M.B. (1999b) Sustainable shrimp farming in the dry tropics: third generation shrimp farming technology. *World Aquaculture,* **30** (4) 34–43 & 62.

New M.B. & Rabanal H.R. (1985) A review of the status of penaeid aquaculture in South East Asia. In: *Second Ausralian National Prawn Seminar* (eds P.C. Rothlisberg, B.J. Hill & D.J. Staples) NPS2, pp. 307–326. Cleveland, Australia.

New M.B. & Wijkstrom U.N. (1990) Feed for thought. *World Aquaculture,* **21** (1) 17–19 & 22–23.

Ninawe A.S. (1999) Coastal aquaculture versus environment: pros and cons. *Infofish International,* (2) 43–47.

NOAA (1988) *Latin American Shrimp Culture Industry, 1986–90*, 79 pp. IFR/88–38, F/IA23:DW. Foreign Fisheries Analysis Branch, US National Marine Fisheries Service.

O'Sullivan D. (1999) Jumping thru hoops for prawn farm licences. *Austasia Aquaculture Magazine,* **13** (2) 44–46.

O'Sullivan D. (2000) New federal regulations on Great Barrier Reef protection to apply to new Queensland aquaculture farms. *Austasia Aquaculture Magazine,* **14** (2) 34–40.

Overton J.L. & Macintosh D.J. (1997) Mud crab culture: prospects for the small-scale Asian farmer. *Infofish International,* (5) 26–32.

Pereira G., Eganez H. & Monente J. (1996) First record of a reproductive wild population of *Macrobrachium rosenbergii* (De Man) (Crustacea, Decapoda, Palaemonidae) from Venezuela. *Acta Biologica Venezuelica,* **16**, 93–95.

Phillips M.J. (1995) Shrimp culture and the environment. In: *Towards Sustainable Aquaculture in South-east Asia and Japan* (eds T.U. Bagarinao & E.E.C. Flores), pp. 37–62. SEAFDEC, Aquaculture Department, Iloilo, Philippines.

Pickering H. (1997) Legal framework governing artificial reefs in the EU. In: *Proceedings of the First Conference of the European artificial reef research network* Ancona, Italy, 26–30 March 1996 (ed. A.C. Jensen), pp. 195–232. Southampton Oceanography Centre, Southampton, UK.

Primavera J.H. (1997) Socioeconomic impacts of shrimp culture. *Aquaculture Research,* **28**, 815–827.

Revenga S., Fernández F., González J.L. & Santaella E. (2000) Artificial reefs in Spain: the regulatory framework. In: *Artificial Reefs in European Seas* (eds A.C. Jensen, K.J. Collins & A.P.M. Lockwood), pp. 185–194. Kluwer Academic, Netherlands.

Rosenberry R. (1984) Cornelius Mock. *Aquaculture Digest,* **9** (11) 23–24.

Rosenberry R. (1985) Corny Mock at large. *Aquaculture Digest,* **10** (11) 5–6.

Rosenberry R. (1990) Understanding shrimp farming in Mexico yesterday, today and tomorrow. *Aquaculture Digest,* **15** (7) 5–7.

Rosenberry R. (1998) *World shrimp farming 1998.* Rosenberry, Shrimp News International, San Diego, USA, **11**, 1–328.

Rothlisberg P.C. (1998) Aspects of penaeid biology and ecology of relevance to aquaculture: a review. *Aquaculture,* **164** (1–4) 49–65.

Ruangpan L., Prapadsorn S. & Sangrungruang K. (1997) Minimal inhibitory concentration of 5 chemotherapeutants against *Vibrio* bacteria and their transfer of R-plasmids. In: *Diseases in Asian Aquaculture III* (eds T.W. Flegal & I.H. MacRae), pp. 349–353. Fish Health section, Asian Fisheries Society, Manila, Philippines.

Sakthivel M. (1985) Shrimp farming – a boon or bane to India. *ICLARM Newsletter*, **8** (3) 9–10.

Sakthivel M. (2000) From India. *Global Aquaculture Advocate,* **3** (3) 9.

Salmon J.E. (1992) Conservation and environmental issues affecting the shrimp industry. In: *Shrimp '92, proceedings of the third global conference on the shrimp industry.* 14–16 September 1992, Hong Kong (eds H. de Saram & T Singh), pp. 30–34. Infofish, Kuala Lumpur, Malaysia.

Santaella E. (1989) Implementation of national aquaculture programmes. In *OECD, Aquaculture: a review of recent experience,* pp. 187–192. OECD, Paris, France.

Schnick R.A. (1988) The impetus to register new therapeutants for aquaculture. *Progressive Fish-Culturist*, **50** (4) 190–196.

Schnick R.A. (1992) Overview of the regulatory aspects of chemotherapy in aquaculture. In: *Chemotherapy in Aquaculture: from theory to reality* (eds C. Michel & D.J. Alderman), pp. 71–9. IOE, Paris, France.

Scott, A. (2000) Crayfish conservation: legislating for non-native species. In: *Crayfish conference*, Leeds (2000), 26–27 April 2000 (eds W.D. Rogers & J. Brickland), pp. 27–31. Leeds, UK.

Secretan P.D. (1986) Risk insurance in aquaculture. In: *Realism in Aquaculture: achievements, constraints, perspectives* (eds M. Bilio, H. Rosenthal & C.J. Sindermann), pp. 536–541. European Aquaculture Society, Bredene, Belgium.

Secretan P.D. (1988) Aquaculture insurance. In: *Shrimp '88, Conference proceedings,* 26–28 January 1988, Bangkok, Thailand, pp. 181–186. Infofish, Kuala Lumpur, Malaysia.

Shang Y.C. (1990) Socio-economic constraints of aquaculture in Asia. *World Aquaculture,* **21** (1) 34–5 & 42–43.

Simonsen B. (1999) Cutting feed plant odours. *Fish Farming International,* **26** (11) 17.

Skladany M. (1992) Conflicts in Southeast Asia: an institutionalist perspective. *World Aquaculture,* **23** (2) 33–35.

Skurdal J. & Taugbøl T. (2001) *Astacus*. In: *Biology of Freshwater Crayfish* (ed. D.M. Holdich), pp. 467–510. Blackwell Science, Oxford, UK.

Smith I.R. (1984) Social feasibility of coastal aquaculture: packaged technology from above or participatory rural development? 42 pp. (mimeo). Discussion document prepared for a consultation on social feasibility of coastal aquaculture, 26 November–1 December 1984. *ICLARM contribution No. 225*. Madras, India.

Smith S.A. (1998) Chemotherapeutics and treatment calculations used in aquaculture. In *Proceedings of Second International Conference on Recirculating Aquaculture,* 16–19 July 1998, pp. 157–162. Virginia Tech, VA, USA.

Söderhäll K. & Cerenius L. (1999) The crayfish plague fungus: history and recent advances. In: *Freshwater Crayfish 12* (eds M. Keller, M.M. Keller, B. Oidtmann, R. Hoffmann & G. Vogt), pp. 11–35. Weltbild Verlag, Germany.

Sze C.P. (1997) The use of mangroves for aquaculture: can it be sustainable? *Infofish International,* (6) 42–44.

Tacon A.G. (1998) Feeding tomorrow's fish. *Infofish International,* (2) 19–25.

Tacon A.G. (2000) Rendered animal by-products. *Global Aquaculture Advocate,* **3** (4) 18–19.

Teichert-Coddington D.R. (1995) Estuarine water quality and sustainable shrimp culture in Honduras. In: *Swimming Through Troubled Water. Proceedings of the special session on shrimp farming* (eds C.L. Browdy & J.S. Hopkins), pp. 144–156. Aquaculture '95. World Aquaculture Society, Baton Rouge, LA, USA.

Tisdell C.A. (1994a) Socioeconomic considerations in the development of aquaculture. In: *Socioeconomics of Aquaculture, Proceedings of international symposium '93* (eds Y.C. Shang, P-S. Leung, C-S. Lee, M-S. Su & I.C. Liao), pp. 1–13. Tungkang Marine Laboratory Conference Proceedings No. 4. Taiwan Fisheries Research Institute, Taiwan.

Tisdell C.A. (1994b) Aquaculture and environmental economics. In: *Socioeconomics of Aquaculture, Proceedings of international symposium '93* (eds Y.C. Shang, P-S. Leung, C-S. Lee, M-S. Su & I.C. Liao), pp. 331–339. Tungkang Marine Laboratory Conference Proceedings No. 4. Taiwan Fisheries Research Institute, Taiwan.

Van Houtte A. (1995) Fundamental techniques of environmental law and aquaculture law. In: *Report on a regional study and workshop on the environmental assessment and management of aquaculture development.* TCP/RAS/2253, 451–7. Food and Agriculture Organization, Network of Aquaculture Centres in Asia-Pacific, April 1995, Bangkok, Thailand.

Veldhuizen T.C. & Stanish S. (1999) *Overview of the life history, distribution, abundance, and impacts of the Chinese mitten crab,* Eriocheir sinensis, 14 pp. (mimeo). California Department of Water Resources, Sacramento, USA.

Weaver A.J. & Green C. (1998) Global climate change: lessons from the past: policy for the future. *Ocean and Coastal Management,* **39**, 73–86.

Weeks P. (1989) *Aquaculture development: an anthropological critique*, 41 pp. (mimeo). Paper presented at the World Aquaculture Society meeting, Aquaculture '89, 12–16 February 1989 Los Angeles, USA.

Wiley K.C. (1992) The application of environmental assessment to the management of risk in shrimp aquaculture in southeast Asia. In: *Shrimp '92, Proceedings of the Third Global Conference on the Shrimp Industry.* 14–16 September 1992, Hong Kong (eds H. de Saram & T. Singh), pp. 146–152. Infofish, Kuala Lumpur, Malaysia.

Wilson G. (1999) "Useless" land may be ideal for aquaculture. *Austasian Aquaculture Magazine,* **13** (3) 18–20.

Xu H-S. (2000) Organic pollution and shrimp diseases in coastal aquaculture area: a review. In: *Abstracts, Aqua 2000, Responsible Aquaculture in the New Millennium* (compiled by R. Flos & L. Creswell), p. 757. European Aquaculture Society, Special Publication No. 28.

Yndestad M. (1992) Public health aspects of residues in animal products: fundamental considerations. In: *Chemotherapy in Aquaculture: from theory to reality* (eds C. Michel & D.J. Alderman), pp. 494–510. IOE, Paris, France.

Chapter 12
The Future for Crustacean Farming

12.1 Introduction

The demand for crustaceans seems set to expand into the twenty-first century, for as long as the economies of consumer nations (Japan, USA and Western Europe) remain buoyant, and with increasing wealth and tourism in the tropical producer countries. Crustacean farming remains, however, a high-risk industry with good prospects for worthwhile profit but also potential for serious loss. It is characterised by high investor appeal and over-optimistic predictions and aspirations. The industry has significantly increased the opportunities for a diversity of employment and trading activities, created high-value exportable products and, in doing so, provided the justification for improvements to national and regional infrastructures. On the other hand, farmed crustaceans seldom contribute to the diets of the poor and their large-scale culture may cause significant changes in the surrounding communities and environment. The potential for detrimental impact in tropical regions has long been recognised, but technologies now exist that could minimise many of the adverse environmental effects. Their application in affected countries nonetheless remains patchy.

Scientific research led to the establishment of today's crustacean hatcheries and much of the ongrowing technology. It is, however, axiomatic that the extraordinary rate of growth and increasing diversity of the crustacean farming industry owes so much to the versatility and innovative skills of technicians, and to managers and technicians alike being driven by commercial pressures to take a number of intuitive technical short cuts. The lack of scientific knowledge underlying many of the practices evolved in this way only becomes of concern when animal survival, growth or reproductive performance is not as expected, when production economics become critical in the face of changes brought about by market forces or, as has become so noticeable in recent years, when environmental degradation, disease and social disquiet force the development of new, more responsible working practices.

The main areas in which technical constraints to progress exist in each of the groups considered in this book are shown in Table 12.1. Compared to other groups, those affecting the advancement of shrimp culture were, until about 1988, few and relatively minor. To those presently in the industry, however, broodstock and postlarvae availability, the spread of diseases and the reper-

Species/group	Broodstock	Larvae	Diet	Disease	Ongrowing	Harvesting
Penaeids	**			***		
Macrobrachium				*	**	**
Crayfish, European			**	***	***	***
Crayfish, Australian	*		*	*	**	*
Crayfish, USA			**		*	***
Lobsters, clawed	**		**		****	
Lobsters, spiny	*	****	**			
Crabs		**	**		**	**

Table 12.1 Areas in which technical constraints seem likely to affect culture progress (**** = major constraint).

cussions arising from public perceptions (real and imaginary) of environmentally harmful farming practices, inequitable socio-economic impacts and adverse impacts on other resource users, are but a few of the negative factors identified which must pose the most serious threats to future expansion. Profligate use of resources (e.g. wild seed and broodstocks, fishmeals and oils in diets, medicants, water supplies) has created additional problems demanding further research and development for their solution.

In response to growing consumer concern for animal welfare issues, research relating to anaesthetisation, slaughter, and the live storage and transport of the larger crustaceans (lobsters, crabs and Australian crayfish) is being increasingly funded. The use of eyestalk ablation techniques, whether for the induction of maturation, to increase growth or to stimulate moulting (in the soft-shell and bait industries), can be anticipated to generate similar concerns. Also, as long as world governments continue to prioritise in favour of economic development rather than the conservation of natural resources, risks will remain that some crustacean farming and stock enhancement activities could, directly or indirectly, cause genetic contamination of isolated populations (potentially useful in selective breeding programmes), destroy habitats and, through competition or disease transfer, even eliminate important species from local ecosystems.

Greater public awareness of such issues, many of which are inherent throughout all aquaculture, has led to widespread apprehension. Considerable, and welcome, progress is undoubtedly being made towards their rectification in several sectors of the industry. However, the indignation generated by NGOs and the media can create disproportionate impediments to further advances, especially when the advantages already gained, for example from increased export earnings, increased variety and quality of food supplies, beneficial infrastructure development and expansion of employment opportunities in remote regions, are deliberately ignored. Appropriate policy development for the amelioration of many of these problems is occurring, but implementation is often slow. Consequent legislative and regulatory changes have been made, some hastily or imprecisely, resulting in further unnecessary hardship and expense. Other changes have been either too difficult or uneconomic to enforce, or implemented unfairly.

The major research programmes and advances in crustacean farming we see today relate primarily to improving the long-term sustainability of crustacean production through the adoption of responsible farming, processing and marketing practices. To these ends, scientific studies during the past 10–15 years have revolutionised understanding as well as the technologies available for crustacean disease management, domestication, broodstock and juvenile supply, feeding practices and waste management. The lessons learned are gradually becoming part of inland and coastal zone management policies and plans.

12.2 Disease management

Despite repeated failures in the shrimp industry due to disease, it would seem that the lessons of the past 10 years are still not being heeded. Jory (2000) reported that, in about 1999, certified high-health stocks of *Litopenaeus vannamei* were imported to Taiwan. Initially they grew well and the industry expanded using further imports of high-health shrimp post-larvae. Then unscrupulous buyers began to import cheaper uncertified stocks, some labelling them falsely as high-health shrimp in order to get a better price. On farms, certified stocks were often mixed with uncertified stocks. Shortly afterwards Taura syndrome virus (TSV) struck and the *L. vannamei* industry collapsed as quickly as it had begun. This virus is one of three shrimp viruses recently classified by Office International des Épizooties as notifiable diseases. The other two are white spot syndrome virus (WSSV) and yellow head virus (YHV). Clearly, effective import controls like those in Europe, the USA and Japan were not enforced, or did not exist. The finding that WSSV is pathogenic to freshwater crayfish has potentially serious implications for the global industry (Evans & Edgerton 2001). Attempts to find internationally acceptable regulations governing imports and exports of aquaculture animals and products are under way in the west (Fegan 2000). At the time of writing the US and EU are finalising an equivalency agreement for the mutual recognition of regulatory systems to protect both public and animal health which, at the same time, guards against protectionism and allows fair and consistent trade access.

Research into shrimp immune systems became a priority only in the late 1990s. A new group of antimicrobial peptides (penaeidins) has been discovered and seems effective against filamentous fungi and Gram-positive bacteria, but not against the important Gram-negative *Vibrio* spp. (section 2.5.1). Examples of short-term resistance to disease are now being reported, especially in shrimp (section 2.5.5), and the general belief that crustaceans do

Plate 12.1 Preparations for the export of certified disease-free shrimp broodstock from a biosecure farm and hatchery complex in Hawaii. Weak and undersized individuals are being excluded and the shipping cartons can be seen stacked in readiness in the background. (Photo courtesy of Greta Martinez, Molokai Sea Farms, Hawaii.)

not have an acquired immunity system has recently been challenged. There is some evidence that an anticipatory-like immune response to some antigens may exist and, if so, this could form the basis for the development of new vaccination strategies (Arala-Chaves & Sequeira 2000).

In Thailand, field observations revealed that very often a period of lethal virus infections and crop losses would be followed a year or so later by a period in which normal survival and growth would occur, even though the virus remained present in equivalent numbers. Unlike bacteria and fungi, viruses usually fail to induce an inflammatory response from the shrimp immune system. In an attempt to explain these phenomena, a new theory of viral accommodation has been proposed (Flegel & Pasharawipas 1998) by which some crustacean cells are thought to actively accommodate (or tolerate) viruses. Although the viruses continue to replicate actively, they do not cause mortalities until the animal becomes critically stressed. This 'tolerine' theory proposes that if a crustacean meets a virus early in life before it can mount a full defence response (perhaps at the first zoeal stage), the virus may become recognised as 'self'. Later, when the virus is next encountered, the cells recognise the virus and bind to it, but not in the same way as they would when binding to engulf and destroy an infective bacterial or fungal agent (Jory 2000). This seems to imply that early rearing in the presence of specific viruses might perhaps produce populations capable of accommodating one or more viruses simultaneously. There is some evidence that the production of stress or shock proteins in response to rapid changes in pond environmental conditions can cause massive, uncontrolled apoptosis (also termed programmed cell death) which might also release the pathogenic viruses bound in the tolerine cells (Owens 1999). Another theory invokes the role of the lymphoid organ and the extent to which it can accumulate and eventually clear infected cells contained in the spherical bodies commonly observed within the interstitial spaces of the organ (Anggraeni & Owens 2000). These spheroids are also associated with tegmental glands during chronic infections and it could be that spheroid formation represents a significant but little-known part of the shrimp cell-mediated, immune response system (Hasson *et al.* 1999). Under any of these scenarios, it is clear that extra attention must be paid to pond management where pathogenic viruses are known to exist. For example, before introducing new stock to ponds, great care must be taken to remove all residual, unharvested shrimp and escapees living in supply channels (even though they may not be causing or showing signs of disease) and to sterilise the ponds.

At present, measurement of a crustacean's osmoregulatory capacity is probably the best way to determine its general level of health or stress (sections 7.2.4 and 8.5) but more specific biological markers are needed to evaluate immune status and hence the ability to resist potential disease challenges: all potentially valuable tools in crustacean health management. Potential markers already identified include haemogram counts (this comprises a total haemocyte count plus a count of three different cell types; the latter being open to misidentification unless monoclonal antibodies are used), measurements of reactive oxygen intermediates produced after cell phagocytosis, phenyloxidase activity, antibacterial activity and plasma protein concentrations (Rodríguez & Le Moullac 2000). Biological markers can also show the effects of contaminants on physiological and biochemical systems, for example, on the cytochrome monoxygenase enzymes that affect moulting and disrupt chitin synthesis. Similarly, heavy metals may cause production of defence metalothioneins while pesticides may inhibit acetylcholinesterase and irreversibly damage the central nervous system. Any of these effects might be measurable and developed into worthwhile diagnostic tools (Bainy 2000).

The development of crustacean cell culture lines is difficult (section 1.4) and remains a major constraint to research on virus diseases, although the development and application of molecular genetic techniques in the management of crustacean disease has come of age. Polymerase chain reaction (PCR) and gene probe methods are used to identify very small amounts of DNA specific to particular micro-organisms. However, a positive reaction may not always mean that a DNA sequence specific to a pathogenic virus has been found. Closely related, but harmless, viruses could conceivably have some sequences in common and identification of additional sequences may be necessary for confirmation if unnecessary destruction of suspect stock is to be avoided (Laramore 2000). Such fingerprinting techniques using genetic markers (RAPD and microsatellites) can identify and map gene loci responsible for growth, disease resistance and disease susceptibility. The results are being used successfully in breeding programmes to locate suitable founding populations and gain faster rates of improvement to captive stocks (section 8.10.1.3). They are also used in developing restocking strategies for crayfish (Schulz & Sypke 1999) and Australian shrimp.

It is well known that stressed animals are vulnerable to disease but there may be circumstances where the degree of pathogenicity of an organism can change. Among the bacteria most associated with crustacean mortalities, the majority are secondary invaders. Recent evidence suggests that strains of at least two species (*Vibrio harveyi* and *V. penaeicida*) may either function as opportunistic pathogens causing secondary infections, or become true primary pathogens. It is important to know the circumstances under which micro-organisms can turn from being benign to causing a disease. Reproducible and standardised tests for pathogenicity are being sought that would help evaluate disease management techniques. These will become increasingly important especially if the use of probiotic bacteria becomes more widespread (Saulnier *et al.* 2000).

There is considerable debate concerning the efficacy of probiotics (section 8.9.4.2) and immunostimulants (sections 2.5.3 and 8.9.4.3) in the management of crustacean diseases. Probiotic bacteria need to be provided with an environment compatible with their biological needs in order to become effective as the dominant microbial population. Knowledge and application of appropriate water and feed management are therefore essential prerequisites. Commercial hatchery trials of probiotic treatments with shrimp and crabs seldom allow critical evaluation and it is not known exactly how the probiotic bacteria are working under any given set of conditions, how they will work under stressful (e.g. intensive) culture regimes or whether they could become pathogenic. While the results are often scientifically inconclusive, the tangible benefits being reported from some hatcheries imply considerable potential (section 7.2.4). Few scientifically controlled experiments have been conducted to critically evaluate these treatments in ongrowing ponds. Those that have, indicate little or no significant beneficial effect, either in conventional or low/zero water exchange pond systems (sections 7.2.6.6, 8.3.6.8 and 8.3.7). Much further research under careful management regimes will be required if credible assessments of probiotic function in ponds are to be obtained (Jory 1998).

Similarly, the results of field trials with immunostimulants are often too variable to instil complete confidence in the efficacy of commercial scale applications. A number of factors may be contributory including inadequate levels of essential dietary components (e.g. vitamin C) and stressful culture conditions. Research is needed to determine the duration of effective administration periods, the dosing frequency for each compound under a range of relevant conditions, and the effects of long-term exposure, especially during sexual maturation. Research on virus disease control in Japan has led to recommendations that include the use of two-step PCR techniques to detect latent pathogenic virus (e.g. WSSV) in asymptomatic shrimp, and the prophylactic administration of an immunostimulant in their diet throughout (and for potential broodstocks for 2–3 months beyond) the normal culture period. A further advance indicated that a sulphated polysaccharide derived from a brown alga when fed to *Marsupenaeus japonicus* gave protection against WSSV, probably by inhibiting the virus's adsorption onto host cell walls (Itami 1999).

The use of high-health, specific pathogen-free (SPF) and resistant (SPR) stocks of shrimp provides considerable benefits to the industry. Nevertheless, a common misconception persists, notably that the shrimp will remain free of disease. In fact, shrimp from any of these stocks may succumb to a 'new' pathogen, that is to say, one from which they are not certified free or selected for having resistance against. Efforts are already under way to select for broader resistance to viruses in general, by genetically improving SPR stocks to the state where they become globally stress-resistant (GSR) stocks (Bédier *et al.* 1998). As a result of the threat of disease and environmental degradation, some authorities believe super-intensive shrimp production with selected, disease-resistant and fast-growing strains could, through the use of water recirculation technologies, expand into temperate regions and thus be closer to markets in Japan, USA and Europe (Kautsky *et al.* 2000). Production of *Cherax* stocks free of particular pathogens (e.g. *Thelohania*) is in the early stages of development.

Crustacean pathology is moving from microscopic to molecular characterisation and probe-based diagnostic aids. These advances, together with research into the characterisation of immune genes, have set the stage for selecting pathogen-resistant strains of shrimp by genetic transformation (sections 8.10.1.3 and 8.10.2).

12.3 Domestication

The potential advantages of current research progress in domestication, particularly the development of genetically engineered crustaceans, seem likely to be constrained by emotional and often ill-informed public opinion for the foreseeable future. Considerable effort will be required to demonstrate scientifically that the strains produced are safe, both ecologically and as food for humans and, if found to be so, to convince consumers at large, the latter being by far the more difficult task. In the future it might be possible to lessen fears of ecological damage due to escapees or from crustaceans released for stock enhancement purposes, by only working with sterile offspring, e.g. polyploids (section 2.6.3; Lutz 1999), although it is possible that, in time, some might revert to their normal or diploid state. There seems to be no internationally agreed definition of what constitutes a genetically modified organism (GMO) and, not surprisingly, few countries have regulatory frameworks in place specifically governing aquatic GMOs (Bartley & Hallerman 1995), although the foundations for national policies are being laid (Hallerman & Kapuscinski 1995). For now, the focus is on the conservation of genetic resources in wild populations (Fetzner *et al.* 1997; Taylor 2001) and much emphasis is placed on containment and safe working practices where GMOs are concerned (Hallerman *et al.* 1999). This latter approach increases the demand for efficient 'biosecure' systems based on minimal water use, including water recirculation technologies (sections 8.3.7 and 8.4.4).

Selective breeding and genetic modification programmes among crustaceans are aimed primarily at improving disease resistance, growth, fecundity and, to a lesser extent, edible meat yield. Increasing cold tolerance and reducing aggressive or cannibalistic behaviour are potentially of interest but have yet to attract significant research funding. Arguably, selection for resistance against pathogenic organisms, i.e. for mechanisms that limit or kill the invader, is likely to be a less stable strategy in evolutionary terms than selection for tolerance, i.e. the ability to survive in the presence of evolving pathogens (Roy & Kirchner 2000). For example, one approach might be to select for animals that produce limited amounts of stress or shock proteins (section 12.2). These proteins are produced in response to pronounced environmental fluctuations during culture and can trigger mass mortalities in the presence of previously accommodated pathogens.

Evidence that significant genetic improvements can be obtained from selective breeding programmes with shrimp, crayfish and possibly freshwater prawns now exists, for example from programmes run over several generations with *Litopenaeus stylirostris* and *L. vannamei* in the USA, with *L. stylirostris, L. vannamei, Fenneropenaeus indicus* and *Penaeus monodon* in Tahiti and New Caledonia. Indeed, in Australia, genetically selected lines of *Marsupenaeus japonicus* and *Cherax* spp. showing improved growth rates are already in production at commercial farms (sections 12.8.1 and 12.8.3), although not all strains developed in one culture environment will necessarily perform as well in another (Coman *et al.* 2000). Among research populations of domesticated broodstock shrimp (*Marsupenaeus japonicus* and *Penaeus monodon*) in Australia, over 40% of shrimp now show comparable fecundity and hatching success to wild broodstock. Further improvements and demonstrations of larvae and post-larvae quality will be required to convince hatchery operators and farmers that such stocks can be consistently as good as wild stocks. National programmes to develop genetic linkage maps for shrimp (*Litopenaeus vannamei, Marsupenaeus japonicus* and *Penaeus monodon)* are already well advanced and will greatly contribute to the breeding programmes designed to supply healthy, genetically improved seed stocks.

Among the promising new technologies that, with development, could advance domestication are the use of retroviral vectors containing viral envelope proteins to transfer genes into crustacean cells (Burns *et al.* 2000). Already firefly luminescence cDNA has been successfully incorporated, and subsequently expressed, in *Litopenaeus stylirostris* (Shimizu *et al.* 2000).

12.4 Reproduction

Wild broodstocks remain under considerable pressure since they are still perceived by many hatchery operators as producing a higher and more consistent quality of larvae than pond-reared or artificially matured stock, even though scientific evidence for this seems equivocal. Also in some, like *Penaeus monodon*, convenience is a major factor since captive stocks must be reared through about four generations before their reproductive performance matches that of their wild counterparts. Proven techniques are available for broodstock production as well as for the control of maturation, artificial impregnation and spawning of most, if not all, cultured species. Yet for 20 years, control over maturation in *Penaeus monodon*, among others, has depended on unilateral eyestalk ablation and little progress has been made in replacing this invasive technique with environmental or hormonal manipulation (Benzie 1997). However, increased egg size *in vitro* and increased fecundity *in vivo* have been induced in shrimp following administration of methyl farnesoate. This hormone is normally produced by the crustacean mandibular organ and may suppress the effects of other hormones that inhibit gonad maturation. In the future, hormones like methyl farnesoate, in combination with other physiological and hormonal treatments, may contribute to a satisfactory solution to the problem (Huberman 2000). Greater endocrinological control over the timing of spawning in *Macrobrachium* could advantageously reduce broodstock facility and management costs, particularly in temperate regions where broodstock held indoors for extended periods tend to show asynchronous egg development (Daniels *et al.* 2000).

Environmental or dietary treatments, sometimes in conjunction with eyestalk ablation, have been used during the past 10 years to rear several commercially important penaeid species (including *Penaeus monodon*) through successive generations, many in support of industries in countries where the shrimp are not endemic. The selection of genetic lines with improved growth and disease resistance afforded by these developments is expected to lead to wider uptake of selective breeding and 'closed cycle' broodstock production (Browdy 1998). Primarily it has been the risk of disease, rather than scarcity or cost of wild females or post-larvae, that has forced many hatchery owners to produce juveniles from their own broodstocks, but extravagant use of the natural resource persists in many countries.

The cryopreservation of sperm (Divan & Joseph 2000) and nauplii would facilitate low-cost preservation of genetic lines, and help to stabilise fluctuations in larvae availability. However, early claims of successful cryopreservation have not been widely substantiated (Benzie 1998), although Oo *et al.* (1998) report that 78% of barnacle nauplii (*Balanus amphitrite*) frozen at −196°C for 28 days recovered, and that 19.8% metamorphosed to the cyprid stage.

While much is known about the reproductive cycles of male and female lobsters, present practical experience is probably insufficient to establish adequate control or accurate cost estimates. Manipulating temperature and photoperiod to control maturation, egg extrusion and incubation period in captive broodstock, in order to produce large and predictable numbers of eggs each month, would be a complex process, and for continuous production under battery conditions will require computer support for implementation. The computers would be needed to manage several independent controlled-environment stock rooms as well as to provide farm managers with stock movement, feeding and mating schedules. For the foreseeable future, production of clawed lobster juveniles will depend upon wild-caught ovigerous females and, for spiny lobsters, the capture of pueruli or juveniles from the wild will provide the main source of seed. Even so, larvae rearing of spiny lobster has made significant advances but much remains to be done before truly commercial-scale mass-rearing techniques can be demonstrated.

12.5 Nutrition

Further research to determine the specific interactions between nutrients, for example, those that affect the digestibility and assimilation of dietary components, is needed for most species. The results will allow the balance of ingredients to be adjusted more coherently and cost-effectively, according to the species and culture method, with the aim of sparing nutrient resources and improving effluent quality (Jory 2000). The search for alternative ingredients, especially protein, and for innovative processing methods that maximise efficient food utilisation is intensifying (section 11.5.1.4). However, at the time of writing (2000), the use of some mammalian ingredients (e.g. meat and bone meals) had already been banned in the EU to combat the spread of 'mad cow' disease. In similar vein, additives to promote diet intake and assimilation, health, growth and reproduction will doubtless be subjected to increasingly rigorous scrutiny.

The potential contribution from pond biota in crustacean nutrition is well recognised (Moss 2000) and, for shrimp, has led to the development of low or zero water exchange systems. One method for enhancing the persistence of nutrient micro-organisms, which in suspension are prone to cycles of bloom and collapse, involves the use of artificial substrates floating in a steady, gentle current of water. These substrates increase the surface area available for microbial colonisation and facilitate conversion of dissolved and some suspended nutrients into edible material (biofiltration) which can be grazed by the cultured crustaceans (McNeil 2000). Another method involves the encouragement of an heterotrophic bacteria dominated community (or floc) in the pond water by increasing the carbon to nitrogen ratio of inputs from below 10 : 1 (typical of many compounded feeds) to about 30 : 1 through the use of carbon-rich fertilisers. Use of flocs has been successfully demonstrated for some *Litopenaeus* species in well-aerated, zero water exchange ponds (McIntosh 2000). These techniques allow the level of expensive protein (mainly fishmeal) in the diet to be reduced. Unfortunately, in some zero water exchange systems, trace elements such as the copper and aluminium that occur in conventional feeds have accumulated to toxic levels in the water by the end of a production cycle. Special dietary formulations may therefore be required to get the best from such systems. Interestingly, this trend towards maximising the contribution of pond biota to the diet in shrimp culture does not seem to be followed by redclaw crayfish researchers, who seem to be placing increased emphasis on the development of nutritionally complete formulations (Lawrence & Jones 2001).

Research has highlighted the nutritional as well as the environmental importance of maintaining balanced bacterial populations in shrimp and spiny lobster larval cultures. Additionally, the benefits of an early feed of live microalgae to stimulate enzyme production, particularly in larvae destined to be fed on microencapsulated diets, are widely appreciated (section 2.4.8).

12.6 Effluents and environmental impacts

Dietary constituents, feeding regimes, pond and water management are all now recognised as integral components in the control and minimisation of crustacean (and especially shrimp) farm effluents. The struggle to reduce the adverse environmental impacts of farms has extended beyond the bounds of the culture ponds and into the surrounding ecosystems where the concept of an 'ecological footprint' reflects the area or resource required not only to neutralise discharges but also to support the farm's structure and provide all its inputs (Kautsky *et al.* 2000). In some situations, settlement ponds coupled with water reuse or treatment lagoons containing filter-feeding bivalves (e.g. mussels) seem to provide a promising solution. The use of constructed wetlands containing salt-tolerant plants suitable for animal fodder is also being investigated. Further research on the assimilative capacity of mangrove plantations is needed (Foster & Robertson 2000) and is being addressed, for example, in North Queensland. There, research to compare the capacity of a range of local mangrove species in artificially constructed wetlands receiving effluent from a commercial farm is under way (Danis & O'Sullivan 2000). Similarly, the development of aquatic macrophyte filter systems in discharge channels has been recommended by US regulatory authorities to reduce silt and organic loads drained from North American crayfish ponds (Huner 2001).

New developments in water treatment technologies suitable for use in controlled-environment, intensive culture systems include floating bead filters that act both as biological filters and as particle traps, and polymeric bead denitrification units. In both systems, active bacterial populations remain advantageously entrapped within the beads throughout backwashing, and in the case of experimental denitrifying beads, an essential carbon source is also retained (section 8.4.5). Regulations governing effluents, particularly those containing antibiotics and other treatment chemicals, already exist and will undoubtedly become more stringent. Indeed the treatment of all effluents may eventually become mandatory, especially in new projects, as environmentally oriented legislation increases (sections 11.4.3 and 11.5.3.2). This will incur non-trivial costs and may come together with some form of charge for environmental restoration or resource usage. In many cases it would be better for the industry to be proactive and become self-regulating than to have edicts imposed from above which, as experience indicates, may not always be appropriate to crustacean farming.

12.7 Stock enhancement

Scientific programmes have shown how hatchery-reared crustaceans released to the wild can enter a fishery and contribute to stocks both as individuals and through breeding (sections 5.7 and 10.6.3.3). In addition to the value of the catch, potential benefits might include sustaining jobs in rural communities and mitigating measures that compensate for other planned, implemented or accidental uses of the seabed. The basic techniques for juvenile production, transportation and release of several species have been demonstrated successfully, and significant recaptures of identifiable animals have been achieved in widely differing localities. Novel species (e.g. *Penaeus esculentus*) are being considered for restocking programmes and the use of reporter genes to enable stock identification, which can be detected by visual inspection (e.g. bioluminescence), or by simple chemical assay, are under investigation (Rothlisberg 1998). The results now available can provide basic data for the cost–benefit analyses to assess different project strategies that need to focus on:

- objectives of the release programme;
- scale of the operation;
- ownership rights;
- policy on population genetics and ecological impact of the released species;
- size and number of hatcheries;
- size at which the crustaceans are to be released;
- social benefits and externalities.

In the future, economic evaluations of releasing for profit will require further experimentation, in particular to study the optimum frequency of releases, release densities, and the frequency of restocking the same ground. The programmes of scientific research needed to clarify these aspects will be inherently expensive to conduct and for clawed lobsters, for example, would best be undertaken in conjunction with pilot or demonstration (extension) schemes (Wickins 1997). In fact a small number of hatcheries have already been built and are releasing juvenile lobsters in support of local fisheries in Great Britain, elsewhere in Europe (Ireland, Norway) and in Atlantic North America. Releases are most likely to show beneficial effects when made in conjunction with appropriate fishery management measures. The most pressing biological or research need now is to know smallest size at release (which has critical cost implications) and the detailed spatial (stocking frequency and density) and habitat needs (food availability, replenishment rate and thus carrying capacity) for lobsters of different sizes to survive and grow within a defined area (Wickins 1999).

While there is considerable enthusiasm in some quarters for stock enhancement projects, little is known of the ecological impacts of large-scale releases of hatchery-reared or transplanted wild-caught, juvenile crustaceans. Likewise, the impacts of creating extensive areas of modified or new habitat on the local hydrodynamic environment and ecology, including biodiversity, have seldom been studied. Truly effective programmes will necessarily be relatively large scale but their implementation, without appropriate research, impact assessments and cost–benefit studies, could cause serious and perhaps irreversible damage as well as significant financial loss.

12.8 Production technologies

12.8.1 Shrimp

With hindsight it is clear that during the boom years of the 1980s and early 1990s the shrimp industry focused too narrowly on expansion, and matters of sustainability were largely ignored. This is understandable since most entrepreneurs could recoup the investment in a farm after just two good crops and were then ready, with additional financial backing from friends and bankers, to build yet more ponds. Only when environmental degradation and viral epidemics started to have a major impact on yields and profits was the industry forced to take notice of the limitations imposed by an overexploited environment. Those harsh experiences have helped drive research and development into more responsible farming techniques, upon which the long-term health of the shrimp farming industry will come to rely more and more.

Zero water exchange systems, for which much research was pioneered by the Waddell Mariculture Center in South Carolina, have key advantages over traditional systems relying on water exchange: they greatly reduce the risk of pathogens gaining entry to the system; they improve the efficiency of nutrient assimilation; and they virtually eliminate effluent discharges (Hopkins *et al.*

1995). On the basis of positive research findings, a privately owned pilot-scale farm was set up in Belize and even greater yields were achieved in a system employing deep plastic-lined ponds and heavy aeration (McIntosh 1999). However, necessity being the mother of invention, the first commercial development of reduced and zero exchange systems for shrimp culture can probably be attributed to the ingenuity of farmers in Thailand, who responded to the immediate need of avoiding the ingress of pathogens in production ponds (Kongkeo 1995).

Another approach to producing healthy crops has involved starting new operations away from disease-ridden areas, either inland or in apparently disease-free countries. In Thailand many shrimp farmers have moved away from infected marine areas and utilise inland freshwater ponds made slightly saline by the addition of hypersaline water trucked in from coastal saltpans. This kind of inland shrimp farming poses some environmental risks if precautions are not taken to prevent the salinisation of surrounding land (section 11.4.1). Researchers in the Harbour Branch Oceanographic Institution in Florida are making progress with the culture of shrimp in water that is virtually fresh (0.5–2‰). This technology holds out prospects for farming in inland areas where pond effluents or sludge can be used as a fertiliser for terrestrial crops with minimal environmental consequences. In Arizona shrimp are being farmed using saline groundwater, further extending the geographical range of the shrimp farming industry (section 7.2.6.6).

Given the usual preference for brackish-water sites for shrimp farming, it is perhaps surprising to see new farms being established in desert areas, for example along the Red Sea coast of Saudi Arabia and elsewhere on the Arabian Peninsula. These developments testify to the ability of certain strains of penaeids to adapt well to a culture environment with salinities that often exceed 40‰. One farm in Saudi Arabia, which has been proposed as a model for sustainable shrimp farming, is notable for its use of round ongrowing ponds and for the inclusion of a very large surface area of ponds (50% of the total) that act as reservoirs for the pre-greening of incoming water and as settling basins for the treatment of pond effluents prior to discharge (Falaise & Boël 1999). Round ponds in conjunction with circular currents, usually generated by paddlewheel aerators or water jets, are a hydro-dynamically efficient way of providing stable water and pond bottom conditions. They were first used commercially in Japan and the concept has been further developed by researchers at the Oceanic Institute in Hawaii.

Hawaii has also become a centre for the development of biosecure, closed-cycle production facilities that, in combination with rigorous quarantine and screening measures, have become important sources of SPF and SPR seedstock and broodstock. In some shrimp farming regions the use of such stocks may currently be the best hope for avoiding disease problems, but many farmers still prefer to use local wild stocks for short-term motives of economy and convenience.

Shrimp production in super-intensive, controlled-environment systems promises to combine biosecurity, low environmental impact and the ability to operate in temperate climates close to important markets (Moss 1998). Such systems attract the attention of researchers and entrepreneurs but have yet to demonstrate commercial viability. Some make use of artificial seagrass-like substrates in the form of buoyant or sinking fronds, which encourage shrimp to occupy the whole water column and make better use of available space. Artificial seagrass substrates are also being used in shrimp nurseries for the same motive, and they can be preseeded with diatoms to provide a nutritionally rich grazing surface (section 7.2.4).

Shrimp farming is spreading into relatively pristine environments where major virus outbreaks have not yet been encountered, for example to Australia and Madagascar. If developments in such countries are regulated and farms are not allowed to become over-concentrated, then there is hope that earlier mistakes can be avoided. Evidence of the willingness of Australian operators to develop a sustainable industry is provided by their Prawn Farming Association's endorsement of a proposal for a compulsory federal levy to fund research and development (Navarro 2001). The potential benefits of public/private co-operation are illustrated by the case of Venezuela, where shrimp farms have been prevented from crowding together and where progress has been made with the domestication of *Litopenaeus vannamei* and *L. stylirostris*. Regulations restrict imports of potentially infected broodstock and seedstock and this has encouraged the closing of the breeding cycle for these two species. So far Venezuela has escaped the epidemics of Taura syndrome virus and white spot syndrome virus that have devastated other parts of Latin America (Jory 2000).

The question as to which of the newly emerging shrimp farming technologies can truly claim to be sustainable, can only really be answered by future generations. All that can be done at this stage is to learn the lessons of the past 20 years and try not to repeat them.

12.8.2 *Macrobrachium*

Three species of *Macrobrachium* now constitute the bulk of farmed freshwater prawns, *Macrobrachium rosenbergii*, *M. nipponense* and *M. malcolmsonii*, but it is the inclusion of production figures from China and Bangladesh since 1996 that has increased awareness of the importance of these crustaceans on world markets. Now that western markets in particular are more familiar with freshwater prawns, other species may also find a niche. Even so, it is widely recognised that the two major constraints to *Macrobrachium* farming centre on its heterogeneous growth and the fact that it yields some 20% less tail meat than penaeid shrimp. The latter presents a marketing and economic constraint while the former limits yield and increases the cost of both ongrowing and harvesting techniques. While the extended larval life is only a minor handicap, the requirement for brackish water throughout larval development can be restrictive in terms of site, water transport and storage requirements or the cost of artificial seawater salts. Arguably, this is less of a handicap than being tied to expensive coastal sites. The culture of *M. nipponense* is becoming popular in temperate China because at least one selected strain can be reared entirely in freshwater (Kutty *et al.* 2000). Attempts to rear larvae in ponds from females held in cages are under way in Thailand and may offer prospects for reducing labour and *Artemia* requirements, provided sufficient control can be exercised over the pond environment (Correia *et al.* 2000). Early larval instars seem to rely on external sources of digestive enzymes which means that most hatcheries will remain dependent on supplies of live foods.

Batch culture with regular harvesting of large individuals has become the most popular method of culture (section 7.3.5.3). It is well suited to the incorporation of new management strategies arising from increased understanding of the factors governing heterogeneous prawn growth. Pond experiments in the USA indicated that culture at slightly lower than normal temperatures (i.e. 25°C) delayed maturation in female *M. rosenbergii,* but temperatures were still sufficiently high to allow rapid growth (Tidwell & D'Abramo 2000). Although male population structure was unaffected, higher yields of marketable animals were obtained due to the better growth of females freed from the demands of reproduction (Karplus *et al.* 2000). These results, together with those from other trials using nursed and graded juveniles cultured in ponds fitted with vertical netting substrates, suggest that farming of *M. rosenbergii* could become viable in cooler regions than those traditionally used (Tidwell & D'Abramo 2000).

Polyculture of *Macrobrachium* spp. with fish, agricultural crops or in rotation with crayfish and crabs is now more widespread than in the 1980s. Such methods suit a variety of economic and environmental circumstances, which has led to their widespread commercial exploitation, particularly in Brazil, China and Vietnam.

12.8.3 Crayfish

The North American crayfish industry has suffered a number of setbacks in the past few years. These included increased competition from cheaper imports, problems from new rice insecticides and droughts that reduced yields and crayfish sizes in the late 1990s, the collapse of the soft-shell market and the imposition of new food safety regulations. All these have combined to forestall or at least limit the widespread introduction of some new developments that could have increased productivity and efficiency. As a result, new markets have been sought (which have required changes to production and grading practices), and the potential of other species has been examined. Among the latter are some of the larger orconectid crayfish (e.g. *Orconectes immunis*, *O. rusticus*) that are predicted to become economically important in the very near future both as food and as bait (Hamr 2001). The apparent high tolerance of waters low in pH and calcium makes two cambarid species (*Cambarus robustus* and *C. bertoni*) superficially attractive for management, and potentially for restocking, in acidified lakes. They have been harvested commercially for bait as well as food but could also be stocked to reduce filamentous algae (which is a common nuisance in acidified lakes) or to control invasive bivalve populations (Guiaşu 2001). However the transplantation (or accidental release by anglers using crayfish as live bait) of any of these crayfish outside their native area could present serious ecological problems. Indeed, non-indigenous crayfish species now threaten biodiversity in large areas of the world (Holdich 1999, 2001).

In Europe, little seems to have changed in crayfish aquaculture over the past 10 years, but the conservation and restoration of native species by restocking hatchery-reared juveniles continues unabated. In support of this, attempts to incubate and hatch eggs artificially (and more cost-effectively) away from the mother are frequently reported but do not seem widely adopted by the industry. Regulations governing the movement, stocking and culture of crayfish are strict but vary considerably between countries and, throughout Europe, introduced species abound (see Gherardi & Holdich 1999). Some have been stocked deliberately to create new fisheries; others like *Astacus leptodactylus* in Britain have escaped to form large, fishable populations. At present, most farming is done in extensive and semi-intensive ponds but it is almost impossible to get meaningful production figures, one reason being the lack of a clear distinction between managed fisheries and extensive cultures. There is a critical lack of information on the precise nutritional requirements of crayfish in general and of European and North American species in particular. An adequate compounded diet will be essential if any of the reported attempts to develop intensive farming systems are to be successful. The latest commercial prototype is based on a computer-controlled, compartmented tray system originally designed for clawed lobsters (section 7.8.9). However, we are not aware of any pilot-scale production data from intensive crayfish rearing systems. It is expected that *Cherax* spp. will increasingly appear in European markets over the next 5 years and, together with imported *Procambarus* products, may challenge marketing opportunities for 'home-grown' crayfish.

The Australian crayfish species hold considerable promise for successful farming enterprises, though not yet anywhere near the scale of shrimp and prawn farming. At present, global production is low, at most a few hundred, possibly a thousand tonnes, and world markets are largely unfamiliar with the product. Until total production volume can be substantially increased, it will be difficult to penetrate key freshwater crayfish markets, to attract further investment and to generate proactive interest from support industries such as feed manufacturers. Nevertheless, the current rate of investment in new farms indicates that production of *Cherax* spp. will gradually increase (Lawrence & Jones 2001). A driving force behind the expansion of *Cherax* culture has been the close and commendable association of researchers with their industry, particularly throughout Australia where considerable emphasis has been put on rapid dissemination and demonstration of research results to farmers at all levels. By applying the best practices currently available, yield increases of up to 70–90% can be achieved (Lawrence & Whisson 2000).

In the case of marron (the most valuable species farmed in Australia) there has been significant investment in well-designed and constructed farms in both Western Australia and South Australia. In the more tropical regions of Australia, redclaw farming is also expanding where construction and operation of semi-intensive farms is clearly benefiting from the experiences gained from marron, and to some extent, shrimp farming practices. The scale of operation is critical, for example, for redclaw farms, 4 ha or more of production area are considered necessary. Redclaw crayfish are now also farmed in Ecuador, New Caledonia and southern China. Ecuadorian production was 200–300 mt in 1998 but, because of marketing and financial problems, only two farms remained in production in 2000; these were expected to produce 80–100 mt (Lawrence & Jones 2001). A total of 21 farms were set up in New Caledonia in 1992 and annual production is expected to rise from the present 3 mt to 50 mt by 2002 (Piroddi & Arrignon 2000). Several million redclaw juveniles were exported to China but we are unaware of any subsequent production figures.

Commercial attempts to farm yabbies more intensively than in traditional farm dams, by using purpose-built ponds, have often been unsuccessful. Such farms, now established in Western Australia, South Australia, Victoria and New South Wales, produce only a small percentage of Australia's total yabby output. One of the major problems has been early breeding in the ongrowing ponds, resulting in detrimental competition between growing stock and new juveniles. Laborious hand selection of males for stocking can increase revenue by up to 70% but the recent exciting discovery of a yabby hybrid (female *Cherax rotundus* × male *C. albidus*) that produces all-male progeny (Lawrence & Morrissy 2000; section 2.6.3) holds great promise for preventing unwanted reproduction in traditional extensive cultures and may open the way for profitable semi-intensive cultures. Early indications are that the all-male hybrids grow faster and give a harvest nearly five times more valuable than the mixed-sex populations currently grown. It is envisaged that quarantined breeding stocks would be made available to farmers who would then breed the hybrids they require from separately held parent stocks. Commercial production of the hybrids is expected to commence in 2001 (Lawrence & Jones 2001).

Further studies of redclaw, marron and yabby populations in Australia have shown that growth rates vary considerably between geographically isolated populations of the same or very closely related species. For example, the fastest-growing yabby strain grew over nine times faster than the slowest, while growth rate differences between marron strains were 30%. These differences indicate potential for selective breeding programmes and at the same time emphasise the importance of keeping these distinct populations pure by preventing the translocation of animals from one region to another. Hybrid vigour is sometimes obtained by cross-breeding; however trial crosses of marron individuals from four different populations have so far failed to show improvements to growth, survival or meat yield. On the other hand, 9 out of 18 yabby hybrids tested grew faster than both parent strains (Lawrence & Morrissy 2000) and a first-generation redclaw cross is currently growing 10% faster than the parent stock (Jones 2000). The potential of selective breeding programmes and use of single-sex populations, combined with recent developments in feeding practices, are together expected to increase the production from these farms markedly over the next few years.

12.8.4 Clawed lobsters

The most important technical constraint to farming clawed lobsters is the need to rear individuals in isolation to prevent fighting and cannibalism. The addition of pacifying agents like lithium, the induction of synchronous moulting, routine periodic claw ablation and genetic selection have all been considered, but no commercially viable advances have yet been achieved. Research into system design, including computer-controlled hardware and redesigned containers, is continuing albeit on a small scale (section 7.8.9). Better prospects for profitability might arise if the lobsters could be sold at a small, 250 g size (section 10.6.3.6) or as a soft-shelled product, but these options would need new market development and possibly new legislation in some countries. Unilateral eyestalk ablation has been found to reduce ongrowing time by as much as 7 months (Peutz *et al.* 1987) but, with current consumer concern for animal welfare issues, marketability cannot be taken for granted. Growth increases of 10–20% have also been achieved in intact lobsters injected with human growth hormone (Charmantier *et al.* 1989) – a useful research procedure but, again, not one likely to be acceptable to consumers.

Prospects for communal culture of juveniles destined for stock enhancement programmes are being reinvestigated in Norway (section 7.8.8) where release trials in a depleted fishery have shown significant returns.

As far as clawed lobsters, spiny lobsters and crabs are concerned, the carrying capacity of new artificial habitat (wrecks, artificial reefs and islands, some submerged coastal protection schemes) seems more likely to be limited by food availability than by the number of crevices. Early seeding of new structures with, for example, mussels has also been considered (Wickins, unpublished) and if resident populations could be established, could confer a number of advantages:

(1) Seeding would reduce the opportunities for colonisation by organisms less nutritious for lobsters;
(2) Mussels would provide a valuable food resource more readily available to lobsters and crabs than to fish;
(3) Lobsters in particular cannot forage in high current speeds and a close supply of food might reduce the need to forage away from the habitat and, at the same time, extend the period of tide over which they can safely feed;
(4) An overall increase in seabed productivity might occur provided the mussel population did not attract undue numbers of predators (e.g. starfish) or suffer recruitment failure as a result of excessive silt or 'mussel mud' deposition (Ardizzone *et al.* 2000).

A fortuitous consequence of UK lobster stock enhancement experiments (section 7.8.11) was that the recapture of micro-tagged individuals of precisely known age, 5–10 years after release, allowed the potential of the age indicator pigment, lipofuscin, to be validated for the first time in clawed lobsters (*Homarus gammarus*; Sheehy *et al.* 1996). This calibration, with supporting data from cultivated signal crayfish (*Pacifastacus leniusculus*), showed that at least seven year-classes could be present in lobsters reaching legal size (Sheehy *et al.* 1999). The results are important for crustacean stock assessment and management, including the evaluation of stock enhancement programmes in natural waters, and are already being extended to other valuable crustaceans whose age has hitherto been difficult to determine.

12.8.5 Spiny lobsters

Despite the creditable progress made from 1988 to 2000 it seems likely to be several years yet before the culture of spiny lobster larvae could be undertaken on a significant, commercially acceptable scale. The rates of development, temperature and salinity tolerances of phyllosoma larvae have been determined for a variety of temperate and tropical species and small numbers have been painstakingly reared to the puerulus stage. In contrast to penaeid larvae, little is known about the digestive capabilities or indeed the prospects for enhancing digestive enzymes in phyllosoma larvae. There is obviously much scope for research in this area. It is now known that the puerulus stage of most species does not feed and, when cultured, suffers high mortality during metamorphosis. Early research experiences with many crustacean species showed that death at or near metamorphosis was common among inadequately nourished larvae, particularly when the levels of DHA and EPA in the diet were inadequate (section 2.4.2). It thus seems likely that as phyllosoma nutrition research advances, much of this source of mortality could be alleviated.

Larvae of the scyllarid or slipper lobsters (e.g. *Scyllarus* spp., *Ibacus* spp.) seem easier to rear than spiny lobster larvae; at least to the non-feeding nisto stage. Now, the commercially valuable Moreton Bay bug, *Thenus orientalis*, has been reared through its four phyllosoma instars and its non-feeding nisto stage (with 48% survival) to become juveniles in about 40 days (Mikami & Greenwood 1997). Commercial applicability of the proprietary culture methodology developed over the last 6 years has been investigated to the stage where animals have been cultured from the egg to marketable size on a pilot scale. Launch of a commercial scale operation in Australia seems likely (S. Mikami, 2001 pers. comm.).

Better understanding of nursery and ongrowing phases now exists for several spiny lobster species. Suitable stocking densities and conditions that may be employed in a variety of culture environments have been assessed but, again, much remains to be done. For example, the high mortalities reported when recirculation systems are used (Kittaka & Booth 2000) may be symptomatic of operational inexperience in a developing technology and, no doubt, they too will soon be overcome. Effort continues on elucidating the further dietary and culture environment needs of the juvenile stages, but it may be some time before a complete, low-cost diet for spiny lobsters is available. Like their clawed counterparts, spiny lobsters fed contemporary compounded diets respond remarkably well to a dietary supplement of live mussel flesh (*Mytilus* spp.) given just once or twice each week. Reported growth rates of the different species in captivity are rapid but variable, no doubt because of the different culture, diet and temperature regimes employed (Booth & Kittaka 2000). Rationalisation will be a prerequisite for credible and comparable economic evaluation.

The key research requirements are now to:

(1) Develop specific diets and feeding strategies for phyllosoma larvae, and cost-effective feeds for juveniles and adults;
(2) Design commercial scale larvae culture vessels and optimise water management protocols, including the controlled use of probiotic bacteria;
(3) Rationalise growth and survival studies and evaluate cost-effectiveness of ongrowing methods (enclosures, cages, tanks, recirculation systems) for each location and species;
(4) Conduct cost–benefit studies and impact assessments for enhancement and habitat modification programmes;
(5) Elucidate critical behavioural and environmental factors for optimising shelter placement for habitat creation;
(6) Develop methods for transportation, release and monitoring of juveniles in enhancement trials;
(7) Investigate possibilities of shortening the duration of the larval phase, through improved culture conditions, diet and possibly hybridisation;
(8) Investigate possibilities of increasing the growth rate of *Palinurus elephas* (one captive individual reached 260 g in 2 years) which has one of the shortest larval phases among spiny lobsters at 65–132 days (Kittaka 2000).

12.8.6 Crabs

Until recently, investors had not paid the same attention to crabs as they had to shrimp, crayfish and lobsters, possibly because crabs lack the solid tail meat so attractive to consumers, because of their cannibalistic tendencies, or perhaps because of the protracted larval life of some species. Nevertheless, new interest has arisen in the farming, and in some countries, the restocking of mud crabs (*Scylla* spp.). In China, production of mitten crab (*Eriocheir sinensis*) has escalated in the past 8–10 years to over 120 000 mt, but growth reduction and early mortality associated with precocious sexual maturity are currently being reported as serious problems (Liu Fengqi, 2001 pers. comm.). In the tropics and northern Australia, hatchery development has been stimulated by concerns arising from extensive overfishing and collection of juvenile mud crabs for fattening. Mud and mitten crab aquaculture is popular because the methods available are appropriate for a variety of economic circumstances and can provide viable alternatives for some farmers when shrimp farms collapse. Some crabs have a short larval life (e.g. *Portunus pelagicus*) and may be more suited to cultivation than many of those presently fished, or farmed from wild-caught juveniles. The full range of cultivable crab species has yet to be determined but prospects for the extensive farming and soft-shell production of some large, valuable species are now being evaluated, for example *Mithrax spinosissimus* in the Caribbean and

Portunus pelagicus in Australia. However, luxury markets for the products will need development.

Collection of wild Dungeness crab megalopae (*Cancer magister*) for restocking programmes may be feasible in some areas (Jamieson & Phillips 1990), but experiments are required to evaluate the prospects for their transplantation to areas of modified seabed habitat for ongrowing. The most promising outlet seems to be the market for soft-shell crabs.

Within the soft-shell crab industry, the most pressing research needs are to develop methods to predict and induce moulting, and gain knowledge of shell hardening rates at different temperatures, sea water calcium levels and salinities so that the hardening can be controlled. Moult-inducing hormones derived from other species have been tested on the blue crab, *Callinectes sapidus*, but it is unlikely that products so treated could be marketed, at least in the USA or on a large scale in the EU. In an alternative approach, research is currently in progress to design and generate compounds based on natural crab hormones that can be used to induce moulting in blue crab (Anon. 1997). The advances so far include the molecular cloning of a cDNA encoding moult-inhibiting hormone (MIH) and the production of anti-MIH antisera. However, because the sequence of MIH is highly conserved among brachyurans, it is possible that the compounds could be effective in other crab species, although they may not be as effective in crayfish and prawns, for which there is also demand for soft-shelled individuals. At present the research results (Lee *et al.* 1998; Umphrey *et al.* 1998) are still a considerable way from commercial applicability, but the underlying principles are applicable and the findings may lead, eventually, to development of methods for inducing moulting in these groups as well (R.D. Watson, 2000 pers. comm.). Similar programmes of research are under way in Australia (Navarro 2001).

12.9 Ornamental shrimp

There are two primary reasons for the growing interest in cultivating ornamental crustaceans. For all, there is of course the commercial profit motive, which can apply to almost any attractive species, e.g. the redclaw crayfish. Beyond this, however, lies a conservation motive (section 11.3.1.3), particularly where tropical marine reef-dwelling and cleaner shrimp are concerned. At present techniques exist for the culture of a few species but they are small-scale and expensive. It may be some while before commercial cultivation becomes tenable and, if it does, it may only be because natural populations have been severely overfished.

12.10 Non-decapod crustaceans

Artemia have long been the main live food used in hatcheries during the larval and early post-larval stages of many fish and crustacean species. Much is known of their nutritional strengths, and deficiencies, but hatcheries suffer most from unpredictable fluctuations in their availability and price. Indeed, the predicted climatic changes are expected to exacerbate the problems. *Artemia* are, however, relatively easy to culture, at least in small quantities, and can be nutritionally enriched to specific standards by a number of means (section 7.11.2.1).

The successes achieved in farming novel species of marine finfish over the past 5–10 years have to a large extent resulted from the ability to culture and use other live, specialist, crustacean feeds. A wide range of cladoceran, copepod and mysid species can now be raised (and enriched) in hatcheries, allowing a choice of prey sizes for young fish and crustaceans as they grow. One frequently encountered problem is the unpredictability with which a population will switch from producing live young to producing resting eggs. This causes continuous cultures to collapse, although the eggs can be harvested and stored either for direct feeding or to start new cultures. Nearly all these live feeds, including *Artemia*, can also be used to convey useful substances into larvae including medicines, essential nutrients and hormones, thereby reducing wastage and discharges to the environment (Abdu *et al.* 1998; Lavens *et al.* 2000).

12.11 References

Abdu U., Takac P., Yehezkel G., Chayoth R. & Sagi A. (1998) Administration of methyl farnesoate through the *Artemia* vector, and its effect on *Macrobrachium rosenbergii* larvae. *Israeli Journal of Aquaculture – Bamidgeh,* **50** (2) 73–81.

Anggraeni M. & Owens L. (2000) Evidence for the haemocytic origin of lymphoid organ spheroid cells in the penaeid prawn *Penaeus monodon. Diseases of Aquatic Organisms,* **40**, 85–92.

Anon. (1997) Novel method to get soft-shelled crab year-round. *Infofish International,* (5) 56.

Arala-Chaves M. & Sequeira T. (2000) Is there any kind of adaptive immunity in invertebrates? *Aquaculture,* **191** (1–3) 247–258.

Ardizzone G., Somaschini A. & Belluscio A. (2000) Prediction of benthic and fish colonisation on the Fregene and other Mediterranean artificial reefs. Artificial reefs in the Adriatic

Sea. In: *Artificial Reefs in European Seas* (eds A.C. Jensen, K.J. Collins & A.P.M. Lockwood), pp. 113–128. Kluwer Academic, Netherlands.
Bainy A.C.D. (2000) Biochemical responses in penaeids caused by contaminants. *Aquaculture*, **191** (1–3) 163–168.
Bartley D.M. & Hallerman E.M. (1995) A global perspective on the utilisation of genetically modified organisms in aquaculture and fisheries. *Aquaculture*, **137** (1–4) 1–7.
Bédier E., Cochard J.C., Le Moullac G., Patrois J. & AQUACOP (1998) Selective breeding and pathology in penaeid shrimp culture: the genetic approach to pathogen resistance. *World Aquaculture*, **29** (2) 46–51.
Benzie J.A.H. (1997) A review of the effects of genetics and environment on the maturation and larval quality of the giant tiger prawn *Penaeus monodon*. *Aquaculture*, **155** (1–4) 69–85.
Benzie J.A.H. (1998) Penaeid genetics and biotechnology. *Aquaculture*, **164** (1–4) 23–47.
Booth J. & Kittaka J. (2000) Spiny lobster growout. In: *Spiny Lobsters: fisheries and culture*, 2nd edn (eds B.F. Phillips & J. Kittaka), pp. 556–585. Fishing News Books, Oxford, UK.
Browdy C.L. (1998) Recent developments in penaeid broodstock and seed production technologies: improving the outlook for superior captive stocks. *Aquaculture*, **164** (1–4) 3–21.
Burns, J.C., Shimizu C. & Shike H. (2000) Pantropic retroviral vectors for gene transfer in aquaculture species. In: *Abstracts, Aqua 2000, Responsible Aquaculture in the New Millennium* (compiled by R. Flos & L. Creswell), p. 101. European Aquaculture Society, Special Publication No. 28.
Charmantier G., Charmantier-Daures M. & Aiken D.E. (1989) La somatotropine humane stimule la croissance de jeunes homards Americains, *Homarus americanus* (Crustacea, Decapoda). *Comptes Rendes Academie Sci. Paris*, **308** (3) 21–26.
Coman G.J., Crocos P.J. & Preston N.P. (2000) Effect of the interaction of genotype and environment on the survival and growth of the kuruma shrimp *Penaeus japonicus*. In: *Abstracts, Aqua 2000, Responsible Aquaculture in the New Millennium* (compiled by R. Flos & L. Creswell), p. 136. European Aquaculture Society, Special Publication No. 28.
Correia E.S., Suwannatous S. & New M.B. (2000) Flow-through hatchery systems and management. In: *Freshwater Prawn Culture: the farming of* Macrobrachium rosenbergii (eds M.B. New & W.C. Valenti), pp. 52–68. Blackwell Science, Oxford, UK.
Daniels W.H., Cavalli R.O. & Smullen R.P. (2000) Broodstock management. In: *Freshwater Prawn Culture: the farming of* Macrobrachium rosenbergii (eds M.B. New & W.C. Valenti), pp. 41–51. Blackwell Science, Oxford, UK.
Danis V. & O'Sullivan D. (2000) Can mangroves solve the problem of effluent discharge? *Austasia Aquaculture Magazine*, **14** (1) 44–46.
Divan A.D. & Joseph S. (2000) Cryopreservation of spermatophores of the marine shrimp *Penaeus indicus* H. Milne Edwards. *Journal of Aquaculture in the Tropics*, **15** (1) 35–43.
Evans L.H. & Edgerton B.F. (2001) Pathogens, parasites and commensals. In: *Biology of Freshwater Crayfish* (ed. D.M. Holdich), pp. 377–438. Blackwell Science, Oxford, UK.
Falaise F. & Boël L. (1999) A new technology for sustainable shrimp farming. *Infofish International*, (3) 33–99.
Fegan D. (2000) International trade in live shrimp: part 1. *Global Aquaculture Advocate*, **3** (6) 22–23.
Fetzner J.W. Jr., Sheehan R.J. & Seeb L.W. (1997) Genetic implications of broodstock selection for crayfish aquaculture in the midwestern United States. *Aquaculture*, **154** (1–2) 39–55.
Flegel T.W. & Pasharawipas T. (1998) Active viral accommodation: a new concept for crustacean response to viral pathogens. In: *Advances in Shrimp Biotechnology* (ed. T.W. Flegel), pp. 245–250. National Centre for Genetic Engineering and Biotechnology, Bangkok, Thailand.
Foster D. & Robertson C. (2000) The development of a constructed mangrove wetland to treat prawn farm effluent in the Australian wet tropics. In: *Abstracts, Aqua 2000, Responsible Aquaculture in the New Millennium* (compiled by R. Flos & L. Creswell), p. 219. European Aquaculture Society, Special Publication No. 28.
Gherardi F. & Holdich D.M. (eds) (1999) *Crayfish in Europe as Alien Species: how to make the best of a bad situation*, 304 pp. Crustacean Issues 11. A.A. Balkema, Rotterdam, Netherlands.
Guiaşu R.C. (2001) *Cambarus*. In: *Biology of Freshwater Crayfish* (ed. D.M. Holdich), pp. 609–34. Blackwell Science, Oxford, UK.
Hallerman E.M. & Kapuscinski A.R. (1995) Incorporating risk assessment and risk management into public policies on genetically modified finfish and shellfish. *Aquaculture*, **137** (1–4) 9–17.
Hallerman E., King D. & Kapuscinski A. (1999) A decision support software for safely conducting research with genetically modified fish and shellfish. *Aquaculture*, **173** (1–4) 309–318.
Hamr P. (2001) *Orconectes*. In: *Biology of Freshwater Crayfish* (ed. D.M. Holdich), pp. 585–608. Blackwell Science, Oxford, UK.
Hasson K.W., Lightner D.V., Mohney L.L., Redman R.M. & White B.M. (1999) Role of lymphoid organ spheroids in chronic Taura syndrome virus (TSV) infections in *Penaeus vannamei*. *Diseases of Aquatic Organisms*, **38**, 93–105.
Holdich D.M. (1999) Negative aspects of crayfish introductions. In: *Crayfish in Europe as Alien Species: how to make the best of a bad situation* (eds F. Gherardi & D.M. Holdich), pp. 31–47. Crustacean Issues 11. A.A. Balkema, Rotterdam, Netherlands.
Holdich D.M. (ed.) (2001) *Biology of Freshwater Crayfish*, 702 pp. Fishing News Books, Oxford, UK.
Hopkins J.S., Sandifer P.A. & Browdy C.L. (1995) A review of water management regimes which abate the environmental impacts of shrimp farming. In: *Swimming Through Troubled Water. Proceedings of the special session on shrimp farming* (eds C.L. Browdy & J.S. Hopkins), pp. 157–166. Aquaculture '95. World Aquaculture Society, Baton Rouge, LA, USA.
Huberman A. (2000) Shrimp endocrinology. A review. *Aquaculture*, **191** (1–3) 191–208.
Huner J.V. (2001) *Procambarus*. In: *Biology of Freshwater Crayfish* (ed. D.M. Holdich), pp. 541–84. Blackwell Science, Oxford, UK.

Itami T. (1999) Prevention and control of white spot syndrome (WSS) in kuruma shrimp in Japan, 22 pp. (mimeo). *Conferencia Regional de Camaronicultura,* Panama, 7–8 July 1999. Groupo FCE, Panama.

Jamieson G.S. & Phillips A.C. (1990) A natural source of megalopae for the culture of Dungeness crab, *Cancer magister* Dana. *Aquaculture,* **86** (1) 7–18.

Jones C. (2000) Recent developments in redclaw research. In: *Proceedings of Australian Crayfish Aquaculture Workshop* (eds C. Lawerence & G. Whisson), pp. 34–36. International Association of Astacology, Curtin University of Technology, Perth, Australia.

Jory D.E (1998) Use of probiotics in penaeid shrimp growout. *Aquaculture Magazine,* **24** (1) 62–67.

Jory D.E. (2000) Status of shrimp aquaculture 2000. *Aquaculture Magazine Buyer's Guide 2000*, 29th Annual Edition, pp. 49–60.

Karplus I., Malecha S.R. & Sagi A. (2000) The biology and management of size variation. In: *Freshwater Prawn Culture: the farming of* Macrobrachium rosenbergii (eds M.B. New & W.C. Valenti), pp. 259–289. Blackwell Science, Oxford, UK.

Kautsky N., Rönnbäck P., Tedenaren M. & Troell M. (2000) Ecosystem perspectives on management of disease in shrimp pond farming. *Aquaculture*, **191** (1–3) 145–161.

Kittaka J. (2000) Culture of larval spiny lobsters. In: *Spiny Lobsters: fisheries and culture,* 2nd edn (eds B.F. Phillips & J. Kittaka), pp. 508–532. Fishing News Books, Oxford, UK.

Kittaka J. & Booth J.D. (2000) Prospectus for aquaculture. In: *Spiny Lobsters: fisheries and culture,* 2nd edn (eds B.F. Phillips & J. Kittaka), pp. 465–473. Fishing News Books, Oxford, UK.

Kongkeo H. (1995) How Thailand made it to the top. *Infofish International*, (1) 25–31.

Kutty M.N., Herman F. & Le Menn H. (2000) Culture of other prawn species. In: *Freshwater Prawn Culture: the farming of* Macrobrachium rosenbergii (eds M.B. New & W.C. Valenti), pp. 393–410. Blackwell Science, Oxford, UK.

Laramore R. (2000) The urgent need for research. *Global Aquaculture Advocate,* **3** (3) 67–69.

Lavens P., Thongrod S. & Sorgeloos P. (2000) Larval prawn feeds and the dietary importance of *Artemia*. In: *Freshwater Prawn Culture: the farming of* Macrobrachium rosenbergii (eds M.B. New & W.C. Valenti), pp. 91–111. Blackwell Science, Oxford, UK.

Lawrence C. & Jones C. (2001) *Cherax*. In: *Biology of Freshwater Crayfish* (ed. D.M. Holdich), pp. 635–69. Blackwell Science, Oxford, UK.

Lawrence C.S. & Morrissy N.M. (2000) Genetic improvement of marron *Cherax tenuimanus* Smith and yabbies *Cherax* spp. in Western Australia. *Aquaculture Research,* **31**, 69–82.

Lawrence C. & Whisson G. (eds) (2000) *Proceedings of Australian Crayfish Aquaculture Workshop,* 44 pp. International Association of Astacology, Curtin University of Technology, Perth, Australia.

Lee K.J., Watson R.D. & Roer R.D. (1998) Molt-inhibiting hormone mRNA levels and ecdysteroid titre during a moult cycle of the blue crab *Callinectes sapidus*. *Biochemical and Biophysical Research Communications*, **249**, 624–627.

Lutz, C.G. (1999) Transgenic organisms in aquaculture: a little bit of this, little bit of that. *Aquaculture Magazine,* **25** (5) 83–85.

McIntosh R.P. (1999) Changing paradigms in shrimp farming: 1. General description. *Global Aquaculture Advocate,* **2** (4–5) 42–47.

McIntosh R.P. (2000) Changing paradigms in shrimp farming: IV. Low protein feeds and feeding strategies. *Global Aquaculture Advocate,* **3** (2) 44–50.

McNeil R. (2000) Zero exchange, aerobic, heterotrophic systems: key considerations. *Global Aquaculture Advocate,* **3** (3) 72–76.

Mikami S. & Greenwood J.G. (1997) Influence of light regimes on phyllosomal growth and timing of moulting in *Thenus orientalis* (Lund) (Decapoda: Scyllaridae). *Marine and Freshwater Research*, **48**, 777–782.

Moss S.M. (ed.) (1998) *Proceedings of the US Marine Shrimp Farming Program Biosecurity Workshop*, 84 pp. 14 February 1998. The Oceanic Institute, Honolulu, HI, USA.

Moss S.M. (2000) Benefits of a microbially dominated intensive shrimp production system: a review of pond water studies at the Oceanic Institute. *Global Aquaculture Advocate*, **3** (2) 53–55.

Navarro R. (ed.) (2001) Major R&D reform heralds new era for Australian prawn sector. *Australian Aquaculture Yearbook,* p. 46. National Aquaculture Council, Executive Media Pty Ltd, Melbourne, Australia.

Oo K-M., Kurokura H., Iwano T., Okamoto K., Kado R. & Hino A. (1998) Cryopreservation of nauplius larvae of the barnacle, *Balanus amphitrite* Darwin. *Fisheries Science,* **64** (6) 857–860.

Owens L. (1999) How environmental changes trigger viral epizootics: theory to control. In: *Book of Abstracts, World Aquaculture '99*, 26 April–2 May 1999, Sydney, Australia, p. 578. World Aquaculture Society, Baton Rouge, LA, USA.

Peutz A.V.H.A., Waddy S.L., Aiken D.E. & Young-Lai W.W. (1987) Accelerated growth of juvenile American lobsters induced by unilateral eyestalk ablation. *Bulletin of the Aquaculture Association of Canada*, **87** (2) 28–29.

Piroddi G. & Arrignon J. (2000) L'élevage de l'écrevisse en Nouvelle Calédonie. *L'Astaciculteur de France*, **64**, 2–7 (*apud* Lawrence & Jones 2001).

Rodríguez J. & Le Moullac G. (2000) State of the art of immunological tools and health control of penaeid shrimp. *Aquaculture*, **191** (1–3) 109–119.

Rothlisberg P.C. (1998) Aspects of penaeid biology and ecology of relevance to aquaculture: a review. *Aquaculture,* **164** (1–4) 49–65.

Roy B.A. & Kirchner J.W. (2000) Evolutionary dynamics of pathogen resistance and tolerance. *Evolution*, **154**, 51–63.

Saulnier D., Haffner P., Goarant C., Levy P. & Ansquer D. (2000) Experimental infection models for shrimp vibriosis studies: a review. *Aquaculture*, **191** (1–3) 133–144.

Schulz R. & Sypke J. (1999) Freshwater crayfish populations *Astacus astacus* (L.) in northeast Brandenburg (Germany): analysis of genetic structure using RAPD–PCR. In: *Freshwater Crayfish 12* (eds M. Keller, M.M. Keller, B. Oidtmann, R. Hoffmann & G. Vogt), pp. 387–395. Weltbild Verlag, Germany.

Sheehy M.R.J., Shelton P.M.J., Wickins J.F., Belchier M. &

Gaten E. (1996) Ageing the European lobster by the lipofuscin in its eyestalk ganglia. *Marine Ecology Progress Series,* **143**, 99–111.

Sheehy M.R.J., Bannister R.C.A., Wickins J.F. & Shelton P.M.J. (1999) New perspectives on the growth and longevity of the European lobster (*Homarus gammarus*). *Canadian Journal of Fisheries and Aquatic Sciences*, **56** (10) 1904–1915.

Shimizu C., Shike H., Dhar A.K., Klimpel K.R. & Burns J.C. (2000) Pantropic retroviral vectors mediate foreign gene expression in shrimp (*Penaeus stylirostris*). In: *Abstracts, Aqua 2000, Responsible Aquaculture in the New Millennium* (compiled by R. Flos & L. Creswell), p. 646. European Aquaculture Society, Special Publication No. 28.

Taylor C. A. (2001) Taxonomy and conservation of native crayfish stocks. In: *Biology of Freshwater Crayfish* (ed. D.M. Holdich), pp. 236–57. Blackwell Science, Oxford, UK.

Tidwell J.H. & D'Abramo L.R. (2000) Grow-out systems – culture in temperate zones. In: *Freshwater Prawn Culture: the farming of* Macrobrachium rosenbergii (eds M.B. New & W.C. Valenti), pp. 177–186. Blackwell Science, Oxford, UK.

Umphrey H.R., Lee K.J., Watson R.D. & Spaziani E. (1998) Molecular cloning of a cDNA encoding molt-inhibiting hormone of the crab, *Cancer magister*. *Molecular and Cellular Endocrinology*, **136**, 145–149.

Wickins J.F. (1997) Strategies for lobster cultivation. *CEFAS Shellfish News,* (4) 6–10.

Wickins J.F. (1999) *Lobster behaviour and stock enhancement*, 4 pp. CEFAS, Lowestoft, UK.

Appendix 1
Summary of Biological Data and Examples of Typical Culture Performance

Selected sources showing major or otherwise pertinent differences from the fishery and aquaculture production statistics given by FAO (2000) are included for interest.

Marine shrimp

Species: *Fenneropenaeus chinensis* (Osbeck, 1765); formerly *Penaeus chinensis* (Osbeck, 1765); *Penaeus orientalis* (Kishinouye, 1918)

Common names: Fleshy shrimp, Chinese white shrimp, pok, taishou-ebi
Home range: Yellow Sea, Gulf of Bohai, Korean Bight
Culture temperature range: 16–28°C
Culture salinity range: 11–38‰
Fished tonnage: 79 595 (1998) (FAO 2000)
Culture tonnage: 143 932 (1998) (FAO 2000); 48 855 (1999) (Rosenberry 1999)
Culture methods and location: Semi-intensive ponds, N. China; Korea
Source of broodstock or seed: 80% wild; 20% captive females

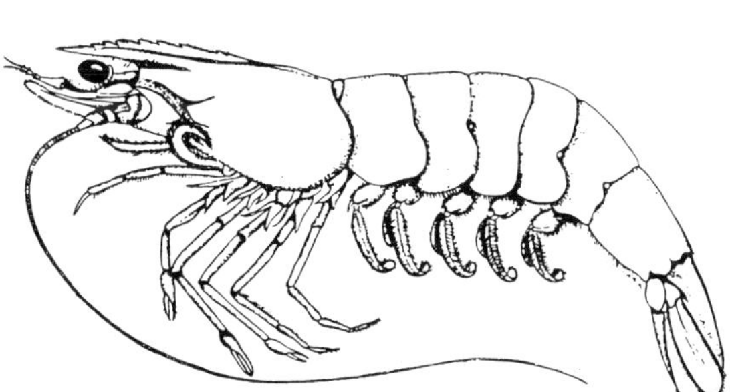

Fig. A1.1 Marine shrimp.

Special culture features: One operation per year due to climate; only 10% use nursery phase
Growth rate (ongrowing): 25 g in under 5 months
Survival: 25–55%
Yield (kg ha^{-1} crop^{-1}): 314–2308

Species: *Fenneropenaeus indicus* (H. Milne Edwards, 1837); formerly *Penaeus indicus* (H. Milne Edwards, 1837)

Common names: Indian white shrimp
Home range: India; South-east Asia
Culture temperature range: 22–33°C
Culture salinity range: 15–25‰
Fished tonnage: 0 (1998) (FAO 2000), 120 000–140 000 (Anon. 1988)
Culture tonnage: 6193 (1998) (FAO 2000), India, Malaysia and Thailand 25 300 (1997) (Rosenberry 1998)
Culture methods and location: Extensive and semi-intensive often in polyculture, India, Indonesia, Philippines
Source of broodstock or seed: Wild or pond-raised seed or broodstock
Special culture features: Non-burrowing
Growth rate (ongrowing): 4–11 g in 70–120 days
Survival: 32–91%
Yield (kg ha^{-1} crop^{-1}): 231–15 000

Species: *Fenneropenaeus merguiensis* (De Man, 1888); formerly *Penaeus merguiensis* (De Man, 1888)

Common names: Banana shrimp
Home range: South-east Asia, Thailand, Indonesia
Culture temperature range: 25–30°C

Culture salinity range: 15–33‰
Fished tonnage: 81 435 (1998) (FAO 2000)
Culture tonnage: 57 731 (1998) (FAO 2000)
Culture methods and location: Extensive and semi-intensive ponds also in polyculture, Indonesia; Thailand
Source of broodstock or seed: Wild seed, some pond-raised broodstock
Special culture features: Often a catch crop
Growth rate (ongrowing): 7–12.5 g in 76–112 days
Survival: 47–73%
Yield (kg ha^{-1} crop^{-1}): 200–5850

Species: *Penaeus monodon* (Fabricius, 1798)

Common names: Jumbo tiger shrimp; black tiger shrimp or prawn; grass prawn (Taiwan); sugpo (Philippines); udang windu (Indonesia)
Home range: Indo-west Pacific, east and southeast Africa, Pakistan to Japan; Malay archipelago and N. Australia
Culture temperature range: 24–34°C
Culture salinity range: 5–25‰
Fished tonnage: 159 208 (1998) (FAO 2000)
Culture tonnage: 577 990 (1998) (FAO 2000); 455 980 (1999) Rosenberry (1999)
Culture methods and location: Thailand, semi-intensive and intensive ponds; Philippines 60% extensive, 25% semi-intensive, 15% intensive; Taiwan intensive and super-intensive ponds
Source of broodstock or seed: Wild-caught and captive (ablated) females
Special culture features: Ablation essential, fastest growing penaeid species
Growth rate (ongrowing): 21–33 g in 80–225 days
Survival: 30–80%
Yield (kg ha^{-1} crop^{-1}): Philippines 250; Thailand 1125; Taiwan 5000–14 500

Species: *Fenneropenaeus penicillatus* (Alcock, 1905); formerly *Penaeus penicillatus* (Alcock, 1905)

Common names: Redtail shrimp; red prawn
Home range: Indo-west Pacific, Pakistan to Taiwan and Indonesia
Culture temperature range: 15–32°C
Culture salinity range: 15–32‰
Fished tonnage: 647 (1998) (FAO 2000)
Culture tonnage: 137 (1998) (FAO 2000), 3500 (1988) (Liao & Chien 1990)
Culture methods and location: Semi-intensive ponds, Taiwan
Source of broodstock or seed: Wild broodstock, hatchery seed
Special culture features: Seasonal operation in Taiwan
Growth rate (ongrowing): 9–21 g in 3–5 months
Survival: 45–90%
Yield (kg ha^{-1} crop^{-1}): 3400–12 300

Species: *Marsupenaeus japonicus* (Bate, 1888); formerly *Penaeus japonicus* (Bate, 1888)

Common names: Kuruma shrimp; Japanese tiger shrimp
Home range: Indo-west Pacific from Red Sea, east and southeast Africa to Japan and Malay archipelago, eastern Mediterranean
Culture temperature range: 18–28°C
Culture salinity range: 35–45‰
Fished tonnage: 5113 (1998) (FAO 2000)
Culture tonnage: 2549 (1998) (FAO 2000), Taiwan 4000; Japan 3020; Korea Rep. 79; Spain 55; France 14; other 3747 (1988) FAO (1990)
Culture methods and location: Semi-intensive ponds in Japan, Brazil, Australia and elsewhere; super-intensive Shigueno tanks in Japan
Source of broodstock or seed: Wild-caught females, hatchery seed
Special culture features: Live sales; one operation per year in Japan, more in the tropics; ranching programmes
Growth rate (ongrowing): 25 g in 6 months
Survival: 40–70%
Yield (kg ha^{-1} crop^{-1}): 300–30 000

Species: *Litopenaeus stylirostris* (Stimpson, 1874); formerly *Penaeus stylirostris* (Stimpson, 1874)

Common names: Blue shrimp
Home range: Eastern Pacific, Mexico to Peru
Culture temperature range: 22–30°C
Culture salinity range: 25–30‰
Fished tonnage: 186 (1988) (FAO 1990)
Culture tonnage: 15 911 (1998) (FAO 2000); 32 570 (1999) (Rosenberry 1999)
Culture methods and location: Extensive and semi-intensive ponds, Ecuador; Peru; semi-intensive ponds, New Caledonia
Source of broodstock or seed: Wild-caught and captive females

Special culture features: Generally performs less well than *L. vannamei* but disease-resistant strains are leading to significant improvements
Growth rate (ongrowing): 28 g in 8 months
Survival: 5–70%
Yield (kg ha^{-1} $crop^{-1}$): 300–2500

Species: *Litopenaeus vanammei* (Boone, 1931); formerly *Penaeus vanammei* (Boone, 1931)

Common names: Whiteleg shrimp, western white shrimp
Home range: Eastern Pacific, Mexico to Peru
Culture temperature range: 26–33°C
Culture salinity range: 5–35‰
Fished tonnage: 6051 (1998) (FAO 2000)
Culture tonnage: 191 009 (1998) (FAO 2000)
Culture methods and location: Extensive and semi-intensive Ecuador; semi-intensive elsewhere
Source of broodstock or seed: Wild and hatchery-reared post-larvae and broodstock
Special culture features: Captive stocks breed readily, acclimation (hardening-off) nurseries
Growth rate (ongrowing): 7–23 g in 2–5 months
Survival: 40–90%
Yield (kg ha^{-1} $crop^{-1}$): Latin America 500–1500; USA 3000

Freshwater prawns

Species: *Macrobrachium malcolmsonii* (H. Milne Edwards, 1844)

Common names: Indian river prawn
Home range: Pakistan, India, Bangladesh
Culture temperature range: 26–30°C
Culture salinity range: 0–5‰ adults, 11–20‰ for larvae
Fished tonnage: Pakistan 100 (1994)
Culture tonnage: Limited
Culture methods and location: Experimental, eastern India
Source of broodstock or seed: Wild seed
Special culture features: Mainly polyculture with carp, will hybridise with *M. rosenbergii*
Growth rate (ongrowing): 20–40 g in 4 months with repeated culls
Survival: 44–57%
Yield (kg ha^{-1} $crop^{-1}$): 475–605

Species: *Macrobrachium nipponense* (de Haan, 1849)

Common names: Oriental river prawn
Home range: Indo-west Pacific, N. China to Japan and Taiwan
Culture temperature range: 23–26.5°C, tolerant to <10°C
Culture salinity range: Freshwater, 0–10‰ for larvae (see below)
Fished tonnage: Unknown
Culture tonnage: China 15 000 (1998) (New 2000)
Culture methods and location: Extensive China.
Source of broodstock or seed: Wild broodstock and hatchery seed
Special culture features: Larvae of some strains can be reared entirely in freshwater
Growth rate (ongrowing): ND, probably >4–6 g in 6 months
Survival: ND
Yield (kg ha^{-1} $crop^{-1}$): 390–1875

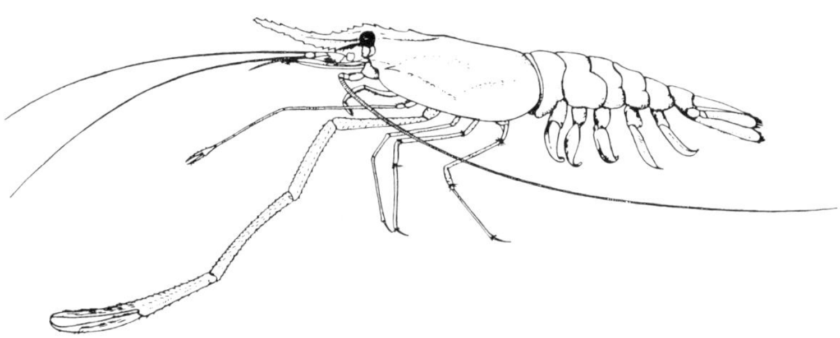

Fig. A1.2 Freshwater prawn.

Species: *Macrobrachium rosenbergii* (De Man, 1879)

Common names: Giant freshwater prawn; udang galah (Malaysia); koong yai (Thailand)
Home range: Indo-west Pacific, NW India to Vietnam, Philippines, N. Australia, Papua New Guinea
Culture temperature range: 26–32°C
Culture salinity range: 0–2‰, 12‰ for larvae
Fished tonnage: 5183 (1998) (FAO 2000)
Culture tonnage: 130 313 (1998) (FAO 2000)
Culture methods and location: Semi-intensive ponds, Thailand, Taiwan, China, Bangladesh, Brazil, Australia
Source of broodstock or seed: Pond-raised broodstock, hatchery seed
Special culture features: Larvae require brackish water, wide variety of stocking/harvesting strategies used
Growth rate (ongrowing): 25–45 g in 3–5 months; batch or repeated harvests
Survival: 40–60%
Yield (kg ha^{-1} yr^{-1}): 1000–4000

Crayfish: USA and Europe

Species: *Astacus astacus* (Linnaeus, 1758)

Common names: Noble crayfish
Home range: Europe and Scandinavia, except Iberian Peninsula.
Culture temperature range: 15–25°C
Culture salinity range: Freshwater
Fished tonnage: 300–500
Culture tonnage: 30–100
Culture methods and location: Extensive and semi-intensive ponds, juveniles reared for restocking
Source of broodstock or seed: Wild or pond-raised
Special culture features: High market value, susceptible to plague fungus
Growth rate (ongrowing): 30–80 g in 2–3 years
Survival: No data
Yield (kg ha^{-1} $crop^{-1}$): Sweden 60–430; Germany 300–600 (Ackefors 2000)

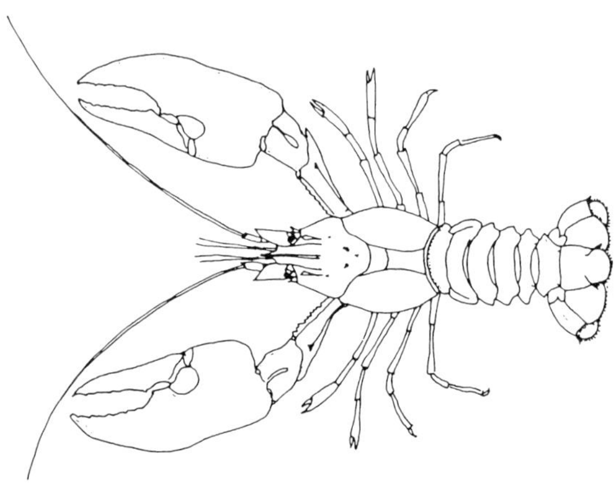

Fig. A1.3 Crayfish: USA and Europe.

Species: *Astacus leptodactylus* (Eschscholtz, 1823)

Common names: Narrow-clawed crayfish; Turkish crayfish
Home range: Eastern Europe, USSR, Turkey
Culture temperature range: 10–18°C
Culture salinity range: Freshwater
Fished tonnage: >1000, previously 8000 from Turkey
Culture tonnage: 15 (1998) (FAO 2000)
Culture methods and location: Extensive, mainly hatchery supported fisheries
Source of broodstock or seed: Wild or pond-raised
Special culture features: Fast-growing, susceptible to plague fungus
Growth rate (ongrowing): 30–80 g in 1–2 years
Survival: 60%
Yield (kg ha^{-1} $crop^{-1}$): 500–1000; Bulgaria 200–500 (Ackefors 2000)

Species: *Pacifastacus leniusculus* (Dana, 1852)

Common names: Signal crayfish
Home range: USA
Culture temperature range: 16–22°C
Culture salinity range: Freshwater
Fished tonnage: Approx. 5000
Culture tonnage: 51 (1998) (FAO 2000), Sweden 42; UK 7
Culture methods and location: Extensive and semi-intensive ponds, canals; juveniles raised for restocking, western Europe
Source of broodstock or seed: Pond-reared broodstock, hatchery juveniles
Special culture features: 3–4 months nursery, 100 m^{-2}
Growth rate (ongrowing): 30–80 g in 1–2 years
Survival: 30–40%

Yield (kg ha^{-1} crop^{-1}): France 900–2400; Spain 500–1000; Sweden and UK 50–680 (Ackefors 2000)

Species: *Procambarus clarkii* (Girard, 1852)

Common names: Red swamp crayfish, crawdad, red swamp crawfish
Home range: Southern USA
Culture temperature range: 18–25°C
Culture salinity range: Freshwater (0–5‰)
Fished tonnage: 2750 (1998) (FAO 2000), 15 000–20 000 (J.V. Huner, 2000 pers. comm. *apud* Ackefors 2000)
Culture tonnage: 17 221 (1998) (FAO 2000), 35 000 (J.V. Huner, 2000 pers. comm. *apud* Ackefors 2000)
Culture methods and location: Extensive ponds China, Southern USA, Spain, Kenya
Source of broodstock or seed: Wild or ponds
Special culture features: Self-sustaining populations, destructive burrower; also some soft-shell production
Growth rate (ongrowing): 17–80 g in 1 year
Survival: 47–87%
Yield (kg ha^{-1} yr^{-1}): USA 200–3000; Spain 350

Species: *Orconectes* (Cope, 1872)

Common names: Rusty crayfish (*O. rusticus)*, spiny cheeked crayfish (*O. limosus)*, virile crayfish (*O. virilis)*
Home range: USA
Culture temperature range: >10°C
Culture salinity range: Freshwater
Fished tonnage: 400–500
Culture tonnage: 175
Culture methods and location: Extensive ponds USA
Source of broodstock or seed: Wild or ponds
Special culture features: Self-sustaining populations, used for bait, possibly for food and soft-shell market, not easily trapped
Growth rate (ongrowing): 25–30 g in 12–14 months
Survival: No data
Yield (kg ha^{-1} yr^{-1}): 323–807

Crayfish: Australia

Species: *Cherax destructor* (Clark 1936); *C. albidus* (Riek, 1951)

Common names: Yabby, yabbie
Home range: S.E. and Central Australia
Culture temperature range: 15–30°C, optimum 28°C
Culture salinity range: Freshwater (0–5‰)
Fished tonnage: <500 (C. Jones, 2001 pers. comm.)
Culture tonnage: 230 (1998) (FAO 2000); Australia 100–300 (Mosig 1999); China ND
Culture methods and location: Semi-intensive ponds, south eastern Australia; Extensive farm dams, Western Australia.
Source of broodstock or seed: Wild or ponds
Special culture features: Destructive burrower; prolific breeder in ponds; all-male offspring produced from specific hybrids
Growth rate (ongrowing): 50–100 g in 4–12 months
Survival: Up to 70%
Yield (kg ha^{-1} crop^{-1}): 300–1500; Australia 700–2000 (Wingfield 2000)

Species: *Cherax tenuimanus* (Smith, 1912)

Common names: Marron
Home range: Western Australia
Culture temperature range: 15–24°C
Culture salinity range: Freshwater (<6‰)
Fished tonnage: <100 (C. Jones 2001 pers. comm.)
Culture tonnage: 59 (1998) (FAO 2000), Australia 49 (Wingfield 2000)
Culture methods and location: Semi-intensive ponds, Western Australia, South Australia
Source of broodstock or seed: Pond-reared broodstock, hatchery juveniles.
Special culture features: Nursery advisable

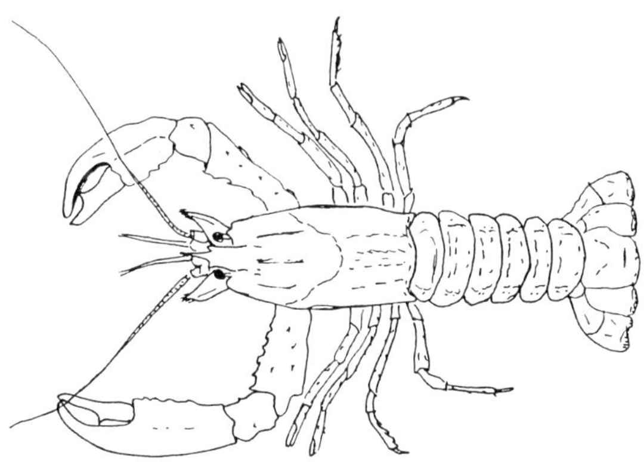

Fig. A1.4 Crayfish: Australia.

Growth rate (ongrowing): 40–120 g in 1–2 years, 200–600 g in 2–4 years
Survival: Up to 60%
Yield (kg ha^{-1} $crop^{-1}$): 1000–4000

Species: *Cherax quadricarinatus* (Von Martens, 1868)

Common names: Redclaw (the name Queensland marron is not correct)
Home range: Queensland, Northern Territory, Papua New Guinea
Culture temperature range: 24–31 °C
Culture salinity range: Freshwater (0–5‰)
Fished tonnage: >200 from reservoirs and dams (C. Jones, 2001 pers. comm.)
Culture tonnage: 113 (1998) (FAO 2000), Australia 79 (Wingfield 2000)
Culture methods and location: Semi-intensive pond culture, Australia, China, Ecuador
Source of broodstock or seed: Wild and captive
Special culture features: Selective breeding practised
Growth rate (ongrowing): 40–200 g in 6–9 months
Survival: 49–94%
Yield (kg ha^{-1} $crop^{-1}$): 1000–6000

Clawed lobsters

Species: *Homarus americanus* (H. Milne Edwards, 1837)

Common names: American lobster, Canadian lobster
Home range: Atlantic Canada, USA
Culture temperature range: 18–23°C
Culture salinity range: 30–35‰
Fished tonnage: 76 211 (1998) (FAO 2000)
Culture tonnage: 0; tonnage fattened unknown; 175 000–500 000 juveniles reared and released annually USA
Culture methods and location: Fattening systems, experimental and pilot hatchery and nursery systems for ranching studies
Source of broodstock or seed: Wild ovigerous females
Special culture features: Individual confinement from metamorphosis
Growth rate (ongrowing): 400 g in 2 years
Survival: 80–90% (post-nursery)
Yield (kg ha^{-1} $crop^{-1}$): Up to 28 534 (extrapolated from pilot studies)

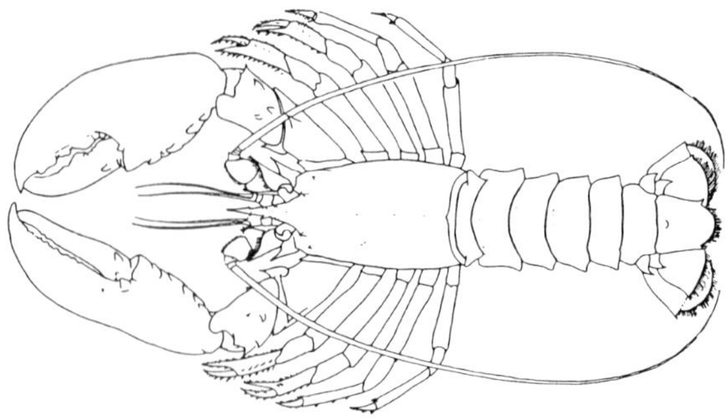

Fig. A1.5 Clawed lobster.

Species: *Homarus gammarus* (Linnaeus, 1758)

Common names: European lobster; hummer (Denmark)
Home range: Mediterranean, France, British Isles, Norway, N. Africa
Culture temperature range: 18–23°C
Culture salinity range: 30–35‰
Fished tonnage: 2935 (1988) (FAO 2000)
Culture tonnage: 0, 10 000–250 000 juveniles reared and released annually in European restocking trials (1980s & 1990s)
Culture methods and location: Pilot hatchery and nursery systems for ranching studies
Source of broodstock or seed: Wild-caught ovigerous females
Special culture features: Individual confinement of juveniles, restocking in Norway, Ireland, UK
Growth rate (experimental battery culture): 350 g in 2 years
Survival: Restocking – UK 30% over 5 years from release to 80 mm CL
Yield: Recapture rate 1–5.5% UK, 6–7% Norway

Spiny lobsters

Species: *Panulirus* (White, 1847); *Jasus* (Parker, 1883); *Palinurus* (Weber, 1795)

Common names: Spiny and rock lobsters, crawfish
Home range: Spiny lobsters, Pacific N. America, Caribbean, Indian Ocean; rock lobsters, Australasia; crawfish Mediterranean, Spain, Atlantic France, North Africa

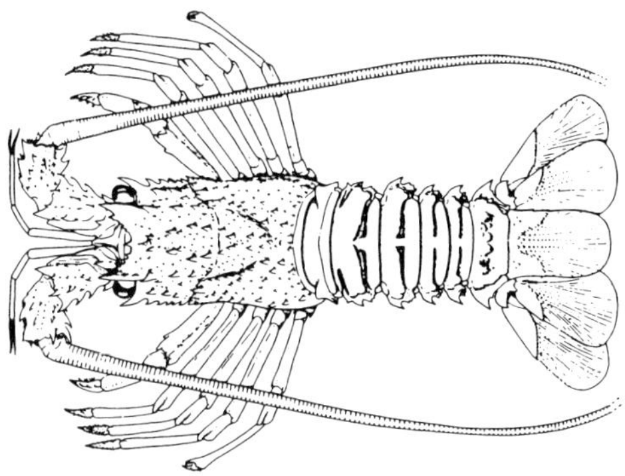

Fig. A1.6 Spiny lobster.

Culture temperature range: 18–30°C depending on species
Culture salinity range: 30–35‰
Fished tonnage: 73 575 (1998) (FAO 2000)
Culture tonnage: 71 (1998) FAO (2000); Taiwan 13.2 (1987) (Chen 1990); Singapore 24 (1988) (Lovatelli 1990)
Culture methods and location: Cage fattening, Singapore, India, Japan, Australia; pond fattening, Taiwan, New Zealand, Australia
Source of broodstock or seed: Wild-caught juveniles
Special culture features: Larvae difficult to culture; ongrowing and fattening in ponds and cages; prospects of ranching by translocation
Growth rate (ongrowing): 350 g in 2–3 years
Survival: 80% Taiwan
Yield: 45 kg m^{-3} in Singapore cages; 10 000 individuals ha^{-1} in Taiwanese ponds; 34 875 (kg ha^{-1} $crop^{-1}$ – extrapolated from pilot studies)

Crabs

Species: *Scylla* (de Haan, 1833)

Common names: Mud crab, mangrove crab
Home range: South-east Asia, Mauritius
Culture temperature range: 23–30°C
Culture salinity range: 18–34‰
Fished tonnage: 14 841 (1998) (FAO 2000)
Culture tonnage: 5883 (1998) (FAO 2000), Japan 56 300 juveniles restocked in 1996 (Imamura 1999).
Culture methods and location: Extensive and polyculture South-east Asia; semi-intensive Philippines and Australia
Source of broodstock or seed: Wild-caught juveniles, Philippines; wild and hatchery Taiwan, Philippines and Australia
Special culture features: Fattening/ripening of maturing females for gourmet Singapore market; pen culture in mangroves
Growth rate (ongrowing): 8–9 cm CW (200+ g) in 3–6 months
Survival: 40–70% Taiwan
Yield (kg ha^{-1} $crop^{-1}$): 340–1800

Species: *Portunus triturbiculatus* (Miers, 1876)

Common names: Blue swimming crab
Home range: Western Pacific, Japan, Korea, China
Culture temperature range: 20–28°C
Culture salinity range: 30–33‰
Fished tonnage: 283 971 (1998) (FAO 2000)
Culture tonnage: *Portunus* spp. 10 (1998) (FAO 2000); Japan, 10–50 × 10^6 post-larvae restocked annually
Culture methods and location: Hatchery juveniles for release to sea, Japan; semi-intensive, monosex trials China
Source of broodstock or seed: Wild-caught females
Special culture features: Fish or shrimp hatcheries used; pre-release crabs, C2–4, held in sea enclosures for 1–3 weeks, 20–40% survival
Growth rate (ongrowing): Stock supplementation (Japan)
Survival: ND

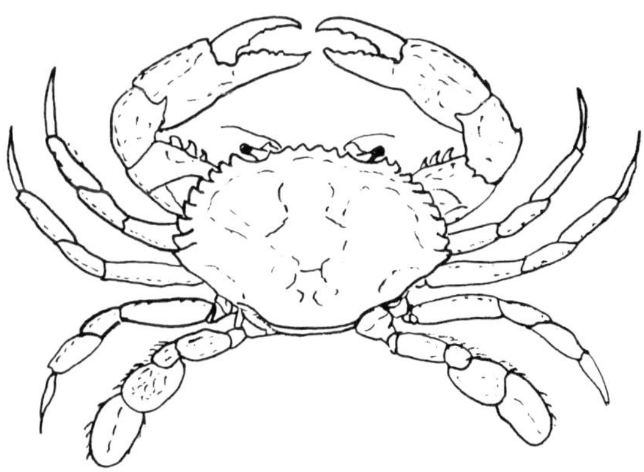

Fig. A1.7 Crab.

Yield: Ranching 3–12% recapture

Species: *Eriocheir sinensis* (H. Milne Edwards, 1854); *E. japonica* (de Haan, 1835)

Common names: Chinese mitten crab, river crab
Home range: Yellow Sea coast, Korea, China
Culture temperature range: 17–31°C
Culture salinity range: Freshwater adults; <15‰ larvae
Fished tonnage: 10 000 (1993) China (Li *et al.* 1993)
Culture tonnage: 123 249 (1998) (FAO 2000)
Culture methods and location: Pond mono- and polyculture of hatchery and wild seed
Source of broodstock or seed: Wild-caught and pond-reared females
Special culture features: Fish or shrimp hatcheries also used; hatchery and wild juveniles for release, nursery rearing megalopae to 10 g crabs in ponds in China
Growth rate (ongrowing): From 5–25 g to 125 g in 6–9 months in polyculture, fattening from 100 g to 250 g in 4–5 months
Survival: 40–60%; 2–4% recapture (ranching)
Yield (kg ha^{-1} $crop^{-1}$): 450–1500 semi-intensive; 300–500 (max. 900) polyculture

ND = No data

References

Ackefors H.E.G. (2000) Freshwater crayfish farming technology in the 1990s: a European and global perspective. *Fish and Fisheries*, **1**, 337–359.

Anon. (1988) The world shrimp industry. In: *Shrimp '88, Conference proceedings,* 26–28 January 1988, Bangkok, Thailand, pp. 1–6. Infofish, Kuala Lumpur, Malaysia.

Chen L.C. (1990) *Aquaculture in Taiwan*, 273 pp. Fishing News Books, Blackwell Scientific Publications, Oxford, UK.

FAO (1990) Fishery Statistics 1988, catches and landings **66**, 1–502. FAO, Rome, Italy.

FAO (2000) http://www.fao.org/waicent/faoinfo/fishery/statist/fisoft/fishplus.htm *(apud* FAO (2000) FAO yearbook, Fishery statistics, Capture production 1998. Vol. 86/1 and FAO (2000) FAO yearbook, Fishery statistics, Aquaculture production 1998. Vol. 86/2).

Imamura K. (1999) The organisation and development of sea farming in Japan. In: *Stock Enhancement and Sea Ranching* (eds B.R. Howell, E. Moksness & T. Svåsand), pp. 91–102. Fishing News Books, Oxford, UK.

Li G., Shen Q. & Xu Z. (1993) Morphometric and biochemical genetic variation of the mitten crab, *Eriocheir*, in southern China. *Aquaculture,* **111** (1–4) 103–115.

Liao I.C. & Chien Y.H. (1990) Evaluation and comparison of culture practices for *Penaeus japonicus, P. penicillatus* and *P. chinensis* in Taiwan. In: *The Culture of Cold-tolerant Shrimp* (eds K.L. Main & W. Fulks), pp. 49–63. Oceanic Institute, Honolulu, HI, USA.

Lovatelli A. (1990) *Regional seafarming resources atlas*, 83 pp. FAO/UNDP Regional seafarming development and demonstration project, RAS/86/024, January 1990. FAO, Rome, Italy.

Mosig J. (1999) Hothouse yabbies keep growing all winter. *Austasia Aquaculture Magazine*, **13** (5) 16–18.

New M.B. (2000) Commercial freshwater prawn farming around the world. In: *Freshwater Prawn Culture: the farming of* Macrobrachium rosenbergii (eds M.B. New & W.C. Valenti), pp. 290–325. Blackwell Science, Oxford, UK.

Rosenberry R. (1998) *World shrimp farming 1998,* 328 pp. Shrimp News International, **11**. Rosenberry, San Diego, USA.

Rosenberry R. (1999) *World shrimp farming 1999*, 320 pp. Shrimp News International, **12**. Rosenberry, San Diego, USA.

Wingfield M. (2000) An overview of the Australian fresh-water crayfish farming industry. In: *The Australian Crayfish Aquaculture Workshop*, Western Australia, 5 August 2000 (eds Whisson & M. Wingfield), pp. 5–13 (mimeo). International Association of Astacology, Aquatic Science Research University of Technology, Perth, Western Australia. (*apud* Ackefors 2000).

Appendix 2
Shrimp Counts

Marine shrimp are customarily graded by size counts. Raw, head-off shrimp or tails, are conventionally, but not universally, counted in pieces to the pound (454 g). Raw, whole head-on shrimp are counted in pieces to the kilogram (1000 g) (ITC 1983).

The count groupings and their tolerances may vary slightly between countries. Equivalent relationships between tail weight and counts of caridean prawns or shrimps will differ from marine shrimp because carideans have a larger head relative to the tail (Figs 2.3a,b).

No. of tails per pound	Approximate weight (g) of Tail	Whole shrimp
over 70	<6	<10
61–70	6–7	10–11
51–60	7–9	12–13
41–50	9–11	14–16
36–40	11–12	17–18
31–35	13–14	19–21
26–30	15–17	22–26
21–25	18–21	27–32
16–20	22–27	33–42
11–15	28–40	43–65
<10	>40	>65

No. of whole shrimp per kg	Approximate weight of whole shrimp (g)
71–90	11–13
61–70	14–16
51–60	17–20
41–50	21–24
31–40	25–32
21–30	33–49
16–20	50–62
11–15	63–91
<15	>91

Reference

ITC (1983) *Shrimps: a survey of the world market*, 273 pp. ITC Publications, International Trade Centre, UNCTAD, GATT, Geneva.

Appendix 3
Glossary

The following descriptions have been freely adapted to explain some of the terms used in this book, and so they are not all-embracing, textbook type definitions. Further guidance on terminology may be obtained from Holmes (1979) and Eleftheriou (1997) and on the Internet: http://www.aquatext.com

Ablation Surgical removal of glands (eyestalk) to stimulate maturation or growth (also called extirpation and enucleation; sections 2.3 and 7.2.2.5).

Acclimation, acclimatisation Gradual exposure to new environmental conditions in order to minimise shock or stress.

Acid sulphate soils Acidic soils typically found where areas of mangrove have been cleared. Generally unsuitable for pond construction.

Activated charcoal, carbon Finely divided carbon material capable of adsorbing organic molecules.

Ad libitum Feeding until individuals or populations seem satisfied.

Aeration The mechanical mixing of air and water, generally to increase oxygen content but also to remove excess carbon dioxide.

Aggression Hostile act or display to protect territory, family or establish dominance.

Algal bloom Rapid increase in unicellular alga population(s), manifested visually as a change in colour or turbidity of the water.

Algicide Chemical that kills unicellular and macroalgae, e.g. copper sulphate.

Alkalinity The concentration of basic minerals (e.g. carbonates) in the water, capable of neutralising excess hydrogen ions (acidity).

Ammonia The main nitrogenous excretion product of crustaceans. High toxicity that increases with pH increase.

Anaerobic Chemical processes occurring in the absence of oxygen. In ponds, an indication of undesirable substrate conditions.

Antenna The second and longer whip-like head appendage of crustaceans used for sensing the environment.

Antibiotic Natural or synthetic compound capable of inhibiting or killing (susceptible) micro-organisms.

Antioxidant Substance added to crustacean feeds to prevent or delay breakdown of fats (lipids). Maintains food value of diet and increases shelf life.

Apoptosis The process of controlled cell destruction in organisms; also termed programmed cell death.

ARR Accounting rate of return (section 10.3.3.1).

Artemia Small brine shrimp that lays drought-resistant cysts. The cysts can be stored for several years and will hatch when placed in seawater to produce a nutritious nauplius larva – an ideal food for many crustacean larvae.

Artificial impregnation Manual transfer of spermatophore from male to female, placing near or inserting spermatophore into genital structures of the female so that sperm contact with spawned eggs is inevitable.

Artificial reef Man-made structure below high-water level.

Artificial seawater A solution of salts, resembling to a greater or lesser extent, that of natural seawater. Often used in inland *Macrobrachium* hatcheries.

Artificial tidelands Intertidal ponds used as shrimp or crab nurseries prior to release.

Atterberg limit The moisture content at which a soil sample changes from one consistency to another. Liquid limit: the percentage moisture content at which a soil changes (with decreasing wetness) from liquid to plastic consistency. Plastic limit: the percentage moisture content at which a soil changes (with decreasing wetness) from plastic to semi-solid consistency.

Automatic feeder A device that dispenses feed pellets at preselected times, usually electrically operated.

Autotroph An organism that manufactures its own food from inorganic constituents, using energy from light or chemical reactions.

Backyard culture Trade press term for Australian 'hobbyist' crayfish growers.

Backyard hatchery Small, family-owned and run hatcheries.

Batch culture Method of culture in which organisms are stocked and grown without grading or culling until harvesting. *Cf.* Continuous culture.

Battery culture Culture of crustaceans in individual compartments or multiple tanks in a controlled indoor environment.

Benthic organisms, benthos Organisms living on or in the bottom sediments.

Bentonite A very fine grained clay with a high shrink/swell potential, often used to seal ponds.

Berried Female crustacean carrying eggs under her abdomen during a period of incubation.

Billion (US) One thousand million, 1×10^9.

Binder Natural or artificial substance added to bind and hold finely ground dietary ingredients together when in water.

Biological filter, biofilter Part of a water treatment system in which there is a large surface area occupied by micro-organisms that oxidise dissolved organic matter, ammonia and nitrite to less harmful products.

Biomass The total quantity of living organisms or specific organisms in a defined body of water.

Biosecure system Culture system for which all precautions are taken to exclude disease organisms; biosecurity.

Bivalve Molluscs with an openable shell – oysters, clams, mussels.

Black mud, sludge A foul-smelling marine sediment, rich in hydrogen sulphide and organic content, typically occurring in poorly managed ponds and uncleaned tanks and pipework.

Blanched Processing term meaning precooked or parboiled, e.g. at 65°C for 15–20 seconds.

Bloom see Algal bloom.

Blue claw male Large, dominant male *Macrobrachium rosenbergii* that has developed blue chelae; see also Orange claw male.

Brackish-water Seawater diluted with freshwater, for example in an estuary.

Breeding cycle A period between hatching and the first spawning of a given generation.

Broodstock Populations of maturing or mature and breeding animals.

Buffer (see also Alkalinity) A substance or substances that resist or counteract changes in the acid or alkali concentrations in water.

Bund Raised embankment separating two bodies of water.

Buster Colloquial term for a crab that has just started to moult.

Butterfly shrimp A form of prepared, value-added shrimp.

Cage culture Growing crustaceans in mesh cages either floating or staked to the bottom.

Canner Small-sized North American clawed lobster suitable for canning, also exported to Europe.

Cannibalism Consumption of one crustacean by others of the same species.

Carapace The one-piece shell structure covering head and thorax of crustaceans.

Caridea Taxonomic group (infraorder) of shrimp and prawns.

Carrying capacity The population of a given species that an area or volume will support without undergoing deterioration.

Cash crop Crustaceans grown for high sale value rather than for use as food locally.

Casitas, casas Cubanas Latin American term for artificial shelters used to attract young (casitas) or adult (casas) spiny lobsters.

Cast net Fine circular throwing net weighted at its circumference, attached at its centre to a thin rope, which is held by the fisherman.

Catch crop A subsidiary population, often of a different species, grown between crops of the main cultured species to maximise revenue.

CBA Cost–benefit analysis (section 10.3.3.3).

Chela(ae) The pincer claws of a crustacean.

Chloramphenicol One of several broad-spectrum antibiotics used, often unwisely, in shrimp hatcheries.

Chlorine, chlorine solution Chemical disinfection agent available in form of powder (calcium hypochlorite) or liquid bleach (sodium hypochlorite solution).

Cholesterol Parent compound for manufacture of many steroids; an essential component of crustacean diets.

CL Carapace length, usually measured from the eye notch to the posterior mid-dorsal margin of the carapace.

Clay Fine grained portion of a soil that can exhibit plasticity within a range of water contents and which exhibits considerable strength when air-dry.

Cobble Rock fragment usually rounded or smooth with average dimensions of 8–30 cm. Also boulder – similar but larger rock.

Colourmorphs Genetically determined, colour variations of a species, useful in identifying individuals.

Compaction Increasing the density and lowering the porosity of a soil by mechanical manipulation, essential in the construction of ponds.

Complete diet A diet capable of supporting good growth and survival throughout a specified culture phase, e.g. larval life or ongrowing.

Continental climate Any climate in which the difference between summer and winter temperatures is greater than the average range for that latitude because of the influence of a large land mass.

Continuous culture, stock and cull Ongrowing method used for farming *Macrobrachium rosenbergii* in which the fastest-growing animals are selectively harvested from the main population at intervals and replaced with new juveniles (see also Batch culture).

Count A measure used to describe a size-graded crustacean product. For example, the number of tails per pound or whole shrimp per kilogram (Appendix 2).

Crackers, shrimp or prawn Shrimp/prawn-flavoured starch-based snack food.

Crawfish Alternative name for marine Palinuridae, also spiny and rock lobsters. Commonly but incorrectly applied to freshwater crayfish in the USA.

Crayfish Freshwater Astacidae. Occasionally but incorrectly applied to marine rock lobsters, e.g. in New Zealand.

Croquettes, shrimp or prawn Peeled and deveined tails of shrimp/prawns prepared with a seasoned coating.

Cryopreservation Specialised process for freezing microscopic organisms or gametes for long-term storage in a dormant condition. Normal life functions are resumed upon subsequent thawing.

Cull Partial harvest of (usually) largest individuals in a population.

Culture To grow an organism or population; a thriving population of micro-organisms.

CW Carapace width, a standard measure of crab size; other farmed crustaceans (shrimps, lobsters) are measured by carapace length or total length.

Cyst Drought-resistant egg-like stage in the life of the brine shrimp *Artemia*. Produced instead of normal eggs in response to drying out of the shrimp's environment.

Dead spot Area or volume of pond bottom or water where circulation is minimal and where sedimentation and anaerobic conditions develop.

Decapsulation Removal of the outer shell of *Artemia* cysts by dissolution in chlorine solution.

Demand function A mathematical expression linking the level of consumer demand for a product to a series of variables including price, income level and individual preference. Used for analysing and quantifying consumer behaviour, it may refer to individuals or to consumers in general, in the latter case to describe aggregate demand (section 3.2.5).

Denitrification The chemical reduction of nitrate to nitrogen by certain micro-organisms (see also Nitrification).

Detritus Fragments of organic matter or other disintegrated material. Often forms a food resource for juvenile crustaceans, particularly crayfish.

Diatoms Single-celled planktonic plants (see also Phytoplankton) covered with two overlapping porous shells of silica. Sometimes form chains.

Dip A treatment or disinfection bath into which one or more animals are placed for a short period of time (1–60 min).

Dip net Small, hand-held net on a wooden or wire frame, used for sampling.

DO Dissolved oxygen.

Domestication The adaptation of an organism for life in intimate association with man. Purposeful selection away from the wild type is implied.

Double cropping Production of two crops from the same pond, simultaneously (polyculture) or alternately (crop rotation). One of the crops need not be an aquaculture product, e.g. rice or salt.

Drop net see Lift net.

Ecdysis The act of casting the external skeleton or shell of crustaceans.

Ecological footprint Area of land and water resources used by a crustacean farm, includes sources of seed, feed and effluent disposal.

Ecosystem The interactions of communities of organisms and their physical environment.

EDTA A chemical chelation agent added to seawater during larvae cultures to favourably adjust the availability of mineral ions.

Effluent Water that is discharged from a hatchery, farm or other industrial unit.
EIRR Economic internal rate of return (section 10.3.3.2)
Electro-fishing, -harvesting Application of an electric pulse to a specially adapted push net that makes shrimp jump out of the substrate and into the water column above the bottom of the pond.
ELISA Enzyme-linked immuno-solvent assay, method for detecting virus and other DNA.
El Niño; La Niña Unseasonable oceanic currents that strongly influence global weather and the occurrence of penaeid post-larvae off the coast of Ecuador. El Niño: warm current setting south along the coast that favours shrimp productivity in the wild and in farms (abundant wild post-larvae; rapid growth due to higher temperatures). La Niña: a cool northbound current with the reverse effects.
Embayment A shoreline indentation that forms an open bay that has been fenced or screened for aquaculture purposes.
Endemic Specific or indigenous to an area; applies both to farmed species and to diseases.
Enhancement see Stock enhancement.
Epibiotic Living organisms infesting the outer covering of an animal or plant, e.g. severe infestations of stalked protozoans that can smother crustacean larvae. Also epifauna, epiphyte.
Epipelagic Inhabiting oceanic water at depths not exceeding *ca* 200 m.
Equity The risk capital used in financing a project.
Estuary The lower reaches of a river influenced by ocean tides and mixing with seawater.
Etang French coastal lagoon.
Etiology (*aetiology*) Assignment or study of the causes of a disease.
Euryhaline Adaptable to a wide range of salinity.
Eutrophication Natural or artificial enrichment (fertilisation) of water, usually characterised by excessive blooms of phytoplankton.
Exoskeleton The external shell or covering of a crustacean.
Expatriate Person from another country; usually an employee, manager or consultant contributing special skills.
Extension service, worker Organisation or person forming the vital two-way link between the farm and the aid or research organisation.
Externalities Indirect costs and benefits that accrue to third parties and fall outside the immediate financial concerns of a project.
Exuvium(a) The cast shell(s) of a crustacean.
FAO Food and Agriculture Organization of the United Nations.
Farm dam Man-made reservoir providing water for cattle or sheep in Australia, sometimes stocked with crayfish.
Fat Processing term sometimes used for mid-gut gland (see also Hepatopancreas).
Fattening Holding wild-caught subadults or adults for a short period to enhance marketable attributes (see also Pound).
Fatty acids A group of straight-chain carbon compounds that form the building blocks of fats, oils, waxes and, in conjunction with other components, cell membranes.
FCR Food conversion ratio, usually measured as the weight of dry food fed to the weight of live animal produced.
Feasibility study Comprehensive evaluation of a proposal to farm crustaceans prior to making an investment decision.
Fecundity The number of eggs produced by a female, commonly but incorrectly used to denote the number of viable larvae produced.
Feeding rate The amount of feed offered to crustaceans in a specified time. The amount may be given in several discrete doses.
Fertiliser A natural (e.g. manure) or chemical material added to water or soil to increase natural productivity.
Filter, biological Part of a water treatment system in which there is a large surface area occupied by microorganisms that oxidise dissolved organic matter, ammonia and nitrite to less harmful products.
Filter, mechanical Part of a water treatment system that mechanically strains or collects suspended particulate material from the water.
Fish meal A dehydrated and often defatted ground/processed fish material, used in animal feed manufacture.
Flagellate alga Single-celled planktonic plant that swims by means of a whip-like flagellum.
Flake A soft form of ice used in processing delicate crustacean flesh.
Flock, floc Fragments of bacterial or other biological growths sloughed off from the surfaces of biological filters and aquaculture tanks. Also suspended bacterial accumulation in ponds.
Foam fractionation, -separation Water treatment methods for the removal of dissolved and colloidal organic material and bacteria from water, usually accomplished by inducing a countercurrent of water (down-

wards) and fine air bubbles (upwards) in a vertical cylinder. The resulting foam (foamate) is discarded.

Food conversion ratio see FCR.

Forage To search for food; the plant material actually consumed by a grazing or detritus feeding crustacean (e.g. red swamp crayfish).

Fouling The assemblage of organisms that attach to and grow on underwater objects; also the deleterious accumulation of dissolved and particulate material in a body of water or pond bottom.

Freeboard The distance between the water surface and the top of the surrounding vessel or bund (pond embankment).

Frequency distribution An arrangement of data (often size measurements) grouped into classes, which shows the number of observations falling within each category.

Fuller's earth A variety of clay or marl containing 50% silica, used in treating batches of water for use in hatcheries.

Future A contract to buy a commodity or security on a future date at a price that is fixed today (section 3.3.1).

Fyke nets Traps made of netting held under tension by a series of hoops.

Gantt chart A project planning schedule relating tasks to time (Fig. 9.3).

Gastrolith 'Stomach stone'; mass of calcium carbonate found each side of the cardiac stomach, especially in heavily calcified crayfish and lobsters. Forms reservoir of calcium for shell remineralisation after moulting.

Gearing ratio The relative proportion of loan capital to risk capital used in financing a project (section 10.3.1.1).

Geothermal water Naturally warm water from below ground.

Gill net see Tangle net.

Glazing A thin layer of ice covering frozen crustaceans, which gives them an attractive shiny appearance.

GMO Genetically modified organism, a living organism whose genetic content has been changed by man.

Gravid Female with eggs.

Groundwater Water that has percolated through the soil into porous bed-rock.

Growout North American term describing the period for which crustaceans are grown from the post-nursery phase to market size. English equivalent is ongrowing.

Habitat The locality, site and particular type of local environment occupied by an organism.

HACCP Hazard analysis, critical control points (section 3.2.2).

haemocytometer Graduated glass microscope slide and cover slip used for counting blood cells, ideal for counting microalgae (unicellular phytoplankton).

Hapas Net cages suspended in the water between poles, not usually in contact with the bottom.

Hardness In practical terms, a measure of the amount of calcium and magnesium ions in water. Frequently expressed as a calcium carbonate equivalent.

Hatchery Building or tanks used for the maintenance and conditioning of broodstock and for the culture of their larvae. A nursery facility may be included.

Heat exchanger Shell and tube or parallel plate type devices for the transfer of heat from one fluid to another, used to recover waste heat or to boost temperature of incoming water.

Heat pump Electrically driven device used to transfer heat from one area or body of water to another; acts like a domestic refrigerator transferring heat from inside to outside.

Hectare Metric unit of area, 10 000 square metres or 2.471 acres.

Hedging Buying one commodity or security and selling another in order to reduce risk.

Hepatopancreas The major digestive gland in Crustacea, also referred to as the mid-gut gland.

Hermaphrodite A species capable of producing male and female sex cells either synchronously or by changing from one sex to the other.

Heterogeneous growth Different rates of growth occurring in the same population of animals, leading to a wide range of sizes at the time of harvesting; especially during ongrowing of *Macrobrachium*.

Heterotrophic organism Organisms that are dependent on organic matter for food.

Hierarchy Behaviourally maintained system of dominance among crustaceans.

High health (shrimp) Mass produced (shrimp) reared for sale to farms from stocks certified free of specific pathogens under strictly controlled, disease-free conditions.

HPV Hepatopancreatic parvo-like virus disease of shrimp.

HUFA Highly unsaturated fatty acid; more than four double bonds.

Husbandry The art of keeping organisms alive and healthy.

Hydraulic load The daily rate of flow of water through a given volume or over a given surface area of a filter, usually expressed as m^3 of water per m^3 or m^2 of filter respectively (e.g. $m^3\ m^{-3}\ d^{-1}$).

Hydrodynamic survey Gathering of information about tides, currents, direction and volume of flow of a body of water.

Hyperbolic bottomed tank Deep, approximately oval larviculture tank whose bottom has a hyperbolic shape over long and short axes to aid even dispersal of larvae and suspended food.

IHHN Infectious hypodermal and haematopoietic necrosis, a viral disease of shrimp.

Immunostimulants Substances (generally from cell wall components of bacteria, yeast or fungus) that enhance an animal's general or non-specific defence mechanisms.

Impregnation In crustaceans, the deposition of a spermatophore on or in the female. Alternatively the treatment of porous soils with other material to prevent leaks.

Inbreeding Mating or crossing of individuals more closely related than average pairs in the population.

Incubation The holding of eggs between spawning and hatching.

Infestation The presence of organisms growing on or in a host species.

Inorganic Chemical compounds not containing carbon as a principal element (except carbonates).

Insolation The amount or duration of sunshine.

Instar (number) Refers to the number of moults a crustacean has passed through since hatching. Often termed 'stage' in literature on *Homarus* (e.g. stage IV larvae) but 'instar' avoids confusion with a stage of development that may consist of several instars e.g. the mysis stage of penaeids that undergoes three moults in producing three instars.

Integrated, vertically integrated Applied to a crustacean farm that maintains or has control over its own support facilities, broodstock and feed supplies, hatchery, nursery, processing, marketing.

Intermoult The period between each ecdysis during which the crustacean is hard-shelled.

IQF Individually quick frozen, a processing step in the preparation of high quality crustaceans.

IRR Internal rate of return (section 10.3.3.1).

Kill-chill To dip in ice water, then precook at 65°C for 15–20 seconds; see Blanched (section 3.3.2).

Kosher Food fulfilling the requirements of Jewish law.

Kreisel A cylindrical vessel with a concave bottom and an upwardly spiralling water flow developed to keep clawed lobster larvae in suspension during culture.

Lab-lab Name used in the Philippines to describe a dense mat of aquatic plant and micro-animal communities that develops on the bottom of ponds.

Larva(ae) Usually the planktonic, free-swimming stage of cultured crustaceans, although some larval stages only exist in incubated eggs.

Levee see Bund.

Lift net Net fixed to a circular, square or cross-shaped frame that is positioned beneath the water and lifted swiftly to catch shrimp and fish that settle on or above it. Sometimes attached to a wooden pole or fixed frame to assist in lifting, and sometimes baited.

Lime Calcium oxide, used as a disinfectant in ponds. Commonly but incorrectly applied to forms of calcium carbonate, such as powdered limestone, used to increase pH levels in ponds.

Liner Clay or plastic sheet applied to a vessel or pond to stop leaking or diffusion of acidic minerals into the water.

Lipid Name given to dietary fats.

Lipofuscin A fluorescent cellular (lysosomal) degradation product found in specific regions of crustacean nerve ganglia, which accumulates with age.

Macrobenthos The larger organisms (from 1 mm upwards) living in or on the pond bottom.

Malachite green Aniline dye effective in the control of external fungal and protozoan infections, banned in Denmark and USA due to human health concerns.

Management Planning and supervision of hatchery and farm operations; assessment and manipulation of water flow to maintain good water quality; assessment and manipulation of fertilisation, feeding and water exchange to control phytoplankton density within desirable limits.

Mangrove A tidal salt marsh community dominated by trees and shrubs, mainly *Rhizophora* spp. mangrove plants; if cleared for pond construction, the underlying soil is usually found to be strongly acidic.

Marl A general term for calcareous clay or calcareous loam.

Maturation Ripening, the cell divisions by which gametes are produced, to enter a phase of reproductive competence.

Maturation facility, unit Part of a farm or hatchery for the production or conditioning of breeding crustaceans.

Megalopa Name given to a late stage crab larva between the zoeal and first crab stage.

Metabolite The product of any chemical change occurring in living organisms; waste metabolites from one species may be used as food by another.

Metamorphosis The marked change in form that occurs between life cycle phases during the development of crustaceans.

Microencapsulated diets Very small particles containing compounded ingredients and surrounded by a digestible coat.

Microparticulate diets Very small particles of food manufactured for larvae.

Microsatellites Regions of DNA containing repeat sequences of short lengths of two to four nucleotide bases e.g. GTGTGT ... These DNA sequences are inherited and vary in length from individual to individual with up to 50 repeats per length. They are amplified by the use of specific primers in combination with PCR. The different segments appear as single bands that can be visualised by fluorescent staining after gel electrophoresis to determine allelic variation.

Microwire tag, microtagged A small length (1 mm × 0.25 mm dia.) of magnetised steel wire, etched with a binary code and injected into crustacean muscle tissue to identify hatchery-reared animals among wild stock during restocking and ranching trials. Detection involves passing captured animals through a sensitive metal detector.

Milkfish (Chanos chanos); an important food fish grown extensively in ponds in South-east Asia.

Moist pellet A compounded diet with a moisture content of around 30%. May also contain fresh ingredients.

Monk A water control structure governing pond depth at a pond exit, and water flow rate at a pond entrance. Screens prevent escape of stock and entry of predators and competitors. Same function as Sluice gate.

Monoclonal antibody An specific antibody derived from a culture of a single clone of cells.

Monoculture Culture of a single species. *Cf.* Polyculture.

Monosex culture The rearing of a single sex in an attempt to reduce size variability among harvested populations; employed in fish culture to avoid uncontrolled reproduction.

Moulting The act of casting the exoskeleton (see Ecdysis).

Mysid A small, swarming crustacean (e.g. *Neomysis integer*) commonly used as fish food in home display aquaria but also a good food for large crustacean larvae such as lobster larvae.

Mysis The stage between protozoea and post-larva in the development of penaeid shrimp larvae.

n-3, n-6 Convention for indicating the position of the first double bond in an unsaturated fatty acid, counting from the carboxyl group.

Natural productivity, primary productivity The development of diverse aquatic communities based on phytoplankton growth in culture ponds. These provide food for grazing crustaceans, especially in extensive cultures, and may be enhanced by the addition of fertiliser.

Nauplius(lii) The first stage of crustacean larval development, often feeding only on internal yolk reserves.

Nisto Post-larval stage in the development of slipper lobsters.

Nitrification The aerobic bacterial oxidation of toxic nitrogenous metabolites, ammonia and nitrite to much less toxic nitrate. Occurs in biological filters, also nitrifying bacteria.

Nitrogen load The amount of nitrogenous waste (usually ammonia) that has to be oxidised to nitrate by a biological filter in a given time in order to maintain acceptable levels in the culture water.

NPK ratio Ratio of nitrogen to phosphorus to potassium in fertilisers.

NPV Net present value (section 10.3.3.1).

Nucleus/plasma schemes Integrated farming operation in which a core company (the 'nucleus') supplies numerous smallholders (the 'plasma') with seed, feed and other resources and subsequently buys back their crop.

Nursed juveniles see Nursery phase.

Nursery phase The culture of post-larvae from the time of metamorphosis to the time they are stocked in the ongrowing ponds or released into the wild.

Ocean ranching see Ranching.

Ongrowing Growth to market size (see also Growout).

Option An option to buy (call option) or sell (put option) an asset at a specified price on or before a specified date (section 3.3.1).

Orange claw male Fast-growing but subdominant male *Macrobrachium rosenbergii*, distinguished by large size and orange coloured claws.

Organic load The amount of dissolved organic material carried in the water.

Osmoregulatory capacity The difference between a crustacean's haemolymph osmotic pressure and that of the external medium. Marine shrimp hyporegulate, i.e. their blood osmotic pressure is held lower than that of seawater. Under stress, the difference cannot be maintained and thus the capacity to osmoregulate becomes reduced.

Overwintering Adults: Stocks of adults held throughout the cold season (sometimes at elevated temperatures) in order to provide broodstock before wild broodstock are available. Juveniles: Populations of juveniles grown at elevated temperatures through the cold season to provide partly grown animals for ongrowing. Useful in areas where there is only a single or short growing season. May also be called nursed juveniles.

Ovigerous Carrying eggs; see also Berried.

Oxidative rancidity see Rancid.

Oxygenation Addition of oxygen to water.

Ozonation Addition of ozone to water to break down refactory organic molecules, oxidize waste metabolites and sterilise the water.

P & D see Peeled and deveined.

Paddle-wheel An electrically driven, floating device used to aerate and circulate water in ponds.

Paper-shell Recently moulted crustacean whose shell has started to harden and has turned leathery.

Parthenogenesis A reproductive strategy in which unfertilised eggs develop to adult females. Environmental change often triggers the development of males and consequent sexual reproduction.

PB Payback period (section 10.3.3.1).

PCR Polymerase chain reaction, the amplification of particular regions of DNA using primers (*q.v.*) that flank the region of DNA to be amplified. Around 30 cycles of sequential heat denaturation and subsequent replication of DNA between the primers by the polymerase enzyme produces large numbers of copies of the DNA segment required for analysis.

Pcs Pieces, term used in the Far East for individual post-larvae or juveniles, e.g. 10 000 pcs means 10 000 juveniles.

Peeled and deveined A processing term describing headless shrimp with the shell and gut removed.

Peeler A crab that is within 1–14 days of moulting. In *Callinectes sapidus* the new shell can be seen forming beneath the old one.

Pellet Compounded feed rations of graded sizes containing about 10% moisture.

Pereopods The walking legs of crustaceans, commonly spelt *pereiopods*.

Permeability, coefficient of Measure of soil permeability to water measured in m s^{-1}.

pH A measure of the hydrogen ion activity (acidity) of water. Strictly the negative logarithm of the hydrogen ion concentration. Seawater has a pH of around 8.0–8.2, freshwater pH 6.0–7.5.

Phagocyte A defensive cell circulating in the haemolymph that is capable of ingesting debris and foreign material.

Phototactic Swims or moves towards light.

Phytoplankton Microscopic plants, usually single-celled, that grow suspended in the water.

PI Plasticity index (section 6.3.3.2).

PL see Post-larva.

Plague Fungal disease to which European and Australasian crayfish are susceptible.

Pleopod The abdominal paddles or swimmerets of a crustacean, also used to provide attachment for incubation of eggs.

Polychaete Marine worms; many are rich in the specific fatty acids and other nutrients required by broodstock crustaceans.

Polyculture The cultivation of two or more species in the same facility (pond), often but not necessarily at the same time.

Polyhedral inclusion body Resistant, crystalline stage/form of the virus *Baculovirus penaei*. Often appears triangular when viewed under light microscope; also known as occlusion body.

Post-larva The stage following the last planktonic larval stage in crustaceans. Frequently the time of transition between planktonic and benthic existence.

Pound, pounding The holding of newly moulted clawed lobsters until their shell hardens to improve market value.

ppm Parts per million, often expressed as mg L^{-1}.

Prawn Natantian decapod crustacean, taxonomically equivalent to shrimp but in use the terms differ between countries. Prawn is increasingly used to describe *Macrobrachium* species while shrimp refers to penaeids.

Prefeasibility study A preliminary assessment of a culture proposal designed to determine the need for or scope of a full feasibility study.

Premix A mixture of vitamins, minerals and other micro-ingredients that is added to a compounded diet during manufacture.

Pre-operating expenses Costs of feasibility studies, construction and legislative requirements.

Primer A short (10–20 bases) single-stranded sequence of DNA that binds to a complementary sequence and initiates the extension of adjacent DNA using PCR. Primers can be designed to target specific regions of repeat sequences of DNA (mini- and microsatellites, *q.v.*).

Probiotics Live microbial preparations added to feed to improve utilisation, or to water to improve chemical quality and/or inhibit growth of pathogenic organisms.

Processing The preparation of crustaceans and crustacean products for the market.

Productivity see Natural productivity.

Propeller-aspirator pump An electrically powered, floating device for drawing air into a stream of water discharged at an angle towards the pond bottom. Serves similar function to a paddle-wheel.

Protanderous hermaphrodite Matures and functions first as a male, later as a female.

Protein skimming see Foam fractionation.

Protozoea The stage between the nauplius and the mysis in the development of penaeid shrimp.

psu Practical salinity units (symbol is ‰) equivalent to ppt or parts per thousand; commonly used to express the salt content or salinity of saline water.

PUD Peeled and un-deveined, see Peeled and deveined.

Pueruli Post-larval stage in the development of spiny lobsters.

PUFA Poly unsaturated fatty acid – two or more double bonds.

Purge The removal of unsightly gut contents by holding the crustaceans without food in clean water for 1–2 days.

Quarantine Enforced isolation of organisms that are, or may be infected, to prevent transmission of diseases to other organisms or the environment.

Raceway, D-ended Rectangular trough with rounded ends, sometimes with a short central barrier to assist creation of water circulation.

Ranching The release of hatchery reared juveniles into the wild, or into enclosed or modified wild habitat.

Rancid The breakdown of lipids and their constituents through poor storage conditions that severely reduces their dietary value.

Recirculation system A culture system incorporating a water treatment unit(s) through which the water continuously passes. A quantity of new water is added periodically or continuously.

Red tides Dense blooms of algae occurring in coastal regions, not always red in colour but often harmful to aquatic life and man.

Refactory organic compound Dissolved organic matter not readily oxidised or broken down by biological filtration.

Refractometer Optical device for measuring the refractive index of a drop of liquid. Some are specifically calibrated for the measurement of salinity.

Release The act of transferring hatchery-reared crustaceans to the wild.

Respiration The acquisition of energy from the oxidation of organic molecules in living cells. During aerobic respiration oxygen and food are consumed, carbon dioxide and ammonia are produced.

Restocking Releasing cultured or wild-caught juveniles into the wild or culture environment (see also Ranching).

Retroviral vector Refers to the use of viral envelope proteins to transfer genes into cells.

Rickettsias Small, rod or coccoid-shaped, bacteria-like, obligate intracellular parasites.

Risk capital Equity (section 10.2.1).

ROI Return on investment (section 10.3.3).

Rostrum The forward pointing spine between the eyes of a crustacean.

Rotating biological contactor, biodisc or biodrum Slowly rotating disks or drums in the water treatment unit of some recirculation systems. They form effective biological filtration units (see Biofilter).

Rotenone Selective fish poison often applied in the form of dry derris root, which contains about 5% active ingredient.

Rotifers A group of microscopic aquatic animals whose anterior end bears tufts of cilia used for feeding and locomotion. Important living food resource for larvae, frequently cultured in crustacean hatcheries.

Salinas Shallow ponds built for the production of sea salts by evaporation of seawater.

Salinity A measurement of the total mineral ions content of seawater (symbol ‰), often expressed as parts per thousand (ppt) or practical salinity units (psu).

Sanitary fish Herbivorous fish introduced into a culture pond to limit the growth of plants.

Saponin see Teaseed cake.

Sashimi Japanese speciality raw seafood.

Secchi disc A circular plate 20 cm in diameter, the upper surface of which is divided into four quadrants painted alternately black and white. When lowered into a pond on a graduated rope the point at which it disappears provides a measure of turbidity or plankton density.

Seed Post-larval or juvenile crustaceans used for ongrowing or release.

Seine net, seining Long fishing net that is suspended in the water by floats and is drawn through the pond to encircle the catch.

Semillero Ecuadorian catcher of post-larval and juvenile shrimp.

Settlement The transition from a planktonic to a benthic existence, usually at or near the first post-larval stage of development.

Shedding trays Shallow trays used to hold adult crustaceans until they moult for the production of soft-shell crustacean delicacies.

Sheepsfoot roller Heavy toothed roller for compacting clay soils during pond construction.

Shellfish Term embracing both aquatic molluscs and crustaceans.

Shell-on Processing term meaning crustaceans with the outer skin (exoskeleton, *q.v.*) left in place, shelled.

Shigueno tanks Circular tanks with a sand-covered, false floor used for the super-intensive culture of kuruma shrimp in Japan.

Shrimp Natantian decapod crustacean, commonly, but not exclusively, penaeid species (see Prawn).

Silviculture Forestry.

Simultaneous hermaphroditism Possessing male and female reproductive systems at the same time.

Skewness Deviations from symmetry. Typically used to describe the distribution of sizes in a population in which the majority of individuals are clustered to the left or right of the mean size. The reason the mean lies outside the bulk of the population is due to the disproportionate influence of individuals that are either extremely large (positive skew) or small (negative skew).

Sludge see Black mud.

Sluice gate A water control structure governing pond depth at a pond exit, and water flow rate at a pond entrance. Screens prevent escape of stock and entry of predators and competitors. Same function as a Monk.

Soft-shelled A newly moulted crustacean prior to remineralisation of the exoskeleton. Also a pathological condition in an intermoult animal.

Sourcing North American term for the catching of broodstock from the wild.

Spat Young (seed) bivalves (oysters, clams, mussels) that have settled.

Spawning The natural extrusion of eggs from a female, not to be confused with the hatching of larvae from eggs being incubated by a female.

Specific surface area The surface area of the material in a biofilter available for bacterial colonisation, usually expressed as m^2 of surface per m^3 of filter volume ($m^2\ m^{-3}$).

Spermatophore A packet or capsule of spermatozoa transferred to the female during mating.

SPF Specific pathogen-free; crustaceans certified free of particular disease-causing organism(s).

SPR Specific pathogen-resistant; crustaceans certified resistant to particular disease-causing organism(s).

Spray-dried algae Larvae food product made by spraying a dense culture of microalgae into a stream of warm air to evaporate the water.

Standing crop Generally the total weight of animals of all sizes on a farm, or in a pond.

Static water tanks Culture tanks in which the water is changed in batches (batch culture, *q.v.*).

Stock enhancement To improve or increase stocks of fishable crustaceans; in the context of this book by release of hatchery-reared juveniles into an existing or new fishery.

Stocking density The numbers of crustaceans stocked into a body of water per unit volume or bottom area.

Stratification The division of a water body into roughly horizontal layers of different temperature, salinity or oxygen content.

Sub-sand filter Water intake point for a hatchery located beneath a sandy beach in order to obtain a filtered/pretreated water supply. If gravity fed then known as a beach well.

Substrate The material constituting the bottom of the pond, lagoon or bay on or in which the farmed species lives.

Summerlings Juvenile crayfish of one summer old.

Surimi Mechanically deboned, washed and stabilised white fish flesh, which is flavoured and extruded to form analogue products such as 'shrimp' tails and 'crab' sticks.

Sushi Japanese speciality dish of raw seafood on flavoured rice.

Tagging Various methods for the identification of individual or cohorts of crustaceans among conspecifics.

Tambak The Indonesian name for coastal brackish-water ponds used in traditional extensive fish culture. Many are being deepened for shrimp culture.

Tangle net Long fine net suspended vertically in the water to trap fish by entangling them; also used to catch spiny lobsters and marine shrimp.

Tax holiday Period of reduced corporate tax liability applied during the early years of a new commercial venture as an incentive to investment.

Teaseed cake Selective fish poison. Contains 10–15% active ingredient, saponin, and is the residue after oil extraction from the seeds of *Camellia*.

TED Turtle exclusion device, device fitted to fishing nets intended to prevent accidentally captured turtles from drowning.

Telson The central, hindmost segment of the crustacean abdomen that, together with the two leaf-like uropods on either side, make up the tail fan.

Tempura Japanese speciality dish comprising small pieces of seafood or other food deep-fried in batter.

Thelycum (plural *Thelyca*) The genital region of female shrimp, may be setose (open thelycum) or set within a pouch (closed thelycum).

Tidal impoundments Intertidal areas enclosed by low walls, used in Japan for shrimp culture (e.g. Amakusa pens) and as nursery areas for juvenile shrimp prior to release into the wild (see also Artificial tidelands).

TL Total length. Length of a crustacean measured from either the tip of the rostrum or from the posterior margin of the orbit to the tip of the telson or, in some cases, the extremities of the uropods. An imprecise measure, rarely properly defined and subject to errors caused by abdominal flexibility.

Tolerance A measure of the ability of a crustacean to survive or grow in the presence of specified adverse conditions.

Total hardness see Hardness.

Transgenic manipulation, see also GMO. Transfer of genetic material from one organism to another.

Transplantation The removal of species from one geographic location to another, often outside the range of natural distribution.

Turnkey operation A complete farm package often including management and investment involvement.

Umbrella net Umbrella-shaped lift-net; see Lift net.

Urea A nitrogenous compound used for the fertilisation of ponds.

Uropod see Telson.

UV sterilisation Ultraviolet irradiation of water to kill or inhibit bacterial development. Only likely to be effective when the water is free of particulate material that the light cannot penetrate.

Vaccine A preparation of non-virulent disease organisms or immunogens that still retain the capacity to stimulate the production of antibodies or resistance.

Value-added Processing or presenting crustacean flesh in a more attractive way to increase its market value.

Venture capital The portion of capital raised by the sale of equity (section 10.2.1).

Venturi An orifice in a pipe that, because of the vacuum created in it by the swift passage of water through the pipe, sucks air into the pipe.

Vivier truck Vehicular transport equipped with facilities for the live transportation of adult crustaceans.

Water exchange rate The partial, exponential replacement of water by a flow of new water into a culture system. Often confused with water flow rate, which only describes the time it would take to fill or empty a vessel.

Water quality A vague but useful term to describe the ability of water to support the cultivated species.

Zeolite Naturally occurring hydrated sodium aluminosilcate mineral. Sometimes applied to intensive shrimp ponds for its ability to bind and remove toxic metabolites (e.g. ammonia) from water. Most effective in freshwater, however.

Zero water exchange systems Pond cultures maintained with little or no new water inflow in which heterotrophic bacteria dominate rather than microalgae (section 8.3.7).

Zoea The free-swimming larval stages of caridean prawns and lobsters.

Zooplankton Microscopic animal life suspended in the water column.

References

Eleftheriou M. (ed.) (1997) *Aqualex: a glossary of aquaculture terms*, 416 pp. Praxis Publishing, UK.

Holmes S. (1979) *Henderson's Dictionary of Biological Terms*, 510 pp. Longman, London.

Index

Tables are denoted in bold, figures in bold italic, plates in brackets.

Respiration rate → how much oxygen taken in?